The HANDBOOK of MECHANICAL ENGINEERING

Staff of Research & Education Association
Dr. M. Fogiel, Director

 *Research & Education Association*
61 Ethel Road West
Piscataway, New Jersey 08854

THE HANDBOOK OF MECHANICAL ENGINEERING

Copyright © 2004, 1999 by Research & Education Association. All rights reserved. No part of this book may be reproduced in any form without permission of the publisher.

Printed in the United States of America

Library of Congress Control Number 2003105332

International Standard Book Number 0-87891-980-5

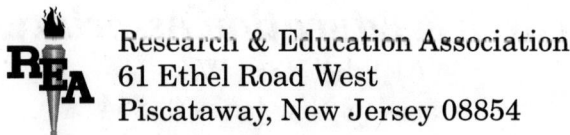

Research & Education Association
61 Ethel Road West
Piscataway, New Jersey 08854

What REA's Mechanical Engineering Handbook Will Do for You

This Mechanical Engineering handbook is a review of the important facts and concepts in mechanical engineering. The handbook is comprehensive and concise.

It is a handy reference source at all times for the professional and the student.

It condenses the vast amount of detail characteristic of the subject matter and summarizes the essentials of the field.

The book provides quick access to the important facts, principles, theorems, and equations in mechanical engineering.

This handbook has been carefully prepared by educators and professionals and was subsequently reviewed by another group of editors to assure accuracy and maximum usefulness.

Dr. Max Fogiel
Chief Editor

CONTENTS

MECHANICS

SECTION A

Chapter No.		Page No.
1	**INTRODUCTION**	A-1
1.1	Fundamental Concepts	1
1.2	Systems of Units	2
1.3	Numerical Accuracy	2
2	**VECTORS**	3
2.1	Scalars and Vectors	3
2.2	Vector Addition	4
2.3	Scalar Product	5
2.4	Vector Product	7
2.5	Mixed Triple Product	8
2.6	Vector Functions	9
2.7	Unit Vector	10
2.8	Transformation Matrix	11
3	**FORCES ON A PARTICLE**	13
3.1	Newton's Laws	13
3.2	Vector Representation	14

3.3	Free-Body Diagram	15
3.4	Rectangular Components	16
3.5	Equilibrium of a Particle	18

4 EQUIVALENT SYSTEMS OF PARTICLES 19

4.1	Rigid Bodies	19
4.2	Principle of Transmissibility	19
4.3	Moment of a Force About a Point	20
4.4	Varignon's Theorem	22
4.5	Moments About a Given Axis	22
4.6	Moment of a Couple	24
4.7	Addition of Couples	24
4.8	Resolution of Forces Into a Force and a Couple	25
4.9	Reduction of Forces	26

5 EQUILIBRIUM OF RIGID BODIES 30

5.1	Introduction	30
5.2	Reactions at Supports and Connections	31
5.3	Equilibrium Conditions	33
5.4	Special Cases in Equilibrium	35

6 CENTROIDS AND CENTER OF GRAVITY 36

6.1	Center of Gravity	36
6.2	Centroids	37
6.3	Composite Bodies	38
6.4	Centroids by Integration	40
6.5	Theorems of Pappus-Guldinus	42
6.6	Distributed Loads on Beams	43

7 ANALYSIS OF STRUCTURES 44

7.1	Introduction	44
7.2	Method of Joints	45
7.3	Special Loading Conditions	45
7.4	Space Trusses	46
7.5	Method of Sections	47
7.6	Multiforce Structures	47
7.7	Analysis of a Frame	48
7.8	Machines	48

8 BEAMS AND CABLES 49

8.1	Types of Loading and Supports	49
8.2	Shear and Bending Moments	50
8.3	Analytical Relationships	51
8.4	Cables— Concentrated Loads	52
8.5	Cables— Distributed Loads	53
8.6	Parabolic Cables	55

9 FRICTION 56

9.1	Dry Friction	56
9.2	Angles of Friction	57
9.3	Wedges	58
9.4	Square-Threaded Screws	59
9.5	Axle Friction	60
9.6	Disk Friction	61
9.7	Rolling Resistance	62
9.8	Belt Friction	63

10 MOMENT OF INERTIA 64

10.1	Integration Method	64
10.2	Polar Moment of Inertia	66
10.3	Radius of Gyration	67
10.4	Perpendicular-Axis Theorem	67
10.5	Parallel-Axis Theorem	68

10.6	Composite Areas and Bodies	69
10.7	Product of Inertia	70
10.8	Principal Axes and Inertias	72
10.9	Mohr's Circle	74
10.10	Thin Plates	76

11   METHOD OF VIRTUAL WORK   78

11.1	Work of a Force	78
11.2	Principle of Virtual Work	79
11.3	Mechanical Efficiency	80
11.4	Finite Displacements	80
11.5	Potential Energy	82
11.6	Stability	83

12   KINEMATICS OF PARTICLES   85

12.1	Basic Definitions	85
12.2	Motion of a Particle	87
12.3	Uniform Motion	89
12.4	Relative Motion	89
12.5	Graphical Solutions	90
12.6	Tangential and Normal Components	93
12.7	Radial and Transverse Components	95
12.8	Angular Velocity	99

13   KINETICS OF PARTICLES   100

13.1	Linear Momentum	100
13.2	Equations of Motion	101
13.3	Dynamic Equilibrium	101
13.4	Velocity-Dependent Force	102
13.5	Angular Momentum	105
13.6	Central Force Fields	106
13.7	Space Mechanics	108
13.8	Kepler's Laws	112
13.9	Inverse-Square Repulsive Fields	113
13.10	Nearly Circular Orbits	115

14 ENERGY AND MOMENTUM METHODS - PARTICLES 118

14.1	Work of a Force	118
14.2	Work-Energy Principle	120
14.3	Power and Efficiency	121
14.4	Conservative Forces	122
14.5	Potential Energy	123
14.6	Gravitational Potential	125
14.7	Conservation of Energy	127
14.8	Application to Space Mechanics	129
14.9	Impulse and Momentum	132
14.10	Impulse Motion	133
14.11	Impact	134
14.12	Methods of Solution	139
14.13	Motion of a Projectile	140
14.14	Motion of a Charged Particle	143
14.15	Constrained Motion of a Particle	145

15 SYSTEMS OF PARTICLES 147

15.1	Systems of Forces	147
15.2	Linear and Angular Momentum	147
15.3	Motion of the Mass Center	148
15.4	Angular Momentum About the Mass Center	149
15.5	Conservation of Momentum	151
15.6	Kinetic Energy	151
15.7	Conservation of Energy	152
15.8	Impulse and Momentum	152
15.9	Continuous System	153
15.10	Variable Mass Systems	155

16 KINEMATICS OF RIGID BODIES 157

16.1	Introduction	157
16.2	Translation	157
16.3	Rotation	158
16.4	Absolute and Relative Velocity	161

16.5	Instantaneous Center of Rotation	162
16.6	Absolute and Relative Acceleration	164
16.7	Inertial Forces	165
16.8	Non-Inertial Reference System	165
16.9	Motion About a Fixed Point	169

17 PLANE MOTION OF RIGID BODIES 170

17.1	Introduction	170
17.2	Center of Mass	170
17.3	Static Equilibrium	173
17.4	Equations of Motion	173
17.5	Angular Momentum	174
17.6	D'Alembert's Principle	177
17.7	Systems of Rigid Bodies	178
17.8	Constrained Motion	178

18 ENERGY AND MOMENTUM METHODS - RIGID BODIES 181

18.1	Principle of Work and Energy	181
18.2	Kinetic Energy in Plane Motion	183
18.3	Conservation of Energy	184
18.4	Power	184
18.5	Impulse and Momentum	185
18.6	Conservation of Angular Momentum	186
18.7	Eccentric Impact	186

19 SPATIAL DYNAMICS 188

19.1	Angular Momentum	188
19.2	Principle Axes	192
19.3	Momental Ellipsoid	193
19.4	Impulse and Momentum	195
19.5	Kinetic Energy	196
19.6	Rigid Body Motion	197
19.7	Euler's Equations of Motion	198

19.8	Motion About a Fixed Point	199
19.9	Motion About a Fixed Axis	199
19.10	Earth's Rotational Effects	200
19.11	Motion of a Gyroscope	204
19.12	Steady Precession	207
19.13	Torque-Free Motion	210

20 LAGRANGIAN MECHANICS 212

20.1	Generalized Coordinates and Forces	212
20.2	Lagrange's Equation	214
20.3	Generalized Momentum	217
20.4	Impulsive Forces	217
20.5	Hamilton's Variation Principle	218
20.6	Hamilton's Equations	219
20.7	Lagrange's Equations with Constraints	220

21 MECHANICAL VIBRATIONS 222

21.1	Introduction	222
21.2	Simple Harmonic Motion	223
21.3	General Harmonic Oscillator	224
21.4	Simple Pendulum	226
21.5	The Spherical Pendulum	228
21.6	The Torsion Pendulum	231
21.7	The Physical Pendulum	232
21.8	Free Vibration	234
21.9	Conservation of Energy	236
21.10	Forced Vibration	236
21.11	Damped Free Vibration	239
21.12	Damped Forced Vibration	241
21.13	Stability	242
21.14	Potential Energy Function	243
21.15	Oscillation with a Single Degree of Freedom	244
21.16	General Theory of Vibrating Systems	245
21.17	Normal Coordinates	248
21.18	Vibration of a Loaded String	250
21.19	Waves	252

STRENGTH OF MATERIALS & MECHANICS OF SOLIDS

SECTION B

Chapter No.		Page No.

1	AXIAL FORCE, SHEAR FORCE AND BENDING MOMENT	B - 1
1.1	Equilibrium of a Solid Body	1
1.1.1	Summation of the Forces Acting on a Solid	1
1.1.2	The Force Vector, $\overline{F}$	1
1.1.3	The Moment Vector, $\overline{M}$	1
1.2	Conventions for Supports and Loadings	2
1.2.1	Link and Roller Supports	2
1.2.2	Pinned Supports	3
1.2.3	Fixed Supports	3
1.2.4	Distributed Loading	3
1.2.5	Hydrostatic Loading	4
1.3	Axial Force, Shear Force and Bending Moment Diagrams	4
1.3.1	Axial Force in Beams	4
1.3.2	Shear	5
1.3.3	Bending Moment	6
1.3.4	Axial, Shear, and Bending Moment Diagrams of Various Beams and Beam Loadings	7

1.3.4.1	Simply Supported Beam	7
1.4	Differential Equations of Equilibrium	9
1.4.1	Basic Equations	9
1.4.2	Shear Diagrams by Summation	9
1.4.3	Moment Diagrams by Summation	11

2 STRESS 12

2.1	Definition of Stress	12
2.1.1	General Definition of Stress at a Point	12
2.1.2	Normal Stress, σ (Sigma)	13
2.1.3	Shearing Stress, τ (Tau)	13
2.1.4	Mathematical Definition of Stress at a Point	13
2.2	Stress Tensor	14
2.3	Stress in Axially Loaded Members	15
2.3.1	Normal Stress (Perpendicular to the Cut)	15
2.3.2	Bearing Stress	15
2.3.3	Average Shearing Stress (Parallel to the Cut)	15
2.4	Design of Axially Loaded Members and Pins	16
2.4.1	Required Area of a Member	16
2.4.2	Example	16

3 STRAIN 19

3.1	Definition of Strain	19
3.1.1	General Definition of Strain	19
3.1.2	Longitudinal Strain, ε (Epsilon)	20

3.1.3	Shear Strain, γ (Gamma)	20
3.1.4	Mathematical Definitions of Strain	20
3.1.4.1	Longitudinal Strain, ε	20
3.1.4.2	Shear Strain, γ	21
3.2	Strain Tensor	21
3.3	Hooke's Law for Isotropic Materials	21
3.3.1	General Equations	22
3.4	Poisson's Ratio	23
3.5	Thermal Strain	23
3.6	Elastic Strain Energy	23
3.7	Deflection of Axially Loaded Members	25
3.8	Stress Concentrations	26

4	**STRESS-STRAIN RELATIONS**	**28**
4.1	Material Properties	28
4.1.1	Definitions	28
4.1.2	Elasticity	30
4.2	Stress-Strain Diagrams	30
4.3	Table of Strength of Materials	32

5	**CENTER OF GRAVITY, CENTROIDS AND MOMENT OF INERTIA**	**35**
5.1	Center of Gravity	35
5.2	Centroids	37

5.2.1	Composite Bodies	37
5.2.2	Centroids of Integration	39
5.3	Theorems of Pappus-Guldinus	41
5.4	Moment of Inertia	43
5.4.1	Polar Moment of Inertia	45
5.4.2	Radius of Gyration	46
5.4.3	Perpendicular-Axis Theorem	47
5.4.4	Parallel-Axis Theorem	47
5.4.5	Composite Areas	49
5.5	Product of Inertia	50
5.6	Principal Axes and Inertias	52
5.7	Properties of Geometric Sections	55

6	**STRESSES IN BEAMS**	**64**
6.1	Bending Stress Distribution	64
6.2	Flexure Formula	65
6.3	Shear Stresses in Beams	68
6.4	Shear Center in Beams	69

7	**DESIGN OF BEAMS**	**73**
7.1	Beam Design Criteria	73
7.1.1	Shear Force	74
7.2	Beam Design Procedures	76
7.3	Plastic Analysis of Beams	77
7.4	Economy in Beam Design	77

7.5	Beam Diagrams and Formulas	78
7.5.1	For Various Static Loading Conditions	78
7.5.2	For Various Concentrated Moving Loads	89

8	**DEFLECTION OF BEAMS**	**91**
8.1	Deflection and Elastic Curve	91
8.2	Differential Equations of Beam	92
8.3	Determination of Deflection	94
8.3.1	The Integration Method	94
8.3.2	Moment Area Method	101
8.3.2.1	Deflection and Slope of a Simply-Supported Beam Due to the Concentrated Load P	101
8.3.2.2	Angles of Rotation	103
8.3.2.3	Simply Supported Beam	105
8.3.2.4	Angle of Rotation θ_b and Deflection δ at the Free End of a Cantilever with a Concentrated Load P	106
8.3.2.5	Deflection of the Free End of the Cantilever Beam	108

9	**STATICALLY INDETERMINATE BEAMS**	**110**
9.1	Introduction	110
9.2	Integration Method	111
9.3	Superposition Method	115
9.4	Moment-Area Method	118
9.5	Three-Moment Theorem	121
9.6	Plastic Analysis	127

10	**COLUMNS**	**132**
10.1	Definition and Types of Columns	132
10.2	Short Columns with Eccentric Load	134
10.3	Euler's Formula for Long Columns	136
10.4	Structural Steel Column Formulas	141
10.5	Design of Columns	143
10.5.1	Steel-Column Design	143
10.5.2	Aluminum-Column Designs	144
10.5.3	Timber-Column Designs	144
10.5.4	Eccentricity Loaded Columns	145

11	**COMPOSITE STRUCTURES**	**147**
11.1	Composite Members in Parallel	147
11.2	Composite Members in Series	148
11.3	Reinforced-Concrete Beams	151

12	**FAILURE CRITERIA IN DESIGN**	**158**
12.1	Introduction	158
12.2	Design for Static Strength	159
12.2.1	Design for Elastic Strength	159
12.2.2	Design for Plastic Strength	162
12.2.3	Design for Brittle Fracture Conditions	162
12.3	Design for Dynamic Strength	163
12.3.1	Definitions	163
12.3.2	Elastic Strength	163
12.3.3	Plastic Strength	166
12.4	Design for Torsional Strength	167
12.4.1	Elastic Strength	167
12.4.2	Plastic Strength	168

12.4.3	Dynamic Torsional Strength	169
12.5	Design of Beam Columns	169
12.6	Fatigue Strength. Elastic Vibration	171
12.6.1	Fatigue Strength	171
12.6.2	Vibration	171

13	**TORSION**	**176**
13.1	Torque	176
13.2	Torsion Formulas	177
13.2.1	Polar Moment of Inertia	177
13.2.2	The Polar Moment of Inertia for a Hollow Circular Shaft	178
13.3	Angle of Twist of Circular Members	178
13.3.1	Twist	179
13.3.2	g_{max} and ϕ	179
13.4	Torque and Twist for Circular Shafts in Inelastic Range	180
13.5	Solid Noncircular Members	181
13.5.1	Torque for Rectangular Shaft in Torsion	181
13.5.2	Angle of Twist for Rectangular Shaft in Torsion	181
13.6	Thin-Walled Hollow Members	182
13.6.1	Shear Flow	182
13.6.2	Torque	182
13.6.3	Angle of Twist	184

14	**BOLTED, RIVETED AND WELDED JOINTS**	**185**
14.1	Bolted Joints	185
14.1.1	Definitions	186

14.1.2	Types of Failures and Stresses	186
14.2	Riveted Joints	187
14.3	Welded Joints	190
14.3.1	Fillet Welds	190
14.3.2	Butt Welds	191
14.4	Thin-Walled Pressure Vessels	194
14.4.1	Stress on Longitudinal Seam	194
14.4.2	Stress on Transverse (Circumferential Seam)	195

15	**ENERGY METHODS**	**197**
15.1	Load-Deformation Relationships	197
15.2	Strain Energy Caused by Stresses	199
15.3	Impact Phenomena	200
15.4	Castigliano's Theorem	202
15.4.1	Energy Stored During Bending	207
15.4.2	Energy Stored During Torsion	211
15.4.3	Shear Strain Energy	214

16	**COMBINED STRESSES**	**216**
16.1	Combined Axial and Bending Stresses	216
16.2	Stresses of Eccentric Loading	217
16.2.1	Eccentric Axial Load	217
16.2.2	Eccentrically Loaded Bolted Joints	218
16.3	Combined Shear and Tension or Compression	219
16.3.1	Shear Stress Due to Tension or Compression	219
16.3.2	Tension or Compression Due to Shear	220
16.3.3	Mohr's Circle for Stress	221
16.4	Combined Torsion and Bending	223

FLUID MECHANICS & DYNAMICS

SECTION C

Chapter No.		Page No.
1	**INTRODUCTION**	C-1
1.1	Definition of a Fluid	1
1.2	Basic Laws in Fluid Mechanics	1
1.3	System and Control Volume	1
1.4	Dimensions and Units	2
2	**FUNDAMENTAL CONCEPTS**	5
2.1	Fluid as a Continuum	5
2.2	Fluid Properties	5
2.3	Two Viewpoints	6
2.4	One, Two, and Three Dimensional Flow	6
2.5	Pathlines, Streaklines, Streamlines, and Streamtubes	7
2.6	Stress Field	8
2.7	Newtonian Fluid Viscosity	9
2.8	Classification of Fluid Motion	11
2.9	Irrotational Flow	11
3	**FLUID STATICS**	13
3.1	The Basic Equation of Fluid Statics	13
3.2	The Standard Atmosphere	15
3.3	Absolute and Gage Pressures	16
3.4	The Simple Manometer	16

3.5	Hydrostatic Forces on Submerged Surfaces	17
3.6	Buoyancy and Stability	20
3.7	Fluids in Rigid-Body Motion	22

4 BASIC LAWS FOR SYSTEMS AND CONTROL VOLUMES 23

4.1	Basic Laws for a System	23
4.2	Reynolds Transport Equation	25
4.3	The Basic Laws Applied to the Control Volume	26

5 INTRODUCTION TO DIFFERENTIAL ANALYSIS OF FLUID MOTION 33

5.1	The Continuity Equation	33
5.2	The Differential Form of Momentum Equation	35
5.3	Euler's Equations	38
5.4	Bernoulli's Equation	39

6 CONSIDERATIONS FOR COMPRESSIBLE FLOW 41

6.1	Review of Thermodynamics	41
6.2	Propagation of Sound Waves	43
6.3	Local Isentropic Stagnation Properites	45
6.4	Critical Condition	45

7 ONE-DIMENSIONAL COMPRESSIBLE FLOW 47

7.1	Basic Equations for Isentropic Flow	47
7.2	Effects of Area Variation on Flow Properties in Isentropic Flow	49
7.3	Isentropic Flow of an Ideal Gas	50
7.4	Isentropic Flow in a Converging Nozzle	51
7.5	Adiabatic Flow in a Constant Area Duct with Friction	52
7.6	The Fanno Line	54
7.7	Frictionless Flow in a Constant Area Duct with Heat Addition	55

7.8	The Rayleigh Line	57
7.9	Normal Shock	58
7.10	Flow in a Converging-Diverging Nozzle	62
7.11	Oblique Shock	63

8 DIMENSIONAL ANALYSIS AND PHYSICAL SIMILARITY 67

8.1	Dimensional Analysis, Physical Similarity	67
8.2	Types of Physical Similarity	68
8.3	The Buckingham's Π Theorem	69
8.4	Dimensional Analysis of a Problem	69

9 INCOMPRESSIBLE VISCOUS FLOW 71

9.1	Internal and External Flows	71
9.2	Laminar and Turbulent Flows	71
9.3	Fully Developed Laminar Flow Between Infinite Parallel Plates	72
9.4	Fully-Developed Laminar Flow in a Pipe	74
9.5	Flow in Pipes and Ducts	76
9.6	Calculation of Head Loss	80
9.7	Solution of Pipe Flow Problems	86

10 BOUNDARY LAYER THEORY 89

10.1	Introduction	89
10.2	Boundary Layer Thickness	90
10.3	Integral Momentum Equation	91
10.4	Flow Over a Flat Plate	92
10.5	The Momentum Integral Equation for Zero Pressure Gradient Flow	93
10.6	Turbulent Boundary-Layer Skin-Friction Coefficient for Smooth Plates	94
10.7	Pressure Gradient in Boundary-Layer Flow Separation	95
10.8	Flow Around Immersed Bodies	96

11 IDEAL-FLUID FLOW 102

11.1	Introduction	102
11.2	Mathematical Considerations	102
11.3	The Stream Function and Important Relations	104
11.4	Basic Equations for Irrotational, Incompressible and Two-Dimensional Flows	105
11.5	Polar Coordinates	106
11.6	Simple Flows	107
11.7	Superposition of Simple Flows	112

12 FLOW IN OPEN CHANNELS 116

12.1	Introduction	116
12.2	Steady Uniform Flow	116
12.3	Surface Waves	119
12.4	Specific Energy	119
12.5	Critical Depth	120
12.6	The Hydraulic Jump	121
12.7	Non-Uniform Flow	122
12.8	The Occurence of Critical Conditions	124

13 TURBOMACHINERY 126

13.1	Introduction	126
13.2	Similarity Relations for Turbomachines	126
13.3	The Basic Laws	129
13.4	Turbines	130
13.5	Cavitation	133
13.6	Radial-Flow Pumps and Blowers	135
13.7	Cavitation in Centrifugal Pumps	137

14 FLOW MEASUREMENT 139

14.1	Introduction	139
14.2	Definition of Coefficients	139
14.3	Flowmeters for Incompressible Internal Flows	140
14.4	Flowmeters for Compressible Internal Flows	143
14.5	Weirs	143

THERMODYNAMICS

SECTION D

Chapter No.		Page No.
1	**BASIC CONCEPTS**	D-1
1.1	Thermodynamic Systems	1
1.2	Properties of Systems	2
1.3	Processes	2
1.4	Thermodynamic Equilibrium	3
1.5	Mutual Equilibrium	3
1.6	Zeroth Law of Thermodynamics	4
1.7	Pressure	4
1.7.1	Units of Pressure	5
1.8	Other Properties	5
1.9	Dimensions and Units	6
2	**PROPERTIES AND STATES OF A PURE SUBSTANCE**	7
2.1	The Pure Substance	7
2.2	P-V-T Behavior of a Pure Substance	7
2.3	P-T Program	8
2.4	T-V Program	9
2.5	P-V Diagram	10
2.6	T-S Diagram	10
2.6.1	Latent Heat of Vaporization	10

2.6.2	Latent Heat of Fusion	10
2.7	H-S Diagram	11
2.8	P-H Diagram	11
2.9	Tables of Thermodynamic Properties	11
2.9.1	Quality	12
2.9.2	Moisture	12

3   WORK AND HEAT   14

3.1	Definition of Work	14
3.2	Thermodynamic Work	14
3.3	Work Done on a Simple Compressible System	15
3.4	Other Modes of Thermodynamic Work	16
3.5	Heat	17
3.6	Comparison of Heat and Work	17
3.7	Units of Work and Heat	18

4   ENERGY AND THE FIRST LAW OF THERMODYNAMICS   19

4.1	Energy of a System	20
4.2	The First Law of Thermodymanics	20
4.3	The First Law for a System Undergoing a Cycle	20
4.4	The First Law for a Change in the State of a System	21
4.5	The Thermodynamic Property, Enthalpy	22
4.6	The Constant— Volume and Constant— Pressure Specific Heats	22
4.7	Control Volume	23
4.8	Law of Conservation of Mass	23
4.9	Conservation of Mass for an Open System	23
4.10	The First Law of Thermodynamics for a Control Volume	24

4.11	The Steady - State Steady - Flow Process (SSSF)	24
4.12	The Joule - Thompson Coefficient and the Trottling Process	25
4.13	The Uniform - State Uniform Flow Process (USUF)	26

5 ENTROPY AND THE SECOND LAW OF THERMODYNAMICS 27

5.1	The Heat Engine	27
5.2	Heat Engine Efficiency	27
5.3	The Second Law	30
5.3.1	Kelvin - Plank Statement	30
5.3.2	Clausius Statement	30
5.4	Perpetual Motion Machines	30
5.4.1	Perpetual Motion Machine of the First Kind (PMM1)	30
5.4.2	Perpetual Motion Machine of the Second Kind (PMM2)	30
5.5	The Carnot Cycle	31
5.6	Thermodynamic Temperature Scales	32
5.7	Entropy	33
5.8	Inequality of Clausius	34
5.9	Principle of the Increase of Entropy	34
5.10	Entropy Change of a System During an Irreversible Process	35
5.11	Two Thermodynamic Relations	35
5.12	The Second Law of Thermodynamics For a Control Volume	36
5.12.1	The Steady - State Steady - Flow Process	36
5.12.2	The Uniform - State Uniform - Flow Process	36
5.12.3	The Reversible Steady - State, Steady Flow Process	37

6	**AVAILABILITY FUNCTIONS**	38
6.1	Reversible Work	38
6.2	Availability	40

7	**GASES**	42
7.1	Gas Constant	42
7.1.1	Universal Gas Constant	42
7.2	Bolzmann's Constant	43
7.3	Compressibility	43
7.4	Semiperfect Gas	44
7.4.1	Properties of a Semiperfect Gas	44
7.5	Perfect Gas (Ideal Gas)	45
7.5.1	Properties of a Perfect Gas	45
7.5.2	Two Diagrams for a Perfect Gas	47
7.6	Real Gas	49
7.6.1	Generalized Enthalpy Chart for Real Gases	50
7.6.2	Generalized Enthropy Chart for Real Gases	52
7.6.3	Residual Volume	53

8	**THERMODYNAMIC RELATIONS**	54
8.1	The Maxwell Relations	54
8.2	Clapeyron Equation	54
8.3	Other Thermodynamic Relations	55
8.4	Equations of State	56
8.5	Law of Corresponding States	59
8.6	Volume Expansion and Isothermal and Adiabatic Compressibility	59

9	**POWER AND REFRIGERATION CYCLES**	61

9.1	Vapor Power Cycles	61
9.1.1	Simple Rankine Cycle	61
9.1.2	Rankine Cycle with Super Heater	62
9.1.3	The Reheat Cycle	63
9.1.4	The Regenerative Cycle with Open Feedwater Heater	64
9.2	Deviation of Actual Cycles from Ideal Cycles	65
9.2.1	Piping Losses	65
9.2.2	Turbine Losses	65
9.2.3	Pump Losses	66
9.2.4	Condenser	66
9.3	Vapor Refrigeration Cycles	66
9.4	Air - Standard Power Cycles	67
9.4.1	The Carnot Cycle	67
9.4.2	The Otto Cycle	68
9.4.3	The Diesel Cycle	69
9.4.4	The Dual Cycle	70
9.4.5	Stirling Cycle	71
9.4.6	Ericsson Cycle	71
9.4.7	The Brayton Cycle	72
9.4.8	Brayton Cycle With Regenerator	73
9.4.9	Gas Refrigeration Cycle	74

10 MIXTURES AND SOLUTIONS 75

10.1	Mole Fraction	75
10.2	Mass Fraction	75
10.3	Dalton's Rule of Partial Pressure	75
10.4	Amagat - Leduc Rule of Partial Volume	76
10.5	Expressions for Perfect Gases	76
10.6	Mixtures Involving Gases and Vapor	77
10.6.1	Dew Point Temperature	78
10.7	Enthalpy and Entropy of a Gas - Vapor Mixture	79
10.8	Adiabatic Saturation Process	79
10.9	Fugacity	80
10.10	Activity	81
10.10.1	Activity Coefficient	81

11 CHEMICAL REACTIONS — 82
11.1 The Combustion Process — 82
11.2 Enthalpy of Reaction — 83
11.3 Enthalpy of Formation (h_f°) — 83
11.4 First Law Analysis of Reacting Systems — 84
11.5 Adiabatic Flame Temperature — 85
11.6 The Third Law of Thermodynamics and Absolute Entropy — 85
11.7 Second Law Analysis of Reacting Systems — 86
11.8 Gibb's Function — 87
11.9 Chemical Potential — 87

12 CHEMICAL EQUILIBRIUM — 89
12.1 Requirements for Chemical Equilibrium — 89
12.2 Equilibrium and Chemical Potential — 89
12.3 Equilibrium Constant of a Reactive Mixture of Ideal Gases — 90
12.4 Equilibrium Between Two Phases of a Pure Substance — 91
12.5 Equilibrium of a Multicomponent, Multiphase System — 91
12.6 Gibb's Phase Rule — 92

13 FLOW THROUGH NOZZLES AND BLADE PASSAGES — 93
13.1 Conservative of Mass for the Control Volume — 93
13.1.1 Special Cases — 93
13.2 Momentum Equation for the Control Volume — 93
13.3 Speed of Sound — 94
13.3.1 Mach Number — 94
13.4 Local Isentropic Stagnation Properties — 94

13.5	Critical Constants	95
13.6	Effects of Area Variation on Flow Properties In Isentropic Flow	95
13.7	Isentropic Flow of an Ideal Gas	96
13.8	Isentropic Flow in a Converging and a Converging/Diverging Nozzle	97
13.8.1	Converging Nozzle	97
13.8.2	Converging/Diverging Nozzle	98
13.9	Normal Shocks	98
13.9.1	Flow in a Converging/Diverging Nozzle	99
13.10	Nozzle and Diffuser Coefficients	100
13.11	Flow Through Blade Passages	101
13.12	Impulse and Reaction Stages for Turbines	103
13.12.1	Impulse Stage	103
13.12.2	Reaction Stage	104

HEAT TRANSFER

SECTION E

Chapter No.		Page No.
1	**INTRODUCTION**	E-1
1.1	Heat Transfer	1
1.2	Modes of Heat Transfer	1

1.2.1	Conduction	1
1.2.2	Convection	2
1.2.3	Radiation	2
1.3	Thermal Diffusivity	2
1.4	Viscosity	3
1.4.1	Dynamic Viscosity	3
1.4.2	Kinematic Viscosity	3
1.5	Units	4

2 STEADY STATE HEAT CONDUCTION IN ONE DIMENSION — 5

2.1	The General Heat Conduction Equation	5
2.1.1	Cartesian Coordinates	5
2.1.2	Cylindrical Coordinates	6
2.1.3	Spherical Coordinates	7
2.2	One Dimensional Governing Equations	7
2.3	The Plane Wall with Specified Surface Temperatures	7
2.4	The Multilayer Wall with Specified Surface Temperatures	8
2.5	The Single Cylinder with Specified Surface Temperatures	9
2.6	The Multilayer Cylinder with Specified Surface Temperatures	10
2.7	The Plane Wall Fluid of Bounded by Specified Temperatures	11
2.8	Cylindrical Surfaces by Fluids of Specified Temperatures	12
2.9	The Plane Wall with Internal Heat Generation	13
2.10	Cylinder with Internal Heat Generation	14
2.11	Fins	15
2.11.1	Uniform Cross-Section	16
2.11.2	Nonuniform Cross-Section	17
2.12	Fin Efficiency	19

3 STEADY STATE HEAT CONDUCTION IN TWO DIMENSIONS — 21

3.1	Introduction	21
3.2	Analytical Method	22
3.2.1	Steady State Conduction in a Rectangular Plate	22

T-27

3.2.2	Steady State Conduction in a Circular Cylinder	25
3.2.3	Two-Dimensional Heat Conduction in Polar Coordinates	26
3.3	Graphical Method	27
3.4	Numerical Method	28

4 UNSTEADY STATE HEAT CONDUCTION 30

4.1	Introduction	30
4.2	Lumped System	30
4.3	One Dimensional Transient Heat Conduction	32
4.3.1	Sudden Temperature Change at the Surface of a Semi-Infinite Body	32
4.3.2	Semi-Infinite Body with Convection Boundary Conditions	34
4.3.3	Charts for Solving Problems with Convection Boundary Conditions	35
4.4	Multiple-Dimensions Systems	40
4.5	Numerical Solution	42

5 FLUID MECHANICS REVIEW 45

5.1	Introduction	45
5.2	Viscous and Invicid Flows	45
5.3	Laminar and Turbulent Flows	46
5.4	Basic Equations	46
5.4.1	Continuity Equation	47
5.4.2	Momentum Equation	47
5.4.3	Energy Equation	47
5.5	Navier-Stokes Equations	47
5.6	Euler Equations	48
5.7	Important Physical Parameters	48

6 FORCED CONVECTION HEAT TRANSFER 50

6.1	Laminar Boundary Layer on a Flat Plate	50
6.2	Solution to the Velocity Boundary Layer	52
6.2.1	Blasius Solution	52
6.2.2	Integral Solution	53
6.3	Solution to the Thermal Boundary Layer	55
6.3.1	Pohlhausen Solution	55
6.3.2	Integral Solution	57

6.4	Laminar Flow Inside Pipes	59
6.4.1	Circular Pipes	59
6.4.2	Noncircular Pipes	60
6.5	Laminar Flow Heat Transfer in Circular Pipes	62
6.5.1	Governing Equations	62
6.5.2	Definitions	63
6.5.3	Dimensionless Heat Transfer Coefficient	63
6.6	Laminar Flow Heat Transfer in Noncircular Pipes	64
6.7	Turbulent Boundary Layer on a Flat Plate	65
6.8	Solution to the Velocity Boundary Layer (Turbulent)	67
6.9	Turbulent Boundary Layer with Heat Transfer (Flat Plate)	68
6.10	Turbulent Flow Inside Pipes	70
6.10.1	Circular Pipes	70
6.10.2	Noncircular Pipes	71
6.11	Flow Inside Pipes with Heat Transfer	72

7 CORRELATIONS FOR FORCED CONVECTION 73

7.1	Definitions	73
7.2	Flow Over a Flat Plate	74
7.2.1	Laminar Flow	74
7.2.2	Turbulent Flow	75
7.3	Flow Inside Cylindrical Pipes and Tubes	76
7.3.1	Laminar Flow	76
7.3.2	Turbulent Flow	77
7.4	Flow Normal to a Circular Cylinder	78
7.5	Flow Normal to Noncircular Cylinders	79
7.6	Flow Over Banks of Tubes	80
7.7	Liquid-Metal Heat Transfer	82
7.7.1	Turbulent Flow on a Flat Plate	82
7.7.2	Turbulent Flow Inside Tubes	82
7.8	Heat Transfer in High Speed Flow	83

8 FREE CONVECTION 84

8.1	Introduction	84
8.2	Boundary Layer Governing Equations	84
8.3	Solution of the Problem of Laminar Free Convection Past a Vertical Surface	85

8.3.1	The Similarity Solution	86
8.3.2	The Integral Solution	88
8.4	Turbulent Flow on a Vertical Flat Surface	89
8.5	Correlations for Free Convections	90
8.5.1	Vertical Flat Surfaces and Vertical Cylinders	90
8.5.2	Horizontal Plates	91
8.5.3	Horizontal Cylinders	91
8.5.4	Spheres	92
8.5.5	Rectangular Blocks	92
8.6	Free Convection in Confined Fluids	92
8.6.1	Between Horizontal Isothermal Plates	92
8.6.2	Between Vertical Plates	93
8.7	Mixed, Free and Forced Convection	94

9 HEAT EXCHANGERS 97

9.1	Types of Heat Exchangers	97
9.1.1	Double-Pipe Arrangement	97
9.1.2	Shell and Tube Arrangement	97
9.1.3	Cross-Flow Heat Exchangers	98
9.2	The Overall Heat Transfer Coefficient	99
9.3	Fouling Factor	100
9.4	The Log-Mean Temperature Difference(LMTD)	100
9.5	Effectiveness-NTU Method	103
9.6	Transfer Units in Heat Exchangers	106

10 RADIATION HEAT TRANSFER 109

10.1	Spectrum of Electromagnetic Radiation	109
10.2	Radiation Properties	110
10.3	Concepts of Thermal Radiation	111
10.3.1	Black Surface	111
10.3.2	Total Hemispheric Emissivity	112
10.3.3	Gray Surface	115
10.4	Intensity of Radiation	115
10.5	Radiation Shape Factor	116
10.6	Radiant Exchange Between Non-Black Surfaces	119
10.6.1	Radiation Between Two Gray Surfaces	121
10.6.2	Infinite Parallel Planes	121
10.6.3	Two Concentric Cylinders	121
10.7	Radiation Shields	122
10.8	Gas Radiation	123
10.9	Heat Exchange Between Gas Volume and Black Enclosure	124

10.10	Heat Exchange Between Gas Volume and Gray Enclosure	125
10.11	Combined Heat Transfer Coefficient: Radiation and Convection	125

11	**BOILING AND CONDENSATION**	**127**
11.1	Boiling Phenomenon	127
11.1.1	Pool Boiling	127
11.1.2	Forced Convection Boiling	127
11.2	Heat Transfer Correlations	129
11.2.1	Regime I: Free Convection	129
11.2.2	Regime II: Nucleate Boiling	129
11.2.3	Regimes III and IV: Film Boiling	131
11.3	Condensation	133
11.4	Film Condensation in Vertical Plates	133

LAPLACE TRANSFORMS

SECTION F

Chapter No.		Page No.
1	**THE LAPLACE TRANSFORM**	**F-1**
1.1	Integral Transforms	1
1.2	Definition of Laplace Transform	2
1.3	Notation	2
1.4	Laplace Transformation of Elementary Functions	3

1.5	Sectionally or Piecewise Continuous Functions	4
1.6	Functions of Exponential Order	4
1.6.1	Sufficient Condition for Exponential Order Functions	5
1.7	Existence of Laplace Transform	6
1.7.1	Behavior of Laplace Transforms at Infinity	6
1.8	Some Important Properties of Laplace Transforms	7
1.8.1	Linearity Property	7
1.8.2	Changes of Scale Property	8
1.8.3	First Shift Property	8
1.8.4	Second Shift Property	9
1.8.5	Laplace Transform of Derivative	10
1.8.6	Derivative of Laplace Transforms	12
1.8.7	Periodic Functions	13
2	**INVERSE LAPLACE TRANSFORM**	**15**
2.1	Definition of Inverse Laplace Transform	15
2.2	Uniqueness of Inverse Laplace Transform	16
2.2.1	Null Functions	16
2.3	Some Inverse Laplace Transforms	17
2.4	Some Properties of Inverse Laplace Transforms	18
2.4.1	Linearity Property	18
2.4.2	First Translation or Shifting Property	18
2.4.3	Second Translation or Shifting Property	19
2.4.4	Change of Scale Property	20
2.4.5	Inverse Laplace Transform of Derivatives	20
2.4.6	Inverse Laplace Transform of Integrals	21
2.5	The Convolution Property	22
2.5.1	Definition	22
2.5.2	Properties of Convolution	23
2.5.3	Inverse Laplace Transform of the Convolution	23
3	**SOME SPECIAL FUNCTIONS**	**25**
3.1	The Gamma Function	25
3.1.1	Properties of the Gamma Function	26

3.2	Bessel Functions	27
3.2.1	Modified Bessel Function	28
3.2.2	Some Properties of Bessel Functions	28
3.3	The Error Function	29
3.3.1	Some Properties of erf(t)	30
3.4	The Complementary Error Function	30
3.4.1	A Property of erfc(t)	31
3.5	The Sine and Cosine Integrals	31
3.6	The Exponential Integral	31
3.7	The Unit Step Function or the Heaviside Function	31
3.7.1	Some Properties of the Unit Step Function	32
3.8	The Unit Impulse Function	35
3.8.1	Some Properties of the Unit Impulse Function	35
3.9	The Beta Function	36
3.10	More Properties of Inverse Laplace Transforms	37
3.10.1	Multiplication by s^n	37
3.10.2	Division by S	38

4 APPLICATION OF ORDINARY LINEAR DIFFERENTIAL EQUATIONS 42

4.1	Ordinary Differential Equation with Constant Coefficients	42
4.2	Ordinary Differential Equations with Variable Coefficients	46
4.3	Systems of Linear Differential Equations	48
4.4	The Vibration of Spring	51
4.4.1	Damped Vibrations	53
4.4.2	Undamped Vibration	54
4.4.3	Free Vibration	54
4.5	Resonance	56
4.6	The Simple Pendulum	58
4.7	Electric Circuits	60
4.8	Beams	63

5	**METHODS OF FINDING LAPLACE TRANS-FORMS AND INVERSE TRANSFORMS**	**65**
5.1	Initial Value Theorem	65
5.2	Final Value Theorem	65
5.3	Methods of Finding Laplace Transforms	65
5.3.1	Direct Method	66
5.3.2	Power Series Method	66
5.3.3	Method of Differential Equations	68
5.3.4	Method of Differentiation with Respect to a Parameter	70
5.3.5	Miscellaneous Methods — Multiplication by T^H Property and Division by T Property	71
5.4	Methods of Finding Inverse Transforms	72
5.4.1	Partial Fraction Method	72
5.4.2	The Heaviside Expansion Formula	76
6	**FOURIER TRANSFORMS**	**77**
6.1	Fourier Series	77
6.2	Fourier Sine and Cosine Series	79
6.2.1	Odd Extension	80
6.2.2	Even Extension	80
6.3	Piecewise Smooth Functions	81
6.4	Periodic Extensions	81
6.5	Theorem 2 (Convergence Theorem)	81
6.5.1	First Criterion for Uniform Convergence	82
6.5.2	Second Criterion for Uniform Convergence	83
6.6	General Criterion for Differentiation	83
6.7	Parseval's Theorem and Mean Square Error	84
6.8	Complex Form of Fourier Series	85
6.9	Parseval's Identity	85
6.10	Fourier Integral Transforms	86
6.10.1	Parseval's Theorem	86
6.10.2	Inversion Theorem for Fourier Transforms	87
6.11	Fourier Cosine Formulas	88

6.12	Fourier Sine Formulas	88
6.13	The Convolution Theorem	89
6.14	Relationship of Fourier and Laplace Transforms	89

7 APPLICATIONS OF LAPLACE TRANSFORMS TO INTEGRAL AND DIFFERENCE EQUATIONS — 91

7.1	Integral Equations	91
7.1.1	Fredholm Integral Equation	91
7.1.2	Volterra Integral Equation	91
7.1.3	Integral Equation of Convolution Type	92
7.2	Integro-Differential Equations	93
7.3	Difference Equations	95
7.4	Differential-Difference Equations	96

8 APPLICATIONS TO BOUNDARY-VALUE PROBLEMS — 98

8.1	Functions of Two Variables	98
8.2	Partial Differential Equation	98
8.3	Some Important Partial Differential Equations	99
8.4	Classification of Second-Order Partial Differential Equations	99
8.5	Boundary Conditions	100
8.5.1	Dirichlet Problem	100
8.5.2	Cauchy Problem	101
8.6	Solution of Boundary-Value Problems by Laplace Transforms	101
8.6.1	One-Dimensional Boundary Value Problems	101
8.6.2	Two-Dimensional Boundary Value Problem	103

9 TABLES — 104

9.1	Table of General Properties	104
9.2	Table of More Common Laplace Transforms	107
9.3	Table of Special Functions	119

AUTOMATIC CONTROL SYSTEMS / ROBOTICS

SECTION G

Chapter No.		Page No.
1	**SYSTEM MODELING: MATHEMATICAL APPROACH**	G-1
1.1	Electric Circuits and Components	1
1.2	Mechanical Translation Systems	3
1.3	Mechanical and Electrical Analogs	5
1.4	Mechanical Rotational Systems	6
1.5	Thermal Systems	7
1.6	Positive-Displacement Rotational Hydraulic Transmission	9
1.7	D-C and A-C Servomotor	10
1.8	Lagrange's Equation	14
2	**SOLUTIONS OF DIFFERENTIAL EQUATIONS: SYSTEM'S RESPONSE**	16
2.1	Standardized Inputs	16
2.2	Steady State Response	16
2.3	Transient Response	20
2.4	First and Second-Order System	24
2.5	Time-Response Specifications	28

T-36

3 APPLICATIONS OF LAPLACE TRANSFORM 29

3.1 Definition of Laplace Transform 29
3.2 Application of Laplace Transform to Differential Equations 32
3.3 Inverse Transform 33
3.4 Frequency Response from the Pole-Zero Diagram 38
3.5 Routh's Stability Criterion 39
3.6 Impulse Function: Laplace-Transform and its Response 41

4 MATRIX ALGEBRA AND Z-TRANSFORM 44

4.1 Fundamentals of Matrix Algebra 44
4.2 Z-Transforms 47

5 SYSTEM'S REPRESENTATION: BLOCK DIAGRAM, TRANSFER FUNCTIONS, AND SIGNAL FLOW GRAPHS 51

5.1 Block Diagram and Transfer Function 51
5.2 Transfer Functions of the Compensating Networks 53
5.3 Signal Flow Graphs 55

6 SERVO CHARACTERISTICS: TIME-DOMAIN ANALYSIS 58

6.1 Time-Domain Analysis Using Typical Test Signals 58
6.2 Types of Feedback Systems 59

6.3	General Approach to Evaluation of Error	60
6.4	Analysis of Systems: Unity Feedback	63

7 ROOT LOCUS 67

7.1	Roots of Characteristic Equations	67
7.2	Important Properties of the Root Loci	69
7.3	Frequency Response	78

8 SPECIAL POLE-ZERO TOPICS: DOMINANT POLES AND THE PARTITION METHOD 80

8.1	Transient Response: Dominant Complex Poles	80
8.2	Pole-Zero Diagram and Frequency and Time Response	85
8.3	Factoring of Polynomials Using Root-Locus	87

9 SYSTEM ANALYSIS IN THE FREQUENCY-DOMAIN: BODE AND POLAR PLOTS 89

9.1	The Frequency Response and the Time Response: Relationship	89
9.2	Frequency Response Plots	90
9.3	Polar Plot	92
9.4	Logarithmic Plots	92
9.5	Log-Magnitude and Phase Diagram: Basic Approach	94
9.6	Relation Between System Type, Gain and Log-Magnitude Curves	97
9.7	Direct Polar Plots	99
9.8	Inverse Polar Plots	101
9.9	Dead Time	105

10 NYQUIST STABILITY CRITERION 107

10.1	Determining and Enhancing the System's Stability	107
10.2	Inverse Polar Plots: Application of Nyquist's Criterion	111
10.3	Phase Margin and Gain Margin: Definitions	113
10.4	System Stability	114
10.5	Stability	115
10.6	Effect of Adding a Pole or a Zero: Effect on the Polar Plots	117

11 PERFORMANCE EVALUATION OF A FEEDBACK CONTROL SYSTEM IN THE FREQUENCY-DOMAIN 118

11.1	Performance Evaluation Using Direct Polar Plot	118
11.2	Resonant Frequency and the Maximum Magnitude of C/R of a Second Order System	120
11.3	Plotting Maximum Magnitude and Resonant Frequency on the Complex Plane	122
11.4	Magnitude and Angle Curves in the Inverse Polar Plane	128
11.5	Gain Adjustment for a Desired Maximum Magnitude Using a Direct Polar Plot	131
11.6	Nichol's Chart	133

12 SYSTEM STABILIZATION: USE OF COMPENSATING NETWORKS AND THE ROOT LOCUS 136

T-39

12.1	Function of a Compensating Network	136
12.2	Types of Compensations	137
12.2.1	PI (Integral and Proportional) Control	137
12.2.2	Lag Compensator	139
12.2.3	Porportional Plus Derivative (PD) Compensator	140
12.2.4	Lead Compensation	141
12.2.5	Lead-Lag Compensation	142
12.2.6	Comparison of Compensators	145

13 FREQUENCY-RESPONSE PLOTS OF CASCADE COMPENSATED SYSTEMS 146

13.1	Selecting a Proper Compensator	146
13.2	Analysis of a Lag Network	149
13.3	Analysis of a Lead Network	151
13.4	Analysis of a Lag-Lead Compensator	153

14 FEEDBACK COMPENSATION: PARALLEL COMPENSATION 155

14.1	Parallel Compensation: Pros and Cons of Selecting a Feedback Compensator	155
14.2	Effects of the Different Types of Feedback on the System's Time Response	157
14.3	Application of Log-Magnitude Curve: for Feedback Compensation	160

15 SYSTEM SIMULATIONS: USE OF ANALOG COMPUTERS 163

| 15.1 | Analog Computer: Basic Components | 163 |

15.2   Simulations Using Analog Computer   166
15.3   Application of Analog Computers for
       System Tuning   172
15.4   Setting of a Control System:
       Controller Setting   174

MATHEMATICS FOR ENGINEERS

SECTION H

Chapter No.		Page No.
1	**VECTORS, MATRICES, AND EQUATION SYSTEMS**	H-1
1.1	Vectors in Three Dimensions	1
1.2	Vectors in N-Dimensions	7
1.3	Matrices and Systems of Equations	9
1.4	Complex Vectors and Matrices	16
2	**ESSENTIALS OF CALCULUS**	18
2.1	Calculus of One Variable	18
2.2	Vector Functions of One Variable	23
2.3	Functions of Two or More Independent Variables	25
2.4	Vector Fields and Divergence	27
2.5	The Double Integral	28

2.6	Line Integrals	32
2.7	Green's Theorem	34
2.8	Surfaces in 3 Dimensions	35
2.9	Volume Integrals	38

3	**COMPLEX FUNCTIONS**	**40**
3.1	Basic Concepts	40
3.2	Sets in the Complex Plane	42
3.3	Functions of a Complex Variable	43
3.4	Limits, Continuity, and Derivatives	45
3.5	Harmonic Functions	46
3.6	The Elementary Complex Functions	47
3.7	Integrals of Complex Functions	51
3.8	Number Sequences and Series	55
3.9	Function Sequences and Series	56
3.10	Poles and Residues	57
3.11	Elementary Mappings and the Mobius Transformation	58
3.12	Conformal Mappings and Harmonic Functions	60
3.13	Zeros and Singular Points of Analytic Functions	62
3.14	Riemann Mapping Theorem	63

4	**ORDINARY DIFFERENTIAL EQUATIONS**	**64**
4.1	Ordinary Differential Equations of First Order	64
4.2	Linear Ordinary Differential Equation	66
4.3	Systems of First Order ODE's	69
4.4	Methods for Solving ODE's	71
4.5	Boundary Value Problems	72

5	**FOURIER ANALYSIS AND INTEGRAL TRANSFORMS**	**74**
5.1	Basic Ideas	74
5.2	Fourier Series	76
5.3	Fourier Series and Vector Space Concepts	79
5.4	Fourier Transforms	82
5.5	Special Functions	85

6	**PARTIAL DIFFERENTIAL EQUATIONS (PDE'S)**	**90**
6.1	Fundamental Ideas	90
6.2	The Laplace Equation	96
6.3	The Heat Equation	100
6.4	The First Order Wave Equation	103
6.5	The Wave Equation	106
7	**CALCULUS OF VARIATIONS**	**110**
7.1	Basic Theory of Maxima and Minima	111
7.2	The Simplest Problem of Variational Calculus	115
7.3	Some Classical Problems	119
7.4	Control	120
7.5	Dynamic Programming	122
7.6	Linear Programming	123
8	**NUMERICAL METHODS**	**125**
8.1	Solutions of Equations	125
8.2	Function Approximation	127
8.3	Numerical Integration	130
8.4	Numerical Linear Algebra	132
8.5	Solving Ordinary Differential Equations	138
8.6	Solving Partial Differential Equations	139
9	**STATISTICS AND PROBABILITY**	**145**
9.1	On Statistics and Probability	145

PHYSICS

SECTION I

Chapter No.		Page No.
1	**VECTORS AND SCALARS**	I-1
1.1	Basic Definitions of Vectors and Scalars	1
1.2	Addition of Vectors $(\bar{a}+\bar{b})$— Geometric Methods	1
1.3	Subtraction of Vectors	2
1.4	The Components of a Vector	3
1.5	The Unit Vector	4
1.6	Adding Vectors Analytically	5
1.7	Multiplication of Vectors	5
2	**ONE DIMENSIONAL MOTION**	7
2.1	Basic Definitions	7
2.2	Motion With Constant Acceleration	8
2.3	Freely Falling Bodies	9
3	**PLANE MOTION**	10
3.1	Displacement, Velocity and Acceleration in General Planar Motion	10

3.2	Motion in a Plane With Constant Acceleration	11
3.3	Projectile Motion	13
3.4	Uniform Circular Motion	16

4 DYNAMICS OF A PARTICLE — 17

4.1	Newton's Second Law	17
4.2	Newton's Third Law	18
4.3	Mass and Weight	18
4.4	Friction	19
4.5	The Dynamics of Uniform Circular Motion—Centripetal Force	20

5 WORK AND ENERGY — 21

5.1	Work Done by a Constant Force	21
5.2	Work Done by a Varying Force	22
5.3	Work Done by a Varying Force-One Dimensional Case	23
5.4	Power	24
5.5	Kinetic Energy	25
5.6	Work-Energy Theorem	25

6 CONSERVATION OF ENERGY — 26

6.1	Conservative Forces	26
6.2	Potential Energy	27
6.3	Conservation of Mechanical Energy	27
6.4	One-Dimensional Conservative Systems	28
6.5	Non-Conservative Forces	29
6.6	The Conservation of Energy	29
6.7	The Relationship Between Mass and Energy	30

7 THE DYNAMICS OF SYSTEMS OF PARTICLES 31

7.1	Center of Mass of a System of Particles	31
7.1.1	Center of Mass in Vector Notation	32
7.2	Motion of the Center of Mass	32
7.3	Work-Energy Theorem for a System of Particles	33
7.4	Linear Momentum of a Particle	33
7.4.1	Newton's Second Law	33
7.5	Linear Momentum of a System of Particles	34
7.6	Conservation of Linear Momentum	34
7.7	Elastic and Inelastic Collisions in One Dimension	35
7.8	Collisions in Two and Three Dimensions	35

8 ROTATIONAL KINEMATICS 37

8.1	Rotational Motion— The Variables	37
8.2	Rotational Motion With Constant Angular Acceleration	38
8.3	Relation Between Linear and Angular Kinematics for a Particle in Circular Motion	39

9 ROTATIONAL DYNAMICS 40

9.1	Torque	40
9.2	Angular Momentum	41
9.3	Kinetic Energy of Rotation and Rotational Inertia	42
9.4	Some Rotational Inertias	43
9.5	Rotational Dynamics of a Rigid Body	43
9.6	Rolling Bodies	44
9.7	Conservation of Angular Momentum	45

10 HARMONIC MOTION 46

10.1 Oscillations—Simple Harmonic Motion (SHM) 46
10.2 Energy Considerations of SHM 49
10.3 Pendulums 50
10.4 Simple Harmonic Motion and Uniform Circular Motion 51

11 SOUND WAVES 53

11.1 Speed of Sound 53
11.2 Intensity of Sound 54
11.3 Allowed Frequencies of a String, Fixed at Both Ends, or an Organ Pipe, Open at Both Ends 54
11.4 The Beat Equation for Particle Displacement 55
11.5 The Beat Frequency 55
11.6 The Doppler Effect 55
11.7 Half Angle of a Shock Wave 56

12 GRAVITATION 57

12.1 Newton's Law of Gravitation 57
12.2 The Motion of Planets and Satellites—Kepler's Laws of Planetary Motion 57
12.3 Angular Momentum of Orbiting Planets 59
12.4 Gravitational Potential Energy, U 60
12.5 Potential Energy for a System of Particles 60
12.6 Kinetic Energy 61
12.7 Total Energy 61

13 EQUILIBRIUM OF RIGID BODIES 62

13.1 Rigid Bodies in Static Equilibrium 62
13.2 Free Body Diagrams 62

14 FLUID STATICS 64
14.1 Pressure in a Static Fluid 64
14.2 Archimedes' Principle (Buoyancy) 65
14.3 Measurement of Pressures 65

15 FLUID DYNAMICS 67
15.1 Equation of Continuity 67
15.2 Bernoulli's Equation 67
15.3 Viscosity 68
15.4 Reynolds Number: N_R 69

16 TEMPERATURE 70
16.1 The Kelvin, Celsius and Fahrenheit Scales 70
16.2 Coefficients of Thermal Expansion 71

17 HEAT AND THE FIRST LAW OF THERMODYNAMICS 73
17.1 Quantity of Heat and Specific Heat 73
17.2 Heat Conduction 74
17.3 Thermal Resistance, R 74
17.4 Convection 75
17.5 Thermal Radiation 75
17.6 Heat and Work 76
17.7 The First Law of Thermodynamics 77
17.8 Applications of the First Law of Thermodynamics 77

18 KINETIC THEORY OF GASES 78
18.1 Ideal Gas Equation 78
18.2 Pressure and Molecular Speed of an Ideal Gas 79

18.3	Kinetic Energy of an Ideal Gas	79
18.4	Internal Energy of an Ideal Monatomic Gas	80
18.5	Specific Heats of an Ideal Gas	80
18.6	Mean Free Path	81

19 ENTROPY AND THE SECOND LAW OF THERMODYNAMICS — 82

19.1	Efficiency of a Heat Engine	82
19.2	Entropy, S	83

Handbook of Mechanical Engineering

SECTION A
Mechanics: Statics and Dynamics

CHAPTER 1

INTRODUCTION

1.1 FUNDAMENTAL CONCEPTS

Mechanics - The science concerned with the conditions of rest or motion of bodies under the action of forces.

Branches of Mechanics

A) Rigid Body Mechanics

    1) Statics - Equilibrium of bodies under the action of forces

    2) Dynamics - Accelerated motion of bodies

        a) Kinematics - Study of geometry of motion

        b) Kinetics - Relate forces to motion

B) Deformable Body Mechanics

C) Fluid Mechanics

D) Relativistic Mechanics

E) Quantum Mechanics

    1) Particle - A dimensionless object that possesses a definite mass.

    2) Rigid Body - Combination of a number of particles but assumed to be not deformable

Fundamental Principles

A) Parallelogram Law for Vector Addition

B) Principle of Transmissibility

C) Newton's Three Fundamental Laws of Motion

D) Newton's Law of Gravitation

1.2 SYSTEMS OF UNITS

Type of System	Name	Length	Mass	Force	Time
Absolute	International System (SI)	Meter (m)	Kilogram (kg)	Newton $\left(N, \frac{kg\text{-}m}{sec^2}\right)$	Second (sec)
Absolute	C.G.S.	Centimeter (cm)	Gram (g)	Dyne $\left(\frac{g\text{-}cm}{sec^2}\right)$	Second (sec)
Gravitational	U.S. Customary units (F.P.S.)	Foot (ft)	Slug $\left(\frac{lb\text{-}sec^2}{ft}\right)$	Pound (lb)	Second (sec)

Absolute system of units are independent of the location where measurements are made.

1.3 NUMERICAL ACCURACY

In engineering problems, the data are seldom known with accuracy greater than 0.2 percent.

General Rule for Expressing a Solution

A) Use four figures to record numbers beginning with a single digit before the decimal point.

B) Use three figures in all other cases.

Examples:

Actual Number	Recorded Number
43.582	43.6
1.37	1.370

A-2

CHAPTER 2

VECTORS

2.1 SCALARS AND VECTORS

Scalar - A quantity possessing magnitude only.

Vector - A quantity possessing magnitude, direction, and sense.
 Common notations of vector A: $\vec{A}, \bar{A}, \hat{A}$ or $\underline{A}$.
  Note: When superscripts or subscripts are omitted, boldface notation is used. The magnitude of a vector A is represented as $|\vec{A}|$.

Graphical Representation

A) Vectors are represented by arrows; the length of the arrow indicates its magnitude. Direction is indicated by its angular orientation.

B) The line of action of a vector is defined as an imaginary straight line of indefinite length passing through the vector.

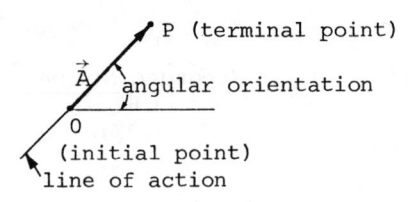

```
              P (terminal point)
           ↗
         $\vec{A}$  angular orientation
         ╱_____
        0
       (initial point)
       line of action
```

C) Two vectors $\vec{A}$ and $\vec{B}$ are equal (e.g., $\vec{A} = \vec{B}$), if they have the same magnitude, direction and sense. They may have different lines of action.

A-3

D) Vectors $\vec{A}$ and $\vec{B}$ are equivalent if $\vec{A} = \vec{B}$ and if they create the same effect(s).

Note: Equal vectors are not necessarily equivalent.

E) The negative vector is defined as the vector having the same magnitude as its positive counterpart but in opposite direction.

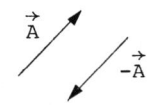

Types of Vectors

A) Fixed vector - Acts at a particular point in space

B) Transmissible or sliding vector - may be applied anywhere along its line of action.

C) Free vector - Can be applied at any point in space, but they should maintain the same magnitude, direction and sense. e.g., Moment vector of a couple.

2.2 VECTOR ADDITION

Graphical Method

Parallelogram Law:

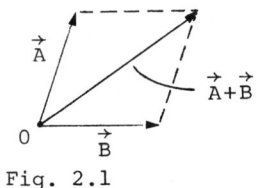

Fig. 2.1

Polygon Rule: (Tip to Tail)

A-4

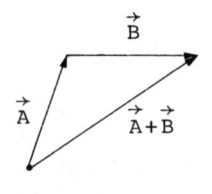

Fig. 2.2

Rectangular Component Method

$$\vec{A} + \vec{B} = [A_x, A_y, A_z] + [B_x, B_y, B_z]$$
$$= [A_x + B_x, A_y + B_y, A_z + B_z] \quad (2.1)$$

$$\vec{A} - \vec{B} = \vec{A} + (-\vec{B}) = [A_x - B_x, A_y - B_y, A_z - B_z] \quad (2.2)$$

Null vector - $\vec{0} = [0,0,0]$ zero magnitude and direction undefined

Commutative Law: $\vec{A} + \vec{B} = \vec{B} + \vec{A}$
Associative Law: $\vec{A} + (\vec{B} + \vec{C}) = (\vec{A} + \vec{B}) + \vec{C}$
Distributive Law: $c(\vec{A} + \vec{B}) = c\vec{A} + c\vec{B}$, where c is a scalar

Useful Trigonometry: $\dfrac{A}{\sin a} = \dfrac{B}{\sin b} = \dfrac{C}{\sin c}$

$$C = \sqrt{A^2 + B^2 - 2AB \cos c}$$

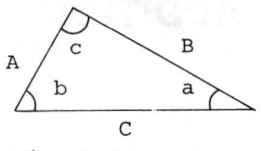

Fig. 2.3

2.3 SCALAR PRODUCT

This is also called the "dot" product. The result is a scalar.

Component Method:

A-5

$$\vec{A} \cdot \vec{B} = A_x B_x + A_y B_y + A_z B_z \qquad (2.3)$$

Graphical Interpretation:

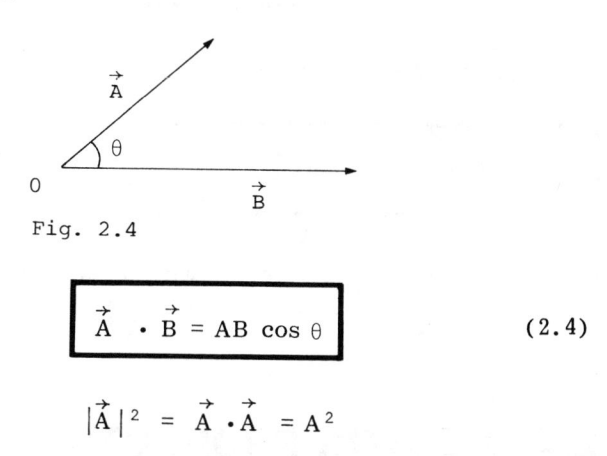

Fig. 2.4

$$\boxed{\vec{A} \cdot \vec{B} = AB \cos \theta} \qquad (2.4)$$

$$|\vec{A}|^2 = \vec{A} \cdot \vec{A} = A^2$$

Applications:

A) To find the angle formed by the two vectors in Figure 2.4:

$$\boxed{\cos \theta = \frac{A_x B_x + A_y B_y + A_z B_z}{AB}} \qquad (2.5)$$

B) Projection of a vector on a given axis:
The projection of $\vec{A}$ on $\overline{OP}$ is

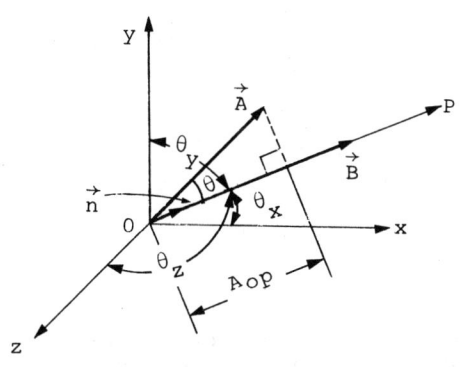

Fig. 2.5 Projection of a vector on an axis.

$$\boxed{A_{OP} = A\cos\theta = \frac{\vec{A}\cdot\vec{B}}{B}}\qquad(2.6)$$

In terms of the unit vector $\vec{n}$:

$$\boxed{A_{OP} = \vec{A}\cdot\vec{n} = A_x\cos\theta_x + A_y\cos\theta_y + A_z\cos\theta_z}\qquad(2.7)$$

where $\vec{n}$ is defined as a vector with a magnitude of one.

2.4 VECTOR PRODUCT

Multiplication of Two Vectors - Known also as the cross product.

Component Method:

$$\vec{A}\times\vec{B} = [A_yB_z - A_zB_y,\ A_zB_x - A_xB_z,\ A_xB_y - A_yB_x]$$

$$= \bar{i}\begin{vmatrix}A_y & A_z\\ B_y & B_z\end{vmatrix} + \bar{j}\begin{vmatrix}A_z & A_x\\ B_z & B_x\end{vmatrix} + \bar{k}\begin{vmatrix}A_x & A_y\\ B_x & B_y\end{vmatrix}$$

$$= \begin{vmatrix}\bar{i} & \bar{j} & \bar{k}\\ A_x & A_y & A_z\\ B_x & B_y & B_z\end{vmatrix}\qquad(2.8)$$

Note: $\bar{i}$, $\bar{j}$ and $\bar{k}$ are unit vectors in the directions x, y and z, respectively. Unit vectors have a magnitude of 1.

Graphical Interpretation:

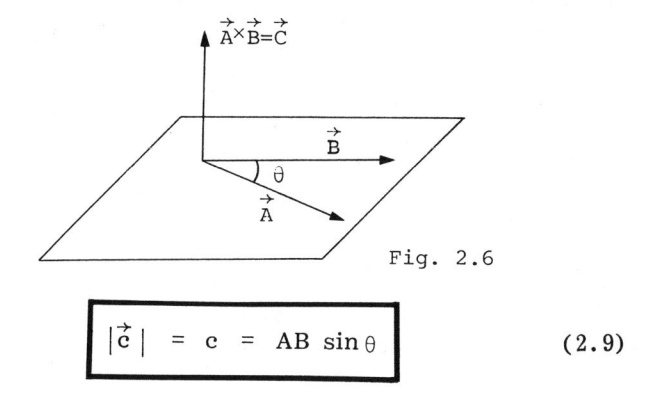

Fig. 2.6

$$|\vec{C}| = c = AB \sin\theta \qquad (2.9)$$

Properties

$$\vec{A} \times \vec{B} = -\vec{B} \times \vec{A}$$

$$\vec{A} \times (\vec{B}+\vec{C}) = (\vec{A}\times\vec{B})+(\vec{A}\times\vec{C})$$

$$c(\vec{A} \times \vec{B}) = (c\vec{A})\times B = A\times(c\vec{B}), \text{ where c is a scalar.}$$

$$|\vec{A} \times \vec{B}|^2 = A^2B^2 - (\vec{A}\cdot\vec{B})^2$$

2.5 MIXED TRIPLE PRODUCT

Triple Scalar Product:

$$\vec{A} \cdot (\vec{B} \times \vec{C})$$

The result is a scalar. It represents the volume of the parallelepiped, in absolute value.

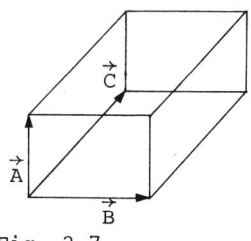

Fig. 2·7

In rectangular components:

$$\vec{A} \cdot (\vec{B} \times \vec{C}) = \begin{vmatrix} A_x & A_y & A_z \\ B_x & B_y & B_z \\ C_x & C_y & C_z \end{vmatrix} \qquad (2.10)$$

$$= A_x(B_y C_z - B_z C_y) + A_y(B_z C_x - B_x C_z)$$

$$+ A_z(B_x C_y - B_y C_x)$$

Property:

$$\vec{A} \cdot (\vec{B} \times \vec{C}) = (\vec{A} \times \vec{B}) \cdot \vec{C}$$

Circular Permutation: The sign of the mixed triple product is conserved if the vectors are permuted in such a way that they are read in counterclockwise direction:

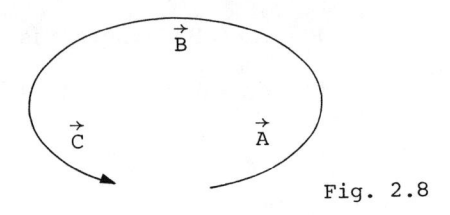

Fig. 2.8

Triple Vector Product

$$\vec{A} \times (\vec{B} \times \vec{C}) = (\vec{A} \cdot \vec{C})\vec{B} - (\vec{A} \cdot \vec{B})\vec{C} \qquad (2.11)$$

2.6 VECTOR FUNCTIONS

Derivative of a Vector

If vector $\vec{A}$ is a function of another parameter u:

i.e., $\vec{A}(u) = \vec{i} A_x(u) + \vec{j} A_y(u) + \vec{k} A_z(u)$

Then

A-9

$$\boxed{\frac{d\vec{A}(u)}{du} = \frac{dA_x}{du}\vec{i} + \frac{dA_y}{dy}\vec{j} + \frac{dA_z}{du}\vec{k}}$$ (2.12)

Vector Integration

$$\boxed{\int \vec{A}(u)dt = \vec{i}\int A_x(u)dt + \vec{j}\int A_y(u)dt + \vec{k}\int A_z(u)dt}$$

For 'u' given as a function of 't'. (2.13)

Derivative Properties:

$$\frac{d(\vec{A} + \vec{B})}{du} = \frac{d\vec{A}}{du} + \frac{d\vec{B}}{du}$$

$$\frac{d(c\vec{A})}{du} = \frac{dc}{du}\vec{A} + c\frac{d\vec{A}}{du}$$

$$\frac{d(\vec{A} \cdot \vec{B})}{du} = \frac{d\vec{A}}{du} \cdot \vec{B} + \vec{A} \cdot \frac{d\vec{B}}{du}$$

$$\frac{d(\vec{A} \times \vec{B})}{du} = \frac{d\vec{A}}{du} \times \vec{B} + \vec{A} \times \frac{d\vec{B}}{du}$$

Note: The scalar 'c' and the vectors A and B are functions of the parameter 'u'.

2.7 UNIT VECTOR

Unit Vector - A vector having a magnitude of unity.

$$\hat{e}_x = [1,0,0]; \quad \hat{e}_y = [0,1,0]; \quad \hat{e}_z = [0,0,1]$$

are unit coordinate vectors. A unit vector is dimensionless.

Expressing a vector in terms of unit vectors:

$$\vec{A} = [A_x, A_y, A_z]$$

$$= \hat{e}_x A_x + \hat{e}_y A_y + \hat{e}_z A_z$$

Cartesian Unit Vectors

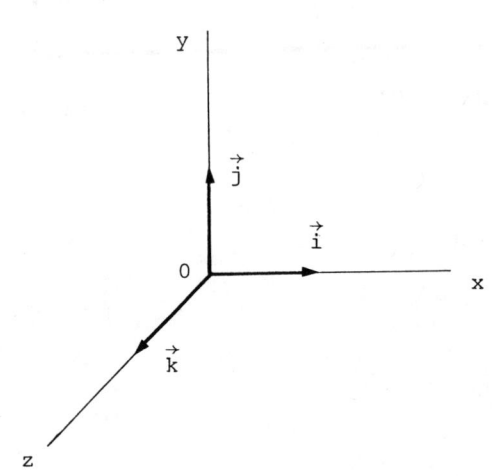

$$\vec{i} = \hat{e}_x \qquad \vec{j} = \hat{e}_y \qquad \vec{k} = \hat{e}_z$$

Right-Hand Coordinate system

Fig. 2.9

2.8 TRANSFORMATION MATRIX

If $\vec{A}$ is expressed in the $\vec{i}\,\vec{j}\,\vec{k}$ triad system:

$$\vec{A} = \vec{i} A_x + \vec{j} A_y + \vec{k} A_z$$

and in a new triad system $\vec{i}'\,\vec{j}'\,\vec{k}'$:

$$\vec{A} = \vec{i}' A_{x'} + \vec{j}' A_{y'} + \vec{k}' A_{z'}$$

The two are related by:

A-11

$$\begin{aligned}
A_{x'} &= \vec{A}\cdot\vec{i}' = (\vec{i}\cdot\vec{i}')A_x + (\vec{j}\cdot\vec{i}')A_y + (\vec{k}\cdot\vec{i}')A_z \\
A_{y'} &= \vec{A}\cdot\vec{j}' = (\vec{i}\cdot\vec{j}')A_x + (\vec{j}\cdot\vec{j}')A_y + (\vec{k}\cdot\vec{j}')A_z \\
A_{z'} &= \vec{A}\cdot\vec{k}' = (\vec{i}\cdot\vec{k}')A_x + (\vec{j}\cdot\vec{k}')A_y + (\vec{k}\cdot\vec{k}')A_z
\end{aligned} \quad (2.14)$$

This may be expressed as:

$$\begin{bmatrix} A_{x'} \\ A_{y'} \\ A_{z'} \end{bmatrix} = \underbrace{\begin{bmatrix} \vec{i}\cdot\vec{i}' & \vec{j}\cdot\vec{i}' & \vec{k}\cdot\vec{i}' \\ \vec{i}\cdot\vec{j}' & \vec{j}\cdot\vec{j}' & \vec{k}\cdot\vec{j}' \\ \vec{i}\cdot\vec{k}' & \vec{j}\cdot\vec{k}' & \vec{k}\cdot\vec{k}' \end{bmatrix}}_{\text{Transformation matrix}} \begin{bmatrix} A_x \\ A_y \\ A_z \end{bmatrix}$$

$(\vec{i}\cdot\vec{i}'), (\vec{j}\cdot\vec{i}')$...are called the coefficients of transformation.

CHAPTER 3

FORCES ON A PARTICLE

3.1 NEWTON'S LAWS

First Law

Every body remains in its state of rest or uniform linear motion, unless a force is applied to change that state.

A) This law describes a common property of matter, called inertia.

B) It defines a reference system called Newtonian or inertial reference system. Rotating or accelerating systems are not inertial.

C) The quantitative measure of inertia is called mass.

Second Law

If the vector sum of the forces $\vec{F}$ acting on a particle of mass m is different from zero, then the particle will have an acceleration, $\vec{a}$, directly proportional to, and in the same direction as $\vec{F}$, but inversely, proportional to mass m. Symbolically

$$\boxed{\vec{F} = m\vec{a}}$$

Third Law

For every action, there exists a corresponding equal and opposing reaction, or the mutual actions of two bodies are always equal and opposing.

$$\boxed{F_A = -F_B}$$

3.2 VECTOR REPRESENTATION

Force:

A) is a push or pull that a body exerts on another.
B) can be represented by a vector.
C) adds and subtracts vectorially.

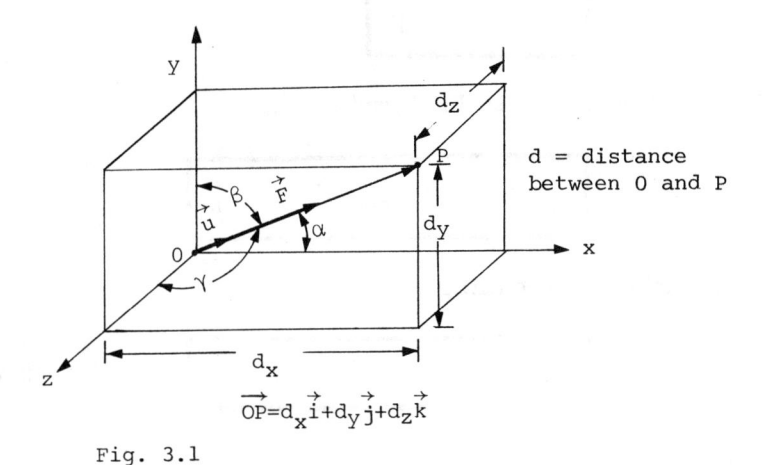

d = distance between O and P

$\overrightarrow{OP} = d_x \vec{i} + d_y \vec{j} + d_z \vec{k}$

Fig. 3.1

Unit Vector:

$$\boxed{\vec{u} = \frac{\overrightarrow{OP}}{OP} = \frac{1}{d}(d_x \vec{i} + d_y \vec{j} + d_z \vec{k})} \quad (3.1)$$

Force F:

$$\boxed{\vec{F} = F\vec{u} = \frac{F}{d}(d_x \vec{i} + d_y \vec{j} + d_z \vec{k})} \quad (3.2)$$

Components:

$$\boxed{F_x = \frac{F d_x}{d}, \quad F_y = \frac{F d_y}{d}, \quad F_z = \frac{F d_z}{d}} \quad (3.3)$$

Distance:

$$d = \sqrt{d_x^2 + d_y^2 + d_z^2} \qquad (3.4)$$

Directional Cosines of F:

$$\alpha = \cos^{-1}\frac{d_x}{d}$$
$$\beta = \cos^{-1}\frac{d_y}{d} \qquad (3.5)$$
$$\gamma = \cos^{-1}\frac{d_z}{d}$$

Unit Vector Expressed In Terms of Angles:

$$\vec{u} = \cos\alpha\,\vec{i} + \cos\beta\,\vec{j} + \cos\gamma\,\vec{k} \qquad (3.6)$$

Relationship Between Angles:

$$\cos^2\alpha + \cos^2\beta + \cos^2\gamma = 1 \qquad (3.7)$$

3.3 FREE-BODY DIAGRAM

Drawing a free-body diagram is the first step in solving a problem. From the free-body diagram, equations of equilibrium are then written.

To construct a free-body diagram:

1) Isolate the body in question.
2) All the forces (active and reactive), acting on the body are labeled and sketched.

Note: Cables carry only tensile forces. These forces are directed along the axes of the cable.

3.4 RECTANGULAR COMPONENTS

Two-Dimensional System

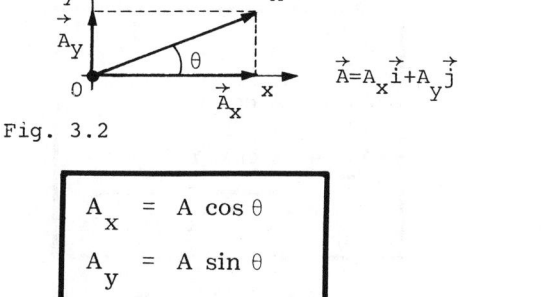

Fig. 3.2

$$\vec{A} = A_x \vec{i} + A_y \vec{j}$$

$$\begin{aligned} A_x &= A \cos \theta \\ A_y &= A \sin \theta \\ \frac{A_y}{A_x} &= \tan \theta \\ A &= \sqrt{A_x^2 + A_y^2} \end{aligned} \qquad (3.8)$$

Three-Dimensional System

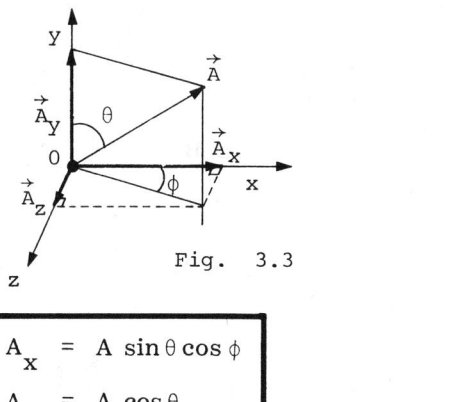

Fig. 3.3

$$\begin{aligned} A_x &= A \sin \theta \cos \phi \\ A_y &= A \cos \theta \\ A_z &= A \sin \theta \sin \phi \end{aligned} \qquad (3.9)$$

Fig. 3.4

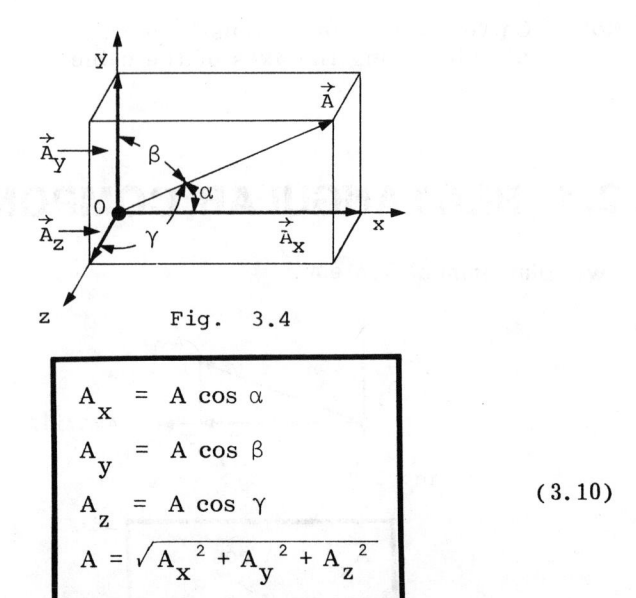

$$A_x = A \cos \alpha$$
$$A_y = A \cos \beta$$
$$A_z = A \cos \gamma \quad (3.10)$$
$$A = \sqrt{A_x^2 + A_y^2 + A_z^2}$$

Concurrent Forces – Forces that pass through the same point.

Addition of Forces by Components:

Given:
$$\vec{A} = A_x \vec{i} + A_y \vec{j} + A_z \vec{k}$$
$$\vec{B} = B_x \vec{i} + B_y \vec{j} + B_z \vec{k}$$

Resultant:
$$\vec{R} = \vec{A} + \vec{B}$$

$$R_x = A_x + B_x$$
$$R_y = A_y + B_y \quad \text{or}$$
$$R_z = A_z + B_z$$

$$R_x = \Sigma F_x$$
$$R_y = \Sigma F_y \quad (3.11)$$
$$R_z = \Sigma F_z$$

A-17

3.5 EQUILIBRIUM OF A PARTICLE

If the resultant force acting on a particle is zero, then the particle is in equilibrium.

Equation of Equilibrium:

$$\Sigma \vec{F} = 0$$
$$\Sigma F_x \vec{i} + \Sigma F_y \vec{j} + \Sigma F_z \vec{k} = 0 \qquad (3.12)$$

ΣF is the vector sum of all the forces acting on the particle, and these forces are resolved into their respective i, j, and k directions.

Component Equations:

$$\Sigma F_x = 0$$
$$\Sigma F_y = 0 \qquad (3.13)$$
$$\Sigma F_z = 0$$

These three equations can solve for no more than three unknowns in a problem.

CHAPTER 4

EQUIVALENT SYSTEMS OF PARTICLES

4.1 RIGID BODIES

A rigid body:

A) is a body that does not deform under a load.

B) can be considered as a combination of many particles which remain at a fixed distance from each other.

Two Groups of Forces Acting on a Rigid Body:

A) External Forces - Actions of other bodies on the rigid body; responsible for external motion of the body.

B) Internal Forces - Forces which hold together parts of the rigid body.

Any external load may be represented by a concentrated force when the area under the load is large compared to the surface area of the body.

4.2 PRINCIPLE OF TRANSMISSIBILITY

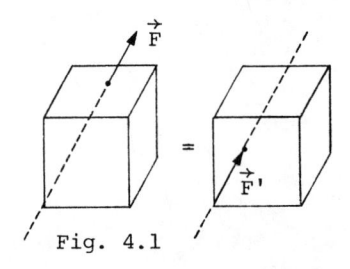

Fig. 4.1

This is used for computing external reactions only.

Note: The two forces have the same line of direction, but act at different points. $\vec{F} = \vec{F}'$ in magnitude and direction. Thus any force acting on a rigid body can be considered as a sliding vector.

Alternate statement:

Two forces are equivalent if

$\vec{F} = \vec{F}'$ (Equal Forces)

$\vec{M}_0 = \vec{M}'_0$ (Equal moment effects about a given point 0)

4.3 MOMENT OF A FORCE ABOUT A POINT

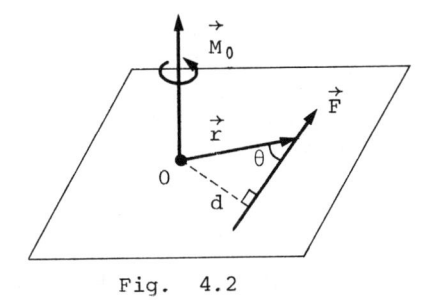

Fig. 4.2

Moment of $\vec{F}$ about point 0:

$$\boxed{\vec{M}_0 = \vec{r} \times \vec{F}} \qquad (4.1)$$

Note: Proper order of multiplication must be maintained. Magnitude of $\vec{M}_0$:

$$\boxed{M_0 = rF\sin\theta = Fd} \quad (\because d = r\sin\theta)$$

$$(4.2)$$

A-20

M_0 measures the tendency of $\vec{F}$ to make the rigid body rotate about a fixed axis directed along $\vec{M}_0$.

Units: SI - Newton-Meters

U.S. - lb · ft or lb · in

Rectangular Components:

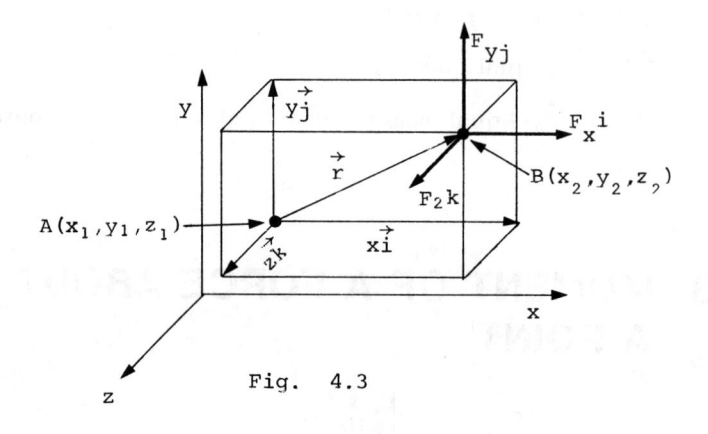

Fig. 4.3

$$\vec{r} = (x_2-x_1)\vec{i} + (y_2-y_1)\vec{j} + (z_2-z_1)\vec{k}$$

$$\vec{F} = F_x\vec{i} + F_y\vec{j} + F_z\vec{k}$$

$$\vec{r} \times \vec{F} = \vec{M}_A = \begin{vmatrix} \vec{i} & \vec{j} & \vec{k} \\ (x_2-x_1) & (y_2-y_1) & (z_2-z_1) \\ F_x & F_y & F_z \end{vmatrix} \quad (4.3)$$

In component form:

$$\vec{M}_A = M_x\vec{i} + M_y\vec{j} + M_z\vec{k}$$

where
$$M_x = yF_z - zF_y$$
$$M_y = zF_x - xF_z \quad (4.4)$$
$$M_z = xF_y - yF_x$$

A-21

where

$$x = x_2 - x_1$$
$$y = y_2 - y_1$$
$$z = z_2 - z_1$$

4.4 VARIGNON'S THEOREM

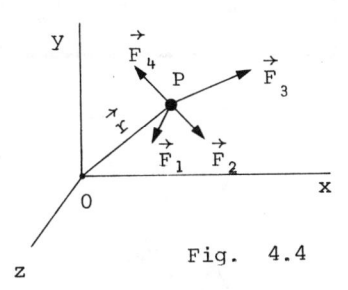

Fig. 4.4

$$\vec{r} \times (\vec{F}_1 + \vec{F}_2 + \vec{F}_3 + \ldots) = (\vec{r} \times \vec{F}_1) + (\vec{r} \times \vec{F}_2) + \ldots \quad (4.5)$$

Statement: About any given point in a concurrent force system the sum of the individual moments of all the forces is equal to the moment of the resultant force of the concurrent force system about the same point.

4.5 MOMENTS ABOUT A GIVEN AXIS

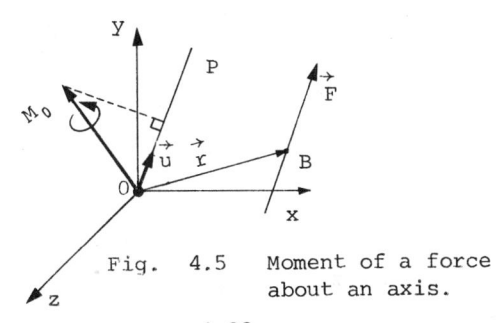

Fig. 4.5 Moment of a force about an axis.

$$M_{OP} = \vec{u} \cdot \vec{M_0} = \vec{u} \cdot (\vec{r} \times \vec{F}) \qquad (4.6)$$

In determinant form:

$$M_{OP} = \begin{vmatrix} u_x & u_y & u_z \\ r_x & r_y & r_z \\ F_x & F_y & F_z \end{vmatrix} \qquad (4.7)$$

More general case:

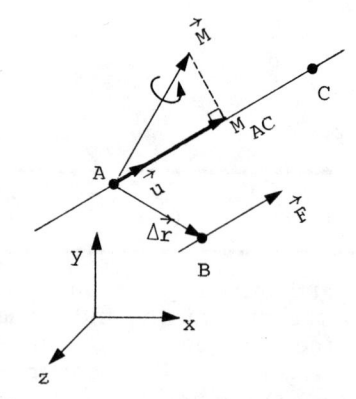

Fig. 4.6 Moment about an axis, (General Case).

$$M_{AC} = \begin{vmatrix} u_x & u_y & u_z \\ \Delta x & \Delta y & \Delta z \\ F_x & F_y & F_z \end{vmatrix} \qquad (4.8)$$

where

$$\Delta x = x_B - x_A$$
$$\Delta y = y_B - y_A$$
$$\Delta z = z_B - z_A$$

4.6 MOMENT OF A COUPLE

Two parallel forces having the same magnitude, opposite in direction and acting at a perpendicular distance apart, constitute a couple.

Couples cause the body to rotate only.

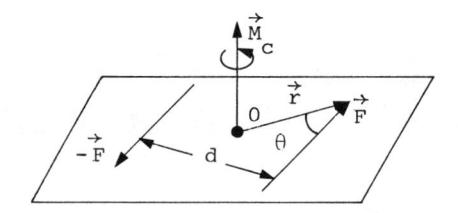

Fig. 4.7 Moment of a couple.

$$M_c = r \times F$$

$$\boxed{M_c = rF\sin\theta = Fd} \qquad (\because d = r\sin\theta) \qquad (4.9)$$

Moment vectors are free vectors.

Two couples are equal if:
1) $F_1 d_1 = F_2 d_2$
2) They lie in the same or parallel planes
3) They have the same sense.

4.7 ADDITION OF COUPLES

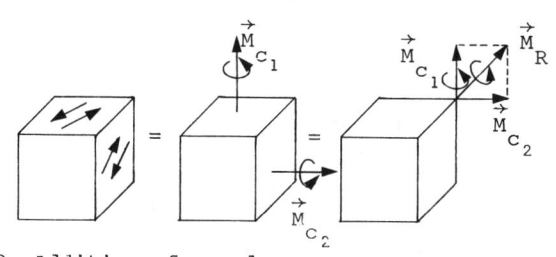

Fig. 4.8 Addition of couples.

A-24

1) Replace all couples by their moment vectors.

2) $$\boxed{\vec{M}_R = \vec{M}_{C_1} + \vec{M}_{C_2} + \ldots}$$ (4.10)

4.8 RESOLUTION OF FORCES INTO A FORCE AND A COUPLE

A) If a force $\vec{F}$ is to be moved to a point '0' not located on its line of action, a couple, of moment equal to the moment of $\vec{F}$ about 0, must be added.

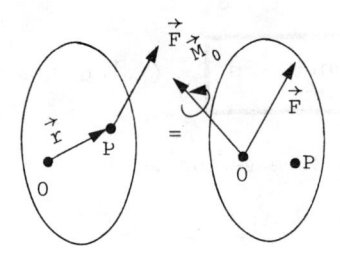

Fig. 4.9 Force-couple system.

B) If $\vec{F}$ was moved to a different point A:

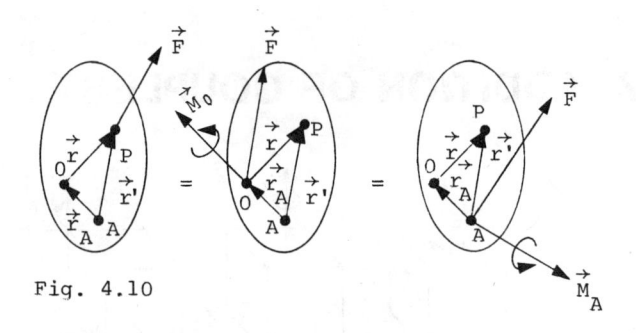

Fig. 4.10

Relation between $\vec{M}_0$ and $\vec{M}_A$:

A-25

$$\boxed{\vec{M}_A = \vec{M}_0 + \vec{r}_A \times \vec{F}}$$ (4.11)

C) A force-couple system may be replaced by a single equivalent force by moving $\vec{F}$ in the plane perpendicular to $\vec{M}_0$ until its moment about 0 equals the vector $\vec{M}_0$ to be eliminated.

4.9 REDUCTION OF FORCES

A) Two systems of forces are equivalent, if:

$$\Sigma F_1 = \Sigma F_2 \quad \text{and} \quad \Sigma M_1 = \Sigma M_2$$

B) When two systems of vectors satisfy the above condition, they are called equipollent.

Equipollent = Equivalent

Equivalent ≠ Equipollent

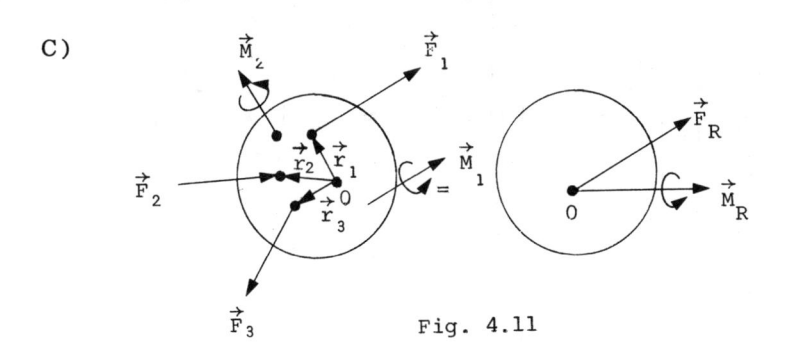

C)

Fig. 4.11

$$\boxed{\begin{aligned}\vec{F}_R &= \Sigma \vec{F} = \vec{F}_1 + \vec{F}_2 + \ldots \\ \vec{M}_R &= \Sigma \vec{M}_0 = \vec{M}_1 + \vec{M}_2 + \ldots\end{aligned}}$$ (4.12)

D) A system of forces may be reduced to a single force if the line of action of $\vec{F}_R$ and $\vec{M}_R$ are mutually perpendicular.

This condition is satisfied by systems of
1) concurrent forces - add directly to give $\vec{F}_R$
2) coplanar forces - forces that act in the same plane.

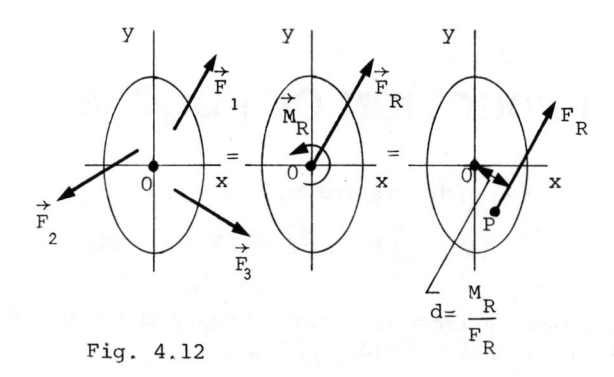

Fig. 4.12

3) parallel forces - forces having parallel lines of action.

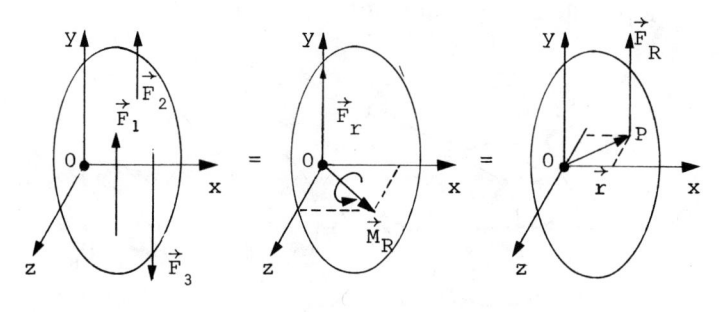

Fig. 4.13

P is on the x-z plane. It coordinates are:

$$\boxed{\begin{array}{l} -z(F_R)_y = (M_R)_x \\ x(F_R)_y = (M_R)_z \end{array}}$$ (4.13)

A-27

$(F_R)_y$ = y-component of $\vec{F}_R$

$(M_R)_x$ = x-component of M_R

$(M_R)_z$ = z-component of M_R

E) General case - For $\vec{F}_R$ and $\vec{M}_R$ not perpendicular to each other:

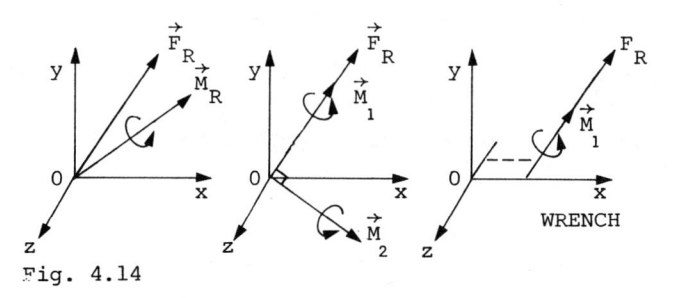

Fig. 4.14

Axis of the Wrench = Line of Action of $\vec{F}_R$

$$\text{Pitch of the Wrench} = \frac{\vec{M_1}}{\vec{F}_R} = \frac{\vec{F}_R \cdot \vec{M}_0}{F_R^2}$$

F) Reduction of Distributed Loading:

The load is usually described by a loading function.
Units: Force/Area or Pressure
Total Resultant Force:

$$F_R = \int_A P(x,y)dA = \int_V dV = V \quad (4.14)$$

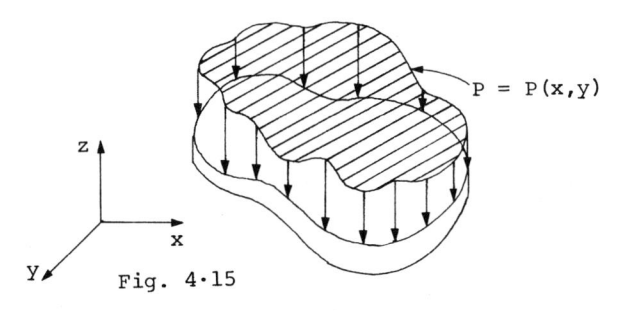

Fig. 4·15

A-28

1) The magnitude of the resultant force is equal to the volume under the distributed loading diagram.

2) The line of action of the resultant force passes through the geometric-center or centroid of this volume.

If the loading function is symmetric, then it can be shown as an area rather than a volume. The loading intensity is then expressed as force/length.

1) The magnitude of the resultant force is equal to the area under the load diagram.

2) The line of action passes through the centroid of this area.

CHAPTER 5

EQUILIBRIUM OF RIGID BODIES

5.1 INTRODUCTION

The necessary and sufficient conditions for equilibrium of a rigid body are that the vector sum of all the forces and couple moments (about any point in space) equal zero:

$$\Sigma \vec{F} = 0$$
$$\Sigma \vec{M}_0 = \Sigma (\vec{r} \times \vec{F}) = 0$$

(5.1)

To solve an equilibrium problem, first draw a free-body diagram of the body in question. Then solve the component equations of (5.1).

Forces and couples usually encountered in a problem. The following are to be considered:

a) External loading

b) Reactions at the support and contact points with other bodies.

c) Weight of the isolated body.

5.2 REACTIONS AT SUPPORTS AND CONNECTIONS

Two-Dimensional Structures

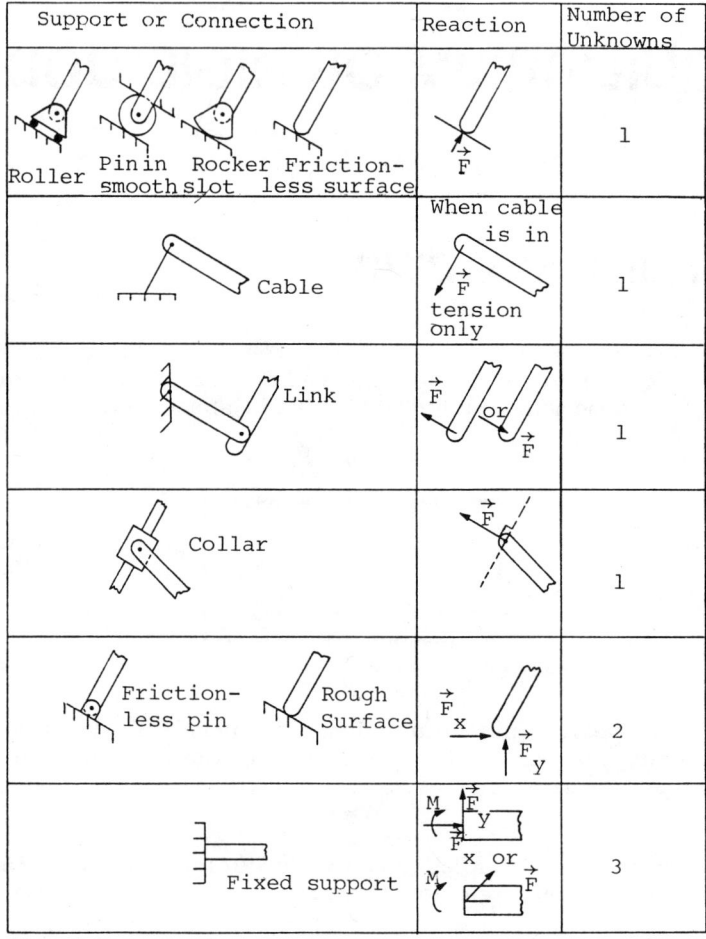

Support or Connection	Reaction	Number of Unknowns
Roller, Pin in smooth slot, Rocker, Frictionless surface	$\vec{F}$	1
Cable	When cable is in tension only, $\vec{F}$	1
Link	$\vec{F}$ or $\vec{F}$	1
Collar	$\vec{F}$	1
Frictionless pin, Rough Surface	$\vec{F}_x$, $\vec{F}_y$	2
Fixed support	M, $\vec{F}_y$, $\vec{F}_x$ or M, $\vec{F}$	3

Fig. 5.1 Reactions at Supports and Connections in two-dimensional structures.

Note:
A) A force is developed at a support if it prevents translation.
B) A moment is developed if the support prevents rotation.
C) When the sense of an unknown force or moment is not apparent, then assume an arbitrary direction. The sign

of the solution will indicate whether the assumption is correct or not.
D) All reactions which can exist at a support must be included in the free-body diagram.
E) A simple way to determine the type of reaction corresponding to a given support or connection and the number of unknowns involved is to find which of the six fundamental motions (translation in x, y, and z directions; rotation about the x, y, and z axes) are allowed and which are prevented.

Three–Dimensional Structures

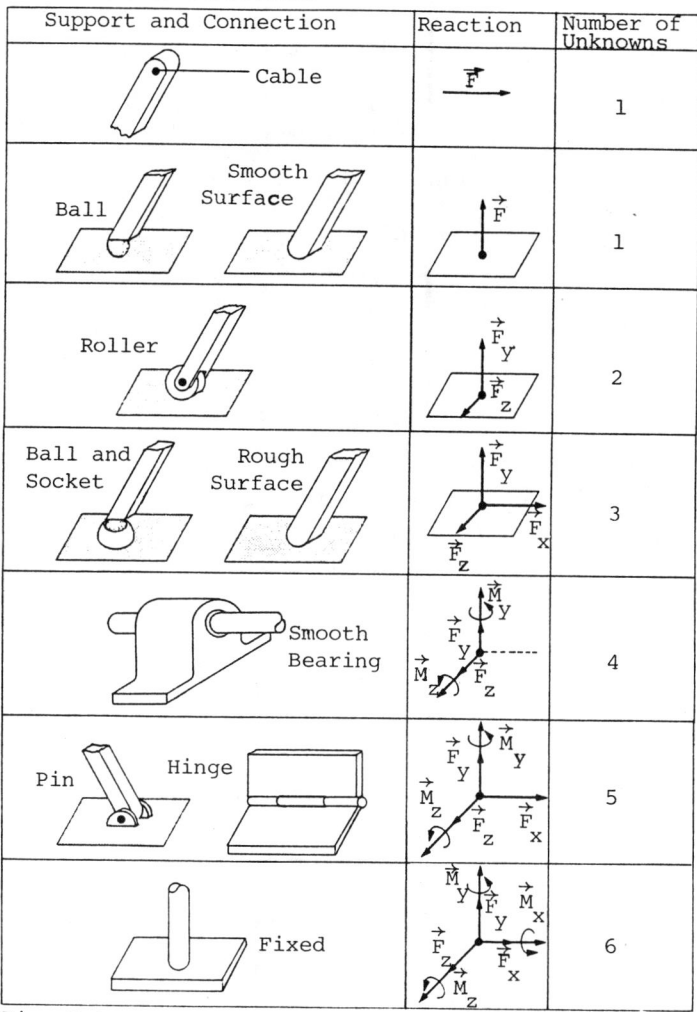

Support and Connection	Reaction	Number of Unknowns
Cable	$\vec{F}$	1
Ball, Smooth Surface	$\vec{F}$	1
Roller	$\vec{F}_y$, $\vec{F}_z$	2
Ball and Socket, Rough Surface	$\vec{F}_x$, $\vec{F}_y$, $\vec{F}_z$	3
Smooth Bearing	$\vec{F}_y$, $\vec{F}_z$, $\vec{M}_y$, $\vec{M}_z$	4
Pin, Hinge	$\vec{F}_x$, $\vec{F}_y$, $\vec{F}_z$, $\vec{M}_y$, $\vec{M}_z$	5
Fixed	$\vec{F}_x$, $\vec{F}_y$, $\vec{F}_z$, $\vec{M}_x$, $\vec{M}_y$, $\vec{M}_z$	6

Fig. 5.2 Reactions at Supports and Connections in three-dimensional structures

5.3 EQUILIBRIUM CONDITIONS

Two-Dimensional Structures

Equations of Equilibrium:

From Eq.(5.1):

$$\boxed{\begin{array}{l} \Sigma F_x = 0 \\ \Sigma F_y = 0 \\ \Sigma M_0 = 0 \end{array}} \quad (5.2)$$

3 independent equations → 3 unknowns are allowed

Alternate sets of equations:

A)
$$\boxed{\begin{array}{l} \Sigma F_y = 0 \\ \Sigma M_A = 0 \\ \Sigma M_B = 0 \end{array}} \quad (5.3)$$

Provided that 'A' and 'B' do not lie on a line perpendicular to the y-axis.

B)
$$\boxed{\begin{array}{l} \Sigma F_x = 0 \\ \Sigma M_A = 0 \\ \Sigma M_B = 0 \end{array}} \quad (5.4)$$

Provided that A and B do not lie on a line perpendicular to the x-axis.

C)
$$\boxed{\begin{array}{l} \Sigma M_A = 0 \\ \Sigma M_B = 0 \\ \Sigma M_C = 0 \end{array}} \quad (5.5)$$

A, B, and C must not lie on the same line.

Note:

A) Orient the x- and y-axes in a direction which will facilitate reduction of forces in components.

B) Moments should be summed about points which lie along the line of action of as many unknown forces as possible.

Three-Dimensional Structures

Equations of Equilibrium:

$$\Sigma F_x = 0 \quad \Sigma M_x = 0$$
$$\Sigma F_y = 0 \quad \Sigma M_y = 0 \quad (5.6)$$
$$\Sigma F_z = 0 \quad \Sigma M_z = 0$$

May be used to solve for no more than six unknowns.

Type of Problems Encountered

A) Statically Indeterminant - If the reactions involve more than six unknowns.

B) Partially Constrained - Reactions involve less than six unknowns.

C) Improperly Constrained - The equations of equilibrium are not satisfied even with six or more unknowns.

To ensure equilibrium, a body should be properly constrained by:

A) At least as many reactive forces as equations of equilibrium.

B) The lines of action of the reactive forces do not intersect points on a common axis.

C) The reactive forces are not parallel to one another.

5.4 SPECIAL CASES IN EQUILIBRIUM

2-Force Body

A rigid body subjected to forces acting at only two points.

If a two-force body is in equilibrium, then the two forces must have the same magnitude, line of action and must act in opposite direction.

3-Force Body

A rigid body subjected to forces acting at three points only.

If a three-force body is in equilbrium, then the forces must either be concurrent or parallel.

CHAPTER 6

CENTROIDS AND CENTER OF GRAVITY

6.1 CENTER OF GRAVITY

For a System of Particles:

$$\bar{x} = \frac{\sum_{i=1}^{n} x_i w_i}{\sum_{i=1}^{n} w_i}, \quad \bar{y} = \frac{\sum_{i=1}^{n} y_i w_i}{\sum_{i=1}^{n} w_i}, \quad \bar{z} = \frac{\sum_{i=1}^{n} z_i w_i}{\sum_{i=1}^{n} w_i} \quad (6.1)$$

$\bar{x}, \bar{y}, \bar{z}$ are the coordinates of the center of gravity of the system.
w_i is the weight of the ith particle.
x_i, y_i, z_i are the algebraic distances to the ith particle from the origin.

Two-Dimensional Conditions (areas and lines)

$$W = \int dw$$
$$\bar{x} = \frac{\int x\, dw}{W} \quad (6.2)$$
$$\bar{y} = \frac{\int y\, dw}{W}$$

W is the sum of the magnitudes of the elementary weights. In the case of a line, the center of cravity generally does not lie on the line.

Three-Dimensional Conditions (volumes)

$$\bar{x} = \frac{\int x\,dw}{W}, \quad \bar{y} = \frac{\int y\,dw}{W}, \quad \bar{z} = \frac{\int z\,dw}{W} \quad (6.3)$$

If the body is made of homogeneous material of specific weight γ then

Weight of the element $(dw) = \gamma\,dV$ where dV = volume of the element and

$$\bar{x} = \frac{\int x\,dv}{V}, \quad \bar{y} = \frac{\int y\,dv}{V}, \quad \bar{z} = \frac{\int z\,dv}{V} \quad (6.4)$$

$\bar{x},\bar{y},\bar{z}$ are the coordinates of the centroid of the volume V of the body.

$\int x\,dv$ is called the first moment of the volume with respect to the yz plane.

6.2 CENTROIDS

When the calculation depends on the geometry of the body only, the point $(\bar{x},\bar{y},\bar{z})$ is called the centroid of the body.

Centroids of Areas and Wires

For an area A of a homogeneous plate

$$\bar{x} = \frac{\int x\,dA}{A}$$
$$\bar{y} = \frac{\int y\,dA}{A}$$
(6.5)

The above equations define the coordinates $\bar{x}$ and $\bar{y}$ of the center of gravity. This point is also known as centroid of the area A.

For a homogeneous wire of length L

$$\bar{x} = \frac{\int x\,dL}{L}$$
$$\bar{y} = \frac{\int y\,dL}{L}$$
(6.6)

Note: If the body is nonhomogeneous, the centroid and the center of gravity will not coincide.

6.3 COMPOSITE BODIES

Areas

A composite body may be divided into parts with common shapes to determine its center of gravity.

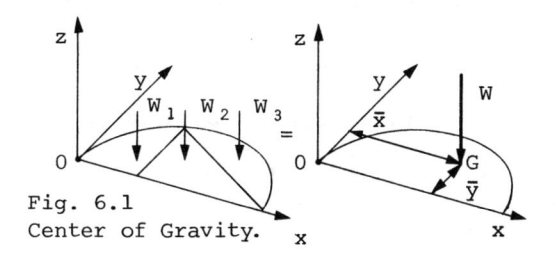

Fig. 6.1 Center of Gravity.

A-38

By equating moments:

$$\boxed{\begin{aligned} \bar{x} &= \bar{x}_1 W_1 + \bar{x}_2 W_2 + \ldots + \bar{x}_n W_n \\ \bar{y} &= \bar{y}_1 W_1 + \bar{y}_2 W_2 + \ldots + \bar{y}_n W_n \end{aligned}} \quad (6.7)$$

To find the centroid - equate moment of areas:

$$\boxed{\begin{aligned} \bar{x} &= \bar{x}_1 A_1 + \bar{x}_2 A_2 + \ldots + \bar{x}_n A_n \\ \bar{y} &= \bar{y}_1 A_1 + \bar{y}_2 A_2 + \ldots + \bar{y}_n A_n \end{aligned}} \quad (6.8)$$

Lines

For composite lines, divide it into simpler segments and add the moments of each segment.

Volumes

A body can be divided into common shapes of revolution.

Center of Gravity:

$$\boxed{\begin{aligned} \bar{x} \sum_{i=1}^{n} W_i &= \sum_{i=1}^{n} \bar{x}_i W_i \\ \bar{y} \sum_{i=1}^{n} W_i &= \sum_{i=1}^{n} \bar{y}_i W_i \\ \bar{z} \sum_{i=1}^{n} W_i &= \sum_{i=1}^{n} \bar{z}_i W_i \end{aligned}} \quad (6.9)$$

If the body is homogeneous, the center of gravity coincides with the centroid:

$$\bar{x}\sum_{i=1}^{n} V_i = \sum_{i=1}^{n} \bar{x}_i V_i$$

$$\bar{y}\sum_{i=1}^{n} V_i = \sum_{i=1}^{n} \bar{y}_i V_i \qquad (6.10)$$

$$\bar{z}\sum_{i=1}^{n} V_i = \sum_{i=1}^{n} \bar{z}_i V_i$$

6.4 CENTROIDS BY INTEGRATION

Area

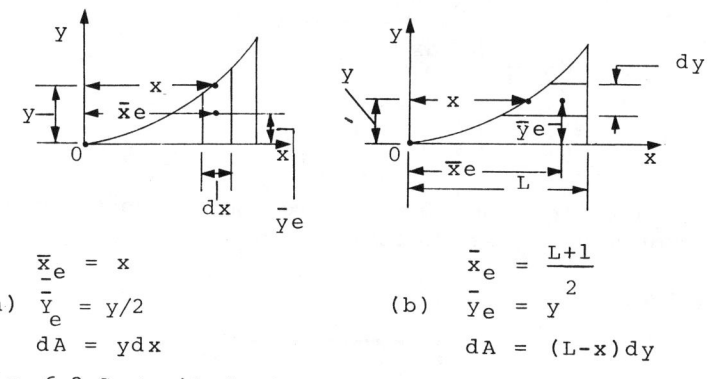

(a) $\bar{x}_e = x$
$\bar{y}_e = y/2$
$dA = y\,dx$

(b) $\bar{x}_e = \dfrac{L+1}{2}$
$\bar{y}_e = y$
$dA = (L-x)\,dy$

Fig. 6.2 Centroids by integration

$$\bar{x}A = \int \bar{x}_e \, dA$$
$$\bar{y}A = \int \bar{y}_e \, dA \qquad (6.11)$$

$\bar{x}_e, \bar{y}_e$ are the coordinates of the centroid of the element dA.

$\bar{x}_e, \bar{y}_e$ should be expressed in terms of the coordinates of a point located on the curve bounding the area.

dA should be expressed in terms of the coordinates of the point and their differentials.

Lines

$$\boxed{\begin{aligned} \bar{x}L &= \int x\,dL \\ \bar{y}L &= \int y\,dL \end{aligned}} \qquad (6.12)$$

dL should be expressed by one of the following:

$$dL = \sqrt{\left(\frac{dy}{dx}\right)^2 + 1}\ dx$$

$$dL = \sqrt{\left(\frac{dx}{dy}\right)^2 + 1}\ dy$$

$$dL = \sqrt{r^2 + \left(\frac{dr}{d\theta}\right)^2}\ d\theta$$

The above equations for dL, are chosen depending on the type of equation used to define the line.

Volume

$$\boxed{\begin{aligned} \bar{x}V &= \int \bar{x}_e\,dV \\ \bar{y}V &= \int \bar{y}_e\,dV \\ \bar{z}V &= \int \bar{z}_e\,dV \end{aligned}} \qquad (6.13)$$

For bodies with two planes of symmetry:

$$\bar{y} = \bar{z} = 0$$

and

$$\bar{x}V = \int \bar{x}_e\,dV$$

A-41

6.5 THEOREMS OF PAPPUS-GULDINUS

Surface of Revolution - Generated by revolving a curve about a fixed axis.

Theorem I

The area of a surface revolution is equal to the product of the length of the generating curve and the distance traveled by the centroid of the curve while the surface is being generated.

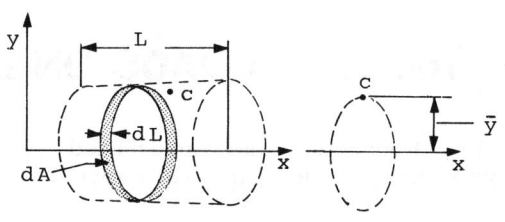

Fig. 6.3

When a differential length dL of line L is revolved about x-axis, a ring is generated having surface area $dA = 2\pi y dL$. Therefore

$$A = 2\pi \int_L y\, dL = 2\pi \bar{y} L \qquad (6.14)$$

Theorem II

The volume of a surface of revolution is equal to the generating area times the distance traveled by the centroid of the area in generating the volume.

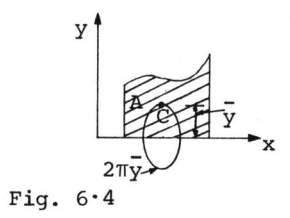

Fig. 6·4

A-42

Volume of Revolution - Generated by revolving an area about a fixed axis. When a differential area dA is revolved about x-axis, it generated a ring of volume dV = $2\pi y dA$. Therefore

$$V = 2\pi \int_A y \, dA = 2\pi \bar{y} A \qquad (6.15)$$

6.6 DISTRIBUTED LOADS ON BEAMS

A distributed load may be represented by plotting the load W supported per unit length (load curve).

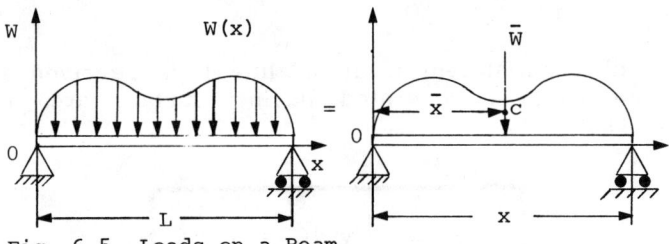

Fig. 6.5 Loads on a Beam.

A distributed load on a beam may thus be replaced by a concentrated load; the magnitude of this single load is equal to the area under the load curve, and its line of action passes through the centroid of that area.

The concentrated load may be used to determine reactions but not internal forces and deflections.

CHAPTER 7

ANALYSIS OF STRUCTURES

7.1 INTRODUCTION

Truss - A structure composed of slender members connected at their ends.

Trusses are designed to carry loads which act in their plane and are considered as two-dimensional structures.

Assumptions:

A) Members are connected by smooth pins; therefore, forces acting at the ends may be reduced to a single force with no couple acting.

B) All loads are applied at the joints.

C) Weight of the members are neglected in the analysis.

Internal Forces - Forces which hold together various part of the structure. Internal forces appear when individual members are considered as a free-body.

Rigid Truss - Will not collapse under load.

Simple Truss - Constructued by adding two members to a basic triangular rigid truss connected to a new joint. The three joints must not be collinear. This process may be repeated as desired. Total number of members:

$$m = 2n - 3 \quad \text{where} \quad n = \text{number of joints.}$$

A-44

7.2 METHOD OF JOINTS

If a truss is in equilibrium, then

$$\Sigma F_x = 0$$
$$\Sigma F_y = 0$$

must be satisfied at each joint. Rotational or moment equilibrium is satisfied as the force system is concurrent. The forces developed in each member is directed along the axis of each member.

Method of Solution:

A) Considering the entire truss as a rigid body, find the reactions at the supports.

B) Begin at a joint with two unknowns and determine the unknowns by writing the two equations of equilibrium.

C) The values of step 2 are transferred to adjacent joints as known quantities.

D) The procedure is repeated until all the unknown quantities are determined.

7.3 SPECIAL LOADING CONDITIONS

Zero Force Members - Members carrying zero load under specific loading conditions of the truss.

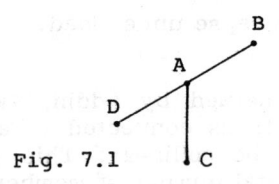

Fig. 7.1

AC is a zero force member if there is no load at joint A.

A-45

Fig. 7.2

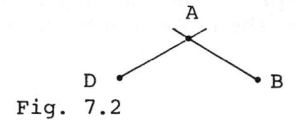

AB and AD are zero force members if no load is applied at A.

Fig. 7.3

Forces in opposite members must be equal:

$$AB = AC$$

7.4 SPACE TRUSSES

Space Truss - Several straight members joined together at their extremities to form a 3-dimensional structure.

The most elementary rigid space truss consists of six members joined at their extremities to form the edges of a tetrahedron.

By adding three members at a time to the basic space truss, attaching them at separate existing joints and connecting them at a new joint, a simple space truss is formed.

Total number of members:

$$m = 3n - 6$$

where n = number of joints.

No couples are applied to the members of the truss. The forces are concurrent.

Equilibrium Condition:

$$\Sigma F_n = 0; \quad \Sigma F_y = 0 \text{ and } \Sigma F_z = 0$$

Must be satisfied at each joint.

A-46

To apply the equilibrium condition to joints, select joints with no more than three unknowns at a time.

7.5 METHOD OF SECTIONS

Procedure:

A) Find external reactions at supports.
B) Pass a section through the truss in which the desired force(s) is one of the external forces acting on that portion.
C) Each portion of the truss may now be used as a free-body.
D) Apply the equations of equilibrium to each portion.

Note:

A) The cutting section may not pass through more than three members in which the forces are unknown.
B) This method is used only when the forces acting in a few members of the truss are to be found.

7.6 MULTIFORCE STRUCTURES

Multiforce Member - Member acted upon by three or more forces.

A) These forces generally will not be directed along the line connecting the joints of members.
B) Their direction is unknown.
C) They should be represented by two unknown components.

Two common types of pin-connected multiforce members are:

A) Frames - Designed to support loads
 - Usually stationary and fully constrained

B) Machines - Designed to transmit and modify loads
- May or may not be stationary
- Always contain moving parts

7.7 ANALYSIS OF A FRAME

Procedure of Analysis

A) Determine the external reactions; solve for as many external reactions as possible.

B) Dismember the frame and draw a free-body diagram for each of its component parts.

C) Consider first, the two force members. Determine their lines of action and assume an arbitrary direction for each force.

D) Consider next the multiforce members. Apply Newton's third law.

E) Solve for the unknowns by applying the equations of equilibrium to each free-body.

7.8 MACHINES

A) Machines are non-rigid structures.

B) In order to determine the output (unknown) forces, one of the component parts must be considered as a free-body.

C) In the case of complicated machines, several free-bodies may be needed.

D) The free-bodies should include input forces, and the reactions to the output forces.

E) Total number of unknown force components should not exceed the number of available independent equations.

CHAPTER 8

BEAMS AND CABLES

8.1 TYPES OF LOADING AND SUPPORTS

Beam - A structural member designed to support loads applied at various points along the member.

Note:
A) Most loads are applied perpendicularly to the member.

B) Axial forces are usually neglected.

Type of Beams

A) Statically Determinate

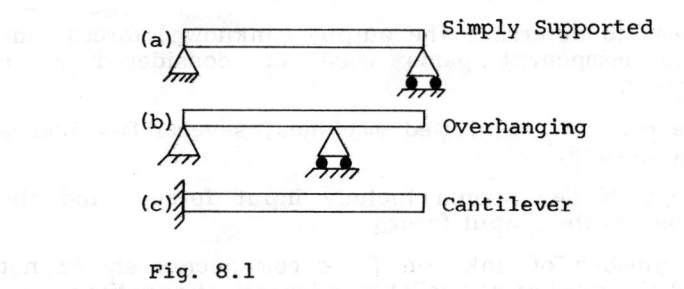

```
(a) ═══════════════════  Simply Supported
(b) ═══════════════════  Overhanging
(c) ═══════════════════  Cantilever
```

Fig. 8.1

B) Statically Indeterminate

A-49

(a) ──────── Continuous

(b) ──────── Fixed at one end and simply supported at the other end.

(c) ──────── Fixed at both ends.

Fig. 8.2

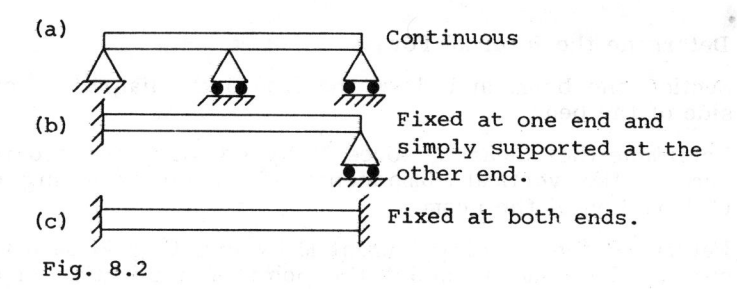

Types of Loading

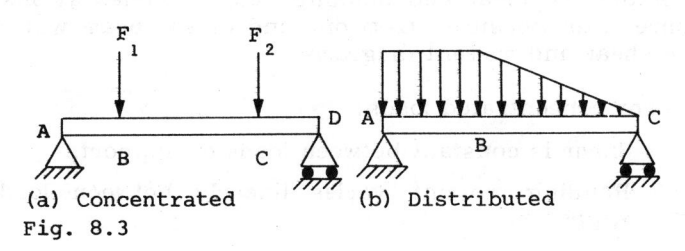

(a) Concentrated   (b) Distributed

Fig. 8.3

8.2 SHEAR AND BENDING MOMENTS

Graphs which show the variation of shear V and bending moment M as functions of length X, are known as shear and moment diagrams.

Sign Convention:

Positive as shown:

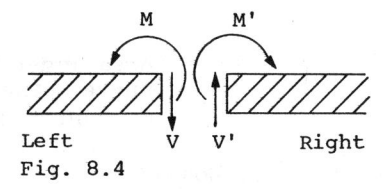

Left   V V'   Right
Fig. 8.4

A good way to remember is:

A) Positive shear tends to rotate the segment clockwise.

B) Positive bending moment tends to bend the segment concave upward.

A-50

To Determine the Internal Forces:

A) Section the beam and draw the free-body diagram of one side of the beam.
B) Determine the shearing force $\vec{V}$ by equating to zero the sum of the vertical components of all forces acting on that portion of the beam.
C) Determine the bending moment $\vec{M}$ by equating to zero the sum of the moments about the point of application of all forces.

A plot of shear and bending moment values against the distance X as measured from one end of the beam will result in the shear and moment diagrams.

Note: For concentrated loads:

- Shear is constant between loads or supports
- Bending moment varies linearly between loads or supports.

8.3 ANALYTICAL RELATIONSHIPS

Between Load and Shear

$$\boxed{\frac{dM}{dx} = -\omega} \quad \text{or} \quad dV = -\omega dx \quad (8.1)$$

V = Shear

ω = Distributed load

$$\therefore \quad V_2 - V_1 = -\int_{x_1}^{x_2} \omega dx = -(\text{Area under the load curve between points } x_1 \text{ and } x_2)$$

Between Shear and Bending Moment

$$\boxed{\frac{dM}{dx} = V} \quad (8.2)$$

$$\therefore \quad M_2 - M_1 = \int_{x_1}^{x_2} V dx = \text{Area under the shear curve between } x_1 \text{ and } x_2.$$

Note: These relationships are invalid where concentrated forces act or concentrated moments are applied.

8.4 CABLES - CONCENTRATED LOADS

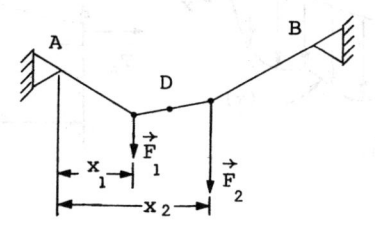

Fig. 8.5 Concentrated loads on a cable.

Assumptions:

A) The cable is flexible and inextensible.
B) Weight of the cable is negligible.
C) Consider any portion of the cable as two-force members in tension.
D) Each load lies in a vertical line.
E) $x_1, x_2, \ldots$ are known.

Method of Solution

A) Draw a free-body diagram of the cable.
B) Represent the reactions at each support by two components.
C) Obtain the moment equation by considering a portion of the cable, say point D: $\Sigma M_D = 0$.
D) Solve for the reactions.
E) The horizontal component of the tension force is the same at any point on the cable.

A-52

8.5 CABLES - DISTRIBUTED LOADS

Assumptions:
A) Perfectly flexible and inextensible cable.
B) Forces are tangential along the cable.

Case I. Externally Loaded

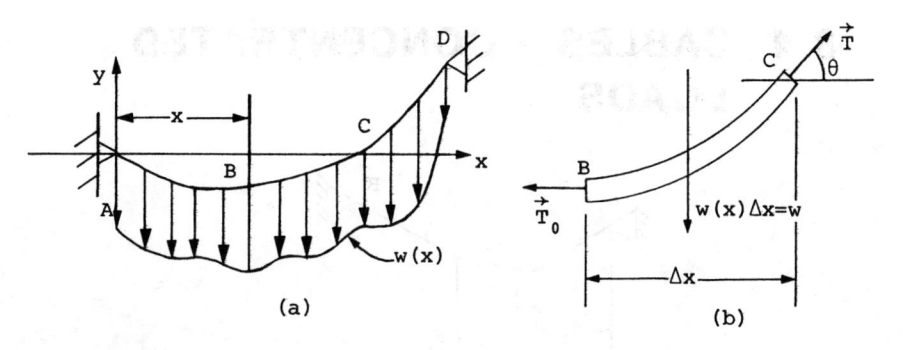

Fig. 8.6

Relationships:

$$T \cos\theta = T_0 \qquad T \sin\theta = W \qquad (8.3)$$
$$T = \sqrt{T_0^2 + W^2} \qquad \tan\theta = W/T_0 \qquad (8.4)$$

Note: T = Cable Tension; $W = W(x) \cdot \Delta x$;
$W(x)$ = loading function

Deflection Formula:

$$y = \frac{1}{T \cos\theta} \int \left[\int w(x)dx \right] dx \qquad (8.5)$$

Note: Tension $\vec{T}$ is minimum at the lowest point and maximum at the supports.

Case II. Loaded by its own weight (catenary)

A-53

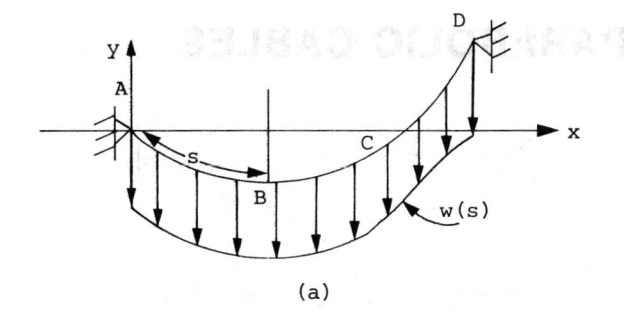

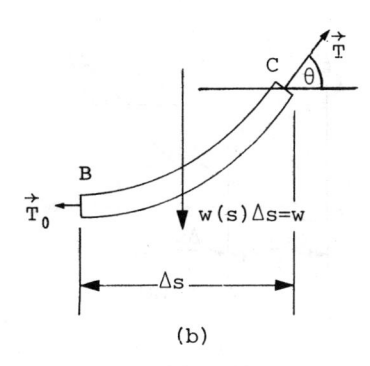

Fig. 8.7

The relationships are the same as in Case I. Equation of the curve:

$$x = \int \frac{ds}{\{1 + \frac{1}{(T\cos\theta)^2}\left[\int w(s)ds\right]^2\}^{\frac{1}{2}}} \quad (8.6)$$

Equation of a catenary with the vertical axis:

$$y = c \cosh \frac{x}{c} \; ; \quad c = \frac{T_0}{W} \quad (8.7)$$

8.6 PARABOLIC CABLES

(a)

(b)

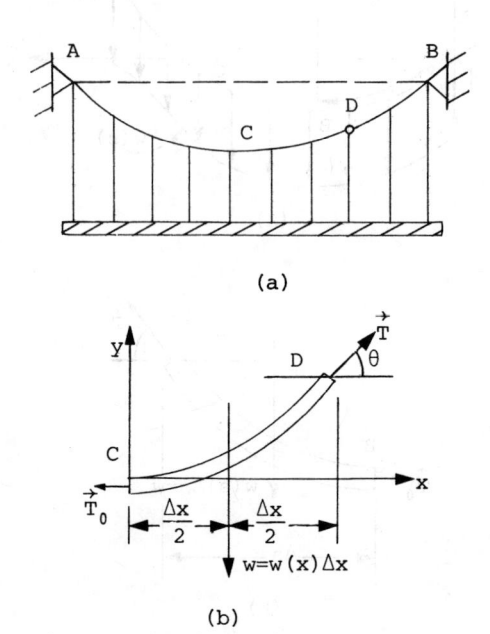

Fig. 8.8 Uniformly distributed load along the horizontal

Relationships:

$$T = \sqrt{T_0^2 + W^2 x^2}$$
$$\tan \theta = \frac{Wx}{T_0}$$
(8.8)

Deflection Curve:

From $\Sigma M_D = 0 \rightarrow$ $\quad y = \dfrac{Wx^2}{2T_0}$ (8.9)

Length of cable from point C to support B, S_B:

$$S_B = \int_0^{x_B} \sqrt{1 + \left(\frac{dy}{dx}\right)^2}\, dx$$
(8.10)

A-55

CHAPTER 9

FRICTION

9.1 DRY FRICTION

Dry Friction - Also called coulomb friction, is friction between bodies in the absence of lubrication.

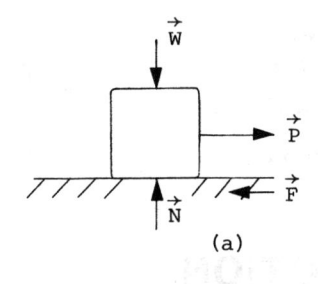

Fig. 9·1

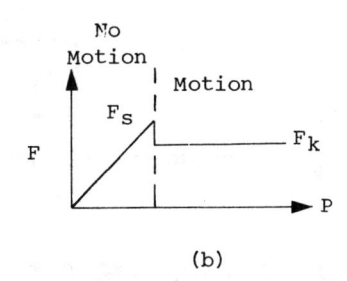

$\vec{N}$ = Normal Force

$\vec{F}$ = Static Friction Force

As $\vec{P}$ increases (refer to Figure 9.1), the body remains in equilibrium until $\vec{F}$ reaches a maximum value $\vec{F}_s$. If $\vec{P}$ continues to increase, then $\vec{F}$ drops to a lower value, $\vec{F}_k$, called kinetic-friction force.

Coefficient of Static Friction

$$\mu_s = \frac{F_s}{N}$$

Coefficient of Kinetic Friction

$$\mu_k = \frac{F_k}{N}$$

A) During equilibrium phase:

$$\boxed{F_s < \mu_s N} \quad (9.1)$$

μ_s = Coefficient of Static Friction

N = Normal Force

B) At the point of impending motion:

$$\boxed{F_s = \mu_s N} \quad (9.2)$$

C) When motion begins:

$$\boxed{F_k = \mu_k N} \quad (9.3)$$

μ_k = Coefficient of Kinetic Friction

9.2 ANGLES OF FRICTION

A) Angle of Static Friction, ϕ_s:

$$\boxed{\phi_s = \tan^{-1} \mu_s} \quad (9.4)$$

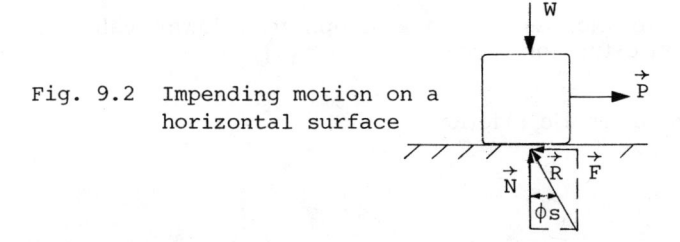

Fig. 9.2 Impending motion on a horizontal surface

A-57

For a body on an inclined surface

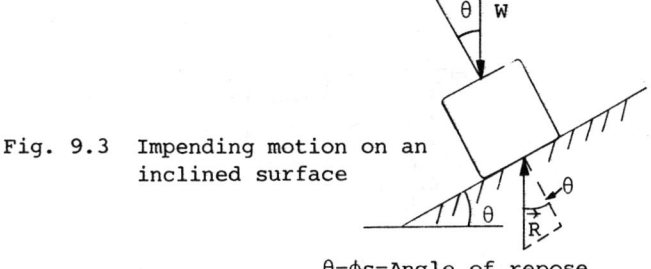

Fig. 9.3 Impending motion on an inclined surface

$\theta = \phi_s =$ Angle of repose

The angle of inclination corresponding to impending motion is the angle of repose.

B) Angle of Kinetic Friction, ϕ_k:

$$\phi_k = \tan^{-1}\mu_k \qquad (9.5)$$

9.3 WEDGES

Wedges are simply machines used to raise heavy loads. They stay in place after being forced under the load because of frictional forces.

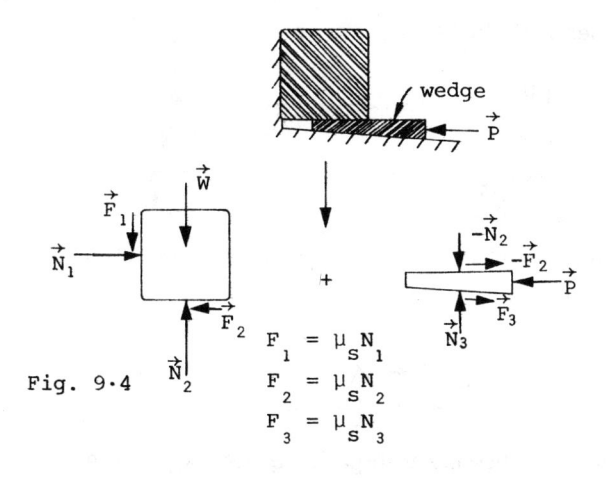

Fig. 9.4

$F_1 = \mu_s N_1$
$F_2 = \mu_s N_2$
$F_3 = \mu_s N_3$

A-58

9.4 SQUARE-THREADED SCREWS

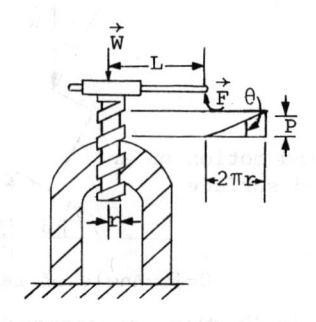

Fig. 9.5

A screw may be thought of as a block on an inclined plane. (Figure 9.6) Let:

r = Mean Radius of the Thread
P = Pitch of the Screw
θ = Pitch Angle, where:

$$\theta = \tan^{-1}\left(\frac{P}{2\pi r}\right)$$

$F_H = \dfrac{FL}{r}$; F_H = Horizontal Force

φ = Angle of Friction, where:

$$\phi = \tan^{-1} \mu$$

L = Lead of the Screw

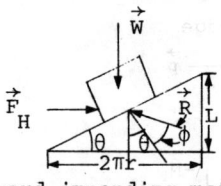

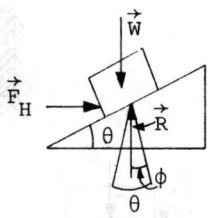

(a) Upward impending motion   (b) Downward impending motion, $\phi < \theta$

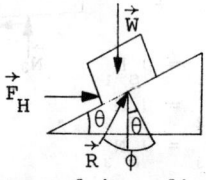

Fig. 9.6   (c) Downward impending motion, $\phi > \theta$

Note:

Lead - Distance through which the screw travels in one rotation

Pitch - Distance between two threads

The force needed to cause upward impending motion of the screw:

$$F = \frac{Wr}{L} \tan(\phi + \theta) \qquad (9.6)$$

If the weight $\vec{W}$ is to be lowered:

$$F = \frac{Wr}{L} \tan(\phi - \theta) \qquad (9.7)$$

9.5 AXLE FRICTION

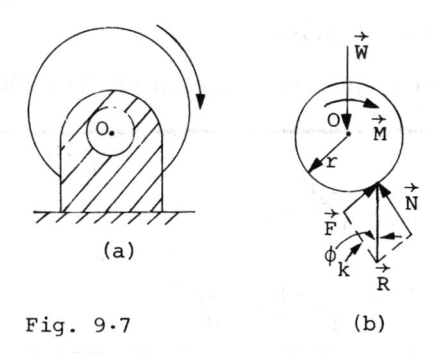

Fig. 9.7 (a) (b)

Moment needed to maintain rotation of the shaft:

$$M = Rr \sin \phi_k \qquad (9.8)$$

A-60

ϕ_k = Angle of Kinetic Friction:

$$\tan \phi_k = \frac{F}{N} = \mu_k$$

If μ_k is small, then M can be approximated:

$$\boxed{M \approx Rr\, \mu_k} \qquad (9.9)$$

9.6 DISK FRICTION

Types of Thrust Bearings:

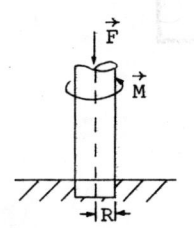

(a) End, or pivot bearing   (b) Collar bearing

Fig. 9·8 Thrust bearings

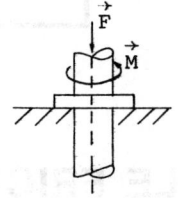

Moment required to overcome friction of the disk:

$$\boxed{\begin{aligned} M &= \frac{\mu_s F}{\pi(R_2^2 - R_1^2)} \int_0^{2\pi}\!\!\int_{R_1}^{R_2} r^2 \, dr\, d\theta \\ &= \frac{2}{3} \mu_s F \left[\frac{R_2^3 - R_1^3}{R_2^2 - R_1^2} \right] \end{aligned}} \qquad (9.10)$$

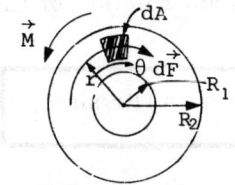

Fig. 9·9

When contact takes place over the whole disk, as in the case of pivot bearing:

$$\boxed{M = \tfrac{2}{3}\mu_s FR}$$
(9.11)

9.7 ROLLING RESISTANCE

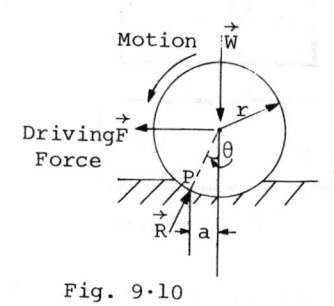

Fig. 9·10

Taking moment about point P:

$$\Sigma M_P = Wa = P(r\cos\theta)$$

For small deformation:

θ is very small and hence $\cos\theta = 1$.

$$\therefore \boxed{P \approx \frac{Wa}{r}}$$
(9.12)

a = coefficient of rolling resistance, given in terms of length.

A-62

9.8 BELT FRICTION

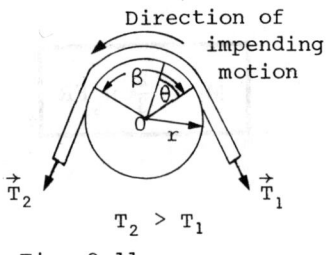

Fig. 9·11

For impending motion only:

$$\ln \frac{T_2}{T_1} = \mu_s \beta \quad (9.13)$$

or

$$T_2 = T_1 e^{\mu_s \beta} \quad (9.14)$$

β is given in radians, and is the angle of belt and surface contact.

μ_s = Coefficient of friction between belt and the surface

Note: Equation (9.13) is used for determining μ_s or β. Equation (9.14) is used to solve for T_1 or T_2. If there is motion, substitute μ_s with μ_k in eqs.(9.13) and (9.14).

V-Belts:

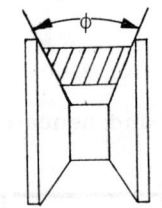

Fig. 9·12

$$\frac{T_2}{T_1} = e^{\mu_s \beta/\sin(\phi/2)} \quad (9.15)$$

ϕ = Angle of the V-Belt

A-63

CHAPTER 10

MOMENT OF INERTIA

10.1 INTEGRATION METHOD

Area

General Formula:

$$I = \int_A s^2 \, dA \qquad (10.1)$$

s = Perpendicular distance from the axis to the area element (Figure 10.1)

In component form:

$$I_x = \int y^2 \, dA$$
$$I_y = \int x^2 \, dA \qquad (10.2)$$

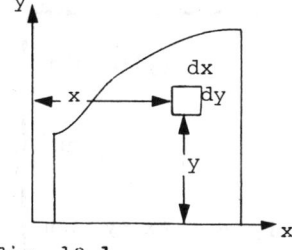

dA = dxdy
$dI_x = y^2 dA$
$dI_y = x^2 dA$

Fig. 10.1

A-64

For a Rectangular Area:

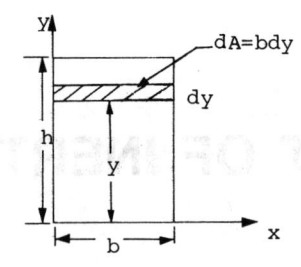

Fig. 10·2

$$I_x = \int_0^h by^2 dy = \frac{1}{3} bh^3 \qquad (10.3)$$

Note: This is the moment of inertia with respect to an axis passing through the base of the rectangle.

Moments of Inertia of Masses

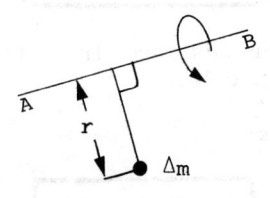

Fig. 10·3

$$I = \int r^2 dm \qquad (10.4)$$

In component form:

$$I_x = \int (y^2+z^2) dm$$
$$I_y = \int (z^2+x^2) dm \qquad (10.5)$$
$$I_z = \int (x^2+y^2) dm$$

A-65

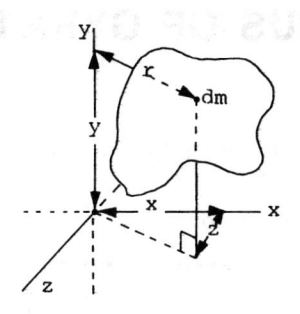

Fig. 10·4

10.2 POLAR MOMENT OF INERTIA

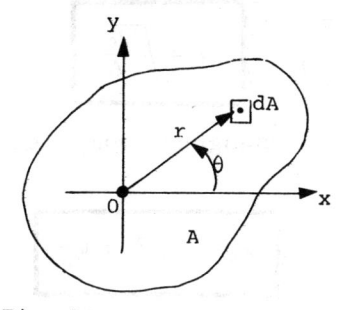

Fig. 10·5

Polar Moment of Inertia:

$$J_0 = \int r^2 dA \qquad (10.6)$$

In terms of rectangular moments of inertia:

$$J_0 = I_x + I_y \qquad (10.7)$$

10.3 RADIUS OF GYRATION

Areas

$$I_x = k_x^2 A; \quad I_y = k_y^2 A$$

Rectangular component form:

$$k_x = \sqrt{\frac{I_x}{A}}$$
$$k_y = \sqrt{\frac{I_y}{A}}$$

(10.8)

Polar form:

$$k_0 = \sqrt{\frac{J_0}{A}}$$

(10.9)

Relation between rectangular component form and polar form:

$$k_0^2 = k_x^2 + k_y^2$$

(10.10)

Masses

$$I = k^2 m$$

$$k = \sqrt{\frac{I}{m}}$$

(10.11)

10.4 PERPENDICULAR-AXIS THEOREM

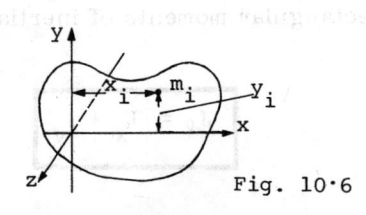

Fig. 10·6

For a thin plane lamina:

$$I_z = \sum_i m_i x_i^2 + \sum_i m_i y_i^2 = I_x + I_y \;;\; i = 1, 2, \ldots \qquad (10.12)$$

Statement

The moment of inertia of any plane lamina about an axis normal to the plane of the lamina is equal to the sum of the moments of inertia about any two mutually perpendicular axes passing through the given axis and lying in the plane of the lamina.

10.5 PARALLEL-AXIS THEOREM

Areas

The theorem states that the moment of inertia of an area about a given axis is equal to the sum of the moment of inertia parallel to the given axis (and passing through the centroid of the area) and the product of the area and the square of the distance between the two parallel axes.

$$I = I_c + Ad^2 \qquad (10.13)$$

I_c = Moment of inertia about an axis through the centroid
d = Distance Between the Axes
A = Area

In terms of the radius of gyration:

$$k_0^2 = k_c^2 + d^2 \qquad (10.14)$$

In polar form:

$$J_0 = J_c + Ad^2 \qquad (10.15)$$

J_c = Polar Moment of Inertia about the Centroid

Masses

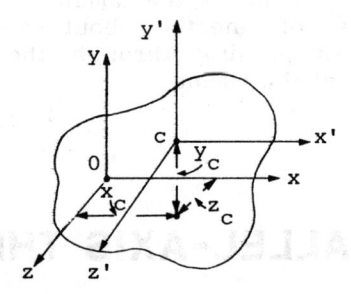

Fig. 10·7

$$I = I_c + md^2 \qquad (10.16)$$

In the component form:

$$\begin{aligned} I_x &= I_{x_c} + m(y_c^2 + z_c^2) \\ I_y &= I_{y_c} + m(z_c^2 + x_c^2) \\ I_z &= I_{z_c} + m(x_c^2 + y_c^2) \end{aligned} \qquad (10.17)$$

10.6 COMPOSITE AREAS AND BODIES

A composite area (or body) consists of several simpler shapes of bodies as components. The moments of inertia of these component shapes (or bodies) about a given axis may be computed separately and then added to give the moment of inertia for the entire area (or body).

The moment of inertia of a region with a hole equals the difference between the moment of inertia of the complete area (neglecting the hole) and the moment of inertia of the hole.

Example: C is the centroid,

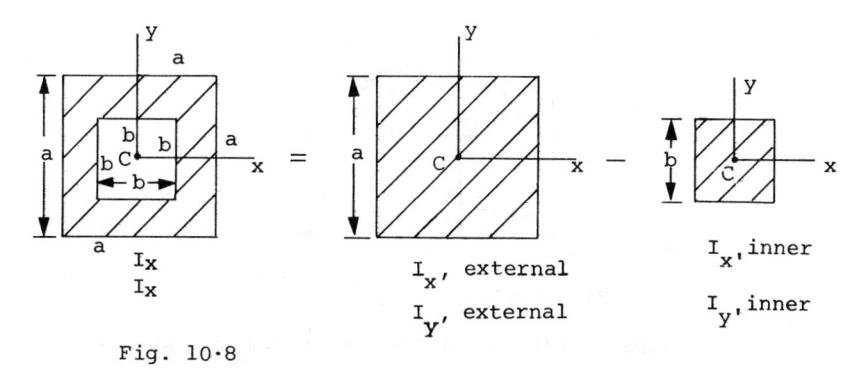

Fig. 10·8

I_x
I_x

$I_{x'}$ external
$I_{y'}$ external

$I_{x'}$ inner
$I_{y'}$ inner

10.7 PRODUCT OF INERTIA

Areas

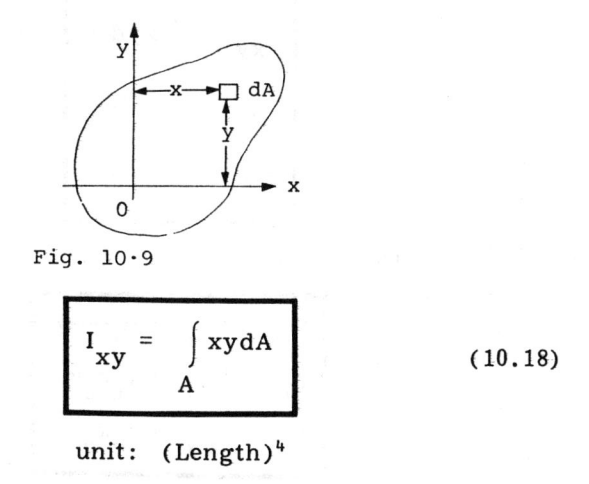

Fig. 10·9

$$I_{xy} = \int_A xy\,dA \qquad (10.18)$$

unit: $(\text{Length})^4$

Note: I_{xy} is zero when x, y, or both x and y axes are axes of symmetry for the region.

Parallel-Axis Theorem:

$$I_{xy} = I_{x'y'} + x_c y_c A \qquad (10.19)$$

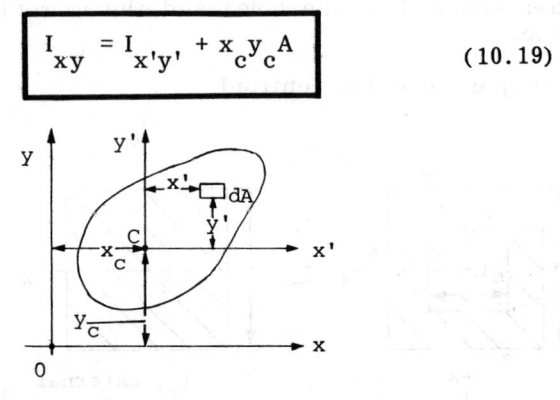

Fig. 10-10

$I_{x'y'}$ = Moment of inertia in the centroidal axis system

x_c, y_c = Coordinates of the centroid in the 'xoy' axis system.

Mass Products of Inertia

Component Equations:

$$I_{xy} = \int xy\,dm$$
$$I_{yz} = \int yz\,dm \qquad (10.20)$$
$$I_{zx} = \int zx\,dm$$

Parallel-Axis Theorem:

$$I_{xy} = I_{x'y'} + x_c y_c m$$
$$I_{yz} = I_{y'z'} + y_c z_c m \qquad (10.21)$$
$$I_{zx} = I_{z'x'} + z_c x_c m$$

m = Total Mass of the Body.

Moment of Inertia with Respect to an Arbitrary Axis:

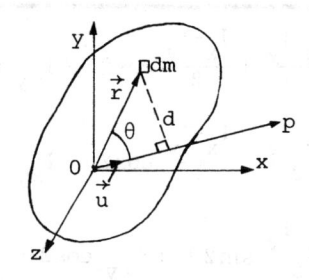

Fig. 10·11

$$I_{OP} = \int d^2 dm = \int (\vec{u} \times \vec{r})^2 \, dm \qquad (10.22)$$

In terms of scalar quantities:

$$I_{OP} = I_x u_x^2 + I_y u_y^2 + I_z u_z^2 - 2I_{xy} u_x u_y - 2I_{yz} u_y u_z - 2I_{zx} u_z u_x$$

$$(10.23)$$

10.8 PRINCIPAL AXES AND INERTIAS

Areas

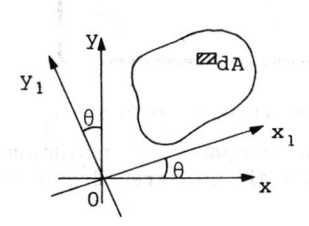

Fig. 10·12

To calculate $I_{x_1}, I_{y_1}, I_{x_1y_1}$ when θ, I_x, I_y, and I_{xy} are known; with respect to the inclined x_1 and y_1 axes:

$$I_{x_1} = \frac{I_x+I_y}{2} + \frac{I_x-I_y}{2}\cos 2\theta - I_{xy}\sin 2\theta$$

$$I_{y_1} = \frac{I_x+I_y}{2} - \frac{I_x-I_y}{2}\cos 2\theta + I_{xy}\sin 2\theta \qquad (10.24)$$

$$I_{x_1y_1} = \frac{I_x-I_y}{2}\sin 2\theta + I_{xy}\cos 2\theta$$

The polar moment of inertia about the z axis is independent of the orientation of x_1 and y_1:

$$J_0 = I_{x_1} + I_{y_1} = I_x + I_y \qquad (10.25)$$

Principal Axes - Axes about which the moments of inertia are maximum or minimum. The corresponding moments of inertia are called principal moments of inertia.

Note: Every point on the body has its own set of principal axes. The important point to consider is the centroid.

Orientation of the principal axes at the centroid; θ_p:

$$\tan 2\theta_p = \frac{-I_{xy}}{\frac{I_x-I_y}{2}} \qquad (10.26)$$

Equation (10.26) has two solutions, θ_{p_1} and θ_{p_2}. They correspond to the maximum and minimum values of the principal moments of inertia and are $90°$ apart.

Principal Moments of Inertia:

A-73

$$I_{\substack{max\\min}} = \frac{I_x + I_y}{2} \pm \sqrt{\left(\frac{I_x - I_y}{2}\right)^2 + I_{xy}^2} \qquad (10.27)$$

Note: The product of inertia with respect to the principal axes is zero. Therefore, axes of symmetry are also principal axes for a given area.

Ellipsoid of Inertia:

Equation of an ellipsoid:

$$I_x x^2 + I_y y^2 + I_z z^2 - 2I_{xy} xy - 2I_{yz} yz - 2I_{zx} zx = 1$$

$$(10.28)$$

The ellipsoid defines the moment of inertia of the body with respect to any axis passing through the origin O, it is also called the ellipsoid of inertia at point O.

If the principal axes, x_1, y_1, z_1 are used as coordinate axes, then the equation of the ellipsoid becomes:

$$I_{x_1} x_1^2 + I_{y_1} y_1^2 + I_{z_1} z_1^2 = 1 \qquad (10.29)$$

Using the principal axes, the moment of inertia with respect to an arbitrary axis through the origin O, equation (10.23), is:

$$I_{OP} = I_{x_1} u_{x_1}^2 + I_{y_1} u_{y_1}^2 + I_{z_1} u_{z_1}^2 \qquad (10.30)$$

10.9 MOHR'S CIRCLE

Purpose – To transform the moment of inertia I_x, I_y, I_{xy}, into the principal moments of inertia I_{max}, I_{min} for a plane area.

A-74

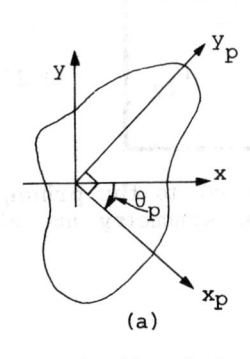

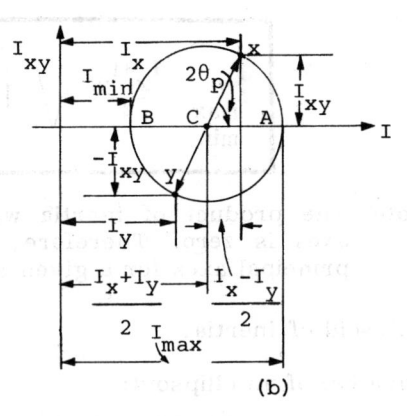

Fig. 10·13 Mohr's Circle

Procedure to Construct Mohr's Circle:

A) Determine I_x, I_y, I_{xy} (Figure 10.13(a)).

B) Construct the coordinate system of I and I_{xy} (Figure 10.13(b)).

C) Determine point C, the center of the circle: $\dfrac{I_x + I_y}{2}$

D) Plot point x: $x = (I_x, I_{xy})$.

E) Draw the diameter $\overline{xy}$. The radius $\overline{cx}$ is $\sqrt{\left(\dfrac{I_x - I_y}{2}\right)^2 + I_{xy}^2}$

F) Draw the circle with the above radius. Points A and B give the values of I_{max} and I_{min}.

G) To find the direction of the principal axes, X_p and Y_p, determine the angle $2\theta_p$ in Figure 10.13(b). Plot θ_p on the area. Both the angle of the area and the angle on the circle must be measured in the same sense.

H) θ_p gives the direction for the maximum moment of inertia. The minimum moment of inertia is perpendicular to this axis.

A-75

10.10 THIN PLATES

Thin Plate Assumptions:
A) Uniform thickness, t.
B) Homogeneous material, with density ρ.

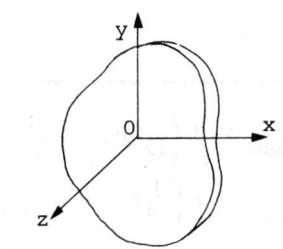

Fig. 10·14

$$(I_x)_{mass} = (I_x)_{area} \, \rho t$$
$$(I_y)_{mass} = (I_y)_{area} \, \rho t \qquad (10.31)$$
$$(I_z)_{mass} = (J_0)_{area} \, \rho t$$

where

$(I_x)_{area}$ = Area moment of inertia about the x-axis.

$(I_x)_{mass}$ = Mass moment of inertia about the x-axis.

$(J_0)_{area}$ = Polar moment of inertia at point O.

Relationship between the mass moments of inertia of a thin plate:

$$(I_z)_{mass} = (I_x)_{mass} + (I_y)_{mass} \qquad (10.32)$$

A-76

Rectangular Plate

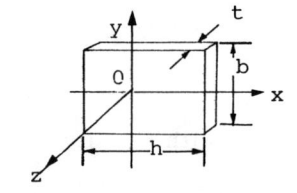

Fig. 10·15

$$(I_x)_{mass} = \rho t \left(\frac{1}{12} b^3 h\right) = \frac{1}{12} mb^2$$

$$(I_y)_{mass} = \rho t \left(\frac{1}{12} bh^3\right) = \frac{1}{12} mb^2 \qquad (10.33)$$

$$(I_z)_{mass} = \frac{1}{12} m(b^2 + h^2)$$

Circular Plate

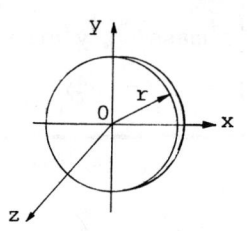

Fig. 10·16

$$(I_x)_{mass} = \rho t \left(\frac{1}{4} \pi r^4\right) = (I_y)_{mass} = \frac{1}{4} mr^2$$

$$(m = \pi r^2 st) \qquad (10.34)$$

$$(I_z)_{mass} = (I_x)_{mass} + (I_y)_{mass} = \frac{1}{2} mr^2$$

CHAPTER 11

METHOD OF VIRTUAL WORK

11.1 WORK OF A FORCE

Work done by a force $\vec{F}$ is defined as:

$$\boxed{d\vec{w} = \vec{F} \cdot d\vec{s}} \qquad (11.1)$$

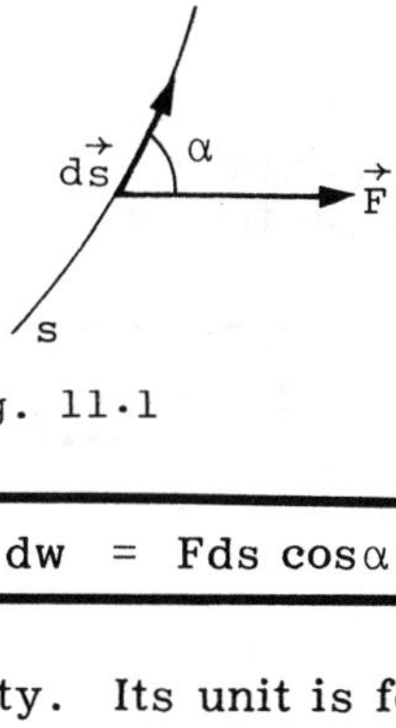

Fig. 11.1

$$\boxed{dw = Fds \cos\alpha} \qquad (11.2)$$

Work is a scalar quantity. Its unit is force-length.

Special Cases:

A) 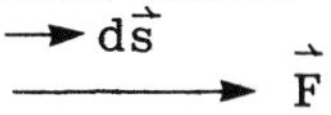 $\xrightarrow{d\vec{s}} \quad \xrightarrow{\vec{F}} \quad dw = Fds$

B) 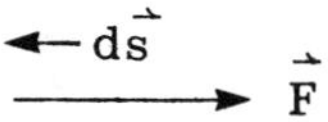 $\xleftarrow{d\vec{s}} \quad \xrightarrow{\vec{F}} \quad dw = -Fds$

C) 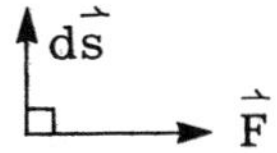 $\uparrow d\vec{s} \quad \xrightarrow{\vec{F}} \quad dw = 0$

A-78

The total work of the internal forces holding particles of a body together, is zero.

Work of a Moment, M:

$$\boxed{dw = Md\theta} \qquad (11.3)$$

θ = Angle the Body has Rotated

11.2 PRINCIPLE OF VIRTUAL WORK

Virtual movement are displacements or rotations which are assumed in a problem under the applied loading conditions, but may not actually take place. They are denoted by δs and $\delta\theta$, respectively.

Virtual work done by:

A Force : $\quad \delta w = F\cos\theta \, \delta s \qquad (11.4)$

A Couple: $\quad \delta w = M\delta\theta \qquad (11.5)$

If the system is in equilibrium, then the virtual work done by every external load on the system is zero:

$$\boxed{\delta w = 0} \qquad (11.6)$$

Note:

A) The equilibrium condition is both necessary and sufficient.

B) This method is most effective when applied to problems which involve series of connected rigid bodies.

C) If the virtual displacements are consistent with the constraints imposed by the supports, then all reactions and internal forces can be eliminated and only the external forces are considered.

11.3 MECHANICAL EFFICIENCY

Ideal Machines - Input work equals output work

Real Machines - Output work is always less than input work because of losses due to friction.

Mechanical efficiency is defined as:

$$\eta = \frac{\text{output work}}{\text{input work}} \qquad (11.7)$$

For an ideal machine: $\eta = 1$.

11.4 FINITE DISPLACEMENTS

For work done between two specific points of displacement:

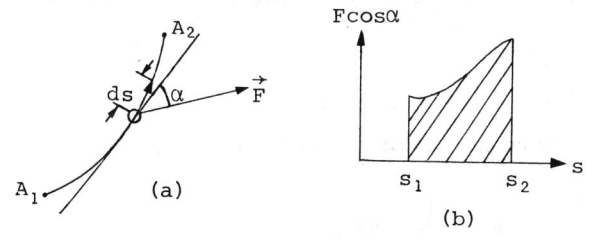

$$W_{1 \to 2} = \int_{A_1}^{A_2} \vec{F} \cdot d\vec{r} \qquad (11.8)$$

Fig. 11·2

In scalar form, equation (11.8) becomes:

A-80

$$\boxed{W_{1\to 2} = \int_{S_1}^{S_2} (F\cos\alpha)\,ds \\ = \text{Area under the curve of } F\cos\alpha \text{ verses } s.}$$ (11.9)

Work of a couple:

$$\boxed{W_{1\to 2} = \int_{\theta_1}^{\theta_2} M\,d\theta}$$ (11.10)

Specific Cases

Weight - For a weight raised or lowered from point 1 to point 2:

$$W_{1\to 2} = -w(y_2 - y_1) = -w\,\Delta y$$

Δy = Vertical Displacement

Work is positive when the body is lowered.

Spring - Spring Force = $F = kx$

where k = Spring Constant

$$W_{1\to 2} = \tfrac{1}{2} kx_1^2 - \tfrac{1}{2} kx_2^2$$

Work is positive when $x_2 < x_1$, or when the spring is returning to its original unstretched position.

$$F = kx$$

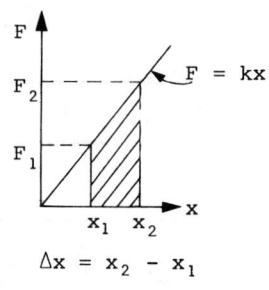

Fig. 11·3

Graphical Representation:

$$W_{1\to 2} = -\frac{1}{2}(F_1+F_2)\Delta x$$

$$= \text{Area under the line } F = kx \text{ between the limits } x_1 \text{ and } x_2.$$

11.5 POTENTIAL ENERGY

If the work done by a force is independent of the path followed and is dependent only upon the initial and final position, the force is said to be conservative.

The work done by a conservative force is stored in the form of an energy called potential energy, denoted by v.

Examples:

A) Gravity:

$$v = wy$$

v increases when $w_{1\to 2}$ is negative. That is, when an object is raised.

B) Spring Force:

$$v = \frac{1}{2}kx^2$$

A-82

In general:

$$\boxed{dw = -dv} \qquad (11.11)$$

or

$$\boxed{w_{1 \to 2} = v_1 - v_2} \qquad (11.12)$$

For a system in equilibrium:

$$\frac{\delta v}{\delta q} = 0 \qquad (11.13)$$

q = Independent Variable Defining Position

Procedure for solving problems:

A) Express the potential energy in terms of displacement.
B) Express the displacement in terms of q.
C) Differentiate the result.
D) Set it equal to zero.
E) Solve for the value(s) of q.

11.6 STABILITY

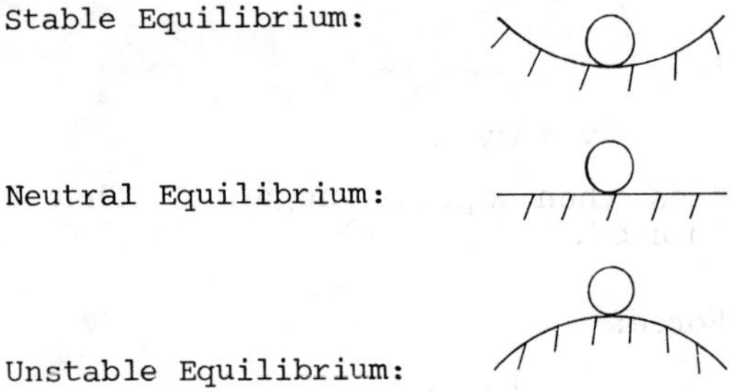

Stable Equilibrium:

Neutral Equilibrium:

Unstable Equilibrium:

Fig. 11·4 Stability

In terms of potential energy, v:

    Stable - Minimum Potential Energy

    Unstable - Maximum Potential Energy

    Neutral - Constant Potential Energy

Mathematical Form:

$$\text{Stable:} \quad \frac{dv}{dq} = 0 \qquad \frac{d^2v}{dq^2} > 0$$

$$\text{Unstable:} \quad \frac{dv}{dq} = 0 \qquad \frac{d^2v}{dq^2} < 0$$

If $\frac{dv}{dq} = 0$, $\frac{d^2v}{dq^2} = 0$, then examine a higher order derivative; the equilibrium is:

    Neutral - If all derivatives are zero

    Stable - If higher order derivatives of $\left(\frac{d^2v}{dq^2}\right)$ are greater than zero.

    Unstable - For all other cases.

CHAPTER 12

KINEMATICS OF PARTICLES

12.1 BASIC DEFINITIONS

Rectilinear Motion - The resulting linear motion of a particle when influenced by the resultant force whose direction and line of action always remain the same.

Position:

Fig. 12·1

The equation of motion is given as a scalar function of time t, x = x(t).

Instantaneous velocity: $v = \dfrac{dx}{dt}$ (12.1)

Instantaneous acceleration:

$$a = \dfrac{dv}{dt} = \dfrac{d}{dt}\left(\dfrac{dx}{dt}\right) = \dfrac{d^2x}{dt^2}$$ (12.2)

Curvilinear Motion - Resulting non-linear motion of a particle when influenced by a variable force.

Position:

A-85

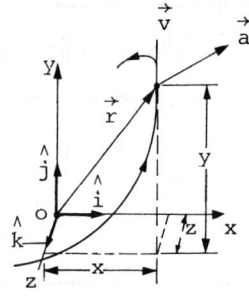

Fig. 12·2

The position is described by vector $\vec{r}$, $\vec{r} = \vec{r}(t)$.

Velocity:

$$\vec{v} = \frac{d\vec{r}}{dt} = \dot{\vec{r}} \tag{12.3}$$

Acceleration:

$$\vec{a} = \frac{d\vec{v}}{dt} = \dot{\vec{v}} \tag{12.4}$$

In terms of rectangular components:

Position:

$$\vec{r} = x\hat{i} + y\hat{j} + z\hat{k} \tag{12.5}$$

Velocity:

$$\vec{v} = \frac{d\vec{r}}{dt} = \dot{x}\hat{i} + \dot{y}\hat{j} + \dot{z}\hat{k} \tag{12.6}$$

Acceleration:

A-86

$$\vec{a} = \frac{d\vec{v}}{dt} = \ddot{x}\hat{i} + \ddot{y}\hat{j} + \ddot{z}\hat{k} \qquad (12.7)$$

Centroidal Motion - Center of gravity remains stationary. Only a moment couple is acting on the body.

Speed - The magnitude of the velocity vector:

$$v = |\vec{v}| = \sqrt{\dot{x}^2 + \dot{y}^2 + \dot{z}^2} \qquad (12.8)$$

Hodograph - A curve generated by the tip of the velocity vector. A chord of the hodograph is the change in velocity during the corresponding time.

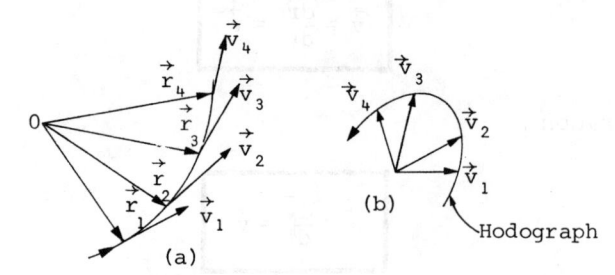

Fig. 12·3 Hodograph

12.2 MOTION OF A PARTICLE

The motion of a particle is known if its position is known for all values of time t.

Case 1: If acceleration is given in terms of t, a = f(t):

Velocity: dv = adt

$$v - v_0 = \int_0^t f(t)dt \qquad (12.9)$$

Gives v in terms of t

Position:

$$x = \int_0^t v(t)dt + x_0 \qquad (12.10)$$

Case 2: If acceleration is a function of velocity, $a = f(v)$:

Time: $f(v) = \dfrac{dv}{dt}$

$$t = \int \dfrac{dv}{f(v)} + t_0 \qquad (12.11)$$

Position: $f(v) = v\dfrac{dv}{dx}$

$$x = \int \dfrac{vdv}{f(v)} + x_0 \qquad (12.12)$$

Case 3: If acceleration is a function of displacement, $a = f(x)$:

Velocity:

$$vdv = adx$$
$$\tfrac{1}{2}v^2 - \tfrac{1}{2}v_0^2 = \int_{x_0}^x f(x)dx \qquad (12.13)$$

Position:

$$x = \int_0^t v\,dt + x_0 \qquad (12.14)$$

A-88

12.3 UNIFORM MOTION

A) For constant velocity:

$$\boxed{x = x_0 + vt}$$ (12.15)

B) For constant acceleration:

$$\boxed{\begin{aligned} v &= v_0 + at \\ x &= x_0 + v_0 t + \tfrac{1}{2} at^2 \\ v^2 &= v_0^2 + 2a(x-x_0) \end{aligned}}$$ (12.16)

x_0, v_0 are initial values for x and v.

12.4 RELATIVE MOTION

Rectilinear Motion

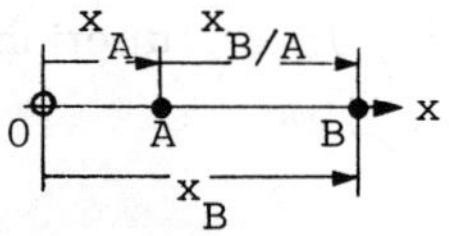

Fig. 12·4

$$\boxed{\begin{aligned} \text{Relative position:} &\quad x_B = x_A + x_{B/A} \\ \text{Relative velocity:} &\quad v_B = v_A + v_{B/A} \\ \text{Relative Acceleration:} &\quad a_B = a_A + a_{B/A} \end{aligned}}$$ (12.17)

Note:
A) All motion is measured from the same origin.
B) Time is recorded at the same instant.

Dependent Motion

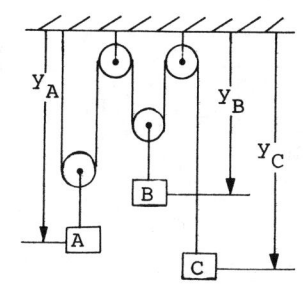

Fig. 12·5

Relative position: $2x_A + 2x_B + x_C$ = Constant

Relative velocity: $2v_A + 2v_B + v_C = 0$

Relative acceleration: $2a_A + 2a_B + a_C = 0$

(12.18)

Curvilinear Motion

Relative position: $\vec{r}_B = \vec{r}_A + \vec{r}_{B/A}$

Relative velocity: $\vec{v}_B = \vec{v}_A + \vec{v}_{B/A}$

Relative acceleration: $\vec{a}_B = \vec{a}_A + \vec{a}_{B/A}$

(12.19)

Absolute Motion — Motion with respect to a fixed frame of reference.

12.5 GRAPHICAL SOLUTIONS

Graphical solutions are useful when:

A) Data is obtained from experiments.
B) When the position, velocity, and acceleration are not analytical functions of time t.

A-90

C) The motion is made up of different parts.

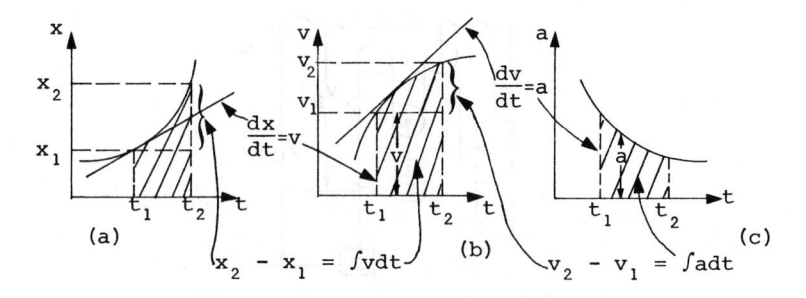

Fig. 12·6 Graphical Solutions

Relationship between the graphs:

A) Slope of the x-t curve represents velocity.

B) Slope of the v-t curve represents acceleration.

C) Area under the a-t curve represents the change in velocity.

D) Area under the v-t curve represents the change in position.

Special Cases:

A) Constant Velocity:

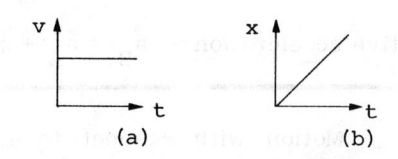

Fig. 12·7 Constant Velocity

B) Constant Acceleration:

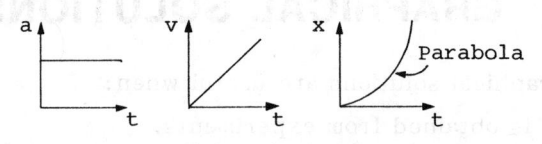

Fig. 12·8 Constant Acceleration

C) Linear Acceleration:

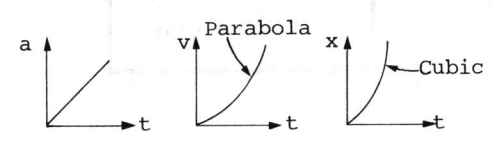

Fig. 12·9 Linear Acceleration

In general, if the acceleration is a polynomial of degree n, then the velocity will be a polynomial of degree n+1, and the position will be a polynomial of degree n+2. These polynomials will be represented by graphs of curves of the same degree.

Moment-Area Method

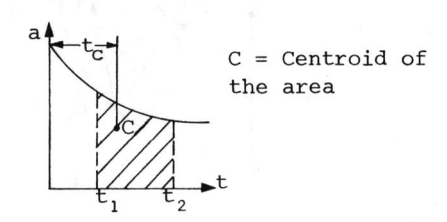

C = Centroid of the area

Fig. 12·10

If the value of t_c is known, then

$$x_2 = x_1 + v_1 t_1 + (\text{area under the a-t curve})(t_2 - t_c)$$

(12.20)

Using the v-x curve to find acceleration:

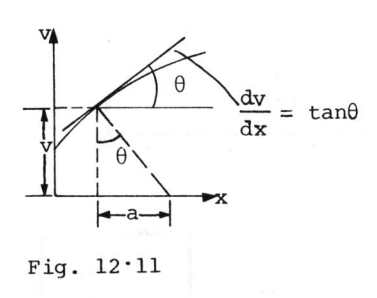

$$\frac{dv}{dx} = \tan\theta$$

Fig. 12·11

A-92

$$a = v\frac{dv}{dx} = v\tan\theta \qquad (12.21)$$

12.6 TANGENTIAL AND NORMAL COMPONENTS

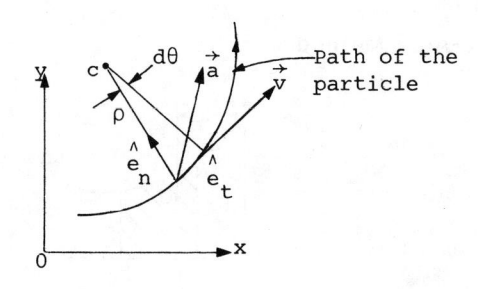

Fig. 12·12

c = Center of curvature of the path

$\hat{e}_t$ = Unit vector in the direction of the velocity vector

$\hat{e}_n$ = Unit vector in the direction normal to the velocity

$$\frac{d\hat{e}_t}{dt} = \dot{\theta}\hat{e}_n = \frac{v}{\rho}\hat{e}_n, \text{ where}$$

$$\rho = \frac{(1+(dy/dx)^2)^{3/2}}{\left|\frac{d^2y}{dx^2}\right|}$$

ρ = radius of curvature of the path

$$(12.22)$$

Velocity:

$$\vec{v} = v\hat{e}_t \qquad (12.23)$$

A-93

Acceleration:

$$\vec{a} = \dot{v}\hat{e}_t + \frac{v^2}{\rho}\hat{e}_n \qquad (12.24)$$

There are two parts to equation (12.24):

Tangential Acceleration:

$$a_t = \dot{v} \qquad (12.25)$$

Centripetal Acceleration:

$$a_n = \frac{v^2}{\rho} \qquad (12.26)$$

Magnitude of Total Acceleration: $|\vec{a}| = \sqrt{\dot{v}^2 + \frac{v^4}{\rho^2}}$ (12.27)

Special cases:

A) If the particle is traveling at constant speed:

$$\vec{a} = \frac{v^2}{\rho}\hat{e}_n; \quad a_t = 0$$

B) If the particle is in rectilinear motion:

$$\vec{a} = \dot{v}\hat{e}_t; \quad a_n = 0$$

Correlation between rectangular components with normal and tangential components of $\vec{a}$:

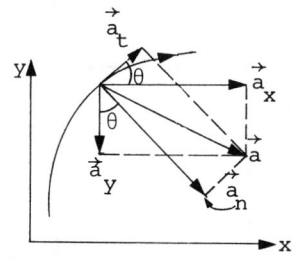

Fig. 12·13

$$\boxed{\begin{aligned} a_n &= a_x \sin\theta + a_y \cos\theta \\ a_t &= a_x \cos\theta - a_y \sin\theta \end{aligned}} \quad (12.28)$$

or

$$\boxed{\begin{aligned} a_x &= a_n \sin\theta + a_t \cos\theta \\ a_y &= a_n \cos\theta - a_t \sin\theta \end{aligned}} \quad (12.29)$$

12.7 RADIAL AND TRANSVERSE COMPONENTS

Polar Coordinates

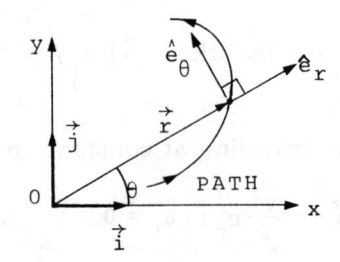

Fig. 12·14

$\hat{e}_r$ = Radial Unit Vector

$\hat{e}_\theta$ = Transverse Unit Vector

Unit Vector Derivatives:

$$\boxed{\begin{aligned} \frac{d\hat{e}_\theta}{dt} &= -\dot{\theta}\,\hat{e}_r \\ \frac{d\hat{e}_r}{dt} &= \dot{\theta}\,\hat{e}_\theta \end{aligned}} \quad (12.30)$$

$$\boxed{\begin{aligned}&\text{Position: } \vec{r} = r\hat{e}_r \\ &\text{Velocity: } \vec{v} = \dot{r}\hat{e}_r + r\dot{\theta}\hat{e}_\theta \\ &\text{Acceleration: } \vec{a} = (\ddot{r} - r\dot{\theta}^2)\hat{e}_r + (r\ddot{\theta} + 2\dot{r}\dot{\theta})\hat{e}_\theta\end{aligned}}$$

(12.31)

Radial Components:	Transverse Components:
Velocity: $v_r = \dot{r}$	Velocity: $v_\theta = r\dot{\theta}$
Acceleration: $a_r = \ddot{r} - r\dot{\theta}^2$	Acceleration: $a_\theta = r\ddot{\theta} + 2\dot{r}\dot{\theta}$
	$= \dfrac{1}{r}\dfrac{d}{dt}(r^2\dot{\theta})$

Transformation into Tangential and Normal Components:

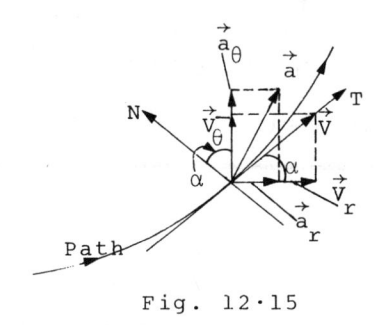

Fig. 12·15

$$\boxed{\begin{aligned}a_t &= a_r \cos\alpha + a_\theta \sin\alpha = \frac{a_r v_r + a_\theta v_\theta}{v} \\ a_n &= a_\theta \cos\alpha - a_r \sin\alpha = \frac{a_\theta v_r - a_r v_\theta}{v}\end{aligned}}$$

(12.32)

A-96

Cylindrical Coordinates

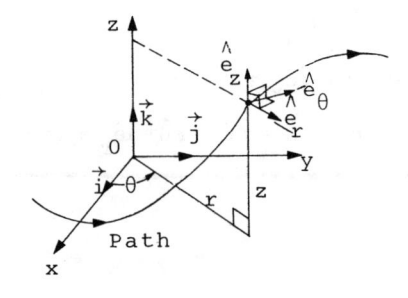

Fig. 12·16

Unit vector derivatives as given by equation (12.30):

$$\frac{d\hat{e}_r}{dt} = \dot{\theta}\hat{e}_\theta$$

$$\frac{d\hat{e}_\theta}{dt} = -\dot{\theta}\hat{e}_r$$

Position: $\vec{r} = r\hat{e}_r + z\hat{e}_z$

Velocity: $\vec{v} = \dot{r}\hat{e}_r + r\dot{\theta}\hat{e}_\theta + \dot{z}\hat{e}_z$

Acceleration: $\vec{a} = (\ddot{r} - r\dot{\theta}^2)\hat{e}_r + (2\dot{r}\dot{\theta} + r\ddot{\theta})\hat{e}_\theta + \ddot{z}\hat{e}_z$

(12.33)

Transformation:

$$\hat{e}_r = \vec{i}\cos\theta + \vec{j}\sin\theta$$
$$\hat{e}_\theta = -\vec{i}\sin\theta + \vec{j}\cos\theta$$
$$\hat{e}_z = \vec{k}$$

(12.34)

Spherical Coordinates

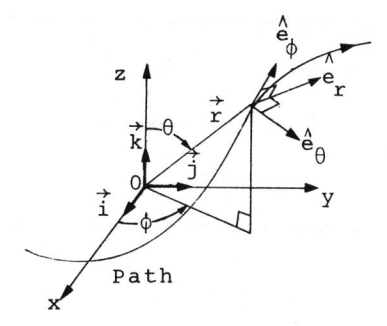

Fig. 12·17

Unit Vector Derivatives:

$$\frac{d\hat{e}_r}{dt} = \dot{\phi}\hat{e}_\phi \sin\theta + \dot{\theta}\hat{e}_\theta$$

$$\frac{d\hat{e}_\theta}{dt} = -\dot{\theta}\hat{e}_r + \dot{\phi}\hat{e}_\phi \cos\theta \qquad (12.35)$$

$$\frac{d\hat{e}_\phi}{dt} = -\dot{\phi}\hat{e}_r \sin\theta - \dot{\phi}\hat{e}_\theta \cos\theta$$

Position: $\vec{r} = r\hat{e}_r$

Velocity: $\vec{v} = \hat{e}_r \dot{r} + \hat{e}_\phi r\dot{\phi}\sin\theta + \hat{e}_\theta r\dot{\theta}$

Acceleration: $\vec{a} = (\ddot{r} - r\dot{\phi}^2\sin^2\theta - r\dot{\theta}^2)\hat{e}_r$

$\qquad + (r\ddot{\theta} + 2\dot{r}\dot{\theta} - r\dot{\phi}^2\sin\theta\cos\theta)\hat{e}_\theta$

$\qquad + (r\ddot{\phi}\sin\theta + 2\dot{r}\dot{\phi}\sin\theta + 2r\dot{\theta}\dot{\phi}\cos\theta)\hat{e}_\phi$

(12.36)

Transformation:

$$\hat{e}_r = \vec{i}\sin\theta\cos\phi + \vec{j}\sin\theta\sin\phi + \vec{k}\cos\theta$$
$$\hat{e}_\theta = \vec{i}\cos\theta\cos\phi + \vec{j}\cos\theta\sin\phi - \vec{k}\sin\theta \quad (12.37)$$
$$\hat{e}_\phi = -\vec{i}\sin\phi + \vec{j}\cos\phi$$

12.8 ANGULAR VELOCITY

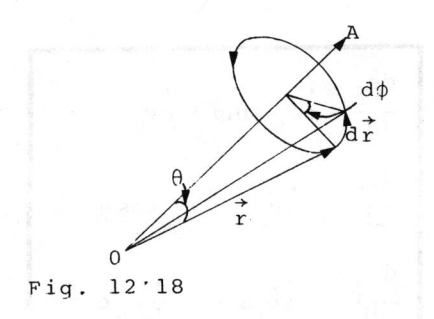

Fig. 12·18

For a particle rotating about an axis $\overline{OA}$ through an angle $d\phi$:

$$\vec{dr} = d\phi(\hat{e}\times\vec{r}) \quad (12.38)$$

The angular velocity is defined by:

$$\vec{\omega} = \dot{\phi}\,\hat{e} \quad (12.39)$$

The velocity of the particle is then expressed as:

$$\vec{v} = \dot{\vec{r}} = \vec{\omega}\times\vec{r} \quad (12.40)$$

Angular velocity vectors obey the rule of vector addition.

CHAPTER 13

KINETICS OF PARTICLES

13.1 LINEAR MOMENTUM

Linear momentum is defined as

$$\boxed{\vec{P} = m\vec{v}}$$ (13.1)

It is directed in the same direction as the vector $\vec{v}$.

Results:

A) Newton's first law can be written as:

If the resultant force acting on a particle is zero, then the linear momentum $\vec{P}$ of the particle is constant. This is the Law of Conservation of Linear Momentum.

B) Newton's second law can be expressed as

$$\Sigma \vec{F} = \frac{d}{dt}(m\vec{v}) = \frac{d\vec{P}}{dt}$$

i.e., the resultant force is equal to the rate of change of linear momentum.

C) Newton's third law:

For two particles A and B,

$$\vec{P}_A + \vec{P}_B = \text{constant}$$

A-100

13.2 EQUATIONS OF MOTION

Rectangular Components

or
$$\Sigma (F_x \vec{i} + F_y \vec{j} + F_z \vec{k}) = m(a_x \vec{i} + a_y \vec{j} + a_z \vec{k})$$

$$\boxed{\begin{aligned} \Sigma F_x &= m\ddot{x} \\ \Sigma F_y &= m\ddot{y} \\ \Sigma F_z &= m\ddot{z} \end{aligned}} \quad (13.2)$$

Tangential and Normal Components (refer to Figure 12.12)

$$\boxed{\begin{aligned} \Sigma F_t &= ma_t = m \frac{dv}{dt} \\ \Sigma F_n &= ma_n = m \frac{v^2}{\rho} \end{aligned}} \quad (13.3)$$

ρ = Radius of curvature of the path

Radial and Transverse Components (refer to Figure 12.14)

$$\boxed{\begin{aligned} \Sigma F_r &= ma_r = m(\ddot{r} - r\dot{\theta}^2) \\ \Sigma F_\theta &= ma_\theta = m(r\ddot{\theta} + 2\dot{r}\dot{\theta}) \end{aligned}} \quad (13.4)$$

13.3 DYNAMIC EQUILIBRIUM

A particle is in dynamic equilibrium if the sum of all forces acting on the particle, including inertial forces, is equal to zero:

$$\boxed{\Sigma \vec{F} - m\vec{a} = 0} \quad (13.5)$$

A-101

The term $-m\vec{a}$ is called the inertia vector.

The first term $\Sigma\vec{F}$ is sometimes called effective forces. i.e., it is the resultant of all external forces.

When expressed in tangential and normal form:

A) The tangential compoment of inertia measures the resistance of the particle to change in speed.

B) The normal component of inertia, also called centrifugal force, measures the resistance of the particle to leave its curved path.

13.4 VELOCITY-DEPENDENT FORCE

The force acting on a particle can be a function of time. Some examples of such forces are:

A) Viscous resistance

B) Air resistance

General equation:

$$\boxed{\vec{F}(v) = m\frac{d\vec{v}}{dt}} \qquad (13.6)$$

Equation 13.6 can be integrated to solve for time:

$$t_2 - t_1 = \int_1^2 \frac{m}{\vec{F}(v)} d(\vec{v}) \qquad (13.7)$$

A second integration yields position:

$$x_2 - x_1 = \int_1^2 v(t)dt \qquad (13.8)$$

Terminal Velocity - When the drag force is equal to the weight of the body there is no acceleration and no increase in velocity.

A) Linear Case - Assume the fluid resistance is proportional to the first power of velocity. The differential equation of motion:

$$-(mg+cv) = m\frac{dv}{dt} \qquad (13.9)$$

where g = gravitational constant and c is a function of viscosity and geometry of the object. If c is constant then,

$$v = -\frac{mg}{c} + \left(\frac{mg}{c} + v_0\right)e^{-ct/m} \qquad (13.10)$$

where v_0 = initial velocity.

As the $t \rightarrow \infty$, equation 13.10 becomes

$$\boxed{v_t = -\frac{mg}{c}} \qquad (13.11)$$

v_t is the terminal velocity.

A second integration of equation 13.9 yields:

$$x-x_0 = \int_0^t v(t)dt = -\frac{mg}{c}t + \left(\frac{m^2g}{c^2} + \frac{mv_0}{c}\right)(1-e^{-ct/m}) \qquad (13.12)$$

Characteristic time is defined as:

$$\tau = \frac{m}{c} \qquad (13.13)$$

By using equation 13.13 and 13.11, equation 13.10 becomes:

$$v = -v_t + (v_t+v_0)e^{-t/\tau} \qquad (13.14)$$

B) Quadratic Case - If the fluid resistance is proportional to the square of velocity.

Differential equation of motion:

$$\boxed{-mg \pm cv^2 - m\frac{dv}{dt}} \qquad (13.15)$$

A-103

Note: A double sign (±) is necessary for any even power of v.

Solving equation 13.15 for time:

$$\text{Upward Motion:} \quad t = \int \frac{m\,dv}{-mg-cv^2} = -\tau \tan^{-1}\frac{v}{v_t} + t_0$$

$$\text{Downward Motion:}$$
$$t = \int \frac{m\,dv}{-mg+cv^2} = -\tau \tanh^{-1}\frac{v}{v_t} + t_0^*$$

(13.16)

where
t_0, t_0^* = Initial time and are not necessarily equal

$$v_t = \sqrt{\frac{mg}{c}} = \text{Terminal Velocity} \quad (13.17)$$

and

$$\tau = \sqrt{\frac{m}{cg}} = \text{Characteristic Time} \quad (13.18)$$

Solve for velocity v:

$$\text{Upward Motion:} \quad v = v_t \tan \frac{t_0-t}{\tau}$$

$$\text{Downward Motion:}$$
$$v = -v_t \tanh \frac{t-t_0^*}{\tau}$$

(13.19)

If $t_0 = 0$, then $t_0^* = 0$ and

$$v = -v_t \tanh \frac{t}{\tau} = -v_t \left(\frac{e^{t/\tau} - e^{-t/\tau}}{e^{t/\tau} + e^{-t/\tau}} \right) \quad (13.20)$$

A-104

According to the definition of the hyperbolic tangent, i.e.,

$$\sinh z = \frac{e^z - e^{-z}}{2} \ ;$$

$$\cosh z = \frac{e^z + e^{-z}}{2} \implies \tanh z = \frac{e^z - e^{-z}}{e^z + e^{-z}}$$

13.5 ANGULAR MOMENTUM

Angular momentum is defined as the moment about the origin O of the angular momentum vector $m\vec{v}$. It is denoted by $\vec{H}_0$:

$$\boxed{\vec{H}_0 = \vec{r} \times m\vec{v}} \qquad (13.21)$$

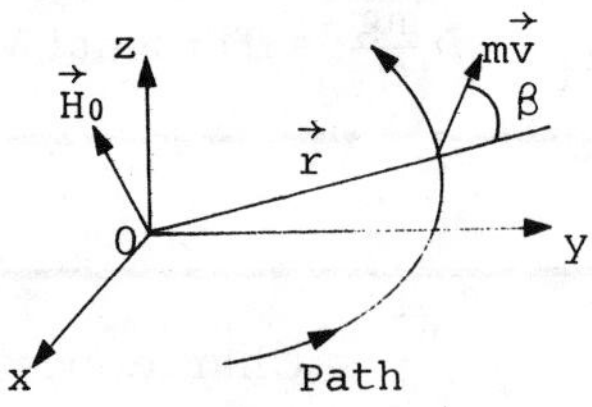

Fig. 13·1

Note: The angular momentum vector acts in a direction perpendicular to the plane containing the position and the linear momentum vectors.

The scalar form of equation 13.21 is

$$\boxed{H_0 = rmv \sin \beta} \qquad (13.22)$$

Determinant form:

$$\boxed{\vec{H}_0 = \begin{vmatrix} \vec{i} & \vec{j} & \vec{k} \\ x & y & z \\ mv_x & mv_y & mv_z \end{vmatrix}} \qquad (13.23)$$

Compoment form by expanding equation 13.23:

$$\boxed{\begin{aligned} H_x &= m(yv_z - zv_y) \\ H_y &= m(zv_x - xv_z) \\ H_z &= m(xv_y - yv_x) \end{aligned}} \qquad (13.24)$$

In terms of polar coordinates:

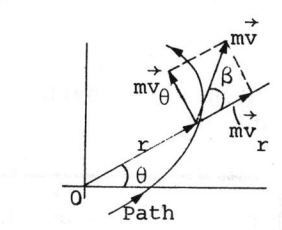

Fig. 13·2

$$\boxed{H_0 = mr^2 \dot\theta} \qquad (13.25)$$

By Newton's second law:

$$\boxed{\Sigma \vec{M}_0 = \dot{\vec{H}}_0,} \qquad (13.26)$$

i.e., the vector sum of the moments about O, acting on the particle equals the time derivative of the moment of momentum, or angular momentum, of the particle about O.

13.6 CENTRAL FORCE FIELDS

Central Force - The force acting on a particle which is directed towards or away from a fixed point O called the center of force. The particle is considered to be moving in a central force field. The magnitude of the force depends on the distance between the particle and the point O.

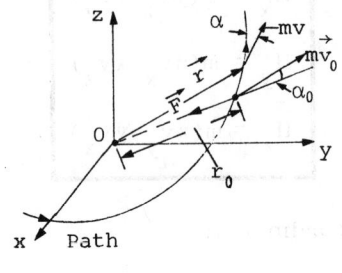

Fig. 13·3

The angular momentum of a particle moving in a central force field is constant:

$$\vec{r} \times m\vec{v} = \vec{H}_0 = \text{constant} \qquad (13.27)$$

Conservation of angular momentum:

$$rmv \sin\alpha = r_0 m v_0 \sin\alpha_0 \qquad (13.28)$$

r_0, v_0, and α_0 are initial parameters.

In terms of polar coordinates (Figure 13.2), equation (13.25), the angular momentum per unit mass is:

$$h = r^2\dot{\theta} \qquad (13.29)$$

Areal Velocity:

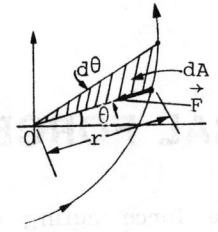

Fig. 13·4

A-107

Areal velocity is defined as:

$$dA/dt$$

Areal Velocity

$$\frac{dA}{dt} = \frac{1}{2} r^2 \frac{d\theta}{dt}$$

For a particle under a central force field, the areal velocity is constant.

Newton's Law of Gravitation

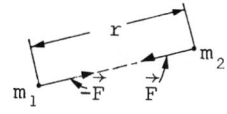

Fig. 13-5

G = universal constant

$$F = G \frac{m_1 m_2}{r} \qquad (13.30)$$

Trajectory of Central Force Motion

Governing equations in radial and transverse compoments:

$$m(\ddot{r} - r\dot{\theta}^2) = -F$$
$$M(r\ddot{\theta} + 2\dot{r}\dot{\theta}) = 0 \qquad (13.31)$$

Using equation (13.29), and introducing a new function $u = \frac{1}{r}$, the differential equation of motion for central force fields is:

$$\frac{d^2u}{d\theta^2} + u = \frac{F}{mh^2 u^2} \qquad (13.32)$$

13.7 SPACE MECHANICS

For satellite motion, equation (13.32) takes the form:

A-108

$$\boxed{\frac{d^2u}{d\theta^2} + u = \frac{GM}{h^2}} \qquad (13.33)$$

where
- M = Mass of the Earth
- m = Mass of the Satellite
- r = Distance from the Earth's Center to the satellite
- $u = \frac{1}{r}$
- G = Gravitational Constant

Solution of equation 13.33:

$$\boxed{\frac{1}{r} = u = \frac{GM}{h^2} + c\cos\theta} \qquad (13.34)$$

c = Constant of Integration

This is an equation of a conic section in polar coordinates.

Eccentricity, e: Ratio of the distance between satellite and earth to the distance between the satellite from a fixed line.

$$\boxed{e = \frac{c}{GM/h^2} = \frac{ch^2}{GM}} \qquad (13.35)$$

Therefore, equation 13.34 can be rewritten as

$$\boxed{\frac{1}{r} = \frac{GM}{h^2}(1+e\cos\theta)} \qquad (13.36)$$

Three cases for orbits

A) $e > 1$:

Two solutions result, θ_1 and $-\theta_1$, the orbit is a hyperbola.

B) $e = 1$.

The radius becomes infinite for $\theta = 180°$, the orbit is a parabola.

C) e < 1:

   The orbit is an ellipse.

   For the special case when e = 0, the orbit is a circle.

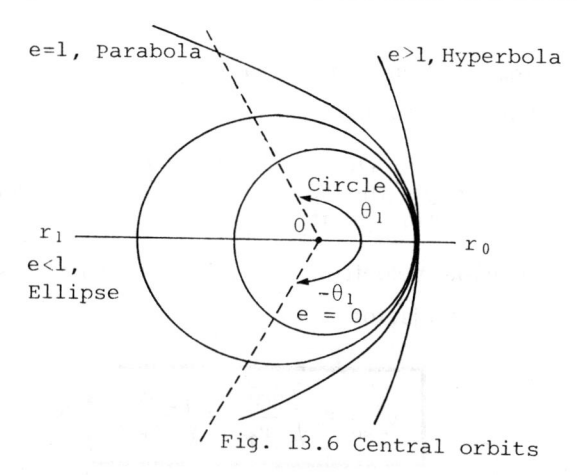

Fig. 13.6 Central orbits

Assume that the satellite begins its free flight at the vertex (r_0) of its orbit, with velocity parallel to the surface of the earth:

Angular Momentum:
$$h = r_0^2 \dot{\theta}_0 = r_0 v_0 \qquad (13.37)$$

Constant c:
$$c = \frac{1}{r_0} - \frac{GM}{h^2} \qquad (13.38)$$

Orbits by Velocity:

For a parabolic trajectory, let $c = GM/h^2$ in equation (13.38), and eliminate h between equations (13.37) and (13.38):

$$v_p = \sqrt{\frac{2GM}{r_0}} \qquad (13.39)$$

For a hyperbolic trajectory:
$$v > v_p$$

For an elliptic trajectory:
$$v < v_p$$

A-110

For a circular trajectory (c=0):

$$v_{cir} = \sqrt{\frac{GM}{r_0}} = \sqrt{\frac{gR^2}{r_0}} \qquad (13.40)$$

where

g = Acceleration due to gravity at earth's surface

R = Earth's Radius

Escape Velocity - The smallest velocity required for a satellite to escape the earth's gravitational field and not return to its starting point.

The escape velocity, v_{esc}, is the same as the velocity for a parabolic trajectory:

$$\boxed{v_{esc} = \sqrt{\frac{2GM}{r_0}} = \sqrt{\frac{2gR^2}{r_0}}} \qquad (13.41)$$

Perigee - A point on the satellite's orbit closest to earth.

Apogee - A point on the satellite's orbit farthest from the earth.

Perihelion - A point on the planet's orbit closest to the sun.

Aphelion - A point on the planet's orbit farthest from the sun.

Orbital Parameters

Let v_0 be the velocity at $\theta = 0$.

$$\boxed{\text{Eccentricity } e = (v_0/v_{cir})^2 - 1} \qquad (13.42)$$

$$\boxed{\text{Equation of the orbit:} \quad r = r_0 \frac{(v_0/v_{cir})^2}{1+[(v_0/v_{cir})^2 - 1]\cos\theta}} \qquad (13.43)$$

$$\boxed{\text{Apogee or Aphelion Distance } r_1 = r_0 \frac{(v_0/v_{cir})^2}{2-(v_0/v_{cir})^2}}$$

(13.44)

Period – The time required for a satellite to complete an orbit. Denoted by τ

$$\boxed{\tau = ca^{3/2}}$$ (13.45)

where $c = 2\pi(GM)^{-\frac{1}{2}}$

a = Semimajor Axis

In terms of Geometry:

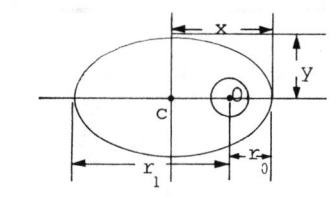

Fig. 13·7

$$\boxed{\tau = \frac{2\pi xy}{h}}$$ (13.46)

13.8 KEPLER'S LAWS

Kepler's Laws of Planetary Motion

First Law:

   Every planet moves in its orbit such that the line joining it to the sun sweeps over equal areas in equal intervals of time regardless of the line's length.

Second Law:

   The orbit of every planet is an ellipse with the sun at one of its foci.

Third Law:

The square of the period of any planet is directly proportional to the cube of the semi-major axis of its orbit.

Notes:

A) The first and third laws result from the inverse-square law.

B) The second is a consequence of the fact that planetary motion is central, and is usually written in the form dA/dt = constant.

Angular Momentum and Kepler's Second Law:

$$\boxed{\frac{dA}{dt} = \frac{H_0}{2m} = \text{constant}}$$ (13.47)

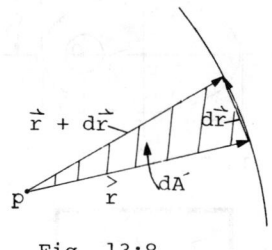

Fig. 13·8

13.9 INVERSE-SQUARE REPULSIVE FIELDS

Inverse-Square Repulsive Field - A central force field in which the direction of the force on a particle is away from the origin. The magnitude of this force is proportional to the inverse square of the distance between the origin and the particle.

The deflection of atomic particles by the nuclei is used as an example of the inverse-square repulsive field:

Coulomb's Law: $F(r) = \frac{Qq}{r^2}$

A-113

where Q = Charge of the Larger Particle
q = Charge of the Smaller Particle
r = Distance Between the Particles.

Let Q be the origin of the system.

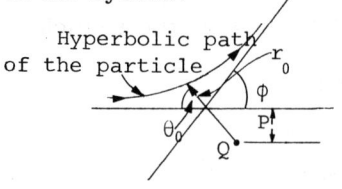

Fig. 13·9

Differential Equation of the Path:

$$\frac{d^2u}{d\theta^2} + u = -\frac{Qq}{mh^2}, \quad u = \frac{1}{r} \qquad (13.48)$$

Solution of Equation (13.48) is:

$$r = \frac{1}{u} = \frac{1}{A\cos(\theta-\theta_0) - Qq/mh^2} \qquad (13.49)$$

This is a hyperbolic orbit.

An alternate form of the solution, i.e., equation (13.49):

$$r = \frac{mh^2 Q^{-1} q^{-1}}{-1 + (1+2Emh^2 Q^{-2} q^{-2})^{\frac{1}{2}} \cos(\theta-\theta_0)} \qquad (13.50)$$

where:

$$E = \text{Energy} = \frac{1}{2}mv^2 + \frac{Qq}{r}$$

From angle relationships, we have the scattering formula:

$$\tan\theta_0 = \sqrt{2Em}\, Q^{-1} q^{-1} h = \cot\frac{\phi}{2} \qquad (13.51)$$

A-114

To express h in terms of the impact parameter P (see Figure 13.9):

$$h = |\vec{r} \times \vec{v}| = Pv_0$$

where
v_0 = Initial Speed of the Particle.

From the fact that:

A) Energy is constant
B) Kinetic energy $E = \frac{1}{2} mv_0^2$
C) Inital potential energy = 0 ($r = \infty$)

The scattering formula, equation (13.51), can be rewritten as:

$$\boxed{\cot \frac{\phi}{2} = \frac{Pmv_0^2}{Qq} = \frac{2PE}{Qq}}\qquad(13.52)$$

13.10 NEARLY CIRCULAR ORBITS

Stability - The tendency of a particle in a central force in its initial circular orbit after sustaining a slight disturbance in its motion.

Radial Equation of Motion:

$$m\ddot{r} - \frac{mh^2}{r^3} = F(r)$$

If the orbit is circular, the equation becomes

$$-mh^2/c^3 = F(c),$$

where c is the radius of the circle.

Let $y = r - c$, and expand the y+c terms in powers of y:

$$\boxed{m\ddot{y} + [-\frac{3}{c}F(c) - F'(c)]y = 0}\qquad(13.53)$$

where higher order terms are taken at zero.

Conditions for Stability:

A) If the coefficient of y is greater than 0, then the motion is stable.
B) If the coefficient of y is less than 0, then the motion is not stable.
C) If the coefficient y equals zero, then higher order terms of the expansion must be considered to determine stability.

Therefore, the circular orbit of radius c is stable if:

$$\boxed{F(c) + \frac{c}{3} F'(c) < 0} \qquad (13.54)$$

Special case: If $F(r) = -kr^a$, then the orbit is stable if $a > -3$.

Apsides and Apsidal Angles

Apsis or Apse - The extremal radius in an orbit.

Note: In planetary orbit, the apsides are known as the aphelion (maximum) and perihelion (minimum).

Apsidal Angle - The angle formed by two consecutive apsides (Figure 13.10).

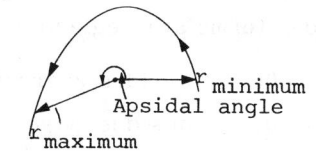

Fig. 13·10

Nearly circular motion (given a stable orbit about r=c):

A) The period is

$$\boxed{\tau = 2\pi \sqrt{\frac{m}{-[\frac{3}{2} f(c) + f'(c)]}}} \qquad (13.55)$$

B) The apsidal angle is

$$\beta = \frac{1}{2}\tau\dot{\phi} = \pi\left[3 + \frac{cF'(c)}{F(c)}\right]^{-\frac{1}{2}} \quad (13.56)$$

C) If $F(r) = -kr^a$,

$$\beta = \pi(a+3)^{-\frac{1}{2}} \quad (13.57)$$

Observations:

A) β is not a function of the orbit size.

B) For $a = -z(\beta = \pi)$ and $a=1(\beta = \pi/2)$, the orbit is said to be repetitive.

C) If there is a slight deviation from the inverse-square law, then, according to the deviation from π in the apsidal angle, the apsides will either move forward or regress.

Special case: If $F(r) = -\left\{\frac{A}{r^2} + \frac{B}{r^4}\right\}$, and B is very small, then:

$$\beta = \pi\left\{1 + \frac{B}{Ac^2}\right\}, \quad (13.58)$$

where higher order terms are neglected.

Results: 1) If $B < 0$, the apsides regress

2) If $B > 0$, the apsides move forward

CHAPTER 14

ENERGY AND MOMENTUM METHODS-PARTICLES

14.1 WORK OF A FORCE

Given a particle on a space curve c at a point R, in an inertial reference frame, and at point S at some instant later, with position vectors $\vec{r}_1$ and $\vec{r}_2$, respectively. Then the work done by a force F through a displacement dr of the particle is defined as

$$dw = \vec{F} \cdot \vec{dr} \quad (14.1)$$

Note: Work in S.I. units is expressed in N·m (Joule); in Figure 14.1 U.S. cutomary units, work is expressed in ft-lb or lb-in.

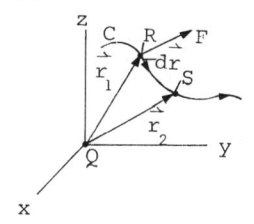

Fig. 14·1

Work for a finite distance from point 1 to point 2 is defined as the line integral:

$$w_{1-2} = \int_1^2 \vec{F} \cdot \vec{dr} \quad (14.2)$$

A-118

Work of a constant force (in linear motion):

$$w_{1-2} = F \cos\theta \cdot \Delta y$$

Fig. 14·2

Work done by a weight:

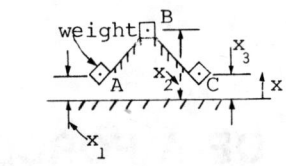

Fig. 14·3

$$\begin{aligned} w_{A-B} &= -\text{weight}(x_2-x_1), \\ w_{B-C} &= -\text{weight}(x_3-x_2), \\ w_{A-C} &= -\text{weight}(x_3-x_1) \end{aligned} \qquad (14.3)$$

If $x_3 = x_1$, then $w_{A-C} = 0$.

Work of Springs

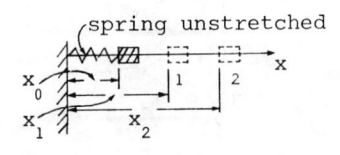

Fig. 14·4

$$w_{1 \to 2} = \frac{1}{2} k x_1^2 - \frac{1}{2} k x_2^2 \qquad (14.4)$$

k = Spring constant or stiffness of the spring force exerted by the spring on the mass: F = kx

A-119

Work of a Gravitational Force - Given two particles of masses M_1 and M_2 such that M_1 is stationary and is at a distance ρ_1 from M_2 at A; while M_2 moves along c to point B, and at a distance ρ_2 from M_1. Then, the work of the gravitational force $\vec{F}$ from A to B is

$$W_{A-B} = -\int_{\rho_1}^{\rho_2} \frac{GM_1M_2}{\rho^2} d\rho = GM_1M_2\left[\frac{1}{\rho_1} - \frac{1}{\rho_2}\right] \quad (14.5)$$

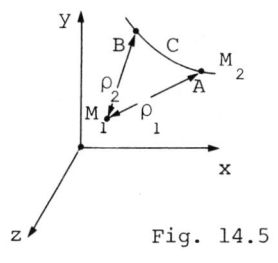

Fig. 14.5

14.2 WORK-ENERGY PRINCIPLE

Kinetic energy for a particle of mass M and velocity v is defined as

$$K.E. = \frac{1}{2} mv^2 \quad (14.6)$$

Kinetic energy is the energy possessed by a particle by virtue of its motion.

Principle of Work and Energy - Given that a particle undergoes a displacement under the influence of a force $\vec{F}$, the work done by $\vec{F}$ equals the change in kinetic energy of the particle.

$$W_{1-2} = (KE)_2 - (KE)_1 \quad (14.7)$$

Note: Since this principle is a consequence of Newton's second law, it is valid only in an inertial reference system.

Results of the Principle of Work and Energy

A) Acceleration is not necessary and may not be obtained directly by this principle.

B) The principle may be applied to a system of particles if each particle is considered separately.

C) Those forces that do not contribute work are eliminated.

Kinetic Energy and Newton's Law:

$$F = mv \frac{dv}{dx} = \frac{d}{dx}(KE),$$

where KE is a function of x.

14.3 POWER AND EFFICIENCY

Power is defined as the time-rate of change of work and is denoted by dw/dt,

$$\boxed{\text{Power} = \frac{dw}{dt} = \vec{F} \cdot \vec{v}} \qquad (14.8)$$

In the SI system of units, power has units J/S or watt (W) (W not to be confused with the lower case w, denoting work). In the U.S. customary system, power is in ft-lb/s or horsepower (H.p.).

Note: 1 h.p. = 550 ft-lb/s and 1 h.p. = 746 W.

Mechanical Efficiency:

$$\boxed{\eta = \frac{\text{Power out}}{\text{Power in}}} \qquad (14.9)$$

where both numertor and denominator must have the same

A-121

units. It is assumed that the rate of work is constant.

Note: Since power out < power in, η is always less than 1.

14.4 CONSERVATIVE FORCES

The del operator - denoted by the symbol $\vec{\nabla}$ is defined as

$$\boxed{\vec{\nabla} = i\frac{\partial \phi}{\partial x} + j\frac{\partial \phi}{\partial y} + k\frac{\partial \phi}{\partial z}} \qquad (14.10)$$

in rectangular coordinates.

The gradient of a scalar function is defined as:

$$\boxed{\vec{\nabla}\phi = \text{grad}\phi = \frac{\partial}{\partial x}i + \frac{\partial}{\partial y}j + \frac{\partial}{\partial z}k} \qquad (14.11)$$

If a force $\vec{F}$ is expressible as the gradient of a scalar function ϕ,

i.e., $\boxed{\vec{F} = -\vec{\nabla}\phi}$ (14.12)

then F is known as a conservative force and ϕ is the potential function.

Note: The gradient represents a forcing action.

Conservative Force - Work done by a particle under the action of this force is independent of the path followed by the particle.

Results:

A) If $\vec{F}$ is conservative, then $\vec{F} = \vec{\nabla}\phi$
B) If $\vec{F} = -\vec{\nabla}\phi$, then

$$\int_1^2 \vec{F} \cdot \vec{dr} = \int_1^2 -\left(\frac{\partial \phi}{\partial x} dx + \frac{\partial \phi}{\partial y} dy + \frac{\partial \phi}{\partial z} dz\right)$$

$$= \int_1^2 -d\phi = \phi(x_1, y_1, z_1) - \phi(x_2, y_2, z_2)$$

(14.13)

where ϕ is a scalar function in cartesian coordinates. Hence, the work of $\vec{F}$ is an exact differential, and independent of the path.

C) If $\vec{F} = -\vec{\nabla}\phi$, then

$$\oint_c \vec{F} \cdot \vec{dr} = 0 \qquad (14.14)$$

i.e., the work of a conservative force for any closed path is zero.

D) If $\vec{F} = -\vec{\nabla}\phi$, then curl $\vec{F} = 0$, or $\vec{\nabla} \times \vec{F} = 0$; the reverse is also true, i.e., if $\vec{\nabla} \times \vec{F} = 0$, then $\vec{F}$ is conservative.

Existance of a Potential Function - The necessary and sufficient conditions for ϕ to exist may be expressed in cartesian coordiantes as

$$\frac{\partial F_x}{\partial y} = \frac{\partial F_y}{\partial x}, \quad \frac{\partial F_z}{\partial x} = \frac{\partial F_x}{\partial z},$$

$$\frac{\partial F_z}{\partial y} = \frac{\partial F_y}{\partial z}$$

(14.15)

14.5 POTENTIAL ENERGY

Potential Energy - In general, it is the stored energy of a

A-123

body or particle in a force field associated with its postiion from a reference frame.

Potential Energy and Work - If potential energy is denoted by PE, then

$$U_{1-2} = (PE)_1 - (PE)_2 \qquad (14.16)$$

where PE = weight (x), see Figure 14.3. A negative value of U_{1-2} is associated with an increase in potential energy.

Potential Energy for Space Vehicles:

$$(PE)_g = -G \frac{M_1 M_2}{r}$$
$$= -\frac{\text{weight}}{r} R^2, \qquad (14.17)$$

where M_1 = Mass of the Earth

M_2 = Mass of the object (space vehicle, satellite, etc.)

R = The Radius of the earth, "weight" as measured on the surface of the earth

r = The distance between centers of M_1 and M_2.

The potential energy and work for a spring are given by:

$$U_{1-2} = PE_1 - PE_2 \qquad (14.18)$$

where PE = $\frac{1}{2} ky^2$, and is known as the potential energy with respect to the elastic force.

Notes:

A) If $U_{1-2} < 0$, then the potential energy increases.

B) The work U_{1-2} of the elastic force is a function of the final and initial deflections of the spring.

Potential energy methods may be utilized when the forces involved are conservative.

14.6 GRAVITATIONAL POTENTIAL

Gravitational potential is defined as

For a system $\phi = G \dfrac{M_1 M_2}{r}$

$$\boxed{\phi = \dfrac{(PE)_g}{M_1} = -G \dfrac{M_2}{r}} \qquad (14.19)$$

For a particle of Mass M_2.

Gravitational Field Intensity – Ratio of the gravitational force on a particle to the mass of that particle:

$$\boxed{\vec{G} = \dfrac{\vec{F}(r)}{m}} \qquad (14.20)$$

Relationship between $\vec{G}$ and ϕ:

$$\boxed{\vec{G} = -\vec{\nabla} \phi} \qquad (14.21)$$

Specific Cases:

A) Uniform Spherical Shell

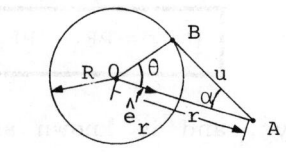

Fig. 14·6

$$\boxed{\phi_A = -G \dfrac{2\pi \rho^2 R}{rR} \int_{r-R}^{r+R} du = \dfrac{-GM}{r}} \qquad (14.22)$$

ρ = Density of the Sphere

Note:

1) The gravitational field outside the shell is the same as if the entire mass was concentrated at the center O.
2) The potential inside the shell is constant and the field is zero.

B) Thin Ring

Using Figure 14.4 again:

$$\Phi_A = -G \int \frac{dM}{u} = -G \int_0^{2\pi} \frac{\mu R d\theta}{u} \qquad (14.23)$$

where μ = Linear Density of the Ring

By use of geometry:

$$\boxed{\Phi_A = -G \frac{4\mu R}{r} k\left(\frac{R}{r}\right)} \qquad (14.24)$$

k = Complete Elliptic Integral

$$\boxed{k\left(\frac{R}{r}\right) = \int_0^{\pi/2} \left[1 - \left(\frac{R}{r}\right)^2 \sin^2\alpha\right]^{-\frac{1}{2}} d\alpha} \qquad (14.25)$$

Equation 14.24 can also be expressed in terms of a series:

$$\boxed{\Phi_A = \frac{-GM}{r}\left[1 + \frac{R^2}{4r^2} + \ldots\right]} \qquad (14.26)$$

Field Intensity:

$$\boxed{\vec{G} = -\frac{\partial \Phi}{\partial r}\hat{e}_r = \left[\frac{-GM}{r^2} - \frac{3GMR^2}{4r^4} - \ldots\right]\hat{e}_r} \qquad (14.27)$$

This field is not inverse-square. But for large distances, the field is predominantly inverse-square. This is true for a finite body of any shape.

Potential in a General Central Field

A general isotropic central field is expressed as:

$$\vec{F} = F(r)\hat{e}_r \qquad (14.28)$$

If the field is conservative, $\vec{\nabla} \times \vec{F} = 0$, and:

$$PE = \int_r^\infty F(r)dr \qquad (14.29)$$

The force function is given by:

$$F(r) = -\frac{\partial v(r)}{\partial r} \qquad (14.30)$$

14.7 CONSERVATION OF ENERGY

A) Conservative Case

For a particle under the action of conservative forces:

$$(KE)_1 + (PE)_1 = (KE)_2 + (PE)_2 = E \qquad (14.31)$$

The sum of kinetic and potential energy at a given point is constant.

Equation (14.31) can be written as:

$$E = \frac{1}{2}mv^2 + (PE) \qquad (14.32)$$

Equation (14.32) can be solved to give the motion of the particle as:

$$\boxed{v = \pm\sqrt{\frac{2}{m}[E - (PE)]}} \qquad (14.33)$$

or in an integral form:

$$\boxed{\int \frac{\pm dx}{\sqrt{\frac{2}{m}[E - v(x)]}} = t} \qquad (14.34)$$

Turning Points:

The potential function (PE) must be less than or equal to the total energy E:

$$(PE) \leq E$$

From equation (14.32), speed v must become zero when (PE) = E. The particle must stop and reverse its direction of motion at this point. These points are called turning points.

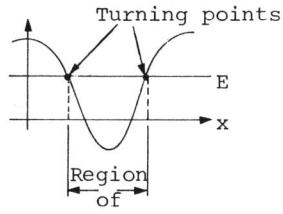

Fig. 14·7

B) Nonconservative Case

When nonconservative forces are present in a system:

Total Force = $\vec{F} + \vec{F}'$

where $\vec{F}$ = Conservative Forces
$\vec{F}'$ = Nonconservative Forces

Relating potential and kinetic energy with the nonconservative force F',

A-128

$$\boxed{d(PE + KE) = \int \vec{F'} \cdot \vec{dr}} \quad (14.35)$$

Note:
1) The direction of $\vec{F'}$ is opposite to that of $\vec{dr}$.
2) Total energy E decreases with motion.
3) Friction forces are nonconservative.

14.8 APPLICATION TO SPACE MECHANICS

Consider the motion of a satellite:

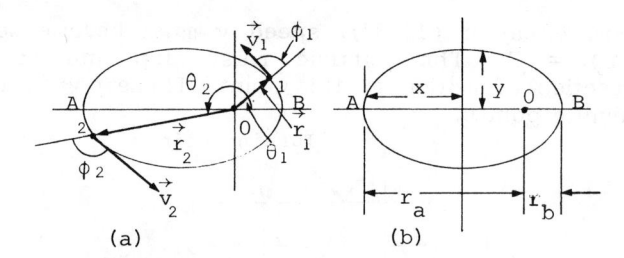

Fig. 14·8

Work done from point 1 to point 2:

$$w_{1 \to 2} = GMm \left(\frac{1}{r_2} - \frac{1}{r_1} \right) \quad (14.36)$$

Change in kinetic energy from 1 to 2:

$$(KE)_{1 \to 2} = \frac{1}{2} m(v_2^2 - v_1^2) \quad (14.37)$$

Conservation of Energy:

$$\boxed{\frac{1}{2} m v_1^2 - \frac{GMm}{r_1} = \frac{1}{2} m v_2^2 - \frac{GMm}{r_2}} \quad (14.38)$$

Conservation of Angular Momentum:

$$r_1 m v_1 \sin \phi_1 = r_2 m v_2 \sin \phi_2 \qquad (14.39)$$

Note: The extremal values of v and r occur when $\phi = 90°$.

Special Case:

If the satellite is moving in a circular orbit with constant velocity:

Gravitational Pull = Centrifugal Inertia Force

or

$$v^2 = \frac{GM}{r} \qquad (14.40)$$

Geometry:

Refer to Figure 14.8(b):

$$x = \frac{1}{2}(r_a + r_b)$$
$$y = \sqrt{r_a r_b} \qquad (14.41)$$

Areal Velocity:

$$\frac{dA}{dt} = \frac{1}{2} rv \sin\phi \qquad (14.42)$$

Polar Coordinates

A) Central Force Field

Energy Equation:

$$\frac{1}{2} mh^2 \left[\left(\frac{du}{d\theta}\right)^2 + u^2 \right] + PE = E \qquad (14.43)$$

where $u = \frac{1}{r}$,

$h = rv$ and PE is a function of $\frac{1}{u}$.

A-130

B) Inverse-Square Field

Potential Function:

$$PE = -\frac{k}{r} = -ku \qquad (14.44)$$

$$K = \text{constant}$$

Energy Equation:

$$\boxed{\frac{1}{2}mh^2\left[\left(\frac{du}{d\theta}\right)^2 + u^2\right] - ku = E} \qquad (14.45)$$

Solution of equation (14.45):

$$\boxed{u = \frac{k}{mh^2}\left[1 + \sqrt{(1+2Emh^2k^{-2})}\cos\theta\right]} \qquad (14.46)$$

or, in terms of r:

$$\boxed{r = \frac{mh^2k^{-1}}{1 + \sqrt{(1+2Emh^2k^{-2})}\cos\theta}} \qquad (14.47)$$

Eccentricity in terms of total energy may be expressed as:

$$\boxed{e = (1 + 2Emh^2k^{-2})^{\frac{1}{2}}} \qquad (14.48)$$

Classification of Orbits:

$E < 0$ or	$(e < 1)$:	Ellipse
$E = 0$ or	$(e = 1)$:	Parabolic
$E > 0$ or	$(e > 1)$:	Hyperbolic

In general:

$KE < |PE|$ for closed orbits

$KE \geq |PE|$ for open orbits

14.9 IMPULSE AND MOMENTUM

Impulse-Momentum Method - An alternate method to solve problems in which forces are expressed as functions of time. It is applicable to situations when forces act over a small interval of time.

Linear Impulse-Momentum Equation:

$$\int_1^2 \vec{F}\, dt = \text{impulse} = m\vec{v}_2 - m\vec{v}_1 \qquad (14.49)$$

Figure (14.9) expresses the idea that the vector sum of the initial momentum and impulse equals the final momentum of the particle.

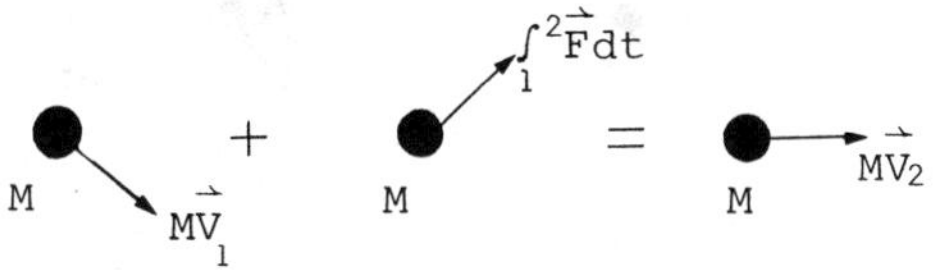

Fig. 14·9

Impulse is a vector quantity and acts in the direction of the force if the force remains constant.

$$\int_1^2 \vec{F}\, dt = \vec{i} \int_1^2 F_x\, dt + \vec{j} \int_1^2 F_y\, dt + \vec{k} \int_1^2 F_z\, dt \qquad (14.50)$$

To obtain solutions, it is necessary to replace equation (14.49) with its compoment equations.

When several forces are involved, the impulse of each force must be considered:

$$m\vec{v}_1 + \Sigma \int_1^2 \vec{F}\, dt = m\vec{v}_2 \qquad (14.51)$$

When a problem involves more than one particle, each particle must be considered separately and then added:

A-132

$$\Sigma m\vec{v}_1 + \Sigma \int_1^2 \vec{F}\, dt = \Sigma m\vec{v}_2 \qquad (14.52)$$

If there are no external forces, conservation of total momentum results:

$$\boxed{\Sigma m\vec{v}_1 = \Sigma m\vec{v}_2} \qquad (14.53)$$

Note: There are cases in which an impulse is exerted by a force which does no work on the particle. Such force should be considered when applying the principle of impulse and momentum in solving problems.

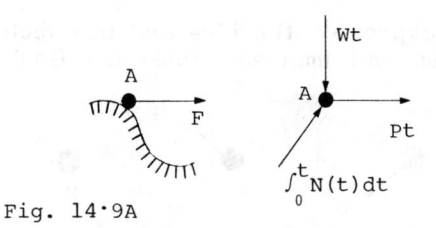

Fig. 14·9A

E.g. Figure (14.9A)

W - Weight of Particle A

F - Horizontal Force

t - Time Taken

N - Normal Force

Here the normal force N does no work but creates an impulse $\int_0^t N(t)\,dt$ on the particle A.

14.10 IMPULSE MOTION

Impulsive Force - A very large force acting on a particle over a very short time interval and produces a significant change in momentum.

Impulse-Momentum equation, equation (14.51), becomes:

$$\boxed{M\vec{v}_1 + \Sigma\vec{F}\,\Delta t = M\vec{v}_2}\qquad(14.54)$$

Note: $\Sigma\vec{F}\,\Delta t$ may also be denoted as $\hat{P}$.

Fig. 14-10

Any force which is non-impulsive may be neglected in equation (14.54), e.g., weight, or small forces.

Unknown forces should be included, e.g., reactive forces (recoil, etc.).

Ideal impulse produces an instantaneous change in momentum and velocity of the particle without producing any displacement.

14.11 IMPACT

Impact - Collision of two bodies.
Types of Impact

A) Central Impact - When the line of impact passes through the center of mass of the colliding bodies.

B) Eccentric Impact - When the line of impact does not pass through the center of mass of the colliding bodies.

Two Types of Central Impacts:

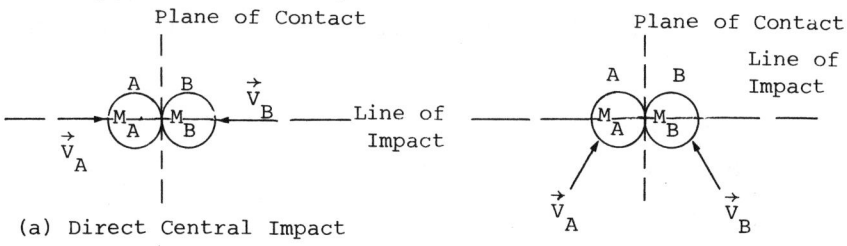

(a) Direct Central Impact

(b) Oblique Central Impact

Fig. 14-11 Types of Central Impacts

A-134

Total Linear Momentum is Conserved Under Impact:

$$\boxed{M_A \vec{v}_A + M_B \vec{v}_B = M_A \vec{v}'_A + M_B \vec{v}'_B}$$ (14.55)

where $\vec{v}_A, \vec{v}_B$ are velocities before impact and $\vec{v}'_A, \vec{v}'_B$ are velocities after impact.

Energy Balance:

$$\boxed{\frac{1}{2} M_A v_A^2 + \frac{1}{2} M_B v_B^2 = \frac{1}{2} M_A {v'_A}^2 + \frac{1}{2} M_B {v'_B}^2 + q}$$

(14.56)

where q = Net Energy Change (deformations, heat, etc.)

Endoergic Impact - When q > 0, representing an energy loss.

Exoergic Impact - When q < 0, representing an energy gain.

Perfectly Elastic Impact - When q = 0, or when e = 1.

A) Direct Central Impact

Coefficient of restitution:

$$\boxed{e = \frac{v'_B - v'_A}{v_B - v_A}}$$ (14.57)

Equations (14.55) and (14.57) are used simultaneously to solve for v'_A and v'_B.

Sign Convention:
 Right-Positive; Left-Negative.

Special Cases:
a) e = 0, perfectly plastic impact:
 Both particles stay together after impact and move

A-135

with the same velocity v'.

Momentum Equation:

$$\boxed{M_A v_A + M_B v_B = (M_A + M_B) v'}\qquad(14.58)$$

b) e = 1, perfectly elastic impact

$$v'_B - v'_A = v_A - v_B \qquad (14.59)$$

Relative velocities are equal before and after impact. Total energy and momentum is conserved.

In most impact cases, e < 1. Total energy is not conserved. The energy loss, q, is expressed as:

$$\boxed{q = \tfrac{1}{2} \lambda\, v_R^2 (1-e^2)},\qquad(14.60)$$

where

$$\lambda = \frac{1}{M_A} + \frac{1}{M_B} \text{ and is known as the reduced mass,}$$

and

$$v_R = |v_B - v_A|, \text{ is the relative speed before impact.}$$

Final Velocities:

$$\boxed{\begin{aligned} v'_A &= \frac{(M_A + M_B e) v_A + (M_B - e M_B) v_B}{M_A + M_B} \\ v'_B &= \frac{(M_A - M_A e) v_A + (M_B + e M_A) v_B}{M_A + M_B} \end{aligned}}\qquad(14.61)$$

B) Oblique Central Impact

Because v'_A and v'_B are unknowns in both directions and magnitude, four independent equations are needed:

a) The sum of the momentum of particles A and B is

conserved in the x direction.

b) The momentum of each particle A and B is conserved in the y direction.

c) The relative velocity of the particles after impact in the x direction equals the product of their relative velocity before impact in the x-direction by the coefficient of restitution.

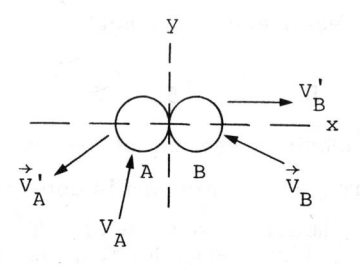

Oblique Central Impact
Fig. 14-12

Center of Mass Coordinates:

Two Systems:

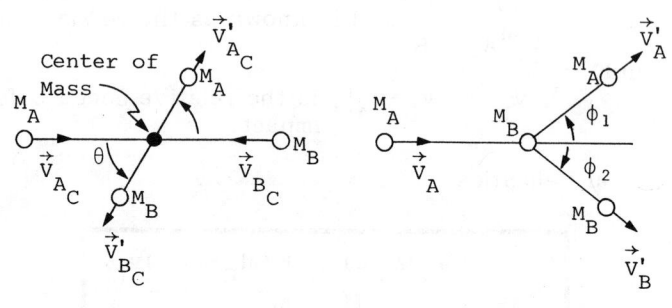

(a) Center of Mass System

(b) Laboratory System

Fig. 14-13

Energy Equation:

$$\frac{\vec{P}_A'^2}{2\lambda} = \frac{\vec{P}_A^2}{2\lambda} + q \qquad (14.62)$$

λ = Reduced Mass

A-137

Velocity Relationships: Refer to Figure 14.14

$$\vec{v}_{A_c} = \vec{v}_A - \vec{v}_{cm} = \frac{M_B \vec{v}_A}{M_A + M_B} \qquad (14.63)$$

$\vec{v}_{cm}$ = Velocity of the Center of Mass

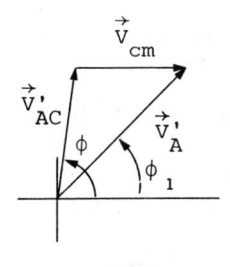

Fig. 14-14

Scattering Angles:

$$\tan \psi_1 = \frac{\sin \phi}{\mu + \cos \phi}$$

$$\mu = \frac{v_{cm}}{v_{A_c}'} = \frac{M_A v_A}{v_{A_c}'(M_A + M_B)} \qquad (14.64)$$

For Elastic Impact:

$$\mu = \frac{M_A}{M_B} \qquad (14.65)$$

Special Cases:

A) If $M_B \gg M_A$, then μ is small

   $\tan \psi_1 \approx \tan \phi$, scattering angles in the laboratory and in the center of mass system are almost equal

B) $M_A = M_B$ then $\mu = 1$

A-138

$$\tan \psi_1 = \tan \frac{\phi}{2} \quad ; \quad \text{or} \quad \psi_1 = \frac{\phi}{2}$$

The particles will leave the point of contact at right angles when seen in the laboratory system.

For the general case of non-elastic impact:

$$\mu = \frac{M_A}{M_B} \left[1 - \frac{q}{KE} \left(1 + \frac{M_A}{M_B} \right) \right]^{-\frac{1}{2}} \qquad (14.66)$$

KE = Kinetic energy of the incident particle in the laboratory system.

Relation Between Impulse and Coefficient of Restitution:

$$e = \frac{\hat{P}_r}{\hat{P}_c} \qquad (14.67)$$

where $\quad \hat{P}_r = M_A \vec{v}_A' - M_A \vec{v}_0$

$$\hat{P}_c = M_A \vec{v}_0 - M_a \vec{v}_a$$

v_0 is the velocity during the time when the two particles are in contact with each other.

14.12 METHODS OF SOLUTION

Thus far, three distinct methods have been presented to solve kinetics problems:

1) Newton's Second Law of Motion: $\Sigma \vec{F} = m\vec{a}$

2) Method of Work and Energy

3) Method of Impulse and Momentum

Except for short impact phases, most problems involve conservative forces. These problems, in general, have three parts:

1) Impact Phase Parts - Use the method of impulse and momentum. (Secton 14.9 - 14.11)

2) Parts that involve the determination of normal forces -

A-139

Use Newton's Laws. (Section 12.1 - 12.8)

3) All Other Parts - Use the method of work and energy. (Section 13.1 - 14.8)

14.13 MOTION OF A PROJECTILE

The projectile motion is a classical problem of particle dynamics.

Two Cases:

A) No Air Resistance

Differential Equation of Motion:

$$M \frac{d^2 \vec{r}}{dt^2} = -Mg\vec{k} \qquad (14.68)$$

with initial velocity, $v_i = 0$.

Energy Equation:

$$\frac{1}{2} M v_i^2 = Mgz + \frac{1}{2} M \left[\left(\frac{dx}{dt}\right)^2 + \left(\frac{dy}{dt}\right)^2 + \left(\frac{dz}{dt}\right)^2 \right]$$

$$v^2 = v_i^2 - 2gz$$

(14.69)

Velocity:

$$\vec{v} = \frac{d\vec{r}}{dt} = -gt\vec{k} + \vec{v}_i \qquad (14.70)$$

Position:

$$\vec{r} = -\frac{1}{2} gt^2 \vec{k} + \vec{v}_i t \qquad (14.71)$$

Trajectory in the x-y plane:

$$y = \frac{\dot{y}_i}{\dot{x}_i} x \qquad (14.72)$$

Equation (14.72) represents a straight line. Value of z in terms of x:

$$z = Ax - Bx^2$$
$$\text{where} \quad A = \dot{z}_i/\dot{x}_i \qquad (14.73)$$
$$B = g/2\dot{x}_i^2$$

Equation (14.73) represents a parabola. The trajectory is shown in Figure 14.15:

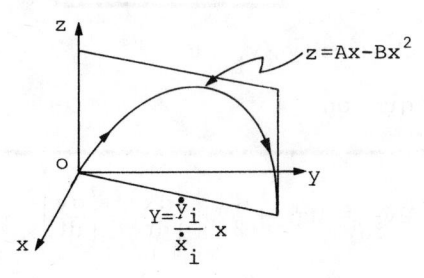

Fig. 14-15 Projectile Motion Without Air Resistance

B) Linear Air Resistance

The motion is now nonconservative.

Force of Resistance = $-Mc\vec{v}$

Differential Equation of Motion:

$$M \frac{d^2\vec{r}}{dt^2} = -Mc\vec{v} - Mg\vec{k} \qquad (14.74)$$

A-141

Component Form of Equation (14.74)

$$\ddot{x} = -c\dot{x}$$
$$\ddot{y} = -c\dot{y}$$
$$\ddot{z} = -c\dot{z} - g$$
(14.75)

Velocity Components:

$$\dot{x} = \dot{x}_i e^{-ct}$$
$$\dot{y} = \dot{y}_i e^{-ct}$$
$$\dot{z} = \dot{z}_i e^{-ct} - \frac{g}{c}(1 - e^{-ct})$$
(14.76)

Position Coordinates:

$$x = \frac{\dot{x}_i}{c}(1 - e^{-ct})$$
$$y = \frac{\dot{y}_i}{c}(1 - e^{-ct})$$
$$z = \left(\frac{\dot{z}_i}{c} + \frac{g}{c^2}\right)(1 - e^{-ct}) - \frac{g}{c}t$$
(14.77)

Vector form of the solution, equation (14.77):

$$\vec{r} = \left(\frac{\vec{V}_0}{c} + \frac{g}{c^2}\vec{k}\right)(1 - e^{-ct}) - \frac{gt}{c}\vec{k}$$
(14.78)

The motion in the x-z plane approaches an asymptote:

A-142

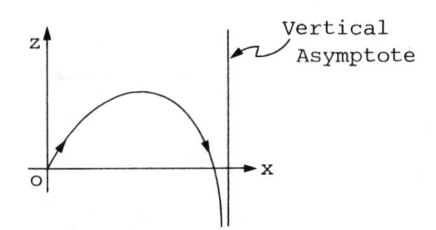

Fig. 14-16 Projectile Motion With Air Resistance

Approximate Solution - When the air resistance is small.

$$\vec{r} = \vec{v}_0 t - \frac{1}{2} gt^2 \vec{k} - \Delta \vec{r} \qquad (14.79)$$

where

$$\Delta \vec{r} = c \left[\vec{v}_0 \left(\frac{t^2}{2!} - \frac{ct^3}{3!} + \ldots \right) - g \left(\frac{t^3}{3!} - \frac{ct^4}{4!} + \ldots \right) \vec{k} \right]$$

$\Delta \vec{r}$ can be considered as a correction factor to the zero-resistance case, i.e., Equation (14.71).

14.14 MOTION OF A CHARGED PARTICLE

The force exerted on the particle is

$$\vec{F} = q\vec{E}_e \qquad (14.80)$$

where q = Charge of the Particle
$\vec{E}_e$ = Electric Field Strength

Differential Equation of Motion:

$$m \frac{d^2 \vec{r}}{dt^2} = q\vec{E}_e \qquad (14.81)$$

A-143

m = Mass of the Charged Particle

Component Form of Equation (14.81)

$$\begin{array}{l} m\ddot{x} = qE_{ex} \\ m\ddot{y} = qE_{ey} \\ m\ddot{z} = qE_{ez} \end{array}$$
(14.82)

Two Cases:

A) $\vec{E}$ is Uniform and Constant

If the field is directed along the x-axis:

$$\begin{array}{l} \ddot{x} = \dfrac{qE_e}{m} = \text{constant} \\ \ddot{y} = \ddot{z} = 0 \end{array}$$
(14.83)

The path is a parabola. The solutions take the same form as those of the projectile in a uniform gravity field, i.e., Equations (14.70) to (14.73).

If $\vec{E}$ is due to static charges:

$$\vec{\nabla} \times \vec{E}_e = 0$$
(14.84)

The field is conservative and the potential function Φ exists,

$$-\vec{\nabla}\Phi = \vec{E}_e$$
(14.85)

Energy Equation:

$$E = \frac{1}{2}mv^2 + q\Phi$$
(14.86)

B) Static Magnetic Field ($\vec{B}$) or $\vec{B}$ = Magnetic Induction Vector

A-144

Force: $$\boxed{\vec{F} = q\vec{v} \times \vec{B}}$$ (14.87)

Differential Equation of Motion:

$$\boxed{M \frac{d^2\vec{r}}{dt^2} = q(\vec{v} \times \vec{B})}$$ (14.88)

Results:
a) Acceleration is always perpendicular to the direction of motion.
b) Tangential compoment of acceleration ($\dot{v}$) is zero (constant speed).

14.15 CONSTRAINED MOTION OF A PARTICLE

Constrained Motion - Motion that is restricted by geometrical constraints. The constraints are assumed to be smooth.

Differential Equation of Motion:

$$\boxed{m \frac{d\vec{v}}{dt} = \vec{F} + \vec{F}_c}$$ (14.89)

where $\vec{F}$ = External Forces
$\vec{F}_c$ = Force of Constraints

If $\vec{F}$ is conservative, then the energy equation, the equation $E = PE + KE$ can be applied.

Motion of a Curve

Let all parameters be expressed in a single coordinate

s, where s = Distance Measured on the Curve from a Fixed Point.

Energy Equation:

$$\frac{1}{2} m\dot{s}^2 + PE = E \qquad (14.90)$$

Differential Equation of Motion:

$$m\ddot{s} - F_s = 0 \qquad (14.91)$$

F_s = Compoment of $\vec{F}$ in the s direction.

Potential Function:

$$F_s = \frac{-d(PE)}{ds} \qquad (14.92)$$

CHAPTER 15

SYSTEMS OF PARTICLES

15.1 SYSTEMS OF FORCES

For a system consisting of many particles:
Newton's Second Law of Motion takes the form:

$$\sum_{j=1}^{n} \vec{F}_j = \sum_{j=1}^{n} M_j \vec{a}_j \qquad (15.1)$$

Moment equation about an origin O:

$$\sum_{j=1}^{n} (\vec{r}_j \times \vec{F}_j) = \sum_{j=1}^{n} (\vec{r}_j \times M_j \vec{a}_j) \qquad (15.2)$$

The system of external forces, $\vec{F}_j$, and the system of effective forces, $m_j \vec{a}_j$, have the same force and moment result, and are therefore equipollent.

15.2 LINEAR AND ANGULAR MOMENTUM

Center of Mass Coordinates:

$$x_{cm} = \frac{\Sigma M_j x_j}{M} \;;\; y_{cm} = \frac{\Sigma M_j y_j}{M} \;;\; z_{cm} = \frac{\Sigma M_j z_j}{M}$$

(15.3)

Linear Momentum

$$\vec{P} = \sum_{j=1}^{n} M_j \vec{v}_j = M\vec{v}_{cm}$$

(15.4)

Angular Momentum (About the origin O)

$$\vec{H}_0 = \sum_{j=1}^{n} (\vec{r}_j \times M_j \vec{v}_j)$$

(15.5)

Conservation of Momentum

If a system is under no external force of moment, then

$$\Sigma \vec{F} = \dot{\vec{P}} = \sum_{j=1}^{n} (M_j \vec{a}_j) = 0$$

(15.6)

and

$$\Sigma \vec{M}_0 = \dot{\vec{H}}_0 = \sum_{j=1}^{n} (\vec{r}_j \times M_j \vec{a}_j) = 0$$

(15.7)

This means that linear momentum and angular momentum are constant and is the princple of conservation of momentum.

15.3 MOTION OF THE MASS CENTER

D'Alembert's Principle:

The vector sum of all external forces acting on a system

of particles is equal to the vector sum of the effective forces acting on all the particles.

$$\sum_{j=1}^{n} \vec{F}_j = \sum_{j=1}^{n} m_j \ddot{\vec{r}}_j = m\vec{a}_{cm} = \dot{\vec{P}} \qquad (15.8)$$

Equation (15.8) defines the motion of the mass center or center of mass.

Properties of the Mass Center:

A) The motion of the mass center is the same as if the total mass of the system and the external forces were concentrated at that point.

B) The center of gravity coincides with the mass center when the weight is taken into consideration.

15.4 ANGULAR MOMENTUM ABOUT THE MASS CENTER

When considering the motion of a system of particles, it is sometimes convenient to use a cartesian centroidal frame of reference x'y'z' which translates with respect to the inertial frame or reference xyz.

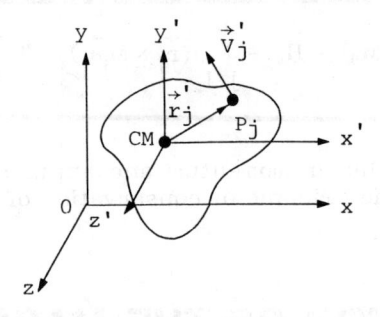

Fig. 15.1

All centroidal coordinates and quantities are denoted by the prime (') notation.

A-149

Angular momentum in the x'y'z' system:

$$\vec{H}'_{cm} = \sum_{j=1}^{n} (\vec{r}'_j \times m_j \vec{v}'_j) \qquad (15.9)$$

Moment resultant in the x'y'z' system:

$$\Sigma \vec{M}_{cm} = \dot{\vec{H}}'_{cm} = \sum_{j=1}^{n} (\vec{r}'_j \times m_j \vec{a}'_j) \qquad (15.10)$$

In the xyz system:

$$\vec{H}_{cm} = \sum_{j=1}^{n} (\vec{r}'_j \times m_j \vec{v}_j) \qquad (15.11)$$

Equation (15.9) and (15.11) are equal:

$$\vec{H}_{cm} = \vec{H}'_{cm} \qquad (15.12)$$

Result:

$$\Sigma \vec{M}_{cm} = \dot{\vec{H}}_{cm}$$

where

$$\vec{H}_{cm} = \sum_{j=1}^{n} (\vec{r}'_j \times m_j \vec{v}_j) = \sum_{j=1}^{n} (\vec{r}'_j \times m_j \vec{v}'_j)$$

(15.13)

Angular momentum may be computed by adding the moments about the center of mass in either the centroidal or inertial reference system.

15.5 CONSERVATION OF MOMENTUM

For the mass center CM, the conservation of momentum, equations (15.6) and (15.7) takes the same form:

$$\dot{\vec{P}}_G = 0 \quad \text{or} \quad \vec{P}_G = \text{constant} \tag{15.14}$$

and

$$\dot{\vec{H}}_G = 0 \quad \text{or} \quad \vec{H}_G = \text{constant} \tag{15.15}$$

under the conditions of zero resultant external force and moment acting on the system.

Equations (15.14) and (15.15) apply to the system as a whole and never to its compoment parts.

15.6 KINETIC ENERGY

Kinetic energy for a system of particles is given by:

$$KE = \frac{1}{2} \sum_{j=1}^{n} m_j v_j^2, \tag{15.16}$$

in an inertial frame of reference.

For a centroidal reference frame,

$$KE_{cm} = \frac{1}{2} mV_{cm}^2 + \frac{1}{2} \sum_{j=1}^{n} m_j v_j^2 \tag{15.17}$$

A-151

Equation (15.17) shows that KE_{cm} consists of two terms:

1) $\frac{1}{2}mV_{cm}^2$ = Kinetic Energy of the Mass Center,

2) $\frac{1}{2} \sum_{j=1}^{n} m_j v_j^2$ = Kinetic Energy of the system's motion relative to the centroidal frame of reference.

15.7 CONSERVATION OF ENERGY

The principle of work and energy is written as follows which is simlar to that of the particle:

$$\boxed{W_{1 \to 2} = KE_2 - KE_1} \qquad (14.8)$$

KE_2 and KE_1 may be computed by either equation (15.16) or (15.17).

$W_{1 \to 2}$ is the work of the entire system. Internal and external forces must be considered.

If all the forces acting on the system are conservative, then equation (14.8) may be replaced by the following equation:

$$\boxed{KE_1 + PE_1 = KE_2 + PE_2 = E} \qquad (14.37)$$

This energy equation, as applied to the particle case, is also applicable to a system of particles.

15.8 IMPULSE AND MOMENTUM

Linear Impulse - Changes the linear momentum of the system.

A-152

Angular Impulse - Changes the angular momentum of the system:

$$\vec{P}_1 + \Sigma \int_{t_1}^{t_2} \vec{F} dt = \vec{P}_2; \qquad \vec{P}_1 = m\vec{V}_{G_1}$$

$$(\vec{H}_0)_1 + \Sigma \int_{t_1}^{t_2} \vec{M}_0 dt = (\vec{H}_0)_2; \qquad \vec{H}_0 = I_G w_1$$

(15.18)

When there are no external forces acting on a system:

$$\vec{P}_1 = \vec{P}_2$$
$$(\vec{H}_0)_1 = (\vec{H}_0)_2$$

(15.19)

Equations (15.19) represent conservation of momentum and are alternate forms of the equations (15.14) and (15.15).

15.9 CONTINUOUS SYSTEM

Consider the flow of fluid through a duct:

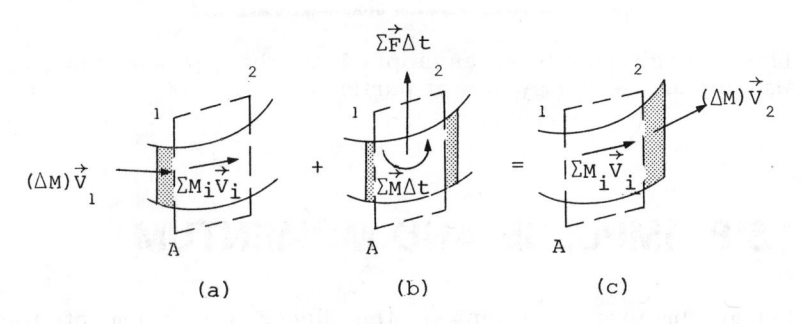

$$(\Delta M)\vec{V}_1 + \Sigma M_i \vec{V}_i \quad + \quad \Sigma \vec{M} \Delta t \quad = \quad \Sigma M_i \vec{V}_i \quad + (\Delta M)\vec{V}_2$$

(a) (b) (c)

Fig. 15.2 Fluid Flow Through A Duct

A-153

Equation Formation:

The momentum of particles entering a constant area 'A' during a time 'Δt' plus the impulses exerted during that time 'Δt' is equipollent to the momentum of the particles leaving 'A' in time Δt:

$$(\Delta m)\vec{v}_1 + \Sigma \vec{F} \Delta t = (\Delta m)\vec{v}_2 \qquad (15.20)$$

Differential Equation of Motion:

$$\boxed{\Sigma \vec{F} = \dot{m}(\vec{v}_2 - \vec{v}_1)} \qquad (15.21)$$

Specific Cases:

A) Fluid Diverted by a Vane

1) Fixed Vane - Direct application of the above relation.
   $\vec{F}$ = Force exerted by the vane on the fluid.

2) Vane Moving at Constant Velocity:
   a) Choose a coordinate system moving with the vane.
   b) $\vec{v}_1$ and $\vec{v}_2$ must be replaced by the relative velocities of the fluid with respect to the moving vane.

B) Fluid Flowing Through a Pipe

In general, the pressure in the flow needs to be considered, therefore, the forces exerted on the area 'A' by the fluid on either side should be included in the analysis.

C) Aircraft Turbine Engine

1) All velocities are relative to the engine.

2) The only external force considered is the force exerted by the engine on the air stream. This force is equal and opposite to the thrust.

D) Aircraft Propeller

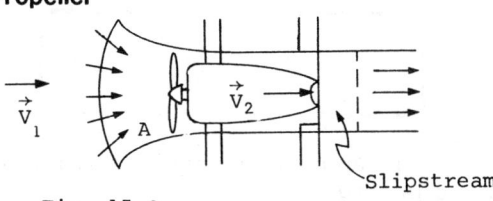

Fig. 15.3

1) Velocities are measured with respect to the aircraft. $\vec{v}_1 \approx 0$

2) The flow rate is the product of $\vec{v}_2$ and the cross sectional area of the slipstream (see Figure 15.3).

3) The only external force considered is the thrust.

15.10 VARIABLE MASS SYSTEMS

From the principle of impulse and momentum:

$$\vec{F} \Delta t = (\vec{P})_{t+\Delta t} - (\vec{P})_t \quad (15.22)$$

General Differential Equation of Motion:

$$\vec{F} = m \frac{d\vec{v}}{dt} - \vec{u} \frac{dm}{dt} \quad (15.23)$$

where
  $\vec{u}$ = Velocity of the Mass Gained (or Lost) Relative to the Moving Body

Specific Cases:

A) Mass Gaining Systems

In this case
$$\vec{u} = -\vec{v},$$

and equation (15.23) becomes:

$$\vec{F} = m \frac{d\vec{v}}{dt} + \vec{v} \frac{dm}{dt} = \frac{d(m\vec{v})}{dt} \quad (15.24)$$

Equation (15.24) is applicable only if the initial velocity of the mass gained is zero.

A-155

In equation (15.24) $v \frac{dm}{dt}$ represents the magnitude of force exerted by the mass gained. $\frac{dm}{dt}$ represents the rate at which the mass is gained.

B) Mass Loosing Systems - Rocket Motion

In this case:

$$\frac{dm}{dt} < 0,$$

Assume that $\vec{F} = 0$, then equation (15.23) becomes

$$\boxed{m \frac{d\vec{v}}{dt} = \vec{u} \frac{dm}{dt}} \qquad (15.25)$$

If u is constant, then equation (15.25) may be solved to obtain the value of v.

$$\boxed{v = v_0 + u \ln \frac{M_0}{M}} \qquad (15.26)$$

CHAPTER 16

KINEMATICS OF RIGID BODIES

16.1 INTRODUCTION

Types of Rigid Body Motion:

A) Translation - All particles of the body move in a straight (rectilinear) or curved (curvilinear) paths.

B) Rotation About a Fixed Axis - All the particles move in circular paths with their centers on a fixed straight line called the axis of rotation.

C) General Plane Motion - All particles in the body remain at constant distance from a fixed reference plane and move in parallel planes. All particles on the same straight line perpendicular to the reference plane have identical values of displacement, velocity and acceleration. Plane in which mass center moves is called plane of motion.

D) Motion About a Fixed Point - Here the distance from a fixed point to any particle of the body is constant. Therefore, the path of motion lies on a sphere and centered at the fixed point.

E) General Motion - All other motion.

16.2 TRANSLATION

For a body in translation:

A) The resultant of all applied forces passes through the mass center of the body.

B) All the particles of the body have the same velocity and acceleration at any given instant.

C) In rectilinear translation, velocity and acceleration have the same direction.

D) In curvilinear translation, velocity and acceleration change in direction and magnitude.

16.3 ROTATION

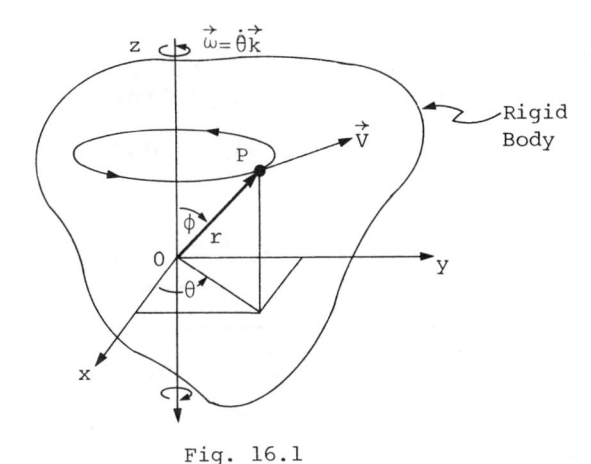

Fig. 16.1

The differential equations of rotation are analogous to the equations of translation:

Translation	Rotation
$v = \dfrac{ds}{dt}$	$\omega = \dfrac{d\theta}{dt}$
$a = \dfrac{dv}{dt}$	$\alpha = \dfrac{d\omega}{dt}$
$ads = vdv$	$\alpha d\theta = \omega d\omega$

A-158

Transformations:

$$s = r\theta$$
$$v = r\omega$$
$$a_t = r\alpha$$
$$a_n = r\omega^2$$

(16.1)

Omega Theorem

The time derivative of a constant length vector fixed in a rotating body is equal to the cross product of the angular velocity with the vector. i.e.,

$$\frac{dr}{dt} = \vec{\omega} \times \vec{r}$$

where $\vec{\omega}$ = Angular Velocity, and
$\vec{r}$ = Constant Length Vector

For a point in the body (see Figure 16.1):

Velocity: $$\boxed{\vec{v} = \frac{d\vec{r}}{dt} = \vec{\omega} \times \vec{r}}$$ (16.2)

Speed: $$\boxed{v = \frac{ds}{dt} = r\dot{\theta}\sin\phi}$$ (16.3)

Acceleration: $$\boxed{\vec{a} = \vec{\alpha} \times \vec{r} + \vec{\omega} \times (\vec{\omega} \times \vec{r})}$$ (16.4)

where

$$\vec{\alpha} = \frac{d\vec{\omega}}{dt} = \text{Angular Acceleration},$$

Also written as $\dot{\omega}\vec{k}$ or $\ddot{\theta}\vec{k}$

Rotation of a Slab

The motion of a slab in a reference plane perpendicular to the axis of rotation is used to define the rotation of a rigid body about a fixed axis.

A-159

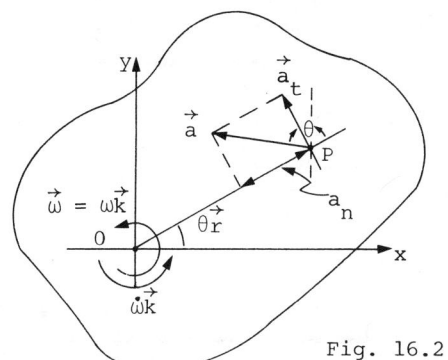

For Point P:

Fig. 16.2

Velocity:
$$\vec{v} = \omega\vec{k} \times \vec{r}$$
(16.3)

Speed:
$$v = r\omega$$
(16.1)

Acceleration:
$$\vec{a} = \dot{\omega}\vec{k} \times \vec{r} - \omega^2 \vec{r}$$
(16.4)

Tangential and Normal Acceleration:

$$\vec{a}_t = \dot{\omega}\vec{k} \times \vec{r}$$
$$\vec{a}_n = -\omega^2 \vec{r}$$
(16.5)

Equations of Motion:

$$\omega = \frac{d\theta}{dt}$$
$$\dot{\omega} = \frac{d^2\theta}{dt^2} = \omega \frac{d\omega}{d\theta}$$
(16.6)

Special Cases:

A) Uniform Rotation where $\vec{\alpha} = 0$

    Angular coordinate is

A-160

$$\theta = \theta_0 + \omega t \qquad (16.7)$$

B) Uniformly Accelerated Rotation where $\bar{\alpha}$ = constant

$$\omega = \omega_0 + \alpha t$$
$$\theta = \theta_0 + \omega_0 t + \tfrac{1}{2}\alpha t^2 \qquad (16.8)$$
$$\omega^2 = \omega_0^2 + 2\alpha(\theta - \theta_0)$$

16.4 ABSOLUTE AND RELATIVE VELOCITY

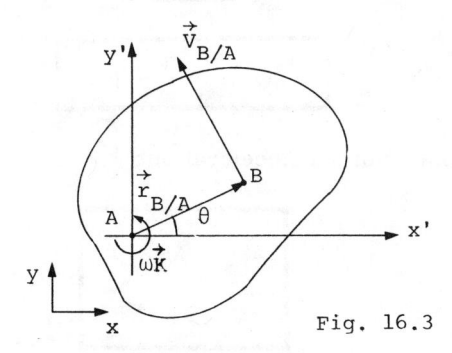

Fig. 16.3

Let $\vec{v}_{B/A}$ denote the velocity of B relative to A.

Relative Velcoity:

$$\vec{v}_{B/A} = \omega \vec{k} \times \vec{r}_{B/A} \qquad (16.9)$$

Relative Speed:

$$v = r\dot{\theta} = r\omega \qquad (16.10)$$

A-161

Absolute velocity of point B:

$$\vec{v}_B = \vec{v}_A + \vec{v}_{B/A} = \vec{v}_A + (\omega \vec{k} \times \vec{r}_{B/A})$$ (16.11)

Absolute speed of point B:

$$v_B = v_A + v_{B/A} = v_A + r\omega$$ (16.12)

16.5 INSTANTANEOUS CENTER OF ROTATION

Given a slab in general plane motion, the velocities of the particles composing the slab are the same as if the slab were rotating about a specific axis normal to the plane of the slab. This axis, known as the instantaneous axis of rotation is positioned at a point known as the instantaneous center of rotation.

Graphical Location of the Instantaneous Center of Rotation

A) General Case

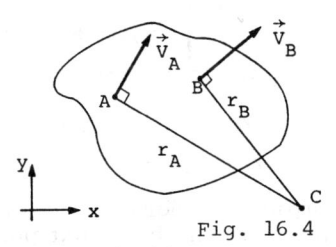

Fig. 16.4

B) Parallel and Unequal Velocities

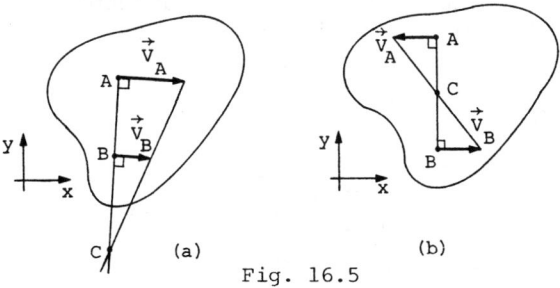

(a)   (b)

Fig. 16.5

A-162

During the instantaneous rotation, all the particles have the same angular velocity about the instantaneous center. The velocities are directed normal to the line connecting the particles to the instantaneous center.

If the linear velocity and the angular velocity are known:

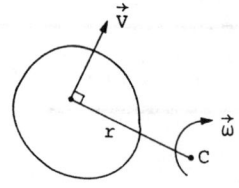

Fig. 16.6

Then C is located at:

$$r = \frac{v}{\omega}$$ (16.13)

For a Wheel or a Sphere:

C is located at distance equal to the radius R from the geometric center.

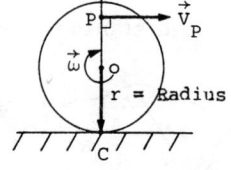

Fig. 16.7

Instantaneous Center of Zero Acceleration

A) It is used to compute acceleration in plane motion as if the body is in pure rotation about that point.

B) It is the same as the instantaneous center of rotation only for a body starting from rest.

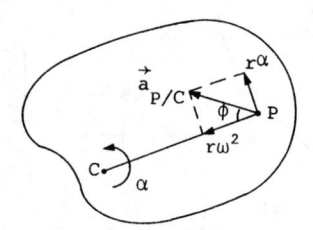

Fig. 16.8

A-163

$$a_p = a_c + a_{p/c}$$

When α and ω are known:

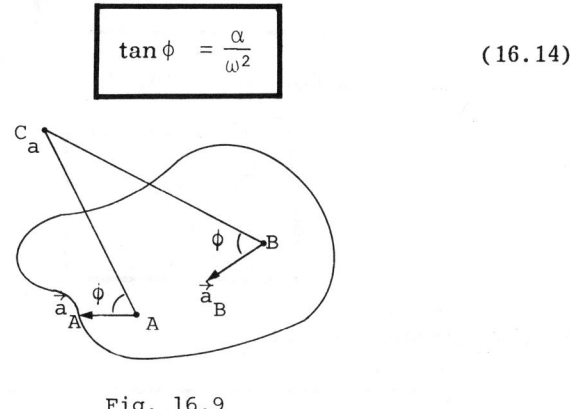

$$\boxed{\tan\phi = \frac{\alpha}{\omega^2}}$$ (16.14)

Fig. 16.9

C_a = Instantaneous Center of Acceleration

16.6 ABSOLUTE AND RELATIVE ACCELERATION

Relative Acceleration Formula:

$$\boxed{\vec{a}_B = \vec{a}_A + \vec{a}_{B/A}}$$ (16.15)

Let $\vec{r}_{B/A}$ = Relative Position of B with Respect to A.

$\omega\vec{k}$ = Angular Velocity

$\alpha\vec{k}$ = Angular Acceleration

Tangential Compoment:

$$\boxed{(\vec{a}_{B/A})_t = \alpha\vec{k} \times \vec{r}_{B/A}}$$ (16.16)

A-164

Normal Compoment:

$$(\vec{a}_{B/A})_n = -\omega^2 \vec{r}_{B/A} \qquad (16.17)$$

Scalar Equations of Equations (16.16) and (16.17):

$$(a_{B/A})_t = r\alpha$$
$$(a_{B/A})_n = r\omega^2 \qquad (16.18)$$

Absolute Acceleration of B:

$$\vec{a}_B = \vec{a}_A + [\alpha \vec{k} \times \vec{r}_{B/A} - \omega^2 \vec{r}_{B/A}] \qquad (16.19)$$

16.7 INERTIAL FORCES

If an accelerated coordinate system is used to describe the motion of a particle, then Newton's second law becomes:

$$\vec{F} - m\vec{a}_0 = m\vec{a} \qquad (16.20)$$

where
$\vec{a}_0$ = Acceleration of the Coordinate System

The term $(-m\vec{a}_0)$ is the inertial or fictitious force term. It is not the result of any interaction with other bodies. It is the result of using an accelerated coordinate system.

16.8 NON-INERTIAL REFERENCE SYSTEM

Non-Inertial Coordinate System - When the reference frame

undergoes translation and rotation.

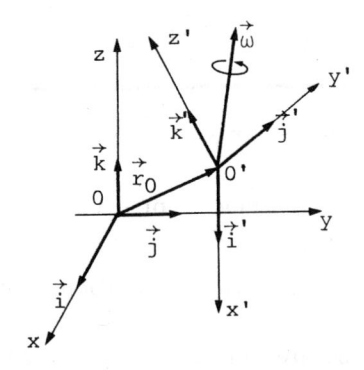

Fig. 16.10 Non Intertial Reference System

In general, the absolute time derivative of any vector A in the rotating frame 0'x'y'z' is:

$$\left(\frac{d\vec{A}}{dt}\right)_{0xyz} = \left(\frac{d\vec{A}}{dt}\right)_{0'x'y'z'} + (\vec{\omega} \times \vec{A}) \qquad (16.21)$$

Equation (16.21) has two parts:

A) $\left(\frac{d\vec{A}}{dt}\right)_{0'x'y'z'}$ = Change of $\vec{A}$ with respect to the rotating system 0'x'y'z'.

B) $\vec{\omega} \times \vec{A}$ = Induced motion of the moving frame 0'x'y'z' with respect to the stationary system 0xyz.

By the principle stated in equation (16.21):

Velocity:

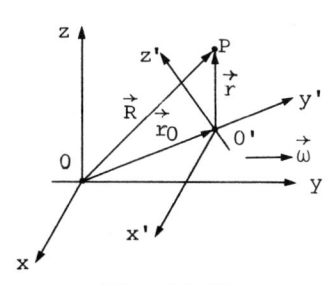

Fig. 16.11

A-166

$$\boxed{\vec{v}_p = \left(\frac{d\vec{R}}{dt}\right)_{0xyz} = (\dot{\vec{r}})_{0'x'y'z'} + \vec{\omega} \times \vec{r} + \vec{v}_0}$$

(16.22)

$\vec{v}_p$ = Absolute velocity of p

$(\dot{\vec{r}})_{0'x'y'z'}$ = Relative velocity of p with respect to the moving system

$\vec{\omega} \times \vec{r}$ = Induced motion due to rotation of 0'x'y'z'

$\vec{v}_0 = \dfrac{d\vec{r}_0}{dt}$ = Absolute velocity of the moving origin 0' with respect to the fixed system 0xyz

Acceleration — Found by differentiating equation (16.21) with $\vec{A}$ equal to $\dfrac{d\vec{R}}{dt} - \vec{v}_0 = \dot{\vec{r}} + \vec{\omega} \times \vec{r}$:

$$\boxed{\begin{aligned}\vec{a}_p &= \left(\frac{d^2 R}{dt^2}\right)_{0xyz} = (\ddot{\vec{r}})_{0'x'y'z'} + [2\vec{\omega} \times (\dot{\vec{r}})_{0xyz}] \\ &\quad + (\dot{\vec{\omega}} \times \vec{r}) + [\vec{\omega} \times (\vec{\omega} \times \vec{r})] + \vec{A}_0\end{aligned}}$$

$\vec{a}_p$ = Absolute Acceleration of p, (16.23)

$(\ddot{\vec{r}})_{0'x'y'z'}$ = Relative acceleration of the particle with respect to the moving coordinate system

$2\vec{\omega} \times (\dot{\vec{r}})$ = Coriolis acceleration

$\dot{\vec{\omega}} \times \vec{r}$ = Transverse acceleration

$\vec{\omega} \times (\vec{\omega} \times \vec{r})$ = Centripetal acceleration

$\vec{A}_0 = \dfrac{d^2\vec{r}_0}{dt^2}$ = Absolute acceleration of the moving origin

Note: Coriolis, transverse, and centripetal acceleration are rotational terms which appear only in the fixed system 0xyz.

Newton's Second Law

Equation (16.20) may now be written as

$$\vec{F} - m\vec{A}_0 - (2m\vec{\omega} \times \dot{\vec{r}}) - (m\dot{\vec{\omega}} \times \vec{r}) - m\vec{\omega} \times (\vec{\omega} \times \vec{r}) = m\ddot{\vec{r}}$$

(16.24)

In equation (16.24), the inertial terms are as follows:

Coriolis Force

$$\vec{F}_{cor} = -2m\vec{\omega} \times \dot{\vec{r}}$$

(16.25)

This force is always perpendicular to the velocity vector.

Transverse Force

$$\vec{F}_{trans} = -m\dot{\vec{\omega}} \times \vec{r}$$

(16.26)

This force is present only if there is an angular acceleration of the rotating reference system and acts normal to the position vector.

Centrifugal Force

$$\vec{F}_{cent} = -m\vec{\omega} \times (\vec{\omega} \times \vec{r})$$

(16.27)

This force acts normal to, and is directed away from the axis of rotation. Its magnitude is $mr\omega^2 \sin\theta$.

Inertial forces from the rotation of the reference system are illustrated as follows:

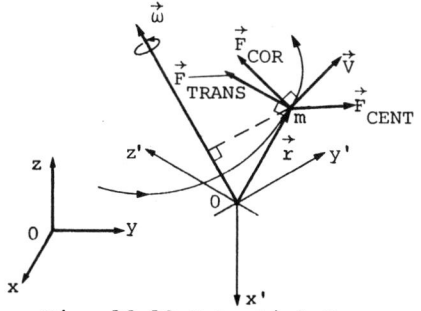

Fig. 16.12 Intertial Forces

16.9 MOTION ABOUT A FIXED POINT

In general, the displacement of a rigid body with respect to a fixed point 0 is equivalent to a rotation of the body about an axis through 0.

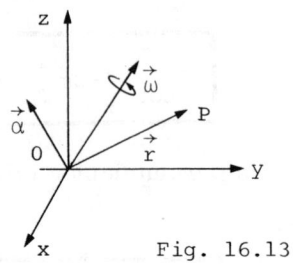

Fig. 16.13

Velocity:
$$\boxed{\vec{v} = \frac{d\vec{r}}{dt} = \vec{\omega} \times \vec{r}}$$
(16.28)

Acceleration:
$$\boxed{\vec{a} = \vec{\alpha} \times \vec{r} + \vec{\omega} \times (\vec{\omega} \times \vec{r})}$$
(16.29)

$\vec{\alpha} = \frac{d\vec{\omega}}{dt}$ — It represents the velocity of the tip of the vector $\vec{\omega}$.

CHAPTER 17

PLANE MOTION OF RIGID BODIES

17.1 INTRODUCTION

Plane Motion - Rigid body motion where the particles composing the body are always a fixed distance from a reference plane.

Plane of Motion - The plane in which the center of mass moves.

General plane motion may be considered to be a sum of a translation and a rotation.

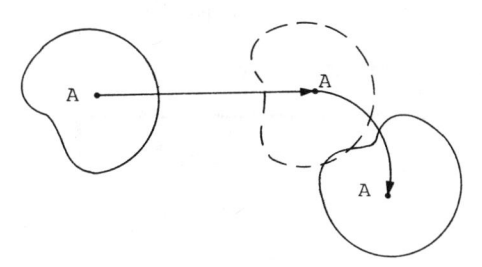

Fig. 17.1 General Plane Motion

17.2 CENTER OF MASS

General Formulas:

$$x_{cm} = \frac{\int_Q PxdQ}{\int_Q PdQ} \quad ; \quad y_{cm} = \frac{\int_Q PydQ}{\int_Q PdQ} \quad ; \quad z_{cm} = \frac{\int_Q PzdQ}{\int_Q PdQ}$$

(17.1)

For a Rigid Body:

Substitute into equation (17.1):

$P = \rho$ = Density of the Material

$Q = V$ = Volume

For a Thin Shell:

Substitute:

$P = m$ = Mass Per Unit Area

$Q = s$ = Surface Area

For a Thin Wire:

Substitute:

$P = n$ = Mass Per Unit Length

$Q = \ell$ = Length

For a Composite Body:

$$x_{cm} = \frac{x_1 m_1 + x_2 m_2 + \ldots}{m_1 + m_2 + \ldots}$$

(17.2)

$x_1, x_2 \ldots$ are center of mass locations for $m_1, m_2 \ldots$ respectively. y_{cm} and z_{cm} are calculated using equations similar to equation (17.2).

Symmetry Considerations – If the body has a plane or axis of symmetry, then the mass center is located on this plane or axis of symmetry.

A) Solid Hemisphere

A-171

Fig. 17.2

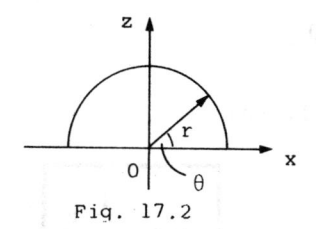

$$z_{cm} = \frac{\int_0^r \rho\pi z(r^2-z^2)dz}{\int_0^r \rho\pi (r^2-z^2)dz} = \frac{3}{8}r \qquad (17.3)$$

$x_{cm} = y_{cm} = 0$

B) Hemispherical Shell
Refer to Figure 17.2:

$$z_{cm} = \frac{\int_0^r \rho\, 2\pi rz\, dz}{\int_0^r \rho\, 2\pi r\, dz} = \frac{1}{2}r \qquad (17.4)$$

$x_{cm} = y_{cm} = 0$

C) Semi-circular Arcs

$$z_{cm} = \frac{\int_0^\pi \rho(r\sin\theta)r\,d\theta}{\int_0^\pi \rho r\,d\theta} = \frac{2}{\pi}r \qquad (17.5)$$

A-172

$x_{cm} = y_{cm} = 0$

D) Semi-Circular Lamina

$$\boxed{z_{cm} = \frac{4}{3\pi} r}$$ (17.6)

$x_{cm} = y_{cm} = 0$

17.3 STATIC EQUILIBRIUM

By Newton's Laws:

Translational Equilibrium:

$$\vec{F}_1 + \vec{F}_2 + \ldots = 0 \qquad (17.7)$$

Rotational Equilibrium:

$$(\vec{r}_1 \times \vec{F}_1) + (\vec{r}_2 \times \vec{F}_2) + \ldots = 0 \qquad (17.8)$$

For a particle in a uniform gravitation field:

Let $m\vec{g}$ = Gravitational Force Acting on the Particle

Translational Equilibrium:

$$\vec{F}_1 + \vec{F}_2 + \ldots + m\vec{g} = 0 \qquad (17.9)$$

Rotational Equilibrium: (17.10)

$$(\vec{r}_1 \times \vec{F}_1) + (\vec{r}_2 \times \vec{F}_2) + \ldots + (\vec{r}_{cm} \times m\vec{g}) = 0$$

17.4 EQUATIONS OF MOTION

Basic Equations:

$$\Sigma \vec{F} = m\vec{a}_{cm}$$
$$\Sigma \vec{M}_G = \dot{\vec{H}}_G = I_G \vec{\alpha}$$

(17.11)

G = Mass Center
a_{cm} = Acceleration of Mass Center
α = Angular Acceleration
M_G = Moment About Mass Center
I_G = Moment of Inertia About G
$\dot{\vec{H}}_G$ = Angular Acceleration

Scalar Form of Equation (17.11):

$$\Sigma F_x = m(a_x)_{cm}; \quad \Sigma F_y = m(a_y)_{cm}; \quad \Sigma M_G = I_G \alpha$$

(17.12)

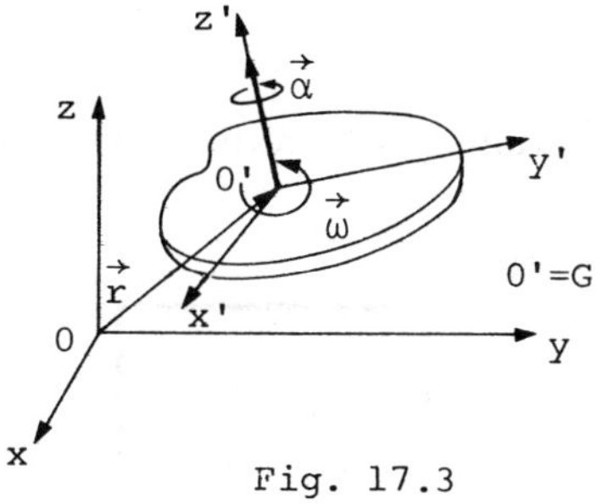

Fig. 17.3

17.5 ANGULAR MOMENTUM

The general theorem for angular momentum is applicable even for cases in which the axis of rotation is not fixed:

For each particle of the body:
$$\vec{r}_i = \vec{r}_{cm} + (\vec{r}_i)_{cm} \qquad (17.13)$$

where
$(\vec{r}_i)_{cm}$ = Position vector of the ith particle relative to the mass center

Also
$$\vec{v}_i = \vec{v}_{cm} + (\vec{v}_i)_{cm} \qquad (17.14)$$

Angular momentum is given by:

$$\boxed{\frac{d}{dt}\left\{\sum_i (\vec{r}_i)_{cm} \times m_i (\vec{v}_i)_{cm}\right\} = \frac{d\vec{H}_G}{dt} = \Sigma \vec{M}_G}$$

(17.15)

Equation (17.15) holds good even if the mass center is accelerating. When using another point as reference the mass center must be stationary in an inertial reference system.

Applications

A) Centroidal Rotation - The axis of rotation passes through the mass center.

Equations of Motion:

$$\boxed{\begin{aligned} \Sigma \vec{F}_x &= 0 \\ \Sigma \vec{F}_y &= 0 \\ \Sigma \vec{M}_G &= I_G \vec{\alpha} \end{aligned}} \qquad (17.16)$$

B) Noncentroidal Rotation

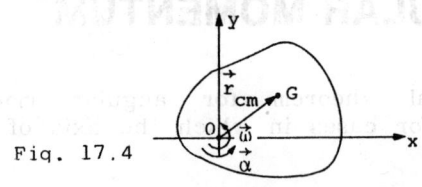

Fig. 17.4

A-175

$$\Sigma F_x = m\, r_{cm}\, \omega^2$$
$$\Sigma F_y = m\, r_{cm}\, \alpha$$
$$\Sigma M_G = I_G\, \alpha \quad \text{or} \quad \Sigma M_0 = I_0\, \alpha$$

(17.17)

C) Rolling Bodies

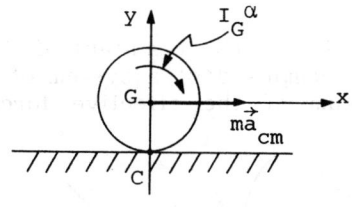

Fig. 17.5

Equations of Motion:

$$\Sigma \vec{F}_x = m\vec{a}_{cm}$$
$$\Sigma \vec{F}_y = 0$$
$$\Sigma \vec{M}_G = I_G\, \vec{\alpha}$$

(17.18)

For rolling without slipping:

$$\Sigma M_c = I_c\, \alpha$$

(17.19)

c = Instant Center of Rotation

If the wheel is unbalanced:

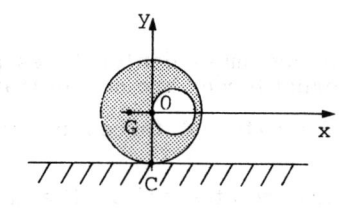

Fig. 17.6

Equation (17.19) is valid at points where the mass center G crosses the line connecting the geometric center O and the instantaneous center C.

17.6 D'ALEMBERT'S PRINCIPLE

D'Alembert's Principle - The resultant of the external forces applied to a body composed of a system of particles is equal to the vector sum of the effective forces acting on all particles.

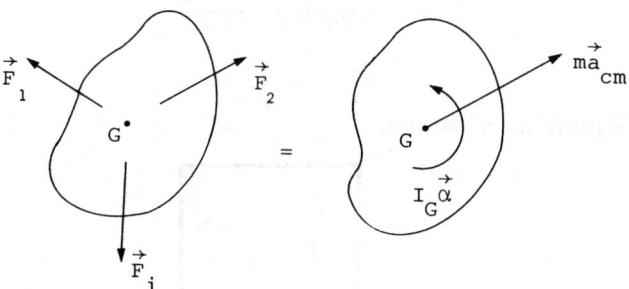

Fig. 17.7

Particular Cases:

A) Translation: $\vec{\alpha} \equiv 0$

Effective forces reduce to a vector $m\vec{a}_{cm}$ acting at the mass center G.

B) Centroidal Rotation: $\vec{a}_{cm} \equiv 0$ as axis of rotation passes through mass center G.

Effective forces reduce to $I_G \vec{\alpha}$ acting at G.

Note:

A) The motion of the mass center takes place as if the total mass and external forces are concentrated at that point.

B) In general, the external forces do not pass through the mass center.

C) The rigid body rotates about the mass center as if it was a fixed point, if the body is in plane motion.

17.7 SYSTEMS OF RIGID BODIES

For a system consisting of several rigid bodies, the method described in the preceding section may also be applied.

Method of Solution:

A) Draw a diagram of the system, including all the external forces and moments acting on the system.
B) Express the external forces as equipollent to a system of effective forces and solve for the unknown quantities.
C) This approach is valid for problems with no more than three unknowns.

17.8 CONSTRAINED MOTION

For rigid bodies moving under constraints, equations (17.11) may be applied to the mass center of the body.

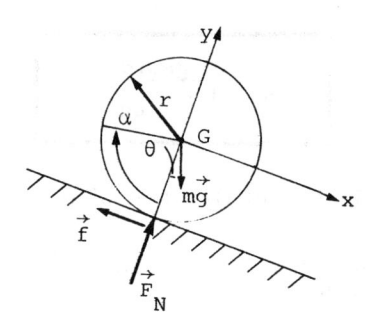

Fig. 17.8

For a body rolling down an incline:

Component equations of translation of the mass center:

$$\boxed{\begin{array}{l} m\ddot{x}_{cm} = mg\sin\theta - f \\ m\ddot{y}_{cm} = -mg\cos\theta + F_N \end{array}}$$

(17.20)

From the constraints of the problem:

$$y_{cm} = \text{constant}$$
$$F_n = mg\cos\theta \tag{17.21}$$

The only force producing a moment is the frictional force $\vec{f}$. Therefore, the equation of rotation is:

$$I_g \dot{\omega} = fr \tag{17.22}$$

Special Cases

A) No Slipping:

$$\dot{x}_{cm} = r\psi \tag{17.23}$$

where ψ is the angle of rotation and equation (17.22) becomes:

$$\frac{I_G}{r^2} \ddot{x}_{cm} = f \tag{17.24}$$

Substituting and solving gives:

$$\boxed{\ddot{x}_{cm} = \frac{g\sin\theta}{1 + (k_{cm}^2/r^2)}} \tag{17.25}$$

where k_{cm} = Radius of Gyration of G

Therefore, the body moves with constant linear and angular acceleration.

Angular Acceleration

Energy Equation:

$$\boxed{\frac{1}{2} m\dot{x}_{cm}^2 + \frac{1}{2} mk_{cm}^2 \frac{\dot{x}_{cm}^2}{r^2} - mgx_{cm}\sin\theta = E} \tag{17.26}$$

B) With Slipping:

Equation of Translation:

$$\boxed{m\ddot{x}_{cm} = mg\sin\theta - \mu mg\cos\theta}\qquad(17.27)$$

μ = Coefficient of Sliding Friction

Equation of Rotation:

$$\boxed{I_G \dot{\omega} = \mu mg\, r\cos\theta}\qquad(17.28)$$

If G has constant acceleration:

$$\ddot{x}_{cm} = g(\sin\theta - \mu\cos\theta)\qquad(17.29)$$

and if angular acceleration is also constant:

$$\dot{\omega} = \frac{\mu g r\cos\theta}{k_{cm}^2}\quad(\therefore\ I_G = mk_{cm}^2)\qquad(17.30)$$

By integration:

$$\boxed{\text{where}\quad \dot{x}_{cm} = Ar\omega \\ A = \frac{k_{cm}^2}{r^2}\left(\frac{\tan\theta}{\mu} - 1\right)}\qquad(17.31)$$

CHAPTER 18

ENERGY AND MOMENTUM METHODS - RIGID BODIES

18.1 PRINCIPLE OF WORK AND ENERGY

The work-energy method is an alternate approach to solving problems in dynamics. Its advantages are:

A) Usually faster and simpler to use.

B) Involves only scalar quantities.

C) Can be applied directly to a system without analysis of its component parts.

D) Directly relates forces, displacements and velocities.

Translation

$$W = \frac{1}{2} m(v^2 - v_0^2) = \Delta kE$$

(18.1)

    W = Work Done

    v_0 = Initial Velocity

    V = Final Velocity

    KE = Kinetic Energy; units: ft-lb

    m = Mass

Fixed Axis Rotation

$$\int_{\theta_0}^{\theta} \Sigma M d\theta = \frac{1}{2} I(\omega^2 - \omega_0^2)$$

(18.2)

If ΣM is variable, integrate equation (18.2), or find the area under the M-Q diagram.

If ΣM is constant, then

$$\Sigma M(\theta) = \frac{1}{2} I(\omega^2 - \omega_0^2)$$

(18.3)

The initial direction of rotation is considered to do positive work.

For rotation under gravity forces:

$$\int_{\theta_0}^{\theta} \Sigma M d\theta = Wh$$

(18.4)

where   h = Change in elevation of the mass center
         = $\bar{r} \sin\theta$

For a system in translation and rotation:

$$W = \Sigma \frac{m}{2}(v^2 - v_0^2) + \Sigma \frac{I}{2}(\omega^2 - \omega_0^2)$$

(18.5)

Work done by a force $\vec{F}$:

$$W = \int \vec{F} \cdot d\vec{r} = \int F(dr\cos\theta)$$

(18.6)

$d\vec{r}$ = Increment in the distance in the direction of the applied force

s = Path of the motion

A-182

θ = Angle between $\vec{F}$ and $\vec{dr}$

For constant forces:

$$W = \Sigma\, Fs = \text{Area under the force-displacement diagram} \quad (18.7)$$

Examples of Forces That Do No Work:

A) Reactions at a frictionless pin when the body rotates about the pin.

B) Reactions at a frictionless surface when the body moves on the surface.

C) Weight of a body when the mass center moves horizontally.

D) No work is done by the frictional force at the point of contact, when a body rolls on a surface without slipping.

18.2 KINETIC ENERGY IN PLANE MOTION

Since plane motion is equal to translation plus rotation, kinetic energy is equal to the work equation as given by equation (18.5):

$$\Delta KE = W = \Sigma \frac{m}{2}(v^2 - v_0^2) + \Sigma \frac{I}{2}(\omega^2 - \omega_0^2) \quad (18.8)$$

In terms of the instant center of rotation:

$$\Delta KE = \frac{1}{2} I_c \omega^2 - \frac{1}{2} I_{c_0} \omega_0^2 \quad (18.9)$$

I_c, I_{c_0} = Moment of inertia about the final and initial positions of the instant center c, respectively

For a homogeneous free-rolling wheel, I_c is constant, and:

$$\Sigma M_c = \frac{1}{2} I_c (\omega^2 - \omega_0^2)$$

(18.10)

Note: The above equation is valid only when there is no slippage. If slippage occurs, then equation (18.8) should be used, i.e., work done by the kinetic friction forces should also be considered.

18.3 CONSERVATION OF ENERGY

If a rigid body, or a system of rigid bodies are in motion under the influence of conservative forces, then the equation of conservation of energy (14.37) for a particle holds good or rigid bodies also.

$$(KE)_1 + (PE)_1 = (KE)_2 + (PE)_2 = E \quad (14.37)$$

For Plane Motion

The kinetic energy term should account for both, translational and rotational effects.

18.4 POWER

Power is the time rate at which work is done.

$$\text{Power} = \frac{dW}{dt}$$

(18.11)

In terms of forces:

$$\text{Power} = \vec{F} \cdot \vec{v}$$

(18.12)

A-184

In terms of moments:

$$\boxed{\text{Power} = M\omega}$$
(18.13)

where
$$\omega = \frac{d\theta}{dt}$$

Efficiency is given by equation (14.9):

$$\boxed{\eta = \frac{\text{Power Output}}{\text{Power Input}} \leq 1}$$
(14.9)

Overall efficiency of machines working in series is equal to the product of their individual efficiencies.

18.5 IMPULSE AND MOMENTUM

Impulse-momentum relationship:

$$\boxed{\begin{array}{c}\text{Initial system momenta} + \text{Resultant external impulse} \\ = \text{Final system momenta}\end{array}}$$

(18.14)

Three separate equations may be derived:

A) Equate the x-components of impulse and momenta.
B) Equate the y-components of impulse and momenta.
C) Equate the moments of the vectors about any given point.

For a system of rigid bodies:

A) Consider each body separately.
B) If the problem contains no more than three unknowns, the principle of impulse and momentum may be applied to the system as a whole.

A-185

18.6 CONSERVATION OF ANGULAR MOMENTUM

If either

A) the lines of action of all external forces passes through a point O; or

B) the sum of the angular impulses of all external forces are zero,

then:

The angular momentum about a given point is conserved:

$$(\vec{H}_0)_1 = (\vec{H}_0)_2$$

(18.15)

Applications:

A) If rotation of the body is constant (uniform angular velocity)

B) Valid only for the system as a whole, never to its compoment parts.

18.7 ECCENTRIC IMPACT

Unconstrained Case

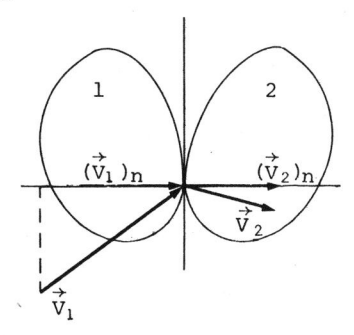

Fig. 18.1

Velocity Relationships:

$$(v_2')_n - (v_1')_n = e[(v_1)_n - (v_2)_n]$$

(18.16)

e = Coefficient of Restitution

Prime (') variables represent parameters after the impact.

Constrained Case

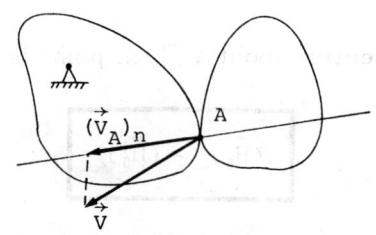

Fig. 18.2

$$e = \frac{(u_A)_n - (v_A')_n}{(v_A)_n - (u_A)_n}$$

(18.17)

where u = Velocity after the period of deformation.

CHAPTER 19

SPATIAL DYNAMICS

19.1 ANGULAR MOMENTUM

Angular momentum of a rigid body about its mass center:

$$H_x = I_{xx}\omega_x - P_{xy}\omega_y - P_{xz}\omega_z$$
$$H_y = -P_{yx}\omega_x + I_{yy}\omega_y - P_{yz}\omega_z$$
$$H_z = -P_{zx}\omega_x - P_{zy}\omega_y + I_{zz}\omega_z$$

(19.1)

where

$$I_{xx} = \int (y^2+z^2)\,dm$$
$$I_{yy} = \int (z^2+x^2)\,dm$$
$$I_{zz} = \int (x^2+y^2)\,dm$$

(19.2)

are the centroidal mass moments of inertia; and

$$P_{xy} = \int xy\,dm$$
$$P_{yz} = \int yz\,dm$$
$$P_{zx} = \int zx\,dm$$

(19.3)

are the centroidal mass products of inertia. From equation (19.1), we get the inertia tensor at the mass center:

$$\begin{pmatrix} I_{xx} & -P_{xy} & -P_{xz} \\ -P_{yx} & I_{yy} & -P_{yz} \\ -P_{zx} & -P_{zy} & I_{zz} \end{pmatrix}$$

(19.4)

Principal Axes of Inertia - A set of body axes for which the products of inertia are zero.

If the principal axes of inertia were used as reference, then:

Angular Momentum:
$$H_{x'} = I_{xx'}\,\omega_{x'}$$
$$H_{y'} = I_{yy'}\,\omega_{y'}$$
$$H_{z'} = I_{zz'}\,\omega_{z'}$$

(19.5)

$\vec{H}_G$ and $\vec{\omega}_G$ have the same direction if, and only if, $\vec{\omega}$ is directed along a principal axis of inertia.

The inertia tensor corresponding to the principal axes of inertia is defined as:
$$\begin{pmatrix} I_{xx'} & 0 & 0 \\ 0 & I_{yy'} & 0 \\ 0 & 0 & I_{zz'} \end{pmatrix}$$

(19.6)

Transfer Formula:

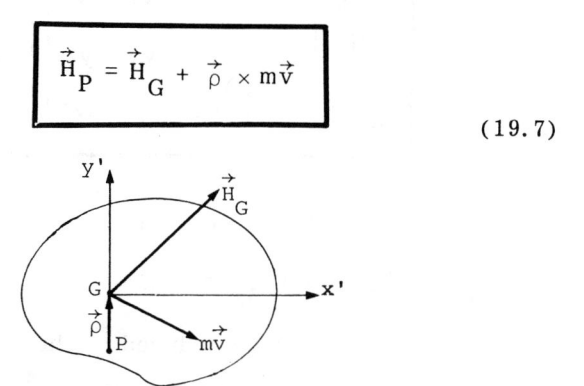

$$\vec{H}_P = \vec{H}_G + \vec{\rho} \times m\vec{v}$$

(19.7)

Fig. 19.1

A standard way of writing expression (19.4) is:

$$\begin{pmatrix} I_{11} & I_{12} & I_{13} \\ I_{21} & I_{22} & I_{23} \\ I_{31} & I_{32} & I_{33} \end{pmatrix}$$

(19.8)

where

$I_{xx} = I_{11}$ $-P_{xy} = I_{12}$

$I_{yy} = I_{22}$ $-P_{xz} = I_{13}$

$I_{zz} = I_{33}$ $-P_{zx} = I_{31}$

$\vdots$

etc.

Note: The moment of inertia terms have repeated subscripts and the product of inertia terms have different subscripts.

The angular momentum expressions, given by (19.1), may be conveniently expressed in matrix notation as:

$$\begin{bmatrix} H_x \\ H_y \\ H_z \end{bmatrix} = \begin{bmatrix} I_{xx} & I_{xy} & I_{xz} \\ I_{yx} & I_{yy} & I_{yz} \\ I_{zx} & I_{zy} & I_{zz} \end{bmatrix} \begin{bmatrix} \omega_x \\ \omega_y \\ \omega_z \end{bmatrix}$$

(19.9)

where $I_{xy} = -P_{xy}$, $I_{xz} = -P_{xz}$, $I_{yx} = -P_{yx}$, $I_{yz} = -P_{yz}$, $I_{zx} = -P_{zx}$, and $I_{zy} = -P_{zy}$.

Equation 19.9 may be reduced to the form,

$$\vec{H} = \bar{I}\,\vec{\omega}$$

(19.10)

where
$$\vec{H} = \begin{bmatrix} H_x \\ H_y \\ H_z \end{bmatrix} \quad \bar{I} = \begin{bmatrix} I_{xx} & I_{xy} & I_{xz} \\ I_{yx} & I_{yy} & I_{yz} \\ I_{zx} & I_{zy} & I_{zz} \end{bmatrix}$$

and
$$\vec{\omega} = \begin{bmatrix} \omega_x \\ \omega_y \\ \omega_z \end{bmatrix}$$

Note: An alternate form of equation (19.10) may be written using equation (19.8), i.e., in tensor notation as

$$H_\alpha = I_{\alpha\beta}\,\omega_\beta \quad \alpha,\beta = 1,2,3$$

where the Einstein summation convention is employed, i.e., repeated subscripts in the same term imply summation over the range of that index.

A-191

19.2 PRINCIPLE AXES

Principal Axes - A set of coordinate axes in which the products of inertia are zero.

The angular momentum in reference to principal axes is given by equation (19.5) and the inertia tensor is given by equation (19.6).

A device such as a rotating shaft is

A) Statically Balanced - If the center of mass lies on the axis of rotation, and

B) Dynamically Balanced - If the products of inertia are zero and the center of mass lies on the axis of rotation.

Finding the principal axes is equivalent to diagonalizing a (3×3) matrix and is accomplished by solving

$$| \bar{I} - \lambda \bar{u} | = 0 \qquad (19.11)$$

where $\bar{u}$ is a unit matrix and $\bar{I}$ is given by equation (19.10). Equation (19.11) may be expanded as

$$\begin{vmatrix} I_{xx}-\lambda & I_{xy} & I_{xz} \\ I_{yx} & I_{yy}-\lambda & I_{yz} \\ I_{zx} & I_{zy} & I_{zz}-\lambda \end{vmatrix} = 0 \qquad (19.12)$$

and reduced to the cubic form as:

$$-\lambda^3 + A\lambda^2 + B\lambda + C = 0 \qquad (19.13)$$

where A, B and C are functions of the I's. The roots of equation (19.13) λ_1, λ_2 and λ_3 are the principal moments of inertia.

To find the orientation, the following equation should be solved:

$$\vec{H} = \lambda \vec{\omega} = \bar{I}\bar{\omega} \qquad (19.14)$$

in which λ is one of the three roots (λ_1, λ_2 or λ_3)

Scalar form of equation (19.14):

$$(I_{xx-\lambda})\cos\alpha + I_{xy}\cos\beta + I_{xz}\cos\gamma = 0$$
$$I_{yx}\cos\alpha + (I_{yy-\lambda})\cos\beta + I_{yz}\cos\gamma = 0 \quad (19.15)$$
$$I_{zx}\cos\alpha + I_{zy}\cos\beta + (I_{zz-\lambda})\cos\gamma = 0$$

α, β and γ are the direction angles of one of the principal axes. The roots of the equations (19.15) must satisfy:

$$\cos^2\alpha + \cos^2\beta + \cos^2\gamma = 1 \quad (19.16)$$

If one principal axis is known, the others can be determined by asuming that the z-axis is the principal axis. Then equations (19.15) reduce to

$$\boxed{\begin{aligned}(I_{xx-\lambda})\cos\alpha + I_{xy}\cos\beta &= 0 \\ I_{xy}\cos\alpha + (I_{yy-\lambda})\cos\beta &= 0\end{aligned}} \quad (19.17)$$

Let η be the angle between the x-axis and one of the unknown principal axes, then:

$$\boxed{\tan 2\eta = \frac{2 I_{xy}}{I_{xx} - I_{yy}}} \quad (19.18)$$

The third axis is normal to this axis: $(\eta + 90°)$.

19.3 MOMENTAL ELLIPSOID

To find the moment of inertia about an arbitrary axis

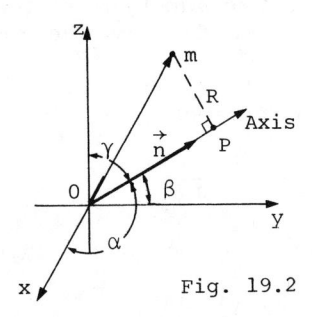

Fig. 19.2

$$I = \Sigma m_j R_j^2$$

(19.19)

Using direction cosines:

$$I = I_{xx}\cos^2\alpha + I_{yy}\cos^2\beta + I_{zz}\cos^2\gamma + 2I_{yz}\cos\gamma\cos\beta$$
$$+ 2I_{zx}\cos\alpha\cos\gamma + 2I_{xy}\cos\alpha\cos\beta$$

(19.20)

If the coordinate axes are also principal axes at the origin, then equation (19.20) becomes:

$$I = I_{xx}\cos^2\alpha + I_{yy}\cos^2\beta + I_{zz}\cos^2\gamma$$

(19.21)

The abbreviation of equation (19.20) is:

$$I = \bar{n}^T \bar{\bar{I}} \bar{n}$$

(19.22)

The distance 'OP' is given by:

$$OP = \frac{1}{\sqrt{I}}$$

(19.23)

By equating the direction cosines in terms of $\sqrt{I}$:

$$\bar{n} = \begin{bmatrix} x\sqrt{I} \\ y\sqrt{I} \\ z\sqrt{I} \end{bmatrix} = \bar{r}\sqrt{I}$$

(19.24)

Substituting into the general formula for the moment of inertia:

$$\bar{r}^T \bar{\bar{I}} \bar{r} = 1$$

(19.25)

A-194

or

$$x^2 I_{xx} + y^2 I_{yy} + z^2 I_{zz} + 2yz I_{yz} + 2zx I_{zx} + 2xy I_{xy} = 1$$

(19.26)

This is the equation of a surface called the momental ellipsoid of the body at O. If the coordinate axes are the principal axes, then equation (19.26) becomes:

$$x^2 I_{xx} + y^2 I_{yy} + z^2 I_{zz} = 1$$

(19.27)

Note:

A) The principal axes of the body are collinear with the principal axes of the ellipsoid.

B) If two principal moments of inertia are equal, then the ellipsoid is a surface of revolution.

C) If three are equal, then it is a sphere, and the moment of inertia is the same for any axis passing through 'O' regardless of its direction.

19.4 IMPULSE AND MOMENTUM

The relationship as described in section 18.5 is applicable to spatial dynamics also:

$$\text{System Momenta}_1 + \text{System External Impulse}_{1 \to 2} = \text{System Momenta}_2$$

(18.14)

Points to Remember

Use figures to write component and moment equations.

Remember that components of $\vec{H}_G$ are related to the components of $\vec{\omega}$.

For motion about a fixed point in rotation, eliminate reactions at that point by writing an equation involving moments of the momenta and impulses about that point.

19.5 KINETIC ENERGY

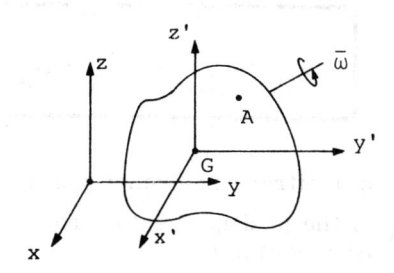

Fig. 19.3

The general expression for kinetic energy is:

$$KE = \frac{1}{2} m V_G^2 + \frac{1}{2} \vec{\omega} \cdot \vec{H}_G + \vec{V}_G \cdot \vec{V}_{A/G} m \qquad (19.28)$$

If G the mass center is used as the reference point, then

$$KE = \frac{1}{2} m V_{cm}^2 + \frac{1}{2} \vec{\omega} \cdot \vec{H}_G \qquad (19.29)$$

In scalar form (of equation (19.29)):

$$KE = \frac{1}{2} m V_{cm}^2 + \frac{1}{2} (I_x \omega_x^2 + I_y \omega_y^2 + I_z \omega_z^2 - 2P_{xy} \omega_x \omega_y \\ - 2P_{yz} \omega_y \omega_z - 2P_{zx} \omega_z \omega_x) \qquad (19.30)$$

If the axes x'y'z' are also the principal axes, then:

$$T = \frac{1}{2} I_x \omega_x^2 + \frac{1}{2} I_y \omega_y^2 + \frac{1}{2} I_z \omega_z^2 \qquad (19.31)$$

Matrix Notation

$$T = \frac{1}{2} \bar{\omega}^T \bar{\bar{I}} \bar{\omega}$$

$$= \frac{1}{2} [\omega_x \; \omega_y \; \omega_z] \begin{bmatrix} I_{xx} & I_{xy} & I_{xz} \\ I_{yx} & I_{yy} & I_{yz} \\ I_{zx} & I_{zy} & I_{zz} \end{bmatrix} \begin{bmatrix} \omega_x \\ \omega_y \\ \omega_z \end{bmatrix} \qquad (19.32)$$

19.6 RIGID BODY MOTION

Moment Equation:

$$\sum \vec{M}_A = \vec{r}_{G/A} \times m\vec{a}_A + \dot{\vec{H}}_A$$

(19.33)

where   A is an arbitrary reference point

$\vec{r}_{G/A}$ is the position vector from A to the mass center G.

General Approach to Solving a Problem:

Reduce a given force system to a resultant force-couple system acting at G. Equation (19.33) becomes:

$$\sum \vec{M}_A = \vec{r}_{G/A} \times m\vec{a}_{cm} + \dot{\vec{H}}_G$$

(19.34)

To evaluate $\dot{\vec{H}}$:

$$\dot{\vec{H}} = \frac{\delta \vec{H}}{\delta t} + \vec{\omega} \times \vec{H}$$

(19.35)

$\dot{\vec{H}}$ = Absolute time rate of change of $\vec{H}$ with respect to the inertial axes

$\frac{\delta \vec{H}}{\delta t} = \dot{\vec{H}}_r$ = Relative time rate of change of $\vec{H}$ with respect to the rotating body axes

$\vec{\omega} \times \vec{H}$ = Rate of change of the direction of $\vec{H}$ due to rotation of the body axes

Component Equations:

$$\sum M_x = \dot{H}_x + H_z \omega_y - H_y \omega_z$$
$$\sum M_y = \dot{H}_y + H_x \omega_z - H_z \omega_x$$
$$\sum M_z = \dot{H}_z + H_y \omega_x - H_x \omega_y$$

(19.36)

A-197

where:

$$H_x = I_x \omega_x - P_{xy} \omega_y - P_{xz} \omega_z$$
$$H_y = I_y \omega_y - P_{yz} \omega_z - P_{yx} \omega_x$$
$$H_z = I_z \omega_z - P_{zx} \omega_x - P_{zy} \omega_y$$

19.7 EULER'S EQUATIONS OF MOTION

By using body axes that coincide with the principal axes of inertia, the moment equation can be simplified to the following known as Euler's equations of motion:

$$\Sigma M_x = I_x \dot{\omega}_x + (I_z - I_y) \omega_y \omega_z$$
$$\Sigma M_y = I_y \dot{\omega}_y + (I_x - I_z) \omega_z \omega_x \qquad (19.37)$$
$$\Sigma M_z = I_z \dot{\omega}_z + (I_y - I_x) \omega_x \omega_y$$

The above equations are used to analyze the motion of a rigid body about its mass center.

$$\Sigma F_x = m a_x$$
$$\Sigma F_y = m a_y \qquad (19.38)$$
$$\Sigma F_z = m a_z$$

Newton's Laws with Euler's equations completely solve the motion of a rigid body in three dimensions.

19.8 MOTION ABOUT A FIXED POINT

Consider the rate of change of $\vec{H}_0$ about the fixed point O:

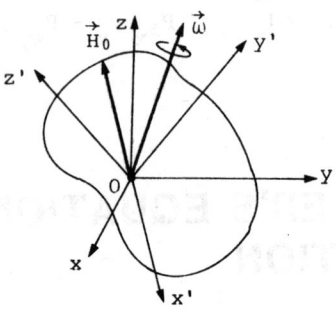

Fig. 19.4

$$\Sigma \vec{M}_0 = \dot{\vec{H}}_0 = (\dot{\vec{H}}_0)_{oxyz} + \vec{\Omega} \times \vec{H}_0$$

(19.39)

$\vec{H}_0$ = Angular momentum with respect to oxyz
$(\dot{\vec{H}}_0)_{oxyz}$ = Rate of change of $\vec{H}_0$ with respect to the rotating frame ox'y'z'
$\vec{\Omega}$ = Angular velocity of ox'y'z'

19.9 MOTION ABOUT A FIXED AXIS

Assume that the z-axis is the axis of rotation:

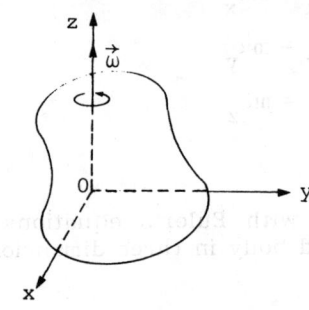

Fig. 19.5

Non-Symmetrical Bodies:

$$\Sigma M_x = -P_{zx}\alpha + P_{yz}\omega^2$$
$$\Sigma M_y = -P_{yz}\alpha - P_{xz}\omega^2 \qquad (19.40)$$
$$\Sigma M_z = I_z\alpha$$

$\alpha = \dot{\omega}_z$ = Angular acceleration in z direction

Symmetrical Bodies:

$$\Sigma M_x = \Sigma M_y = 0$$
$$\Sigma M_z = I_z\dot{\omega}_z = I_z\alpha \qquad (19.41)$$

Application:

A) When the forces applied are known, use ΣM_z to find α.
B) Determine ω by integration and substitute it into the ΣM_x and ΣM_y equations.
C) Together with the compoment equations of $\Sigma \vec{F} = m\vec{a}$, these equations define the motion of the mass center.

19.10 EARTH'S ROTATIONAL EFFECTS

Static Effects - A study of the plum line. Choose a coordinate system with the origin at the bob.

Equation of motion, equation (16.24) becomes:

$$\boxed{\vec{F} - m\vec{A}_0 = 0} \qquad (19.42)$$

where $\vec{F}$ includes:

1) $m\vec{G}$ = Gravitational Attraction
2) $-m\vec{g}$ = Vertical Tension of the Plum Line and

A-200

$\vec{A}_0$ = Centripetal Acceleration of the Moving Origin with a magnitude of $\rho \omega^2$

where $\rho = r\cos \lambda$ (see Fig. 19.6)

The plum line does not point to the earth's center, but deviates by an amount e:

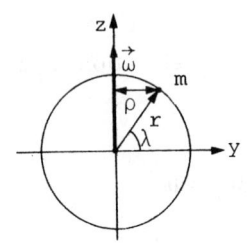

Fig. 19.6

$$\frac{\sin e}{mr\omega^2 \cos\lambda} = \frac{\sin\lambda}{mg}$$

(19.43)

Since e is a small quantity, equation (19.43) can be approximated as:

$$\sin e \doteq e = \frac{r\omega^2}{g} \sin\lambda \cos\lambda = \frac{r\omega^2}{2g} \sin 2\lambda$$

(19.44)

Maximum deviation occurs at $\lambda = 45°$ where

$$e_{max} = \frac{1}{10} \text{ degree}$$

(19.45)

Fig. 19.7

Dynamic Effects - Projectile motion.

Equation of motion:

$$\vec{F} + m\vec{g} - 2m\vec{\omega} \times \dot{\vec{r}} - m\vec{\omega} \times (\vec{\omega} \times \vec{r}) = m\ddot{\vec{r}}$$

(19.46)

Neglect air resistance $\vec{F} = 0$ in above equation.
Neglect the term $-m\vec{\omega} \times (\vec{\omega} \times \vec{r})$ as it is very small.

Resulting equation is:

$$m\ddot{\vec{r}} = m\vec{g} - 2m\vec{\omega} \times \dot{\vec{r}}$$

(19.47)

Coordinate System:

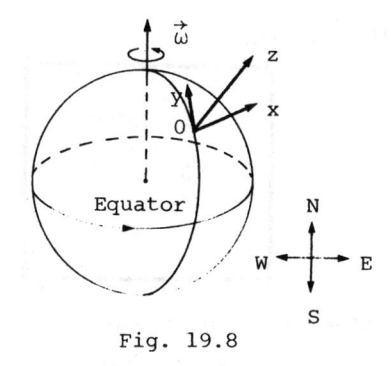

Fig. 19.8

With this system:

$$\vec{g} = -g\vec{k}$$
$$\vec{\omega} = (\omega \cos \lambda)\vec{j} + (\omega \sin \lambda)\vec{k}$$

(19.48)

Component Equations of Motion:

$$\ddot{x} = -2\omega(\dot{z}\cos\lambda - \dot{y}\sin\lambda)$$
$$\ddot{y} = -2\omega(\dot{x}\sin\lambda)$$
$$\ddot{z} = -g + 2\omega\dot{x}\cos\lambda$$

(19.49)

A-202

These equations are solved simultaneously by integration and eliminating ω^2.

$$x = \frac{1}{3}\omega g t^3 \cos\lambda - \omega t^2(\dot{z}_0 \cos\lambda - \dot{y}_0 \sin\lambda) + \dot{x}_0 t$$
$$y = \dot{y}_0 t - \omega \dot{x}_0 t^2 \sin\lambda$$
$$z = -\frac{1}{2}g t^2 + \dot{z}_0 t + \omega \dot{x}_0 t^2 \cos\lambda$$

(19.50)

where $\dot{x}_0, \dot{y}_0, \dot{z}_0$ are the initial conditions.

Special Cases:

A) If the projectile is released from rest in mid air:

Initial Conditions: $\dot{x}_0 = \dot{y}_0 = \dot{z}_0 = 0$

The solution is then

$$x = \frac{1}{3}\omega g t^3 \cos\lambda$$
$$y = 0$$
$$z = -\frac{1}{2}g t^2$$

(19.51)

The projectile drifts to the east.

B) If the projectile is moving with high velocity in a horizontal direction initially:

$$\dot{x}_0 = v_0, \quad \dot{y}_0 = \dot{z}_0 = 0$$

Equation of Motion:

$$y = -\omega v_0 t^2 \sin\lambda$$

(19.52)

The projectile drifts to the right.

The amount of drift is given by:

$$\text{Drift} = \frac{\omega H^2}{v_0}\sin\lambda$$

(19.53)

where $H \simeq v_0 t$, t = Time of Flight

Note: The amount of drift is the same regardless of the initial direction, provided that the trajectory is planar.

19.11 MOTION OF A GYROSCOPE

Consider a body rotating about two axes simultaneously.

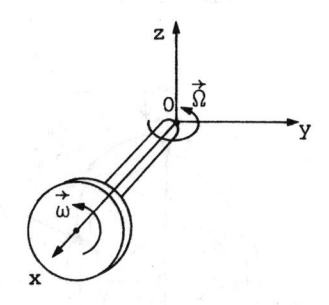

Fig. 19.9 Simple Gyroscope

Nomenclature:

x-Axis = Spin Axis

y-Axis = Torque Axis

z-Axis = Precession Axis

If all the axes are mutually perpendicular to each other, then:

$$\vec{M}_0 = \dot{\vec{H}}_0$$

and $\quad \vec{M}_0 = \vec{I}_{cm} \omega \Omega \qquad (19.54)$

Ω = Rate of Precession

Note: A force applied in the x-y plane trying to increase the precession will result in making the disk and axle rise.

A-204

Reference Position of a Gyroscope:

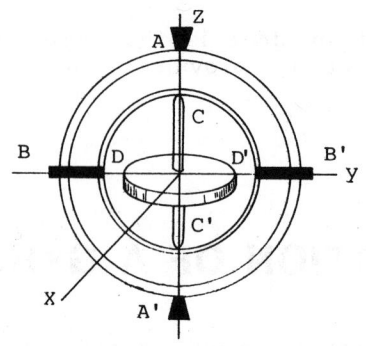

Fig. 19.10

Reference Angles:

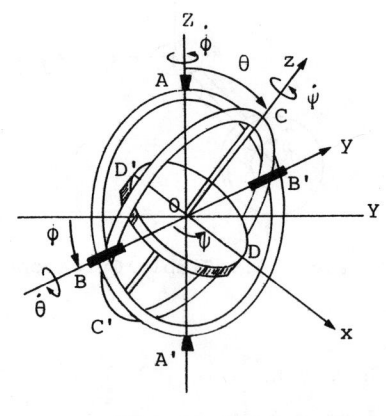

Fig. 19.11

Note:

A) The angles θ, ϕ, ψ and   are Eulerian angles. They describe the position of the gyroscope.

B) $\dot{\phi}$ = Rate of Precession

$\dot{\theta}$ = Rate of Nutation

$\dot{\psi}$ = Rate of Spin

C) Angular displacements are not vectors; time derivatives of the angles are vectors.

D) Eulerian angles may be used for any rigid body, in defining its orientation with respect to axes situated at any point of the body.

E) Oxyz are the coordinate axes and are attached to the inner gimbal ring. They are the principal axes of inertia for the gyroscope.

Angular velocity with respect to the fixed axes oxyz:

$$\boxed{\vec{\omega} = -\dot{\phi}\sin\theta\,\vec{i} + \dot{\theta}\,\vec{j} + (\dot{\psi} + \dot{\phi}\cos\theta)\vec{k}}$$
(19.55)

Angular Momentum:

$$\boxed{\vec{H}_0 = -I'\dot{\phi}\sin\theta\,\vec{i} + I'\dot{\theta}\,\vec{j} + I(\dot{\psi} + \dot{\phi}\cos\theta)\vec{k}}$$
(19.56)

where $I = I_z$

$I' = I_x$

Angular velocity with respect to the rotating axes O_{xyz}:

$$\boxed{\vec{\Omega} = -\dot{\phi}\sin\theta\,\vec{i} + \dot{\theta}\,\vec{j} + \dot{\phi}\cos\theta\,\vec{k}}$$
(19.57)

From equation (19.39)

$$\boxed{\begin{aligned} \Sigma M_x &= -I'(\ddot{\phi}\sin\theta + 2\dot{\theta}\dot{\phi}\cos\theta) + I\dot{\theta}(\dot{\psi} + \dot{\phi}\cos\theta) \\ \Sigma M_y &= I'(\ddot{\theta} - \dot{\phi}^2\sin\theta\cos\theta) + I\dot{\phi}\sin\theta(\dot{\psi} + \dot{\phi}\cos\theta) \\ \Sigma M_z &= I\frac{d}{dt}(\dot{\psi} + \dot{\phi}\cos\theta) \end{aligned}}$$
(19.58)

Solutions of equation (19.58) may be found by numerical methods and digital computers. Analytical solutions have not yet been obtained.

A-206

19.12 STEADY PRECESSION

Steady Precession - Special condition in which θ, $\dot{\phi}$, and $\dot{\psi}$ are all constants.

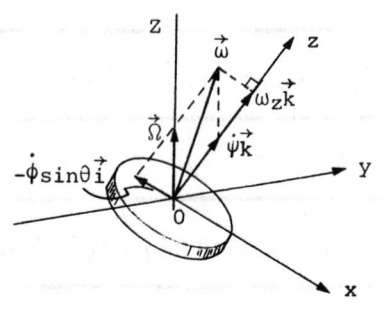

Fig. 19.12

$$\vec{\omega} = -\dot{\phi} \sin\theta \, \vec{i} + \omega_z \vec{k} \tag{19.59}$$

$$\vec{H}_0 = -I' \dot{\phi} \sin\theta \, \vec{i} + I \omega_z \vec{k} \tag{19.60}$$

and

$$\vec{\Omega} = -\dot{\phi}\sin\theta \, \vec{i} + \dot{\phi}\cos\theta \, \vec{k} \tag{19.61}$$

$$\omega_z = \dot{\psi} + \dot{\phi}\cos\theta \tag{19.62}$$

The couple which must be applied to maintain precession:

$$\boxed{\Sigma\vec{M}_0 = (I\omega_z - I'\dot{\phi}\cos\theta)\dot{\phi}\sin\theta \, \vec{j}} \tag{19.63}$$

This couple must act about an axis normal to the precession axis and the spin axis of the gyroscope:

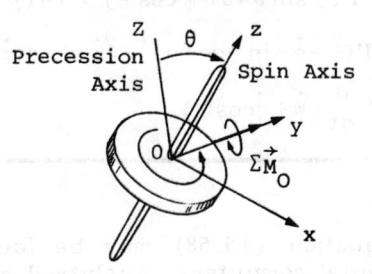

Fig. 19.13

A-207

Special Case: When $\theta = 90°$

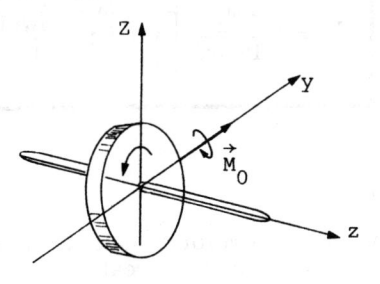

Fig. 19.14

$$\Sigma \vec{M}_0 = I\dot{\psi}\dot{\phi}\vec{j}$$

(19.64)

Spinning Top

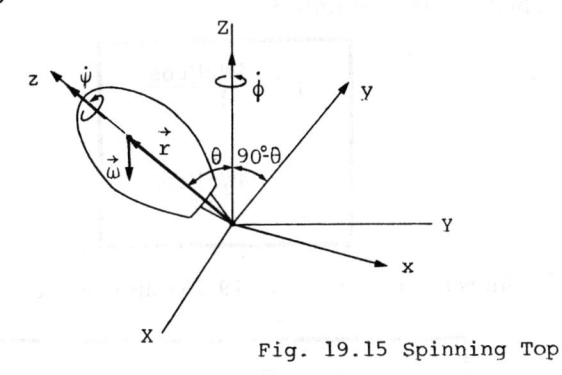

Fig. 19.15 Spinning Top

Equations of Motion:

$$\Sigma M_x = I[(\dot{\psi} + \dot{\phi}\cos\theta)\dot{\phi} - I'\dot{\phi}^2\cos\theta]\sin\theta$$
$$\Sigma M_y = \Sigma M_z = 0$$

(19.65)

The applied moment is due to weight:

$$\Sigma M_x = Wr\sin\theta = [I(\dot{\psi} + \dot{\phi}\cos\theta)\dot{\phi} - I'\dot{\phi}^2\cos\theta]\sin\theta \quad (19.66)$$

Rate of Precession:

$$\dot{\phi} = \frac{I\omega_z}{2I'\cos\theta}\left[1 \pm \sqrt{1 - \frac{4WrI\cos\theta}{I^2\omega_z^2}}\right] \qquad (19.67)$$

Two Cases:

A) If $\theta > 90°$:

The top or gyro rotor is called a pendulous gyroscope with ($\cos\theta < 0$) and $\dot{\phi}$ is real.

There are two possible values of $\dot{\phi}$:
1) Positive
2) Negative-Retrograde Precession

B) If $\theta < 90°$:

Then $\dot{\phi}$ is real only if

$$1 \geq \frac{4WrI'\cos\theta}{I^2\omega_z^2}$$

or

$$\omega_z^2 \geq \frac{4WrI'\cos\theta}{I^2} \qquad (19.68)$$

As ω_z increases, equation (19.65) approaches:

$$\dot{\phi} = \frac{I\omega_z}{2I'\cos\theta}\left[1 \pm \left(1 - \frac{2WrI'\cos\theta}{I^2\omega_z^2}\right)\right] \qquad (19.69)$$

The two roots of equation (19.69) approach the limiting values:

1) Rapid Rate

$$\dot{\phi} = \frac{I\omega_z}{I'\cos\theta} \qquad (19.70)$$

2) Slow Rate - Occurs usually

A-209

$$\boxed{\dot{\phi} = \frac{Wr}{I\omega_z}}$$

(19.71)

Note: To achieve steady precession, the top must be started with $\dot{\theta} = 0$ and $\dot{\phi}$ specified by equation (19.67).

19.13 TORQUE - FREE MOTION

Motion of a spinning body moving with no external moment about its mass center.

i.e. $\Sigma M_G = 0$.

Typical Examples

Orbiting space vehicles, gimbal supported gyroscope with frictionless bearings, etc.

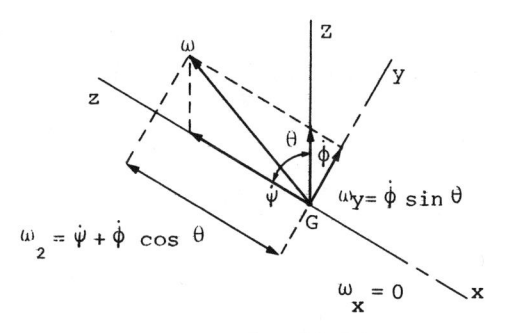

Fig. 19.16

Angular Momentum H_G = Constant as $\Sigma M_G = H_G$

$\omega \cdot H_G$ = Constant

Body spin is stabilized if 'ω' and 'H_G' are collinear but does not happen usually (Fig. 19.16).

For steady precession about z-axis

A-210

$$\Sigma M_x = 0 \quad \text{and} \quad \dot{\phi} = \frac{I\dot{\psi}}{(I'-I)\cos\theta} \qquad (19.72)$$

Relative magnitudes of I and I' determine the direction of precession.

For elongated body where I' > I

The positive value of $\dot{\phi}$ produces direct or regular precession.

For a flattened body where I' < I

The resulting negative value of $\dot{\phi}$ produces retrograde precession.

CHAPTER 20

LAGRANGIAN MECHANICS

20.1 GENERALIZED COORDINATES AND FORCES

The position of a particle is described by employing the concept of a coordinate system. Given, for example, a coordinate system such as the spherical or the oblate spherical coordinates, etc., a particle in space may be characterized as an ordered triple of numbers called coordinates.

A constrained particle in motion on a surface requires two coordinates, and a constrained particle on a curve, requires one coordinates to characterize its location.

Given a system of 'm' particles, 3M coordinates are required to describe the location of each particle. This is the configuration of the system. (If constraints are imposed on the system, fewer coordinates are required.)

A rigid body requires six coordinates - three for orientation and three for the reference point, to completely locate its position.

Generalized coordinates - A set of coordinates, $q_1, q_2, \ldots, q_m$, equal to the number of degrees of freedom of the system.

If each q_i is independent of the others, then it is known as holonomic.

The rectangular coordinates for a particle expressed in generalized coordinates:

$x = x(q)$ Motion on a curve (one degree of freedom),

$x = x(q_1, q_2)$
$y = y(q_1, q_2)$ motion on a surface (two degrees of freedom),

$x = x(q_1, q_2, q_3)$
$y = y(q_1, q_2, q_3)$ Spatial motion (three degrees of freedom).
$z = z(q_1, q_2, q_3)$

Small Changes in Coordinates:

$$\delta x = \frac{\partial x}{\partial q_1} \delta q_1 + \frac{\partial x}{\partial q_2} \delta q_2 + \frac{\partial x}{\partial q_3} \delta q_3$$

$$\delta y = \frac{\partial y}{\partial q_1} \delta q_1 + \frac{\partial y}{\partial q_2} \delta q_2 + \frac{\partial y}{\partial q_3} \delta q_3 \qquad (20.1)$$

$$\delta z = \frac{\partial z}{\partial q_1} \delta q_1 + \frac{\partial z}{\partial q_2} \delta q_2 + \frac{\partial z}{\partial q_3} \delta q_3$$

For a system of 'm' particles in generalized coordinates:

$$\delta x_i = \frac{\partial x_i}{\partial q_k} \delta q_k \qquad \begin{array}{l} k = 1, 2, \ldots, m \\ 1 < i < m \end{array}$$

$$\delta y_i = \frac{\partial y_i}{\partial q_k} \delta q_k \qquad \begin{array}{l} k = 1, 2, \ldots, m \\ 1 < i < m \end{array} \qquad (20.2)$$

$$\delta z_i = \frac{\partial z_i}{\partial q_k} \delta q_k \qquad \begin{array}{l} k = 1, 2, \ldots, m \\ 1 < i < m \end{array}$$

expressed in tensor notation.

Generalized Forces:

Work done $\delta w = \bar{F} \cdot \delta \bar{r} = F_i \cdot \delta x_i$

For one particle, $1 < i < 3$ and
for 'm' particles, $1 < i < 3m$

In terms of generalized coordinates:

$$\delta w = F_i \frac{\partial x_i}{\partial q_k} \delta q_k \qquad (20.3)$$

or $\delta w = Q_k \delta q_k$, where $Q_k = F_i \frac{\partial x_i}{\partial q_k}$ and is known as the generalized force.

Conservative Systems

Forces expressed in terms of the potential energy function:

$$\boxed{F_i = \frac{-\partial v}{\partial x_i},} \qquad (20.4)$$

where v is the potential energy.

In terms of the generalized force,

$$\boxed{Q_k = -\frac{\partial v}{\partial x_i} \frac{\partial x_i}{\partial q_k} = -\frac{\partial v}{\partial q_k}} \qquad (20.5)$$

20.2 LAGRANGE'S EQUATION

For a system, kinetic energy KE is

$$T = KE = \frac{1}{2} m_i \dot{x}_i^2 \qquad i = 1, 2, \ldots, 3M \qquad (20.6)$$

where $\dot{x}_i = \frac{\partial x_i}{\partial q_k} \dot{q}_k + \frac{\partial x_i}{\partial t} \qquad k = 1, 2, \ldots, M \qquad (20.7)$

The Lagrange equation of motion using equations (20.6) and (20.7) is

$$\frac{d}{dt} \frac{\partial T}{\partial \dot{q}_k} = Q_k + \frac{\partial T}{\partial q_k} \qquad k = 1, 2, \ldots, M \qquad (20.8)$$

A-214

or if the motion is conservative and if the potential energy is a function of generalized coordinates, then equation (20.8) becomes

$$\frac{d}{dt}\left(\frac{\partial T}{\partial \dot{q}_k}\right) = \frac{\partial T}{\partial q_k} - \frac{\partial V}{\partial q_k} \qquad k = 1, 2, \ldots, M \qquad (20.9)$$

Lagrange's Function (L)

L = T-V where T and V are in terms of generalized coordinates.

Lagrange's Equation in Terms of L

$$\boxed{\frac{d}{dt}\left(\frac{\partial L}{\partial \dot{q}_k}\right) = \frac{\partial L}{\partial q_k} \qquad k = 1, 2, \ldots, M} \qquad (20.10)$$

Lagrange's equation for non-conservative generalized forces:

If $Q_k = Q' - \frac{\partial V}{\partial q_k}$, where Q' is non-conservative then Lagrange's equation becomes

$$\boxed{\frac{d}{dt}\frac{\partial L}{\partial \dot{q}_k} = Q_k' + \frac{\partial L}{\partial q_k},}$$

(20.11)

and is useful, for example, when frictional forces are present.

General Procedure for Obtaining the Equation of Motion:

A) Choose a coordinate system.

B) Write the kinetic energy equation as a function of these coordinates.

C) Find the potential energy, if the system is conservative.

D) Combining these terms in Lagrange's equation results in the equation of motion.

Applications

A) The harmonic oscillator (a non-conservative system), using rectangular coordinates,

Lagrange Function: $L = T - V = \frac{1}{2} m\dot{x}^2 - \frac{1}{2} kx^2$

and
$$\frac{\partial L}{\partial \dot{x}} = m\dot{x}$$
$$\frac{\partial L}{\partial x} = -kx$$

Using the non-conservative version of Lagrange's equation with $Q' = -L\dot{x}$,

$$\frac{d}{dt}(m\dot{x}) = -c\dot{x} + (-kx)$$

and the equation of motion becomes

$$\boxed{m\ddot{x} + c\dot{x} + kx = 0}$$

(20.12)

B) Particle in a Central Field

In terms of polar coordinates: $q_1 = r$; $q_2 = \theta$

$$T = \frac{1}{2} mv^2 = \frac{1}{2} m(\dot{r}^2 + r^2\dot{\theta}^2)$$

$$V = v(r)$$

$$L = \frac{1}{2} m(\dot{r}^2 + r^2\dot{\theta}^2) - v(r)$$

The required partial derivatives are:

$$\frac{\partial L}{\partial \dot{r}} = m\dot{r}; \quad \frac{\partial L}{\partial r} = mr\dot{\theta}^2 - \frac{\partial v}{\partial r} = mr\dot{\theta}^2 + F(r)$$

$$\frac{\partial L}{\partial \theta} = 0 \quad \text{and} \quad \frac{\partial L}{\partial \dot{\theta}} = mr^2\dot{\theta}$$

where $F(r)$ is a function of r equal to $-\partial v/\partial r$.
Substituting in Lagrange's equation, the equations of motion become

$$\boxed{\begin{array}{l} m\ddot{r} = mr\dot{\theta}^2 + F(r) \\ \frac{d}{dt}(mr^2\dot{\theta}) = 0 \end{array}}$$

(20.13)

20.3 GENERALIZED MOMENTUM

Momentum P in terms of kinetic energy T:

$$P = \frac{\partial T}{\partial \dot{x}} = m\dot{x}$$

(20.14)

Generalized momentun is defined as the momentum P in terms of the generalized coordinates:

$$P_k = \frac{\partial L}{\partial \dot{q}_k}$$

(20.15)

For a conservative system equation (20.15) becomes:

$$\dot{P}_k = \frac{\partial L}{\partial q_k}$$

(20.16)

20.4 IMPULSIVE FORCES

Given a system with all generalized forces being non-zero for a very small time interval Δt, then the generalized impulse equation is

$$\lim_{\Delta t \to 0} \int_0^{\Delta t} Q_k \, dt = \hat{P}_k$$

and

$$\int_0^{\Delta t} \left(\frac{\partial T}{\partial q_k}\right) dt \quad \text{approaches zero.}$$

The impulse equation may be written as

A-217

$$\Delta \left[\frac{\partial T}{\partial \dot{q}_k} \right] = \hat{P}_k$$

(20.17)

If the potential energy does not involve q's explicitly,

$$\Delta P_k = \hat{P}_k$$

(20.18)

$\hat{P}_K$ is calculated from the impulsive work ($\delta \hat{w}$) equation

$$\delta \hat{w} = \hat{P} \, \delta q_1 + \hat{P}_2 \, \delta q_2 + \ldots$$
$$= \hat{P}_k \, \delta q_k$$

(20.19)

20.5 HAMILTON'S VARIATION PRINCIPLE

Hamilton's Variation Principle - The motion of any system occurs such that

$$\int_{t_a}^{t_b} L \, dt$$

(20.20)

takes an extremal value.

More precisely,

$$\delta \int_{t_a}^{t_b} L \, dt = 0$$

(20.21)

A-218

where δ = a variation due to different paths and

L = the Lagrangian function.

20.6 HAMILTON'S EQUATIONS

The Hamiltonian function is defined as

$$H = P_k \dot{q}_k - L$$

(20.22)

H will always be a function of P_k, q_k and t. From equation (20.22) we get

$$dH = P_k d\dot{q}_k + \dot{q}_k dP_k - \frac{\partial L}{\partial q_k} dq_k - \frac{\partial L}{\partial \dot{q}_k} d\dot{q}_k - \frac{\partial L}{\partial t} dt$$

Since $P_k = \frac{\partial L}{\partial \dot{q}_k}$

$$dH = \dot{q}_k dP_k - \frac{\partial L}{\partial q_k} dq_k - \frac{\partial L}{\partial t} dt$$

(20.23)

Because $H = H(P_k, q_k, t)$,

$$dH = \frac{\partial H}{\partial P_k} dP_k + \frac{\partial H}{\partial q_k} dq_k + \frac{\partial H}{\partial t} dt$$

(20.24)

Equating (20.24) and (20.23) and since $\dot{P}_k = \frac{\partial L}{\partial q_k}$

$$\dot{q}_k = \frac{\partial H}{\partial P_k} \; ; \quad \dot{P}_k = -\frac{\partial H}{\partial q_k}$$

(20.25)

Equations (20.25) are known as Hamilton's Canonical Equations of motion.

20.7 LAGRANGE'S EQUATIONS WITH CONSTRAINTS

Holonomic Constraint - Constraints of the form

$$\boxed{\frac{\partial q}{\partial q_k} \delta q = 0}$$

(20.26)

Non-holonomic Constraint - Constraints of the form

$$\boxed{h_k \delta q_k = 0}$$

(20.27)

Differential equations of motion by the method of undetermined multipliers:

(The Non-Holonomic Case)

Multiply the equation by a constant λ and add the result to the integrand of

$$\int_{t_a}^{t_b} \left[\frac{\partial L}{\partial q_k} - \frac{d}{dt} \frac{\partial L}{\partial \dot{q}_k} \right] \delta q_k \, dt = 0$$

Select λ such that the term in brackets equals zero,

$$\boxed{\begin{array}{c} \dfrac{\partial L}{\partial q_k} - \dfrac{d}{dt} \dfrac{\partial L}{\partial \dot{q}_k} + \lambda h_k = 0 \\[2mm] h_k \dot{q}_k = 0 \end{array}} \quad (k=1,2,\ldots,m)$$

(20.28)

There now exist m+1 equations to obtain m+1 unknowns, i.e., $q_1, q_2, \ldots, q_n$, λ.

This technique may be employed with moving constraints or with several constraints by having corresponding undetermined coefficients with corresponding h's in the Lagrangian equations.

CHAPTER 21

MECHANICAL VIBRATIONS

21.1 INTRODUCTION

Nomenclature and Definitions

Mechanical Vibration - The oscillation of a body about an equilibrium position.

Period - Time taken by the system to complete one cycle.

Frequency - The number of cyles per unit time.

Amplitude - Maximum displacement from the equilibrium position.

Free Vibration - Motion is sustained by restoring forces only.

Forced Vibration - Caused by periodic forces applied external to the system.

Undamped Vibration - Friction forces are neglected.

Damped Vibration - Internal and external frictional forces and other damping forces are included.

A-222

21.2 SIMPLE HARMONIC MOTION

Simple Harmonic Motion - Linear motion of a body where the acceleration is proportional to the displacement from a fixed origin and is always directed towards the origin. The direction of acceleration is always opposite to that of the displacement.

Equation of Motion

$$m\ddot{x} + kx = 0$$
or
$$\ddot{x} + p^2 x = 0 \quad (21.1)$$

where $p^2 = k/m$.

General Solution of Equation (21.1)

$$x = c_1 \sin pt + c_2 \cos pt \quad (21.2)$$

where c_1 and c_2 may be obtained from initial conditions.

An alternate form of equation (21.2)

$$x = x_m \sin(pt + \phi) \quad (21.3)$$

where x_m = the amplitude, and ϕ is the phase angle.

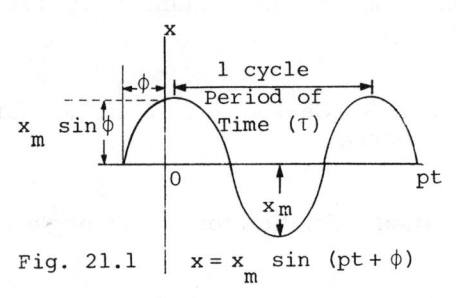

Fig. 21.1 $x = x_m \sin(pt + \phi)$

Period = $\tau = 2\pi/P$

Frequency, $f = 1/\tau = P/2\pi$

A-223

Maximum Values:

$$\boxed{\begin{array}{l} V_m = X_{mp} \\ a_m = X_m p^2 \end{array}}$$

(21.4)

21.3 GENERAL HARMONIC OSCILLATOR

Two-Dimensional Oscillator

Component equation of motion:

$$\boxed{\begin{array}{l} m\ddot{x} = -kx \\ m\ddot{y} = -ky \end{array}}$$

(21.5)

Solution of equation (21.5) yields

$$\boxed{\begin{array}{l} x = c_1 \cos(\omega t + \theta_1) \\ y = c_2 \cos(\omega t + \theta_2) \end{array}} \quad \omega = \sqrt{\frac{k}{m}}$$

(21.6)

The constants c_1, c_2, θ_1 and θ_2 are determined from initial conditions.

Elimination of 't' between x and y yields the path equation:

$$\frac{x^2}{c_1^2} - xy\frac{2\cos\phi}{c_1 c_2} + \frac{y^2}{c_2^2} = \sin^2\phi \quad \text{where} \quad \phi = \theta_2 - \theta_1$$

Because the discriminant is less that zero, the path is an ellipse.

If $\phi = \pi/2$, then the path is an ellipse whose axes coincide with the coordinate axes and the equation is given by

$$\frac{x^2}{c_1^2} + \frac{y^2}{c_2^2} = 1$$

If $\phi = 0$, $y = B/A\, X$

If $\phi = \pi$, $y = -B/A\, X$

Than angle of inclination ψ can be obtained from the relation:

$$\boxed{\tan 2\psi = \frac{2c_1 c_2 \cos\phi}{c_1^2 - c_2^2}}$$
(21.7)

Three-Dimensional Oscillator

Equations of Motion are:

$$\boxed{\begin{array}{c} m\ddot{x} = -kx \\ m\ddot{y} = -ky \\ m\ddot{z} = -kz \end{array}}$$
(21.8)

with corresponding solutions as below:

$$\boxed{\begin{array}{l} x = C_1 \sin \omega t + D_1 \cos \omega t \\ y = C_2 \sin \omega t + D_2 \cos \omega t \\ z = C_3 \sin \omega t + D_3 \cos \omega t \end{array}}$$
(21.9)

or in vector notation,

$$\vec{r} = \vec{C} \sin \omega t + \vec{D} \cos \omega t$$

Motion is in the plane of the vectors $\vec{C}$ and $\vec{D}$ and the path is an ellipse.

Nonisotropic Oscillator

The restoring force is a function of the direction of displacement.

A-225

Equations of Motion:

$$\boxed{\begin{array}{l} m\ddot{x} = -k_a x \\ m\ddot{y} = -k_b y \\ m\ddot{z} = -k_c z \end{array}}$$

(21.10)

With three corresponding frequencies being

$$\omega_a = \sqrt{k_a/m}$$
$$\omega_b = \sqrt{k_b/m}$$
$$\omega_c = \sqrt{k_c/m}$$

and three corresponding solutions

$$\boxed{\begin{array}{l} x = c_1 \cos(\omega_a t + \theta_1) \\ y = c_2 \cos(\omega_b t + \theta_2) \\ z = c_3 \cos(\omega_c t + \theta_3) \end{array}}$$

(21.11)

The motion is confined to a cuboid with sides $2c_1$, $2c_2$, and $2c_3$ situated at the origin.

21.4 SIMPLE PENDULUM

The simple pendulum consists of a weightless cord fixed at one end and has a particle at the other end that oscillates in an arc of a circle. The forces involved are gravity and tension. (See Fig. 21.2)

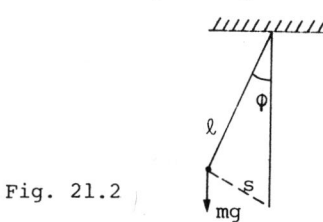

Fig. 21.2

For small angles of vibration

$$\phi = s/\ell$$

Equation of Motion in Terms of ϕ:

$$\ddot{\phi} + \frac{g}{\ell} \sin \phi = 0 \qquad (21.12)$$

If ϕ (in radians) is small, then $\sin \phi \approx \phi$. Equation (21.12) then becomes

$$\ddot{\phi} + \frac{g}{\ell} \phi = 0 \qquad (12.13)$$

The solution for equation (21.13) is given by

$$\phi = \phi_0 \cos(\omega_0 t + \alpha_0)$$

where $\omega_0 = \sqrt{g/\ell}$; ϕ_0 = Max. amplitude of oscillation and α_0 = Phase factor.

The period of oscillation is given by

$$\tau_0 = \frac{2\pi}{\omega_0} = 2\pi \sqrt{\frac{\ell}{g}}$$

Exact Solution

$$\dot{\alpha}^2 = \frac{g}{\ell}(1 - \sin^2 \frac{\alpha_0}{2} \sin^2 \alpha) \qquad (21.14)$$

Let $K = \sin^2 \frac{\alpha_0}{2}$

$$\dot{\alpha}^2 = \frac{g}{\ell}(1 - K^2 \sin^2 \alpha)$$

$$\dot{\alpha} = \sqrt{\frac{g}{\ell}} \sqrt{(1 - K^2 \sin^2 \alpha)} \qquad (21.15)$$

Solving for τ

$$\tau = \sqrt{\frac{\ell}{g}} \int_0^\phi \frac{d\phi}{\sqrt{1 - K^2 \sin^2 \alpha}} = \sqrt{\frac{\ell}{g}} F(K, \alpha) \qquad (21.16)$$

where $F(k, \alpha)$ is known as the incomplete elliptic integral of the first kind.

Period

$$\tau = 4\sqrt{\frac{\ell}{g}} \int_0^{\pi/2} \frac{d\phi}{\sqrt{1-K^2\sin^2\alpha}} = 4\sqrt{\frac{\ell}{g}} \; K(k) \qquad (21.17)$$

where $K(k) = F(K,\pi/2)$ known as the complete elliptic integral of the first kind.

Numerical values of elliptic integrals can be obtained from tables.

21.5 THE SPHERICAL PENDULUM

The spherical pendulum may oscillate in space as shown in Fig. 21.3.

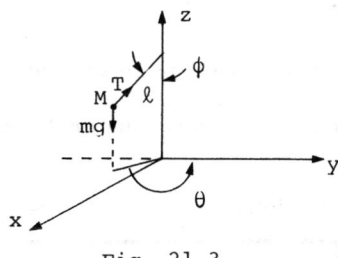

Fig. 21.3

Approximate solution in cartesian coordinates:

Equation of motion:

$$\boxed{M\ddot{\vec{r}} = M\vec{g} + \vec{T}}$$

(21.18)

or in compoment form:

$$m\ddot{x} = T_x$$
$$m\ddot{y} = T_y \qquad (21.19)$$
$$m\ddot{z} = T_z - mg$$

Assuming that the displacement from equilibrium is very small:

$$T_x \approx -Mg\, x/\ell$$
$$T_y \approx -Mg\, y/\ell$$

The equations of motion become:

$$\ddot{x} + (g/\ell)x = 0$$
$$\ddot{y} + (g/\ell)y = 0$$

(21.20)

with solutions:

$$x = A\cos(\omega t + r)$$
$$y = B\cos(\omega t + \delta)$$

(21.21)

where $\omega = \sqrt{g/\ell}$

On the x-y plane, the motion is an ellipse.

Spherical Coordinates - More accurate than the previous solution.

Equations of Motion:

$$ma_r = F_r = Mg\cos\theta - T$$
$$ma_\phi = F_\phi = -Mg\sin\phi$$
$$ma_{\phi\theta} = F_{\phi\theta} = 0$$

(21.22)

In spherical coordinates:

$$\ddot{\phi} - \dot{\psi}^2 \sin\phi\cos\phi + g\sin\phi = 0$$

$$\frac{1}{\sin\phi}\frac{d}{dt}(\dot{\alpha}\sin^2\phi) = 0$$

(21.23)

The ($\dot{\alpha}\sin^2\phi$) term is constant and is known as the angular momentum 'h':

$$\dot{\alpha} = h/\sin^2\phi$$

Equation in ϕ direction:

$$\boxed{\ddot{\phi} + g/\ell \sin\phi - h^2 \frac{\cos\phi}{\sin^2\phi} = 0}$$

(21.24)

Special Cases:

A) α = Constant:

Equation of motion becomes:

$$\boxed{\ddot{\phi} + g/\ell \sin\phi = 0}$$

(21.25)

which is the case of the simple pendulum discussed in the previous section.

B) $\phi = \phi_0$ = Constant: known as the conical pendulum

Equation of motion:

$$\boxed{h^2 = g/\ell \sin^4\phi_0 \ \sec\phi_0}$$

(21.26)

Condition for Conical Motion of the Pendulum:

$$\dot{\alpha}_0^2 = \frac{g}{\ell} \sec\phi_0$$

For nearly conical motion ϕ is close to ϕ_0.

We define μ as $\mu = \phi - \phi_0$

Expanding in series and neglecting higher order terms,

$$\ddot{\mu} + \frac{g}{\ell}(\sec\phi_0 + 3\cos\phi)\mu = 0$$

The resulting motion is then defined by the following equation:

$$\boxed{\mu = \phi - \phi_0 = \mu_0 \cos\left(\sqrt{\frac{2A}{\ell}}\, t + t\right)}$$

(21.27)

A-230

with period:

$$T = 2\pi \sqrt{\frac{\ell}{gA}}$$

(21.28)

where $A = \sec\phi_0 + 3\cos\phi_0$.

Energy Considerations

$$V = -Mg\ell\cos\phi$$
$$T = \frac{1}{2}Mv^2 = \frac{1}{2}M(\ell^2\dot\phi^2 + \ell\dot\alpha^2\sin^2\phi)$$

(21.29)

Energy Equation:

$$E = \frac{M\ell^2}{2}(\dot\phi^2 + \dot\alpha^2\sin^2\phi) - Mg\ell\cos\phi$$

(21.30)

Solving for $\dot\phi^2$:

$$\dot\phi^2 = \frac{2E}{M\ell^2} + \frac{2g}{\ell}\cos\phi - \frac{h^2}{1-\cos^2\phi} = f(\cos\phi)$$

(21.31)

The values for which $f(\cos\phi) = 0$ sets the limits of oscillation in ϕ, and is limited for the values of ϕ such that $f(\cos\phi) > 0$. The vertical oscillation is confined therefore between two horizontal circles.

21.6 THE TORSION PENDULUM

The torsion pendulum consists of an elastic rod or wire clamped at one end and having a body attached to its other end such that the axis of the rod passes through the center of gravity of the body (Fig. 12.4).

External moment applied:

$$M = \frac{JG}{\ell}\phi \quad (21.32)$$

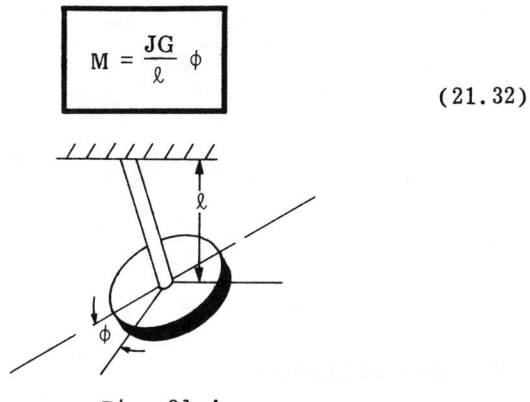

Fig. 21.4

where J = polar moment of inertia; for a rod of diameter α, $J = \pi d^4/32$; G = shear modulus of elasticity; ℓ = length of the rod; and JG/ℓ = Torsional spring constant.

Equation of motion:

$$\alpha = -\frac{JG}{I\ell}\phi \quad (21.33)$$

and with period:

$$\tau = 2\pi\sqrt{\frac{I\ell}{JG}} \quad (21.34)$$

and I is the mass moment of inertia of the body about the axis of rotation.

21.7 THE PHYSICAL PENDULUM

The physical or compound pendulum consists of a rigid body that rotates freely about a fixed horizontal axis and also swings under gravitational influence. Point Q in Fig. 21.5 is known as the center of oscillation, Point C, the center of suspension, and G is the center of gravity.

Fig. 21.5

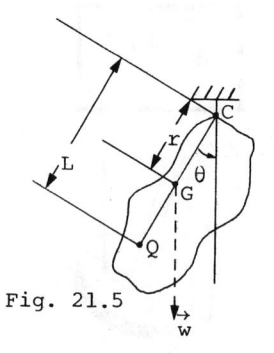

Equation of Motion:

Applying $\Sigma M_x = I_x \alpha$,

or

$$-wr\sin\theta = I_x \alpha$$

$$-wr\sin\theta = \frac{w}{g} K_x^2 \alpha$$

(21.35)

where k_x is the radius of gyration.

If θ is small, the $\sin\theta \approx \theta$, and

$$\alpha = -\frac{gr}{K_x^2}\theta$$

(21.36)

period τ is given by

$$\tau = 2\pi \sqrt{\frac{K_x^2}{gr}} = 2\pi\sqrt{\frac{L}{g}}$$

(21.37)

Note: The period given here equals that of a simple pendulum having a length L equal to K_x^2/r.

With the transfer relation

$$K_x^2 = \overline{K}^2 + r^2$$

A-233

the length L may now be expressed as

$$L = \frac{\overline{K}^2}{r} + r$$

(21.38)

Therefore, this result implies that for any parallel axis of suspension situated at the same distance from the center of gravity, the period remains the same.

If a body can be made to swing as a pendulum, then equation (21.37) may be employed to experimentally determine its radius of gyration or moment of inertia.

21.8 FREE VIBRATION

Consider the spring-mass system in Fig. 21.6 in which the mass, after being displaced, vibrates freely and indefinitely, neglecting all resistances. Such a condition is known as free vibration.

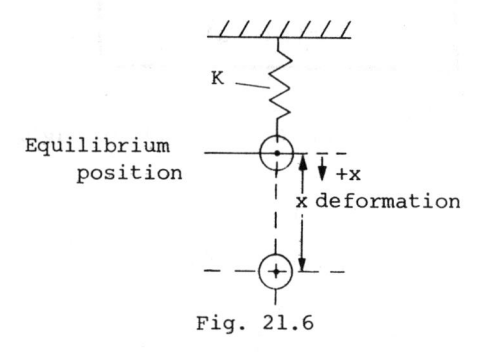

Fig. 21.6

Using Newton's second law, the equation of motion becomes

$$\frac{d^2x}{dt^2} = -\omega^2 x,$$

(21.39)

where
$$\omega = \sqrt{\frac{Kg}{W}}$$

General solution of equation (21.39)

$$\boxed{x = A \sin \omega t + B \cos \omega t} \qquad (21.40)$$

Constants A and B are determined from initial conditions.

Period
$$\tau = 2\pi \sqrt{\frac{W}{Kg}} \qquad (21.41)$$

W = Weight of Body

Note: τ depends only on the weight of the body and the spring constant K.

Frequency:
$$f = \frac{1}{2\pi} \sqrt{\frac{Kg}{W}} \qquad (21.42)$$

Specifying initial conditions as x_0 and v_0 at $t = 0$, equation (21.40) becomes,

$$\boxed{\begin{aligned} x &= \frac{v_0}{\omega} \sin \omega t + x_0 \cos \omega t \\ &= C \cos(\omega t - \alpha), \end{aligned}} \qquad (21.43)$$

where
$$C = \sqrt{x_0^2 + \left(\frac{v_0}{\omega}\right)^2} \quad \text{and} \quad \alpha \text{ is the phase angle,}$$

equal to $\tan^{-1}(v_0/\omega x_0)$.

Equation (21.43) may be graphically interpreted as in Fig. (21.7).

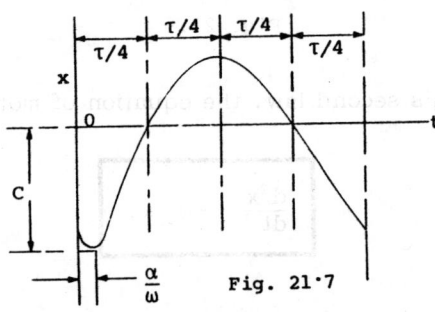

Fig. 21·7

21.9 CONSERVATION OF ENERGY

The method of conservation of energy involves the determination of two quantities:

A) The kinetic energy at the static equilibrium position, and
B) The stored elastic energy at the moment of maximum displacement.

These two quantities are then equated.

For the spring-mass system discussed in section (21.8) the work-energy relation becomes:

$$\frac{1}{2} K x_{max}^2 = \frac{W}{2g}(v^2 + v_{max}^2)$$

(21.44)

and is reduced further since $v = 0$ at the maximum displacement,

$$\frac{1}{2} K x_{max}^2 = \frac{W}{2g}(v_{max}^2)$$

(21.45)

$x = x_m \sin\omega t$

and $v = \omega x_m \cos\omega t$

which gives $v_m = \omega x_m$

From the above relations and substituting them in equation (21.45) we get

$$\omega^2 = \frac{Kg}{W}$$

(21.46)

21.10 FORCED VIBRATION

Forced vibrations occur when an external disturbing force acts on the body continuously and periodically (see Fig. 21.8).

Equation of motion (from Newton's second law):

$$\frac{d^2x}{dt^2} = -\omega^2 x + \frac{F_0}{m}\cos\omega_f t \quad (21.47)$$

where ω_F = Forcing frequency, F_0 which is the maximum magnitude of the external force, and x is the distance taken from the equilibrium position.

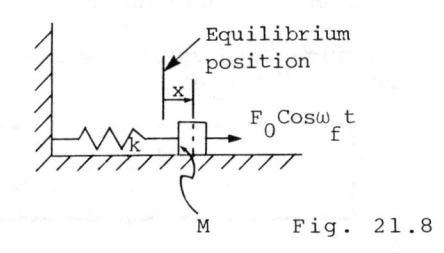

Fig. 21.8

The solution of equation (21.47) consists of the sum of two parts:

A) The complementary solution, X_c and
B) The particular solution, X_p:

The complementary solution is obtained by setting the $\frac{F_0}{m}\cos\omega_f t$ term to zero and

$$x_c = A\cos\omega t + B\sin\omega t \quad (21.48)$$

The particular solution is found by assuming that $x_p = D\cos\omega_f t$ and substituting it in equation (21.47) to obtain the unkown constant D:

$$D = \frac{F_0}{m} \cdot \frac{1}{\omega^2 - \omega_f^2}$$

Therefore, the solution of equation (21.47) is

$$X = A\cos\omega t + B\sin\omega t + \frac{F_0/m}{\omega^2 - \omega_f^2}\cos\omega_f t \quad (21.49)$$

Note: The forcing function may be any combination of periodic functions, not neccessarily as simple as $F_0 \cos \omega_f t$.

The magnification factor is defined as the ratio of the amplitude of the steady state vibration D to the static deflection F_0/K (see Fig. 21.9),

$$\text{Magnification Factor} = \frac{1}{1 - \left(\dfrac{\omega_f}{\omega}\right)^2}$$

(21.50)

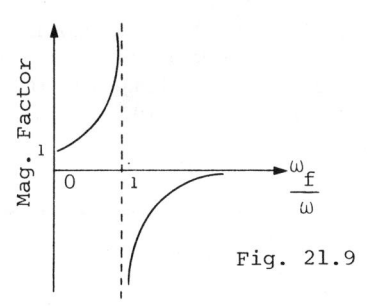

Fig. 21.9

When $\omega_f \approx \omega$, the condition is called resonance and the amplitude of vibration is very large causing quick failure of structures such as bridges or mechanical elements.

When $\omega_f > \omega$, the magnification factor is negative and the disturbing force is in opposite direction to the displacement.

When ω_f is significantly larger than ω, the magnification factor is close to zero and the body is nearly stationary.

A-238

21.11 DAMPED FREE VIBRATION

In general, damping is caused by

A) Internal friction,

B) Dry friction between two surfaces,

C) Viscous damping.

The third is the most common in vibrating systems, and is caused by fluid friction. It may be expressed as

$$\vec{F} = c \frac{dx}{dt}$$

(21.51)

It is assumed that the body is moving at low velocities (the constant c is the coefficient of viscous damping). An example is the viscous damping provided by the dashpot device as shown in Fig. 21.10.

The equation of motion is:

$$M \frac{d^2 x}{dt^2} + c \frac{dx}{dt} + kx = 0$$

(21.52)

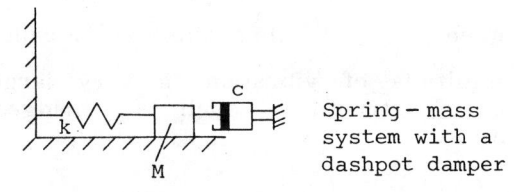

Fig. 21.10 Spring-mass system with a dashpot damper

Equation (21.52) has exponential solutions of the form $x = e^{at}$. Substituting this in equation (21.52) yields two values for 'a' given by:

$$a_1 = \frac{-c}{2m} - \sqrt{\left(\frac{c}{2m}\right)^2 - \frac{K}{M}}$$

$$a_2 = \frac{-c}{2m} + \sqrt{\left(\frac{c}{2m}\right)^2 - \frac{K}{M}}$$

The critical damping coefficient is defined as

$$C_{critical} = 2M\sqrt{\frac{K}{M}}$$

(21.53)

Now, three cases must be considered with respect to $C_{critical}$:

A) If $c > C_{critical}$, a_1 and a_2 are both real, the motion is nonoscillating, and the system is overdamped. The general solution is given by

$$x = Ae^{a_1 t} + Be^{a_2 t}$$

(21.54)

B) If $c = C_{critical}$, $a_1 = a_2$ and the system is critically damped with general solution:

$$x = (A+Bt)e^{-\left(\frac{C_{critical}}{2m}\right)t}$$

(21.55)

C) If $c < C_{critical}$ a_1 and a_2 are complex and imaginary and the system is underdamped with the solution given by:

$$X = E[e^{-(c/2M)t} \sin\left(\sqrt{\frac{K}{M} - \left(\frac{C}{2M}\right)^2}\, t + \psi\right)]$$

(21.56)

where constants A, B, E and ψ are determined from initial conditions. The graph representing equation (21.56) is shown in Figure 21.11.

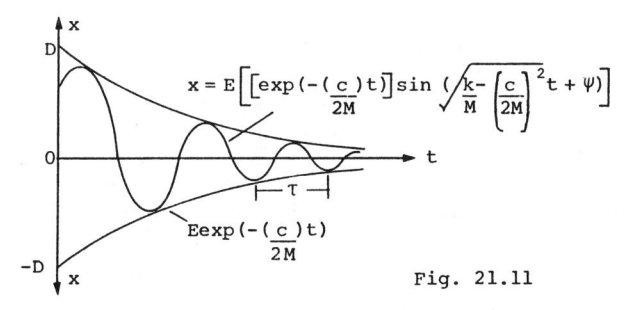

Fig. 21.11

A-240

Note: Period of damped vibration 'τ' is given by

$$\tau = \frac{2\pi}{\sqrt{\frac{K}{M} - \left(\frac{c}{2M}\right)^2}}$$

(21.57)

21.12 DAMPED FORCED VIBRATION

Damped forced vibrations is the most general case of single degree of freedom vibrating motion.

The equation of motion becomes:

$$M\frac{d^2x}{dt^2} + c\frac{dx}{dt} + kx = F\sin\omega t$$

(21.58)

The corresponding spring-mass system is shown in Figure 21.12.

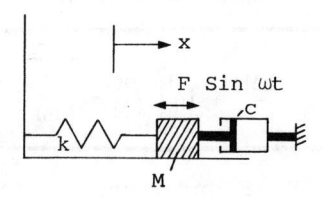

Fig. 21.12

Since equation (21.58) is a second-order, ordinary non-homogeneous differential equation, its solution will consist of the sum of the complementary solution and the particular solution. The homogeneous solution will therefore be identical to equation (21.54), (21.55) or (21.56), and will dampen-out, leaving only the particular or steady-state solution. Assuming the particular solution to be of the form $x_p = A\sin(\omega t - \psi)$, the constants A and ψ are given by

A-241

$$A = \frac{F/K}{\sqrt{\left[1 - \frac{4m^2\omega^2}{C^2_{critical}}\right]^2 + \left[\frac{4m\omega C}{C^2_{critical}}\right]^2}} \quad (21.59)$$

$$\psi = \tan^{-1}\left[\frac{c\omega/k}{\left(1 - \frac{m\omega^2}{K}\right)}\right] \quad (21.60)$$

Note: Angle ψ is the phase difference between the resulting steady-state vibration and the applied force.

The magnification factor is defined as:

$$\text{Magnification factor} = \frac{A}{F/K}$$

$$= \frac{1}{\sqrt{\left[1 - \frac{4m^2\omega^2}{C^2_{critical}}\right]^2 + \left[\frac{4m\omega c}{C^2_{critical}}\right]^2}} \quad (21.61)$$

and the graph is shown in Fig. 21.13. Resonance occurs only when the damping is zero and the frequency ratio is one.

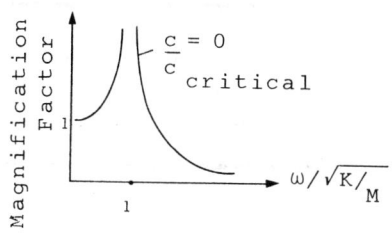

Fig. 21.13

21.13 STABILITY

Taking the equilibirum position of a system as the reference and if a disturbance is imposed, then the system is

A) Neutral if it moves either towards or away from equilibrium.

A-242

B) Stable if it comes back to equilibrium, or

C) Unstable if it moves away from equilibrium.

Considering the potential energy v:

If $\frac{dv}{dq} = 0$ at equilibrium then

A) if $\frac{d^2v}{dq^2} > 0$, it is stable

B) if $\frac{d^2v}{dq^2} < 0$, it is unstable and

C) if $\frac{d^2v}{dq^2} = 0$, then higher order terms must be considered.

21.14 POTENTIAL ENERGY FUNCTION

Given a one degree of freedom system, the potential energy function is:

$$v(q) = k_0 + k_1(q-b) + \frac{1}{2!}k_2(q-b)^2 + \ldots$$
$$\ldots + \frac{1}{n!}k_n(q-b)^n$$

(21.62)

where
$$k_n = \left(\frac{d^n v}{dq^n}\right)_{q=b}$$

If a point of equilibrium is $q = b$, then

$$v(q) = k_0 + \frac{1}{2!}k_2(q-b)^2 + \ldots$$

(21.63)

The stability of $q = b$ depends on the first non-zero term after K_0 of the expansion:

A) If n is even then it is stable if $d^n v/dq^n$ is positive.

B) If n is odd or if $d^n v/dq^n$ is negative, then it is unstable.

For a system with n degrees of freedom, the potential energy function is given by

$$v(q_1, q_2, \ldots q_n) = \frac{1}{2}(k_{11} q_1^2 + 2 k_{12} q_1 q_2 + k_{22} q_2^2 + \ldots)$$

(21.64)

where

$$k_{11} = \left(\frac{\partial^2 v}{\partial q_1^2}\right)_{q_1 = q_2 = \ldots = q_n = 0}$$

$$k_{12} = \left(\frac{\partial^2 v}{\partial q_1 \partial q_2}\right)_{q_1 = q_2 = \ldots = q_n = 0}$$

Equation (21.64) is in quadratic form and if it is zero for all q, then the configuration $q_1 = q_2 = \ldots = q_n = 0$ is stable.

21.15 OSCILLATION WITH A SINGLE DEGREE OF FREEDOM

The kinetic energy for a one degree of freedom oscillation is given by

$$T = \frac{1}{2} \mu \dot{q}^2$$

(21.65)

where μ may be expanded as,

$$\mu = \mu(0) + \left(\frac{d\mu}{dq}\right)_{q=0} q + \ldots$$

(21.66)

A-244

$\mu = \mu(0)$ = constant if $q = 0$ is an equilibrium position.

The Lagrangian Function:

$$L = T - V = \frac{1}{2}\mu \dot{q}^2 - \frac{1}{2}kq^2$$

(21.67)

where $\quad k = k_2 = \left(\dfrac{d^2 v}{dq^2}\right)_{q=0}$

Lagrange's Equation of Motion:

$$\mu \ddot{q} + kq = 0$$

(21.68)

If $q = 0$ is stable:

$$k > 0$$
$$\omega = \sqrt{k/\mu} \quad \text{and}$$
$$q = q_0 \cos(\omega t + \psi)$$

(21.69)

where
q_0 = Amplitude and ψ = Phase angle

21.16 GENERAL THEORY OF VIBRATING SYSTEMS

Consider a system with 'n' degrees of freedom.

The kinetic energy function is given by:

$$T = \sum_i \sum_j \frac{1}{2} \mu_{ij} \dot{q}_i \dot{q}_j$$

(21.70)

A-245

where μ_{ij} is constant and equal to the value at the equilibrium configuration, i.e., with $q_1 = q_2 = \ldots = q_n = 0$.

Potential Energy:

$$V = \sum_i \sum_j \frac{1}{2} K_{ij} q_i q_j \qquad (21.71)$$

The Lagrangian Function:

$$L = \sum_i \sum_j \frac{1}{2} (\mu_{ij} \dot{q}_i \dot{q}_j - K_{ij} q_i q_j) \qquad (21.72)$$

The Equation of Motion:

$$\sum_i \mu_{ij} \ddot{q}_i = -\sum_i K_{ij} q_i \qquad (j=1,2,\ldots) \qquad (21.73)$$

In matrix notation:

$$M\ddot{q} = -Kq \qquad (21.74)$$

where

$$M = \begin{bmatrix} \mu_{11} & \mu_{12} & \cdots \\ \mu_{21} & \mu_{22} & \cdots \\ \vdots & & \end{bmatrix} = [\mu_{ij}]$$

$$K = \begin{bmatrix} K_{11} & K_{12} & \cdots \\ K_{21} & K_{22} & \cdots \\ \vdots & & \end{bmatrix} = [K_{ij}]$$

The generalized displacement vector:

$$q = \begin{bmatrix} q_1 \\ q_2 \\ \vdots \\ q_n \end{bmatrix}$$

If a solution exists of the form:

$$q_j = B_j \cos(\omega t + \phi_j)$$

then the equation in matrix notation becomes

$$\boxed{(K - M\omega^2)q = 0} \quad (21.75)$$

with the requirement (for a nontrivial solution) that the determinant:

$$|K_{ij} - \mu_{ij}\omega^2| = 0$$

The roots give the normal frequencies and their corresponding eigenvectors give the normal modes.

By methods of linear algebra:

If M^{-1} exists, then

$$\boxed{(M^{-1}K - I\omega^2)q = 0} \quad (21.76)$$

where I is the identity matrix, and the equation becomes an eigenvalue problem.

The secular equation is

$$\boxed{\text{Det}(M^{-1}K - I\omega^2) = 0} \quad (21.77)$$

and the same results are obtained.

A-247

21.17 NORMAL COORDINATES

Kinetic Energy:

$$T = \frac{1}{2} \dot{q}^T M \dot{q}$$

(21.78)

Potential Energy:

$$V = \frac{1}{2} q^T K q$$

(21.79)

These two may be written in a simpler form when expressed in terms of coordinates such that M and K are diagonal (the secular determinant equation also becomes diagonal). By quating to zero each diagonal term, the roots are determined.

Normal coordinates are determined by first defining the matrix B:

$$q = Bq'$$

$$q' = B^{-1}q$$

Then T and v become

$$T' = \frac{1}{2} \dot{q}'^T B^T M B \dot{q}'$$
$$V' = \frac{1}{2} q'^T B^T K B q'$$

and

It is required that the matrices be diagonal

$$B^T M B = M' = \begin{bmatrix} \mu_1' & 0 & 0 & \ldots & 0 \\ 0 & \mu_2' & 0 & \ldots & 0 \\ 0 & 0 & \mu_3' & \ldots & 0 \\ \vdots & \vdots & \vdots & & \\ 0 & 0 & 0 & \ldots & \end{bmatrix}$$

and $B^T KB = K' = \begin{bmatrix} K'_1 & 0 & 0 & \cdots & 0 \\ 0 & K'_2 & 0 & \cdots & 0 \\ 0 & 0 & K'_3 & \cdots & 0 \\ \vdots & \vdots & \vdots & & \\ 0 & 0 & 0 & \cdots & \end{bmatrix}$

To obtain the required transformation, first find a_K:

$$Ma_K = \mu'_K a_K \quad (K=1,2,\ldots,n)$$
$$Ka_K = K'_K a_K \tag{21.80}$$

Subtracting these we get

$$(K - M\lambda_K)a_K = 0 \qquad \lambda_K = K'_K/\mu'_K \tag{21.81}$$

with eigenvalue $\lambda_K = \omega_K^2$

To obtain B: Find the roots of $|K - M\omega^2| = 0$ and then for each ω_K, solve the corresponding $(K-M\omega_k^2)a_k = 0$ to obtain the eigenvector a_k.

Damped Motion: Given that the damping forces are proportional to the velocities

Lagrange's Equation:

$$\frac{d}{dt}\left(\frac{\partial L}{\partial \dot{q}_k}\right) = \frac{\partial L}{\partial q_k} + P_k \tag{21.82}$$

where P_k is the generalized damping force given in tensor notation as

$$P_k = -c_{nk}\dot{q}_n \qquad k = 1,2,\ldots,n \tag{21.83}$$

A normal coordinate transformation may be determined in some cases to yield the following governing equation:

$$\mu'_k \ddot{q}'_k + C'_k \dot{q}'_k + K'_k q'_k = 0 \tag{21.84}$$

Motion with outside driving force: In normalized coordinates:

$$\mu'_k \ddot{q}'_k + C'_k \dot{q}'_k + K'_k q'_k = R_k e^{i\omega t} \qquad (21.85)$$

21.18 VIBRATION OF A LOADED STRING

Kinetic Energy:

$$\boxed{T = \frac{M}{2}(\dot{q}_1^2 + \dot{q}_2^2 + \ldots + \dot{q}_n^2)} \qquad (21.86)$$

Potential Energy (for a section of the string)

$$\boxed{V = \frac{1}{2} K(q_{a+1} - q_a)^2} \qquad (21.87)$$

'a' is any arbitrary particle
and k = Elastic stiffness.

Potential Energy (For the system):

$$\boxed{V = \frac{K}{2}[q_1^2 + (q_2-q_1)^2 + (q_n-q_{n-1})^2 + q_n^2]} \qquad (21.88)$$

where K = s/h and h is the equilibrium distance between two points.

Lagrangian Function:

$$\boxed{L = \frac{1}{2} \sum_a [M\dot{q}_a^2 - K(q_{a+1} - q_a)^2]} \qquad (21.89)$$

A-250

Lagrangian Equation of Motion becomes:

$$\boxed{M\ddot{q}_a = -K(q_a - q_{a-1}) + K(q_{a+1} - q_a)}$$
(21.90)

$a = 1, 2, \ldots, n$

Using a trial solution, $q_a = b_a e^{i\omega t}$ where b_a = the amplitude of the ath particle we obtain the following recursion formula:

$$\boxed{-m\omega^2 b_a = K(b_{a-1} - 2b_a + b_{a+1})}$$
(21.91)

where the amplitude of the end points of the string correspond to $b_0 = b_{n+1} = 0$

The secular determinant:

$$\begin{vmatrix} -M\omega^2 + 2K & -K & 0 & \cdots & 0 \\ -K & -M\omega^2 + 2K & -K & \cdots & 0 \\ 0 & -K & -M\omega^2 + 2K & \cdots & 0 \\ \vdots & \vdots & \vdots & & \vdots \\ 0 & \cdots & 0 & \cdots & -M\omega^2 + 2K \end{vmatrix} = 0$$
(21.92)

An alternate method consists of substituting $b_a = c \sin(a\phi)$ into the recursion formula resulting in

$$\boxed{\omega = 2\omega_0 \sin \frac{\phi}{2}}$$
(21.93)

where $\omega_0 = \left(\dfrac{K}{M}\right)^{\frac{1}{2}}$

To determine ϕ, let $(n+1)\phi = N\pi$, N = integer

Normal frequencies are given by

A-251

$$\boxed{\omega_N = 2\omega_0 \sin\left(\frac{N\pi}{2n+2}\right)} \qquad (21.94)$$

Amplitude
$$\boxed{b_a = c \sin\left(\frac{N\pi a}{n+1}\right)} \qquad (21.95)$$

where $a = 1,\ldots,n$ denotes a particle and $N = 1,2,\ldots,n$ denotes a normal mode.

21.19 WAVES

The typical wave may be written as

$$\boxed{\frac{\partial^2 q}{\partial t^2} = \nu^2 \frac{\partial^2 q}{\partial y^2}} \qquad (21.96)$$

where

ν is the wave speed equal to $\sqrt{\frac{Ka^2}{m}}$ or $\sqrt{s/\rho}$ where $K = s/a$, s = the tension of the string, a = the distance between particles at equilibrium and ρ = the density of the string.

The general solution of equation (21.96) may be expressed as

$$\boxed{q = \alpha(y+\nu t) + \phi(y-\nu t)} \qquad (21.97)$$

Sinusoidal Waves: Given that $q = q(y,t)$ then

$$\boxed{q = c_1 g\left\{\frac{2\pi}{\lambda}(y+\nu t)\right\} + c_2 g\left\{\frac{2\pi}{\lambda}(y-\nu t)\right\}} \qquad (21.98)$$

where 'g' represents either sin or cos and 'λ' is the wavelength.

Frequency

$$f = \nu/\lambda = \omega/2\pi$$

Note: Because the wave equation is a linear partial differential equation, lienar combinations of solutions also satisfy the equation.

Standing Waves - The sum of two waves traveling in opposing directions with equal amplitudes results in standing waves which are represented by the following equation:

$$\boxed{q = c \, \sin\left(\frac{2\pi y}{\lambda}\right) \cos \omega t}$$

(21.99)

where $\omega = 2\pi\nu/\lambda$

Handbook of Mechanical Engineering

SECTION B
Strength of Materials and Mechanics of Solids

CHAPTER 1

AXIAL FORCE, SHEAR FORCE AND BENDING MOMENT

1.1 EQUILIBRIUM OF A SOLID BODY

1.1.1 SUMMATION OF THE FORCES ACTING ON A SOLID

$$\sum F_x = 0, \; \sum F_y = 0, \; \sum F_z = 0 \qquad (1\text{-}1)$$

$$\sum M_x = 0, \; \sum M_y = 0, \; \sum M_z = 0 \qquad (1\text{-}2)$$

1.1.2 THE FORCE VECTOR, $\overline{F}$

$$\overline{F} = F_x \overline{i} + F_y \overline{j} + F_z \overline{k} \qquad (1\text{-}3)$$

1.1.3 THE MOMENT VECTOR, $\overline{M}$

$$\overline{M} = M_x \overline{i} + M_y \overline{j} + M_z \overline{k} \qquad (1\text{-}4)$$

Where F_x, F_y, F_z are components of the force, F, and $\overline{i}, \overline{j}, \overline{k}$ are the unit vectors along the *x, y* and *z* axes, respectively.

Concurrently, M_x, M_y, M_z are components of *M* along the coordinate axes.

1.2 CONVENTIONS FOR SUPPORTS AND LOADINGS

1.2.1 LINK AND ROLLER SUPPORTS

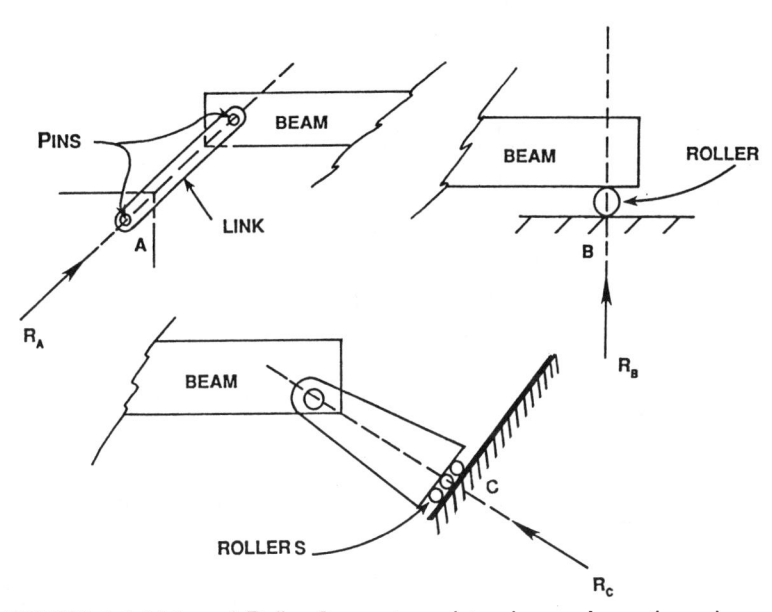

FIGURE 1.1–Link and Roller Supports resist only one force through a specific Line of Action (the Line of Action is indicated with dashes).

1.2.2 PINNED SUPPORTS

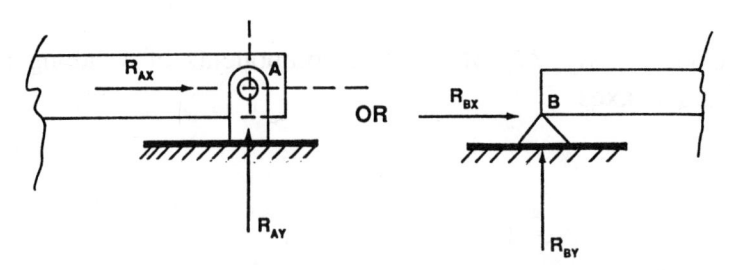

FIGURE 1.2–A Pinned Support is capable of resisting a force in any direction of the Plane. In general, it is resolved into two forces — Horizontal and Vertical.

1.2.3 FIXED SUPPORTS

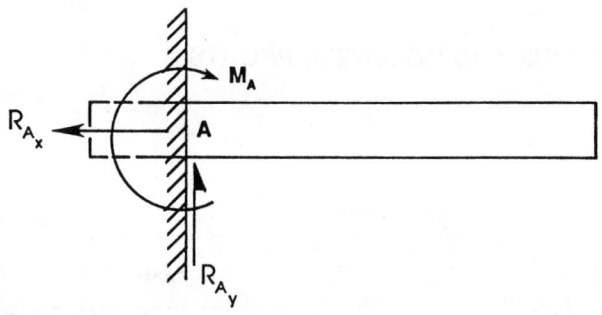

FIGURE 1.3–Fixed Supports can resist a force in any direction and are also capable of resisting a Couple or a Moment.

1.2.4 DISTRIBUTED LOADING

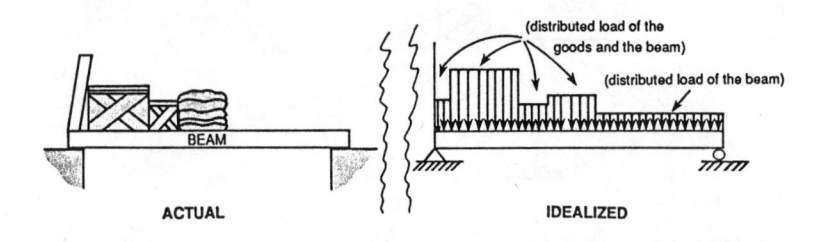

FIGURE 1.4–Simply supported Beam with uniformly distributed loads.

B-3

1.2.5 HYDROSTATIC LOADING

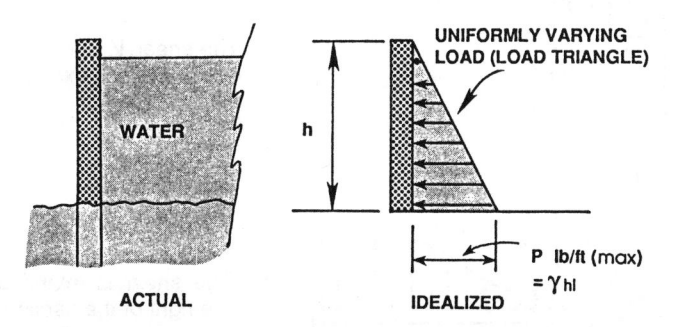

FIGURE 1.5–The force due to Hydrostatic Loading is uniformly varying. Here γ is the unit weight of the liquid, and P is the pressure (maximum) at the base.

NOTE: For a uniformly varying loading, the load may be resolved into a single load applied at 1/3 the distance from the base of the "Load Triangle" and equal to F_T, where

$$F_T = \frac{1}{2} Ph = \frac{1}{2} \gamma h^2 l \qquad (1\text{-}5)$$

1.3 AXIAL FORCE, SHEAR FORCE AND BENDING MOMENT DIAGRAMS

1.3.1 AXIAL FORCE IN BEAMS

The **axial force** is a force, P, acting through the beam's cross-sectional area, along the direction of the longitudinal axis.

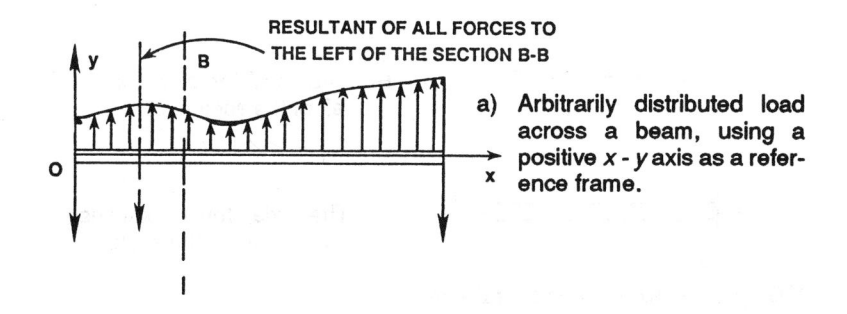

a) Arbitrarily distributed load across a beam, using a positive x - y axis as a reference frame.

B-4

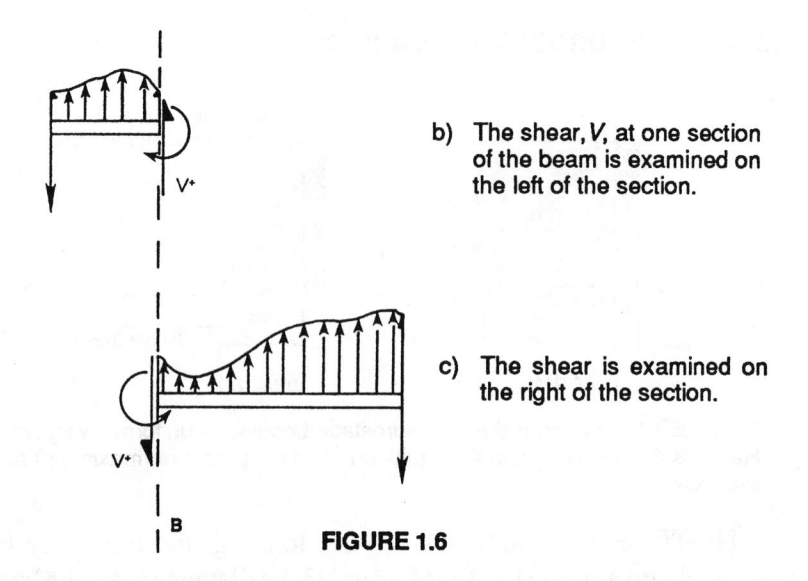

b) The shear, V, at one section of the beam is examined on the left of the section.

c) The shear is examined on the right of the section.

FIGURE 1.6

1.3.2 SHEAR

The **shear** or **shearing force** is the internal force, V, acting at right angles to the axis of the beam. It is numerically equal to the algebraic sum of all the external vertical forces acting on the isolated segment, but it is opposite in direction.

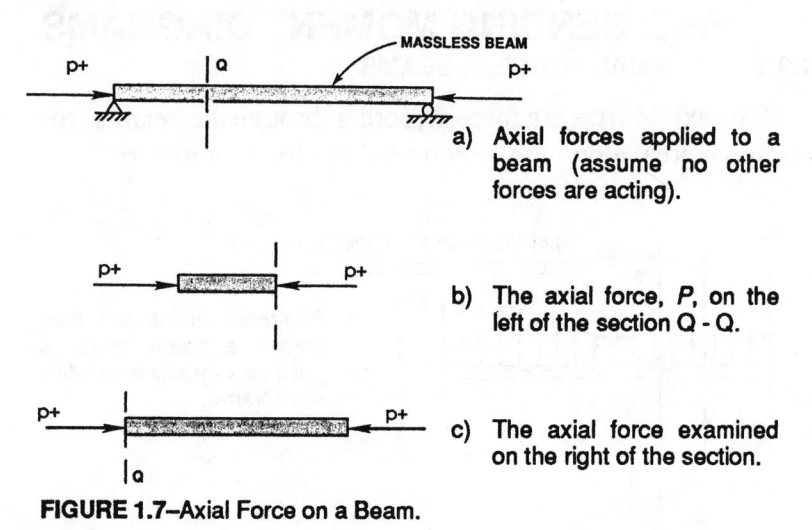

a) Axial forces applied to a beam (assume no other forces are acting).

b) The axial force, P, on the left of the section Q - Q.

c) The axial force examined on the right of the section.

FIGURE 1.7–Axial Force on a Beam.

1.3.3 BENDING MOMENT

The **bending moment** or the **internal resisting moment**, M, acts in the direction opposite to the external moment (or force couple) and it is equal in magnitude.

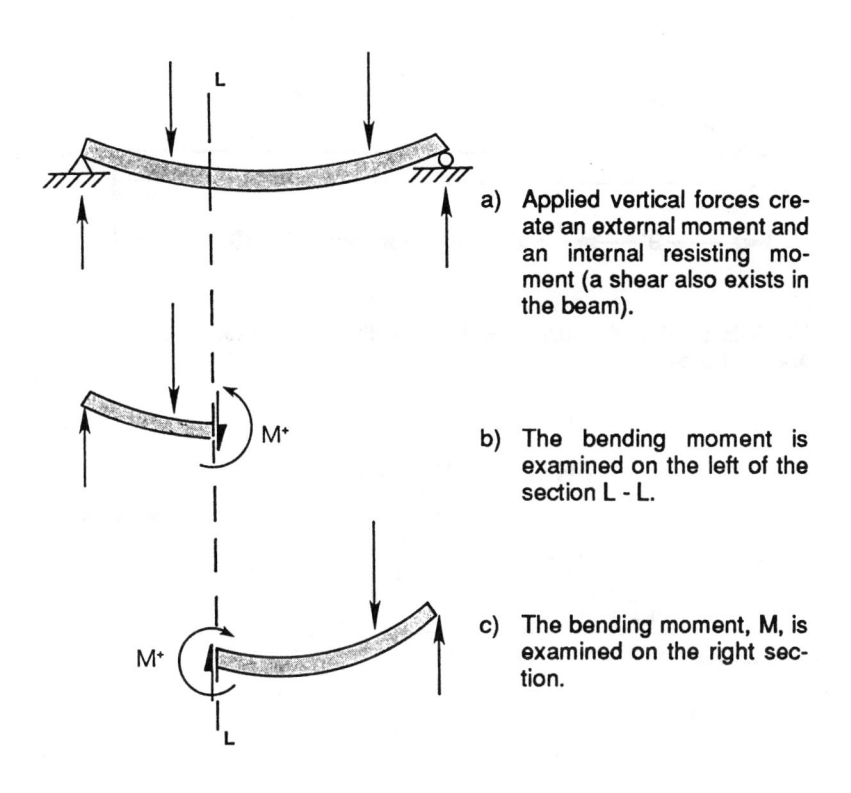

FIGURE 1.8–Bending Moment in a Beam

a) Applied vertical forces create an external moment and an internal resisting moment (a shear also exists in the beam).

b) The bending moment is examined on the left of the section L - L.

c) The bending moment, M, is examined on the right section.

1.3.4 AXIAL, SHEAR, AND BENDING MOMENT DIAGRAMS OF VARIOUS BEAMS AND BEAM LOADINGS (METHOD OF SECTIONS)

1.3.4.1 SIMPLY SUPPORTED BEAM

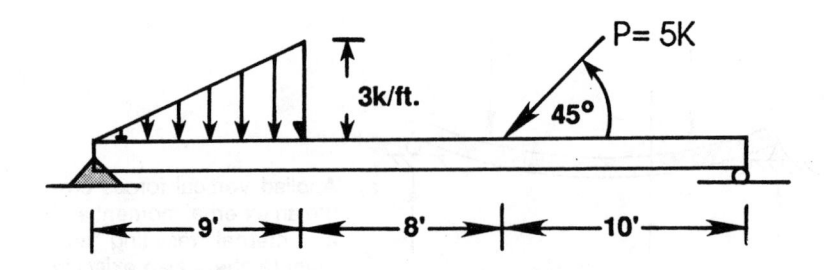

FIGURE 1.9–Simply Supported Beam with a Distributed Load and an Applied Force, P.

FREE BODY DIAGRAM

From equilibrium, we find the external forces:

$$F_H = 3.5k \quad F_V = 8k \quad P_V = 3.5k \quad P_H = 3.5k$$
$$Q = 9k \quad R_V = 4.5$$

We can then analyze the beam using the method of sections.

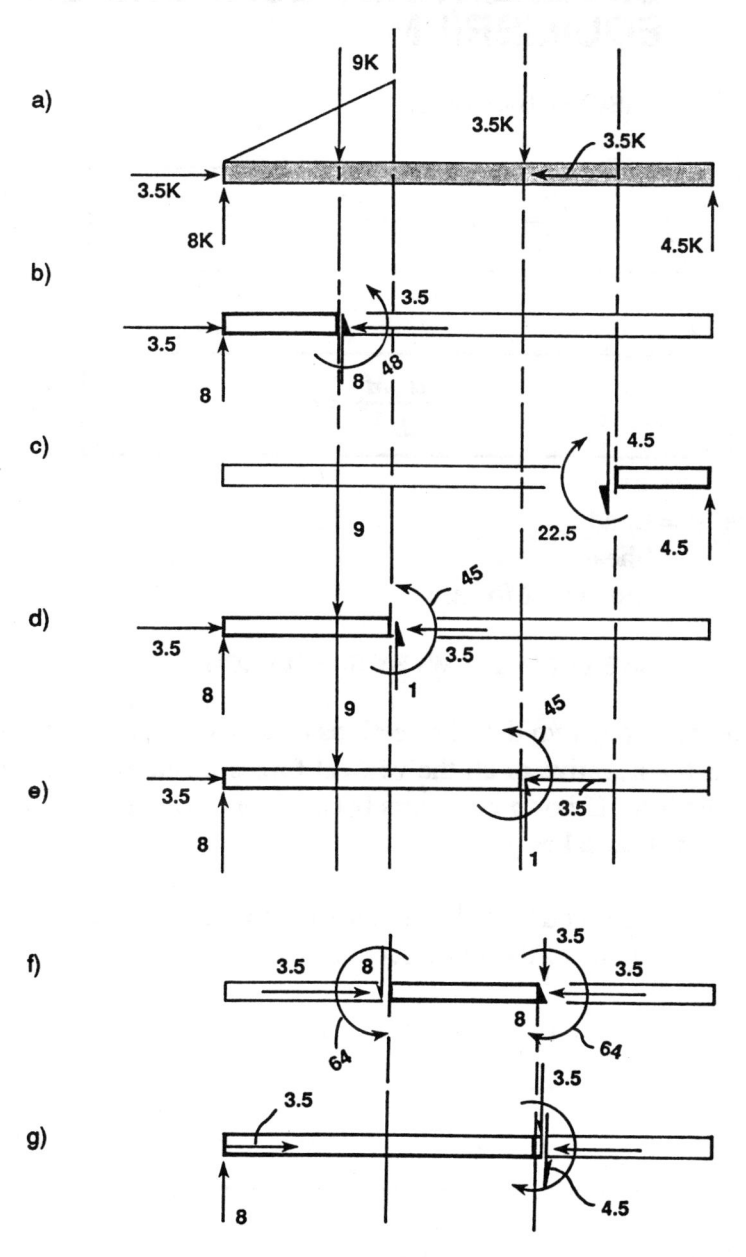

1.4 DIFFERENTIAL EQUATIONS OF EQUILIBRIUM

1.4.1 BASIC EQUATIONS:

$$\frac{dV}{dx} = -P \text{ and } \frac{dM}{dx} = -V \qquad (1\text{-}6)$$

$$\frac{d^2M}{dx^2} = P \qquad (1\text{-}7)$$

where P = Load
V = Shear
M = Bending Moment

1.4.2 SHEAR DIAGRAMS BY SUMMATION

Between any two definite sections of a beam, the change in shear is the negative of all the vertical forces included between these sections. If no force occurs between any two sections, no change in shear takes place.

If a concentrated force comes into the summation, a discontinuity in the value of the shear occurs (see Fig. 1.10).

The basic relation for shear, V:

$$V(x) = -\int_0^x p\,dx + C \qquad (1\text{-}8)$$

where C = constant of integration.

FIGURE 1.10

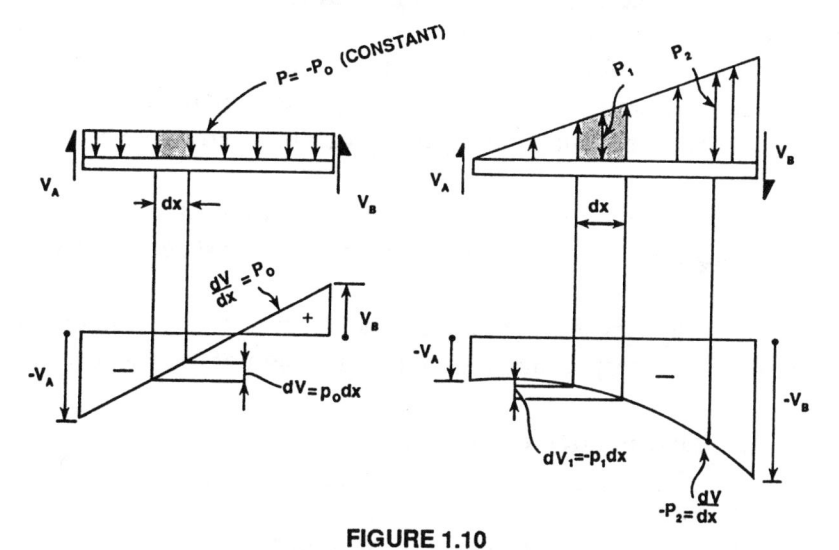

(a) Shear Diagram for a beam with a constant distributed load.

(b) Shear Diagram for a beam with a uniformly varying distributed load.

The change in shear depends entirely on whether the forces act up or down. If they act **upward**, the change is **negative**.

The **slope** of the shear diagram $\left(\frac{dv}{dx} = -p\right)$ determines the **rate of change in shear**. For a load of constant magnitude, the shear curve is a straight line. For a variable load, the shear curve is concave downward for an upward force and concave upward for a downward force.

The rate of change of the slope of the shear diagram equals the negative of the rate of change of the load.

1.4.3 MOMENT DIAGRAMS BY SUMMATION

If the ends of a beam are on rollers, the starting and terminal moments are zero. If the end is built in, the end moment is known from the reaction calculations (for statically determinate beams).

The change in moment in a given segment of a beam is equal to the negative of the area of the corresponding shear diagram.

The slope of the bending moment curve is determined by noting the corresponding magnitude and sign of the shear.

The basic relation for the moment, M:

$$M(x) = -\int_0^x V dx + C_1 \quad \text{(1-9)}$$

where C_1 = constant of integration.

A constant shear yields a uniform rate of change in the bending moment, resulting in a straight line in the moment diagram. If no shear occurs along a certain portion of the beam, no change in moment occurs.

The maximum or minimum moment occurs at a point where the shear is zero, as the derivative of M is then zero. This occurs at a point where the shear changes sign.

CHAPTER 2

STRESS

2.1 DEFINITION OF STRESS

2.1.1 GENERAL DEFINITION OF STRESS AT A POINT

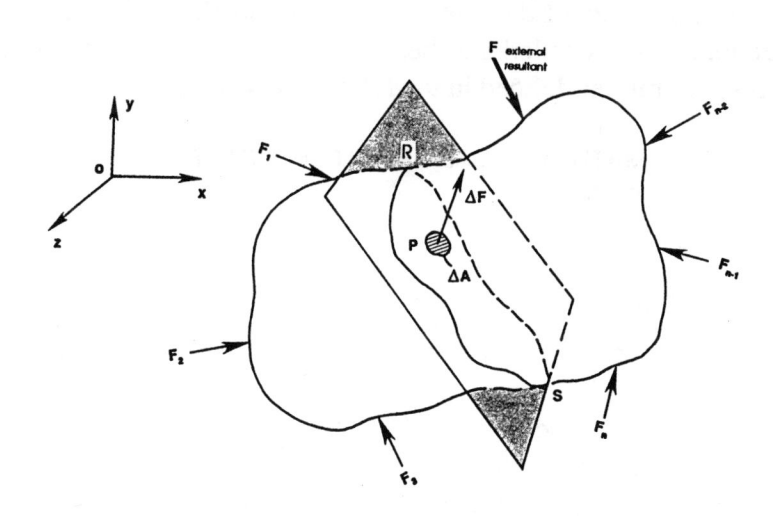

FIGURE 2.1–Arbitrary Solid Being Acted on by External Forces

Consider the elemental area ΔA at a point P located in the cross-sectional plane, RS, with an internal resultant force ΔF acting on the area (See Fig. 2.1).

The stress, or force intensity, at P is the limiting value of the ratio $\Delta F/\Delta A$ as the elemental area approaches zero. It is written as

$$\text{Stress at point } P = \left(\frac{\Delta F}{\Delta A}\right)_{\Delta A \to 0} \quad (2\text{-}1)$$

2.1.2 NORMAL STRESS, σ (SIGMA)

The intensity of the forces perpendicular to the cross-sectional area is called the normal stress (tensile or compressive stress).

2.1.3 SHEARING STRESS, τ (TAU)

The intensity of the forces parallel to the plane of the cross-sectional area is called the shearing stress. (The components of stress are further defined in the following sections.)

2.1.4 MATHEMATICAL DEFINITION OF STRESS AT A POINT

$$\underbrace{\tau_{xx} = \lim_{\Delta A \to 0} \frac{\Delta F_x}{\Delta A}}_{\text{NORMAL STRESS}}, \underbrace{\tau_{xy} = \lim_{\Delta A \to 0} \frac{\Delta F_y}{\Delta A} \text{ and } \tau_{xz} = \lim_{\Delta A \to 0} \frac{\Delta F_z}{\Delta A}}_{\text{SHEARING STRESSES}}$$

(**NOTE**: The subscripts for the stresses indicate the plane in which they act. τ_{xx} is often written as σ_x, denoting a normal stress in the $x-x$ plane. It follows that $\tau_{yy} \equiv \sigma_y$ and $\tau_{zz} \equiv \sigma_z$.)

2.2 STRESS TENSOR

When considering the most general state of stress acting on an element, we isolate an infinitesimal cubic slice of the body. This cube may be called the stress tensor, and it may be represented in a matrix form as follows:

$$\begin{pmatrix} \tau_{xx} & \tau_{xy} & \tau_{xz} \\ \tau_{yx} & \tau_{yy} & \tau_{yz} \\ \tau_{zx} & \tau_{zy} & \tau_{zz} \end{pmatrix} \equiv \begin{pmatrix} \sigma_x & \tau_{xy} & \tau_{xz} \\ \tau_{yx} & \sigma_y & \tau_{yz} \\ \tau_{zx} & \tau_{zy} & \sigma_z \end{pmatrix} \qquad (2\text{-}2)$$

An examination of the stress symbols shown in Figure 2.2 shows that there are three normal stresses

$$\tau_{xx} \equiv \sigma_x, \ \tau_{yy} \equiv \sigma_y, \ \tau_{zz} \equiv \sigma_z,$$

and six shearing stresses

$$\tau_{xy}, \tau_{xz}, \ \tau_{yx}, \tau_{yz}, \tau_{zx}, \tau_{zy}.$$

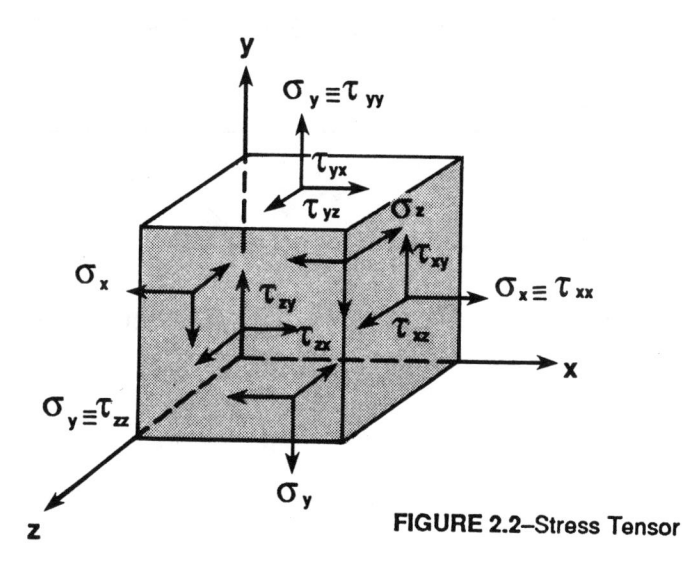

FIGURE 2.2–Stress Tensor

2.3 STRESS IN AXIALLY LOADED MEMBERS

2.3.1 NORMAL STRESS (PERPENDICULAR TO THE CUT)

$$\sigma = \frac{P}{A} \qquad (2\text{-}3)$$

$\sigma =$ Normal stress uniformly distributed over the cross-sectional area
$P =$ Force
$A =$ Area

2.3.2 BEARING STRESS

If the resultant of the applied forces coincides with the centroid of the contact area between the two bodies, the normal stress, σ, is called a **bearing stress**. One body supported by another creates a bearing stress.

2.3.3 AVERAGE SHEARING STRESS (PARALLEL TO THE CUT)

$$\tau = \frac{P}{A} \qquad (2\text{-}4)$$

$\tau =$ Shearing stress uniformly distributed across the cross-sectional area
$P =$ Force
$A =$ Area

2.4 DESIGN OF AXIALLY LOADED MEMBERS AND PINS

2.4.1 REQUIRED AREA OF A MEMBER

$$A = \frac{P}{\sigma_a}$$ (2-5)

$A =$ Required Area
$P =$ Axial Force
$\sigma_a =$ Allowable Stress *

* The allowable stress is determined by dividing the normal stress by a given safety factor,

$$\sigma_a = \frac{\sigma}{S.F.}$$ (2-6)

2.4.2 EXAMPLE

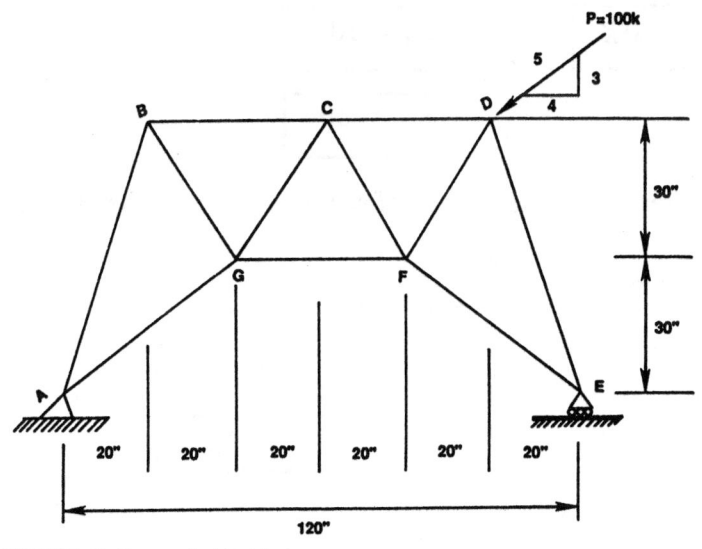

FIGURE 2.3–Example Problem

B-16

Choose members *BC* and *CF* in the truss of Figure 2.3 to carry a force *P* of 100 Kips, acting as shown. The allowable tensile stress is 20 Kpsi. Find the cross-sectional areas of the members.

For overall equilibrium,

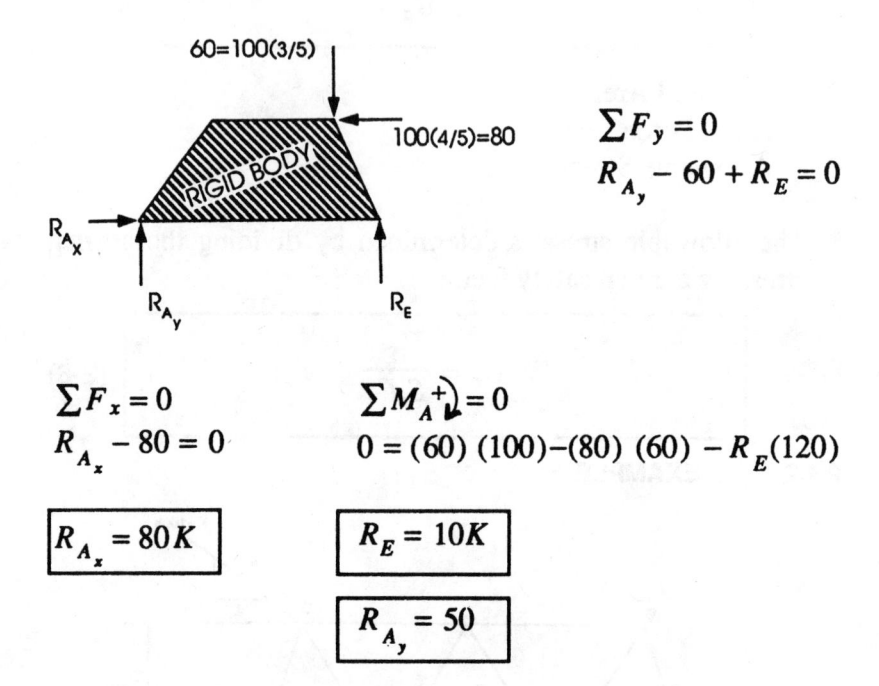

$\sum F_y = 0$
$R_{A_y} - 60 + R_E = 0$

$\sum F_x = 0$
$R_{A_x} - 80 = 0$

$\sum M_A{}^+ = 0$
$0 = (60)(100) - (80)(60) - R_E(120)$

$\boxed{R_{A_x} = 80K}$

$\boxed{R_E = 10K}$

$\boxed{R_{A_y} = 50}$

Then, using the method of sections, we solve for *BC* and *CF*.

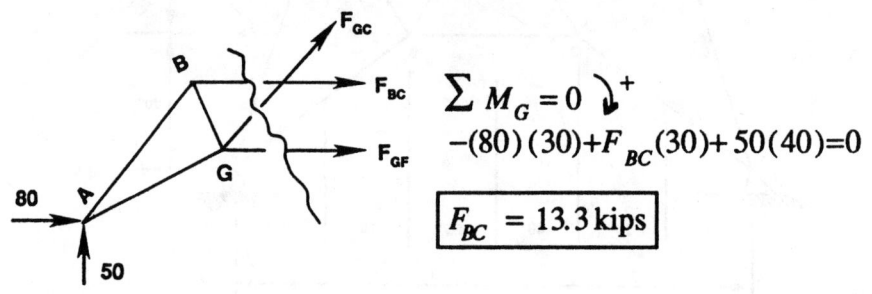

$\sum M_G = 0 \;\;\downarrow^+$
$-(80)(30) + F_{BC}(30) + 50(40) = 0$

$\boxed{F_{BC} = 13.3 \text{ kips}}$

B-17

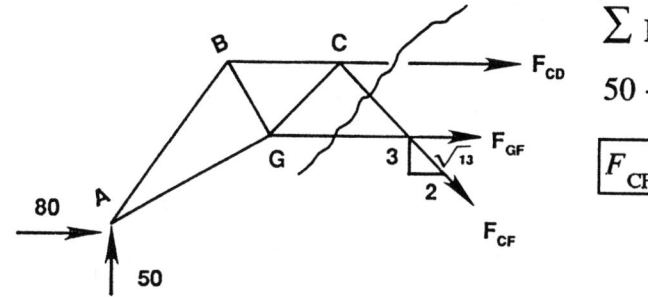

$$\sum F_y = 0$$
$$50 - F_{CF}\left(\frac{\sqrt{13}}{3}\right) = 0$$

$$\boxed{F_{CF} = 41.6 \text{ kips}}$$

$A_{BC} = F_{BC}/\sigma_{\text{allowable}}$

$A_{BC} = 13.3/20 = .667 \text{ in}^2$

$A_{CF} = F_{CF}/\sigma_{\text{allowable}}$

$A_{CF} = 41.6/20 = 2.08 \text{ in}^2$

CHAPTER 3

STRAIN

3.1 DEFINITION OF STRAIN

3.1.1 GENERAL DEFINITION OF STRAIN

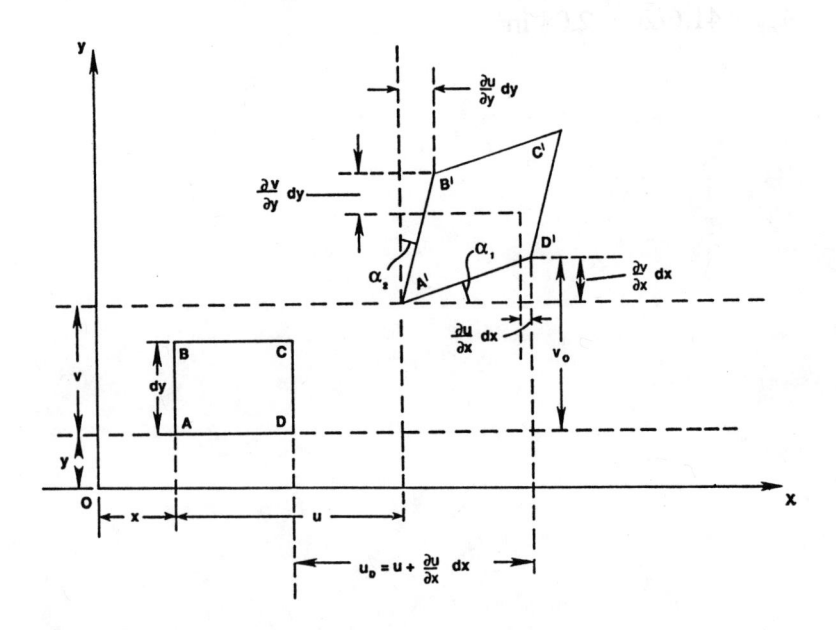

FIGURE 3.1–Rectangular Element Before and After Deformation.

Consider a small rectangular element *ABCD* in a homogeneous, isotropic body, with sides *dx* and *dy* in the plane before deformation. External forces are applied to the body, and the element is displaced to a final position *A'B'C'D'* (See Figure 3.1).

The displacement consists of two basic geometric deformations of the element:

- A change in length of an initial straight line in a certain direction

- A change in the value of the given angle.

These deformations are classified, respectively, as the longitudinal strain and the shear strain.

3.1.2 LONGITUDINAL STRAIN, ε (EPSILON)

The ratio of the change in length, Δ*l*, to the initial length, *l*, of a straight-line element.

3.1.3 SHEAR STRAIN, γ (GAMMA)

The change in value of the initial right angle between the straight-line elements *AB* and *AD* (See Figure 3.1).

3.1.4 MATHEMATICAL DEFINITIONS OF STRAIN

3.1.4.1 LONGITUDINAL STRAIN, ε

$$\varepsilon_x = \frac{\partial u}{\partial x}, \varepsilon_y = \frac{\partial v}{\partial y}, \varepsilon_z = \frac{\partial w}{\partial z} \qquad (3\text{-}1)$$

where *u*, *v*, and *w* are the three displacement components, at a point of the body, occurring, respectively, in the *x*, *y*, and *z* directions of the coordinate axes.

3.1.4.2 SHEAR STRAIN, γ

$$\gamma_{xy} = \gamma_{yx} = \frac{\partial u}{\partial x} + \frac{\partial u}{\partial y} \qquad (3\text{-}2)$$

$$\gamma_{xz} = \gamma_{zx} = \frac{\partial w}{\partial x} + \frac{\partial u}{\partial z} \qquad (3\text{-}3)$$

$$\gamma_{yz} = \gamma_{zy} = \frac{\partial w}{\partial y} + \frac{\partial v}{\partial z} \qquad (3\text{-}4)$$

where u, v, and w are the three displacement components, at a point of the body, occurring, respectively, in the *x, y,* and *z* directions of the coordinate axes.

3.2 STRAIN TENSOR

The general state of strain, like that of stress, may be represented in a matrix form as follows:

$$\begin{pmatrix} \varepsilon_x & \frac{\gamma_{xy}}{2} & \frac{\gamma_{xz}}{2} \\ \frac{\gamma_{yx}}{2} & \varepsilon_y & \frac{\gamma_{yz}}{2} \\ \frac{\gamma_{zx}}{2} & \frac{\gamma_{zy}}{2} & \varepsilon_z \end{pmatrix} \equiv \begin{pmatrix} \varepsilon_{xx} & \varepsilon_{xy} & \varepsilon_{xz} \\ \varepsilon_{yx} & \varepsilon_{yy} & \varepsilon_{yz} \\ \varepsilon_{zx} & \varepsilon_{zy} & \varepsilon_{zz} \end{pmatrix} \qquad (3\text{-}5)$$

3.3 HOOKE'S LAW FOR ISOTROPIC MATERIALS

The linear relationships between the six components of stress and the six components of strain may be generalized and simplified, as stated in Hooke's Laws for homogeneous, isotropic materials.

3.3.1 GENERAL EQUATIONS

$$\varepsilon_x = \frac{\sigma_x}{E} - \nu\frac{\sigma_y}{E} - \nu\frac{\sigma_z}{E} \tag{3-6}$$

$$\varepsilon_y = -\frac{\nu\sigma_x}{E} + \frac{\sigma_y}{E} - \nu\frac{\sigma_z}{E} \tag{3-7}$$

$$\varepsilon_z = -\frac{\nu\sigma_x}{E} - \frac{\nu\sigma_y}{E} + \frac{\sigma_z}{E} \tag{3-8}$$

and

$$\gamma_{xy} = \frac{\tau_{xy}}{G}, \; \gamma_{yz} = \frac{\tau_{yz}}{G}, \; \gamma_{zx} = \frac{\tau_{zx}}{G} \tag{3-9}$$

where E = a constant called Young's Modulus or the Modulus of Elasticity
 G = a constant called the Shearing Modulus of Elasticity or the Modulus of Rigidity
 ν = Poisson's Ratio (a strain relation)

The equation relating all three constants is:

$$G = \frac{E}{2(1+\nu)} \tag{3-10}$$

3.4 POISSON'S RATIO

$$\nu = -\frac{\varepsilon_y}{\varepsilon_x} = -\frac{\varepsilon_z}{\varepsilon_x} \qquad (3\text{-}11)$$

$\nu =$ Poisson's Ratio
$\varepsilon_y =$ Lateral Strain
$\varepsilon_z =$ Lateral Strain
$\varepsilon_x =$ Axial Strain

3.5 THERMAL STRAIN

Strains due to the temperature change

$$\varepsilon_x = \varepsilon_y = \varepsilon_z = \alpha \delta T \qquad (3\text{-}12)$$

$\varepsilon_x, \varepsilon_y, \varepsilon_z$ = Thermal strains
α = Coefficient of linear thermal expansion
δT = Change in temperature

NOTE: For an increase in temperature, δT is taken positive. For example,

$$\varepsilon_z = -\frac{\nu \sigma_x}{E} - \frac{\nu \sigma_y}{E} + \frac{\sigma_z}{E} + \alpha \delta T \qquad (3\text{-}13)$$

3.6 ELASTIC STRAIN ENERGY

The elastic strain energy is the internal work done in a body due to externally applied forces.

3.6.1

$$dU = \tfrac{1}{2}\sigma_x\varepsilon_x dV \qquad (3\text{-}14)$$

dU = The elemental, internal elastic strain energy for uniaxial stress
σ_x = Normal stress
ε_x = Linear (longitudinal) strain
dV = Volume of the element

The strain energy density is the strain energy stored in an elastic body per unit volume of the material.

3.6.2

$$\frac{dU}{dV} = U_0 = \frac{\sigma_x \varepsilon_x}{2} \qquad (3\text{-}15)$$

where U_0 = strain-energy density for uniaxial stress.

3.6.3

$$dU_s = \tfrac{1}{2}\tau_{xy}\gamma_{xy}dV \qquad (3\text{-}16)$$

dU_s = The elemental, internal elastic strain energy for shearing stresses
τ_{xy} = Shear stress
γ_{xy} = Displacement due to deformation
dV = Volume of the element

3.6.4

$$\left(\frac{dU}{dV}\right)_s = (U_0)_s = \frac{\tau_{xy}\gamma_{xy}}{2} \qquad (3\text{-}17)$$

where $(U_0)_s$ = Strain-energy density for shearing stresses.

3.6.5 General equation for strain-energy density having multiaxial states of stress.

$$U_0 = \frac{1}{2E}\left(\sigma_x^2 + \sigma_y^2 + \sigma_z^2\right) - \frac{\nu}{E}\left(\sigma_x\sigma_y + \sigma_y\sigma_z + \sigma_z\sigma_x\right)$$
$$+ \frac{1}{2G}\left(\tau_{xy}^2 + \tau_{yz}^2 + \tau_{zx}^2\right) \tag{3-18}$$

3.7 DEFLECTION OF AXIALLY LOADED MEMBERS

With applied forces only, the deflection of a beam, axially loaded, may be represented as:

$$u = \sum \frac{PL}{AE} \tag{3-19}$$

u = Total deflection (displacement)
P = Applied force
L = Length over which the force is acting
A = Cross-sectional area of the member
E = Modulus of elasticity

For a fixed-supported rod free at one end, the deflection at the free end caused by an applied force is

$$u(L) = \frac{PL}{AE} \tag{3-20}$$

where $u(L)$ is the deflection as a function of the length of the rod.

The deflection caused by the applied force **and** the weight of the rod may be written as:

$$|u| = \frac{PL}{AE} + \frac{WL}{2AE} = \frac{\left[P + (\frac{W}{2})\right]L}{AE} \quad (3\text{-}21)$$

where W is the weight of the rod.

STIFFNESS

$$k = \frac{AE}{L} \quad (3\text{-}22)$$

k = Stiffness influence coefficient
A = Cross-sectional area
E = Modulus of elasticity
L = Length of the rod

(The reciprocal of k defines the flexibility coefficient $f = k^{-1}$).

3.8 STRESS CONCENTRATIONS

When a rigid body is undergoing stress, no matter how irregular the stress distribution is at a given section, the sum of the stress must be equal to the applied force. However, calculations for peak (maximum) stresses may be computed by the following equation:

$$\sigma_{max} = k\frac{P}{A} \quad (3\text{-}23)$$

where K is a calculate constant varying from material to

material. *K* may also be calculated knowing the maximum stress, σ_{max}, and the average stress, σ_{avg}.

$$K = \frac{\sigma_{max}}{\sigma_{avg}}$$

CHAPTER 4

STRESS-STRAIN RELATIONS

4.1 MATERIAL PROPERTIES

4.1.1 DEFINITIONS

Ductile materials are capable of withstanding large strains. The converse applies to **brittle materials.**

True stress is obtained by dividing the applied force by the corresponding actual area of a specimen at the same instant. This differs from the **conventional** or **engineering stresses**, which are computed on the basis of the original area of a specimen.

The **proportional limit** of a material may be roughly defined as the point on the stress-strain diagram where the equation for the slope of the $\sigma - \varepsilon$ curve, due to an applied load, is no longer accurately approximated as being linear.

The slope of the line from 0 (the unloaded starting point of the diagram) to A (the proportional limit) is the elastic modulus, E. This is also a measure of stiffness of the material, due to an imposed load.

The highest point in the curve on a $\sigma - \varepsilon$ diagram is said to correspond with the **ultimate strength** of a material.

The **yield point** of a material exists only in ductile substances. It is the point where a large amount of deformation occurs due to an essentially constant stress.

Often, the yield point is nearly indistinguishable from the proportional limit, and for experimental purposes, the **offset method** is used to make this point more distinct. (See Figure 4.1.) Here, a line offset an aribtrary amount (0.2 percent) of strain is drawn parallel to the straight-line portion of the initial stress-strain diagram. The point Q is then taken to be the yield point of the material at 0.2 percent offset.

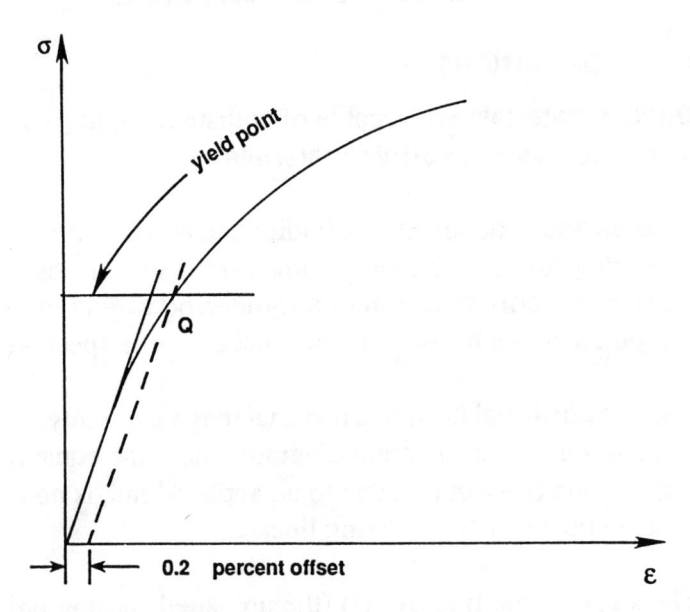

FIGURE 4.1—Offset Method of Determining the Yield Point of a Material.

The **critical point** of a material is the point (approximately) where a certain substance ruptures, under specific experimental conditions, due to an applied tensile loading.

4.1.2 ELASTICITY

An elastic material is one that is able to **completely** regain its original dimensions upon the removal of the applied forces. This implies that no permanent deformation occurs under any given conditions.

The point at which **permanent set** or **permanent deformation** occurs in a material is called the **elastic limit**. Non-elastic reactions are called **plastic**.

4.2 STRESS-STRAIN DIAGRAMS

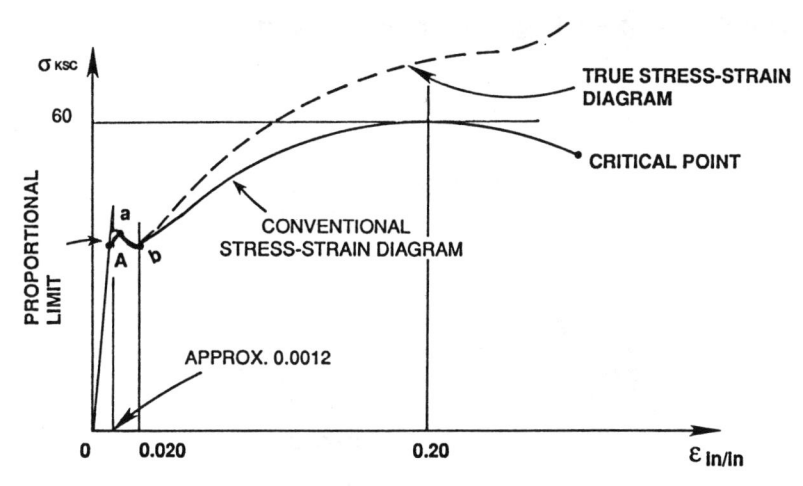

FIGURE 4.2—Stress-Strain Diagram for Mild Steel.

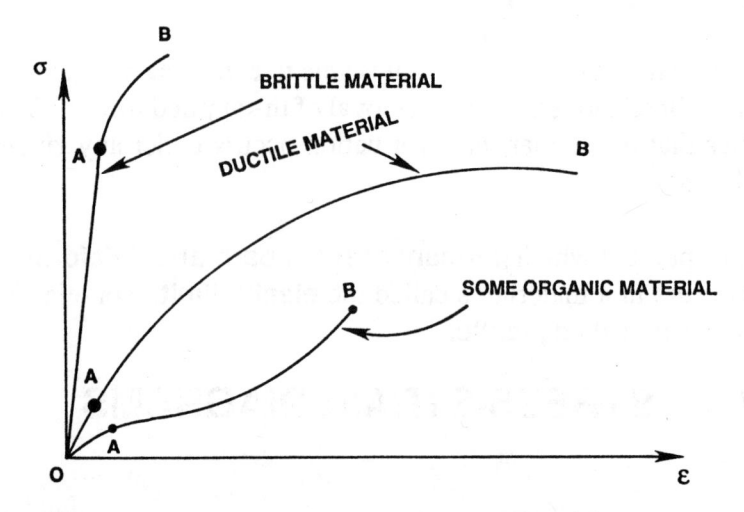

FIGURE 4.3—$\sigma - \varepsilon$ Diagram of Three Types of Materials.

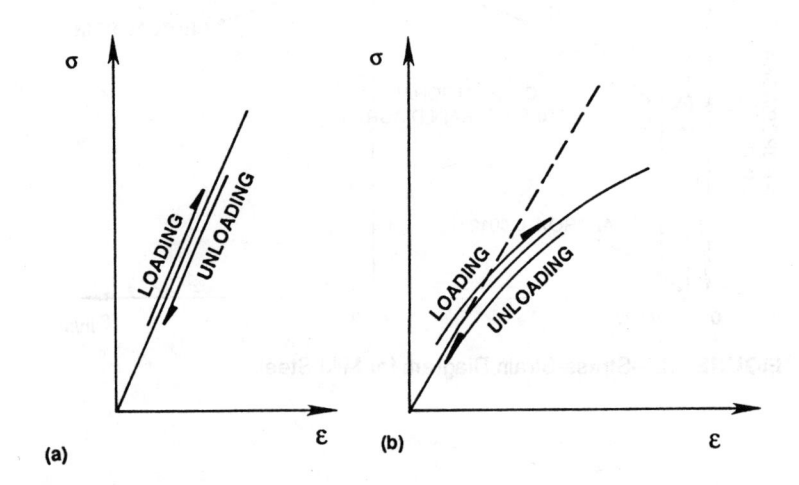

FIGURE 4.4—(a) Linearly Elastic Materials and (b) Nonlinearly Elastic Materials.

4.3 TABLE OF STRENGTH OF MATERIALS

METALS AND ALLOYS

Material	Tension Ultimate	Elastic Limit	Compression Ultimate	Bending Ultimate	Shearing Ultimate	Modulus of Elasticity Pounds per Sq. In.	Elongation Percent
Aluminum, bars, sheets	24-28	12-14					
" wire, annealed	20-35	14					5.0
Brass, 50% Zn	31	17.9	117	33.5	—	9,000,000	
" cast, common	18-24	6	30	20	36		
" wire, hard	80	16	—	—	—	14,000,000	
" " annealed	50	40	120				
Bronze, aluminum 5 to 7½%	75						
" Tobin, cast 38% Zn	66	40	—	—	—	14,500,000	
" " rolled 1½% Sn	80						
" " C. " ⅓% Pb	100						
Copper, plates, rods, bolts	32-35	10	32				
Iron, cast, gray	18-24	—	—	25-33	40	28,000,000	
" " malleable	27-35	15-20	46	30			
" wrought, shapes	48	26	Tensile	Tensile	⅚ Tens.	29,000,000	*1,500,000 Tensile Strength
Steel, plates for cold pressing	48-58	½ Tens.	Tensile	Tensile	¾ Tens.	29,000,000	
" cars	50-65	½ Tens.	Tensile	Tensile	¾ Tens.		

B-32

" locos., stat. boilers	55-65	½ Tens.	Tensile	Tensile	¾ Tens.	29,000,000	
" bridges and bldgs., ships	60-72	33	Tensile	Tensile	¾ Tens.	29,000,000	
" structural silicon	80-95	45	Tensile	Tensile	¾ Tens.	29,000,000	
" struc. nickel (3.25% Ni)	85-100	50	Tensile	Tensile	¾ Tens.	29,000,000	*1,500,000
Steel, rivet, boiler	45-55	½ Tens.	Tensile	Tensile	¾ Tens.	29,000,000	Tensile Strength
" " br.,bldg.,loco.,cars	52-62	28	Tensile	Tensile	¾ Tens.	29,000,000	
" " ships	55-65	30	Tensile	Tensile	¾ Tens.	29,000,000	
" " high-tensile	70-85	38	Tensile	Tensile	¾ Tens.	29,000,000	
Steel, cast, soft	60	27	Tensile	Tensile	¾ Tens.	29,000,000	†24
" " medium	70	31.5	Tensile	Tensile	¾ Tens.	29,000,000	†20
" " hard	80	36	Tensile	Tensile	¾ Tens.	29,000,000	†17
Steel wire, unannealed	120	60					
" " annealed	80	40					* 8" gage length
" " bridge cable	215	95					† 2" gage length

B-33

BUILDING MATERIALS

Material	Average Ultimate Stress Pounds per Square Inch			Safe Working Stress Pounds per Square Inch			Modulus of Elasticity Pounds Per Sq. In.
	Compression	Tension	Bending	Compression	Bearing	Shearing	
Masonry, granite	—	—	—	420	600		
" limestone, bluestone	—	—	—	350	500		
" sandstone	—	—	—	280	400		
" rubble	—	—	—	140	250		
" brick, common	10,000	200	600				
Ropes, cast steel hoisting	—	80,000					
" standing, derrick	—	70,000					
" manila	—	8,000					
Stone, bluestone	12,000	1,200	2,500	1,200	1,200	200	7,000,000
" granite, gneiss	12,000	1,200	1,600	1,200	1,200	200	7,000,000
" limestone, marble	8,000	800	1,500	800	800	150	7,000,000
" sandstone	5,000	150	1,200	500	500	150	3,000,000
" slate	10,000	3,000	5,000	1,000	1,000	175	14,000,000

CHAPTER 5

CENTER OF GRAVITY, CENTROIDS AND MOMENT OF INERTIA

5.1 CENTER OF GRAVITY

For a System of Particles:

$$\bar{x} = \frac{\sum_{i=1}^{n} x_i w_i}{\sum_{i=1}^{n} w_i}, \quad \bar{y} = \frac{\sum_{i=1}^{n} y_i w_i}{\sum_{i=1}^{n} w_i}, \quad \bar{z} = \frac{\sum_{i=1}^{n} z_i w_i}{\sum_{i=1}^{n} w_i} \quad (5\text{-}1)$$

$\bar{x}, \bar{y}, \bar{z}$ are the coordinates of the center of gravity of the system.

w_i is the weight of the ith particle.

x_i, y_i, z_i are the algebraic distances to the ith particle from the origin.

TWO-DIMENSIONAL CONDITIONS (AREAS AND LINES)

$$W = \int dw$$

$$\bar{x} = \frac{\int x\,dw}{W}$$

$$\bar{y} = \frac{\int y\,dw}{W}$$

(5-2)

W is the sum of the magnitudes of the elementary weights. In the case of a line, the center of gravity generally does not lie on the line.

THREE-DIMENSIONAL CONDITIONS (VOLUMES)

$$\bar{x} = \frac{\int x\,dw}{W},\ \bar{y} = \frac{\int y\,dw}{W},\ \bar{z} = \frac{\int z\,dw}{W}$$

(5-3)

If the body is made of homogeneous material of specific weight γ, then

Weight of the element $(dw) = \gamma\,dV$ where dV = volume of the element and

$$\bar{x} = \frac{\int x\,dv}{V},\ \bar{y} = \frac{\int y\,dv}{V},\ \bar{z} = \frac{\int z\,dv}{V}$$

(5-4)

$\bar{x}, \bar{y}, \bar{z}$ are the coordinates of the centroid of the volume V of the body.

$\int x\,dv$ is called the first moment of the volume with respect to the yz plane.

5.2 CENTROIDS

When the calculation depends on the geometry of the body only, the point $(\bar{x}, \bar{y}, \bar{z})$ is called the centroid of the body.

For an area A of a homogeneous plate:

$$\bar{x} = \frac{\int x\,dA}{A}$$

$$\bar{y} = \frac{\int y\,dA}{A}$$

(5-5)

The above equations define the coordinates $\bar{x}$ and $\bar{y}$ of the center of gravity. This point is also known as centroid of the area A.

For a homogeneous wire of length L:

$$\bar{x} = \frac{\int x\,dL}{L}$$

$$\bar{y} = \frac{\int y\,dL}{L}$$

(5-6)

NOTE: If the body is nonhomogeneous, the centroid and the center of gravity will not coincide.

5.2.1 COMPOSITE BODIES

A composite body may be divided into parts with common shapes to determine its center of gravity.

FIGURE 5.1–Center of Gravity.

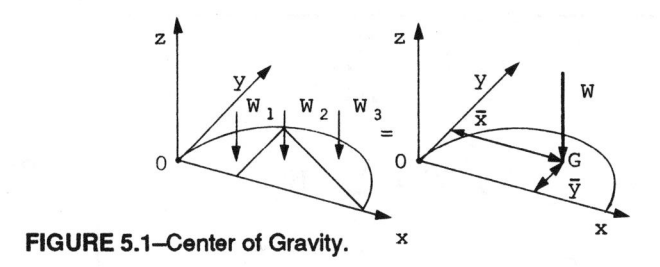

By equating moments:

$$\bar{x} = \frac{\bar{x}_1 W_1 + \bar{x}_2 W_2 + \ldots + \bar{x}_n W_n}{W_1 + W_2 + \ldots + W_n}$$

$$\bar{y} = \frac{\bar{y}_1 W_1 + \bar{y}_2 W_2 + \ldots + \bar{y}_n W_n}{W_1 + W_2 + \ldots + W_n}$$

(5-7)

To find the centroid, equate moment of areas:

$$\bar{x} = \frac{\bar{x}_1 A_1 + \bar{x}_2 A_2 + \ldots + \bar{x}_n A_n}{A_1 + A_2 + \ldots + A_n}$$

$$\bar{y} = \frac{\bar{y}_1 A_1 + \bar{y}_2 A_2 + \ldots + \bar{y}_n A_n}{A_1 + A_2 + \ldots + A_n}$$

(5-8)

LINES

For composite lines, divide it into simpler segments and add the moments of each segment.

VOLUMES

A body can be divided into common shapes of revolution.

Center of Gravity:

$$\bar{x}\sum_{i=1}^{n} W_i = \sum_{i=1}^{n} \bar{x}_i W_i$$

$$\bar{y}\sum_{i=1}^{n} W_i = \sum_{i=1}^{n} \bar{y}_i W_i \qquad (5\text{-}9)$$

$$\bar{z}\sum_{i=1}^{n} W_i = \sum_{i=1}^{n} \bar{z}_i W_i$$

If the body is homogeneous, the center of gravity coincides with the centroid:

$$\bar{x}\sum_{i=1}^{n} V_i = \sum_{i=1}^{n} \bar{x}_i V_i$$

$$\bar{y}\sum_{i=1}^{n} V_i = \sum_{i=1}^{n} \bar{y}_i V_i \qquad (5\text{-}10)$$

$$\bar{z}\sum_{i=1}^{n} V_i = \sum_{i=1}^{n} \bar{z}_i V_i$$

5.2.2 CENTROIDS BY INTEGRATION

$$\bar{x} A = \int \bar{x}_e \, dA$$

$$\bar{y} A = \int \bar{y}_e \, dA \qquad (5\text{-}11)$$

x_e, y_e are the coordinates of the centroid of the element dA.

B-39

x_e, y_e should be expressed in terms of the coordinates of a point located on the curve bounding the area.

dA should be expressed in terms of the coordinates of the point and their differentials.

(a) $\overline{x}_e = x$
$\overline{Y}_e = y/2$
$dA = ydx$

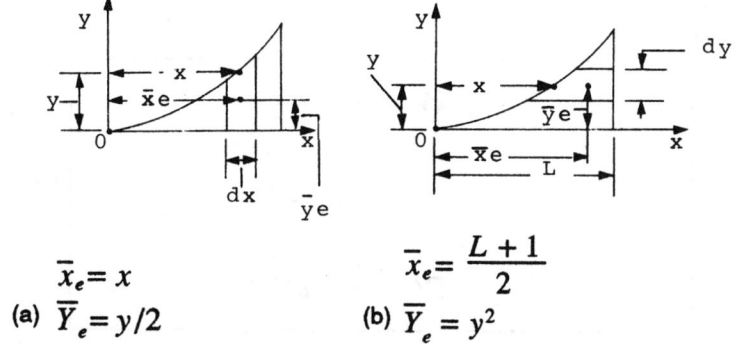

(b) $\overline{x}_e = \dfrac{L+1}{2}$
$\overline{Y}_e = y^2$
$dA = (L - x)dy$

FIGURE 5.2–Centroids by Integration.

LINES

$$\overline{x}L = \int xdL$$

$$\overline{y}L = \int ydL$$

(5-12)

dL should be expressed by one of the following:

$$dL = \sqrt{\left(\dfrac{dy}{dx}\right)^2 + 1}\ dx$$

$$dL = \sqrt{\left(\dfrac{dx}{dy}\right)^2 + 1}\ dy$$

$$dL = \sqrt{r^2 + \left(\frac{dr}{d\theta}\right)^2}\, d\theta$$

The above equations for dL, are chosen depending on the type of equation used to define the line.

VOLUME

$$\bar{x}V = \int \bar{x}_e\, dV$$

$$\bar{y}V = \int \bar{y}_e\, dV \qquad (5\text{-}13)$$

$$\bar{z}V = \int \bar{z}_e\, dV$$

For bodies with two planes of symmetry:

$$\bar{y} = \bar{z} = 0$$

and

$$\bar{x}V = \int \bar{x}_e\, dV$$

5.3 THEOREMS OF PAPPUS-GULDINUS

Surface of Revolution – Generated by revolving a curve about a fixed axis.

THEOREM I

The area of a surface revolution is equal to the product of the length of the generating curve and the distance traveled by the centroid of the curve while the surface is being generated.

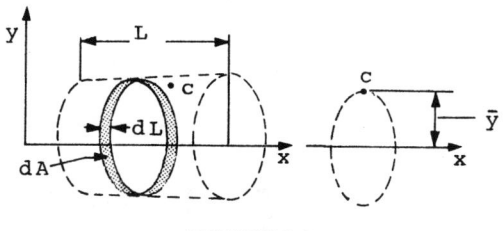

FIGURE 5.3

When a differential length dL of line L is revolved about x-axis, a ring is generated having surface area $dA = 2\pi y dL$. Therefore

$$A = 2\pi \int_L y \, dL = 2\pi \bar{y} L \qquad (5\text{-}14)$$

THEOREM II

The volume of a surface of revolution is equal to the generating area times the distance traveled by the centroid of the area in generating the volume.

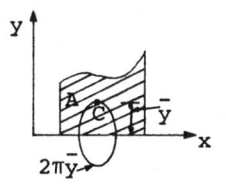

Volume of Revolution – Generated by revolving an area about a fixed axis. When a differential area dA is revolved about x-axis, it generated a ring of volume $dV = 2\pi y dA$. Therefore

$$v = 2\pi \int_A y \, dA = 2\pi \overline{y} A \qquad (5\text{-}15)$$

5.4 MOMENT OF INERTIA

AREA

General Formula:

$$I = \int_A s^2 \, dA \qquad (5\text{-}16)$$

s = Perpendicular distance from the axis to the area element (Figure 5.4).

In component form:

$$I_x = \int y^2 \, dA$$
$$I_y = \int x^2 \, dA \qquad (5\text{-}17)$$

$dA = dxdy$
$dI_x = y^2 \, dA$
$dI_y = x^2 \, dA$

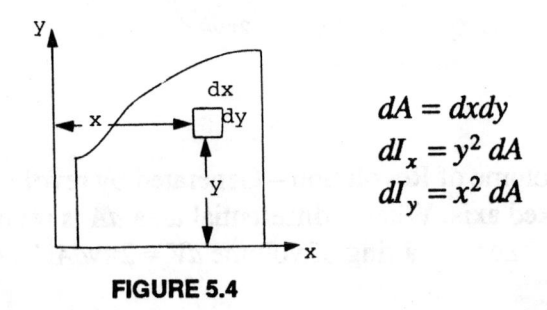

FIGURE 5.4

For a Rectangular Area:

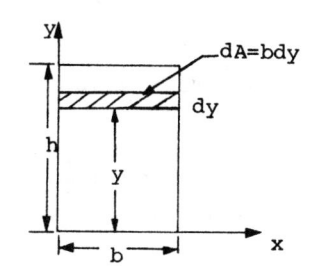

FIGURE 5.5

$$I_x = \int_0^h by^2 dy = \frac{1}{3} bh^3 \qquad (5\text{-}18)$$

NOTE: This is the moment of inertia with respect to an axis passing through the base of the rectangle.

Moments of Inertia of Masses

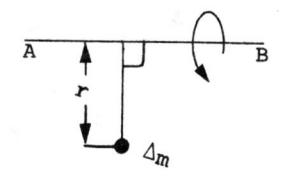

FIGURE 5.6

$$I = \int r^2 dm \qquad (5\text{-}19)$$

In component form:

B-44

$$I_x = \int (y^2 + z^2)\, dm$$

$$I_y = \int (z^2 + x^2)\, dm \qquad (5\text{-}20)$$

$$I_z = \int (x^2 + y^2)\, dm$$

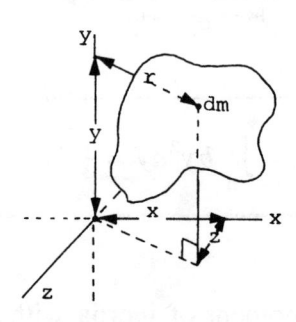

FIGURE 5.7

5.4.1 POLAR MOMENT OF INERTIA

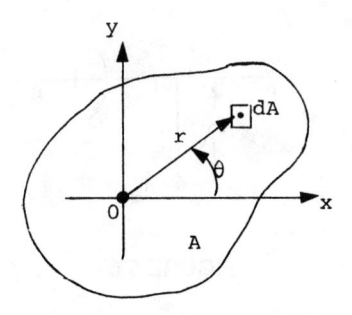

FIGURE 5.8

Polar Moment of Inertia:

$$J_0 = \int r^2\, dA \qquad (5\text{-}21)$$

In terms of rectangular moments of inertia:

$$J_0 = I_x + I_y \qquad (5\text{-}22)$$

5.4.2 RADIUS OF GYRATION
AREAS

$$I_x = k_x^2 A \; , \; I_y = k_y^2 A$$

Rectangular component form:

$$k_x = \sqrt{\frac{I_x}{A}}$$

$$k_y = \sqrt{\frac{I_y}{A}} \qquad (5\text{-}23)$$

POLAR FORM

$$k_0 = \sqrt{\frac{J_0}{A}} \qquad (5\text{-}24)$$

Relation between rectangular component form and polar form:

$$k_0^2 = k_x^2 + k_y^2 \qquad (5\text{-}25)$$

MASSES

$$I = k^2 m$$

$$k = \sqrt{\frac{I}{m}}$$ (5-26)

5.4.3 PERPENDICULAR-AXIS THEOREM

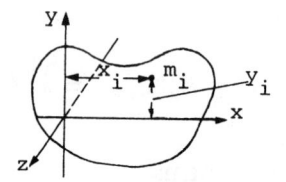

FIGURE 5.9

For a thin plane lamina:

$$I_z = \sum_i m_i x_i^2 + \sum_i m_i y_i^2 = I_x + I_y \; ; \; i = 1, 2, \ldots$$ (5-27)

STATEMENT

The moment of inertia of any plane lamina about an axis normal to the plane of the lamina is equal to the sum of the moments of inertia about any two mutually perpendicular axes passing through the given axis and lying in the plane of the lamina.

5.4.4 PARALLEL-AXIS THEOREM

AREAS

The theorem states that the moment of inertia of an area about a given axis is equal to the sum of the moment of inertia parallel to the given axis (and passing through the centroid of the area) and the product of the area and the square of the distance between the two parallel axes.

$$I = I_c + Ad^2 \qquad (5\text{-}28)$$

I_c = Moment of inertia about an axis through the centroid
d = Distance Between the Axes
A = Area

In terms of the radius of gyration:

$$k_0^2 = k_c^2 + d^2 \qquad (5\text{-}29)$$

In polar form:

$$J_0 = J_c + Ad^2 \qquad (5\text{-}30)$$

J_c = Polar Moment of Inertia about the Centroid

MASSES

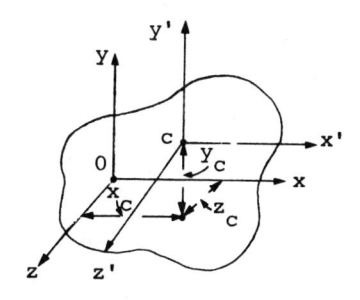

FIGURE 5.10

$$I = I_c + md^2 \qquad (5\text{-}31)$$

In the component form:

$$I_x = I_{x_c} + m(y_c^2 + z_c^2)$$
$$I_y = I_{y_c} + m(z_c^2 + x_c^2) \qquad (5\text{-}32)$$
$$I_z = I_{z_c} + m(x_c^2 + y_c^2)$$

5.4.5 COMPOSITE AREAS

A composite area consists of several regular shapes of bodies as components. The moments of inertia of these component shapes about a given axis may be computed separately and then added to give the moment of inertia for the entire area (or body).

The moment of inertia of a region with a hole equals the difference between the moment of inertia of the complete area (neglecting the hole) and the moment of inertia of the hole.

EXAMPLE

C is the centroid,

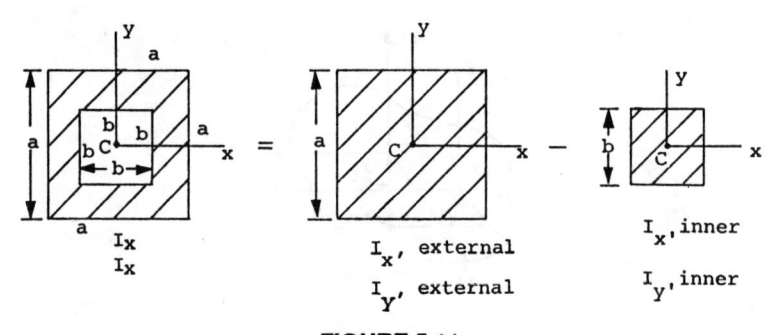

FIGURE 5.11

B-49

5.5 PRODUCT OF INERTIA

AREAS

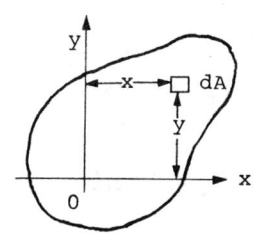

FIGURE 5.12

$$I_{xy} = \int_A xy\, dA \qquad (5\text{-}33)$$

unit: (Length)^4

NOTE: I_{xy} is zero when x, y, or both x and y axes are axes of symmetry for the region.

Parallel-Axis Theorem:

$$I_{xy} = I_{x'y'} + x_c y_c A \qquad (5\text{-}34)$$

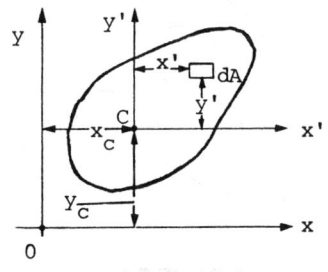

FIGURE 5.13

B-50

$I_{x'y'}$ = Moment of inertia in the centroidal axis system.
x_c, y_c = Coordinates of the centroid in the 'xoy' axis system.

MASS PRODUCTS OF INERTIA

Component Equations:

$$I_{xy} = \int xy\,dm$$
$$I_{yz} = \int yz\,dm \qquad (5\text{-}35)$$
$$I_{zx} = \int zx\,dm$$

Parallel-Axis Theorem:

$$I_{xy} = I_{x'y'} + x_c y_c m$$
$$I_{yz} = I_{y'z'} + y_c z_c m \qquad (5\text{-}36)$$
$$I_{zx} = I_{z'x'} + z_c x_c m$$

m = Total Mass of the Body.

Moment of Inertia with Respect to an Arbitrary Axis:

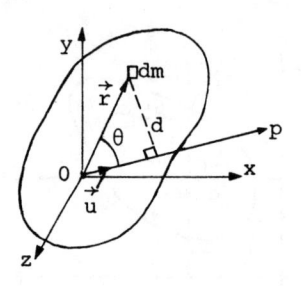

FIGURE 5.14

B-51

$$I_{OP} = \int d^2 dm = \int (\overline{u} \times \overline{r})^2 dm \qquad (5\text{-}37)$$

In terms of scalar quantities:

$$I_{OP} = I_x u_x^2 + I_y u_y^2 + I_z u_z^2 - 2I_{xy}u_x u_y \\ - 2I_{yz}u_y u_z - 2I_{zx}u_z u_x \qquad (5\text{-}38)$$

5.6 PRINCIPAL AXES AND INERTIAS

AREAS

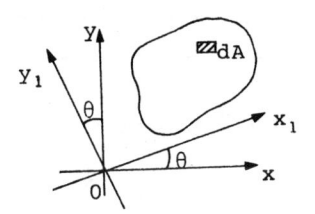

FIGURE 5.15

To calculate I_{x_1}, I_{y_1}, $I_{x_1 y_1}$ when θ, I_x, I_y, and I_{xy} are known; with respect to the inclined x_1 and y_1 axes:

$$I_{x_1} = \frac{I_x + I_y}{2} + \frac{I_x - I_y}{2} \cos 2\theta - I_{xy} \sin 2\theta$$

$$I_{y_1} = \frac{I_x + I_y}{2} - \frac{I_x - I_y}{2} \cos 2\theta - I_{xy} \sin 2\theta \qquad (5\text{-}39)$$

$$I_{x_1 y_1} = \frac{I_x - I_y}{2} \sin 2\theta + I_{xy} \cos 2\theta$$

The polar moment of inertia about the z axis is independent of the orientation of x_1 and y_1:

$$J_0 = I_{x_1} + I_{y_1} = I_x + I_y \tag{5-40}$$

Principal Axes – Axes about which the moments of inertial are maximum or minimum. The corresponding moments of inertia are called principal moments of inertia.

NOTE: Every point on the body has its own set of principal axes. The important point to consider is the centroid.

Orientation of the principal axes at the centroid; θ_p:

$$\tan 2\theta_p = \frac{-I_{xy}}{\dfrac{I_x - I_y}{2}} \tag{5-41}$$

Equation (5-41) has two solutions, θ_{p_1} and θ_{p_2}. They correspond to the maximum and minimum values of the principal moments of inertia and are 90° apart.

PRINCIPAL MOMENTS OF INERTIA:

$$I_{\substack{max \\ min}} = \frac{I_x + I_y}{2} \pm \sqrt{\left(\frac{I_x - I_y}{2}\right)^2 + I_{xy}^2} \tag{5-42}$$

NOTE: The product of inertia with respect to the principal axes is zero. Therefore, axes of symmetry are also principal axes for a given area.

Ellipsoid of Inertia:

Equation of an ellipsoid:

$$I_x x^2 + I_y y^2 + I_z z^2 - 2I_{xy} xy - 2I_{yz} yz - 2I_{zx} zx = 1 \quad (5\text{-}43)$$

The ellipsoid defines the moment of inertia of the body with respect to any axis passing through the origin O. It is also called the ellipsoid of inertia at point O.

If the principal axes, x_1, y_1, z_1 are used as coordinate axes, then the equation of the ellipsoid becomes:

$$I_{x_1} x_1^2 + I_{y_1} y_1^2 + I_{z_1} z_1^2 = 1 \quad (5\text{-}44)$$

Using the principal axes, the moment of inertia with respect to an arbitrary axis through the origin O, equation (5-38), is:

$$I_{OP} = I_{x_1} u_{x_1}^2 + I_{y_1} u_{y_1}^2 + I_{z_1} u_{z_1}^2 \quad (5\text{-}45)$$

5.7 PROPERTIES OF GEOMETRIC SECTIONS

SQUARE
Axis of moments through center

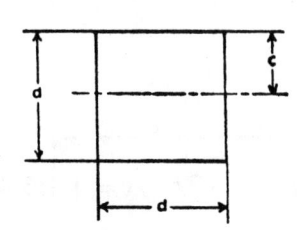

$A = d^2$

$c = \dfrac{d}{2}$

$I = \dfrac{d^4}{12}$

$S = \dfrac{d^3}{6}$

$r = \dfrac{d}{\sqrt{12}}$

SQUARE
Axis of moments on base

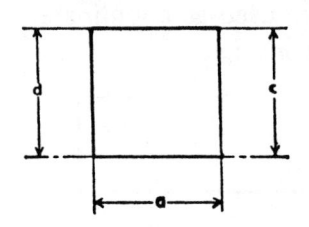

$A = d^2$

$c = d$

$I = \dfrac{d^4}{3}$

$S = \dfrac{d^3}{3}$

$r = \dfrac{d}{\sqrt{3}}$

SQUARE
Axis of moments on diagonal

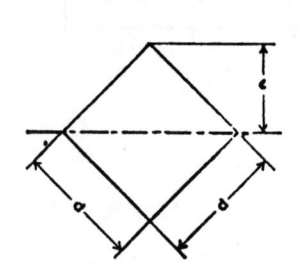

$A = d^2$

$c = \dfrac{d}{\sqrt{2}}$

$I = \dfrac{d^4}{12}$

$S = \dfrac{d^3}{6\sqrt{2}}$

$r = \dfrac{d}{\sqrt{12}}$

RECTANGLE
Axis of moments through center

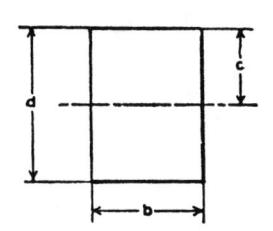

$A = bd$

$c = \dfrac{d}{2}$

$I = \dfrac{bd^3}{12}$

$S = \dfrac{bd^2}{6}$

$r = \dfrac{d}{\sqrt{12}}$

RECTANGLE
Axis of moments on base

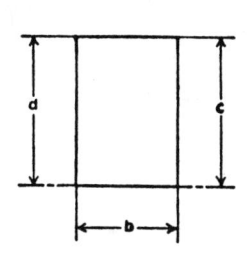

$A = bd$

$c = d$

$I = \dfrac{bd^3}{3}$

$S = \dfrac{bd^2}{3}$

$r = \dfrac{d}{\sqrt{3}}$

RECTANGLE
Axis of moments on diagonal

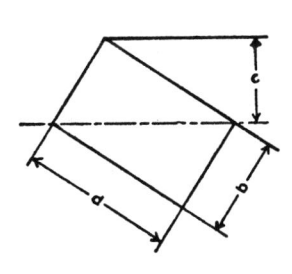

$A = bd$

$c = \dfrac{bd}{\sqrt{b^2 + d^2}}$

$I = \dfrac{b^3 d^3}{6(b^2 + d^2)}$

$S = \dfrac{b^2 d^2}{6\sqrt{b^2 + d^2}}$

$r = \dfrac{bd}{\sqrt{6(b^2 + d^2)}}$

RECTANGLE
Axis of moments on any line through center of gravity

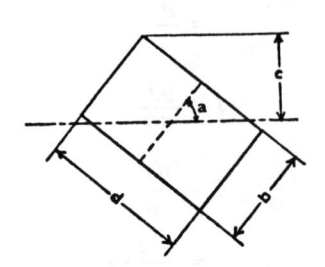

$$A = bd$$

$$c = \frac{b\sin a + d\cos a}{2}$$

$$I = \frac{bd(b^2 \sin^2 a + d^2 \cos^2 a)}{12}$$

$$S = \frac{bd(b^2 \sin^2 a + d^2 \cos^2 a)}{6(b\sin a + d\cos a)}$$

$$r = \sqrt{\frac{b^2 \sin^2 a + d^2 \cos^2 a}{12}}$$

HOLLOW RECTANGLE
Axis of moments through center

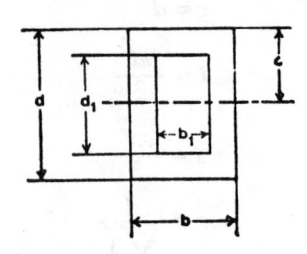

$$A = bd - b_1 d_1$$

$$c = \frac{d}{2}$$

$$I = \frac{bd^3 - b_1 d_1^3}{12}$$

$$S = \frac{bd^3 - b_1 d_1^3}{6d}$$

$$r = \sqrt{\frac{bd^3 - b_1 d_1^3}{12A}}$$

EQUAL TRIANGLES
Axis of moments through center of gravity

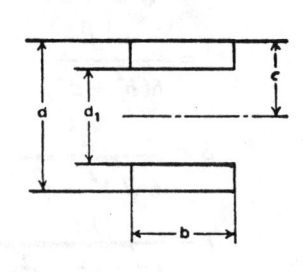

$$A = b(d - d_1)$$

$$c = \frac{d}{2}$$

$$I = \frac{b(d^3 - d_1^3)}{12}$$

$$S = \frac{b(d^3 - d_1^3)}{6d}$$

$$r = \sqrt{\frac{d^3 - d_1^3}{12(d - d_1)}}$$

UNEQUAL RECTANGLES
Axis of moments through center of gravity

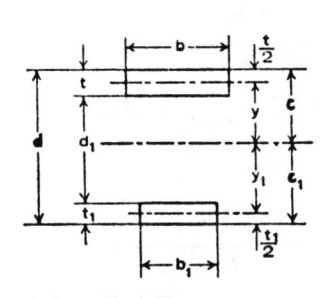

$A = bt + b_1 t_1$

$c = \dfrac{\frac{1}{2}bt^2 + b_1 t_1 (d - \frac{1}{2}t_1)}{A}$

$I = \dfrac{bt^3}{12} + bty^2 + \dfrac{b_1 t_1^3}{12} + b_1 t_1 y_1^2$

$S = \dfrac{I}{c} \quad S_1 = \dfrac{I}{c_1}$

$r = \sqrt{\dfrac{I}{A}}$

TRIANGLE
Axis of moments through center of gravity

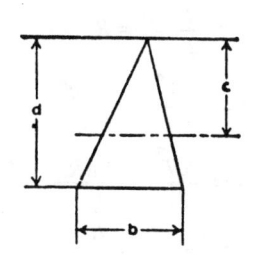

$A = \dfrac{bd}{2}$

$c = \dfrac{2d}{3}$

$I = \dfrac{bd^3}{36}$

$S = \dfrac{bd^2}{24}$

$r = \dfrac{d}{\sqrt{18}}$

TRIANGLE
Axis of moments on base

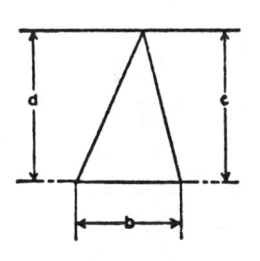

$A = \dfrac{bd}{2}$

$c = d$

$I = \dfrac{bd^3}{12}$

$S = \dfrac{bd^2}{12}$

$r = \dfrac{d}{\sqrt{6}}$

TRAPEZOID
Axis of moments through center of gravity

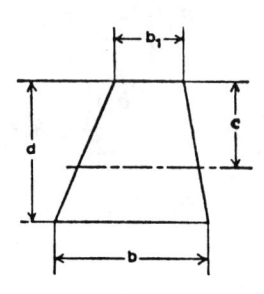

$$A = \frac{d(b+b_1)}{2}$$

$$c = \frac{d(2b+b_1)}{3(b+b_1)}$$

$$I = \frac{d^3(b^2 + 4bb_1 + b_1^2)}{36(b+b_1)}$$

$$S = \frac{d^2(b^2 + 4bb_1 + b_1^2)}{12(2b+b_1)}$$

$$r = \frac{d}{6(b+b_1)}\sqrt{2(b^2 + 4bb_1 + b_1^2)}$$

CIRCLE
Axis of moments through center

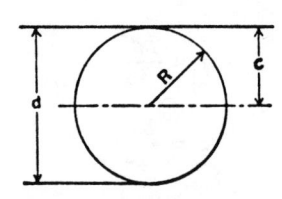

$$A = \frac{\pi d^2}{4} = \pi R^2$$

$$c = \frac{d}{2} = R$$

$$I = \frac{\pi d^4}{64} = \frac{\pi R^2}{4}$$

$$S = \frac{\pi d^3}{32} = \frac{\pi R^3}{4}$$

$$r = \frac{d}{4} = \frac{R}{2}$$

HOLLOW CIRCLE
Axis of moments through center

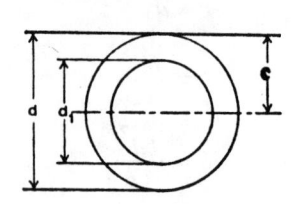

$$A = \frac{\pi(d^2 - d_1^2)}{4}$$

$$c = \frac{d}{2}$$

$$I = \frac{\pi(d^4 - d_1^4)}{64}$$

$$S = \frac{\pi(d^4 - d_1^4)}{32d}$$

$$r = \frac{\sqrt{d^2 + d_1^2}}{4}$$

HALF CIRCLE
Axis of moments through
center of gravity

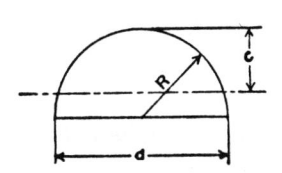

$$A = \frac{\pi R^2}{2}$$

$$c = R\left(1 - \frac{4}{3\pi}\right)$$

$$I = R^4\left(\frac{\pi}{8} - \frac{8}{9\pi}\right)$$

$$S = \frac{R^3}{24} \frac{(9\pi^2 - 64)}{(3\pi - 4)}$$

$$r = R \frac{\sqrt{9\pi^2 - 64}}{6\pi}$$

PARABOLA

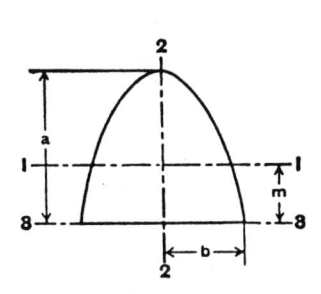

$$A = \frac{4}{3} ab$$

$$m = \frac{2}{5} a$$

$$I_1 = \frac{16}{175} a^3 b$$

$$I_2 = \frac{4}{15} ab^3$$

$$I_3 = \frac{32}{105} a^3 b$$

HALF PARABOLA

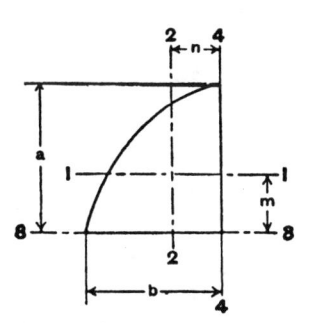

$$A = \frac{2}{3} ab$$

$$m = \frac{2}{5} a$$

$$n = \frac{3}{8} b$$

$$I_1 = \frac{8}{175} a^3 b, \quad I_2 = \frac{19}{480} ab^3$$

$$I_3 = \frac{16}{105} a^3 b, \quad I_4 = \frac{2}{15} ab^3$$

COMPLEMENT OF HALF PARABOLA

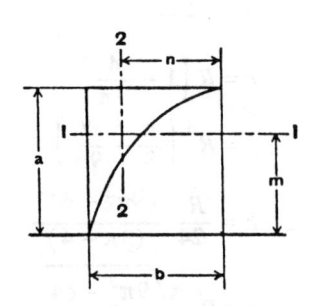

$A = \dfrac{1}{3} ab$

$m = \dfrac{7}{10} a$

$n = \dfrac{3}{4} b$

$I_1 = \dfrac{37}{2100} a^3 b$

$I_2 = \dfrac{1}{80} ab^3$

PARABOLIC FILLET IN RIGHT ANGLE

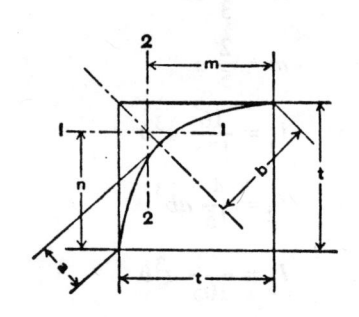

$a = \dfrac{t}{2\sqrt{2}}$

$b = \dfrac{t}{\sqrt{2}}$

$A = \dfrac{1}{6} t^2$

$m = n = \dfrac{4}{5} t$

$I_1 = I_2 = \dfrac{11}{2100} t^4$

* HALF ELLIPSE

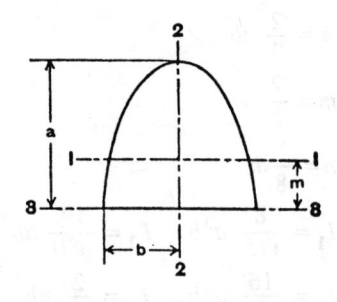

$A = \dfrac{1}{2} \pi ab$

$m = \dfrac{4a}{3\pi}$

$I_1 = a^3 b \left(\dfrac{\pi}{8} - \dfrac{8}{9\pi} \right)$

$I_2 = \dfrac{1}{8} \pi ab^3$

$I_3 = \dfrac{1}{8} \pi a^3 b$

* To obtain properties of half circle, quarter circle and circular complement substitute $a = b = R$.

*QUARTER ELLIPSE

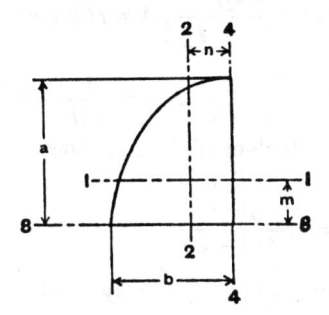

$$A = \frac{1}{4}\pi ab$$

$$m = \frac{4a}{3\pi}$$

$$n = \frac{4b}{3\pi}$$

$$I_1 = a^3 b\left(\frac{\pi}{16} - \frac{4}{9\pi}\right), \quad I_3 = \frac{1}{16}\pi a^3 b$$

$$I_2 = ab^3\left(\frac{\pi}{16} - \frac{4}{9\pi}\right), \quad I_4 = \frac{1}{16}\pi ab^3$$

*ELLIPTIC COMPLEMENT

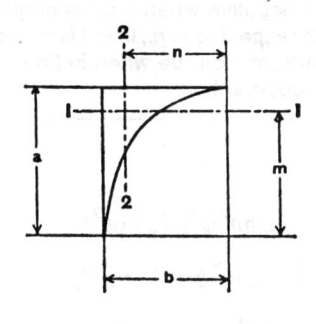

$$A = ab\left(1 - \frac{\pi}{4}\right)$$

$$m = \frac{a}{6\left(1 - \frac{\pi}{4}\right)}$$

$$n = \frac{b}{6\left(1 - \frac{\pi}{4}\right)}$$

$$I_1 = a^3 b\left(\frac{1}{3} - \frac{\pi}{16} - \frac{1}{36\left(1 - \frac{\pi}{4}\right)}\right)$$

$$I_2 = ab^3\left(\frac{1}{3} - \frac{\pi}{16} - \frac{1}{36\left(1 - \frac{\pi}{4}\right)}\right)$$

REGULAR POLYGON
Axis of moments through center

$n = $ Number of sides, $\quad \phi = \frac{180°}{n}$

$$a = 2\sqrt{R^2 - R_1^2}$$

$$R = \frac{a}{2\sin\phi}, \quad R_1 = \frac{a}{2\tan\phi}$$

$$A = \frac{1}{4}na^2 \cot\phi = \frac{1}{2}nR^2 \sin 2\phi = nR_1^2 \tan\phi$$

$$I_1 = I_2 = \frac{A(6R^2 - a^2)}{24} = \frac{A(12R_1^2 + a^2)}{48}$$

$$r_1 = r_2 = \sqrt{\frac{6R^2 - a^2}{24}} = \sqrt{\frac{12R_1^2 + a^2}{48}}$$

---

* To obtain properties of half circle, quarter circle and circular complement substitute $a = b = R$.

ANGLE
Axis of moments through center of gravity

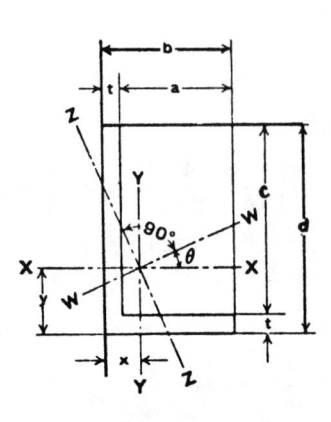

$$\tan 2\phi = \frac{2K}{I_y - I_x}, \quad A = t(b+c)$$

$$x = \frac{b^2 + ct}{2(b+c)}, \quad y = \frac{d^2 + at}{2(b+c)}$$

K = Product of Inertia about $X - X$ & $Y - Y$

$$= \pm \frac{abcdt}{4(b+c)}$$

$$I_x = \frac{1}{3}\left(t(d-y)^3 + by^3 - a(y-t)^3\right)$$

$$I_y = \frac{1}{3}\left(t(b-x)^3 + dx^3 - c(x-t)^3\right)$$

$$I_z = I_x \sin^2\theta + I_y \cos^2\theta + K \sin 2\theta$$

$$I_w = I_x \cos^2\theta + I_y \sin^2\theta - K \sin 2\theta$$

K is negative when heel of angle, with respect to *c. g.*, is in 1st or 3rd quadrant, positive when in 2nd or 4th quadrant.

Z–Z is axis of minimum I

BEAMS AND CHANNELS
Transverse force oblique through center of gravity

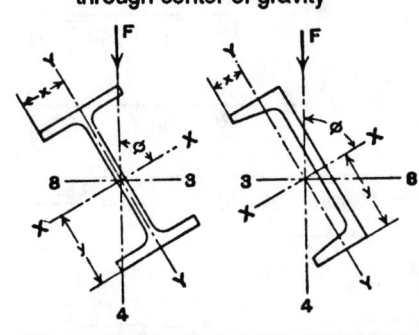

$$I_3 = I_x \sin^2\phi + I_Y \cos^2\phi$$

$$I_4 = I_x \cos^2\phi + I_Y \sin^2\phi$$

$$f = M\left(\frac{Y}{I_x}\sin\phi + \frac{x}{I_Y}\cos\phi\right)$$

where M is bending moment due to force F.

CHAPTER 6

STRESSES IN BEAMS

6.1 BENDING STRESS DISTRIBUTION

A beam deforms as a result of the action of a bending moment. During the deformed condition, the fibers on the outer surface (convex) of the beam are in tension and the fibers on the inner surface (concave) of the beam are in compression. The stress distribution across the cross-section of a deformed beam under bending is shown in Figure 6.1. The locus of all those points inside the beam at which the stress is zero, is called the neutral axis.

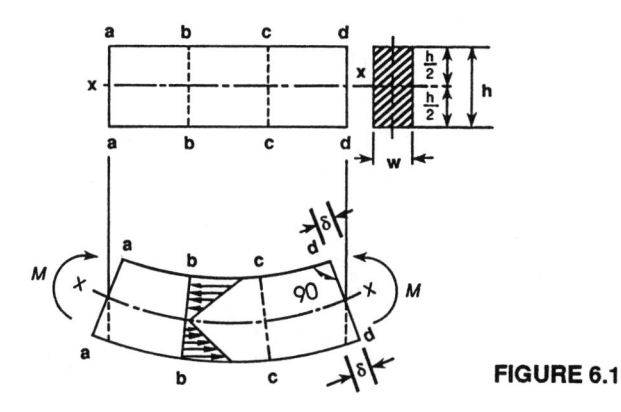

FIGURE 6.1

By the stress distribution diagram, it is clear that the bending stresses increase linearly from zero at the neutral plane to a maximum at the outer fibers. The compressive stress in the top fibers are equal to the tensile stress in the bottom fibers following Hooke's law, that the stress is proportional to deformation.

6.2 FLEXURE FORMULA

The following assumptions are based on the relationship between bending moment M and the resulting bending stress σ_b, also known as flexure formula.

1. Transverse planes remain transverse before and after the bending, i.e. no warping occurs.

2. The beam has a homogenous and isotropic material which obeys Hooke's law. E is the same for tension and compression.

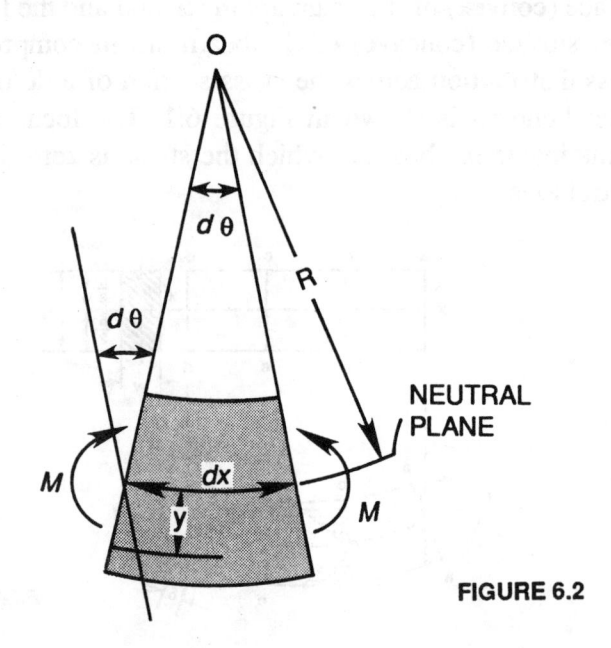

FIGURE 6.2

3. The beam is subjected to a pure bending moment i.e., there is no twisting or buckling load on the beam.

4. The beam is straight with a rectangular cross-section.

From Figure 6.2, the strain in a fiber located at a depth Y from the neutral axis is,

$$\varepsilon_x = \frac{dl}{dx} \qquad (6\text{-}1)$$

where the bending deformation $dl = yd\theta$

$$\therefore \; \varepsilon_x = \frac{yd\theta}{dx} \qquad (6\text{-}2)$$

Using similar triangles approach

$$\frac{yd\theta}{dx} = \frac{y}{R} \Rightarrow \frac{d\theta}{dx} = \frac{1}{R} \qquad (6\text{-}3)$$

$$\therefore \; \varepsilon_x = \frac{y}{R} \qquad (6\text{-}4)$$

Using Hooke's law,

$$\sigma_x = \varepsilon_x E = \frac{E}{R} y \qquad (6\text{-}5)$$

Now, if dF is the small increment of force acting on an elemental area dA, then

$$dF = \sigma_x dA \qquad (6\text{-}6)$$

The corresponding infinitesimal moment is

$$dM = dF \cdot y = (\sigma_x dA) \cdot y$$

or

$$dM = \left(\frac{E}{R}\right)y^2 \, dA \qquad (6\text{-}7)$$

on integration, the total resisting moment is,

$$M = \int \left(\frac{E}{R}\right)y^2 \, dA = \frac{E}{R}\int y^2 \, dA \qquad (6\text{-}8)$$

where $\int y^2 \, dA = I$ is called the second moment of area or moment of inertia of the area of cross-section with respect to the neutral axis.

Thus

$$M = \frac{E}{R} I \quad \text{or} \quad \frac{M}{I} = \frac{E}{R} \qquad (6\text{-}9)$$

since,

$$\sigma_x = \left(\frac{E}{R}\right) y \Rightarrow \frac{\sigma_x}{y} = \frac{E}{R}$$

Therefore,

$$\boxed{\frac{M}{I} = \frac{\sigma_x}{y} = \frac{E}{R}} \qquad (6\text{-}10)$$

This equation is known as flexure formula and is extensively used in the design of structural members.

When c is the distance of the extreme fiber from the neutral axis, $\sigma_x = \sigma_{max}$

$$\boxed{\therefore \quad M = \sigma_{max} \frac{I}{c}} \qquad (6\text{-}11)$$

where I/c is called section modulus of the beam.

6.3 SHEAR STRESSES IN BEAMS

In a beam of rectangular cross-section, both the horizontal and the vertical shearing stresses vary parabolically. The general equation for shear stress in a beam is,

$$\tau = \frac{VQ}{Ib} \quad (6\text{-}12)$$

Shear Formula

where τ = Shear stress
V = Total transverse shear
I = Second moment of area
Q = First moment of area

Now

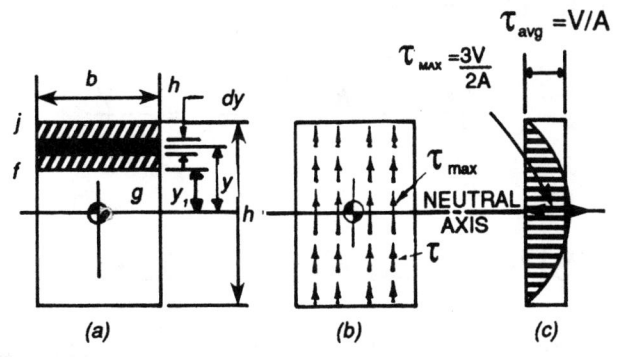

$$Q = \int_A y\, dA \quad (6\text{-}13)$$

FIGURE 6.3—(a) beam cross-sectional area, (b) longitudinal cut, (c) shear stress distribution.

From Figure 6.3 $dA = b\,dy$

B-68

$$\therefore Q = \int y\, dA = \int bydy = b\int_{y_1}^{h/2} y\, dy = \left.\frac{by^2}{2}\right|_{y_1}^{h/2}$$

$$Q = \frac{b}{2}\left[\left(\frac{h}{2}\right)^2 - y_1^2\right] \qquad (6\text{-}14)$$

Therefore, in a beam of rectangular cross-section, the shear stress varies parabolically. The maximum value of a shearing stress $\tau = \tau_{max}$ is obtained when $y_1 = 0$ or at the neutral axis. At the extreme fibers, $y_1 = \pm h/2$, therefore $Q = 0$ thus making $\tau = 0$. At increasing distances from the neutral axis, shear stress gradually diminishes as shown in Figure 6.3(b).

The direction of the shearing stresses at the section through a beam is the same as that of the shearing force V. This fact may be used to determine the sense of the shearing stresses.

As noted above, the maximum shearing stress in a rectangular beam occurs at the neutral axis, and for this case the general expression for τ_{max} may be simplified since $y_1 = 0$.

$$\frac{Vh^2}{8I} = \frac{Vh^2}{\frac{8bh^3}{12}} = \frac{3}{2}\frac{V}{bh} = \frac{3}{2}\frac{V}{A} \qquad (6\text{-}15)$$

6.4 SHEAR CENTER IN BEAMS

If a structural member does not have a longitudinal plane of symmetry or if the loads are not acting in the plane of symmetry, the member will twist. Bending without twisting (pure bending) can be made to occur if the loads are applied in the same plane in which the resultant of the shear stresses acts. In other words, a pure bending is achieved only when the resultant shear force passes through the shear center. Therefore, a shear center is defined as a point in the cross-section of a member through which the resultant of the transverse shearing stresses must act, regardless of the plane of transverse loads so that pure bending could result.

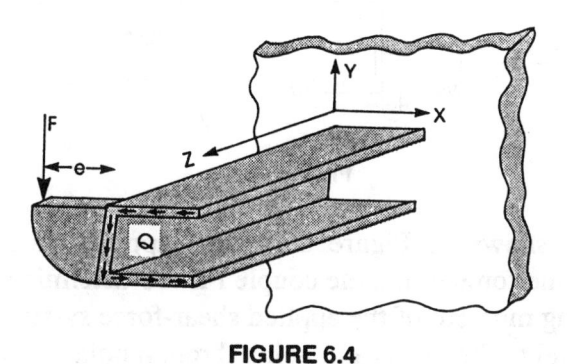

FIGURE 6.4

A shear force in any portion of a cross section can be found from $dF = \tau \, dA$. The internal shear forces, Q, in the flanges of the channel of Figure 6.4 tend to produce clockwise torsion in the channel cross-section. To avoid torsion, the applied load F should intersect through a point distance e from the channel. Generally speaking, this point is known as the **shear center.** The distance e is called the eccentricity of the load.

The following illustrations will explain how the shear center is determined.

Given a cantilever beam with a channel cross-section (Figure 6.5(a)) and a force applied as shown, it is required to compute e.

In Figure 6.5(b) we have shown a free body exposing a section at x having a normal in the plus x coordinate direction. Finally, in Figure 6.5(c) this section is shown with a shear-stress distribution corresponding to a positive shear force V_y. For the

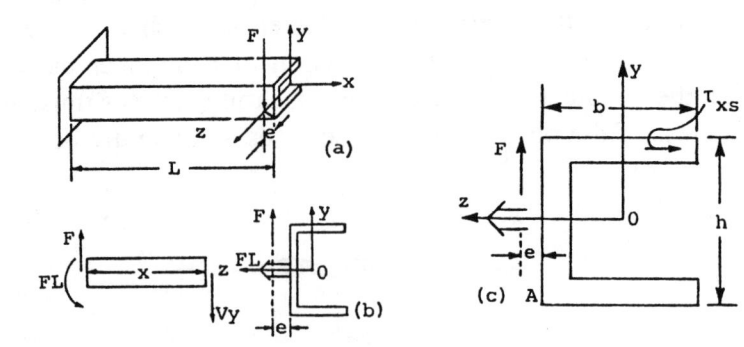

FIGURE 6.5

free body shown in Figure 6.5(b) the applied shear loading system is the force *F* and the couple *FL*. To determine *e*, we set the twisting moment of the applied shear-force system about an axis parallel to the *x* axis and going through point *A* (see Figure 6.5(c)) equal and opposite to the moment about this axis of the shear-stress distribution of the section at position *x*. For such an axis only the stresses in the upper flange contribute moment, because the other stresses have no moment arm with respect to *A*. Accordingly, we can say, using the proper directional signs, for the moments that

$$[-(Fe)] = -\left[\int_0^b \{[-(h-t_1)\tau_{xs}\} t_1 \, ds\right] \quad (6\text{-}16)$$

Note that the couple *FL* has no moment about the axis at *A* since it is orthogonal to this axis. Employing

$$(\tau_{xs})_{\text{flange}} = \frac{V_y s \dfrac{h-t_1}{2}}{I_{zz}}$$

for τ_{xs} we obtain

$$-[Fe] = \int_0^b t_1 \frac{V_y (h-t_1)^2}{2 I_{zz}} s \, ds$$

B-71

$$= \frac{t_1 V_y (h-t_1)^2}{2 I_{zz}} s_2^2 \Big|_0^b = \frac{t_1 V_y (h-t_1)^2 b^2}{4 I_{zz}} \quad (6\text{-}17)$$

Since $V_y = -F$, for e we obtain

$$e = \frac{t_1 (h-t_1)^2 b^2}{4 I_{zz}} \quad (6\text{-}18)$$

Since t_1 and t_2 are small compared to b, we shall approximate I_{zz} employing

$$I_{zz} = b t_1 (h-t_1)\left(\frac{h}{2}-\frac{t_1}{2}\right) + \frac{1}{12} t_2 (h-t_1)^3 + \frac{1}{6} b t_1^3$$

for this purpose. We may thus say that for $t \ll h$ that

$$I_{zz} \approx \frac{b t_1 h^2}{2} + \frac{1}{12} t_2 h^3 \quad (6\text{-}19)$$

$$\therefore \quad e = \frac{t_1 (h-t_1)^2 b^2}{4} \cdot \frac{12}{6 b t_1 h^2 + t_2 h^3}$$

dropping t_1 in the squared bracket

$$e = \frac{t_1 h^2 b^2}{1} \times \frac{3}{6 b t_1 h^2 + t_2 h^3} = \frac{t_1 b^2}{2 b t_1 + \dfrac{t_2 h}{3}} \quad (6\text{-}20)$$

$$\therefore \quad e = \frac{t_1 b^2}{2 b t_1 + \dfrac{t_2 h}{3}}$$

B-72

CHAPTER 7

DESIGN OF BEAMS

7.1 BEAM DESIGN CRITERIA

DESIGN FACTORS

— Loading (arrangement and magnitude)

— Span of beams

— Type of support

— Permissible stress

— Permissible deflection

— Size and shape of beam

DESIGN STEPS

— Analyze the loads (dynamic and static) on beams, unsupported span and type of support, in detail.

— Estimate bending moment, shearing force, twisting moment, etc.

— Choose the factor of safety, depending on the accuracy of assumptions made for the above estimations.

- Select the material of beam having the allowable stress and deflection within permissible limits.

- Select the size and shape of the beam which will given an optimum section modulus.

In addition to the above factors, we have to keep fatigue, creep and vibrational criteria for design in mind. Moreover, the design should be economical.

Rupture strength of the material in bending is found by Equation (7-1)

$$\sigma_r = \frac{M_r \bar{y}}{I} \tag{7-1}$$

where σ_r = modulus of rupture
M_r = moment at the breaking point
I = moment of inertia of beam sections.

7.1.1   SHEAR FORCE

VERTICAL SHEAR FORCE

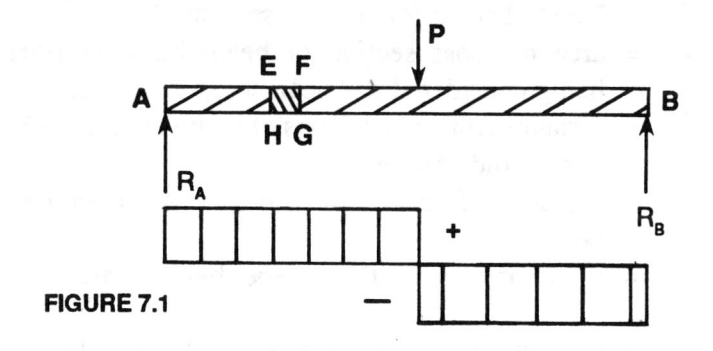

FIGURE 7.1

Vertical shear force at a section is the algebraic sum of all vertical forces acting on the beam to one end of that section.

B-74

HORIZONTAL SHEAR FORCE

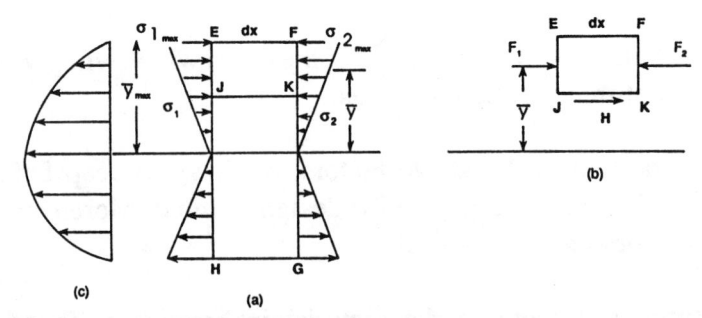

FIGURE 7.2—(a) Enlarged section *EFGH*, (b) horizontal force on section *EFGH*, (c) horizontal shear stress distribution.

From Figure 7.2(b), Horizontal Shear force:

$$H = F_2 - F_1$$

Shear stress

$$\tau_H = \frac{Va\overline{y}}{Ib} \qquad (7\text{-}2)$$

where τ_H = Horizontal shear stress
V = Vertical shear force at the section
a = area of cross section of beam between fiber being considered and nearest extreme fiber
$\overline{y}$ = distance from neutral axis of entire cross section to centroid of area, a.
I = moment of inertia of entire cross section of beam
b = width of cross section at fiber being considered

NOTE: At any point in a member subjected to shearing forces, there exists equal shearing stresses in planes mutually perpendicular.

Beam diagrams and formulas for various static and moving loads are given in Section 7.5.

7.2 BEAM DESIGN PROCEDURES

The most general procedures of beam design are given as follows:

DESIGN PROCEDURE 1:

By trying different values of b to solve the section modulus equation, $z = \dfrac{bh^2}{6}$, a table of results could be formed. Through this procedure, only a few of the suitable beams, which can satisfy the requirements of the strength, are picked up out of several beam sizes available. Further rational constraints cause the final selection of the most suitable beam size.

DESIGN PROCEDURE 2:

Using a pre-determined width-depth ratio (b/t), generally $1/3$, $1/2$, $2/3$ or $3/4$, and selecting the most appropriate ratio, one can solve the equation by having two unknown variables, b and h. The selection of b/t ratio is totally up to the designer's discretion. Any such pre-calculated ratio can vary in size, in either variable b or t, in the beam finally selected.

DESIGN PROCEDURE 3:

The values of the section modulus and the area are determined first. Thereafter, a suitable selection of a beam is made from the tables of beam sizes and beam properties such as moment of inertia, section modulus, area and weight.

7.3 PLASTIC ANALYSIS OF BEAMS

Materials that deform without fracture on further applications of load beyond the yield point are called plastic materials. Plasticity, therefore, is the ability of a material to deform nonelastically without having a rupture. (Refer to Section 9.6 for more details.)

7.4 ECONOMY IN BEAM DESIGN

A beam with a minimum cross-sectional area and satisfying all strength requirements within the specified constraints, is the most economic. Generally, for the same strength criteria, a beam of greater depth is more economical than one with lesser depth. Higher beam depth increases the moment of inertia and hence, the section modulus which in turn reduces the stress developed in the material at the same loading conditions.

The arrangement of supports, selection of material, and proper analysis of design criteria make a design economically sound.

7.5 BEAM DIAGRAMS AND FORMULAS

7.5.1 FOR VARIOUS STATIC LOADING CONDITIONS

1. SIMPLE BEAM—UNIFORMLY DISTRIBUTED LOAD

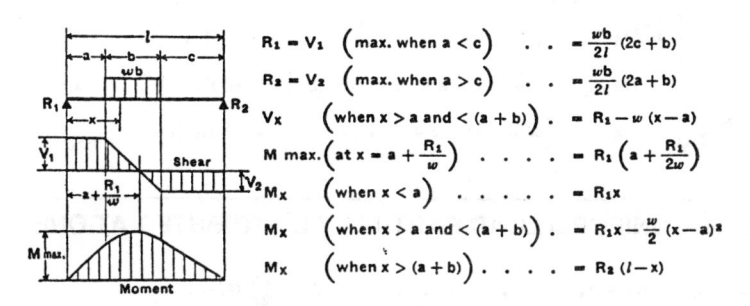

$R_1 = V_1$ (max. when $a < c$) $\quad \ldots \quad = \frac{wb}{2l}(2c + b)$

$R_2 = V_2$ (max. when $a > c$) $\quad \ldots \quad = \frac{wb}{2l}(2a + b)$

V_x (when $x > a$ and $< (a + b)$) $. = R_1 - w(x - a)$

M max. (at $x = a + \frac{R_1}{w}$) $\ldots \ldots = R_1 \left(a + \frac{R_1}{2w}\right)$

M_x (when $x < a$) $\ldots \ldots = R_1 x$

M_x (when $x > a$ and $< (a + b)$) $. = R_1 x - \frac{w}{2}(x - a)^2$

M_x (when $x > (a + b)$) $\ldots \ldots = R_2 (l - x)$

2. SIMPLE BEAM—LOAD INCREASING UNIFORMLY TO ONE END

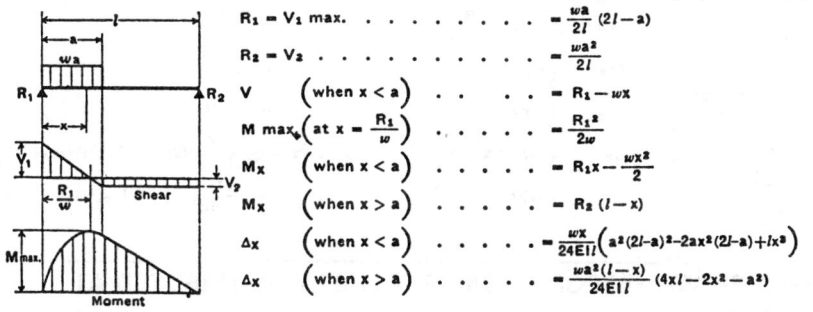

$R_1 = V_1$ max. $\ldots \ldots \ldots = \frac{wa}{2l}(2l - a)$

$R_2 = V_2 \ldots \ldots \ldots \ldots = \frac{wa^2}{2l}$

V (when $x < a$) $\ldots \ldots = R_1 - wx$

M max. (at $x = \frac{R_1}{w}$) $\ldots \ldots = \frac{R_1^2}{2w}$

M_x (when $x < a$) $\ldots \ldots = R_1 x - \frac{wx^2}{2}$

M_x (when $x > a$) $\ldots \ldots = R_2(l - x)$

Δ_x (when $x < a$) $\ldots \ldots = \frac{wx}{24Ell}\left(a^2(2l-a)^2 - 2ax^2(2l-a) + lx^3\right)$

Δ_x (when $x > a$) $\ldots \ldots = \frac{wa^2(l-x)}{24Ell}(4xl - 2x^2 - a^2)$

3. SIMPLE BEAM—LOAD INCREASING UNIFORMLY TO CENTER

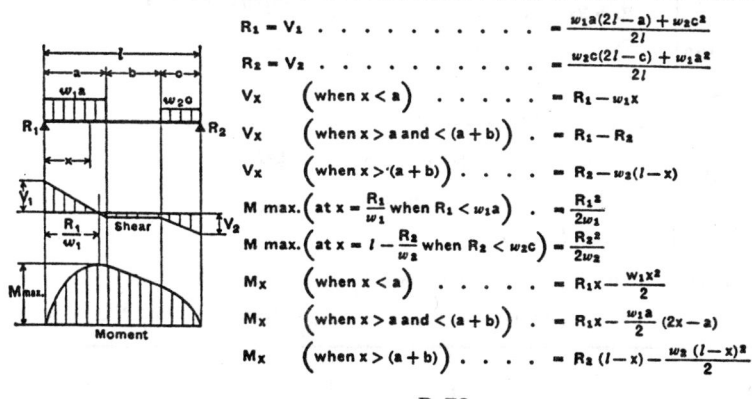

$R_1 = V_1 \ldots \ldots \ldots \ldots = \frac{w_1 a(2l - a) + w_2 c^2}{2l}$

$R_2 = V_2 \ldots \ldots \ldots \ldots = \frac{w_2 c(2l - c) + w_1 a^2}{2l}$

V_x (when $x < a$) $\ldots \ldots = R_1 - w_1 x$

V_x (when $x > a$ and $< (a + b)$) $. = R_1 - R_2$

V_x (when $x > (a + b)$) $\ldots \ldots = R_2 - w_2(l - x)$

M max. (at $x = \frac{R_1}{w_1}$ when $R_1 < w_1 a$) $. = \frac{R_1^2}{2w_1}$

M max. (at $x = l - \frac{R_2}{w_2}$ when $R_2 < w_2 c$) $= \frac{R_2^2}{2w_2}$

M_x (when $x < a$) $\ldots \ldots = R_1 x - \frac{w_1 x^2}{2}$

M_x (when $x > a$ and $< (a + b)$) $. = R_1 x - \frac{w_1 a}{2}(2x - a)$

M_x (when $x > (a + b)$) $\ldots \ldots = R_2(l - x) - \frac{w_2(l - x)^2}{2}$

B-78

4. SIMPLE BEAM—UNIFORM LOAD PARTIALLY DISTRIBUTED

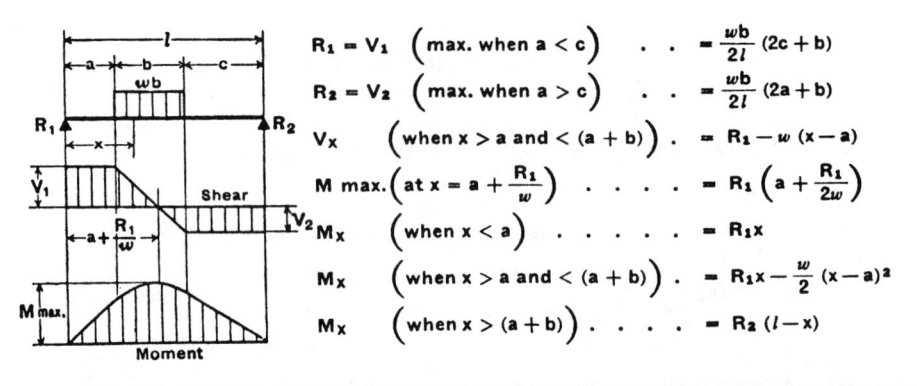

$R_1 = V_1$ (max. when $a < c$) . . $= \dfrac{wb}{2l}(2c+b)$

$R_2 = V_2$ (max. when $a > c$) . . $= \dfrac{wb}{2l}(2a+b)$

V_x (when $x > a$ and $< (a+b)$) . $= R_1 - w(x-a)$

M max. (at $x = a + \dfrac{R_1}{w}$) . . . . $= R_1\left(a + \dfrac{R_1}{2w}\right)$

M_x (when $x < a$) . . . . . $= R_1 x$

M_x (when $x > a$ and $< (a+b)$) . $= R_1 x - \dfrac{w}{2}(x-a)^2$

M_x (when $x > (a+b)$) . . . . $= R_2(l-x)$

5. SIMPLE BEAM—UNIFORM LOAD PARTIALLY DISTRIBUTED AT ONE END

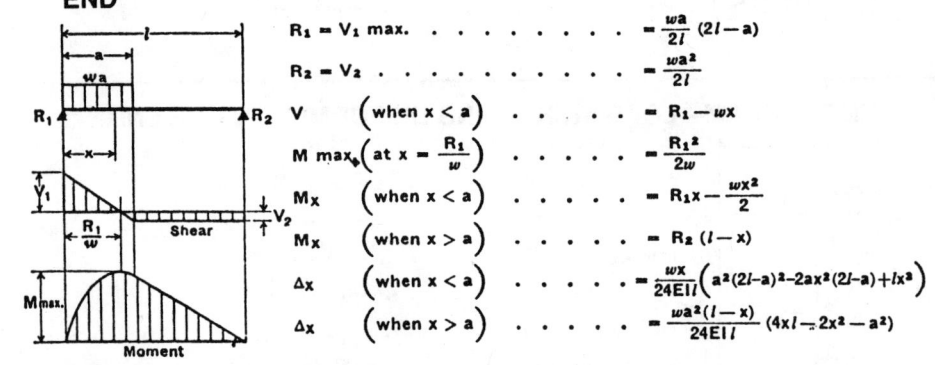

$R_1 = V_1$ max. . . . . $= \dfrac{wa}{2l}(2l-a)$

$R_2 = V_2$ . . . . $= \dfrac{wa^2}{2l}$

V (when $x < a$) . . . . $= R_1 - wx$

M max. (at $x = \dfrac{R_1}{w}$) . . . . $= \dfrac{R_1^2}{2w}$

M_x (when $x < a$) . . . . $= R_1 x - \dfrac{wx^2}{2}$

M_x (when $x > a$) . . . . $= R_2(l-x)$

Δ_x (when $x < a$) . . . . $= \dfrac{wx}{24EIl}\left(a^2(2l-a)^2 - 2ax^2(2l-a) + lx^3\right)$

Δ_x (when $x > a$) . . . . $= \dfrac{wa^2(l-x)}{24EIl}(4xl - 2x^2 - a^2)$

6. SIMPLE BEAM—UNIFORM LOAD PARTIALLY DISTRIBUTED AT EACH END

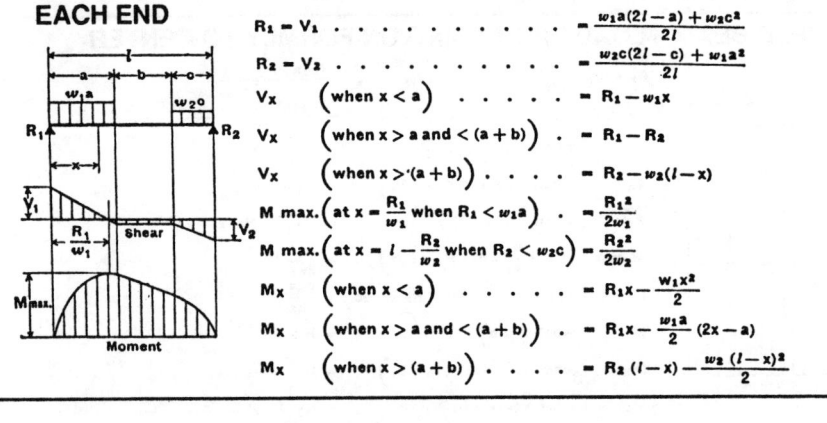

$R_1 = V_1$ . . . . $= \dfrac{w_1 a(2l-a) + w_2 c^2}{2l}$

$R_2 = V_2$ . . . . $= \dfrac{w_2 c(2l-c) + w_1 a^2}{2l}$

V_x (when $x < a$) . . . . $= R_1 - w_1 x$

V_x (when $x > a$ and $< (a+b)$) . $= R_1 - R_2$

V_x (when $x > (a+b)$) . . . . $= R_2 - w_2(l-x)$

M max. (at $x = \dfrac{R_1}{w_1}$ when $R_1 < w_1 a$) . $= \dfrac{R_1^2}{2w_1}$

M max. (at $x = l - \dfrac{R_2}{w_2}$ when $R_2 < w_2 c$) $= \dfrac{R_2^2}{2w_2}$

M_x (when $x < a$) . . . . . $= R_1 x - \dfrac{w_1 x^2}{2}$

M_x (when $x > a$ and $< (a+b)$) . $= R_1 x - \dfrac{w_1 a}{2}(2x-a)$

M_x (when $x > (a+b)$) . . . . $= R_2(l-x) - \dfrac{w_2(l-x)^2}{2}$

7. SIMPLE BEAM—CONCENTRATED LOAD AT CENTER

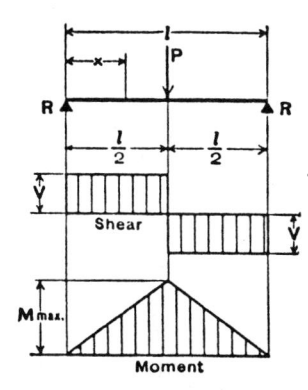

Equivalent Tabular Load . . . . . . $= 2P$

$R = V$ . . . . $= \dfrac{P}{2}$

M max. (at point of load) . . . . . $= \dfrac{Pl}{4}$

M_x (when $x < \dfrac{l}{2}$) . . . . . . $= \dfrac{Px}{2}$

Δmax. (at point of load) . . . . . $= \dfrac{Pl^3}{48EI}$

Δ_x (when $x < \dfrac{l}{2}$) . . . . . $= \dfrac{Px}{48EI}(3l^2 - 4x^2)$

8. SIMPLE BEAM—CONCENTRATED LOAD AT ANY POINT

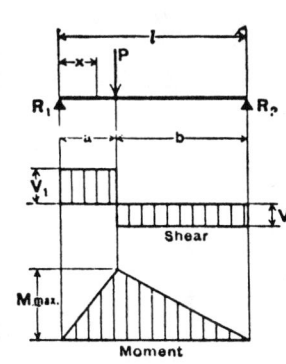

Equivalent Tabular Load . . . . . . $= \dfrac{8 Pab}{l^2}$

$R_1 = V_1$ (max. when $a < b$) . . . . $= \dfrac{Pb}{l}$

$R_2 = V_2$ (max. when $a > b$) . . . . $= \dfrac{Pa}{l}$

M max. (at point of load) . . . . $= \dfrac{Pab}{l}$

M_x (when $x < a$) . . . . . $= \dfrac{Pbx}{l}$

Δmax. (at $x = \sqrt{\dfrac{a(a+2b)}{3}}$ when $a > b$) $= \dfrac{Pab(a+2b)\sqrt{3a(a+2b)}}{27\,EI\,l}$

Δa (at point of load) . . . . . $= \dfrac{Pa^2b^2}{3EI\,l}$

Δ_x (when $x < a$) . . . . . $= \dfrac{Pbx}{6EI\,l}(l^2 - b^2 - x^2)$

9. SIMPLE BEAM—TWO EQUAL CONCENTRATED LOADS SYMMETRICALLY PLACED

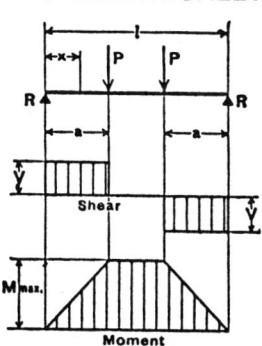

Equivalent Tabular Load . . . . . . $= \dfrac{8 Pa}{l}$

$R = V$ . . . . $= P$

M max. (between loads) . . . . . $= Pa$

M_x (when $x < a$) . . . . . . $= Px$

Δmax. (at center) . . . . $= \dfrac{Pa}{24EI}(3l^2 - 4a^2)$

Δ_x (when $x < a$) . . . . . $= \dfrac{Px}{6EI}(3la - 3a^2 - x^2)$

Δ_x (when $x > a$ and $< (l-a)$) . . $= \dfrac{Pa}{6EI}(3lx - 3x^2 - a^2)$

10. SIMPLE BEAM—TWO EQUAL CONCENTRATED LOADS UNSYMMETRICALLY PLACED

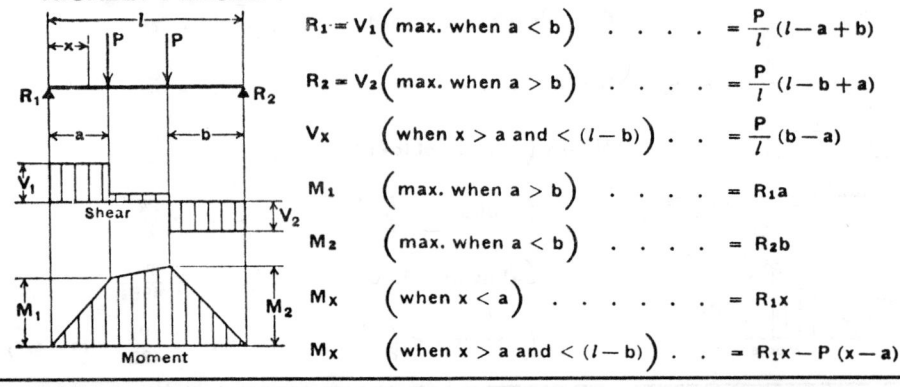

$R_1 = V_1$ (max. when $a < b$) . . . . $= \dfrac{P}{l}(l - a + b)$

$R_2 = V_2$ (max. when $a > b$) . . . . $= \dfrac{P}{l}(l - b + a)$

V_x (when $x > a$ and $< (l - b)$) . . $= \dfrac{P}{l}(b - a)$

M_1 (max. when $a > b$) . . . . $= R_1 a$

M_2 (max. when $a < b$) . . . . $= R_2 b$

M_x (when $x < a$) . . . . . . $= R_1 x$

M_x (when $x > a$ and $< (l - b)$) . . $= R_1 x - P(x - a)$

11. SIMPLE BEAM—TWO UNEQUAL CONCENTRATED LOADS UNSYMMETRICALLY PLACED

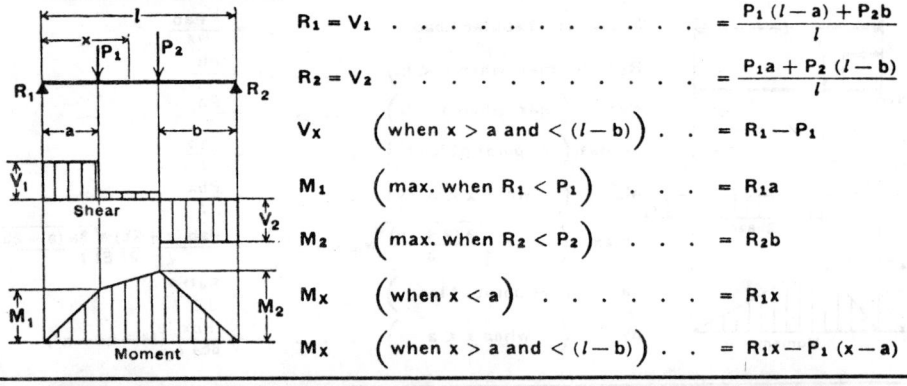

$R_1 = V_1$ . . . . $= \dfrac{P_1(l - a) + P_2 b}{l}$

$R_2 = V_2$ . . . . $= \dfrac{P_1 a + P_2(l - b)}{l}$

V_x (when $x > a$ and $< (l - b)$) . . $= R_1 - P_1$

M_1 (max. when $R_1 < P_1$) . . . $= R_1 a$

M_2 (max. when $R_2 < P_2$) . . . $= R_2 b$

M_x (when $x < a$) . . . . . . $= R_1 x$

M_x (when $x > a$ and $< (l - b)$) . . $= R_1 x - P_1(x - a)$

12. BEAM FIXED AT ONE END, SUPPORTED AT OTHER—UNIFORMLY DISTRIBUTED LOAD

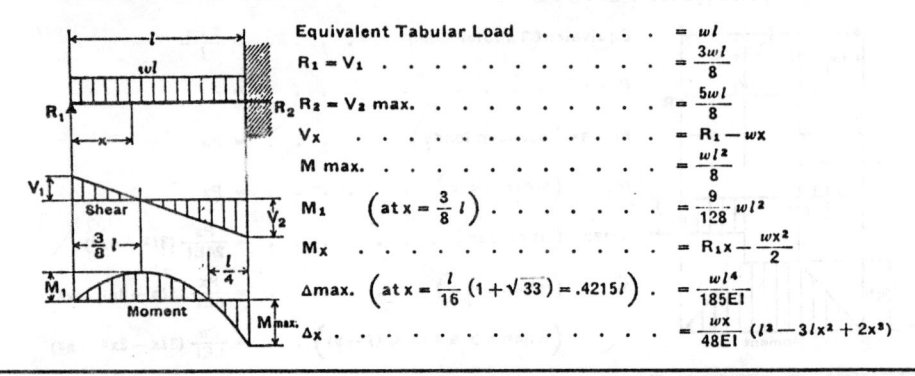

Equivalent Tabular Load . . . . . $= wl$

$R_1 = V_1$ . . . . $= \dfrac{3wl}{8}$

$R_2 = V_2$ max. . . . . $= \dfrac{5wl}{8}$

V_x . . . . $= R_1 - wx$

M max. . . . . $= \dfrac{wl^2}{8}$

M_1 (at $x = \dfrac{3}{8} l$) . . . . $= \dfrac{9}{128} wl^2$

M_x . . . . $= R_1 x - \dfrac{wx^2}{2}$

Δ max. (at $x = \dfrac{l}{16}(1 + \sqrt{33}) = .4215 l$) . $= \dfrac{wl^4}{185 EI}$

Δ_x . . . . $= \dfrac{wx}{48 EI}(l^3 - 3lx^2 + 2x^3)$

B-81

13. BEAM FIXED AT ONE END, SUPPORTED AT OTHER—CONCENTRATED LOAD AT CENTER

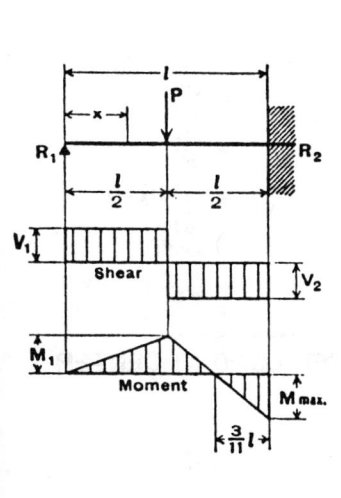

Equivalent Tabular Load $= \dfrac{3P}{2}$

$R_1 = V_1 \ldots = \dfrac{5P}{16}$

$R_2 = V_2$ max. $\ldots = \dfrac{11P}{16}$

M max. (at fixed end) $\ldots = \dfrac{3Pl}{16}$

M_1 (at point of load) $\ldots = \dfrac{5Pl}{32}$

M_x $\left(\text{when } x < \dfrac{l}{2}\right) \ldots = \dfrac{5Px}{16}$

M_x $\left(\text{when } x > \dfrac{l}{2}\right) \ldots = P\left(\dfrac{l}{2} - \dfrac{11x}{16}\right)$

Δmax. $\left(\text{at } x = l\sqrt{\dfrac{1}{5}} = .4472l\right) = \dfrac{Pl^3}{48EI\sqrt{5}} = .009317\dfrac{Pl^3}{EI}$

Δ_x (at point of load) $\ldots = \dfrac{7Pl^3}{768EI}$

Δ_x $\left(\text{when } x < \dfrac{l}{2}\right) \ldots = \dfrac{Px}{96EI}(3l^2 - 5x^2)$

Δ_x $\left(\text{when } x > \dfrac{l}{2}\right) \ldots = \dfrac{P}{96EI}(x-l)^2(11x - 2l)$

14. BEAM FIXED AT ONE END, SUPPORTED AT OTHER—CONCENTRATED LOAD AT ANY POINT

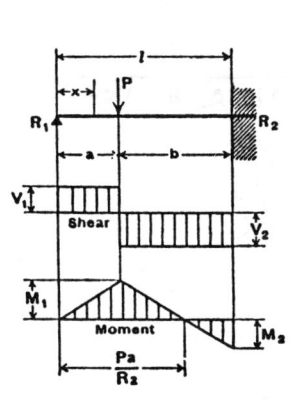

$R_1 = V_1 \ldots = \dfrac{Pb^2}{2l^3}(a + 2l)$

$R_2 = V_2 \ldots = \dfrac{Pa}{2l^3}(3l^2 - a^2)$

M_1 (at point of load) $\ldots = R_1 a$

M_2 (at fixed end) $\ldots = \dfrac{Pab}{2l^2}(a + l)$

M_x (when $x < a$) $\ldots = R_1 x$

M_x (when $x > a$) $\ldots = R_1 x - P(x-a)$

Δmax. $\left(\text{when } a < .414l \text{ at } x = l\dfrac{l^2 + a^2}{3l^2 - a^2}\right) = \dfrac{Pa}{3EI} \cdot \dfrac{(l^2 - a^2)^3}{(3l^2 - a^2)^2}$

Δmax. $\left(\text{when } a > .414l \text{ at } x = l\sqrt{\dfrac{a}{2l+a}}\right) = \dfrac{Pab^2}{6EI}\sqrt{\dfrac{a}{2l+a}}$

Δ_a (at point of load) $\ldots = \dfrac{Pa^2 b^3}{12EIl^3}(3l + a)$

Δ_x (when $x < a$) $\ldots = \dfrac{Pb^2 x}{12EIl^3}(3al^2 - 2lx^2 - ax^2)$

Δ_x (when $x > a$) $\ldots = \dfrac{Pa}{12EIl^3}(l-x)^2(3l^2x - a^2x - 2a^2l)$

15. BEAM FIXED AT BOTH ENDS—UNIFORMLY DISTRIBUTED LOADS

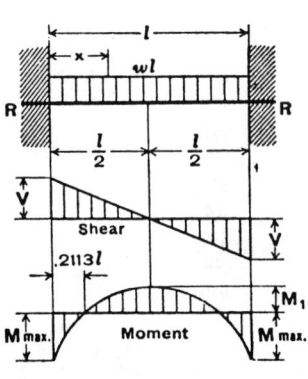

Equivalent Tabular Load $= \dfrac{2wl}{3}$

$R = V = \dfrac{wl}{2}$

$V_x = w\left(\dfrac{l}{2} - x\right)$

M max. (at ends) $= \dfrac{wl^2}{12}$

M_1 (at center) $= \dfrac{wl^2}{24}$

$M_x = \dfrac{w}{12}(6lx - l^2 - 6x^2)$

Δ max. (at center) $= \dfrac{wl^4}{384EI}$

$\Delta_x = \dfrac{wx^2}{24EI}(l - x)^2$

16. BEAM FIXED AT BOTH ENDS—CONCENTRATED LOAD AT CENTER

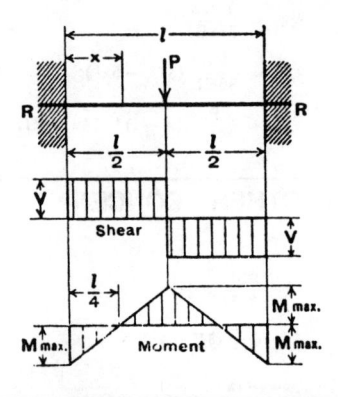

Equivalent Tabular Load $= P$

$R = V = \dfrac{P}{2}$

M max. (at center and ends) $= \dfrac{Pl}{8}$

M_x (when $x < \dfrac{l}{2}$) $= \dfrac{P}{8}(4x - l)$

Δ max. (at center) $= \dfrac{Pl^3}{192EI}$

$\Delta_x = \dfrac{Px^2}{48EI}(3l - 4x)$

17. BEAM FIXED AT BOTH ENDS—CONCENTRATED LOAD AT ANY POINT

$R_1 = V_1$ (max. when $a < b$) $= \dfrac{Pb^2}{l^3}(3a + b)$

$R_2 = V_2$ (max. when $a > b$) $= \dfrac{Pa^2}{l^3}(a + 3b)$

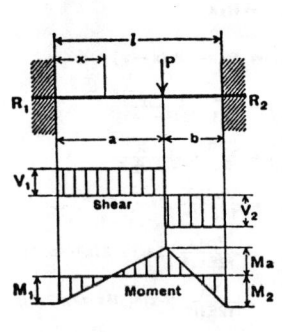

M_1 (max. when $a < b$) $= \dfrac{Pab^2}{l^2}$

M_2 (max. when $a > b$) $= \dfrac{Pa^2b}{l^2}$

M_a (at point of load) $= \dfrac{2Pa^2b^2}{l^3}$

M_x (when $x < a$) $= R_1 x - \dfrac{Pab^2}{l^2}$

Δ max. (when $a > b$ at $x = \dfrac{2al}{3a+b}$) $= \dfrac{2Pa^3b^2}{3EI(3a+b)^2}$

Δ_a (at point of load) $= \dfrac{Pa^3b^3}{3EIl^3}$

Δ_x (when $x < a$) $= \dfrac{Pb^2x^2}{6EIl^3}(3al - 3ax - bx)$

B-83

18. CANTILEVER BEAM—LOAD INCREASING UNIFORMLY TO FIXED END

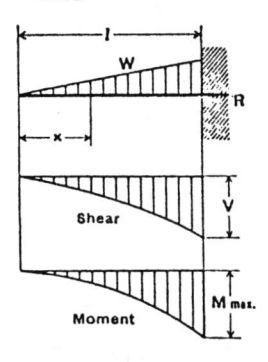

Equivalent Tabular Load $= \frac{8}{3} W$

$R = V = W$

$V_x = W \frac{x^2}{l^2}$

M max. (at fixed end) $= \frac{Wl}{3}$

$M_x = \frac{Wx^3}{3l^2}$

Δmax. (at free end) $= \frac{Wl^3}{15EI}$

$\Delta_x = \frac{W}{60EIl^2}(x^5 - 5l^4 x + 4l^5)$

19. CANTILEVER BEAM—UNIFORMLY DISTRIBUTED LOAD

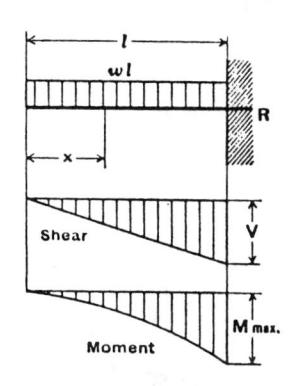

Equivalent Tabular Load $= 4wl$

$R = V = wl$

$V_x = wx$

M max. (at fixed end) $= \frac{wl^2}{2}$

$M_x = \frac{wx^2}{2}$

Δmax. (at free end) $= \frac{wl^4}{8EI}$

$\Delta_x = \frac{w}{24EI}(x^4 - 4l^3 x + 3l^4)$

20. BEAM FIXED AT ONE END, FREE BUT GUIDED AT OTHER—UNIFORMLY DISTRIBUTED LOAD

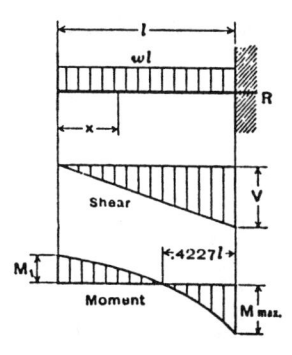

The deflection at the guided end is assumed to be in a vertical plane.

Equivalent Tabular Load $= \frac{8}{3} wl$

$R = V = wl$

$V_x = wx$

M max. (at fixed end) $= \frac{wl^2}{3}$

M_1 (at guided end) $= \frac{wl^2}{6}$

$M_x = \frac{w}{6}(l^2 - 3x^2)$

Δmax. (at guided end) $= \frac{wl^4}{24EI}$

$\Delta_x = \frac{w(l^2 - x^2)^2}{24EI}$

B-84

21. CANTILEVER BEAM—CONCENTRATED LOAD AT ANY POINT

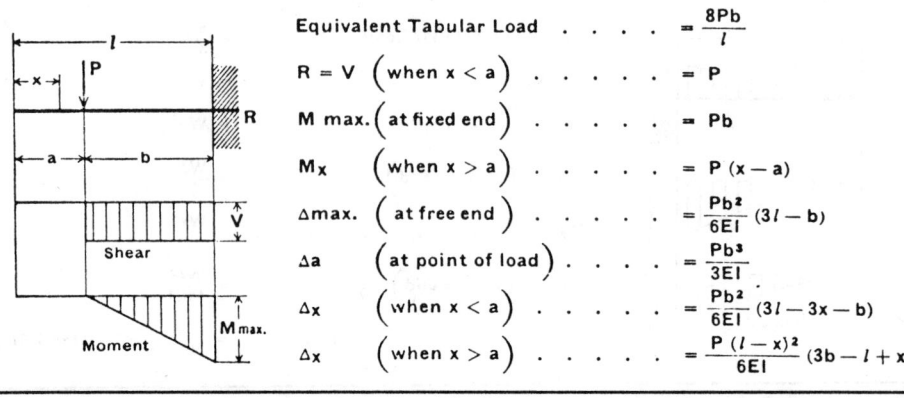

Equivalent Tabular Load $= \dfrac{8Pb}{l}$

$R = V$ (when $x < a$) $= P$

M max. (at fixed end) $= Pb$

M_x (when $x > a$) $= P(x - a)$

Δmax. (at free end) $= \dfrac{Pb^2}{6EI}(3l - b)$

Δa (at point of load) $= \dfrac{Pb^3}{3EI}$

Δ_x (when $x < a$) $= \dfrac{Pb^2}{6EI}(3l - 3x - b)$

Δ_x (when $x > a$) $= \dfrac{P(l-x)^2}{6EI}(3b - l + x)$

22. CANTILEVER BEAM—CONCENTRATED LOAD AT FREE END

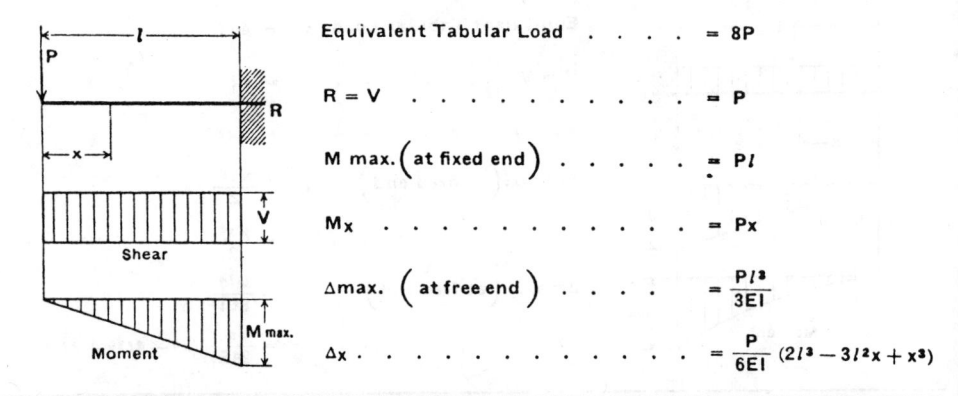

Equivalent Tabular Load $= 8P$

$R = V = P$

M max. (at fixed end) $= Pl$

$M_x = Px$

Δmax. (at free end) $= \dfrac{Pl^3}{3EI}$

$\Delta_x = \dfrac{P}{6EI}(2l^3 - 3l^2 x + x^3)$

23. BEAM FIXED AT ONE END, FREE BUT GUIDED AT OTHER—CONCENTRATED LOAD AT GUIDED END

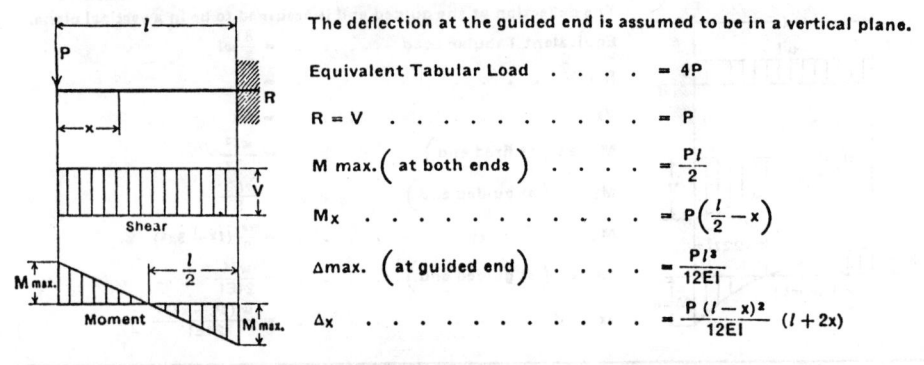

The deflection at the guided end is assumed to be in a vertical plane.

Equivalent Tabular Load $= 4P$

$R = V = P$

M max. (at both ends) $= \dfrac{Pl}{2}$

$M_x = P\left(\dfrac{l}{2} - x\right)$

Δmax. (at guided end) $= \dfrac{Pl^3}{12EI}$

$\Delta_x = \dfrac{P(l-x)^2}{12EI}(l + 2x)$

24. BEAM OVERHANGING ONE SUPPORT—UNIFORMLY DISTRIBUTED LOAD

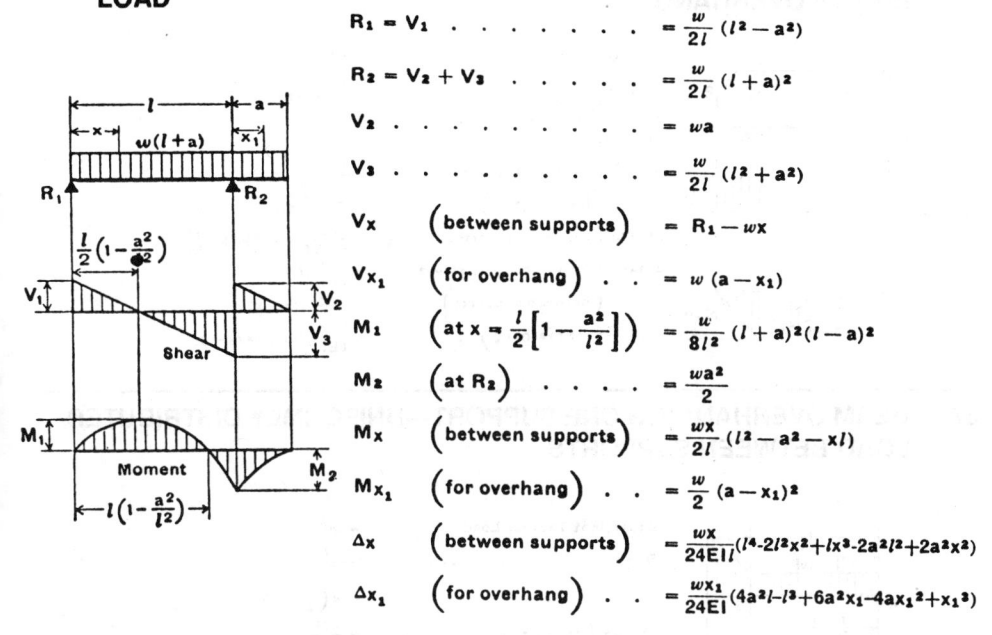

$R_1 = V_1$ . . . . $= \dfrac{w}{2l}(l^2 - a^2)$

$R_2 = V_2 + V_3$ . . . . . $= \dfrac{w}{2l}(l + a)^2$

V_2 . . . . $= wa$

V_3 . . . . $= \dfrac{w}{2l}(l^2 + a^2)$

V_x (between supports) $= R_1 - wx$

V_{x_1} (for overhang) . . $= w(a - x_1)$

$M_1 \left(\text{at } x = \dfrac{l}{2}\left[1 - \dfrac{a^2}{l^2}\right]\right) = \dfrac{w}{8l^2}(l + a)^2(l - a)^2$

M_2 (at R_2) . . . . $= \dfrac{wa^2}{2}$

M_x (between supports) $= \dfrac{wx}{2l}(l^2 - a^2 - xl)$

M_{x_1} (for overhang) . . $= \dfrac{w}{2}(a - x_1)^2$

Δ_x (between supports) $= \dfrac{wx}{24EIl}(l^4 - 2l^2x^2 + lx^3 - 2a^2l^2 + 2a^2x^2)$

Δ_{x_1} (for overhang) . . $= \dfrac{wx_1}{24EI}(4a^2l - l^3 + 6a^2x_1 - 4ax_1^2 + x_1^3)$

25. BEAM OVERHANGING ONE SUPPORT—UNIFORMLY DISTRIBUTED LOAD ON OVERHANG

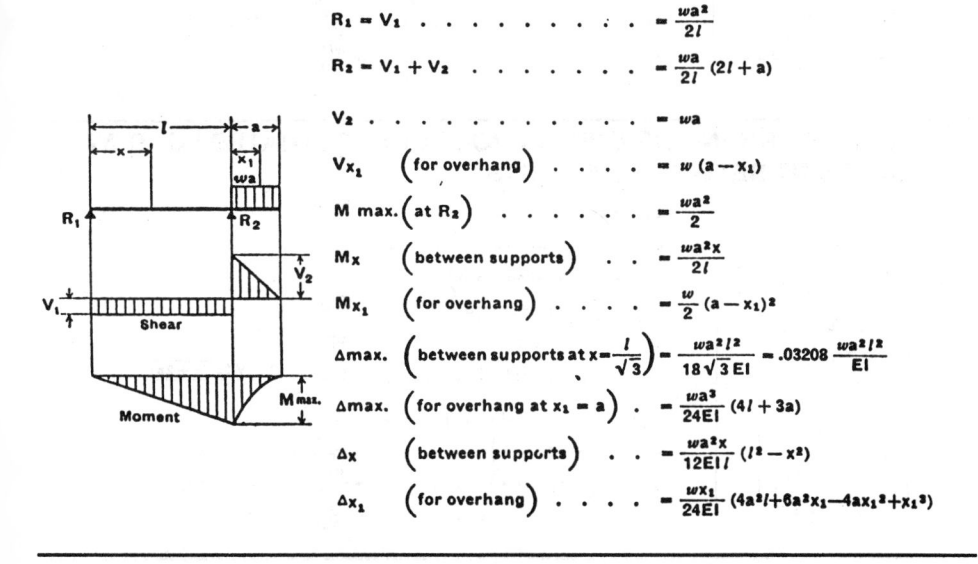

$R_1 = V_1$ . . . . $= \dfrac{wa^2}{2l}$

$R_2 = V_1 + V_2$ . . . . . . $= \dfrac{wa}{2l}(2l + a)$

V_2 . . . . $= wa$

V_{x_1} (for overhang) . . . $= w(a - x_1)$

M max. (at R_2) . . . . . $= \dfrac{wa^2}{2}$

M_x (between supports) . . $= \dfrac{wa^2 x}{2l}$

M_{x_1} (for overhang) . . . . $= \dfrac{w}{2}(a - x_1)^2$

Δmax. $\left(\text{between supports at } x = \dfrac{l}{\sqrt{3}}\right) = \dfrac{wa^2 l^2}{18\sqrt{3}\,EI} = .03208 \dfrac{wa^2 l^2}{EI}$

Δmax. (for overhang at $x_1 = a$) . $= \dfrac{wa^3}{24EI}(4l + 3a)$

Δ_x (between supports) . . $= \dfrac{wa^2 x}{12EIl}(l^2 - x^2)$

Δ_{x_1} (for overhang) . . . . $= \dfrac{wx_1}{24EI}(4a^2l + 6a^2 x_1 - 4ax_1^2 + x_1^3)$

26. BEAM OVERHANGING ONE SUPPORT—CONCENTRATED LOAD AT END OF OVERHANG

$R_1 = V_1 \ldots \ldots \ldots = \dfrac{Pa}{l}$

$R_2 = V_1 + V_2 \ldots \ldots = \dfrac{P}{l}(l+a)$

$V_2 \ldots \ldots \ldots \ldots = P$

M max. (at R_2) $\ldots \ldots = Pa$

M_x (between supports) $\ldots = \dfrac{Pax}{l}$

M_{x_1} (for overhang) $\ldots \ldots = P(a - x_1)$

Δmax. (between supports at $x = \dfrac{l}{\sqrt{3}}$) $= \dfrac{Pal^2}{9\sqrt{3}EI} = .06415 \dfrac{Pal^2}{EI}$

Δmax. (for overhang at $x_1 = a$) $= \dfrac{Pa^2}{3EI}(l+a)$

Δ_x (between supports) $\ldots = \dfrac{Pax}{6EIl}(l^2 - x^2)$

Δ_{x_1} (for overhang) $\ldots \ldots = \dfrac{Px_1}{6EI}(2al + 3ax_1 - x_1^2)$

27. BEAM OVERHANGING ONE SUPPORT—UNIFORMLY DISTRIBUTED LOAD BETWEEN SUPPORTS

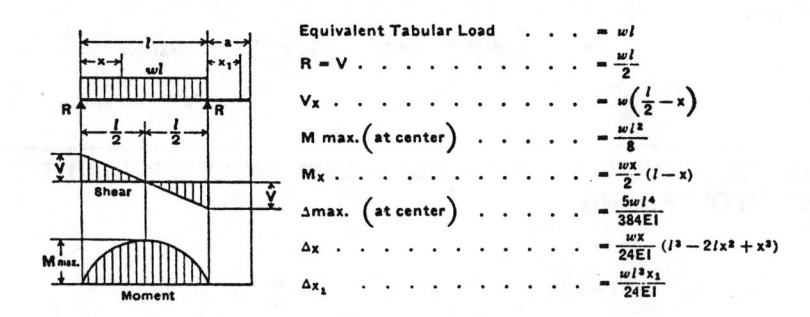

Equivalent Tabular Load $\ldots = wl$

$R = V \ldots \ldots \ldots \ldots = \dfrac{wl}{2}$

$V_x \ldots \ldots \ldots \ldots \ldots = w\left(\dfrac{l}{2} - x\right)$

M max. (at center) $\ldots \ldots = \dfrac{wl^2}{8}$

$M_x \ldots \ldots \ldots \ldots \ldots = \dfrac{wx}{2}(l - x)$

Δmax. (at center) $\ldots \ldots = \dfrac{5wl^4}{384EI}$

$\Delta_x \ldots \ldots \ldots \ldots \ldots = \dfrac{wx}{24EI}(l^3 - 2lx^2 + x^3)$

$\Delta_{x_1} \ldots \ldots \ldots \ldots \ldots = \dfrac{wl^3 x_1}{24EI}$

28. BEAM OVERHANGING ONE SUPPORT-CONCENTRATED LOAD AT ANY POINT BETWEEN SUPPORTS

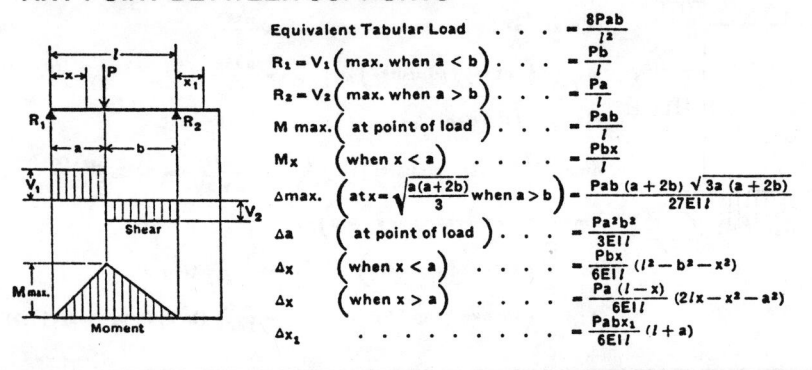

Equivalent Tabular Load $\ldots = \dfrac{8Pab}{l^2}$

$R_1 = V_1$ (max. when $a < b$) $\ldots = \dfrac{Pb}{l}$

$R_2 = V_2$ (max. when $a > b$) $\ldots = \dfrac{Pa}{l}$

M max. (at point of load) $\ldots = \dfrac{Pab}{l}$

M_x (when $x < a$) $\ldots \ldots = \dfrac{Pbx}{l}$

Δmax. $\left(\text{at } x = \sqrt{\dfrac{a(a+2b)}{3}} \text{ when } a > b\right) = \dfrac{Pab(a+2b)\sqrt{3a(a+2b)}}{27EIl}$

Δa (at point of load) $\ldots = \dfrac{Pa^2b^2}{3EIl}$

Δ_x (when $x < a$) $\ldots \ldots = \dfrac{Pbx}{6EIl}(l^2 - b^2 - x^2)$

Δ_x (when $x > a$) $\ldots \ldots = \dfrac{Pa(l-x)}{6EIl}(2lx - x^2 - a^2)$

$\Delta_{x_1} \ldots \ldots \ldots \ldots \ldots = \dfrac{Pabx_1}{6EIl}(l+a)$

B-87

29. CONTINUOUS BEAM—TWO EQUAL SPANS—UNIFORM LOAD ON ONE SPAN

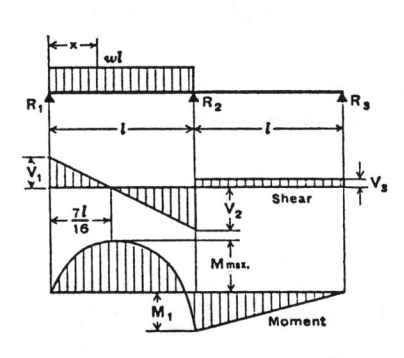

Equivalent Tabular Load $= \frac{49}{64}wl$

$R_1 = V_1 = \frac{7}{16}wl$

$R_2 = V_2 + V_3 = \frac{5}{8}wl$

$R_3 = V_3 = -\frac{1}{16}wl$

$V_2 = \frac{9}{16}wl$

M Max. $\left(\text{at } x = \frac{7}{16}l\right) = \frac{49}{512}wl^2$

M_1 (at support R_2) $= \frac{1}{16}wl^2$

M_x (when $x < l$) $= \frac{wx}{16}(7l - 8x)$

30. CONTINUOUS BEAM—TWO EQUAL SPANS—CONCENTRATED LOAD AT CENTER OF ONE SPAN

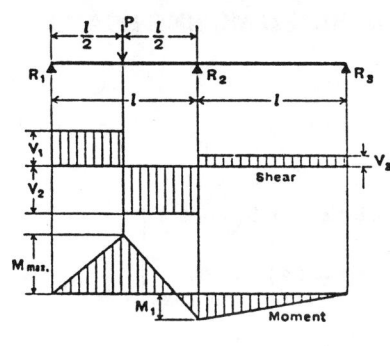

Equivalent Tabular Load $= \frac{13}{8}P$

$R_1 = V_1 = \frac{13}{32}P$

$R_2 = V_2 + V_3 = \frac{11}{16}P$

$R_3 = V_3 = -\frac{3}{32}P$

$V_2 = \frac{19}{32}P$

M Max. (at point of load) $= \frac{13}{64}Pl$

M_1 (at support R_2) $= \frac{3}{32}Pl$

31. CONTINUOUS BEAM—TWO EQUAL SPANS—CONCENTRATED LOAD AT ANY POINT

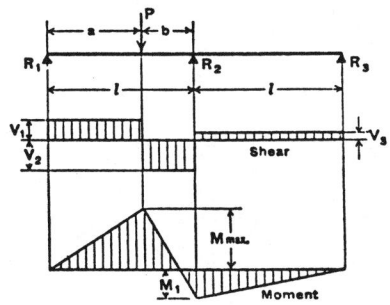

$R_1 = V_1 = \frac{Pb}{4l^3}\left(4l^2 - a(l+a)\right)$

$R_2 = V_2 + V_3 = \frac{Pa}{2l^3}\left(2l^2 + b(l+a)\right)$

$R_3 = V_3 = -\frac{Pab}{4l^3}(l+a)$

$V_2 = \frac{Pa}{4l^3}\left(4l^2 + b(l+a)\right)$

M max. (at point of load) $= \frac{Pab}{4l^3}\left(4l^2 - a(l+a)\right)$

M_1 (at support R_2) $= \frac{Pab}{4l^2}(l+a)$

B-88

7.5.2 FOR VARIOUS CONCENTRATED MOVING LOADS

The values given in these formulas do not include impact which varies according to the requirements of each case.

---

32. SIMPLE BEAM—ONE CONCENTRATED MOVING LOAD

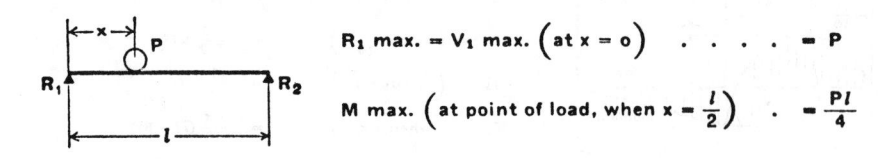

R_1 max. $= V_1$ max. $\left(\text{at } x = 0\right)$ . . . . $= P$

M max. $\left(\text{at point of load, when } x = \dfrac{l}{2}\right)$. $= \dfrac{Pl}{4}$

---

33. SIMPLE BEAM—TWO EQUAL CONCENTRATED MOVING LOADS

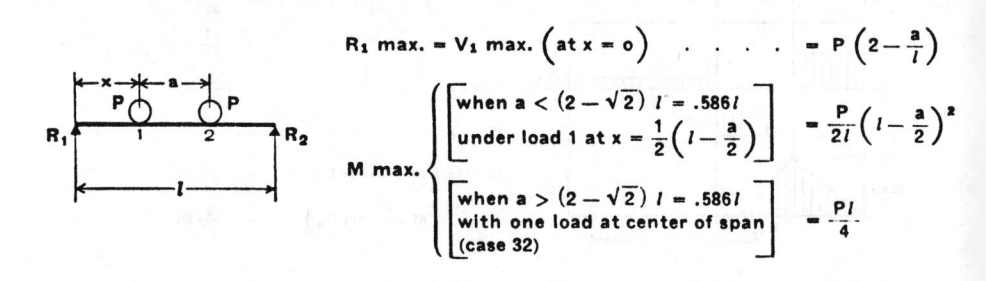

R_1 max. $= V_1$ max. $\left(\text{at } x = 0\right)$ . . . . $= P\left(2 - \dfrac{a}{l}\right)$

M max. $\begin{cases} \left[\begin{array}{l}\text{when } a < (2-\sqrt{2})\, l = .586\, l \\ \text{under load 1 at } x = \dfrac{1}{2}\left(l - \dfrac{a}{2}\right)\end{array}\right] = \dfrac{P}{2l}\left(l - \dfrac{a}{2}\right)^2 \\[2em] \left[\begin{array}{l}\text{when } a > (2-\sqrt{2})\, l = .586\, l \\ \text{with one load at center of span} \\ \text{(case 32)}\end{array}\right] = \dfrac{Pl}{4} \end{cases}$

---

34. SIMPLE BEAM—TWO EQUAL CONCENTRATED MOVING LOADS

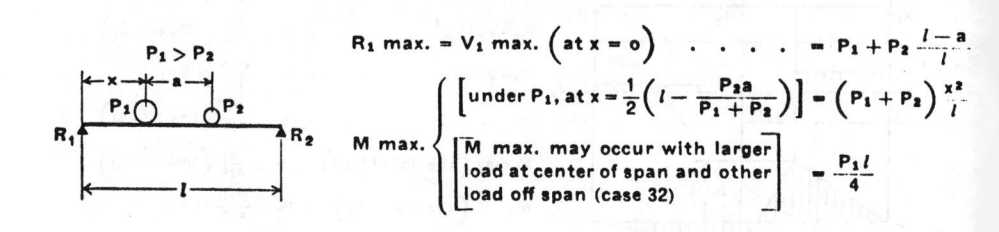

$P_1 > P_2$

R_1 max. $= V_1$ max. $\left(\text{at } x = 0\right)$ . . . . $= P_1 + P_2 \dfrac{l-a}{l}$

M max. $\begin{cases} \left[\text{under } P_1, \text{at } x = \dfrac{1}{2}\left(l - \dfrac{P_2 a}{P_1 + P_2}\right)\right] = (P_1 + P_2)\dfrac{x^2}{l} \\[1em] \left[\begin{array}{l}\text{M max. may occur with larger} \\ \text{load at center of span and other} \\ \text{load off span (case 32)}\end{array}\right] = \dfrac{P_1 l}{4} \end{cases}$

GENERAL RULES FOR SIMPLE BEAMS CARRYING MOVING CONCENTRATED LOADS

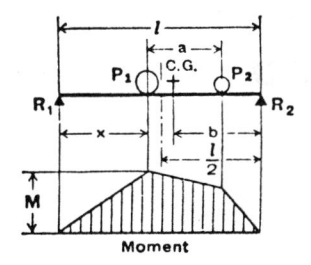

The maximum shear due to moving concentrated loads occurs at one support when one of the loads is at that support. With several moving loads, the location that will produce maximum shear must be determined by trial.

The maximum bending moment produced by moving concentrated loads occurs under one of the loads when that load is as far from one support as the center of gravity of all the moving loads on the beam is from the other support.

In the accompanying diagram, the maximum bending moment occurs under load P1 when $x = b$. It should also be noted that this condition occurs when the center line of the span is midway between the center of gravity of loads and the nearest concentrated load.

CHAPTER 8

DEFLECTION OF BEAMS

8.1 DEFLECTION AND ELASTIC CURVE

The distance traversed by a point on the neutral plane of an already straight beam in the same plane of loading is called the **deflection of the point.**

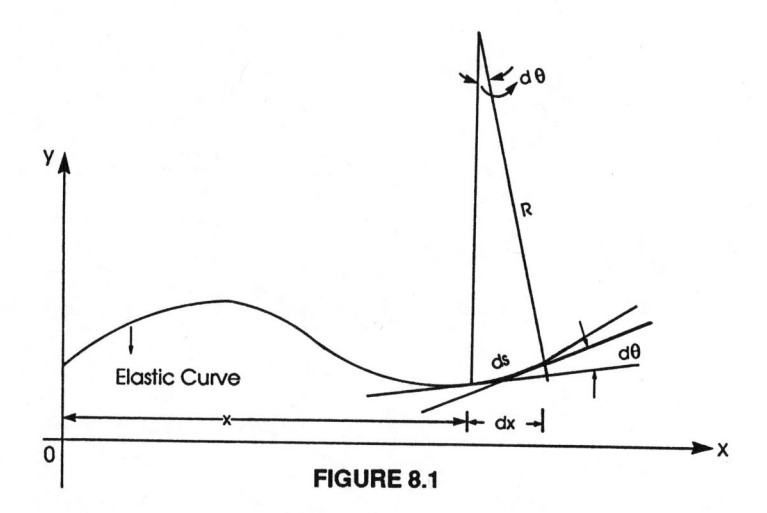

FIGURE 8.1

A deflection curve or an elastic curve is a line in the plane of loading collinear with the neutral axis in a bent shape.

8.2 DIFFERENTIAL EQUATIONS OF BEAM

In the elastic curve in Figure 8.1, consider the incremental arc dS which will cause an infinitesimal change in the slope, $d\theta$, so that

$$dS = R d\theta$$

$$\frac{d\theta}{dS} = \frac{1}{R} \qquad (8\text{-}1)$$

where R is the radius of curvature and $1/R$ is called curvature.

For small deflections (small angles),

$$\frac{dy}{dx} \approx \tan\theta \approx \theta, \quad dS \approx dx$$

Therefore,

$$\frac{d\theta}{dS} \approx \frac{d\theta}{dx} \approx \frac{d^2 y}{dx^2} \qquad (8\text{-}2)$$

Comparing equations (8-1) and (8-2)

$$\frac{d^2 y}{dx^2} = \frac{1}{R} \qquad (8\text{-}3)$$

Now, using the flexure formula:

$$\frac{M}{I} = \frac{\sigma}{Y} = \frac{E}{R}$$

or

$$\frac{E}{R} = \frac{M}{I}$$

$$\frac{1}{R} = \frac{M}{EI} \qquad (8\text{-}4)$$

Comparing equations (8-3) and (8-4)

$$\frac{M}{EI} = \frac{d^2y}{dx^2}$$

or

$$M = EI \frac{d^2y}{dx^2} \quad (8\text{-}5)$$

This differential equation together with the boundary conditions are used to determine the deflection y of the beam at any point, of distant x, from the origin. We can also find the shear load Q by further differentiating the moment M, i.e.,

$$Q = \frac{dM}{dx} = \frac{d}{dx}\left(EI \frac{d^2y}{dx^2}\right)$$

$$\therefore \quad Q = EI \frac{d^3y}{dx^3} \quad (8\text{-}6)$$

The above facts can be summarized as follows:

Deflection: $\quad \Delta = y$

Slope: $\quad \tan\theta \approx \theta \approx \frac{dy}{dx}$

where θ is very small.

Bending Moment: $\quad M = EI \frac{d^2y}{dx^2}$

Curvature: $\quad \frac{1}{R} = \frac{d^2y}{dx^2}$

Bending Moment: $\quad M = \frac{EI}{R} = EI \frac{d^2y}{dx^2}$

Shear force: $$Q = \frac{dM}{dx} = EI\frac{d^3y}{dx^3}$$

Load: $$W = \int Q\,dx$$

In a standard calculus text, the expression for curvature is also given as:

$$\frac{1}{R} = \frac{d^2y/dx^2}{\left[1 + \left(\frac{dy}{dx}\right)^2\right]^{\frac{3}{2}}} \qquad (8\text{-}7)$$

The sign convention follows that the moment is positive for a beam bent downward or when the elastic curve is concave upward. Since curvature corresponds to the moment, thus a positive moment yields a positive curvature, thus giving a positive value of $\frac{d^2y}{dx^2}$. The slope, $\frac{dy}{dx}$ is also positive, corresponding to a positive moment.

8.3 DETERMINATION OF DEFLECTION

The deflection in the beams can be determined by the following three methods:

8.3.1 THE INTEGRATION METHOD

According to this method, the deflection of a beam can be determined by integrating the moment equation twice, directly from the free body diagram analysis. The constants of integrations are determined by using appropriate boundary conditions. Some interesting examples of calculating deflections in beams that are usually encountered in structural mechanics are given as follows:

SIMPLY SUPPORTED BEAM

Considering the free body diagram in Figure 8.2(a) and writing the equation for equilibrium of moments,

$$M - \frac{wL}{2}x + \frac{wx^2}{2} = 0$$

where M is the resisting moment.

Now,

$$M = \frac{wL}{2}x - \frac{wx^2}{2}$$

$$M = \frac{w}{2}x(L - x)$$

@ $x = 0, \quad M = 0$

$x = L, \quad M = \frac{w}{2}L(L - L) = 0$

$x = \frac{L}{2}, \quad M = \frac{w}{2} \cdot \frac{L}{2}\left(L - \frac{L}{2}\right)$

or $M = \frac{wL^2}{8}$

Thus the bending moment is zero at both ends and has a greatest value at the center with a parabolic variation along the length of the beam. (Refer to Figure 8.2(b).)

From equation (8-5),

$$M = EI\frac{d^2y}{dx^2}$$

FIGURE 8.2

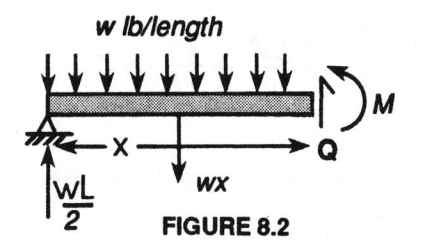

or

$$\frac{d^2y}{dx^2} = \frac{M}{EI} = \frac{1}{EI} \frac{w}{2} (Lx - x^2)$$

On integration,

slope, $$\frac{dy}{dx} = \frac{1}{EI} \int \frac{w}{2} (Lx - x^2) \, dx + C_1$$

or

$$\frac{dy}{dx} = \frac{w}{2EI} \left(\frac{Lx^2}{2} - \frac{x^3}{3} \right) + C_1$$

Further integration yields,

$$y = \frac{w}{2EI} \int \left(\frac{Lx^2}{2} - \frac{x^3}{3} \right) dx + C_1 x + C_2$$

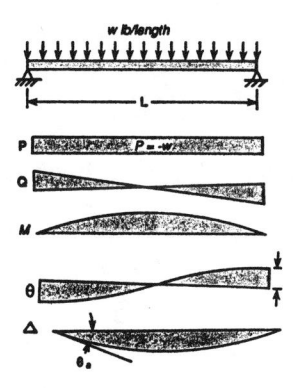

FIGURE 8.3

$$y = \frac{w}{2EI}\left(\frac{Lx^3}{6} - \frac{x^4}{12}\right) + C_1 x + C_2 \qquad (8\text{-}8)$$

The following boundary conditions are used to eliminate the constants of integration,

(i) at $x = 0$, $y = 0$

(ii) $x = L$, $y = 0$.

For the first B.C., $C_2 = 0$.

For the second B.C., we get,

$$0 = \frac{w}{2EI}\left[\frac{L^4}{6} - \frac{L^4}{12}\right] + C_1 L$$

or

$$0 = \frac{w}{12EI}\left[L^3 - \frac{L^3}{2}\right] + C_1$$

$$0 = \frac{w}{12EI}\left[\frac{2L^3 - L^3}{2}\right] + C_1$$

$$\therefore \quad C_1 = -\frac{wL^3}{24EI}$$

Substituting the values of C_1 and C_2 back into equation (8-8) we obtain

$$\boxed{y = \frac{w}{2EI}\left(\frac{Lx^3}{6} - \frac{x^4}{12} - \frac{L^3 x}{12}\right)} \qquad (8\text{-}9)$$

The deflection is maximum at $x = L/2$, hence

$$\Delta = \frac{w}{2EI}\left[\frac{L}{6}\left(\frac{L}{2}\right)^3 - \frac{1}{12}\left(\frac{L}{2}\right)^4 - \frac{L^3}{12}\left(\frac{L}{2}\right)\right]$$

$$= \frac{w}{2EI}\left[\frac{L^4}{48} - \frac{L^4}{48\times 4} - \frac{L^4}{24}\right]$$

$$\therefore \boxed{\Delta = -\frac{5wL^4}{384EI}} \qquad (8\text{-}10)$$

CANTILEVER BEAM SUPPORTING A TRIANGULARLY DISTRIBUTED LOAD OF MAXIMUM INTENSITY

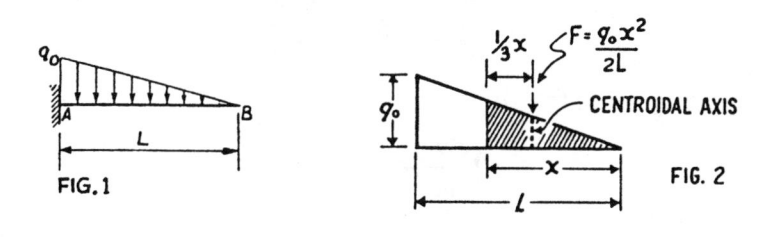

FIGURE 8.4

Again, as in the previous case, this problem will be solved by using the differential equation of curvature for elastic beams. To use this equation, the moment as a function of x is needed.

To find the moment an imaginary section cut is made at x (See Figure 8.4(b)). The internal moment acting in the beam at the cut is found by moment equilibrium. The intensity of the distributed load at x is $q_0 x/L$. The average magnitude of the distributed load in the shaded section is $q_0 x/2L$. The weight of the shaded portion of the distributed loading is the average intensity times the length.

$$F = \frac{q_0 x^2}{2L}$$

The moment about point x is the force times the distance to the centroidal axis of the distributed loading from x. For a triangle the centroidal axis is one-third of the way from the base to the apex. The moment is thus

$$\sum M_x = 0 = M - \frac{q_0 x^2}{2L}\left(\frac{x}{3}\right)$$

$$M = \frac{q_0 x^3}{6L}$$

The curvature equation can now be employed.

$$EI\frac{d^2 y}{dx^2} = M = \frac{q_0 x^3}{6L}$$

Integrating gives the slope.

$$EI\frac{dy}{dx} = \frac{q_0 x^4}{24L} + c_1$$

The integration constant can be found using a boundary condition. At the wall, the slope is zero. Therefore at $x = L$, $dy/dx = 0$. Substituting into the slope equation

$$0 = \frac{q_0 L^4}{24L} + c_1$$

Solving for c_1

$$c_1 = -\frac{q_0 L^3}{24}$$

The equation of the slope is thus

$$EI\frac{dy}{dx} = \frac{q_0 x^4}{24L} - \frac{q_0 L^3}{24}$$

Integrating the slope gives the deflection

$$EIy = \frac{q_0 x^5}{120L} - \frac{q_0 L^3 x}{24} + c_2$$

The integration constant, c_2, can also be found using boundary conditions. At the wall the deflection is zero. Therefore at $x = L$, $y = 0$. Substituting

$$0 = \frac{q_0 L^5}{120L} - \frac{q_0 L^3 (L)}{24} + c_2$$

$$c_2 = \frac{4 q_0 L^4}{120} = \frac{q_0 L^4}{30}$$

The deflection equation is thus

$$EIy = \frac{q_0 x^5}{120L} - \frac{q_0 L^3 x}{24} + \frac{q_0 L^4}{30}$$

The statement of the problem asks for formulas for the slope and the deflection at the free end. At the free end $x = 0$. Substituting into the equations for slope

$$EI \frac{dy}{dx} = -\frac{q_0 L^3}{24}$$

$$\frac{dy}{dx} = -\frac{q_0 L^3}{24 EI} \qquad \text{(slope)}$$

and for the deflection

$$EIy = \frac{q_0 L^4}{30}$$

$$y = \frac{q_0 L^4}{30 EI} \qquad \text{(deflection)}$$

8.3.2 MOMENT AREA METHOD

This method is more helpful for the beams with concentrated loads, in which case discontinuities occur in the M, V, θ, and Δ diagrams. Refer to Section 9.4 for details.

Following are some illustrations of the application of the moment-area method.

8.3.2.1 DEFLECTION AND SLOPE OF A SIMPLY-SUPPORTED BEAM DUE TO THE CONCENTRATED LOAD P

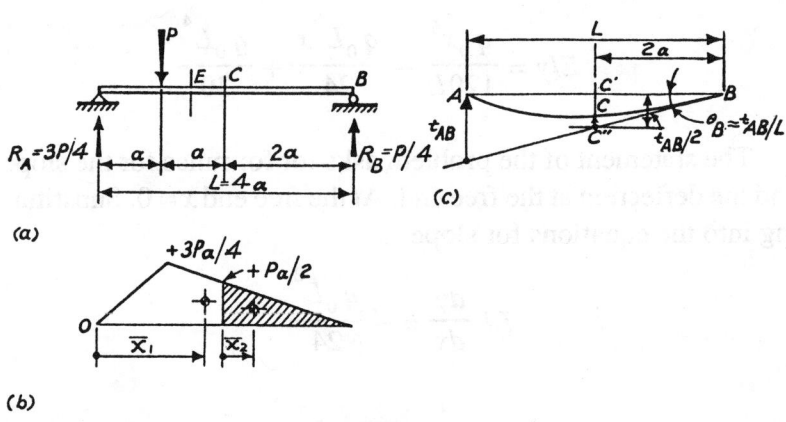

FIGURE 8.5

The bending-moment diagram is in Figure 8.5(a). Since EI is constant, the $M/(EI)$ diagram need not be made, as the areas of the bending-moment diagram divided by EI give the necessary quantities for use in the moment-area theorems. The elastic curve is in Figure 8.5(c). It is concave upward throughout its length as the bending moments are positive. This curve must pass through the points of the support at A and B.

B-101

It is apparent from the sketch of the elastic curve that the desired quantity is represented by the distance CC'. Moreover, from purely geometrical or kinematic considerations, $CC' = C'C'' - C''C$, where the distance $C''C$ is measured from a tangent to the elastic curve passing through the point of support B. However, since the deviation of a support point from a tangent to the elastic curve at the other support may always be computed by the second moment-area theorem, a distance such as $C'C''$ may be found by proportion from the geometry of the figure. In this case, t_{AB} follows by taking the whole $M/(EI)$ area between A and B and multiplying it by its $\overline{x}$ measured from a vertical through A, whence $C'C'' = \frac{1}{2} t_{AB}$. By another application of the second theorem, t_{CB}, which is equal to $C''C$, is determined. For this case, the $M/(EI)$ area is shaded in Figure 8.5(b), and, for it, the $\overline{x}$ is measured from C. Since the right reaction is $P/4$ and the distance $CB = 2a$, the maximum ordinate for the shaded triangle is $+Pa/2$.

$$v_C = C'C'' - C''C = (t_{AB}/2) - t_{CB}$$

$$t_{AB} = \Phi_1 \overline{x}_1 = \frac{1}{EI}\left(\frac{4a}{2} \cdot \frac{3Pa}{4}\right)\frac{(a+4a)}{3} = +\frac{5Pa^3}{2EI}$$

$$t_{CB} = \Phi_2 \overline{x}_2 = \frac{1}{EI}\left(\frac{2a}{2} \cdot \frac{Pa}{2}\right)\frac{(2a)}{3} = +\frac{Pa^3}{3EI}$$

$$v_C = \frac{t_{AB}}{2} - t_{CB} = \frac{5Pa^3}{4EI} - \frac{Pa^3}{3EI} = \frac{11Pa^3}{12EI}$$

The positive signs of t_{AB} and t_{CB} indicate that points A and C lie above the tangent through B. As may be seen from Figure 8.5(a), the deflection at the center of the beam is in a downward direction.

The slope of the elastic curve at C can be found from the slope at one of the ends. For point B on the right

$$\theta_B = \theta_C + \Delta\theta_{BC} \text{ or } \theta_C = \theta_B - \Delta\theta_{BC}$$

$$\theta_C = \frac{t_{AB}}{L} - \Phi_2 = \frac{5Pa^2}{8EI} - \frac{Pa^2}{2EI} = \frac{Pa^2}{8EI}$$

radians counterclockwise.

8.3.2.2 ANGLES OF ROTATION

Angles of rotation, θ_a and θ_b, at the ends of the beam and the deflection δ at the middle of a simple beam acted upon by couples M_0 at the ends:

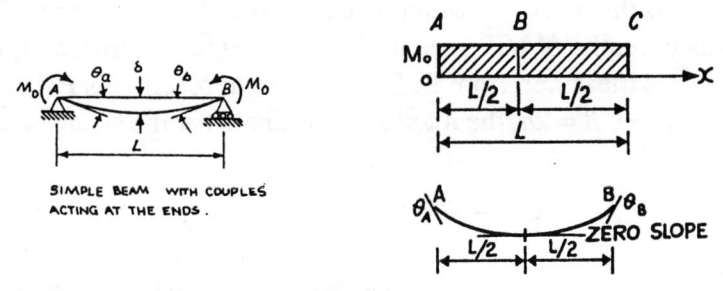

FIGURE 8.6

SOLUTION: This problem will be solved using the first and second moment area theorems.

The moment diagram is given in Figure 8.6(b). The slope at the mid-point of the beam, $x = L/2$, is zero (See Figure 8.6(c).

By the first moment-area theorem, the difference in slope between points A and B is

$$\theta_B - \theta_A = \int_A^B \frac{M}{EI} dx$$

Since θ_B is zero and the value of the integral is the area of the moment diagram (Figure 8.6(b)) between A and B ($M_0 L/2$) divided by EI the slope at A is

$$\theta_A = -\frac{M_0 L}{2EI}$$

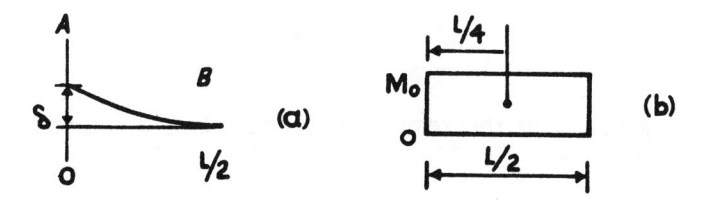

FIGURE 8.7

Again by the first moment-area theorem the difference in slope between points B and C is

$$\theta_C - \theta_B = \int_B^C \frac{M}{EI}\, dx$$

The area of the moment diagram Mx from B to C is $M_0 L/2$ and θ_B is zero. Substituting yields

$$\theta_C = \frac{M_0 L}{2EI}$$

By the second moment-area theorem, the deflection is from Figure 8.7.

$$\delta = \frac{A\bar{x}}{EI} = \frac{M_0\left(\frac{L}{2}\right)\left(\frac{L}{4}\right)}{EI} = \frac{M_0 L^2}{8EI}$$

The deflection at the middle of the beam is $M_0 L^2/8EI$ downwards.

The results obtained can be checked by referring to a table of beam deflections and slopes. If the case of a simply supported beam with couples acting at both ends is not listed, the formulas for a beam with a couple acting at only one end can be modified using superposition. This results in

$$-\theta_a = \theta_b = \frac{M_0 L}{3EI} + \frac{M_0 L}{6EI} = \frac{M_0 L}{2EI}$$

$$\delta = (2)\frac{M_0 L^2}{16EI} = \frac{M_0 L^2}{8EI}$$

This agrees with the results obtained using the moment-area theorems.

8.3.2.3 SIMPLY SUPPORTED BEAM

A simply supported beam having maximum deflection and rotation of the elastic curve at the ends caused by the application of a uniformly distributed load of P_0 lb per foot, Figure 8.8(a). Assume constant EI.

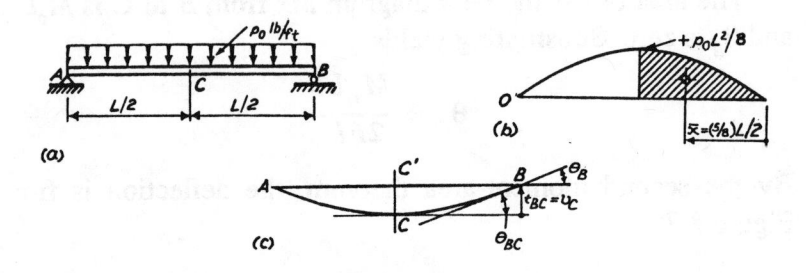

FIGURE 8.8

The bending-moment diagram is in Figure 8.8(a). It is a second-degree parabola with a maximum value at the vertex of $P_0 L^2/8$. The elastic curve passing through the points of the support A and B is shown in Figure 8.8(a).

In this case, the $M/(EI)$ diagram is symmetrical about a

vertical line passing through the center. Therefore the elastic curve must be symmetrical, and the tangent to this curve at the center of the beam is horizontal. From the figure, it is seen that $\Delta\theta_{BC}$ is equal to θ_B, and the rotation of the end B is equal to one-half the area of the whole $M/(EI)$ diagram. The distance CC' is the desired deflection, and from the geometry of the figure it is seen to be equal to t_{BC}.

$$\Phi = \frac{1}{EI}\left(\frac{2}{3}\frac{L}{2}\frac{P_0 L^2}{8}\right) = \frac{P_0 L^3}{24 EI}$$

$$\theta_B = \Delta\theta_{BC} = \Phi = +\frac{P_0 L^3}{24 EI}$$

$$v_C = V_{max} = t_{BC} = \Phi\bar{x} = \frac{P_0 L^3}{24 EI}\frac{5L}{16} = \frac{5 P_0 L^4}{384 EI}$$

Since the point B is above the tangent through C, the sign of v_C is positive.

8.3.2.4 ANGLE OF ROTATION θ_b AND DEFLECTION δ AT THE FREE END OF A CANTILEVER WITH A CONCENTRATED LOAD P (SEE FIGURE 8.9)

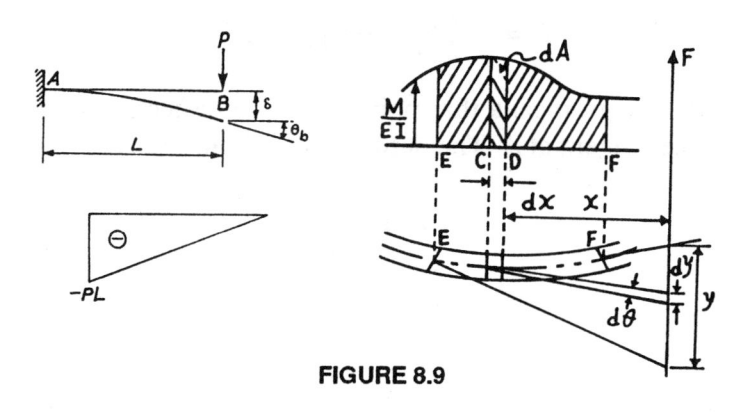

FIGURE 8.9

This problem can be solved using the first and second moment area theorems (bending moment diagram is triangular in shape and is shown in the lower part of the figure).

SECOND MOMENT-AREA THEOREM

On any vertical line, the distance y, between the intersections made by the tangents at points E and F is equal to the first moment about that line of the area of the $M/(EI)$ graph between the points E and F.

From the first moment-area theorem we observe that the difference in angles between points A and B is equal to the area of the bending moment diagram divided by EI; this is $-PL^2/2EI$. The minus sign means that the tangent at B is rotated clockwise from the tangent at A, which is horizontal. Therefore, the angle θ_b, positive as shown in the figure, is

$$\theta_b = \frac{PL^2}{2EI}$$

The deflection δ at the end of the beam can be obtained by applying the second theorem. The distance Δ of point B on the deflection curve from the tangent at A is equal to the first moment of the bending moment area about B divided by EI:

$$\Delta = -\frac{PL^2}{2EI}\left(\frac{2L}{3}\right) = -\frac{PL^3}{3EI}$$

The minus sign means that point B on the deflection curve is below the tangent at A. The deflection δ, therefore is

$$\delta = \frac{PL^3}{3EI}$$

8.3.2.5 DEFLECTION OF THE FREE END OF THE CANTILEVER BEAM OF FIGURE 8.10(a) IN TERMS OF w, L, E AND I

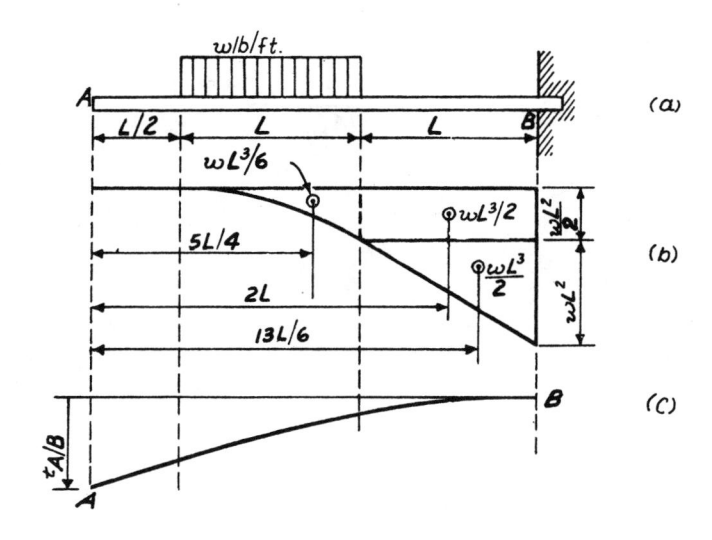

FIGURE 8.10

Since E and I are constant, a bending moment diagram is used instead of an $M/(EI)$ diagram. The moment diagram is shown in Figure 8.10(b), and the area under it has been divided into rectangular, triangular, and parabolic parts for ease in calculating areas and moments. The elastic curve is shown in Figure 8.10(c) with the deflection greatly exaggerated. Points A and B are selected at the ends of the beam because the beam has a horizontal tangent at B and the deflection at A is required. The vertical distance to A from the tangent at B, $t_{A/B}$ in Figure 8.10(c), equals the deflection of the free end of the beam, y_A. For this reason, the second area-moment theorem can be used directly to obtain the required deflection. The area of each of the three portions of the area under the moment diagram is shown in

Figure 8.10(b) along with the distance from the centroid of each part to the moment axis at the free end A. The second area-moment theorem gives

$$EIt_{A/B} = -\frac{wL^3}{6}\left(\frac{5L}{4}\right) - \frac{wL^3}{2}(2L) - \frac{wL^3}{2}\left(\frac{13L}{6}\right),$$

from which

$$y_A = t_{A/B} = -\frac{55wL^4}{24EI} = \frac{55wL^4}{24EI}$$

downward.

CHAPTER 9

STATICALLY INDETERMINATE BEAMS

9.1 INTRODUCTION

A beam is a structural member subject to loading, which is perpendicular to its longitudinal axis. Basically, a beam is classified in two ways:

a) Statically determinate beam and

b) Statically indeterminate beam.

A beam is said to be statically determinate if the unknown reactions and moment can be solved by the conditions of static equilibrium, i.e., $\Sigma F_x = 0$, $\Sigma F_y = 0$ and $\Sigma M_z = 0$. If the member of statics equations becomes insufficient to solve the reactions and moment on a loaded beam, the beam is said to be statically indeterminate.

Statically indeterminate beam problems can be solved by the following methods:

1. Integration method

2. Superposition method

3. Moment-Area method

4. Three-Moment Theorem

5. Plastic Analysis

9.2 INTEGRATION METHOD

The differential equations of different parameters of a beam can be written as

$$\delta = y \tag{9-1}$$

$$\theta = \frac{dy}{dx} \tag{9-2}$$

$$M = EI \frac{d^2 y}{dx^2} \tag{9-3}$$

$$V = EI \frac{d^3 y}{dx^3} \tag{9-4}$$

$$w = EI \frac{d^4 y}{dx^4} \tag{9-5}$$

where y = deflection
 θ = slope
 M = bending moment
 V = shear force
 w = loading
 E = modulus of elasticity
 I = moment of inertia of beam sections
 EI = flexural rigidity (product of E and I)

The bending moment equation can be derived from the loading and shearing functions. Integrating it once gives the slope equation, and twice gives the deflection equation. Equations 9.1 to 9.5 can be written as

$$V = \int w\, dx + C_1 \quad (9\text{-}6)$$

$$M = \int V\, dx + C_2 \quad (9\text{-}7)$$

$$\theta = \int \frac{M}{EI}\, dx + C_3 \quad (9\text{-}8)$$

$$\delta = \int \theta\, dx + C_4 \quad (9\text{-}9)$$

By applying the beam's boundary conditions to the slope and deflection equations, we will get adequate equations to solve the unknowns of the indeterminate beams. The following example illustrates this method.

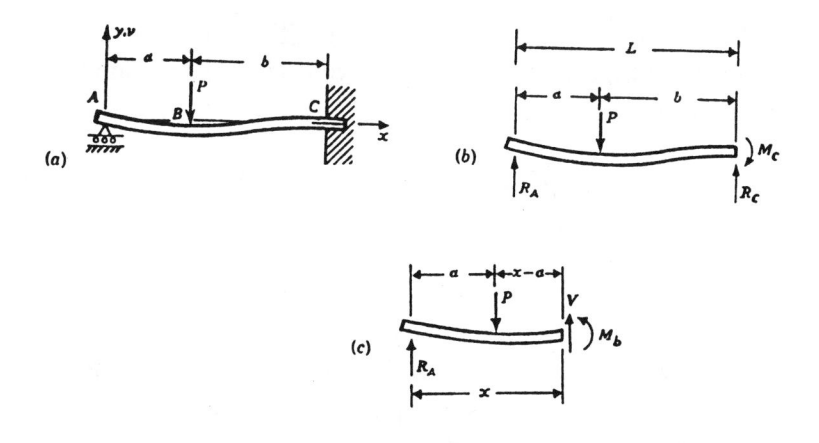

FIGURE 9.1

Figure 9.1(a) shows a beam whose neutral axis coincided with the x axis before the load P was applied. The beam has a simple support at A and a clamped or built-in support at C. The bending modulus EI is constant along the length of the beam.

The moment curvature relation will be used to analyze this statically indeterminate beam. Figure 9.1(a) shows a free-body diagram of the entire beam. Since P is given as vertical and R_A because of the rollers, can only be vertical, the reaction at C can only consist of a vertical force R_C and a clamping moment M_C. There are no horizontal forces, and hence there are only two independent equilibrium requirements, but there are three unknowns: R_A, R_C, and M_C. The equilibrium conditions furnish only two relations between three quantities. The best we can do by considering only equilibrium is to take one of the reactions as an unknown and express the other two in terms of this unknown. For example, taking R_A as unknown, the conditions of equilibrium applied to Figure 9.1(b) yield

$$R_C = P - R_A$$
$$M_C = Pb - R_A L$$
(9-10)

Similarly, applying the conditions of equilibrium to the segment of length x in Figure 9.1(c) gives the following expression for the bending moment:

$$M_b = R_A x - P <x-a>^1 \quad (9\text{-}11)$$

which is valid for $0 < x < L$.

Turning to the geometrical requirements for the deformed beam, we see that now we have three compatibility conditions

$v = 0$ at $x = 0$

$v = 0$ at $x = L$

$$\frac{dv}{dx} = 0 \quad \text{at} \quad x = L \quad (9\text{-}12)$$

Thus, if we integrate the moment-curvature relation, we have enough conditions not only to evaluate the two constants of integration but also to evaluate the unknown reaction R_A which appears in (9-11). Setting up the moment-curvature relation and carrying out one integration gives

$$EI \frac{d^2 v}{dx^2} = M_b = R_A x - P<x-a>^1$$

$$EI \frac{dv}{dx} = R_A \frac{x^2}{2} - P \frac{<x-a>^2}{2} + c_1 \quad (9\text{-}13)$$

In order for the third of (9-12) to be satisfied, we must have

$$c_1 = \frac{Pb^2}{2} - \frac{R_A L^2}{2} \quad (9\text{-}14)$$

Inserting (9-14) in (9-13) and carrying out one more integration gives

$$EIv = R_A \frac{x^3}{6} - P \frac{<x-a>^3}{6} + \frac{Pb^2 x}{2} - \frac{R_A L^2 x}{2} + c_2 \quad (9\text{-}15)$$

In order for the first of (9-12) to be satisfied we must have $c_2 = 0$. Finally, to satisfy the second of (9-12) we must have

$$0 = R_A \frac{L^3}{6} - P \frac{b^3}{6} + \frac{Pb^2 L}{2} - \frac{R_A L^3}{2} \quad (9\text{-}16)$$

from which we find

$$R_A = \frac{Pb^2}{2L^3}(3L - b) \qquad (9\text{-}17)$$

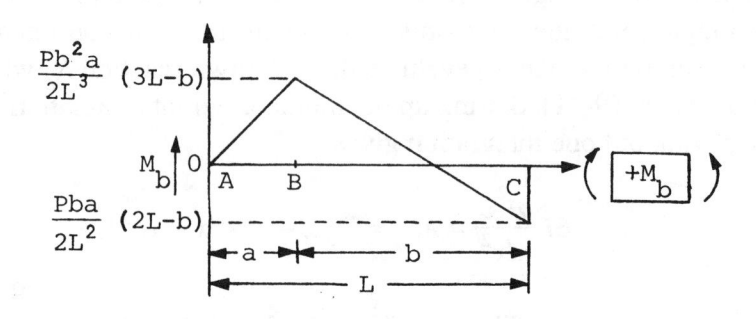

FIGURE 9.2

Thus, to complete the force analysis, we had to bring in the geometric restrictions and the moment-curvature relation. Now we can return to (9-10) with the value (9-17) to obtain explicit results for the reactions. With these values it is an easy matter to sketch the bending-moment diagram shown in Figure 9.2.

9.3 SUPERPOSITION METHOD

PRINCIPLE OF SUPERPOSITION:

If several causes act simultaneously on a system and each effect is directly proportional to its cause, the total effect is the sum of the individual effects when considered separately, provided that Hooke's law is valid within the range of these effects, either individually or combined.

That is, the total deflection, slope, bending moment, or shear force, at a point in a beam, is the algebraic sum of the corresponding parameters produced by each load under the limitations that it does not exceed the proportional limit of the material.

The example given below is self-explanatory. The slope at end A should be zero, since it is a fixed end. The indeterminate structure given in Figure 9.3(a) is split into two determinate structures as in Figure 9.3(b) and Figure 9.3(c). The slope at A is estimated for Figures 9.3(b) and (c) and superimposing them (i.e., equating the algebraic sum to zero), we get the moment M_A. Once we get that, the rest may be treated as a determinate structure and, hence, we may solve for the unknown reactions.

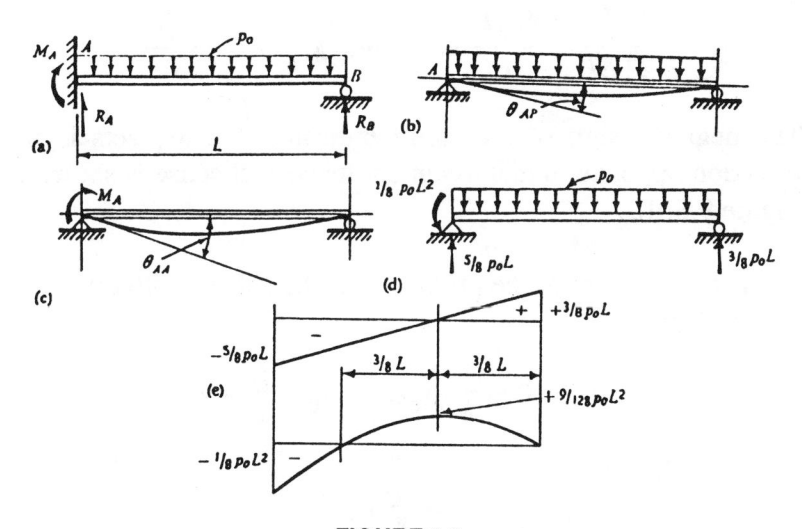

FIGURE 9.3

This beam is indeterminate to the first degree, but it can be reduced to determinacy by removing M_A as in Figure 9.3(b). A positive moment M_A acting at A on the same structure is shown in Figure 9.3(c). The rotations at A for the two determinate cases can be found from the Table of beam deflection and slopes. The requirement of zero rotation at A in the original structure provides the necessary equation for determining M_A.

$$\theta_{AP} = \frac{p_0 L^3}{(24EI)} \qquad \text{(clockwise)}$$

and

$$\theta_{AA} = \frac{M_A L}{(3EI)} \qquad \text{(clockwise)}$$

$$\theta_A = \theta_{AP} + \theta_{AA} = 0$$

Taking clockwise rotations as positive,

$$\frac{P_0 L^3}{24 EI} + \frac{M_A L}{3 EI} = 0 \text{ and } M_A = -\frac{P_0 L^2}{8}$$

The negative sign of the result indicates that M_A acts in the direction opposite to that assumed. Its correct sense is shown in Figure 9.3(d).

The remainder of the problem may be solved with the aid of statics. Now,

$$\sum M_B = 0 = \frac{1}{8} p_0 L^2 + p_0 L \frac{L}{2} - R_A L$$

from which

$$R_A = \frac{5}{8} p_0 L .$$

Applying equation $\Sigma F_y = 0$, one obtains,

$$p_0 L - R_A - R_B = 0$$

or

$$R_B = p_0 L - \frac{5}{8} p_0 L = \frac{3}{8} p_0 L .$$

Reactions, shear diagram, and moment diagram are in Figures 9.3(d), (e), and (f) respectively.

This problem may also be analyzed by treating R_B as the redundant.

9.4 MOMENT-AREA METHOD

This is a semi-graphical method that relates the M/EI diagram to the deflection and the slopes of the elastic curve, which in turn is a tool to solve many structural problems, including indeterminate beams.

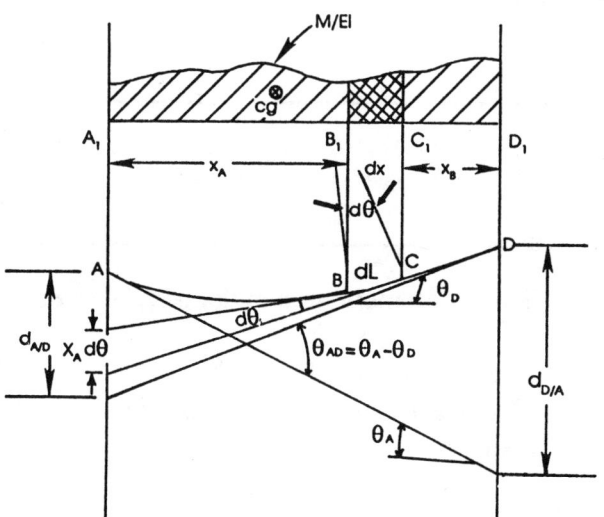

FIGURE 9.4—M/EI, slope and deflection curve segment.

Integrating M/EI diagram between A_1 and D_1 gives the angle between the tangents to the elastic curve at A and D (Figure 9.4).

$$\theta_{AD} = \theta_D - \theta_A = \int_A^D \frac{M}{EI}\, dx \qquad (9\text{-}18)$$

THEOREM 1

The area under the curve M/EI between A and D is equal to the change in slopes between the tangents at A and D.

In Figure 19.3, $d_{A/D}$ and $d_{D/A}$ are deviations of tangents. The deviations contributed by curve length dL is $x_A d\theta$.

$$d_{A/D} = \int_A^D x_A d\theta = \int_A^D \frac{M}{EI} x_A \, dx_A \qquad (9\text{-}19)$$

Since $d\theta = M/EI \, dx_A$.

Similarly

$$d_{D/A} = \int_D^A \frac{M}{EI} x_D \, dx_D \qquad (9\text{-}20)$$

$M/EI \, dx$ is the increment area under M/EI curve.

$M/EI \, x \, dx$ is the moment of the above area. For Equation (9-19) it is about A and for Equation (9-20) it is about D.

THEOREM 2

The tangential deviation of a point D, located on the elastic curve to the tangent at another point A, is equal to the moment of area under the M/EI curve, between points A and B, about B. The following example illustrates the method.

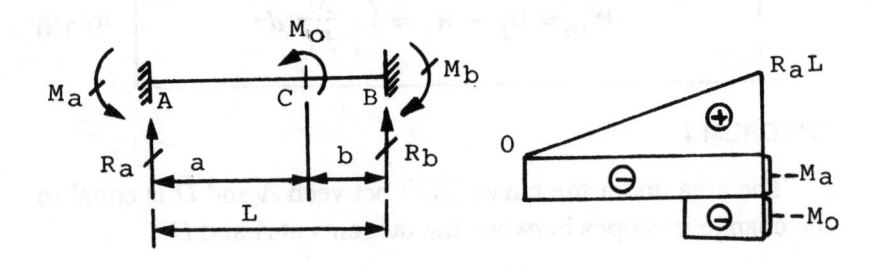

FIGURE 9.5

The choice of redundants must be made first; the possibilities are R_a and M_a, or R_b and M_b, or M_a and M_b. Let us make the first choice and take the reactions at support A as the redundants. Then we have a cantilever beam fixed at B as the released structure, and we can easily draw the bending moment diagrams produced by R_a, M_a, and the load M_0 (See Figure 9.5(b)).

Two conditions concerning the deflections of the beam are required for finding the two redundants. As a first condition, we note that both ends of the beam have zero slopes; hence the change in slope between A and B is zero. It follows from the first moment-area theorem that the area of the M/EI diagram between A and B must be zero; thus,

$$\frac{1}{2}\left(\frac{R_a L}{EI}\right)(L) - \frac{M_a}{EI}(L) - \frac{M_0}{EI}(b) = 0$$

or

$$R_a L^2 - 2M_a L = 2M_0 b \tag{9-21}$$

The second condition is obtained from the fact that the tangent to the deflection curve at A passes through point B, which means that the first moment of the area of the M/EI diagram between A and B, taken about B, is zero. The resulting equation is

$$\frac{1}{2}\left(\frac{R_a L}{EI}\right)(L)\left(\frac{L}{3}\right) - \frac{M_a}{EI}(L)\left(\frac{L}{3}\right) - \frac{M_0}{EI}(b)\left(\frac{b}{2}\right) = 0$$

or

$$R_a L^3 - 3M_a L^2 = 3M_0 b^2. \tag{9-22}$$

Now we can solve simultaneously Equations 9-21 and 9-22 for the redundants:

$$R_a = \frac{6M_0 ab}{L^3} \qquad M_a = \frac{M_0 b}{L^2}(2a - b) \tag{9-23}$$

The other two reactions are

$$R_b = -R_a \qquad M_b = \frac{M_0 a}{L^2}(a - 2b)$$

as found from static equilibrium considerations.

The deflection δ_c at the point where the load is applied can be found from the second moment-area theorem. This deflection is equal to the first moment of the area of the M/EI diagram between A and C, taken about point C. Referring to Figure 9.5(b), we see that this deflection is

$$\delta_c = \frac{1}{2}\left(\frac{a}{L}\right)\left(\frac{R_a L}{EI}\right)(a)\left(\frac{a}{3}\right) - \frac{M_a}{EI}(a)\left(\frac{a}{2}\right) = \frac{R_a a^3}{6EI} - \frac{M_a a^2}{2EI}$$

Substituting the expressions for R_a and M_a (See Equation 9-23) we get

$$\delta_c = \frac{M_0 a^2 b^2 (b - a)}{2L^3 EI} \qquad (9\text{-}24)$$

for the deflection under the load.

When the couple M_0 acts as the midpoint of the span, the reactions for the beam are

$$M_a = -M_b = \frac{M_0}{4} \qquad R_a = -R_b = \frac{3M_0}{2L}$$

and the deflection at the middle (Equation 9-24) becomes zero.

9.5 THREE-MOMENT THEOREM

Two adjacent spans of a continuous beam are considered, as in Figure 9.6(a), with spans l_1 and l_2. The bending moment due to the loads are taken as positive, (shown in Figure 9.6(b)) by considering each span as an independent simply-supported beam. From the deflection curve (Figure 9.6(c) at supports, it is obvious that the top fibers experience tension, which indicates that the moments M_1, M_2 and M_3 respectively at supports 1, 2 and 3 are negative (See Figure 9.6(b)).

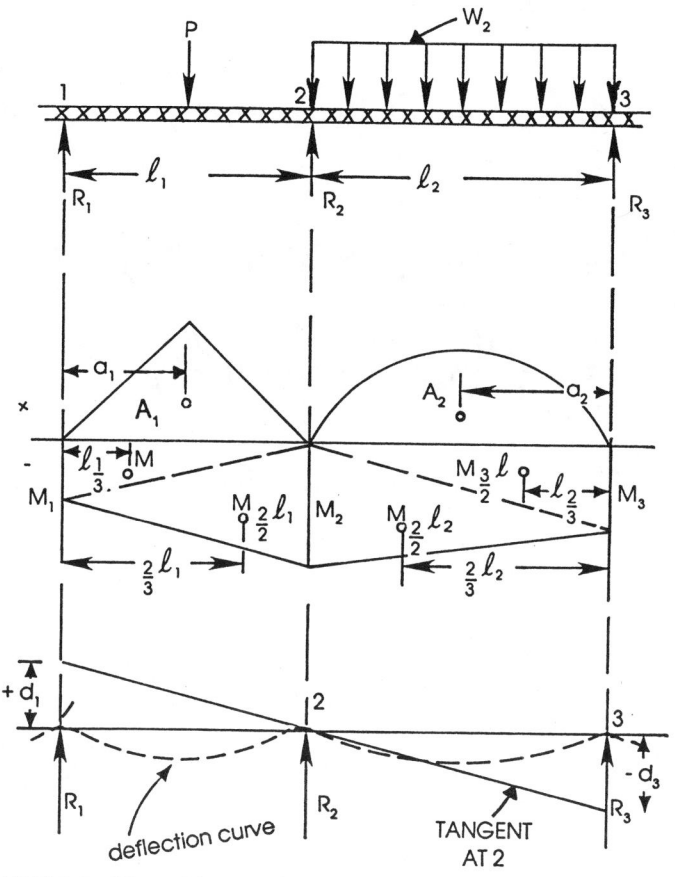

FIGURE 9.6—Three-Moment Theorem: (a) span of a continuous beam between three supports, (b) bending moment and (c) deflection.

A tangent to the deflection curve at two of the loaded beams would slope and its displacement at supports 1 and 3 are d_1 and $-d_3$, as shown in Figure 9.6(c).

By applying Theorem 2 of the Moment-Area Method, we will obtain two expressions for d_1 and d_3 as shown in Equations 9-25 and 9-26, respectively.

$$d_1 = \frac{1}{EI}\left(A_1 a_1 + M_1 \frac{l_1}{2}\frac{l_1}{3} + M_2 \frac{l_1}{2}\frac{2l_1}{3}\right)$$

$$\therefore \quad d_1 = \frac{1}{EI}\left(A_1 a_1 + M_1 \frac{l_1^2}{6} + M_2 \frac{l_1^2}{3}\right) \qquad (9\text{-}25)$$

$$d_3 = \frac{1}{EI}\left(A_2 a_2 + M_2 \frac{l_2}{2}\frac{2l_2}{3} + M_3 \frac{l_2}{2}\frac{l_2}{3}\right)$$

$$\therefore \quad d_3 = \frac{1}{EI}\left(A_2 a_2 + M_2 \frac{l_2^2}{3} + M_3 \frac{l_2^2}{6}\right) \qquad (9\text{-}26)$$

From Figure 9.6(c), by using a similar triangle principle, we have

$$\frac{d_1}{l_1} = -\frac{d_3}{l_2} \qquad (9\text{-}27)$$

Substituting Equations 9-25 and 9-26 in Equation 9-27,

$$\frac{A_1 a_1}{l_1} + \frac{M_1 l_1}{6} + \frac{M_2 l_1}{3} = -\left(\frac{A_2 a_2}{l_2} + \frac{M_2 l_2}{3} + \frac{M_3 l_2}{6}\right)$$

Multiplying each term by 6 and rearranging the term, we obtain

$$\boxed{M_1 l_1 + 2M_2 (l_1 + l_2) + M_3 l_2 = -\frac{6A_1 a_1}{l_1} - \frac{6A_2 a_2}{l_2}} \qquad (9\text{-}28)$$

Equation 9-28 is the Three-Moment Theorem.

Thus, we take two adjacent spans and write a three-moment equation for each pair until the whole beam is covered. The solution of these equations gives the moments and hence we can solve for the reactions.

The assumptions made in this theorem are:

i) The beam is straight before loading

ii) The supports remain in a straight line after loading

iii) E and I are constant throughout its length.

For varying moment of inertia, I we can write the Three-Moment Theorem as in Equation 9-29.

$$M_1 l_1 + 2M_2 \left(l_1 + \frac{I_1}{I_2} l_2 \right) + M_3 l_2 \left(\frac{I_1}{I_2} \right)$$
$$= -\frac{6A_1 a_1}{l_1} - \frac{6A_2 a_2}{l_2} \left(\frac{I_1}{I_2} \right) \quad (9\text{-}29)$$

LOAD TERMS

The terms on the right side of Equations 9-28 and 9-29 are called load terms. These are determined by taking the moment of the bending moment diagram of the corresponding span. For the left span moment it is about its left support and for the right span, it is about its right support. Table 9.1 gives the load terms of left and right spans for different loading conditions.

TABLE 9.1—LOAD TERMS

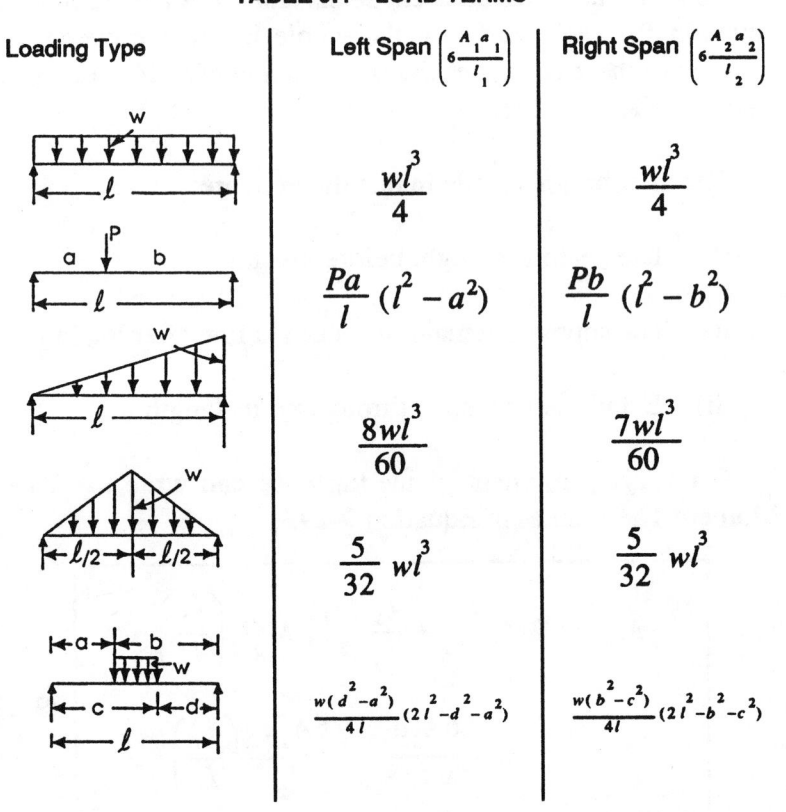

Loading Type	Left Span $\left(6\dfrac{A_1 a_1}{l_1}\right)$	Right Span $\left(6\dfrac{A_2 a_2}{l_2}\right)$
Uniform load w over span l	$\dfrac{wl^3}{4}$	$\dfrac{wl^3}{4}$
Concentrated load P at distance a from left, b from right, span l	$\dfrac{Pa}{l}(l^2 - a^2)$	$\dfrac{Pb}{l}(l^2 - b^2)$
Triangular load (increasing), span l	$\dfrac{8wl^3}{60}$	$\dfrac{7wl^3}{60}$
Triangular (symmetric) load, peak w at center, half-spans $l/2$	$\dfrac{5}{32} wl^3$	$\dfrac{5}{32} wl^3$
Partial uniform load w over segment b (distances a, c, d), span l	$\dfrac{w(d^2-a^2)}{4l}(2l^2-d^2-a^2)$	$\dfrac{w(b^2-c^2)}{4l}(2l^2-b^2-c^2)$

The example given below illustrates a typical application of the theorem.

By using the three-moment equation for the spans AB and BC, one equation is written. From statics, the beam convention being used for signs, $M_A = -10(5) = -50$ kip-ft. The moments of inertia I_L and I_R are equal.

$$12M_A + 2(12 + 20)M_B + 20M_C$$

$$= -\dfrac{2(12)^3}{4} - 8(15)5\left(1 + \dfrac{15}{20}\right) - 12(10)10\left(1 + \dfrac{10}{20}\right)$$

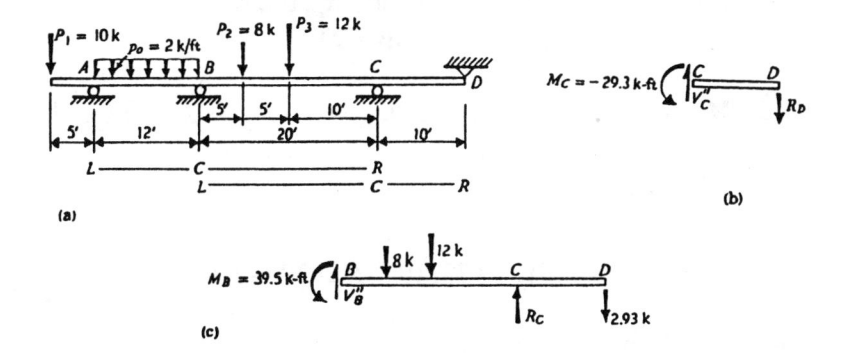

FIGURE 9.7

Substituting $M_A = -50$ kip-ft and simplifying gives

$$64 M_B + 20 M_C = -3,114 \qquad (9\text{-}30)$$

Next, the three-moment equation is applied again for the spans *BC* and *CD*. No constant terms are contributed to the right side of the three-moment equation by the unloaded span *CD*. At the pinned end, $M_D = 0$.

$$20 M_B + 2(20 + 10) M_C + 10 M_D$$

$$= -8(5)15\left(1 + \frac{5}{20}\right) - 12(10)10\left(1 + \frac{10}{20}\right)$$

or

$$20 M_B + 60 M_C = -2,550 \qquad (9\text{-}31)$$

Solving the Equations 9-30 and 9-31 simultaneously gives

$$M_B = -39.6 \text{ kip}-\text{ft} \quad \text{and} \quad M_C = -29.3 \text{ kip}-\text{ft}$$

By isolating the span *CD* as in Figure 9.7(b), one obtains the reaction R_D from statics. Instead of isolating the span BC and computing V'_C to add to V''_C to find R_C, the free body shown in Figure 9.7(c) is used. For free body CD:

$$\Sigma M_C = 0 \circlearrowright \quad +, \quad 29.3 - 10R_D = 0, \quad R_D = 2.93 \text{ kips} \downarrow$$

For free body BD:

$$\Sigma M_B = 0, \circlearrowleft \quad +, \quad 8(5) + 12(10) - R_C(20) + 2.93(30) - 39.5 = 0,$$

$R_D = 10.42$ kips $\uparrow$

9.6 PLASTIC ANALYSIS

A material stressed beyond its proportional or elastic limit is said to be in the plastic region, as shown in Figure 9.8. From O to A, the material is in proportional limit. B and C are the upper and lower yield stress points. Portion A to E is the plastic region. D is the point of ultimate stress.

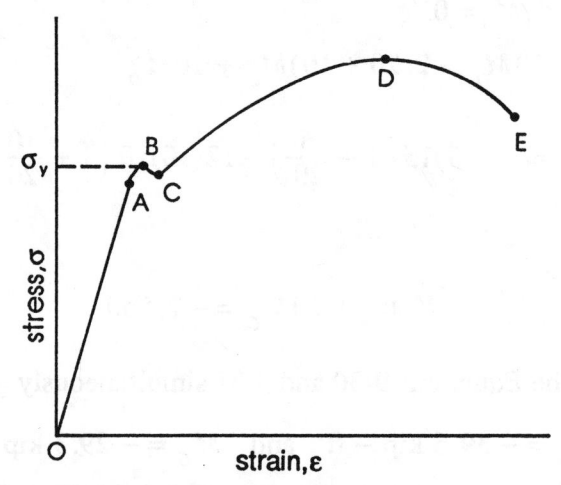

FIGURE 9.8—Stress-strain diagram of structural steel.

B-127

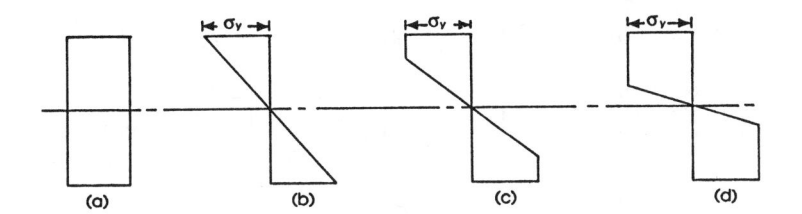

FIGURE 9.9—Stress-Strain relation: (a) rectangular section of a beam, (b) stress within the elastic limit, (c) and (d) stress in the plastic region.

The stress distribution of a beam with rectangular cross-section is shown in Figure 9.9. The stress distribution becomes almost rectangular (See Figure 9.9(d)) when the moment is increased further. The moment at this limiting strength position is called the plastic moment, M_p.

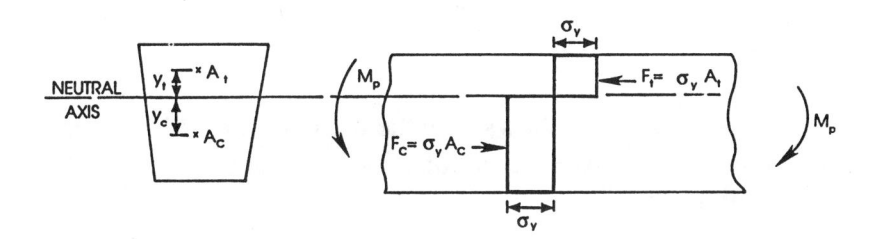

FIGURE 9.10—Plastic moment on a beam.

A beam is subject to a plastic moment as shown in Figure 9.10. The top fiber is in tension and the bottom is in compression.

For $\Sigma M = 0$, we get

$$M_p = \sigma_y A_t y_t + \sigma_y A_c y_c = \sigma_y (A_t y_t + A_c y_c) \quad (9\text{-}32)$$

Table 9-2 gives the comparison between the elastic moment, M_E and plastic moment, M_p for some common section shapes.

TABLE 9.2—PLASTIC MOMENT/ELASTIC MOMENT

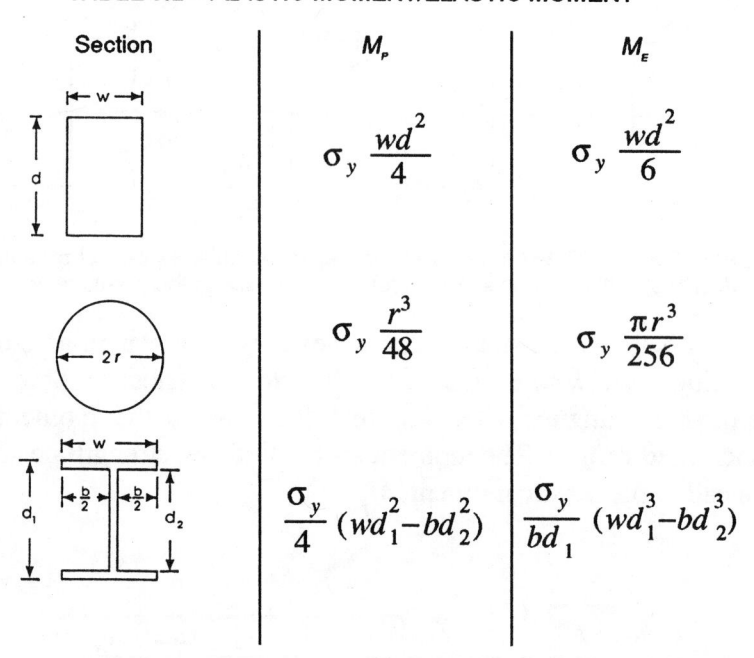

Section	M_P	M_E
Rectangle (width w, depth d)	$\sigma_y \dfrac{wd^2}{4}$	$\sigma_y \dfrac{wd^2}{6}$
Circle (radius r)	$\sigma_y \dfrac{r^3}{48}$	$\sigma_y \dfrac{\pi r^3}{256}$
I-section	$\dfrac{\sigma_y}{4}(wd_1^2 - bd_2^2)$	$\dfrac{\sigma_y}{bd_1}(wd_1^3 - bd_2^3)$

The next example shows how the method of plastic analysis can be used to solve the reactions and moments of a statically indeterminate beam.

EXAMPLE

Find the ultimate load that the beam, shown below, can carry.

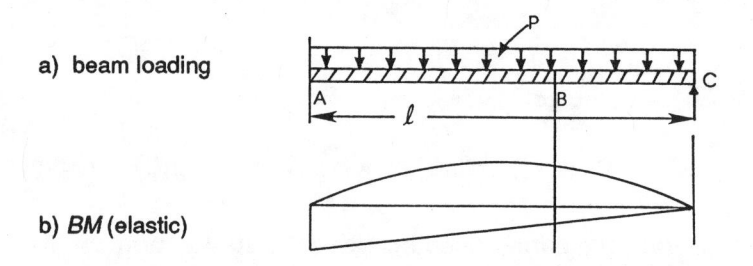

a) beam loading

b) *BM* (elastic)

B-129

c) Deflection Curves:
   (1) Elastic
   (2) Plastic at A
   (3) Plastic at A and B

d) BM (plastic)

e) Plastic hinge analysis

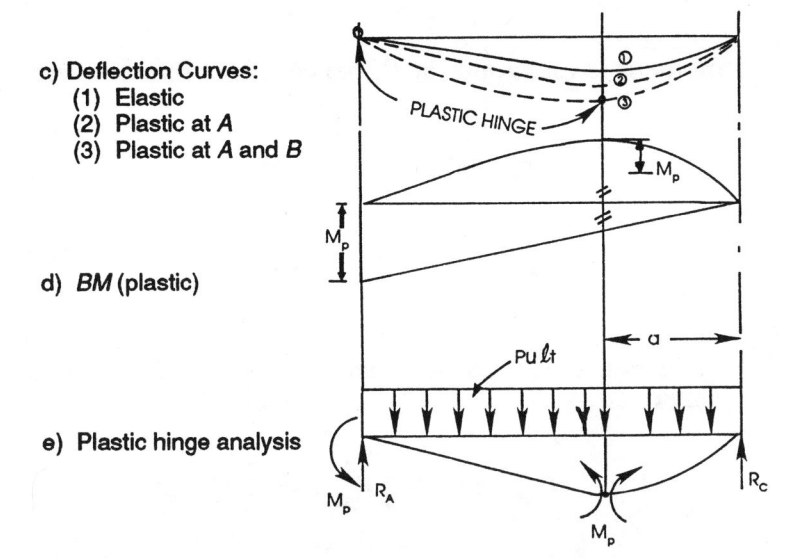

FIGURE 9-11—Plastic Analysis of Statically indeterminate beam.

A and B are points of maximum negative and positive moment. M_A is greater than M_B. By increasing the load, A reaches its plastic state first, forming a plastic hinge. By further increasing the load, C also becomes a plastic hinge. The situations when both A and B become plastic hinges is called **mechanism**. The load which collapses the beam is the **ultimate load**.

$\Sigma M_A = 0$, gives

$$R_c l = \frac{P_{ult} l^2}{2} - M_p$$

$$\therefore \quad R_c = \frac{P_{ult} l}{2} - \frac{M_p}{l} \tag{9-33}$$

$\Sigma M_c = 0$,

$$M_p = R_c a - \frac{P_{ult} a^2}{2} \tag{9-34}$$

The moment is maximum at B; hence, shear force is zero there.

$$R_c - P_{ult} \cdot a = 0$$

$$\therefore \quad a = \frac{R_c}{P_{ult}} \tag{9-35}$$

Substitute Equation 9-35 in Equation 9-34 and those from Equation 9-33, we obtain

$$M_p^2 - 3P_{ult}\, l^2 M_p + \frac{1}{4}P_{ult}^2\, l^4 = 0$$

$$\therefore \quad M_p = P_{ult}\, l^2 (1 \cdot 5 \pm 1 \cdot 414)$$

$$\therefore \quad P_{ult} = \frac{M_p}{(1 \cdot 5 - 1 \cdot 414)l^2} = \frac{11 \cdot 6 M_p}{l^2} \tag{9-36}$$

M_p can easily be determined by using Equation 9-32. Thus, from Equation 9-34 we can find out R_c. Once we know R_c, M_A and R_A can be solved using equations of static equilibrium ($\Sigma M_B = 0$, $\Sigma F_y = 0$).

CHAPTER 10

COLUMNS

10.1 DEFINITION AND TYPES OF COLUMNS

The Column, a structural member used in the construction of buildings, bridges, roof trusses and similar structures, is defined as a slender compression member which is subjected to axial loads and has the tendency to buckle at loads much smaller than those which would cause crushing failure.

TYPES OF COLUMNS

There are three principal classes of columns, viz.

i) Short columns or struts,

ii) Intermediate columns, and

iii) Long columns.

The factor determining the proper group of columns is known as the slenderness ratio, defined as the ratio of length to radius of gyration of the column section in the direction of impending bending. The slenderness ratios of metal columns, e.g., pipes, rolled structural steel or aluminum sections etc, are measured by L/K, k being the radius of gyration. The slenderness ratio for timber columns is given by L/d, where d is the least dimension of a square or rectangular section.

TYPES OF COLUMN CROSS-SECTIONS

The selection of the appropriate type of column cross-section varies according to the application. For example, in mechanical engineering, the piston rod of an engine, air-craft wing struts and connecting rods in locomotives, all compression members, come under the definition of columns, In civil engineering works, the most familiar columns used in buildings and bridges are that of structural steel, timber and reinforced concrete.

The following are the types of cross-sections used in structural steel work:

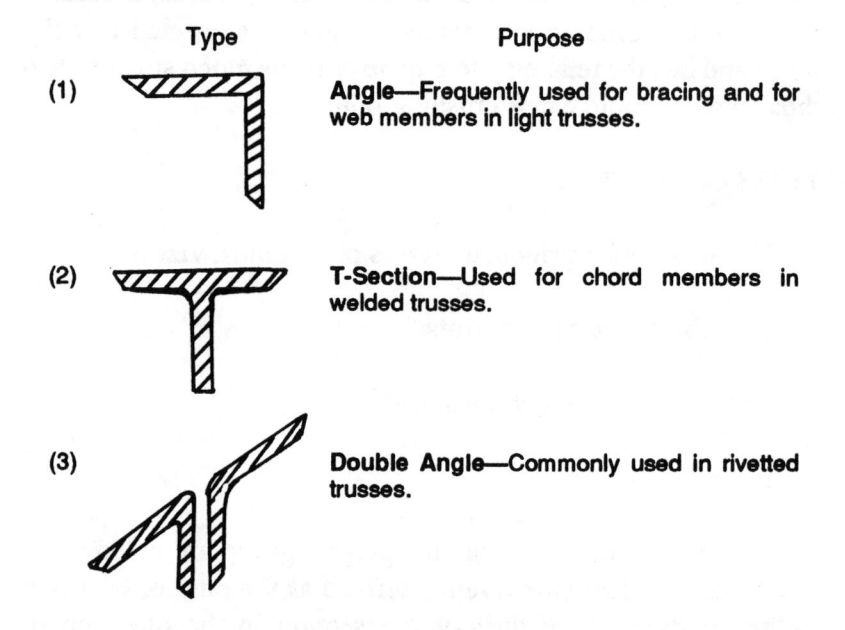

	Type		Purpose
(1)			**Angle**—Frequently used for bracing and for web members in light trusses.
(2)			**T-Section**—Used for chord members in welded trusses.
(3)			**Double Angle**—Commonly used in rivetted trusses.

(4) **I-Section or Wide Flange Section**—a standard cross-section for columns used in structural steel frames of buildings. Also used in trusses with heavier weights.

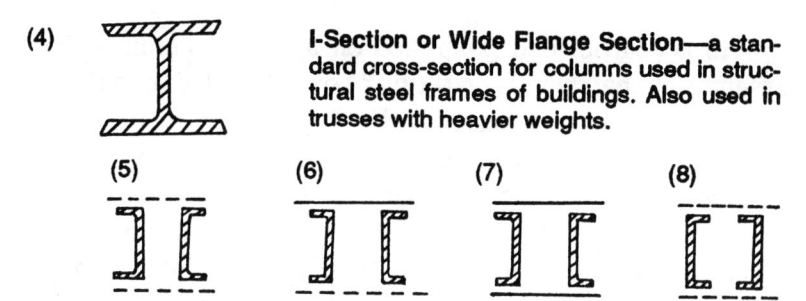

(5) (6) (7) (8)

These four types are generally used in the construction of bridges and buildings

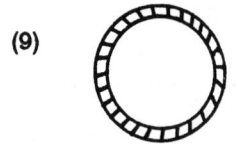

(9) **Hollow Circular Tube**—Heavily used in structural frames of air-crafts. Also used commonly as pipe columns in small buildings.

10.2 SHORT COLUMNS WITH ECCENTRIC LOAD

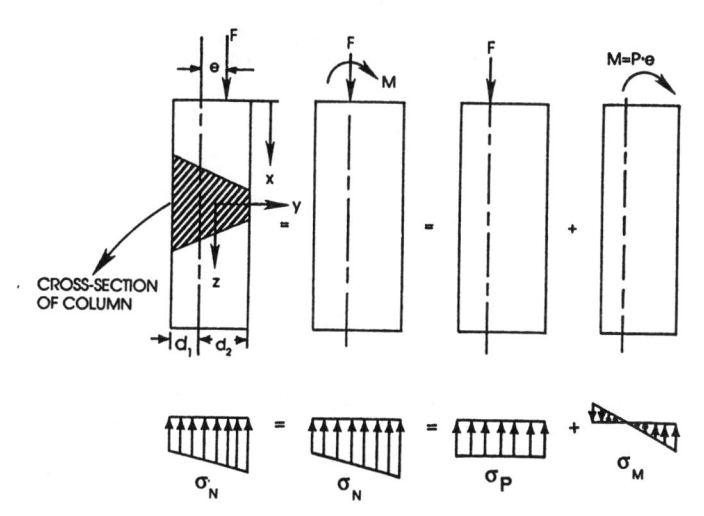

FIGURE 10.1—Column with applied eccentric load and stress distribution.

B-134

An eccentric load is applied on a short column under the assumption that there is no tilting of one end relative to the other, and that the column has an axis of symmetry in the plane of buckling (bending). The buckling and compression of the column occur due to the load, F. The net stress σ_N in the column can be found by superimposing the compression and bending effects. Like the elastic bending of beams, the variation of bending stress is linear with the centroidal distance of the column. Maximum and minimum compression stress is produced in the surface fibers of the column (Figure 10.2). The net stress σ_N is given by the expression

$$\sigma_N = \frac{F}{A} \pm Fe\left(\frac{d_1}{I}\right) \qquad (10\text{-}1)$$

Plus and minus signs indicate maximum and minimum net stresses in the outermost fibers of the column.

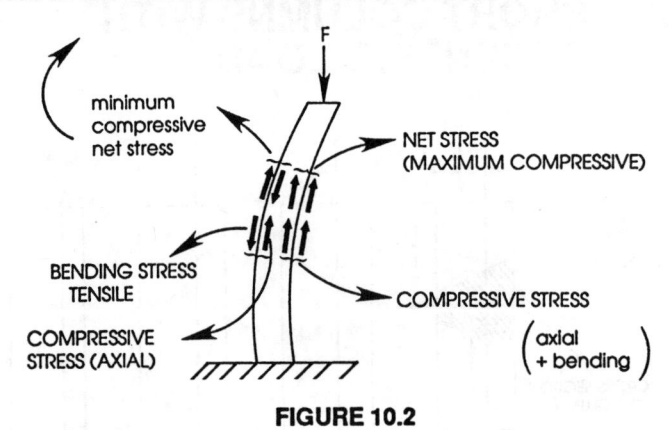

FIGURE 10.2

Now consider the relation for maximum stress,

$$\sigma_N = \frac{F}{A} + Fe\left(\frac{d_1}{I}\right)$$

For maximum load, $P = P_{max}$ for the maximum allowable

stress σ_{max} so that

$$\sigma_{max} = \frac{F_{max}}{A} + Fe\left(\frac{d_1}{I}\right) \quad (10\text{-}2)$$

Multiplying the whole equation by A:

$$A\sigma_{max} = F_{max} + F_{max} e A\left(\frac{d_1}{I}\right)$$

$$= F_{max}\left[1 + \frac{Aed_1}{I}\right]$$

Therefore,

$$F_{max} = \frac{A\sigma_{max}}{\left[1 + \frac{Aed_1}{I}\right]} \quad (10\text{-}3)$$

For no tension on the left face or for net stress to be zero, we have from the minimum stress relation,

$$\sigma_N = 0 = \frac{F}{A}\left[1 + \frac{Aed_2}{I}\right] \quad (10\text{-}4)$$

This yields the maximum eccentricity equal to

$$l_{max} = \frac{I}{Ad_2} \quad (10\text{-}5)$$

10.3 EULER'S FORMULA FOR LONG COLUMNS

In long columns buckling failure is caused by lateral forces much smaller than the ones causing failure of short columns.

Thus for long columns, lateral deflection or buckling cannot be ignored.

Suppose that an appreciably small load P is applied on one end of a long slender column whose other end is pinned jointed. The column shows an elastic behavior within the range of the load.

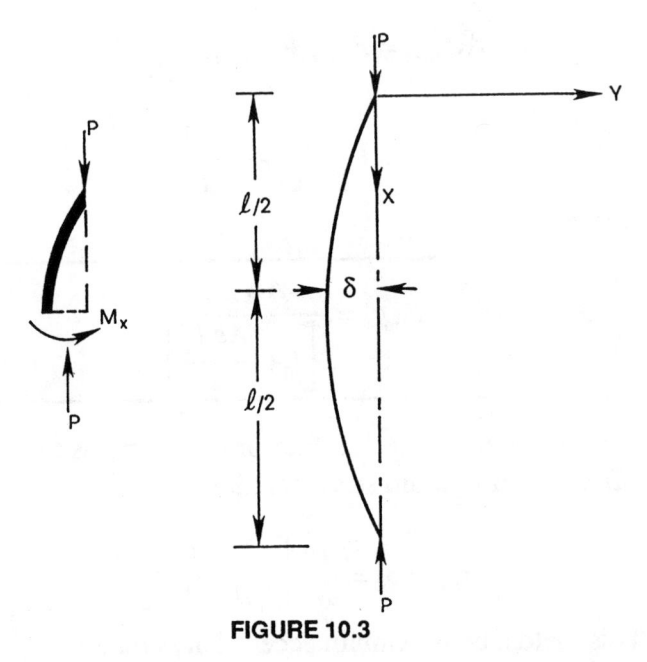

FIGURE 10.3

The bending moment at a certain section of a buckled column is

$$M_x = -P_y$$

The second order differential equation for the bending of columns can be written as,

$$EI \frac{d^2 y}{dx^2} = M_x = -P_y$$

where M has a negative sign with the given y, as shown in Figure 10-3.

Rearranging,

$$\frac{d^2y}{dx^2} + k^2 y = 0 \qquad (10\text{-}6)$$

where $k = \left(\sqrt{\frac{P}{EI}}\right)$

The solution of this linear, second order, homogeneous equation is,

$$y = A \cos kx + B \sin kx$$

A and B are arbitrary constants and can be determined from the end conditions, which conform to the constraints and deflection of the column.

Figure 10.3 shows that,

(i) $y = 0$ at $x = 0, l$

(ii) $\frac{dy}{dx} = 0$ and $y = \delta$ at $x = \frac{l}{2}$

A substitution of these conditions yields,

$$y = 0 = A\cos(0)x + B\sin(0)x$$

$$\therefore A = 0 \quad \text{and} \quad y = B \sin kx$$

Now,

$$\frac{dy}{dx} = Bk \cos kx$$

at

$$x = \frac{l}{2}, \quad \frac{dy}{dx} = 0 = Bk \cos \frac{kl}{2} \quad B \neq 0, \quad k \neq 0,$$

B-138

therefore,

$$\cos \frac{kl}{2} = 0 = \cos \frac{\pi}{2} \quad \text{or} \quad \frac{kl}{2} = \frac{\pi}{2}$$

From condition (ii)

$$y = \delta = B \sin \frac{kl}{2}$$

But

$$\frac{kl}{2} = \frac{\pi}{2},$$

Therefore,

$$B \sin \frac{\pi}{2} = \delta \quad \text{or} \quad B = \delta$$

Hence,

$$y = \delta \sin kx$$

Again, from condition (i), $y = 0$, at $x = l$

$$\therefore \quad y = 0 = \delta \sin kl$$

But $\delta \neq 0$, thus $\sin kl = 0 = \sin n\pi$ or

$$kl = n\pi \Rightarrow k = \frac{n\pi}{l}$$

where $n = 1, 2, 3, \ldots$

For a column with no lateral support, $P = P_{er}$ at $n = 1$. We have $k = \sqrt{\dfrac{P}{EI}}$,

$$\therefore \quad \frac{n\pi}{l} = \sqrt{\frac{P}{EI}}$$

For a critical load, $P = P_{er}$ and $n = 1$, hence,

$$\frac{\pi}{l} = \sqrt{\frac{P_{cr}}{EI}}$$

Squaring both sides.

$$P_{cr} = \frac{\pi^2 EI}{l^2} \qquad (10\text{-}7)$$

The above relation is known as Euler's formula for a column with a pinned-end.

CRITICAL LOAD ON THE COLUMNS

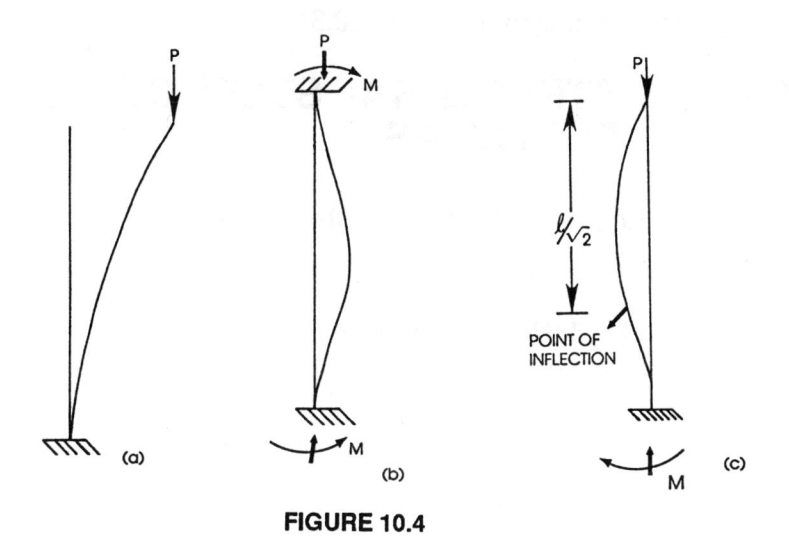

FIGURE 10.4

In Figure 10.4 formulas (i), (ii) and (iii) show the critical loads for Figures 10(a), (b) and (c) respectively.

$$P_{cr} = \frac{\pi^2 EI}{(2l^2)} = \frac{\pi^2 EI}{4l^2} \qquad (10\text{-}8a)$$

$$P_{cr} = \frac{\pi^2 EI}{(l/2)^2} = \frac{4\pi^2 EI}{l^2} \qquad (10\text{-}8b)$$

$$P_{cr} = \frac{\pi^2 EI}{\left(\frac{1}{\sqrt{2}} l\right)^2} = \frac{2\pi^2 EI}{l^2} \qquad (10\text{-}8c)$$

These formulas are obtained by modifying the length l in Euler's equation of critical load [defined in Section 10.3] when any segment of one of the columns shown in Figure 10.4 corresponds to the column in Figure 10.3.

10.4 STRUCTURAL STEEL COLUMN FORMULAS

Empirical formulas used in the design of columns are as follows:

(1) GORDAN-RANKINE FORMULA

$$\frac{P}{A} = \frac{18{,}000}{1 + (1/18{,}000)(L/k)^2} \qquad (10\text{-}9)$$

(2) AISC (American Institute of Steel Construction) RECOMMENDED FORMULAS

(a) For the main and secondary members

$$\frac{P}{A} = \frac{\left[1 - \dfrac{(KL/k)^2}{2C_c^2}\right]}{F.S.} \sigma_y \qquad (10\text{-}10)$$

where

$$C_c = \sqrt{\frac{2\pi^2 E}{\sigma_y}}$$

$$F.S. = \frac{5}{3} + \frac{3(KL/k)}{8C_c} - \frac{(KL/k)^3}{8C_c^3}$$

where KL/k = the slenderness ratio

(b) For the main members with $\frac{KL}{k} > C_c$ AISI recommends

$$\frac{P}{A} = \frac{12\pi^2 E}{23(KL/k)^2} \qquad (10\text{-}11)$$

(c) For secondary members, $L/k > 120$,

$$\frac{P}{A} = \frac{\left[1 - \frac{(KL/k)^2}{2C_c^2}\right]^{FS}_{\sigma_y} \text{ or } \frac{12\pi^2 E}{23(KL/k)^2}}{1 \cdot 6 - L/200k} \qquad (10\text{-}12)$$

NOTE: Main members (columns) are those subjected to permanent, calculated, known or approximately estimated loads. Secondary members (columns) are those subjected to temporary load, e.g., braces which resist temporary load conditions due to winds, earthquakes, etc.

(3) For the main compression members in railway bridges, the following formula has been adopted by the American Railway Engineering Association (AREA).

$$\frac{P}{A} = 15,000 - \frac{1}{3}\left(\frac{L}{k}\right)^2 \qquad (10\text{-}13)$$

For hinged or pinned ends.

$$\frac{P}{A} = 15,000 - \frac{1}{4}\left(\frac{L}{k}\right)^2 \qquad (10\text{-}14)$$

For rivetted ends.

The same formulas are used for compression members in highway bridges by the American Association of State Highway Officials.

10.5 DESIGN OF COLUMNS

Columns allow slightly high stress if its ends are rivetted, particularly in structural work, where the columns are secured by concrete footings and intermediate floor level beams. The steel columns are restrained to some degree, which adds to the factor of safety, even though it is not used to allow higher stresses.

10.5.1 STEEL-COLUMN DESIGN

Column design is an iterative process. Choose the upper and lower L/K values in which the column L/K falls. Take the mean L/K and estimate the allowable stress from the equations in Section 10.4. A design procedure is as follows:

— From the allowable stress σ_a, calculate the required area $A = P_a/\sigma_a$.

— Find a section, with least radius of gyration k and economical area A by iteration.

— Find L/K and hence actual σ_a.

— Calculate $P_a = \sigma_a A$ and

$$\text{Percent stressed} = \frac{\text{actual load}}{\text{allowable load}} \times 100 = \frac{P}{P_a} \times 100$$

— Repeat the above 3 steps if the stressed percent is greater than 102 or less than 95.

10.5.2 ALUMINUM-COLUMN DESIGNS

Aluminum and its alloys ($11,000 < \sigma_y < 53,000$ psi) are commonly used in structures where lightweight materials are required. ASCE (American Society of Civil Engineers) provides different formulas, graphs and other specifications of various Aluminum column designs.

10.5.3 TIMBER-COLUMN DESIGNS

The National Forest Products Association's design formula and recommendations are as stated below:

1) SHORT COLUMNS (L/d ≤ 11.0)

where L = column length
          d = least cross-sectional dimension
          σ_{ac} = allowable compressive stress parallel to the grains (given in design manuals)

2) LONG COLUMNS (K < L/D < 50)

where $K = 0.671\sqrt{E/\sigma_{a_c}}$
          E = modulus of elasticity

EULER'S EQUATION

$$\frac{P}{A} = \frac{\pi^2 E}{(L/K)^2}$$

where $K = \sqrt{I/A} = \sqrt{bd^3/(12bd)} = \frac{d}{\sqrt{12}}$

$$\boxed{\frac{P}{A} = \frac{\pi^2 E}{n(L\sqrt{12}/d)^2}} \quad (10\text{-}15a)$$

$$\therefore \quad \sigma_{AC} = 0.3 \frac{E}{(L/d)^2} \quad (10\text{-}15b)$$

where n = factor of safety (= 2.74)

3) INTERMEDIATE COLUMNS (11.0 < L/d < K)

(10-16)
$$\sigma_{AC} = \left[1 - \frac{1}{3}\left(\frac{L/d}{k}\right)^4\right]\begin{bmatrix}\text{Allowable compressive stress}\\ \text{parallel to the grain}\end{bmatrix}$$

All the formulas above are only for dry conditions.

10.5.4 ECCENTRICITY LOADED COLUMNS

The total combined stress,

$$\sigma = \frac{P}{A} + \frac{Pe}{Z} + \frac{P\Delta}{Z} \quad (10\text{-}17)$$

where P = load
 A = cross-sectional area
 e = eccentricity of load
 Z = section modulus in the direction of bending
 Δ = lateral deflection of the longitudinal axis.

1) ECCENTRICITY LOADED STEEL COLUMNS

$$\sigma = \frac{\Sigma P}{A} + \frac{M}{Z} = \sigma_a + \sigma_b \quad (10\text{-}18)$$

or

$$\frac{\sigma_a}{\sigma} + \frac{\sigma_b}{\sigma} = 1$$

AISC (American Institute of Steel Construction) Formula:

$$\frac{\sigma_a}{\sigma_A} + \frac{\sigma_b}{\sigma_B} \leq 1.0$$

where M = bending moment
σ_a = axial stress (actual)
σ_b = bending stress in column (actual)
σ_A = axial stress allowable if only axial columns stress exists
σ_B = bending stress allowable if only bending stress exists.

2) ECCENTRICITY LOADED TIMBER COLUMNS

Formula (Forest Product Laboratory – U.S.)

$$\boxed{\frac{P/A}{\sigma_{ac}} + \frac{M/Z + P/A(6 + 1.5J)(e/d)}{\sigma_B - J(P/A)} \leq 1.0} \quad (10\text{-}19)$$

where σ_{ac} = allowable compressive stress
σ_B = allowable bending stress
J = $[(L/d) - 11.0]/(k - 11.0)$ and $\not< 0$ or $\not> 1$
e = eccentricity of load
d = side dimensions of rectangular columns in direction of eccentricity.

This formula is valid for rectangular or square eccentricity loaded columns, whose ends can be assumed to be pin-connected.

CHAPTER 11

COMPOSITE STRUCTURES

11.1 COMPOSITE MEMBERS IN PARALLEL

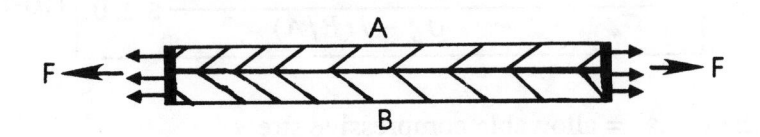

FIGURE 11.1—A composite beam (parallel).

Even though A and B are of two materials arranged in parallel (See Figure 11.1), they act as a single member.

The elongations δ_A and δ_B are the same. The strains ε_A and ε_B are the same, since $l_A = l_B$.

$$\varepsilon_A = \frac{\sigma_A}{E_A}, \varepsilon_B = \frac{\sigma_B}{E_B} \qquad (11\text{-}1)$$

$$\therefore \quad \frac{\sigma_A}{E_A} = \frac{\sigma_B}{E_B} \qquad (11\text{-}2)$$

Total force = force in $A(F_A)$ + force in $B(F_B)$

$$\therefore F = a_A \sigma_A + a_B \sigma_B \qquad (11\text{-}3)$$

Where a_A and a_B are cross-sectional areas of A and B respectively.

11.2 COMPOSITE MEMBERS IN SERIES

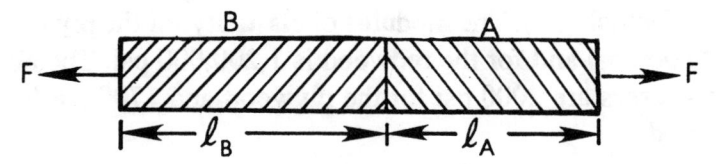

FIGURE 11.2—A composite beam (series).

Figure 11.2 shows a beam composed of two materials, arranged in a series, subjected to a tensile force F.

$$\therefore \sigma_A = \sigma_B$$

Total elongation, $\delta = \delta_A + \delta_B$

$$\delta_A = \frac{F l_A}{a_A E_A} \qquad (11\text{-}4)$$

$$\delta_B = \frac{F l_B}{a_B E_B} \qquad (11\text{-}5)$$

$$\therefore \delta = F \left(\frac{l_A}{a_A E_A} + \frac{l_B}{a_B E_B} \right) \qquad (11\text{-}6)$$

where

$$F = a_A \sigma_A = a_B \sigma_B \qquad (11\text{-}7)$$

The following example is a typical one for composite structures.

A box beam is constructed with webs of Douglas fir plywood and with flanges of redwood, as shown in the figure. The plywood is 1 in. thick and 12 in. wide. The redwood flanges are 2 in. x 4 in. (actual size). The modulus of elasticity for the plywood is 1,600,000 psi and for the redwood is 1,200,000 psi. The allowable stresses are 2000 psi for the plywood and 1700 psi for the redwood.

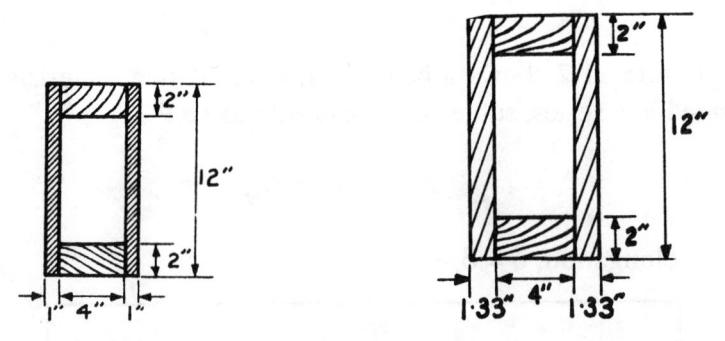

FIGURE 11.3

In solving this problem, use the equivalent section method. Convert the plywood to the redwood equivalent by multiplying the width by

$$n = \frac{E_{\text{plywood}}}{E_{\text{redwood}}} = \frac{1600000}{1200000} = 1.333 \ .$$

The resulting section is shown in Figure 11.3(b). The moment of

B-149

inertia of the equivalent section is the moment of inertia of a solid beam minus the moment of inertia of the hollow inside.

$$I_{eq} = \frac{b_1 h_1^3}{12} - \frac{b_2 h_2^3}{12}$$

$$I_{eq} = \frac{1}{12} \times 6.667 \times 12^3 - \frac{1}{12} \times 4 \times 8^3$$

$$= 789.33 \text{ in.}^4$$

The expression relating stress with moment is

$$\sigma = \frac{My}{I}$$

The beam will fail first at the external fibers, but which of the woods will fail first is not obvious.

Assuming the redwood fails first,

$$\therefore \ 1700 = \frac{M \times 6}{789.33}$$

Bending moment, $M = 223{,}643.5$ in-lb. Assuming the plywood fails first,

$$\therefore \ 2000 = \frac{M \times 6}{789.33} \times 1.33$$

Bending moment, $M = 197{,}827.07$ in-lb.

The assumption that the plywood fails first is the correct one because the maximum moment it can sustain is lower. The allowable bending moment is 197,827.07 in-lb.

11.3 REINFORCED-CONCRETE BEAMS

For a concrete beam, reinforcement with steel bars is done near the tensile side since concrete is weak in tension and relatively strong in compression.

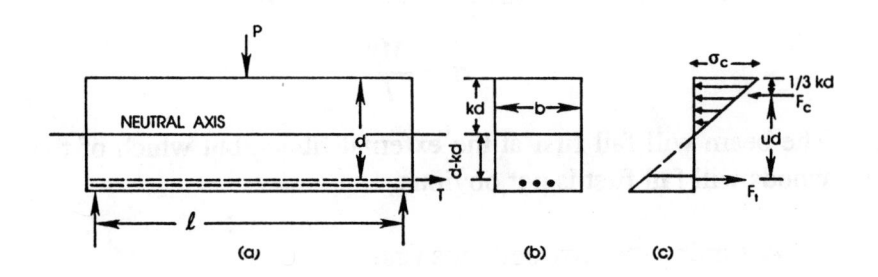

FIGURE 11.4—Stress in a reinforced concrete (a) beam, (b) section, (c) stress.

The tensile force of concrete is negligible.

Effective cross section area of beam = bd

$$V = A_s/bd \quad \text{and} \quad f = E_{ss}/E_c$$

Total compressive force (F_c) = Total tensile force (F_t).

$$A_s \sigma_s = bkd \tfrac{1}{2}\sigma_c \quad (11\text{-}8)$$

where $\sigma_c/2$ is the average stress in concrete.

When the stress in the steel governs the strength of the beam,

$$M = M_r = F_t \cdot ud = A_s \sigma_s \cdot ud = uv\, bd^2 \sigma_s$$

$$\boxed{\sigma_s = \frac{M}{uv\, bd^2}} \qquad (11\text{-}9)$$

where M = bending moment,
M_r = resisting moment.

When the beam strength is governed by the stress in the concrete,

$$M = M_r = F_c \cdot ud = bkd\,\tfrac{1}{2}\sigma_s \cdot ud = \tfrac{1}{2} ku\, bd^2 \sigma_c$$

$$\boxed{\sigma_c = \frac{M}{\tfrac{1}{2}ku\, bd^2}} \qquad (11\text{-}10)$$

we have,

$$ud = d - \tfrac{1}{3}kd$$

$$\boxed{u = 1 - \tfrac{k}{3}} \qquad (11\text{-}11)$$

How to find values of k and u:

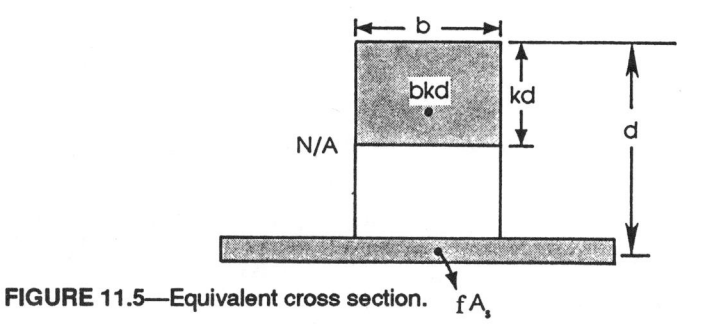

FIGURE 11.5—Equivalent cross section.

Figure 11.5 shows the equivalent section of concrete. The bottom steel section is equivalent to fA_s of concrete.

Taking moment about the top of the section,

$$M_A = bkd \cdot \tfrac{1}{2} kd + f A_s \cdot d$$

$$= \tfrac{b}{2} k^2 d^2 + fv\, bd^2$$

The area of section, $A_E = bkd + fA_s = bkd + fvbd$ ∴ neutral axis from top.

$$kd = \frac{M_A}{A_E}$$

$$\therefore \quad k^2 + 2kfv - 2fv = 0$$

$$\boxed{k = \sqrt{f^2 v^2 + 2fv} - fv} \quad (11\text{-}12)$$

where $f = E_s/E_c$ and the area of a steel bar section, A_s. Hence, v is known.

The working stress of steel = f × working times of concrete.

$$\boxed{\sigma_s = f\sigma_c} \quad (11\text{-}13)$$

from equations (11-9) and (11-10)

$$\frac{M}{uv\, bd^2} = f \cdot \frac{M}{\tfrac{1}{2} ku\, bd^2}$$

$$\boxed{\therefore \quad V = \frac{k}{2f} = \frac{kE_c}{2E_s} = \frac{k\sigma_c}{2\sigma_s}} \quad (11\text{-}14)$$

B-153

∴ *vbd* gives the amount of reinforcement required to develop, given working stresses in the concrete and steel.

The following example is a typical one for reinforced concrete analysis.

The reinforced-concrete beam with the section shown in Figure 11.6 is subjected to a positive bending moment of 50,000 ft-lb. The reinforcement consists of two #9 steel bars. (These bars are $1\frac{1}{8}$ in. in diameter and have a cross-sectional area of 1 in.^{2}). The ratio of E for steel to that of concrete to be 15, i.e., $n = 15$.

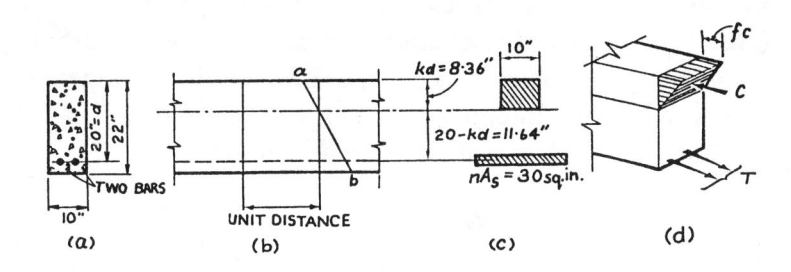

FIGURE 11.6

Plane sections are assumed to remain plane in a reinforced-concrete beam. Strains vary linearly from the neutral axis as shown in Figure 11.6(b) by the line *ab*. A transformed section in terms of concrete is used to solve this problem. However, concrete is so weak in tension that there is no assurance that minute cracks will not occur in the tension zone of the beam. For this reason no credit is given to concrete for resisting tension. On the basis of this assumption, concrete in the tension zone of a beam only holds the reinforcing steel in place. Hence, in this analysis, it virtually does not exist at all, and the transformed section assumes the form shown in Figure 11.6(c). The cross-section of concrete has its own shape above the neutral axis; below it no concrete is shown.

Steel, of course, can resist tension, so it is shown as the transformed concrete area. For computation purposes, the steel is located by a single dimension from the neutral axis to its centroid. There is a negligible difference between this distance and the true distances to the various steel fibers.

So far the idea of the neutral axis has been used, but its location is unknown. However, it is known that this axis coincides with the axis through the centroid of the transformed section. It is further known that the first (or statical) moment of the area on one side of a centroidal axis is equal to the first moment of the area on the other side. Thus, let kd be the distance from the top of the beam to the centroidal axis as shown in Figure 11.6(c), where k is an unknown ratio and d is the distance from the top of the beam to the center of the steel. An algebraic restatement of the foregoing locates the neutral axis, about which I is computed and stresses are determined.

$$\underbrace{10(kd)}_{\text{concrete area}} \times \underbrace{(kd/2)}_{\text{arm}} = \underbrace{30}_{\substack{\text{transformed} \\ \text{steel area}}} \times \underbrace{(20 - kd)}_{\text{arm}}$$

$$5(kd)^2 = 600 - 30(kd)$$
$$(kd)^2 + 6(kd) - 120 = 0$$
$$kd = \frac{-6 \pm \sqrt{36 + 480}}{2}$$

Hence $kd = 8.36$ in. and $20 - kd = 11.64$ in.

$$I = \Sigma(I_0 + Ay^2)$$

where I is the total moment of inertia of the cross section, I_0 is the moment of inertia of one section of the total about its own centroid and Ay^2 is the transfer term from the parallel axis theorem which when added to I_0 gives the moment of inertia about the centroidal axis of the total cross section.

$$I = \frac{10(8.36)^3}{12} + 10(8.36)\left(\frac{8.36}{2}\right)^2 + 0 + 30(11.64)^2 = 6,020 \text{ in.}^4$$

$$(\sigma_c)_{max} = \frac{Mc}{I} = \frac{(50,000)12(8.36)}{6,020} = 833 \text{ psi}$$

$$\sigma_s = n\frac{Mc}{I} = \frac{15(50,000)12(11.64)}{6,020} = 17,400 \text{ psi}$$

ALTERNATE APPROACH

After kd is determined, instead of computing I, a procedure evident from Figure 11.6(d) may be used. The resultant force developed by the stresses acting in a "hydrostatic" manner on the compression side of the beam must be located $kd/3$ below the top of the beam. Moreover, if b is the width of the beam, this resultant force $C = \frac{1}{2}(\sigma_c)_{max} b(kd)$ (average stress times area). The resultant tensile force T acts at the center of the steel and is equal to $A_s\sigma_s$, where A_s is the cross-sectional area of the steel. Then if jd is the distance between T and C, and since $T = C$, the applied moment M is resisted by a couple equal to Tjd or Cjd.

$$jd = d - \frac{kd}{3} = 20 - \left(\frac{8.36}{3}\right) = 17.21 \text{ in.}$$

$$M = Cjd = \tfrac{1}{2} b(kd)(\sigma_c)_{max}(jd)$$

$$(\sigma_c)_{max} = \frac{2M}{b(kd)(jd)} = \frac{2(50,000)12}{10(8.36)(17.21)} = 833 \text{ psi}$$

$$M = T(jd) = A_s \sigma_s jd$$

$$\sigma_s = \frac{M}{A_s(jd)} = \frac{(50,000)12}{2(17.21)} = 17,400 \text{ psi}$$

Both methods naturally give the same answer. The second method is more convenient in practical applications. Since steel and concrete have different allowable stresses, the beam is said to have balanced reinforcement when it is designed so that the respective stresses are at their allowable level simultaneously. Note that the beam shown would become virtually worthless if the bending moments were applied in the opposite direction.

CHAPTER 12

FAILURE CRITERIA IN DESIGN

12.1 INTRODUCTION

The prime consideration in the design of a member should be the determination of its mode of failure.

MODES OF FAILURE

(a) Fracture—brittle material

(b) Fracture—ductile material

(c) Excessive plastic deformation by overloading

(d) Excessive elastic deformation by overloading

(e) Fatigue fracture—due to cyclic/repeated loads

(f) Creep—continuing deformation over long periods

(g) Bearing—between two members

(h) Tearing—due to inadequate clearance

(i) Shearing—due to shearing, torsional, impact loads

The material of the member should have good quality, suitable for the particular mode of failure. (See Section 4.1 for material properties.)

THE FACTOR OF SAFETY

$$n = \frac{\sigma_c}{\sigma_w} \quad (12\text{-}1)$$

where n = factor of safety
  σ_w = working stress or allowable stress, used for design
  σ_c = critical stress at the condition of failure
   = ultimate strength for fracture of brittle and ductile material
   = yield strength for a small inelastic deformation of ductile material
   = stress as the maximum deflection positions for elastic deflection

WHY FACTORS OF SAFETY?

This is needed because of

(a) Uncertainties about the material properties

(b) Uncertainties due to the assumptions made in the development of the method.

(c) Uncertainties about the loads.

12.2 DESIGN FOR STATIC STRENGTH

12.2.1 DESIGN FOR ELASTIC STRENGTH

Failure is assumed to occur when the stress in the outer fiber reaches the yield point or the shear stress reaches the shearing yield point

$$\sigma_y = \frac{M_y}{I/\overline{y}} \qquad (12\text{-}2)$$

where σ_y = yield stress
M_y = bending moment of failure
I = moment of inertia
$\overline{y}$ = distance of outer fiber from neutral axis.

WORKING STRESS

$$\sigma_w = \frac{\sigma_y}{n_y} = \frac{M_y/n_y}{I/\overline{y}} = \frac{M_w}{I/\overline{y}} \qquad (12\text{-}3)$$

where σ_w = working or allowable stress
M_w = corresponding bending moment
n_y = factor of safety (See Equation 12-1)

$$\tau_y = \frac{V_y Q}{It} \qquad (12\text{-}4)$$

where τ_y = shearing yield stress
V_y = shear force at failure
Q = moment of area above the point of consideration (See Figure 12.1) about the neutral axis

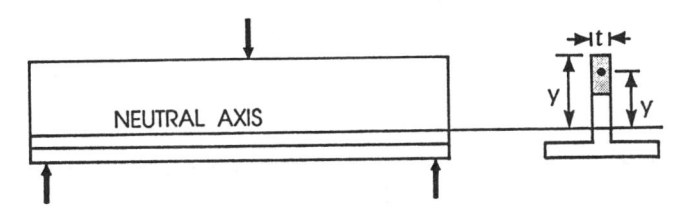

FIGURE 12.1—Shear stress in a beam.

where t = thickness of the section (See Figure 12.1)
       I = moment of inertia of the whole section

WORKING SHEARING STRESS

$$\tau_w = \frac{\tau_y}{n_y} = \frac{V_y}{n_y}\frac{Q}{It} = \frac{V_w Q}{It} \qquad (12\text{-}5)$$

where V_w = working shear force

ECONOMICAL SELECTION OF BEAM SECTIONS FOR ELASTIC STRENGTH

(a) Choose a material suitable for the use of the member.

(b) Choose the section shape and distribution of material along the length of the member to get the maximum utilization from the least quantity of material. (Subject to the cost and strength involved in the above procedure).

From Equation 12-2, the same M_y stress will be low for higher section modulus ($I/\bar{y}$). Figure 12.2 shows the values of $I/\bar{y}$ for different sections having the same cross-sectional area.

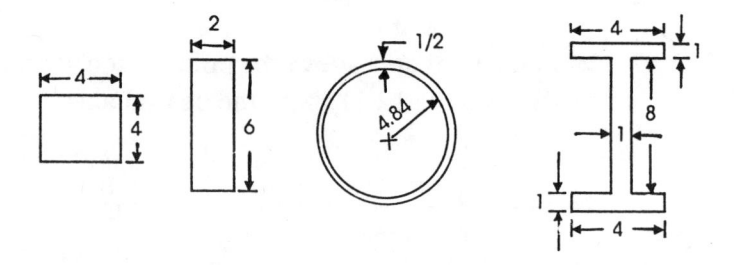

FIGURE 12.2—Strength behavior of different sections with same area.

12.2.2 DESIGN FOR PLASTIC STRENGTH

Failure is assumed to occur at the formation of a plastic hinge, which destroys the elasticity at the section and yields, and leads to the failure of the beam by large deflections.

$$\sigma_y = \frac{M_{fp}/K}{I/\overline{y}} \qquad (12\text{-}6)$$

where M_{fp} = moment of fully plastic condition
K = ratio of M_{fp} to elastic-limit moment

$$\sigma_{wp} = \frac{\sigma_y}{n_p} = \frac{(M_{fp}/K)/n_p}{I/\overline{y}} = \frac{M_{wp}/K}{I/\overline{y}} \qquad (12\text{-}7)$$

where σ_{wp} = working stress in plastic condition
M_{wp} = working bending moment in plastic condition
n_p = factor of safety for plastic design

12.2.3 DESIGN FOR BRITTLE FRACTURE CONDITIONS

It is assumed that the material acts elastically into the load that causes fracture to the beam.

$$\sigma_{wf} = \frac{\sigma_r}{n_f} = \frac{M_f/n_f}{I/\overline{y}} = \frac{M_{wf}}{I/\overline{y}} \qquad (12\text{-}8)$$

where σ_{wp} = working stress in fracture condition
σ_r = modulus of rupture = M_f/Iy
M_f = bending moment at fracture condition
n_f = factor of safety for brittle fracture design
M_{wf} = working M_f

Usually, $n_f > n_p > n_y$.

12.3 DESIGN FOR DYNAMIC STRENGTH

12.3.1 DEFINITIONS

Dynamic load—load caused by a moving body.

Equivalent static load—Static load required to produce a stress or deformation equivalent to those by the dynamic load (= dead load + live load + impact load).

Dead load—weight of a resisting member or structure.

Live load—weight of a moving body.

Impact load—increase in the static load required to cause the same stress that was produced by the dynamic load.

Impact factor—the factor to be multiplied with the live load to get the impact load.

Modulus of Resilience—the energy absorbed per unit volume in stressing the material to the elastic limit.

12.3.2 ELASTIC STRENGTH

AXIAL DYNAMIC LOAD

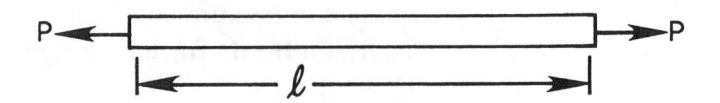

FIGURE 12.3

$$\sigma = \frac{P}{a}$$

$$U = \frac{\sigma^2}{2E} al$$

$$\sigma = \sqrt{\frac{2UE}{al}} \quad (12\text{-}9)$$

where σ = stress, which should not exceed the proportional limit of the material.
U = elastic energy
E = modulus of elasticity
a = cross-section area of the bar
l = length of bar
P = low-velocity impact load

TRANSVERSE DYNAMIC LOAD

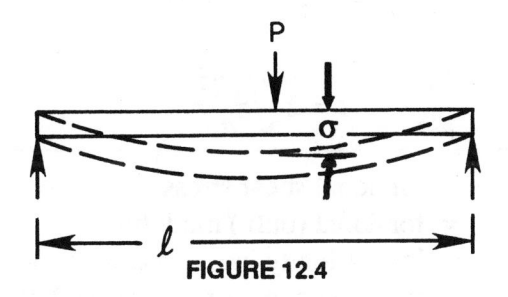

FIGURE 12.4

$$U = \frac{P}{2} \delta$$

$$\delta = \frac{Pl^3}{48 EI}$$

$$M = \frac{Pl}{4} = \frac{\sigma I}{\bar{y}}$$

$$\sigma = \bar{y} \sqrt{\frac{6EU}{Il}} \quad (12\text{-}10)$$

where $\overline{y}$ = distance of the outer fiber from neutral axis
I = moment of inertia of the section about neutral axis

TORSIONAL DYNAMIC LOAD

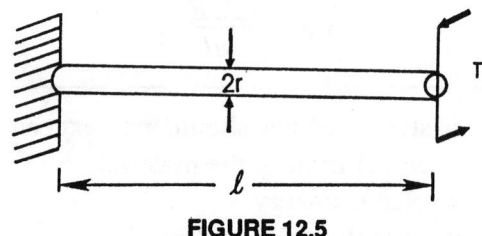

FIGURE 12.5

$$U = \frac{1}{2}T\theta, \ \theta = \frac{Tl}{GJ}, \ T = \tau \frac{J}{r}$$

$$J = \frac{\pi r^4}{2}, \ a = \pi r^2$$

$$\tau = 2\sqrt{\frac{UG}{al}}$$

(12-11)

where τ = torsional shear stress
G = torsional (bulk) modulus
T = low-velocity impact twisting moment
θ = the twist angle of the free end of shaft
J = polar moment of inertia
r = shaft radius

IMPACT LOAD DUE TO FALLING WEIGHT

See Figure 12.4. P is taken as a load falling from a height, h.

$$\sigma = \sigma_s + \sigma_s \sqrt{1 + \frac{2h}{\delta_s}}$$

(12-12)

where σ = stress at the falling point
σ_s and δ_s = stress and deflection, respectively, due to a static load equal to the weight of the falling body.

$$\text{Impact factor} = \sqrt{1 + \frac{2h}{\delta_s}} \quad \text{(12-12a)}$$

For sudden load, h = 0

$$\sigma = 2\sigma_s \quad \text{(12-12b)}$$

12.3.3 PLASTIC STRENGTH

Ductile materials offer more resistance to energy loads than brittle materials, since the modulus of resilience of the former one is higher. Even though a member is usually designed within the elastic limit, the plastic region acts as a reserve energy of the material in resisting total collapse of the member.

TOUGHNESS

This is the property of a material which can be defined as the energy absorbed per unit volume in stressing the material to fracture. This energy is equal to the work done on the material in stressing it to its ultimate strength. That is, toughness is the area under the stress-strain diagram.

$$\text{Toughness} = \frac{\sigma_y + \sigma_u}{2} \varepsilon_u \quad \text{(12-13)}$$

∴ Energy required for the fracture of the member,

$$U = \frac{\sigma_y + \sigma_u}{2} \varepsilon_u \cdot V_m \quad \text{(12-14)}$$

where σ_y = yield stress of the material
σ_u = ultimate stress
ε_u = ultimate strain
V_m = volume of member
= $a\,l$

12.4 DESIGN FOR TORSIONAL STRENGTH

Maximum shear stress in a cylindrical bar

$$\tau_{max} = \frac{T}{J/r} \qquad (12\text{-}15)$$

Shear flow in thin-walled tube,

$$q = \tau t = \frac{T}{2A} \qquad (12\text{-}16)$$

Angle of Twist,

$$\theta = \frac{T\,l}{GJ} \qquad (12\text{-}17)$$

where τ = wall thickness
A = area enclosed by the median line of the shell

12.4.1 ELASTIC STRENGTH

DUCTILE MATERIAL

$$T_w = \frac{\tau_w J}{r} = \frac{\tau_y J}{n\,r} \qquad (12\text{-}18)$$

where n = factor of safety
T_w = working torque
τ_w = working torsional shear stress

B-167

BRITTLE MATERIAL

The torque reaches its limit when σ_t reaches σ_u (i.e., $\tau = \sigma_u$)

$$T_w = \frac{\sigma_w}{r} J = \frac{\sigma_u J}{n r}$$ (12-19)

ELASTIC STIFFNESS

For specified angle of twist, θ_s

$$T_w = \frac{T}{n} = \frac{GJ\theta_s}{n r}$$ (12-20)

For specified working stress

$$\tau_s = G\gamma_s = G\frac{r\theta_s}{l}$$

$$\tau_w = \frac{\tau_s}{n}$$ (12-21)

$$T_w = \tau_w \frac{J}{r}$$

where γ_s = specified shear strain

12.4.2 PLASTIC STRENGTH

Fully plastic torque,

$$T_{fp} = \frac{4}{3}\frac{\tau_y J}{r}$$ (12-22)

B-168

Working torque,

$$T_w = \frac{T_{fp}}{n} = \frac{4}{3} \tau_w \frac{J}{r}$$ (12-23)

where τ_y = yield torsional shear stress

For ultimate or fracture torque, T_u

$$T_w = \frac{4}{3} \frac{\tau_u J}{n r} = \frac{4}{3} \tau_w \frac{J}{r}$$ (12-24)

where τ_u = ultimate torsional shear stress

12.4.3 DYNAMIC TORSIONAL STRENGTH

See Section 12.3.2 (Torsional Dynamic Load).

12.5 DESIGN OF BEAM COLUMNS

Refer to Chapter 10 also.

A bar subject to combined axial compressive and bending loads is called a beam column (See Figure 12.6). The maximum values of combined loads, P and M, are determined by the combination of two limiting cases, such as (i) $P = 0$ and (ii) $M = 0$. But this condition is difficult to satisfy, and hence we use the interactions curve method (See Figure 12.7 and Equation 12-25.)

$$\frac{P}{P_{cr}} + \frac{M}{M_{fp}} = 1$$ (12-25)

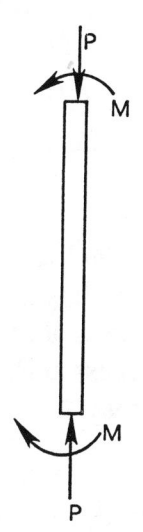

FIGURE 12.6—Beam column.

where P_{cr} = critical value of P
M_{fp} = fully plastic moment
P = $\sigma_1 a$
P_{cr} = $\sigma_{cr} a$
M = $\sigma_2 I/\bar{y}$
M_{fp} = $K\sigma_y I/\bar{y}$

∴ from Equation 12-25 we can write the interaction equation in terms of stresses

$$\frac{\sigma_1}{\sigma_{cr}} + \frac{\sigma_2}{K\sigma_y} = 1 \quad (12\text{-}26)$$

Considering the factor of safety,

$$\frac{\sigma_1/n_1}{\sigma_{cr}/n_1} + \frac{\sigma_2/n_2}{K\sigma_y/n_2} = 1$$

$$\frac{\sigma_{a_1}}{\sigma_{a_2}} + \frac{\sigma_{b_1}}{K\sigma_{b_2}} = 1 \qquad (12\text{-}27)$$

where σ_{a1} = P_w/a and $\sigma_{b1} = M_w \bar{y}/I$
P_w = working axial load
M_w = working bending moment
σ_{a2} = working stress at $M = 0$
σ_{b2} = working stress at $P = 0$
σ_{cr} = critical stress
σ_y = yield stress

12.6 FATIGUE STRENGTH. ELASTIC VIBRATION.

12.6.1 FATIGUE STRENGTH

The fatigue strength of a material is its strength to resist repeated stress.

Mode of failure is gradual or progressive fracture.

Localized stresses due to internal and external discontinuities imitates fatigue fracture.

12.6.2 VIBRATION

Vibration—periodic motion of a particle or member

Period—time interval for one cycle of motion

Frequency—number of cycles per unit of time

Amplitude—maximum displacement of the body from the equilibrium position.

Effects of vibration—Excessive wear in the structural members or machine parts.

— Repeated stresses that will cause fatigue fracture

— Creates objectionable noise

FREE VIBRATIONS

Free vibration—without any damping.

Let a body of weight W, hanging at the end of a spring, be given a small displacement. The body oscillates up and down. The period of the oscillating motion,

$$T = 2\pi \sqrt{\frac{W}{kg}} \quad (12\text{-}28)$$

$$\therefore \quad \text{Frequency} \cdot f = \frac{1}{T} = \frac{1}{2\pi} \sqrt{\frac{kg}{W}} \quad (12\text{-}29)$$

where k = spring constant (the force required to displace the spring through the unit of distance)
g = acceleration due to gravity = 386 in./sec²

FREE VIBRATION OF A BEAM

Deflections,

$$\delta = \frac{W l^3}{48 EI}$$

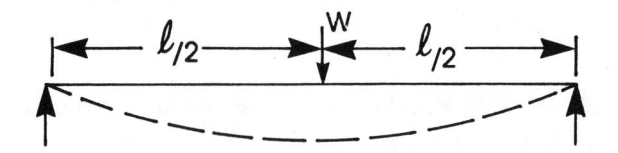

FIGURE 12.7—Simply supported beam (centrally loaded).

From Equation (12-29),

Frequency,
$$f = \frac{1}{2\pi}\sqrt{\frac{g}{W/k}} = \frac{1}{2\pi}\sqrt{\frac{g}{\delta}} = \frac{1}{2\pi}\sqrt{\frac{48EIg}{Wl^3}}$$

Period,
$$T = \frac{1}{f} = 2\pi\sqrt{\frac{Wl^3}{48EIg}}$$

(12-29a)

FREE TORSIONAL VIBRATION

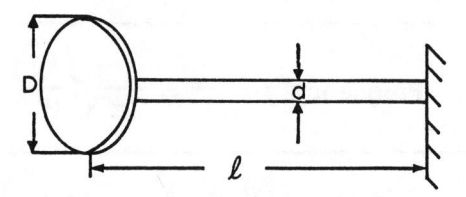

FIGURE 12.8—Torsional vibration.

Give an angular displacement θ to the disc and release it. Then the disc oscillates freely. By neglecting the moment of inertia of the shaft, we can write the period of oscillation,

$$T_t = 2\pi\sqrt{\frac{I_D}{K_t}}$$

(12-30)

where k_t = torsional spring constant
            = T/θ
       T = torque applied to have θ displacement

We know that, $\theta = T\, l/GJ$ and $I_D = \dfrac{WD^2}{8g}$

$$\therefore \quad k = \frac{T}{\theta} = \frac{G\pi d^4}{32\, l} \tag{12-31}$$

$$\therefore \quad T_t = 2\pi \sqrt{\frac{4WD^2 l}{\pi g d^4 G}} \tag{12-32}$$

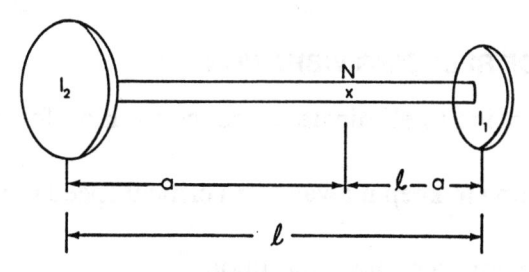

FIGURE 12.9—Nodal point system.

An initial torque is applied to the discs in opposite directions and then released from that. The two discs rotate in opposite directions about the nodal point N. The periods of oscillation for both discs are the same.

$$T = 2\pi \sqrt{\frac{I_1}{K_1}} = 2\pi \sqrt{\frac{I_2}{K_2}} \tag{12-33}$$

where k_1, k_2 = torsional spring constants
       I_1, I_2 = moment of inertia of discs

From Equations 12-31 and 12-33, we obtain

$$\frac{a}{l-a} = \frac{I_1}{I_2}$$

$$\therefore \quad a = \frac{lI_1}{I_1 + I_2}$$

Substituting this back to Equation 12-31 and then value of k to Equation 12-33, we obtain

$$T = 2\pi \sqrt{\frac{32 l I_1 I_2}{\pi d^4 G (I_1 + I_2)}} = 2\pi \sqrt{\frac{I_1 I_2 l}{JG(I_1 + I_2)}} \quad (12\text{-}34)$$

METHODS OF REDUCING VIBRATION

— Balancing to eliminate or reduce exciting force.

— Turning to keep it away from critical speed or resonance.

— Damping to reduce amplitude.

— Isolating (by giving elastic support) to change the resonance region.

CHAPTER 13

TORSION

13.1 TORQUE

From equilibrium, it is known that the sum of the externally applied torques on a body must equal the sum of the internal torques. For this, $\Sigma M_x = 0$ is the equation of statics.

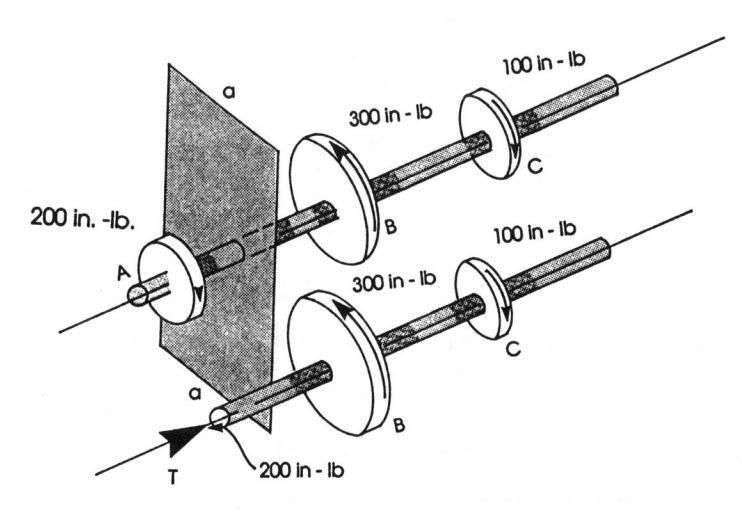

FIGURE 13.1—(a) A shaft in equilibrium has external torque applied. (b) To determine the torque at any section of the shaft, total equilibrium applies.

By applying this equation to an isolated section of the body, one finds the internal resisting torque developed within the member necessary to balance the externally applied torques. (See Figure 13.1).

13.2 TORSION FORMULAS

$$T = \frac{\tau_{max}}{c} \int_A \rho^2 dA \qquad (13\text{-}1)$$

where T = resisting torque
 τ_{max} = maximum shearing stress
 c = radius of the member
 ρ = distance from the center of the member to some arbitrary point
 A = cross-sectional area of the member

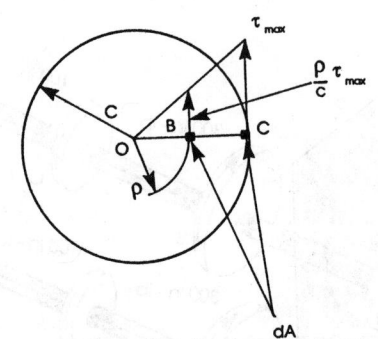

FIGURE 13.2—Cross-section of a circular member and the stress variation.

13.2.1 POLAR MOMENT OF INERTIA

$$J = \int \rho^2 dA \qquad (13\text{-}2)$$

where J = the polar moment of inertia

For a circular section,

$$J = \int_A \rho^2 dA = \int_\sigma^C 2\pi\rho^3 d\rho = \frac{\pi c^4}{2} = \frac{\pi d^4}{32} \quad (13\text{-}3)$$

where d = the diameter of a solid circular shaft. Hence, the equation for resisting torque may be written as:

$$T = \tau_{max} \frac{J}{C} \quad (13\text{-}4)$$

13.2.2 THE POLAR MOMENT OF INERTIA FOR A HOLLOW CIRCULAR SHAFT

$$J = \int_A \rho^2 dA = \int_b^C 2\pi\rho^3 d\rho = \frac{\pi c^4}{2} - \frac{\pi b^4}{2} \quad (13\text{-}5)$$

where c = the outer radius
       b = is the inner radius.

13.3 ANGLE OF TWIST OF CIRCULAR MEMBERS

Knowing the angular twist, ϕ, the torque and the polar moment of inertia can be found. There is also a geometric relation between the maximum shearing strain, γ_{max}, and ϕ (See Figure 13.3).

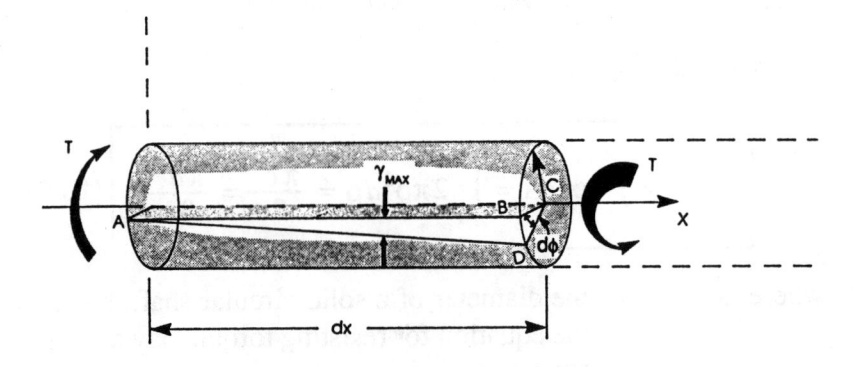

FIGURE 13.3—Element of circular shaft subjected to a torque.

13.3.1 TWIST

$$\phi = \int_B^x \frac{T(x)}{J(x)G} dx + C_1 \quad (13\text{-}6)$$

where ϕ = angle of twist (in radians)
$T(x)$ = torque on the element
$J(x)$ = polar moment of inertia
G = shearing modulus of elasticity

Where C_1 is the initial angle of twist measured in radians.

13.3.2 γ_{max} AND ϕ

$$\gamma_{max} dx = d\phi c \quad (13\text{-}7)$$

where c = the radius of the circular element.

13.4 TORQUE AND TWIST FOR CIRCULAR SHAFTS IN INELASTIC RANGE

$$\phi = \int_0^x \frac{\gamma_a}{\rho_a} dx \qquad (13\text{-}8)$$

where ϕ = angle of twist
x = length of shaft
γ_a = shearing strain at point a that is not on the outer annulus of the shaft
ρ_a = distance from the axis of the circular shaft to some point, a, not on the outer annulus of the shaft

$$T = \int_A [\tau_a(dA)] \rho_a \qquad (13\text{-}9)$$

where T = torque
A = area
τ_a = stress at point a which is in the inelastic range
ρ_a = distance from the axis of the shaft to some point a.

Note: These equations are especially useful when analyzing thin-walled shafts using stress-strain diagrams. They are used to analyze the properties of materials that undergo both elastic and plastic deformations.

13.5 SOLID NONCIRCULAR MEMBERS

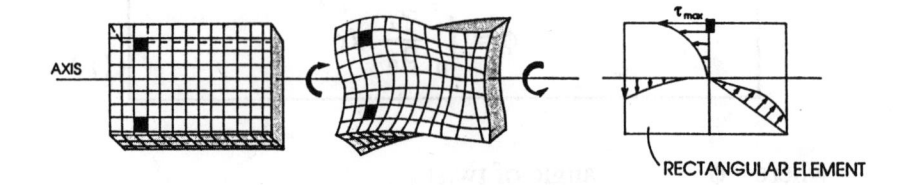

FIGURE 13.4—Effects on a rectangular shaft under a twisting torque. (a) A rectangular shaft before deformation. (b) After deformation. (c) Shearing Stress – Distribution due to applied torque on a rectangular element.

13.5.1 TORQUE FOR RECTANGULAR SHAFT IN TORSION

$$T = \tau_{max} \alpha b c^2 \quad (13\text{-}10)$$

where T = torque
τ_{max} = maximum shearing stress
b = long side of the rectangle
c = short side of the rectangle
α = calculated coefficient determined by the ratio of b/c

13.5.2 ANGLE OF TWIST FOR RECTANGULAR SHAFT IN TORSION

$$\phi = \frac{TL}{\beta b c^3 G} \quad (13\text{-}11)$$

where φ = angle of twist
       T = torque
       L = length of the member
       b = long side of the rectangular cross-section
       c = short side of the rectangular cross-section
       β = calculated coefficient determined by the ratio of b/c
       G = shearing modulus of elasticity.

Note: α and β are obtained from tables. For thin sections, when b is much greater than c, the values of α and β approach $1/3$.

13.6 THIN-WALLED HOLLOW MEMBERS

13.6.1 SHEAR FLOW

In any thin-walled member of variable thickness, the product of the shearing stress and the wall thickness, at any section of the member, is a constant, q, called the shear flow.

$$q = \tau t \qquad (13\text{-}12)$$

where q = shear flow at a point
       τ = shear stress at that point
       t = wall thickness at that point

13.6.2 TORQUE

The general equation for torque acting on a section of tubing may be calculated using the formula

$$T = \oint rq \, ds \qquad (13\text{-}13)$$

Where T is the torque, $q\,ds$ is the force per differential length,

q is the shear flow and ds is the length of the perimeter, and r is the perpendicular distance from an axial line passing through the center of the hollow tube, to the center line, which runs through the center of the area of material itself (See Figure 13.5).

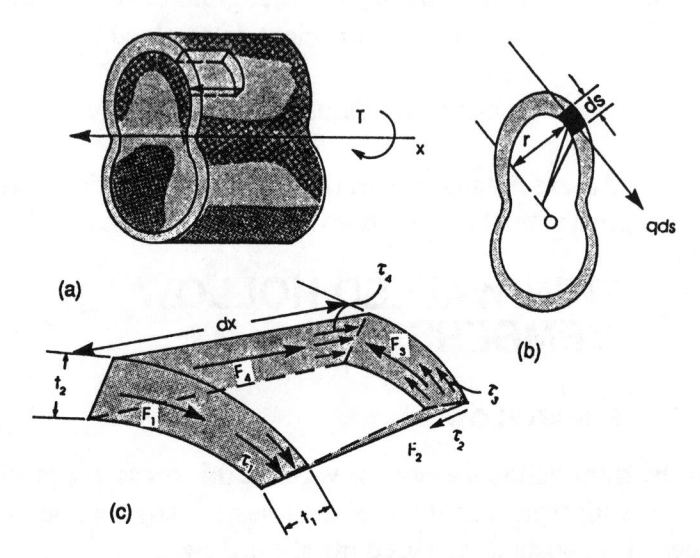

FIGURE 13.5—Thin-walled member of variable thickness. (a) Cross-section (b) and an element in that face, showing the forces and shear stresses.

This equation for torque in thin-walled tubes may be approximated.*

$$T = 2Aq \quad (13\text{-}14)$$

where T = torque

---

* This equation, Brent's Formula, is only applicable to thin-walled tubes. Here, the equation proves valid because the area, A, approximately an average of the two areas enclosed by the inside and outside surfaces of the tube. This formula is not applicable if the tube is a slit.

A	=	area bounded by the center line of the perimeter of the tube
q	=	shear flow

13.6.3 ANGLE OF TWIST

For a hollow tube made of linearly elastic material, the angle of twist may be written as,

$$\theta = \frac{d\phi}{dx} = \frac{T}{4A^2 G} \oint \frac{ds}{t} \quad (13\text{-}15)$$

where
θ	=	angle of twist in a unit distance
T	=	torque
A	=	cross-sectional area of the section
G	=	shearing modulus of elasticity
ds	=	differential length of the perimeter of the section
t	=	thickness of the section

CHAPTER 14

BOLTED, RIVETED AND WELDED JOINTS

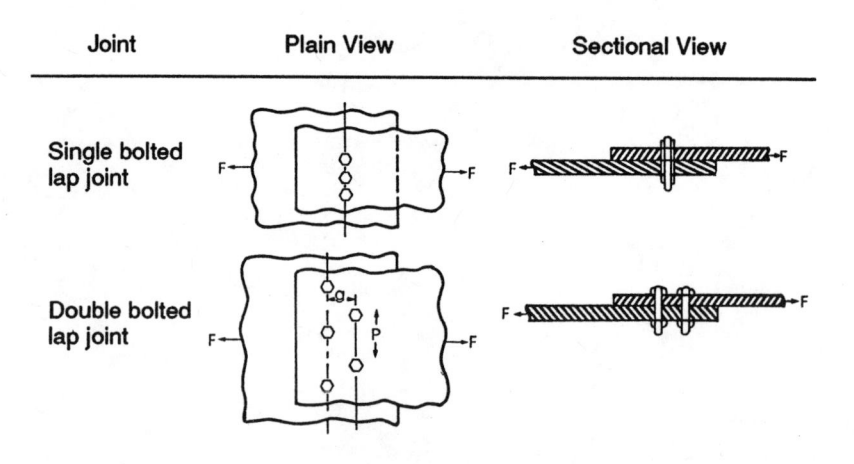

14.1 BOLTED JOINTS

Types of joints (See Figure 14.1).

i) Lap joints

ii) Butt joins

Joint	Plain View	Sectional View
Single bolted lap joint		
Double bolted lap joint		

B-185

Single bolted
butt joint

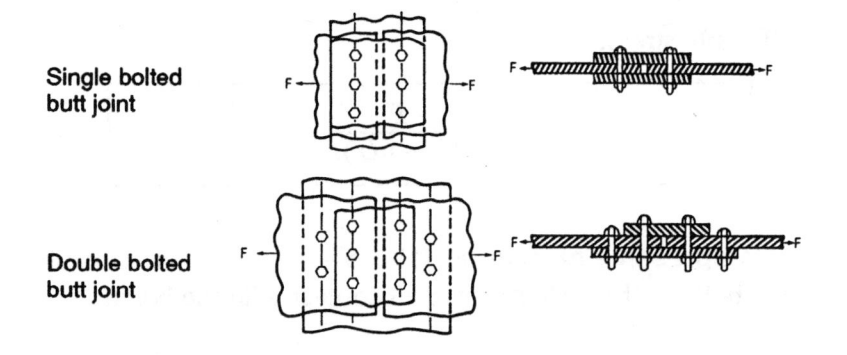

Double bolted
butt joint

FIGURE 14.1—Bolted Joints.

14.1.1   DEFINITIONS

Pitch—distance p between centers of adjacent rows of rivets or bolts.

Gage—distance g between centerline of parallel rows of rivets or bolts.

Cover plate—longitudinal plate connecting the two main plates, as in a butt joint.

Edge distance—the distance from the center of a bolt or rivet to the edge of the plate.

14.1.2   TYPES OF FAILURES AND STRESSES

Shearing failure—joint may fail by shearing the bolts.

Shear stress,

$$\sigma_S = \frac{F}{n\left(\frac{\pi d^2}{4}\right)} \qquad (14.1)$$

Tension failure—plate may pull apart at its weakest section.

Tensile stress,

$$\sigma_t = \frac{F}{(b - nD)t} \quad (14\text{-}2)$$

Compressive or bearing failure—occurs by the crushing of the bolts or the plate material in contact with the bolts.

Compressive or bearing stress,

$$\sigma_c = \frac{F}{ntd} \quad (14\text{-}3)$$

where F = external force applied to the joint, lb.
 n = number of bolts in the joint
 d = diameter of bolt (inches)
 t = thickness of plate (inches)
 b = width of plate (inches)
 D = diameter of bolt hold (inches)

Efficiency of a bolted joint

$$\eta = \frac{\text{strength of the joint}}{\text{strength of the unpunched plate}} \quad (14\text{-}4)$$

Strength of the joint should be the least among the shearing, tensile and bearing strengths.

14.2 RIVETED JOINTS

The procedures discussed above for the bolted joints are

applicable to riveted joints too; the type of joints and other behaviors are the same as well.

EFFECT OF ECCENTRIC LOADS

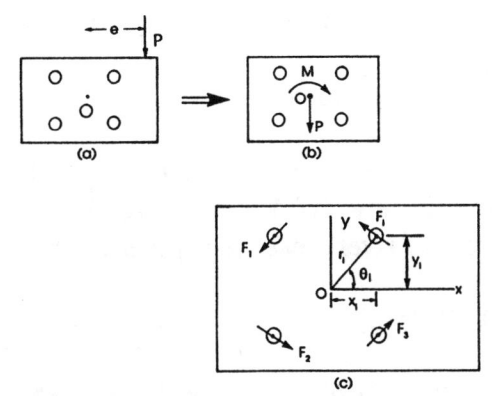

FIGURE 14.2—Eccentric load effect.

A load, P, acts on an eccentric distance, e, from the centroid O of the rivet group. The effect is a force, P, evenly distributed to all n rivets, and a clockwise moment, M. The rivet group sets up an anti-clockwise moment to oppose the clockwise external moment, as shown in Figure 14.2(c).

Shear stress in a rivet is proportional to the radial distance

$$\tau_i = kr_i \qquad (14\text{-}5)$$

Resisting force

$$F_i = kr_i A_i \qquad (14\text{-}6)$$

At equilibrium

$$\Sigma F_x = 0 \,, \Sigma F_y = 0 \text{ and } \Sigma M = 0 \qquad (14\text{-}7)$$

$$\Sigma F_y = F_1 \cos \theta_1 + F_2 \cos \theta_2 + \ldots = \sum_{i=1}^{n} F_i \cos \theta_i = 0 \quad (14\text{-}8)$$

From equation 14-6

$$\Sigma F_x = \sum_{i=1}^{n}(kr_i A_i)\cos \theta_i = k\sum_{i=1}^{n} x_i A_1 = k\overline{x}_i A \quad (14\text{-}9)$$

where A = total area of n rivets.

From Equation 14-9, it is clear that $\overline{x}_i = 0$ since $kA \neq 0$. Similarly, we can show that $\overline{y}_i = 0$. Hence, the center of rotation and the centroid of rivet areas are at the same point 0.

$\Sigma M_0 = 0$ yields

$$Pe = \Sigma r_i F_i = \Sigma r_i (kr_i A_i) = k\Sigma r_i^2 A_i \quad (14\text{-}10)$$

For same size rivets of area a

$$F_i = kar_i \quad (14\text{-}11)$$

Therefore, Equation 14-10 yields

$$Pe = ka\Sigma r_i^2 = \left(\frac{F_i}{r_i}\right)\Sigma r_i^2 \quad (14\text{-}12)$$

where $r_i = \sqrt{x_i^2 + y_i^2}$ \quad (14-13)

$$Pe = \frac{F_i \Sigma(x^2 + y^2)}{\sqrt{x_i^2 + y_i^2}} \quad (14\text{-}14)$$

Therefore

$$\boxed{F_i = \frac{Pe\sqrt{x_i^2 + y_i^2}}{\Sigma x^2 + \Sigma y^2}} \quad (14\text{-}15)$$

F_i is the force in rivet caused by rotation alone, so the total force in a rivet is

$$R_i = F_i + \frac{P}{n} \qquad (14\text{-}16)$$

where P/n = the component of translation force

14.3 WELDED JOINTS

Types of Welds

14.3.1 FILLET WELDS

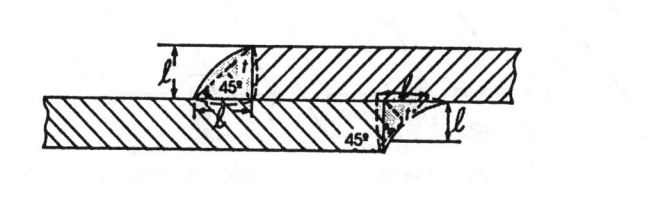

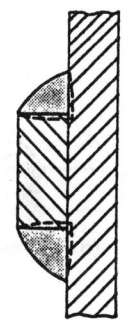

FIGURE 14.3—Types of fillet welds: (a) End fillet weld (b) Side fillet weld.

The size of the fillet weld is determined by the length of the leg, l. The throat area, (the throat thickness, (t) x effective length of weld (L)) is considered the main design parameter of the weld.

Throat
$$t = l \sin 45° = 0.707 l \qquad (14\text{-}17)$$

The allowable load per inch of effective length of weld is

$$P_a = \text{throat area} \times \frac{\sigma_s}{L} \quad (14\text{-}18)$$

$$P_a = 0.707 \, l \, \sigma_s \quad (14\text{-}19)$$

where σ_s = allowable shearing stress
= 18,000 psi for A36 steel Class E 60 series electrodes
= 21,000 psi for A36, A242, A441, A572 and A588 steels with Class E 70 series electrodes

14.3.2 BUTT WELDS

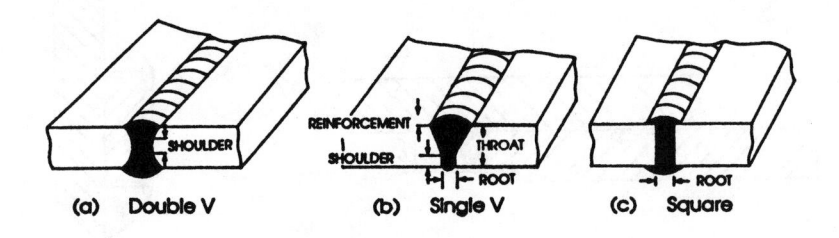

FIGURE 14.4

(a) Double V (b) Single V (c) Square

A square butt weld is used for plate thickness (t_p) up to $1/4$ in. single-V for $1/4 \leq t_p \leq 1/2$ in. and double-V for $t_p > 1/2$ in. Throat, $t = t_p$ or thinner plate.

The strength of a complete-penetration weld can be considered to be the same as the strength of the plate material.

The effective weld length for a given load, P, is

$$L = \frac{P}{P_a} \quad (14\text{-}20)$$

EXAMPLE

An angle of size 9 in × 3 in × $1/2$ is welded to a plate with the longer side parallel to the plate. A load of 63,630 lb is applied through the centroid of the angle, as in Figure 14.5. Find the effective length of fillet welds if (i) only the sides of angles are welded and (ii) the sides and end are welded. Assume working shearing stress is 12,000 psi.

The leg of the weld is assumed to be equal to the thickness of the angle (i.e.; $l = 1/2$ in).

Throat, $t = 0.707l = 0.3535$

Force per inch of weld = t · 1· 12,000 = 4242 lb.

Total length of weld required to withstand the load = 63,630/ 4242 = 15 in.

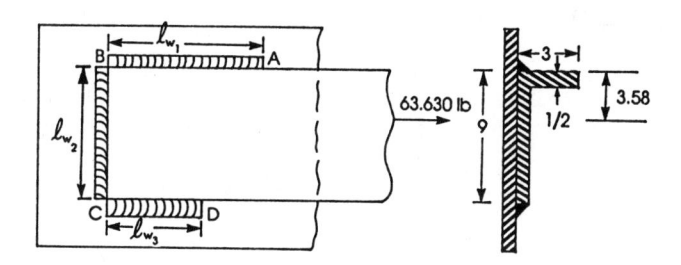

FIGURE 14.5—Fillet weld arrangement.

Centroid of angle from top of flange

$$= \frac{\frac{5}{2} \times \frac{1}{2} \times \frac{1}{4} + 9 \times \frac{1}{2} \times \frac{9}{2}}{\frac{5}{2} \times \frac{1}{2} + 9 \times \frac{1}{2}} = 3.58 \text{ in.}$$

SIDE WELDS ALONE

For equilibrium, moment about B is zero

$$\left(4242 \cdot l_{w_3}\right) \cdot 9 = 63{,}630 \times 3.58$$

$$l_{w_3} = 5.97 \text{ in.}$$

We have $l_{w_1} + l_{w_3} = 15$

$$\therefore \quad l_{w_1} = 15 - l_{w_3} = 9.03 \text{ in.}$$

WELDS AT SIDES AND END

Moment about B gives

$$4242\, l_{w_3} \cdot 9 + 4242\, l_{w_2} \cdot \frac{l_{w_2}}{2} = 63{,}630 \times 3.58$$

But $l_{w_2} = 9$ in.

$$\therefore \quad l_{w_3} = \frac{\left(63{,}630 \times 3.58 - 4242 \cdot 9 \cdot \frac{9}{2}\right)}{(4242 \cdot 9)}$$

$$= 1.47 \text{ in.}$$

We have

$$l_{w_1} + l_{w_2} + l_{w_3} = 15 \text{ in.}$$

$$\therefore \quad l_{w_1} = 15 - l_{w_2} - l_{w_3} = 3.53 \text{ in.}$$

14.4 THIN-WALLED PRESSURE VESSELS

Pressure vessels are mostly cylindrical or spherical in shape. The fluid contained in a vessel exerts pressure normal to the surface of the container. The thickness of a wall less than 10% of the vessel diameter would come under the category of thin-walled pressure vessels.

14.4.1 STRESS ON LONGITUDINAL SEAM

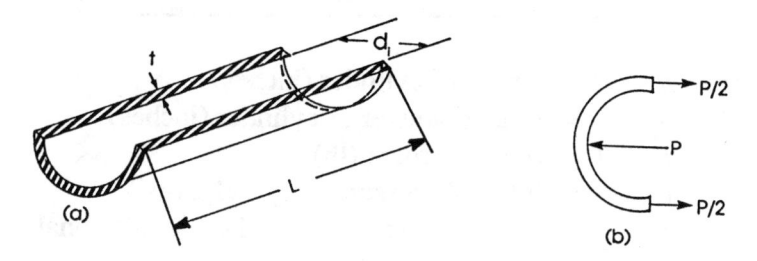

FIGURE 14.6—(a) Longitudinal section of a cylinder (thin-walled). (b) Resultant horizontal components.

The pressure components normal to the longitudinal plane are the cause of failure. Let P be the resultant of all these components.

Force = Pressure × projected area in the longitudinal plane.

$$P = pd_i L \qquad (14\text{-}21)$$

This pressure force is held to equilibrium by the reaction force of the material (See Figure 14.4(b)), sharing equally by both sides.

B-194

Therefore, $\frac{P}{2} = Lt\sigma_t$

$$P = 2Lt\sigma_t \quad (14\text{-}22)$$

For equilibrium, equating Equations 14-21 and 14-22

$$2Lt\sigma_t = pd_i L$$

$$\sigma_t = \frac{pd_i}{2t} \quad (14\text{-}23)$$

where L = length of cylinder (inches)
d_i = inner diameter of cylinder (inches)
P = pressure force (lb)
p = internal pressure in cylinder (psi)
σ_t = tensile stress in material on longitudinal section (psi)
t = plate thickness (inches)

Force on one longitudinal seam of length l

$$F = \frac{pd_i L}{2} \quad (14\text{-}24)$$

14.4.2 STRESS ON TRANSVERSE (CIRCUMFERENTIAL) SEAM

Pressure force

$$P = pA = p\frac{\pi d_i^5}{4} \quad (14\text{-}25)$$

This force should be resisted by the material in the transverse ring. Therefore, the resistance force is

$$P_r = \pi d_i t \sigma_t \qquad (14\text{-}26)$$

But $P = P_R$

$$p \frac{\pi d_i^2}{4} = \pi d_i t \sigma_t$$

$$\sigma_t = \frac{p d_i}{4t} \qquad (14\text{-}27)$$

Force on a transverse seam of length l,

$$F = A_t \sigma_t = \frac{p d_i l}{4} \qquad (14\text{-}28)$$

where A_t = area of the transverse seam = $l \cdot t$
  l = length of transverse seam (need not be the whole circumference)

CHAPTER 15

ENERGY METHODS

15.1 LOAD-DEFORMATION RELATIONSHIPS

An elastic body deforms under the action of an external load and the work done on the body due to the external load, which is stored in the body in the form of potential energy. This internal stored energy is termed strain energy and it can be completely

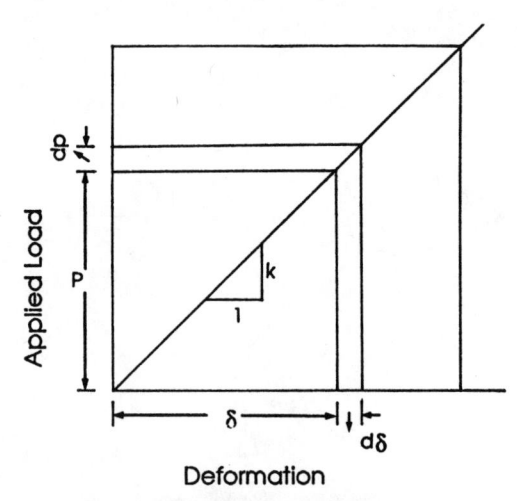

FIGURE 15.1—Load-deformation relationship.

restored for materials obeying Hooke's law. The load-deformation relation for such materials is represented by a straight line as shown in Figure 15.1.

The mathematical expression for strain energy can be obtained as follows:

If a load, P, is initially applied on an elastic structural member resulting in the deformation, δ, in the direction of a load, then a small increment in P, i.e, dP causes a small increment in δ, i.e. $d\delta$ thus yielding a small increase in the stored strain energy du of the material.

Mathematically,

$$du = Pd\delta$$

$$\boxed{P = k\delta} \quad (15\text{-}1)$$

k is called the stiffness of the structural member.

$$du = k\delta d\delta$$

$$\therefore \quad U = \int du = \int k\delta d\delta \quad (15\text{-}2)$$

$$= k\int \delta d\delta$$

$$= \frac{k\delta^2}{2}$$

$$\boxed{\therefore \quad U = \frac{(k\delta)\delta}{2} = \frac{1}{2}P\delta} \quad (15\text{-}3)$$

15.2 STRAIN ENERGY CAUSED BY STRESSES

Consider a steel bar of a length l and uniform cross sectional area A, (Figure 15.2), subject to axial load P (tension or compression). Let δ be the elongation in the bar. The total energy stored in the bar due to the deformation of the bar can be expressed as,

$$U = \frac{1}{2}P\delta$$

But $\sigma = E\varepsilon$ and $\varepsilon = \delta$

$$\therefore \quad \sigma = \frac{E\delta}{l} \qquad (15\text{-}4)$$

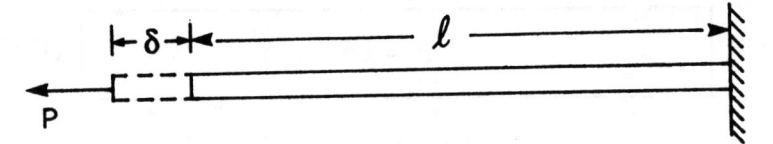

FIGURE 15.2

Also, $\sigma = \dfrac{P}{A}$, therefore

$$\frac{E\delta}{l} = \frac{P}{A} \Rightarrow \boxed{\delta = \frac{Pl}{EA}} \qquad (15\text{-}5)$$

Substituting for δ in Equation 15-3,

$$\boxed{U = \frac{1}{2}P\left(\frac{Pl}{EA}\right) = \frac{P^2 l}{2EA}} \qquad (15\text{-}6)$$

Again, $P = \sigma A = \dfrac{E\delta A}{l}$.

Substituting for P in Equation 15-6, therefore,

$$U = \frac{1}{2}\frac{\delta^2 EA}{l} \qquad (15\text{-}7)$$

where u is the strain energy or energy of deformation. U is given in terms of a load by Equation 15-6 and in terms of deformation by Equation 15-7.

15.3 IMPACT PHENOMENA

Consider a vertically hung rod, (Figure 15.3) along which a load W is suddenly allowed to drop freely from a height, h. The load strikes a collar at the end of the rod and elongates the rod. Assume the bar to be weightless.

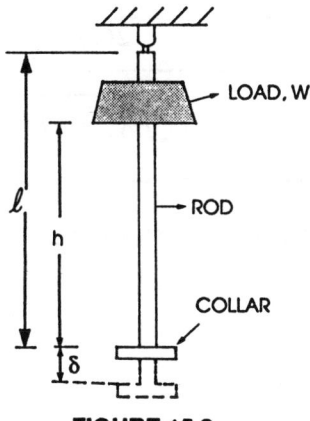

FIGURE 15.3

The energy stored in the rod during the elongation δ comes from the work done by the load through a vertical distance ($h + \delta$). This face can be expressed using Equation 15-7 as follows:

$$\frac{\delta^2 AE}{2l} = W(h + \delta)$$

or

$$\frac{\delta^2}{2(h+\delta)} = \frac{Wl}{AE} = \delta_s \qquad (15\text{-}8)$$

where δ_s = static elongation of rod due to load W.
δ = instantaneous elongation of rod.

From Equation 15-8, therefore,

$$\delta^2 = 2(h+\delta)\delta_s$$

or

$$\delta^2 - 2\delta_s\delta - 2h\delta_s = 0$$

Applying the quadratic formula,

$$\delta = -\frac{(-2\delta_s) \pm \sqrt{4\delta_s^2 - 4(1)(-2h\delta_s)}}{2(1)}$$

$$\boxed{\delta = \delta_s + \sqrt{\delta_s^2 + 2h\delta_s}} \qquad (15\text{-}9)$$

where, in ± signs, + sign indicates the elongation.

From Equation 15-4,

$$\delta_s = \frac{\sigma_s l}{E}$$

$$\frac{\sigma l}{E} = \frac{\sigma_s l}{E} + \sqrt{\frac{\sigma_s^2 l^2}{E^2} + \frac{2h\sigma_s l}{E}}$$

$$\frac{\sigma l}{E} = \sigma_s \frac{l}{E} + \frac{l}{E}\sqrt{\sigma_s^2 + \frac{2h\sigma_s E}{l}}$$

Finally,

$$\sigma = \sigma_s + \sqrt{\sigma_s^2 + \left(\frac{2hE}{l}\right)\sigma_s} \qquad (15\text{-}10)$$

15.4 CASTIGLIANO'S THEOREM

Castigliano's theorem states that the rate of change of the strain energy for a body with respect to any statically independent force P gives the deflection component of the point of application of this force in the direction of the force, i.e. $\delta = \dfrac{\partial U}{\partial P}$. A similar statement is made for moments and rotations: The rate of change of strain energy with respect to a statically independent point couple M, gives the amount of rotation at the point of application of the point couple, about an axis collinear with the complete moment,

$$\frac{\partial U}{\partial P} = \theta$$

In most beams, the bending stress, $\sigma = M_b y/I$, is the only stress that causes significant deflection. The bending stress acts only in the direction of the axis of the beam; therefore, the equation for the strain energy due to stress that acts in only one direction can be used:

$$U = \int \frac{\sigma^2}{2E}\, dV \qquad (15\text{-}11)$$

The integral is taken over the whole volume, V, of the beam. Substituting $M_b y/I$ for σ and $(\int dA)dx$ for dV results in

$$U = \int_0^L \frac{M_b^2}{2EI^2} \left(\int y^2 dA\right) dx$$

Since I is defined as $\int y^2 dA$, the integral cancels with one of the I's.

$$U = \int_0^L \frac{M_b^2}{2EI} dx \qquad (15\text{-}12)$$

The terminal deflection, δ, is in line with the load P, and so according to Castigliano's theorem, $\delta = \frac{\partial U}{\partial P}$, where σ is positive downward, the direction of the applied load P. By Leibnitz's rule for differentiation under integral sign:

$$\frac{\partial U}{\partial P} = \frac{\partial \int_0^L \frac{M_b^2}{2EI} dx}{\partial P} = \int_0^L \frac{1}{2EI} \frac{\partial (M_b^2)}{\partial P} dx \qquad (15\text{-}13)$$

$$= \int_0^L \frac{1}{2EI} 2 M_b \frac{\partial M_b}{\partial P} dx = \int_0^L \frac{M_b}{EI} \frac{\partial M_b}{\partial P} dx$$

The following examples will clear up the application of Castigliano's theorem.

EXAMPLE I

Deflection in an elastic beam due to bending caused by applied force, P, at the center:

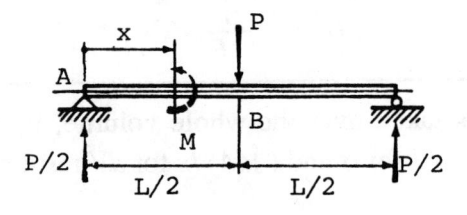

FIGURE 15.4

B-203

The expression of the internal strain energy in bending is

$$U = \int \frac{M^2}{(2EI)} dx$$

Since, according to Castigliano's theorem, the required deflection is a derivative of this function, it is advantageous to differentiate the expression for U before integrating. In problems where M is a complex function, this scheme is particularly useful. In this case the following relation becomes applicable:

$$\Delta = \frac{\partial U}{\partial P} = \int_0^L \frac{M}{EI} \frac{\partial M}{\partial P} dx \qquad (15\text{-}14)$$

Proceeding on this basis, one has from A to B:

$$M = +\frac{P}{2} x \quad \text{and} \quad \frac{\partial M}{\partial P} = \frac{x}{2}$$

On substituting these relations into Equation 15-14 and observing the symmetry of the problem,

$$\Delta = 2 \int_0^{L/2} \frac{Px^2}{4EI} dx \qquad (15\text{-}15)$$

Integration yields

$$\Delta = \frac{2P}{4EI} \frac{x^3}{3} \bigg|_0^{L/2} = \frac{2P}{4EI} \frac{\left(\frac{L}{2}\right)^3}{3} = +\frac{PL^3}{48EI} \qquad (15\text{-}16)$$

The positive sign indicates that the deflection takes place in the direction of the applied force P.

EXAMPLE II

Deflection and angular rotation of the end of a uniformly loaded cantilever, Figure 15.5(a) (constant EI):

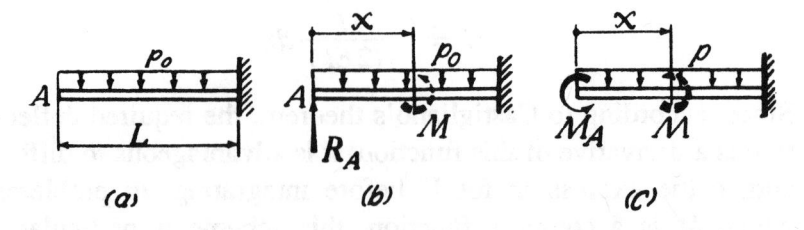

FIGURE 15.5

No forces were applied at the end of the cantilever where the displacements are to be found. Therefore, in order to be able to apply Castigliano's theorem, a fictitious force must be added corresponding to the displacement sought. Thus, as shown in Figure 15.5(b), in addition to the specified loading, a force R_A has been introduced. This permits determining $\dfrac{\partial U}{\partial R_A}$, which with $R_A = 0$ gives the vertical deflection of point A.

The expression for the internal strain energy in bending is

$$U = \int \frac{M^2}{2EI} dx \qquad (15\text{-}17)$$

Then the vertical deflection of point A is

$$\Delta_A = \frac{\partial U}{\partial R_A} = \int_0^L \frac{M}{EI} \frac{\partial M}{\partial R_A} dx \qquad (15\text{-}18)$$

Now,

$$M = \frac{-p_0 x^2}{2} + R_A x$$

Hence, $\dfrac{\partial M}{\partial R_A} = x$

B-205

Substitution of these relations into Equation 15-18 yields

$$\Delta_A = \frac{1}{EI} \int_0^L \left(\frac{-p_0 x^2}{2} + R_A x \right) x\, dx$$

Since $R_A = 0$, the above equation becomes

$$\Delta_A = \frac{1}{2EI} \int_0^L (-p_0 x^3)\, dx$$

Integration gives

$$\Delta_A = \frac{-p_0 L^4}{8EI} \tag{15-19}$$

where the negative sign shows that the deflection is in the opposite direction to that assumed for force R_A. If R_A in the above integration were not set equal to zero, the end deflection due to p_0 and R_A would be found.

The angular rotation of the beam at A can be found in a manner analogous to the above. A fictitious moment M_A is applied to the end, Figure 15-5(c), and the calculations are made in much the same manner as before:

$$M = -\frac{p_0 x^2}{2} - M_A \quad \text{and} \quad \frac{\partial M}{\partial M_A} - 1$$

$$\Delta_A = \frac{\partial U}{\partial M_A} = \frac{1}{EI} \int_0^L \left(-\frac{p_0 x^2}{2} - \overset{0}{\cancel{M_A}} \right)(-1)\, dx \tag{15-20}$$

$$= +\frac{p_0 L^3}{6EI}$$

where the sign indicates that the sense of the rotation of the end coincides with the assumed sense of fictitious moment M_A.

15.4.1 ENERGY STORED DURING BENDING

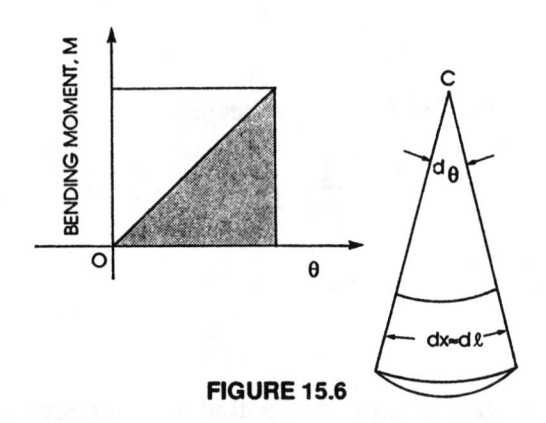

FIGURE 15.6

Consider a short piece of a beam, (Figure 15.6), under pure bending. Using flexure formula,

$$\frac{M}{I} = \frac{\sigma}{y} = \frac{E}{R} \Rightarrow \frac{1}{R} = \frac{M}{EI}$$

Let dx be the length of beam, then $dx \cong dl = Rd\theta$ so that,

$$d\theta = \frac{1}{R} dx = \frac{M}{EI} dx$$

Integration yields,

$$\theta = \int_0^l \frac{M}{EI} dx \quad \text{or} \quad \theta = \frac{Ml}{EI} \qquad (15\text{-}21)$$

This shows a linear relation between bending moment M and the curvature angle θ. As we saw previously in the cases of axial and torsional deformation, the strain energy during bending is the area below the M-θ curve and is derived by,

$$U_b = \frac{1}{2} M\theta \qquad (15\text{-}22)$$

where U_b = strain energy during bending
       $M/2$ = average moment (or couple)

$$U_b = \frac{1}{2} M \left(\frac{Ml}{EI}\right) = \frac{1}{2} \frac{M^2 l}{EI} \qquad (15\text{-}23)$$

In terms of curvature angle θ,

$$U_b = \frac{1}{2}\left(\frac{EI\theta}{l}\right)\theta = \frac{EI}{2l}\theta^2 \qquad (15\text{-}24)$$

The general equation of strain energy due to bending is given by,

$$U_b = \int_l \frac{M^2}{2EI}\, dx \qquad (15\text{-}25)$$

where M = bending moment at any point x in a beam or bar
       l = length of beam over which integration is carried out; dx is the elemental length.

EXAMPLE I

Simply supported beam with a central point load, F. Strain energy caused by bending:

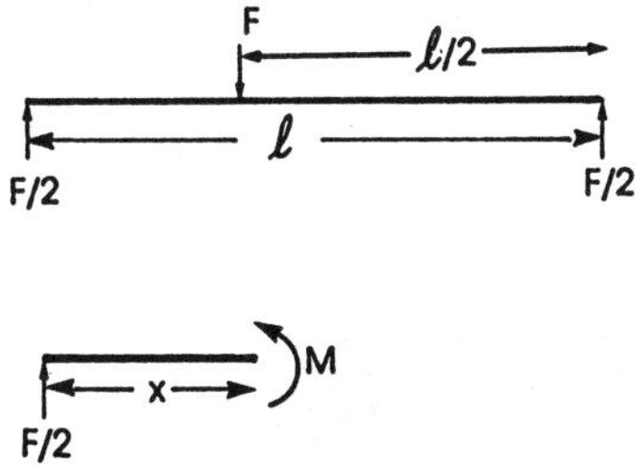

FIGURE 15.7

The total strain energy of the beam can be found by first finding the energy for half of the beam, and then doubling it. Thus for $0 < l < x$,

$$M - \frac{F}{2}x = 0 \Rightarrow M = \frac{F}{2}x \qquad (15\text{-}26)$$

Equation for total strain energy caused by bending:

$$U_b = \int_l \frac{M^2}{2EI}\, dx \qquad (15\text{-}27)$$

Therefore, substituting for M

$$U_b = \int_0^{l/2} \frac{F^2 x^2}{8EI}\, dx \qquad (15\text{-}28)$$

$$= \frac{F^2}{8EI}\left[\frac{x^3}{3}\right]_0^{l/2} = \frac{F^2}{8EI}\left[\frac{l^3}{24}\right]$$

$$= \frac{F^2 l^3}{8 \times 24 EI}$$

For total length of beam,

$$U_m = 2 \times \frac{F^2 l^3}{8 \times 24 EI}$$

$$\therefore \quad U_b = \frac{F^2 l^3}{96 EI} \qquad (15\text{-}29)$$

$$U_b = \frac{F^2}{2EI}\int_0^l x^2\, dx$$

$$= \frac{F^2}{2EI}\left[\frac{x^3}{3}\right]_0^l$$

$$U_b = \frac{F^2 l^3}{6EI} \tag{15-30}$$

Maximum deflection of the beam occurs at the end, therefore

Work causing maximum deflection
= strain energy

$$\frac{1}{2} F \delta_{max} = \frac{F^2 l^3}{6EI}$$

$$\therefore \quad \delta_{max} = \frac{Fl^3}{3EI} \tag{15-31}$$

Maximum deflection of the beam occurs mid-span, where the work done on the beam is stored in the form of strain energy. This can be expressed symbolically as,

$$\frac{1}{2} F \delta_{max} = \frac{F^2 l^3}{96EI} \quad \text{or} \quad \delta_{max} = \frac{Fl^3}{48EI} \tag{15-31}$$

EXAMPLE II

Cantilever with a point load at the end, Figure 15.8(a). The same procedure will be followed as in Example I.

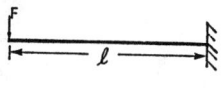

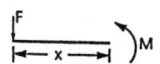

FIGURE 15.8—Free-body diagram.

Equation of moment for free-body diagram, Figure 158(b):

$$M + Fx = 0 \Rightarrow M = -Fx \qquad (15\text{-}32)$$

Equation of strain energy:

$$U_b = \int_l \frac{M^2}{2EI}\, dx$$

$$U_b = \int_o^l \frac{F^2 x^2}{2EI}\, dx \qquad (15\text{-}33)$$

15.4.2 ENERGY STORED DURING TORSION

Consider a circular shaft whose torsional deformation has a linear relationship with the twisting moment. Figure 15.9 shows this relationship. This is analogous to linearity between elongation and axial load in an axial deformation. Therefore the work done during twisting, is strain energy stored in the shaft

$$= U_t = \left(\frac{0+T}{2}\right)\theta = \frac{T\theta}{2} \qquad (15\text{-}34)$$

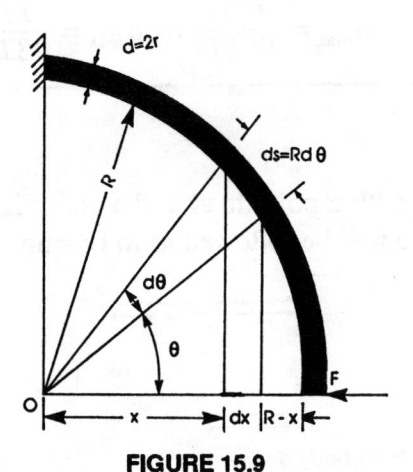

FIGURE 15.9

B-211

where T = twisting moment or torque
 θ = torsional deformation or angle of twist.

Now, (flexure formula for torsion)

$$\frac{T}{J} = \frac{\tau}{r} = \frac{G\theta}{l}$$

$$\therefore \quad \theta = \frac{Tl}{GJ} \quad \text{or} \quad T = G\frac{J\theta}{l}$$

Therefore,

$$U_t = \frac{T}{2}\left(\frac{Tl}{GJ}\right) = \frac{T^2 l}{2GJ} \tag{15-35}$$

In terms of θ,

$$U_t = G\frac{J\theta}{2l} \times \theta = G\frac{J\theta^2}{2l} \tag{15-36}$$

Let τ be the maximum shear stress in the shaft. From torsion formula,

$$\frac{I}{J} = \frac{\tau}{r} \Rightarrow \boxed{T = \frac{\tau J}{r}}$$

Also,

$$\frac{\tau}{r} = \frac{G\theta}{l} \Rightarrow \boxed{\theta = \frac{\tau l}{Gr}}$$

(15-37)

Therefore, from Equation 15-34, in terms of τ

$$U_t = \frac{1}{2}\left(\frac{\tau J}{r}\right)\left(\frac{\tau l}{Gr}\right)$$

$$\therefore \quad \boxed{U_t = \frac{1}{2}\left(\frac{\tau J}{G}\right)\left(\frac{\tau}{r}\right)^2} \tag{15-38}$$

EXAMPLE

Consider a rod of radius, r, bent to form a curve of radius, R, Figure 15.9.

The same procedure that is used for bending is followed, except that instead of bending, torque T is to be determined.

From Figure 15.9, it is clear that the torque varies with x; therefore,

$$T = F(R - x) \text{ where } x = R \cos \theta$$

$$\therefore \quad T = F(R - R\cos \theta) = FR(1 - \cos \theta) \quad (15\text{-}39)$$

For an elemental length dl, the torsional strain energy is,

$$dU = \frac{T^2 ds}{2GJ} = \frac{T^2 R d\theta}{2GJ}$$

Substitution for T yields

$$dU = \frac{F^2 R^3}{2GJ}(1 - \cos \theta)^2 d\theta \quad (15\text{-}40)$$

Integration yields

$$U = \int_0^{\pi/2} \frac{F^2 R^3}{2GJ}(1 - \cos \theta)^2 d\theta$$

$$\boxed{U = \frac{F^2 R^3}{2GJ} \int_0^{\pi/2} (1 - \cos \theta)^2 d\theta} \quad (15\text{-}41)$$

Now

$$J = \frac{\pi d^4}{32} = \frac{\pi (2r)^4}{32} = \frac{\pi r^4}{2}$$

Evaluating the integral and substituting for J gives,

$$U = 0.113 \frac{F^2}{Gr}\left(\frac{R}{r}\right)^3 \quad (15\text{-}42)$$

15.4.3 SHEAR STRAIN ENERGY

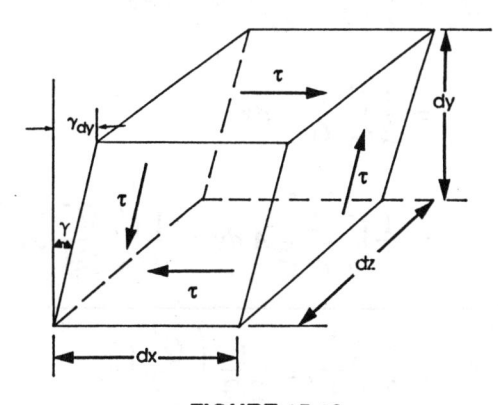

FIGURE 15.10

Consider a tiny element subjected to pure shear. The shear stress τ, shown in Figure 15.10, is numerically equal on all the faces of the element. Now, the shear force on the upper surface is $\tau dx\, d$. If γ is the shear strain, then the work done by the average force $1/2\, \tau\, dx\, dy$ in moving through a shear displacement γdy yields

$$dU_S = \frac{1}{2}(\tau dx\, dz)(\gamma dy)$$

Therefore,

$$U_S = \frac{1}{2}\iiint_{\text{vol}} \tau\gamma\, dx\, dy\, dz \quad (15\text{-}43)$$

Now, $\tau = \gamma G$ gives $\gamma = \tau/G$.

B-214

Let the area of cross-section be constant

$$\therefore \text{ shear force } Q = \tau A \text{ or } \tau = Q/A$$

Thus $\gamma = \tau/G = Q/GA$. Hence Equation changes to,

$$U_s = \frac{1}{2} \iiint \frac{\tau^2}{G} (dx\, dz)\, dy$$

$$= \frac{1}{2} \int_0^l \frac{\tau^2}{G} A\, dy$$

$$= \frac{1}{2} \int_0^l \frac{Q^2}{GA^2} A\, dy = \frac{1}{2} \int_0^l \frac{Q^2}{GA}\, dy$$

$$\boxed{U_s = \frac{1}{2} \frac{Q^2}{GA} l = \frac{1}{2G} \left(\frac{Q^2}{A^2}\right) Al} \quad (15\text{-}44)$$

$\therefore$ shear strain energy per unit volume equals

$$\boxed{U_s = \frac{1}{2G} \tau^2} \quad (15\text{-}45)$$

CHAPTER 16

COMBINED STRESSES

16.1 COMBINED AXIAL AND BENDING STRESSES

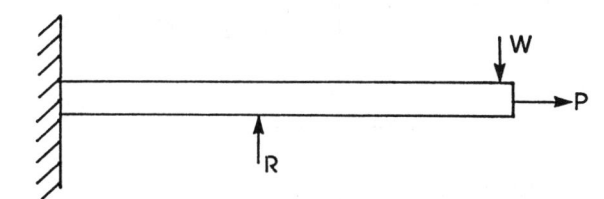

FIGURE 16.1—Combined axial and bending stress.

Axial stress,

$$\sigma_a = \frac{P}{A} \qquad (16\text{-}1)$$

Bending stress,

$$\sigma_b = \frac{M}{I/y} \qquad (16.2)$$

By method of superposition, combined stress, $\sigma = \sigma_a + \sigma_b$
(+ tensile, − compressive)

$$\boxed{\sigma = \frac{P}{A} \pm \frac{M}{I/\overline{y}}} \qquad (16\text{-}3)$$

B-216

where P = axial force
A = cross-section area of beam
M = maximum bending moment
I = moment of inertia
$\overline{y}$ = outer fiber distance from neutral axis

16.2 STRESSES OF ECCENTRIC LOADING

16.2.1 ECCENTRIC AXIAL LOAD

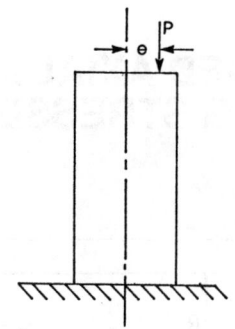

FIGURE 16.2—Eccentric axial loading.

Axial stress,

$$\sigma_a = P/A$$

Bending Moment $= Pe$ (16-4)

From Equation 16-2, Bending stress

$$\sigma_b = \frac{Pe}{I/\overline{y}}$$

Combined stress,

$$\sigma = \frac{P}{A} \pm \frac{Pe}{I/\overline{y}}$$ (16-5)

where e = eccentricity of the load

16.2.2 ECCENTRICALLY LOADED BOLTED JOINTS

Refer to Section 14.2, Equations 14-5 to 14-6 and Figure 14.2.

The direct shear stress in one bolt,

$$\tau_{d_i} = \frac{P/n}{a} \quad (16\text{-}6)$$

From Equation 14-15, force due to eccentricity,

$$F_i = \frac{Pe\sqrt{x_i^2 + y_i^2}}{\Sigma x^2 + \Sigma y^2}$$

∴ resulting shear stress due to eccentricity,

$$\tau_{r_i} = \frac{F_i}{a} \quad (16\text{-}7)$$

∴ resultant shear stress

$$\tau = \tau_{d_i} + \tau_{r_i}$$

$$\tau = \frac{P}{na} \pm \frac{F_i}{a} \quad (16\text{-}8)$$

where n = number of bolts
a = cross-section area of one bolt
x_i, y_i = ordinates to the bolt from reference origins (see Figure 14.2)

16.3 COMBINED SHEAR AND TENSION OR COMPRESSION

16.3.1 SHEAR STRESS DUE TO TENSION OR COMPRESSION

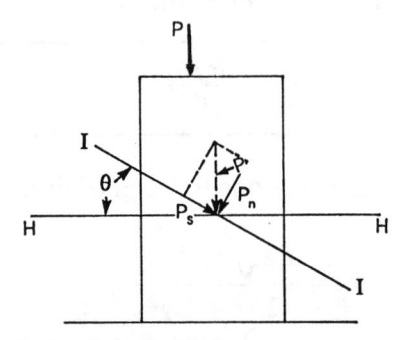

FIGURE 16.3—Forces in inclined plane.

where H-H = horizontal plane
I-I = inclined plane

Normal force,

$$P_n = P \cos \theta \qquad (16\text{-}8)$$

Shear force

$$P_s = P \sin \theta \qquad (16\text{-}9)$$

Area of inclined plane,

$$A_n = \frac{A}{\cos \theta} \qquad (16\text{-}10)$$

where A = area of H-H section

Normal stress,

$$\sigma = \frac{P_n}{A_n} = \frac{P}{A} \cos^2 \theta \qquad (16\text{-}11)$$

Shear stress,

$$\tau = \frac{P_s}{A_s} = \frac{P \sin \theta}{A/\cos \theta} = \frac{P}{2a} \sin 2\theta \quad (16\text{-}12)$$

$\sigma = \sigma_{max}$ for $\theta = 0°$
$\tau = \tau_{max}$ for $\theta = 45°$

16.3.2 TENSION OR COMPRESSION DUE TO SHEAR

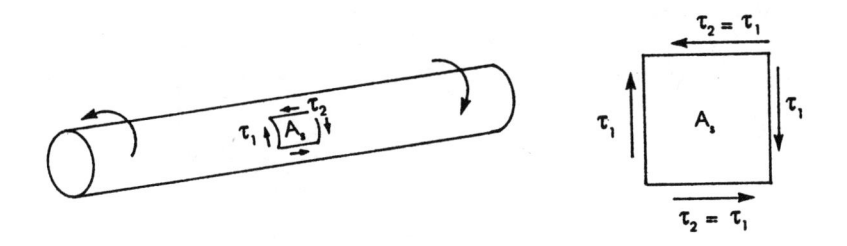

FIGURE 16.4—Shaft in torsion and shear stress on the surface.

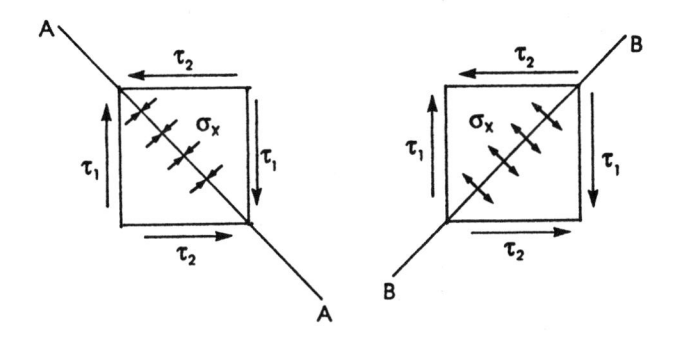

FIGURE 16.5—Internal stresses in diagonal planes.

B-220

Figure 16.4 shows a cylinder in pure torsion and the shear stresses acting on an element. From Figure 16.5 we see that the diagonal plane A - A is under compressive stress and diagonal plane B - B is under tensile stress due to the torsional shear stress. But no shear exists on these diagonal (45°) planes. In any other plane it would be noted that combinations of shear and tension, or compression, exist. The analysis of these stresses can be performed like the analysis in the above sections.

16.3.3 MOHR'S CIRCLE FOR STRESS

Graphic representation of combined normal and shear stresses in a section.

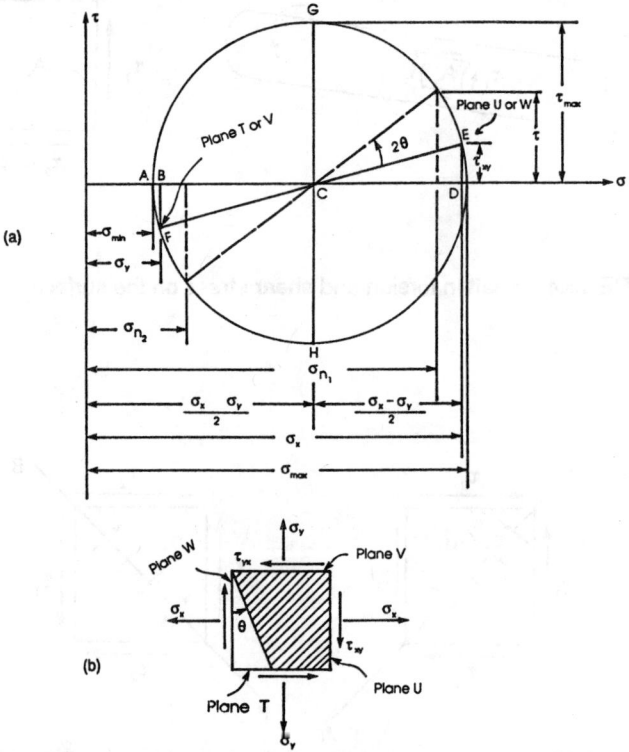

FIGURE 16.6—Graphic method of combined stresses: (a) Mohr's circle (b) section.

B-221

Normal stress,

$$\sigma_n = \left(\frac{\sigma_x + \sigma_y}{2}\right) + \left(\frac{\sigma_x - \sigma_y}{2}\right)\cos 2\theta - \tau_{xy}\sin 2\theta \quad (16\text{-}13)$$

Shear stress,

$$\tau = (\sigma_x - \sigma_y)\sin\theta\cos\theta + \tau_{xy}(\cos^2\theta - \sin^2\theta) \quad (16\text{-}14)$$

$$\tau_{max} = \pm\sqrt{\left(\frac{\sigma_x - \sigma_y}{2}\right)^2 + \tau_{xy}^2} \quad (16\text{-}15)$$

$$\sigma_{max} = \frac{\sigma_x + \sigma_y}{2} + \tau_{max} \quad (16\text{-}16)$$

$$\sigma_{min} = \frac{\sigma_x + \sigma_y}{2} - \tau_{max} \quad (16\text{-}17)$$

The equation of Mohr's circle, with origins at $\{(\sigma_x + \sigma_y)/2, 0\}$, is

$$(\sigma_n - k)^2 + \tau^2 = R^2 \quad (16\text{-}18)$$

where
$$k = \frac{\sigma_x + \sigma_y}{2} \quad (16\text{-}19)$$

$$R = \sqrt{\left(\frac{\sigma_x - \sigma_y}{2}\right)^2 + \tau_{xy}^2} \quad (16\text{-}20)$$

σ_n = normal stress
τ = shear stress
σ_{max} = maximum normal stress
σ_{min} = minimum normal stress

Principal stresses — σ_{max} and σ_{min} (where $\tau = 0$)
Principal planes — the planes on which $\tau = 0$.

16.4 COMBINED TORSION AND BENDING

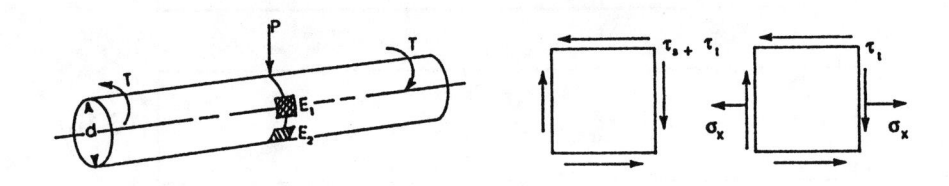

FIGURE 16.7—Stresses due to bending and torsion (a) cylinder (b) element E_1 and (c) element E_2.

For a solid cylindrical shaft, $I = \pi d^4/64$ and $J = 2I$. Shear stress due to shear force V is

$$\tau_s = \frac{VQ}{Ib} \quad (16\text{-}21)$$

Shear stress due to torque T is

$$\tau_t = \frac{Td/2}{J} = \frac{16T}{\pi d^3} \quad (16\text{-}22)$$

Bending stress due to bending moment M is

$$\sigma_x = \frac{My}{I} = \frac{32M}{\pi d^3} \qquad (16\text{-}23)$$

τ_s is negligible compared to τ_t. From Equations 16-15 to 16-17, by substituting $\sigma_y = 0$ and $\tau_{xy} = \tau_t$ we obtain

$$\tau_{max} = \frac{16}{\pi d^3} \sqrt{M^2 + T^2} \qquad (16\text{-}24)$$

$$\sigma_{max,min} = \frac{16}{\pi d^3} (M \pm \sqrt{M^2 + T^2}) \qquad (16\text{-}25)$$

where Q = first moment of area, under consideration, about the neutral axis
I = moment of inertia
b = breadth of sections
d = shaft diameter
J = polar moment of inertia
$\overline{y}$ = distance of fiber from neutral axis

Handbook of Mechanical Engineering

SECTION C
Fluid Dynamics/ Mechanics

CHAPTER 1

INTRODUCTION

1.1 DEFINITION OF A FLUID

A fluid is a substance that changes its shape continuously under the application of a shear stress no matter how slight the shear stress.

1.2 BASIC LAWS IN FLUID MECHANICS

Five basic laws are used for the solution of any problem in fluid mechanics. These basic laws are:

A) Conservation of mass

B) Newton's second law of motion

C) Conservation of momentum

D) The first law of thermodynamics

E) The second law of thermodynamics

1.3 SYSTEM AND CONTROL VOLUME

A) System

A system is a fixed, identifiable quantity of matter,

that may change in shape, position, and thermal condition. The system boundary separates the system from the surroundings (Fig. 1.1).

B) Control Volume

A control volume is a fixed, arbitrary volume in space through which fluid flows. The boundary of the control volume, is called the control surface (Fig. 1.2).

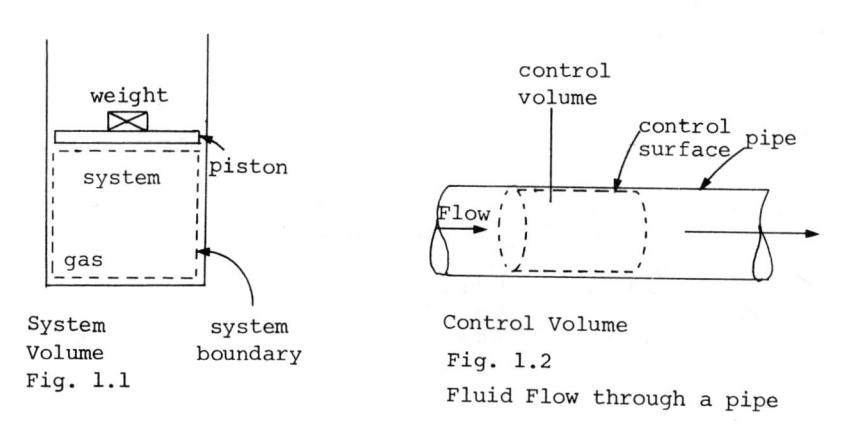

System Volume
Fig. 1.1

Control Volume
Fig. 1.2
Fluid Flow through a pipe

1.4 DIMENSIONS AND UNITS

Dimensions are names given to measurable quantities, such as time, length, velocity, mass, volume, and force.

Primary dimensions are those dimensions which are independent of other dimensions.

Secondary dimensions are those dimensions which are expressed in terms of the primary dimensions.

A unit is a definite standard or measure of a dimension.

The four basic systems of units are shown in table 1.1. The physical equivalents between some of the units are listed in table 1.2.

C-2

Table 1.1 Systems of Units

	Dimensions	Units	Definition of a secondary unit
I	International System		
Primary Dimensions	Mass, M Length, L Time, t Temperature, T	Kilogram, Kg Meter, m Second, sec Kelvin, K	
Secondary Dimensions	Force, F	Newton, N	$1N = \dfrac{1 \text{ Kg} \cdot \text{m}}{\text{sec}^2}$
II	Absolute Metric System		
Primary Dimensions	Mass, M Length, L Time, t Temperature, T	Gram, gm Centimeter, cm Second, sec Kelvin, K	
Secondary Dimensions	Force, F	dyne	$1 \text{ dyne} = \dfrac{1 \text{ gm} \cdot \text{cm}}{\text{sec}^2}$
III	British Gravitational System		
Primary Dimensions	Force, F Length, L Time, t Temperature, T	Pound-force, lbf Foot, ft Second, sec Rankine, R	
Secondary Dimensions	Mass, M	slug	$1 \text{ slug} = \dfrac{1 \text{ lbf} \cdot \text{sec}^2}{\text{ft}}$
IV	English Engineering System		
Primary Dimensions	Force, F Mass, M Length, L Time, t Temperature, T	Pound-force, lbf Pound-mass, lbm Foot, ft Second, sec Rankine, R	
Secondary Dimensions			

Table 1.2

Equivalence Relations Between Units	
1 in. ≡ 2.54 cm 1 ft. ≡ 30.5 cm 1 ft ≡ 0.305 m 5,280 ft ≡ 1 mile 1 gal (U.S.) ≡ 0.1337 ft^3 1 lbf ≡ 16 oz 1 newton ≡ 10^5 dynes 1 lbf ≡ 445,000 dynes	1 slug ≡ 32.2 lbm 1 g ≡ 2.20 × 10^{-3} lbm 1 g ≡ 0.685× 10^{-4} slug 1 ft-lb ≡ 0.001285 Btu 1 joule ≡ 9.48 × 10^{-4} Btu 1 hp ≡ 550 ft-lb/sec 1°K ≡ 1.8°R (where °K= °C + 273° and °R = °F + 460°)

NOTES

In the solution of problems, the following logical steps should be taken:

A) State the information given.

B) State the unknowns to be found.

C) Select and draw the system or control volume to be used. Be sure to label the boundaries of the system or control volume, and use appropriate coordinate directions.

D) Write the basic equations and laws that are necessary to solve the problem.

E) List the appropriate assumptions.

F) Using a set of units, substitute numerical values into equations or laws.

G) Check the answer to make sure that it is reasonable with the assumptions made.

CHAPTER 2

FUNDAMENTAL CONCEPTS

2.1 FLUID AS A CONTINUUM

When a fluid is treated as an infinitely divisible substance, one which is composed of many molecules in constant motion and in collision with each other, it is said to be in continuum.

2.2 FLUID PROPERTIES

As a consequence of the continuum assumption, the fluid's properties are considered to be continuous functions of position and time. Thus, the representation of fluid properties are:

A) Density field:

The density field is represented by

$$\rho = \rho(x,y,z,t) \qquad (2.1)$$

where x, y, z are the space coordinates at any instant of time.

B) Velocity field:

The velocity field is represented by

$$\vec{v} = \vec{v}(x,y,z,t) \qquad (2.2)$$

If the properties and flow characteristics at each position in space remain invariant with time, the flow is called steady flow.

For steady flow

$$\rho = \rho(x,y,z) \quad \text{or}, \quad \frac{\partial \rho}{\partial t} = 0 \qquad (2.3)$$

and

$$\vec{v} = \vec{v}(x,y,z) \quad \text{or}, \quad \frac{\partial \vec{v}}{\partial t} = 0 \qquad (2.4)$$

A time dependent flow, on the other hand, is designated as an unsteady flow.

Uniform flow is a flow in which the magnitude and direction of the velocity vector do not change with time.

2.3 TWO VIEWPOINTS

There are two possibilities to describe the velocity field in computations involving the motion of fluid particles. The velocity of particles at any time can be expressed using fixed coordinates x, y, z in the flow field. Mathematically it may be given by $\vec{v} = \vec{v}(x,y,z,t)$. This procedure is called the Eulerian viewpoint.

Now, to study any one particle in the flow, one must follow the particle as it moves. This is obtained by using the corresponding time functions for each particle. Mathematically it may be expressed by $\vec{v} = \vec{v}[x(t),y(t),z(t)]$, this procedure is called the Lagrangian viewpoint.

2.4 ONE, TWO, AND THREE DIMENSIONAL FLOW

A one-dimensional flow is a flow in which the parameters are expressed as functions of one space coordinate and time.

C-6

A two-dimensional flow is a flow in which the parameters are expressed as functions of two space coordinates and time.

A three-dimensional flow is a flow in which the parameters are expressed as functions of three space coordinates and time.

2.5 PATHLINES, STREAKLINES, STREAMLINES, AND STREAMTUBES

Pathline is a line traced out by a moving particle in time.

Streakline is a line traced out by all fluid particles which at some time passed through one fixed location in space.

Streamlines are imaginary lines drawn in the flow whose tangents are parallel to the velocity vector at a given instant of time.

Streamtube is an elementary region of the flow whose walls are made up of streamlines (Fig. 2.1).

In the case of steady flow $\left(\frac{\partial v}{\partial t} = 0\right)$, pathlines, streaklines, and streamlines all coincide.

Streamlines

Stream-tube

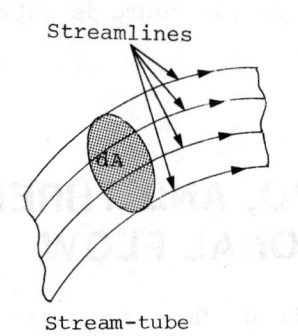

Fig. 2.1

2.6 STRESS FIELD

A) Surface and Body Forces

Surface forces are all of the forces that act upon the boundaries of a medium by direct contact.

Forces which develop without physical contact and are distributed over a volume of the fluid, are called body forces. Two examples are gravitational and electromagnetic forces.

B) Stress at a Point

The stress at a point is defined as

$$\text{stress} = \lim_{\vec{\delta A} \to 0} \frac{\vec{\delta F}}{\vec{\delta A}} \qquad (2.5)$$

where
$\vec{\delta F} \equiv$ The force acting at the point B, as shown in Fig. 2.2.

$\vec{\delta A} \equiv$ The area element.

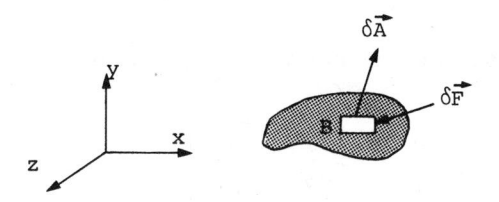

Definition of stress

Fig. 2.2

We can write the definition of stress in terms of components as

$$T_{ij} = \lim_{\delta A_i \to 0} \frac{\delta F_j}{\delta A_i} \qquad (2.6)$$

where T_{ij} denotes the stress acting on an i plane in the j direction and where i and j each can stand for x, y, or z. For example, $T_{xz} = \lim\limits_{\delta A_x \to 0} \dfrac{\delta F_z}{\delta A_x}$ defines the stress on an x plane in the y direction (Fig. 2.3).

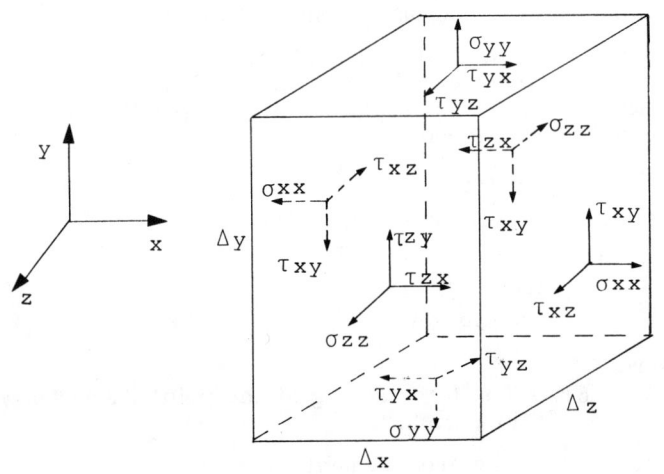

Fig. 2.3 Diagram showing stresses

The stress at a point is specified by the nine compoments

$$\begin{bmatrix} \sigma_{xx} & \tau_{xy} & \tau_{xz} \\ \tau_{yx} & \sigma_{yy} & \tau_{yz} \\ \tau_{zx} & \tau_{zy} & \sigma_{zz} \end{bmatrix} \qquad (2.7)$$

where σ has been used to denote a normal stress and τ to denote a shear stress.

2.7 NEWTONIAN FLUID VISCOSITY

Fluids may be classified according to the relation between the applied shear stress and the rate of deformation of the fluid. Fluids in which the shear stress is proportional to the rate of deformation are called Newtonian fluids (Fig. 2.5), or

C-9

$$\tau_{yx} \, \alpha \, \frac{du}{dy} \qquad (2.8)$$

where

$u \equiv$ The velocity of the upper plate (Fig. 2.4)

$\frac{du}{dy} \equiv$ Rate of deformation

$\tau_{yx} \equiv$ The shear stress

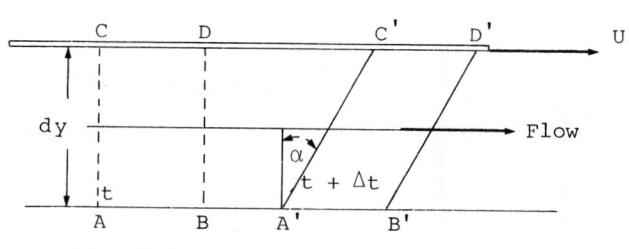

Fig. 2.4

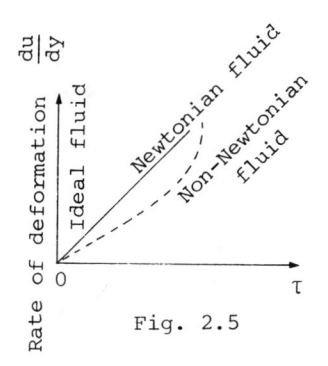

Fig. 2.5

The constant of proportionality in equation 2.8 is the coefficient of viscosity μ. Thus, equation 2.8 may be written as

$$\tau_{yx} = \mu \frac{du}{dy} \qquad (2.9)$$

which is the Newton's law of viscosity.

The ratio of the absolute viscosity, μ, to the density, ρ, of a fluid is called kinematic viscosity and is represented by the symbol ν,

therefore,

$$\nu = \frac{\mu}{\rho} \qquad (2.10)$$

2.8 CLASSIFICATION OF FLUID MOTION

One possible classification of fluid mechanics on the basis of observable physical characteristics of flow fields is shown in Fig. 2.6

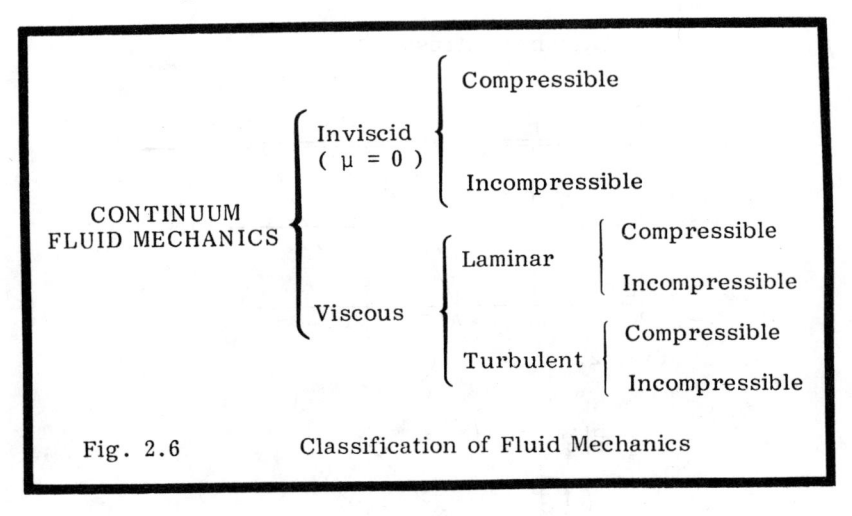

```
                      ┌ Inviscid    ┌ Compressible
                      │ (μ = 0)     │
CONTINUUM             │             └ Incompressible
FLUID MECHANICS  ─────┤
                      │             ┌ Laminar    ┌ Compressible
                      │             │            └ Incompressible
                      └ Viscous ────┤
                                    └ Turbulent  ┌ Compressible
                                                 └ Incompressible
```

Fig. 2.6   Classification of Fluid Mechanics

A) Laminar Flow is a flow pattern characterized by motion in layers or laminae.

B) Turbulent Flow is a flow pattern characterized by random movements of fluid particles in all directions.

C) Incompressible Flow is a flow in which variations in density can be assumed to be constant. (ρ = const.)

D) Compressible Flow is a flow in which variations in density cannot be neglected. ($\rho \neq$ const.)

E) Inviscid fluids are fluids in which the fluid viscosity, μ, is assumed to be zero, ($\mu = 0$)

F) Viscous fluids are fluids in which the fluid viscosity, μ, cannot be neglected, ($\mu \neq 0$)

2.9 IRROTATIONAL FLOW

For a fluid element, the angular velocity, $\vec{w}$, is given by the relation

C-11

$$\boxed{\vec{w} = \frac{1}{2} \nabla \times \vec{v}} \qquad (2.11)$$

or
$$\vec{w} = \frac{1}{2}\left[\left(\frac{\partial w}{\partial y} - \frac{\partial v}{\partial z}\right)\vec{i} + \left(\frac{\partial u}{\partial z} - \frac{\partial w}{\partial x}\right)\vec{j} + \left(\frac{\partial v}{\partial x} - \frac{\partial u}{\partial y}\right)\vec{k}\right] \quad (2.12)$$

where, $\boxed{\nabla \times \vec{v} = \text{curl } \vec{v}}$

$\nabla \stackrel{\Delta}{=}$ vector operator del $\stackrel{\Delta}{=} \hat{i}\frac{\partial}{\partial x} + \hat{j}\frac{\partial}{\partial y} + \hat{k}\frac{\partial}{\partial z}$

$\vec{v} = u\hat{i} + v\hat{j} + w\hat{k}$

u, v, w = The velocity compoments in the x, y, z directions, respectively.

When the angular velocity vector, $\vec{w}$, is equal to zero, the flow is called irrotational or,

$$\boxed{\vec{w} = \frac{1}{2} \nabla \times \vec{v} = 0} \qquad (2.13)$$

or
$$\frac{\partial w}{\partial y} - \frac{\partial v}{\partial z} = \frac{\partial u}{\partial z} - \frac{\partial w}{\partial x} = \frac{\partial v}{\partial x} - \frac{\partial u}{\partial y} = 0 \qquad (2.14)$$

or,
$$\frac{\partial r V_\theta}{\partial r} - \frac{\partial V_r}{\partial \theta} = 0 \quad \text{(polar coordinate)} \qquad (2.15)$$

The vorticity vector ($\vec{\zeta}$) for a fluid element is defined as:

$$\boxed{\vec{\zeta} = 2\vec{w} = \nabla \times \vec{v}} \qquad (2.16)$$

In cylindrical coordinates the vorticity vector is:

$$\boxed{\begin{aligned}\nabla \times \vec{v} = \hat{i}_r &\left(\frac{1}{r}\frac{\partial V_z}{\partial \theta} - \frac{\partial V_\theta}{\partial z}\right) + \hat{i}_\theta \left(\frac{\partial V_r}{\partial z} - \frac{\partial V_z}{\partial r}\right) \\ &+ \hat{i}_z \left(\frac{1}{r}\frac{\partial rV_\theta}{\partial r} - \frac{1}{r}\frac{\partial V_z}{\partial \theta}\right) \\ \text{where} \quad \vec{V} &= \hat{i}_r V_r + \hat{i}_\theta v_\theta + \hat{i}_z v_z\end{aligned}}$$

C-12

CHAPTER 3

FLUID STATICS

3.1 THE BASIC EQUATION OF FLUID STATICS

To determine the pressure field within the fluid a differential element shown in Figure 3.1 is used.

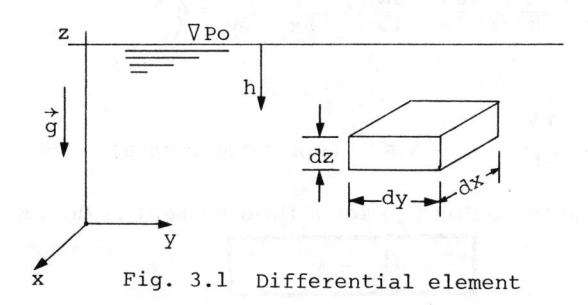

Fig. 3.1 Differential element

For this differential fluid element the following equation applies:

$$-\text{grad } p + \rho \vec{g} = 0 \qquad (3.1)$$

where:

$$\text{grad } p \equiv \nabla p \equiv \left(\hat{i} \frac{\partial p}{\partial x} + \hat{j} \frac{\partial p}{\partial y} + \hat{k} \frac{\partial p}{\partial z} \right)$$

$$\equiv \left(\hat{i} \frac{\partial}{\partial x} + \hat{j} \frac{\partial}{\partial y} + \hat{k} \frac{\partial}{\partial z} \right) p \qquad (3.2)$$

The term in parentheses is called the gradient of the pressure.

The physical significance of the ∇p is the pressure force per unit volume at a point.

$\rho \vec{g} \equiv$ body force per unit volume

$\rho \equiv$ density

$\vec{g} \equiv$ the local gravity vector

By expanding equation (3.1) into components, we find

$$x\text{-direction} \quad -\frac{\partial p}{\partial x} + \rho g_x = 0 \qquad (3.3)$$

$$y\text{-direction} \quad -\frac{\partial p}{\partial y} + \rho g_y = 0 \qquad (3.4)$$

$$z\text{-direction} \quad -\frac{\partial p}{\partial z} + \rho g_z = 0 \qquad (3.5)$$

If the coordinate system is chosen such that the z-axis is directed vertically, then

$$g_x = 0, \quad g_y = 0, \quad \text{and} \quad g_z = -g$$

Because the pressure P varies only in the z direction and because it is not a function of x and y, we may use a total derivative in equation (3.5). Then the equation (3.5) becomes

$$\frac{dp}{dz} = -\rho g = -\gamma \qquad (3.6)$$

where γ = specific weight

This equation is the basic pressure-height relation of fluid statics. It holds under the following restrictions:

1) Static fluid
2) Gravity is the only body force
3) The z-axis is vertical

A) Incompressible Fluid

In an equation for an incompressible fluid, ρ = constant (Fig. 3.1). Therefore, for constant gravity, the equation (3.6) becomes

C-14

$$p = p_0 + \rho g h \qquad (3.7)$$

where

$p_0 \equiv$ the pressure p at the reference level

$h \equiv z-z_0$ (h measured positive downward)

$\rho \equiv$ density

B) Compressible Fluid

If we know the manner in which the specific weight, γ, varies, the pressure variation in a compressible fluid can be evaluated using equation (3.6).

3.2 THE STANDARD ATMOSPHERE

The internationally accepted properties of the standard atmosphere are summarized in Table 3.1:

Table 3.1 U.S. Standard Atmosphere (Conditions here are at sea level)		
Property	English Units	SI Units
Temperature, T	59°F	288 K
Pressure, p	14.696 psia	101.3k Pa(abs)
Viscosity, μ	$3.719 \cdot 10^{-7}$ lbf·sec/ft^2	$1.781 \cdot 10^{-5}$ kg/m sec
Density, ρ	0.002377 slug/ft^3	1.225 kg/m^3
Specific weight, γ	0.07651 lbf/ft^3	-

The temperature is assumed to decrease linearly with height according to the relation,

$$T = (519 - 0.00357z) \ °R \qquad (3.8)$$

where

$z \equiv$ The elevation above sea level in feet.

3.3 ABSOLUTE AND GAGE PRESSURES

Pressure values are stated with respect to a reference pressure level. If the reference pressure level is the vacuum then the pressure is called absolute. The difference between the absolute pressure and the surrounding pressure (mainly atmospheric pressure) is called gage pressure

$$P_{gage} = P_{absolute} - P_{atm} \qquad (3.9)$$

or, $\qquad P_{absolute} = P_{atm} + P_{gage}$

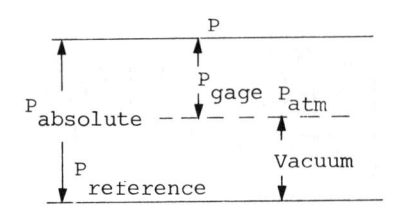

Fig. 3.2 Absolute and gage pressure

3.4 THE SIMPLE MANOMETER

Manometers are instruments that use liquid to determine differences in pressure between two points, or between a certain point and the atmosphere.

A manometer consisting of n different substances with different specific gravities $\gamma_1, \gamma_2, \ldots, \gamma_n$ is shown in Fig. 3.3.

To determine the pressure difference, $(P_A - P_n)$, (Fig. 3.3), the following equation is used.

$$(P_A - P_n) = (P_A - P_{A_1}) + (P_{A_1} - P_{A_2}) + (P_{A_2} - P_{A_3}) + (P_{A_3} - P_{A_4}) +$$

C-16

$$+\ldots+(P_{A_{n-1}}-P_n)$$
$$= -\gamma_1 z_1 - \gamma_2 z_2 + \gamma_3 z_3 + \gamma_4 z_4 \ldots -\gamma_n z_n \quad (3.10)$$

where

$\gamma_1, \gamma_2, \gamma_3, \gamma_4, \ldots, \gamma_n$ = the specific gravities of the substances

$z_1, z_2, z_3, \ldots, z_n$ = the distances between two successive points in the columns of the manometer.

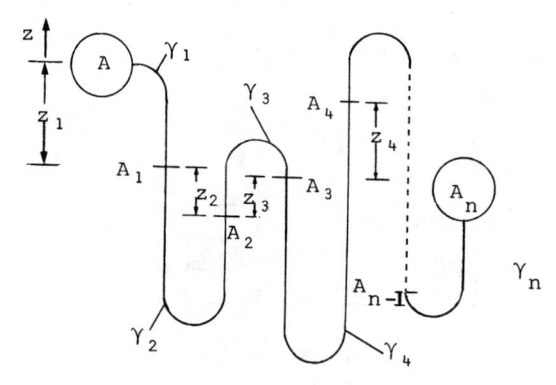

Fig. 3.3 Simple manometer

Note: The distances $z_1, z_2, z_3, \ldots, z_n$ are considered as positive if the end point, A_3, is at a higher position than the start point, A_2, (Distance z_3), or as negative if the end points, A_1, is at lower position than the start point, A (Distance z_1).

3.5 HYDROSTATIC FORCES ON SUBMERGED SURFACES

In order to determine the force acting on a submerged surface we must specify the magnitude, the direction and the line of action of the resultant force ($\vec{F}_R$).

A) Hydrostatic Force on a Plane Surface Submerged in a Static Incompressible Fluid

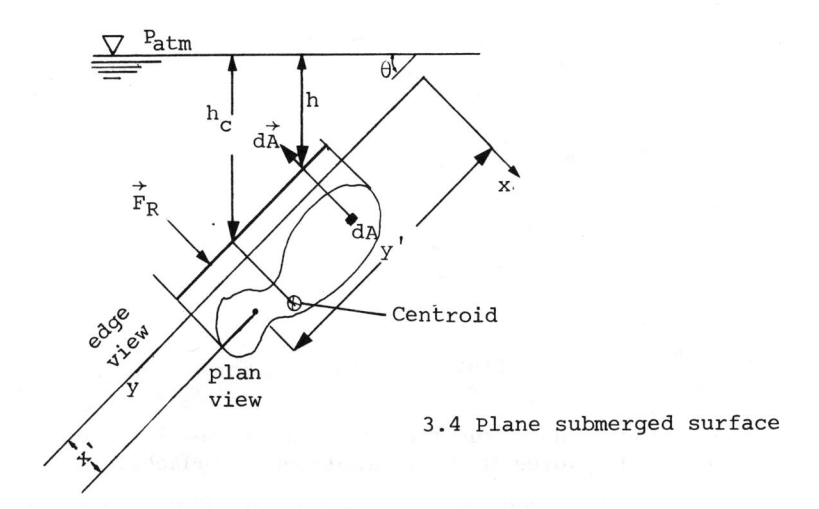

3.4 Plane submerged surface

The magnitude of the resultant force acting on the submerged suface is

$$F_R = |\vec{F}_R| = \int p \, dA$$

where A is the surface area or,

$$\boxed{F_R = \gamma h_c A} \qquad (3.11)$$

The direction of $\vec{F}_R$ is normal to the surface and the line of action passes through the points x', y', which can be located by

$$\boxed{y' = y_c + \frac{I_c}{A y_c} \qquad \frac{I_c}{A y_c} > 0} \qquad (3.12)$$

$$\boxed{x' = x_c + \frac{(I_{xy})_c}{A y_c}} \qquad (3.13)$$

where I_c, $(I_{xy})_c$ is the moment of inertia about its center of gravity axes, and x_c, y_c is the center of gravity coordinates.

C-18

B) Hydrostatic Force on Curved Submerged Surfaces

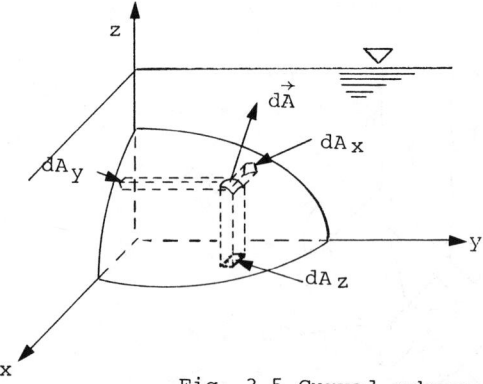

Fig. 3.5 Curved submerged surface

The methods used on plane surfaces can be used to determine the force on curve submerged surfaces.

For the submerged curved surface of Figure 3.5 the pressure force acting on the element of area $\vec{dA}$ is

$$\vec{F}_R = \hat{i}\, F_{Rx} + \hat{j}\, F_{Ry} + \hat{k}\, F_{Rz} = -\int_A p\, \vec{dA} \qquad (3.14)$$

where

$$p = p_0 + \rho g h$$

The components of the force in the x and y direction are

$$F_{Rx} = \pm \int_{Ax} p\, dA_x \qquad (3.15)$$

and

$$F_{Ry} = \pm \int_{Ay} p\, dA_y \qquad (3.16)$$

where $dA_x = dA \cos\theta_x$, and is the projection of the area element dA on a plane perpendicular to the x-axis.

$dA_y = dA \cos\theta_y$, and is the projection of the area element dA on a plane perpendicular to the y-axis.

C-19

The component of the force in the z-direction is

$$F_z = - \int_{A_z} \int_{z'}^{z_0} \rho g \, dz \, dA_z \qquad (3.17)$$

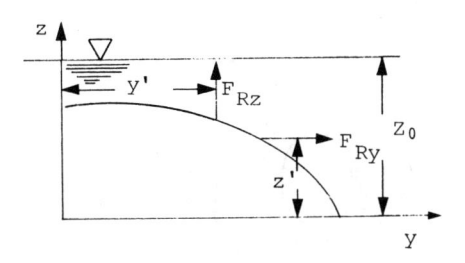

Fig.3.6 Two-dimensional curved submerged surface.

The lines of action of the components for a curved submerged surface are given by

$$z' = \frac{1}{F_{Ry}} \int_{A_y} z p \, dA_y \qquad (3.18)$$

$$y' = - \frac{1}{F_{Rz}} \int_{A_z} y p \, dA_z \qquad (3.19)$$

3.6 BUOYANCY AND STABILITY

A) Buoyancy of a Body

The resultant vertical force exerted on a body by a static fluid in which it is submerged or floating, is called the buoyant force.

The magnitude of this force is

C-20

$$\boxed{F_z = \int \rho g(z_2 - z_1)\,dA = \rho g \forall} \qquad (3.20)$$

where ρ = density of the fluid

g = gravitational constant

$\forall$ = volume of the body

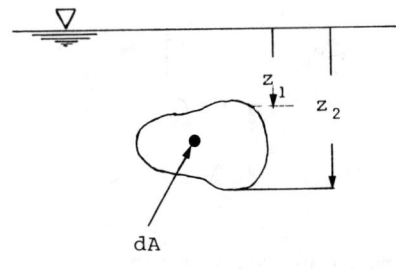

Fig. 3.7

The buoyant force for an incompressible fluid goes through the centroid of the volume displaced by the body.

B) Stability

For a completely submerged body, the condition for stability (stable equilibrium) is that the center of gravity, G, must be directly below the center of buoyancy.

The vertical alignment of B and G is important for stability. If G and B coincide, neutral equilibrium is obtained.

In a floating body (Fig. 3.8) stable equilibrium can be achieved even when G is above B. The magnitude of the length GA serves as a measure of the stability of a floating body.

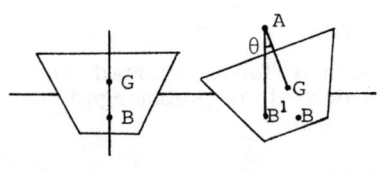

Fig. 3.8

Note: When a body undergoes a displacement and then returns to its original position, the body is in stable equilibrium. If the body does not return to its original position, but moves further away from it, the body is in unstable equilibrium. If the body adopts a new position the body is in neutral equilibrium.

3.7 FLUIDS IN RIGID-BODY MOTION

A) Accelerating Fluid

The equation of motion for a uniformly accelerating fluid is given by

$$\boxed{-\text{grad } p + \rho \vec{g} = \rho \vec{a}} \qquad (3.21)$$

where

$\text{grad } p$ = pressure force per unit volume

$\rho \vec{g}$ = body force per unit volume at a point

ρ = density

$\vec{g}$ = gravitational constant

$\vec{a}$ = acceleration of the fluid

The vector equation (3.20) has three components which are

$$\frac{\partial p}{\partial x} + \rho g_x = \rho a_x \quad \text{(x-direction)} \qquad (3.22)$$

$$\frac{\partial p}{\partial y} + \rho g_y = \rho a_y \quad \text{(y-direction)} \qquad (3.23)$$

$$\frac{\partial p}{\partial z} + \rho g_z = \rho a_z \quad \text{(z-direction)} \qquad (3.24)$$

B) Rotating Fluid

For a fluid which rotates with a constant angular velocity, ω, about an axis z, equation (3.21) is reduced to

$$-\left.\frac{\partial p}{\partial z}\right|_z + \rho g_z = 0 \qquad (3.25)$$

CHAPTER 4

BASIC LAWS FOR SYSTEMS AND CONTROL VOLUMES

4.1 BASIC LAWS FOR A SYSTEM

A) Newton's Second Law

This law states that the total external force acting upon a system which is moving relative to an inertial reference is equal to the rate of change of the linear momentum of the system. Thus,

$$\boxed{\vec{F} = \vec{F}_S + \vec{F}_B = \frac{d\vec{P}}{dt}} \qquad (4.1)$$

where

$\vec{F}_S = \int_{CS} \vec{T} d\vec{A}$ = Surface forces acting on the control volume.

$\vec{F}_B = \int_{CV} \vec{B} \rho d\mathcal{V}$ = Body forces acting inside the control volume

$\vec{P}$ = Linear momentum of the system

$\phantom{\vec{P}} = \int_{\mathcal{V}} V \rho d\mathcal{V}$

V = Velocity of the system

C-23

ρ = Density

V = Volume

B) Conservation of Mass

The conservation of mass simply states that the mass M of a system is always constant, or

$$\left.\frac{dM}{dt}\right|_{system} = 0 \qquad (4.2)$$

where

$$M = \int_{M, system} dm = \int_{V, system} \rho \, dV$$

C) Moment of Momentum

This law states that the total torque acting on the system is equal to the rate of change of angular momentum, or

$$\vec{T} = \frac{d\vec{H}}{dt} \qquad (4.3)$$

where

$\vec{H}$ = The angular momentum of the system

$$= \int_M \vec{r} \times \vec{V} dm = \int_V \vec{r} \times \vec{V} d V$$

$\vec{T}$ = The total torque of the system

$$= \vec{r} \times \vec{F}_S + \int_M \vec{r} \times \vec{g} dm + \vec{T}_{shaft}$$

D) The First Law of Thermodynamics

This law states that

$$\delta Q + \delta W = dE \qquad (4.4)$$

where

Q = Heat transferred to the system

W = Work done on the system

E = Total energy of the system

$$= \int_M e\,dm = \int_V e\rho\,d\forall$$

$$e = u + \frac{V^2}{2} + gz$$

In rate form, equation (4.4) can be written as:

$$\boxed{\dot{Q} + \dot{W} = \left.\frac{dE}{dt}\right|_{system}} \quad (4.5)$$

E) The Second Law of Thermodynamics

This law states that

$$dS \geq \frac{\delta Q}{T} \quad (4.6)$$

where

δQ = Heat transferred to a system
T = Temperature
dS = Change in entropy, S, of the system

On a rate basis, equation (4.6) can be written as

$$\boxed{\left.\frac{dS}{dt}\right|_{system} \geq \frac{\dot{Q}}{T}} \quad (4.7)$$

where

$$S_{system} = \int_M s\,dm = \int_V s\rho\,d\forall$$

4.2 REYNOLDS TRANSPORT EQUATION

A general relation between the rate of change of any arbitrary extensive property, N, of a system and the time variations of this property associated with the control volume is given by

C-25

$$\boxed{\left.\frac{dN}{dt}\right|_{system} = \frac{\partial}{\partial t}\int_{CV} n\rho\, d\forall + \int_{CS} n\rho\, \vec{V}d\vec{A}} \qquad (4.8)$$

$\left.\dfrac{dN}{dt}\right|_{system} \equiv$ Total rate of change of any extensive property of the system

$\dfrac{\partial}{\partial t}\displaystyle\int_{CV} n\rho\, d\forall \quad$ The time rate of change of the extensive property N in the control volume

n = N per unit mass (Intensive property)

$\rho\, d\forall$ = An element of mass in the control volume

$\displaystyle\int_{CV} n\rho\, d\forall$ = The total amount of the extensive property N in the control volume

$\displaystyle\int_{CS} n\rho\, \vec{V}d\vec{A} \equiv$ Net rate of efflux of the extensive property N, crossing the control surface

$\rho\vec{V}d\vec{A} \equiv$ The rate of mass efflux crossing the element of area $d\vec{A}$ per unit of time

$n\rho\vec{V}d\vec{A} \equiv$ The rate of efflux of the extensive property N crossing $d\vec{A}$

4.3 THE BASIC LAWS APPLIED TO THE CONTROL VOLUME

A) Continuity Equation

The control volume formulation of the conservation of mass is given by the relation

$$\boxed{\frac{\partial}{\partial t}\int_{CV} \rho\, d\forall + \int_{CS} \rho\vec{V}d\vec{A} = 0} \qquad (4.9)$$

where

$$\frac{\partial}{\partial t} \int_{CV} \rho \, dV = \text{The rate of change of mass within the control volume}$$

$$\int_{CS} \rho \vec{V} d\vec{A} = \text{The net rate of mass efflux through the control surface}$$

For steady or unsteady incompressible flow, the continuity equation is reduced to

$$\int_{CS} \vec{VdA} = 0 \qquad (4.10)$$

or, $\qquad Q = A_1 V_1 = A_2 V_2$

where

Q = Volumetric flow rate

A_1, A_2 = Cross-sectional areas at the entrance and exit of the control volume, respectively

V_1, V_2 = Fluid velocities at the entrance and exit of the control volume, respectively

For any steady flow, the continuity equation is reduced to

$$\int_{CV} \rho \vec{V} d\vec{A} = 0 \qquad (4.11)$$

B) Linear Momentum Equation for the Inertial Control Volume

For an inertial control volume the linear momentum equation is given by the relation

$$\vec{F} = \vec{F}_S + \vec{F}_B = \int_{CS} \vec{V} \rho \vec{V} dA + \frac{\partial}{\partial t} \int_{CV} \vec{V} \rho \, dV \qquad (4.12)$$

The vector form of the equation (4.12) may be written as

$$F_i = \int_{CS} T_i dA + \int_{CV} B_i \rho \, dV = \int_{CS} V_i \rho \vec{V} d\vec{A} + \frac{\partial}{\partial t} \int_{CV} V_i \rho \, dV \qquad (14.13)$$

C-27

for i = x,y,z

Note: The sign of the scalar quantity $\rho\vec{V}d\vec{A}$ depends on the dirction of flow. $\rho\vec{V}d\vec{A}$ is positive where flow is out through the control volume, and negative where flow is in through the control volume.

C) Linear Momentum Equation for a Control Volume Moving with Constant Velocity

This equation can be written as

$$\vec{F} = \vec{F}_S + \vec{F}_B = \int_{CS} \vec{V}_{xyz}\, \rho \vec{V}_{xyz} \cdot d\vec{A} + \frac{\partial}{\partial t}\int_{CV} \vec{V}_{xyz}\, \rho\, d\forall \quad (4.14)$$

where the xyz subscript shows that the velocities must be measured relative to a reference coordinate system xyz (see Fig. 4.1).

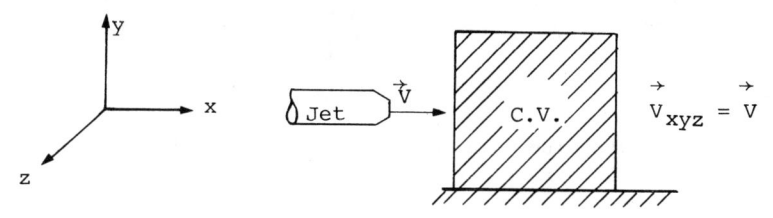

Stationary Control Volume

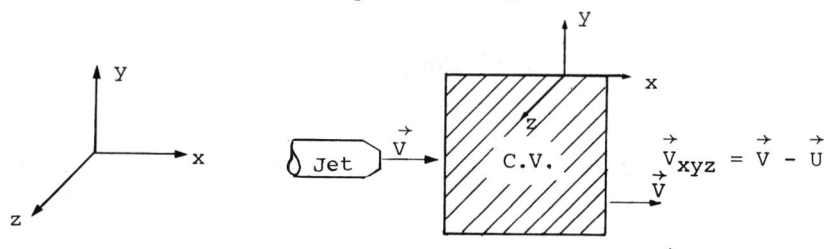

Moving control Volume with a constant velocity $\vec{U}$
Fig. 4.1

In the case of a steady one-dimensional flow shown in Figure 4.2, the equation is reduced to

$$F_x = \rho Q(V_{X2} - V_{X1}) \quad (4.15)$$

The vector form of the equation (4.14) may be written as

C-28

$$F_i = \int_{CS} T_i dA + \int_{CV} B_i \rho dV$$

$$= \int_{CS} (\vec{V}_{xyz})_i \rho \vec{V}_{xyz} dA + \frac{\partial}{\partial t} \int_{CV} (\vec{V}_{xyz})_i \rho dV \quad (4.16)$$

for $i = x, y, z$

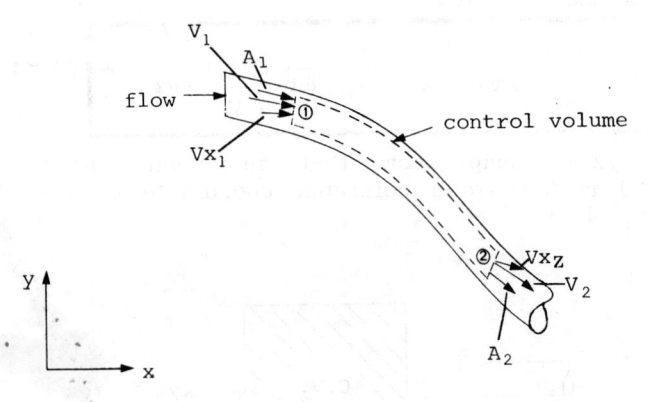

Fig 4.2 Control volume for a uniform flow normal to the control surface.

D) Linear Momentum Equation for Control Volume with Rectilinear Acceleration

This equation can be written as

$$\vec{F}_S + \vec{F}_B - \int_{CV} \vec{a}_{rf} \rho dV = \frac{\partial}{\partial t} \int_{CV} \vec{V}_{xyz} \rho dV$$

$$+ \int_{CS} \vec{V}_{xyz} \rho \vec{V}_{xyz} dA \quad (4.17)$$

where

a_{rf} = The acceleration of the control volume relative to a fixed reference coordinate xyz.

The vector form of the equation (4.17) may be written as

C-29

$$F_i = \int_{CS} T_i dA + \int_{CV} B_i \rho dV - \int_{CV} (\vec{a}_{rf})_i \rho dV$$

$$= \frac{\partial}{\partial t} \int_{CV} (v_{xyz})_i \rho dV + \int_{CS} (V_{xyz})_i \rho \vec{V}_{xyz} \cdot d\vec{A}$$
(4.18)

for $i = x, y, z$

Note: For an accelerating control volume one must label a coordinate system xyz on the control volume and a fixed reference coordinate xyz.

E) Linear Momentum Equation for Control Volume with Arbitrary Acceleration

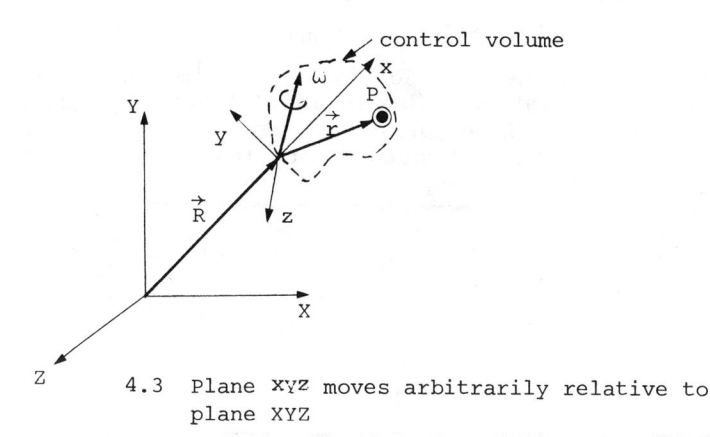

4.3 Plane xyz moves arbitrarily relative to plane XYZ

The most general control volume formulation of Newton's second law is

$$\boxed{\begin{array}{l} \vec{F}_S + \vec{F}_B - \int_{CV} [\vec{a}_{rf} + 2\vec{\omega} \times \vec{V}_{xyz} + \vec{\omega} \times (\vec{\omega} \times \vec{r}) + \dot{\vec{\omega}} \times \vec{r}] \rho dV \\[6pt] = \frac{\partial}{\partial t} \int_{CV} \vec{V}_{xyz} \rho dV + \int_{CS} \vec{V}_{xyz} \rho \vec{V}_{xyz} dA \end{array}}$$

(4.19)

where

$\frac{\partial}{\partial t} \int \vec{V}_{xyz} \rho dV$ = The sum of the rate of change of momentum in the control volume

$$\int_{CS} \vec{V}_{xyz} \rho \vec{V}_{xyz} \cdot d\vec{A} = \text{The net rate of efflux of momentum through the control volume}$$

$\vec{a}_{rf}$ = Absolute rectilinear acceleration of a moving reference xyz relative to a fixed coordinate (XYZ).

$2\vec{\omega} \times \vec{V}_{xyz}$ = Coriolis acceleration due to motion of the particle in the moving coordinate.

$\vec{\omega} \times (\vec{\omega} \times \vec{r})$ = Centrifugal acceleration due to rotation of the moving coordinate.

$\vec{\dot{\omega}} \times \vec{r}$ = Tangential acceleration due to angular acceleration of the moving reference.

F) Moment of Momentum Equation for Control Volume

a) Equation for Fixed Control Volume

The total torque that acts on the control volume is equal to the rate of efflux of angular momentum from the control volume plus the rate of change of angular momentum inside the control volume. The general vector equation is

$$\vec{r} \times \vec{F}_s + \int_{CV} \vec{r} \times \vec{g} \rho dV + \vec{T}_{shaft} = \int_{CS} \vec{r} \times \vec{V} \rho \vec{V} \cdot d\vec{A} + \frac{\partial}{\partial t} \int_{CV} \vec{r} \times \vec{V} \rho dV \quad (4.20)$$

b) Equation for Rotating Control Volume

The general vector equation for the moment of momentum for a non-inertial control volume is

$$\vec{r} \times \vec{F}_s + \int_{CV} \vec{r} \times \vec{g} \rho dV + \vec{T}_{shaft} - \int_{CV} \vec{r} \times [2\vec{\omega} \times \vec{V}_{xyz} + \vec{\omega} \times (\vec{\omega} \times \vec{r}) + \vec{\dot{\omega}} \times \vec{r}] \rho dV$$

$$= \frac{\partial}{\partial t} \int_{CV} \vec{r} \times \vec{V}_{xyz} \rho dV + \int_{CS} \vec{r} \times \vec{V}_{xyz} \rho \vec{V}_{xyz} d\vec{A} \quad (4.21)$$

C-31

G) The First Law of Thermodynamics

The control volume formulation of the first law of thermodynamics is

$$\dot{Q} + \dot{W}_s + \dot{W}_{shear} + \dot{W}_{other} = \frac{\partial}{\partial t} \int_{CV} e\rho d\forall + \int_{CS} (u + pv + \frac{v^2}{2} + gz)\rho \vec{v} \cdot d\vec{A}$$ (4.22)

where

$\dot{Q}$ = Heat transferred to the control volume

$\dot{W}_s$ = Rate of work crossing the control surface by shaft work

$\dot{W}_{shear}$ = Rate of work done by shear stresses

$\dot{W}_{other}$ = Any other work that can be added to the control volume (electrical energy, electromagnetic energy, etc.).

$\frac{\partial}{\partial t} \int_{CV} e\rho d\forall$ = The rate of change of energy in the control volume

$$e = \left(u + \frac{v^2}{2} + gz \right)$$

$\int_{CS} (u + pv + \frac{v^2}{2} + gz)\rho \vec{V} \cdot d\vec{A}$ = The net rate of energy through the control surface

H) The Second Law of Thermodynamics

The control volume formulation of the second law of thermodynamics is

$$\frac{\partial}{\partial t} \int_{CV} s\rho d\forall + \int_{CS} s\rho \vec{V} \cdot d\vec{A} \geq \int_{CS} \frac{1}{T} \left(\frac{\dot{Q}}{A} \right) dA$$ (4.23)

CHAPTER 5

INTRODUCTION TO DIFFERENTIAL ANALYSIS OF FLUID MOTION

In Chapter 4 the basic equations in integral form for a control volume were developed. To obtain a detailed knowledge of the flow field the equations must also be developed in a differential form.

5.1 THE CONTINUITY EQUATION

A) In Rectangular Coordinates

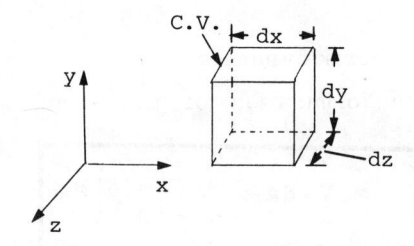

Fig. 5.1 Differential control volume in rectangular coordinates

The differential form of the continuity equation is given by

$$\frac{\partial(\rho u)}{\partial x} + \frac{\partial(\rho v)}{\partial y} + \frac{\partial(\rho w)}{\partial z} = -\frac{\partial \rho}{\partial t} \qquad (5.1)$$

where u, v, and w are the velocity components in the x, y, and z directions, respectively. The continuity equation may be written more compactly in vector notation as

$$\nabla \cdot \rho \vec{v} = -\frac{\partial \rho}{\partial t} \qquad (5.2)$$

where

$$\nabla = \hat{i}\frac{\partial}{\partial x} + \hat{j}\frac{\partial}{\partial y} + \hat{k}\frac{\partial}{\partial z}$$

For steady incompressible flow, ρ constant, and $\partial \rho / \partial t = 0$. Thus the continuity equation is reduced to

$$\frac{\partial u}{\partial x} + \frac{\partial v}{\partial y} + \frac{\partial w}{\partial z} = 0 \qquad (5.3)$$

or

$$\nabla \cdot \vec{v} = 0 \qquad (5.4)$$

and for steady compressible flow,

$$\frac{\partial(\rho u)}{\partial x} + \frac{\partial(\rho v)}{\partial y} + \frac{\partial(\rho w)}{\partial z} = 0 \qquad (5.5)$$

or,

$$\nabla \cdot \rho \vec{v} = 0 \qquad (5.6)$$

B) In a Cylindrical Coordinate System

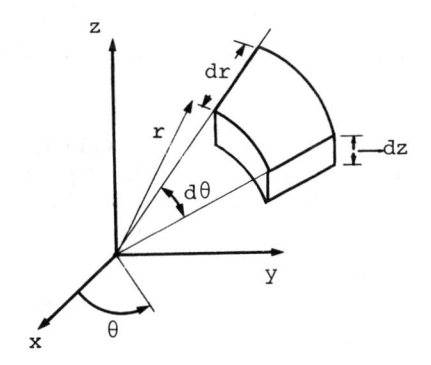

Fig. 5.2 Cylindrical coordinates

The differential form of the continuity equation in cylindrical coordinates is given by

$$\frac{1}{r}\frac{\partial r \rho V_r}{\partial r} + \frac{1}{r}\frac{\partial \rho V_\theta}{\partial \theta} + \frac{\partial \rho V_z}{\partial z} = -\frac{\partial \rho}{\partial t} \qquad (5.7)$$

C-34

or,

$$\nabla \cdot \rho \vec{v} = -\frac{\partial \rho}{\partial t} \qquad (5.8)$$

For steady incompressible flow, ρ = constant, and the continuity equation is reduced to

$$\boxed{\frac{1}{r}\frac{\partial r V_r}{\partial r} + \frac{1}{r}\frac{\partial V_\theta}{\partial \theta} + \frac{\partial V_z}{\partial z} = 0} \qquad (5.9)$$

and for steady compressible flow,

$$\frac{1}{r}\frac{\partial r\rho V_r}{\partial r} + \frac{1}{r}\frac{\partial \rho V_\theta}{\partial \theta} + \frac{\partial \rho V_z}{\partial z} = 0 \qquad (5.10)$$

5.2 THE DIFFERENTIAL FORM OF MOMENTUM EQUATION

A) Acceleration of a Fluid Particle in a Velocity Field

For a velocity field $\vec{v} = \vec{v}(x,y,z,t)$, the acceleration $\left(\frac{D\vec{v}}{Dt}\right)$ of a fluid particle is given by

$$\boxed{a = \frac{D\vec{v}}{Dt} = \frac{d}{dt}\vec{v}(x,y,z,t) = u\frac{\partial \vec{v}}{\partial x} + v\frac{\partial \vec{v}}{\partial y} + w\frac{\partial \vec{v}}{\partial z} + \frac{\partial \vec{v}}{\partial t}} \qquad (5.11)$$

where

$u\frac{\partial \vec{v}}{\partial x} + v\frac{\partial \vec{v}}{\partial y} + w\frac{\partial \vec{v}}{\partial z}$ = Time rate of change of velocity of the particle, due to its changing position in the field (acceleration of transport or convective acceleration)

u,v,w = Scalar velocity components of the particle in the x, y, and z directions respectively

$\frac{\partial \vec{v}}{\partial t}$ = Rate of change of the velocity field itself at the position occupied by the particle at time t (local acceleration)

B) Forces Acting on a Fluid Particle

Forces acting on a fluid element may be classified as

body forces and surface forces. Surface forces include both normal forces and tangential (shear) forces (Fig. 5.3).

The net force in the x-direction is given by:

(5.12)
$$dF_x = dF_{sx} + dF_{Bx} = \left(\rho B_x + \frac{\partial \sigma_{xx}}{\partial x} + \frac{\partial \tau_{yx}}{\partial y} + \frac{\partial \tau_{zx}}{\partial z}\right) dxdydz$$

In the y-direction

(5.13)
$$dF_y = dF_{sy} + dF_{By} = \left(\rho B_y + \frac{\partial \tau_{xy}}{\partial x} + \frac{\partial \sigma_{yy}}{\partial y} + \frac{\partial \tau_{zy}}{\partial z}\right) dxdydz$$

In the z-direction

(5.14)
$$dF_z = dF_{sz} + dF_{Bz} = \left(\rho B_z + \frac{\partial \tau_{xz}}{\partial x} + \frac{\partial \tau_{yz}}{\partial y} + \frac{\partial \sigma_{zz}}{\partial z}\right) dxdydz$$

where B is the body force per unit of mass.

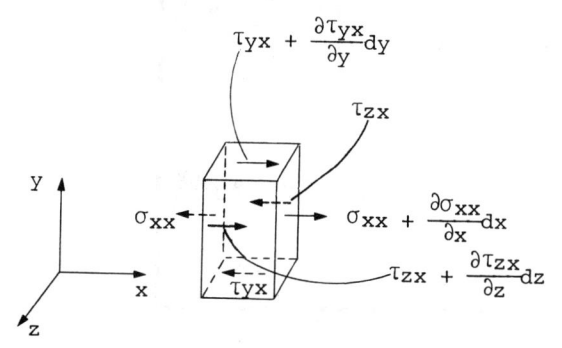

Fig. 5.3 Stresses in the x direction

C) Differential Momentum Equation

For a system of mass, dm, Newton's second law can be written as

$$d\vec{F} = dm \frac{d\vec{v}}{dt} \qquad (5.15)$$

By substituting the expressions of acceleration $\left(\frac{d\vec{V}}{dt}\right)$ and the expressions of force ($d\vec{F}$) acting on the element of mass (dm) in the above equation, we can obtain the differential equations of motion which are:

C-36

$$\rho B_x + \frac{\partial \sigma_{xx}}{\partial x} + \frac{\partial \tau_{yx}}{\partial y} + \frac{\partial \tau_{zx}}{\partial z} = \rho\left(\frac{\partial u}{\partial t} + u\frac{\partial u}{\partial x} + \sigma\frac{\partial u}{\partial y} + w\frac{\partial u}{\partial z}\right) \quad (5.16)$$

$$\rho B_y + \frac{\partial \tau_{xy}}{\partial x} + \frac{\partial \sigma_{yy}}{\partial y} + \frac{\partial \tau_{zy}}{\partial z} = \rho\left(\frac{\partial v}{\partial t} + u\frac{\partial v}{\partial x} + \sigma\frac{\partial \sigma}{\partial y} + w\frac{\partial \sigma}{\partial z}\right) \quad (5.17)$$

$$\rho B_z + \frac{\partial \tau_{xz}}{\partial x} + \frac{\partial \tau_{yz}}{\partial y} + \frac{\partial \sigma_{zz}}{\partial z} = \rho\left(\frac{\partial w}{\partial t} + u\frac{\partial w}{\partial x} + v\frac{\partial w}{\partial y} + w\frac{\partial w}{\partial z}\right) \quad (5.18)$$

For Newtonian fluids the stresses may be expressed in terms of velocity gradients and fluid properties as

$$\tau_{xy} = \tau_{yx} = \mu\left(\frac{\partial \sigma}{\partial x} + \frac{\partial u}{\partial y}\right) \quad (5.19)$$

$$\tau_{yz} = \tau_{zy} = \mu\left(\frac{\partial w}{\partial y} + \frac{\partial v}{\partial z}\right) \quad (5.20)$$

$$\tau_{zx} = \tau_{xz} = \mu\left(\frac{\partial u}{\partial z} + \frac{\partial w}{\partial x}\right) \quad (5.21)$$

$$\sigma_{xx} = -p - \frac{2}{3}\mu\nabla\cdot\vec{V} + 2\mu\frac{\partial u}{\partial x} \quad (5.22)$$

$$\sigma_{xy} = -p - \frac{2}{3}\mu\nabla\cdot\vec{V} + 2\mu\frac{\partial v}{\partial y} \quad (5.23)$$

$$\sigma_{zz} = -p - \frac{2}{3}\mu\nabla\cdot\vec{V} + 2\mu\frac{\partial w}{\partial z} \quad (5.24)$$

where p is the local pressure.

If these equations are introduced into the differential equations of motion, we can obtain the Navier-Stokes equations, which are:

$$\rho\left(\frac{\partial u}{\partial t} + u\frac{\partial u}{\partial x} + v\frac{\partial u}{\partial y} + w\frac{\partial u}{\partial z}\right) = \rho B_x - \frac{\partial p}{\partial x} + \mu\left(\frac{\partial^2 u}{\partial x^2} + \frac{\partial^2 u}{\partial y^2} + \frac{\partial^2 u}{\partial z^2}\right) \quad (5.25)$$

$$\rho\left(\frac{\partial v}{\partial t} + u\frac{\partial v}{\partial x} + v\frac{\partial v}{\partial y} + w\frac{\partial v}{\partial z}\right) = \rho B_y - \frac{\partial p}{\partial y} + \mu\left(\frac{\partial^2 v}{\partial x^2} + \frac{\partial^2 v}{\partial y^2} + \frac{\partial^2 v}{\partial z^2}\right) \quad (5.26)$$

$$\rho \left(\frac{\partial w}{\partial t} + u \frac{\partial w}{\partial x} + v \frac{\partial w}{\partial y} + w \frac{\partial w}{\partial z} \right) = \rho B_z - \frac{\partial p}{\partial z} + \mu \left(\frac{\partial^2 w}{\partial x^2} + \frac{\partial^2 w}{\partial y^2} + \frac{\partial^2 w}{\partial z^2} \right)$$

(5.27)

5.3 EULER'S EQUATIONS

Euler's equations are obtained from the equations of motion asuming a frictionless flow. Since, in a frictionless flow, there can be no shear stress present, the surface forces are due to pressure.

Thus, the Euler's equations are

$$\rho B_x - \frac{\partial p}{\partial x} = \rho \left(\frac{\partial u}{\partial t} + u \frac{\partial u}{\partial x} + v \frac{\partial u}{\partial y} + w \frac{\partial u}{\partial z} \right) \quad (5.28)$$

$$\rho B_y - \frac{\partial p}{\partial y} = \rho \left(\frac{\partial v}{\partial t} + u \frac{\partial v}{\partial x} + v \frac{\partial v}{\partial y} + w \frac{\partial v}{\partial z} \right) \quad (5.29)$$

$$\rho B_z - \frac{\partial p}{\partial z} = \rho \left(\frac{\partial w}{\partial t} + u \frac{\partial w}{\partial x} + v \frac{\partial w}{\partial y} + w \frac{\partial w}{\partial z} \right) \quad (5.30)$$

These equations can also be expressed in a vectorial form as

$$\rho \vec{B} - \nabla p = \rho \left(\frac{\partial \vec{V}}{\partial t} + u \frac{\partial \vec{V}}{\partial x} + v \frac{\partial \vec{V}}{\partial y} + w \frac{\partial \vec{V}}{\partial z} \right)$$

or,

$$\rho \vec{B} - \nabla p = \rho \frac{D\vec{V}}{Dt} \quad (5.31)$$

If the only body force is due to gravity, Euler's equation becomes

$$\rho \vec{g} - \nabla p = \rho \frac{D\vec{V}}{Dt} \quad (5.32)$$

For frictionless flow along a streamline the Euler's equation is reduced to

$$\frac{dp}{\rho} + gdz + vdv = 0$$

In cylindrical coordinates, Euler's equation (with gravity the only force) becomes

$$g_r - \frac{1}{\rho}\frac{\partial p}{\partial r} = a_r = \frac{\partial v_r}{\partial t} + v_r\frac{\partial v_r}{\partial r} + \frac{v_\theta}{r}\frac{\partial v_r}{\partial \theta} + v_z\frac{\partial v_r}{\partial z} - \frac{v_\theta^2}{r} \quad (5.33)$$

$$g_\theta - \frac{1}{\rho r}\frac{\partial p}{\partial \theta} = a_\theta = \frac{\partial v_\theta}{\partial t} + v_r\frac{\partial v_\theta}{\partial r} + \frac{v_\theta}{r}\frac{\partial v_\theta}{\partial \theta} + v_z\frac{\partial v_\theta}{\partial z} + \frac{v_r v_\theta}{r} \quad (5.34)$$

$$g_z - \frac{1}{\rho}\frac{\partial p}{\partial z} = a_z = \frac{\partial v_z}{\partial t} + v_r\frac{\partial v_z}{\partial r} + \frac{v_\theta}{r}\frac{\partial v_z}{\partial \theta} + v_z\frac{\partial v_z}{\partial z} \quad (5.35)$$

5.4 BERNOULLI'S EQUATION

By integrating Euler's equation we get the Bernoulli equation:

$$\frac{p}{\rho} + gz + \frac{v^2}{2} = \text{constant} \quad (5.36)$$

(along a streamline)

The assumptions by which the Bernoulli equation holds, are:

A) Steady flow
B) Incompressible flow
C) Frictionless flow
D) Flow along a streamline

Applied between any two points on a streamline the Bernoulli equation becomes

$$\frac{p_1}{\rho} + \frac{v_1^2}{2} + gz_1 = \frac{p_2}{\rho} + \frac{v_2^2}{2} + gz_2$$

or

$$\frac{p_1}{\gamma} + \frac{v_1^2}{2g} + z_1 = \frac{p_2}{\gamma} + \frac{v_2^2}{2g} + z_2 \quad (5.37)$$

The Bernoulli equation can also be derived by using the first law of thermodynamics.

The first law of thermodynamics is reduced to Bernoulli's equation under the following restrictions:

A) $\dot{W}_s = \dot{W}_{shear} = \dot{W}_{other} = 0$
B) Steady flow
C) Uniform flow at each section
D) Incompressible flow
E) $u_2 - u_1 = \dot{Q}/\dot{m}$

CHAPTER 6

CONSIDERATIONS FOR COMPRESSIBLE FLOW

6.1 REVIEW FOR THERMODYNAMICS

A) Equation of State

The pressure, density, and temperature of a substance may be related by the equation of state:

$$p = \rho RT \qquad (6.1)$$

where

$$R = \frac{R_u}{M}$$

R_u = Universal gas constant
   = 8314 Nm/kg mole K

M = Molecular mass of gas

Substances that satisfy this equation are known as perfect gases.

B) Internal Energy

For any substance that follows the equation of state $p = \rho RT$, the internal energy, 'u', is given by the equation $du = c_v dT$, or by integrating this equation with constant specific heat

C-41

$$u_2 - u_1 = \int_{u_1}^{u_2} du = \int_{T_1}^{T_2} c_v dT = c_v(T_2 - T_1) \quad (6.2)$$

C) Enthalpy

The enthalpy of a substance is defined as $h = u+pv$, and for an ideal gas, $dh = c_p dT$, and when integrating with constant specific heat,

$$h_2 - h_1 = \int_{h_1}^{h_2} dh = \int_{T_1}^{T_2} c_p dT = c_p(T_2 - T_1) \quad (6.3)$$

D) Entropy

Entropy is defined by the equation

$$\Delta s = \int_{rev} \frac{\delta Q}{T} \quad \text{or,} \quad \Delta s = \left(\frac{\delta Q}{T}\right)_{rev} \quad (6.4)$$

The inequality of Clausius states that

$$\oint \frac{\delta Q}{T} \leq 0 \quad (6.5)$$

As a consequence of the second law the results can be extended to

$$Tds \geq \delta Q \quad (6.6)$$

For reversible processes,

$$Tds = \frac{\delta Q}{dm} \quad (6.7)$$

For irreversible processes,

$$Tds > \frac{\delta Q}{dm} \quad (6.8)$$

For an adiabatic process, $\delta Q/\delta m = 0$. Thus,

$\quad ds = 0 \quad$ (reversible adiabatic process)

$\quad ds > 0 \quad$ (irreversible adiabatic process)

For an ideal gas with constant specific heats, we have

$$s_2 - s_1 = c_v \ln \frac{T_2}{T_1} + R\ln \frac{V_2}{V_1} \quad (6.9)$$

$$= c_p \ln \frac{T_2}{T_1} - R\ln \frac{p_2}{p_1} \quad (6.10)$$

E) Specific Heats

The constant pressure of specific heat, c_p, is defined as

$$c_p = \left(\frac{\partial h}{\partial T}\right)_p \quad (6.11)$$

The constant volume of specific heat, c_v, is defined as

$$c_v = \left(\frac{\partial u}{\partial T}\right)_v \quad (6.12)$$

Also, for an ideal gas we have the following relations:

$$c_p - c_v = R \quad (6.13)$$

$$K = \frac{c_p}{c_v} \quad (6.14)$$

$$c_p = \frac{KR}{K-1} \quad (6.15)$$

$$c_v = \frac{R}{K-1} \quad (6.16)$$

6.2 PROPAGATION OF SOUND WAVES

A) Speed of Sound; the Mach Number

The Mach number is defined as:

$$M = \frac{V}{c} \quad (6.17)$$

where   V = Local flow speed

c = Local speed of sound

For an ideal gas the speed of sound is given by the relation

$$c = \sqrt{KRT} \qquad (6.18)$$

where
R = A constant for each gas
T = Temperature of the gas
$K = \dfrac{c_p}{c_v}$

B) Types of Flow; The Mach Cone

The types of flow are shown in the following table

$M < 1$	subsonic
$M = 1$	sonic
$M > 1$	supersonic
$M \gtrsim 5$	hypersonic

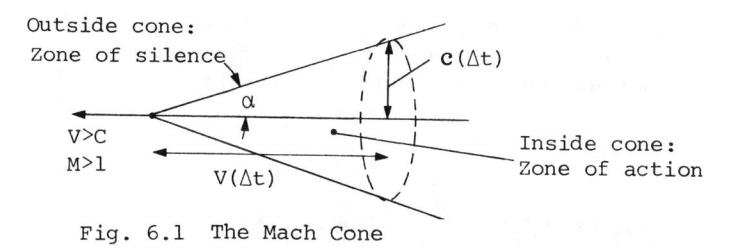

Fig. 6.1 The Mach Cone

The cone angle, α, can be related to the Mach number at which the source moves. From the geometry of (Fig. 6.1), we conclude that

$$\sin \alpha = \frac{c}{V} = \frac{1}{M} \quad \text{or,}$$

$$\alpha = \sin^{-1}\left(\frac{1}{M}\right) \qquad (6.19)$$

where α is the half-angle of the Mach cone.

The regions inside and outside the cone are sometimes called the zone of action and the zone of silence, respectively.

C-44

6.3 LOCAL ISENTROPIC STAGNATION PROPERTIES

Local isentropic stagnation properties are properties that would be obtained at any point in a flow field if the fluid at that point were decelerated from local conditions to zero velocity following an isentropic process.

The equations for the isentropic stagnation properties of an ideal gas are:

$$\frac{p_0}{p} = \left[1 + \frac{K-1}{2} M^2\right]^{K/(K-1)} \qquad (6.20)$$

$$\frac{T_0}{T} = 1 + \frac{K-1}{2} M^2$$

$$\frac{\rho_0}{\rho} = \left[1 + \frac{K-1}{2} M^2\right]^{1/(K-1)} \qquad (6.21)$$

For ideal gases, the ratios of local isentropic stagnation properties to corresponding static properties for an ideal gas can be obtained from standard textbooks.

6.4 CRITICAL CONDITION

Critical conditions are those conditions where the Mach number is equal to unity (sonic conditions).

Sonic conditions are marked with an asterisk:

$$V^* \equiv C^*$$

At critical conditions the stagnation properties become

$$\frac{p_0^*}{p^*} = \left[1 + \frac{K-1}{2}\right]^{K/(K-1)} \qquad (6.22)$$

C-45

$$\frac{T_0^*}{T^*} = 1 + \frac{K-1}{2} \qquad (6.23)$$

$$\frac{\rho_0^*}{\rho^*} = \left[1 + \frac{K-1}{2}\right]^{1/(K-1)} \qquad (6.24)$$

The critical speed may be written in terms of either the critical temperature, T^*, or the critical stagnation temperature, T_0^*.

$$V^* = c^* = \sqrt{\frac{2K}{K+1} R T_0^*} \qquad (6.25)$$

$$V^* = c^* = \sqrt{KRT^*} \qquad (6.26)$$

CHAPTER 7

ONE-DIMENSIONAL COMPRESSIBLE FLOW

7.1 BASIC EQUATIONS FOR ISENTROPIC FLOW

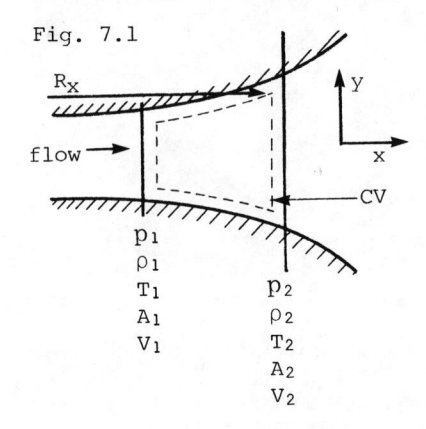

Fig. 7.1

p_1
ρ_1
T_1
A_1
V_1

p_2
ρ_2
T_2
A_2
V_2

The basic equations applied to the control volume shown in Figure 7.1 are

A) Continuity Equation

$$\frac{\partial}{\partial t} \int_{cv} \rho \, d\forall \overset{0}{} + \int_{cs} \rho \vec{V} \cdot d\vec{A} = 0$$

or
$$-\rho_1 V_1 A_1 + \rho_2 V_2 A_2 = 0$$

or
$$\boxed{\rho_1 V_1 A_1 = \rho_2 V_2 A_2 = \rho V A = \dot{m} = \text{constant}} \quad (7.1)$$

B) Momentum Equation

$$\vec{F}_{S_x} + \vec{F}_{B_x} = \int_{cs} \vec{V} \rho \vec{V} \cdot d\vec{A} + \frac{\partial}{\partial t} \int_{cv} \vec{V} \rho \, d\cancel{V}^{\,0}$$

or
$$R_x + p_1 A_1 - p_2 A_2 = -V_1^2 \rho_1 A_1 + V_2^2 \rho_2 A_2$$

or
$$R_x + p_1 A_1 - p_2 A_2 = \dot{m} V_2 - \dot{m} V_1 \quad (7.2)$$

where

R_x = External force acting on the control volume

C) First Law of Thermodynamics

$$\cancel{\dot{Q}}^{\,0} + \cancel{\dot{W}_s}^{\,0} + \cancel{\dot{W}_{shear}}^{\,0} + \dot{W}_{other} = \frac{\partial}{\partial t} \int_{cv} e\rho \, d\cancel{V}^{\,0}$$

$$+ \int_{cs} (e + pv) \rho \vec{v} \cdot d\vec{A}$$

where
$$e = u + \frac{V^2}{2} + \cancel{gz}^{\,0}$$

Then the first law of thermodynamics becomes

$$\boxed{h_1 + \frac{V_1^2}{2} = h_2 + \frac{V_2^2}{2} = \text{constant}} \quad (7.3)$$

where
$$h = u + pv$$

D) Second Law of Thermodynamics

$$\int_{cs} \frac{1}{T} \frac{\dot{Q}}{A} \, dA \leq \frac{\partial}{\partial t} \int_{cv} S\rho \, d\cancel{V} + \int_{cs} S\rho \vec{V} \cdot d\vec{A}$$

or
$$s_1[-\rho_1 V_1 A_1] + s_2[\rho_2 V_2 A_2] = 0$$

or
$$\boxed{s_1 = s_2 = s = \text{constant}} \quad (7.4)$$

E) Equation of State
$$p = \rho RT \quad (7.5)$$

7.2 EFFECTS OF AREA VARIATION ON FLOW PROPERTIES IN ISENTROPIC FLOW

In order to study the effects of area variation on flow properties in isentropic flow, the following equations are considered:

$$\boxed{\begin{aligned}\frac{dA}{A} &= \frac{dP}{\rho V^2}(1-M^2) \\ \frac{dA}{A} &= \frac{-dV}{V}(1-M^2)\end{aligned}} \quad \begin{aligned}(7.6a)\\ \\ (7.6b)\end{aligned}$$

From these equations, any area variation will result in a positive or a negative change of velocity and pressure depending upon the M value:

for $M^2 < 1$, $dP > 0$ and $dV < 0$

for $M^2 > 1$, $dP < 0$ and $dV > 0$

The results are summarized in Figure 7.2.

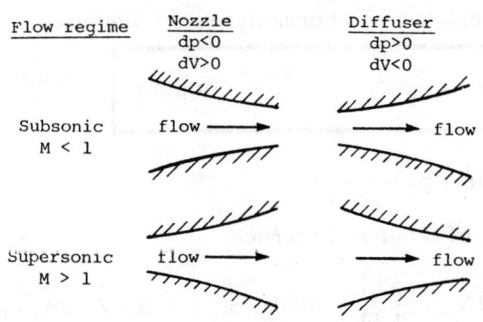

Flow regime	Nozzle dp<0 dV>0	Diffuser dp>0 dV<0
Subsonic M < 1	flow ⟶	⟶ flow
Supersonic M > 1	flow ⟶	⟶ flow

Fig. 7.2 Nozzle and diffuser shapes as a function of initial Mach number.

7.3 ISENTROPIC FLOW OF AN IDEAL GAS

A) Basic Equations

Continuity:	$\rho_1 V_1 A_1 = \rho_2 V_2 A_2 = \dot{m}$	(7.7)
Momentum:	$R_x + p_1 A_1 - p_2 A_2 = \dot{m} V_2 - \dot{m} V_1$	(7.8)
First Law:	$h_1 + \dfrac{V_1^2}{2} = h_2 + \dfrac{V_2^2}{2} = h + \dfrac{V^2}{2}$	(7.9)
Second Law:	$s_1 = s_2 = s$	(7.10)
Equation of State:	$p = \rho R T$	(7.11)
Process Equation:	$p/\rho^k = \text{constant}$	(7.12)

B) Stagnation Properties and Critical Properties for Isentropic Flow of an Ideal Gas

Stagnation pressure:	$\dfrac{p_0}{p} = \left[1 + \dfrac{K-1}{2} M^2\right]^{K/(K-1)}$	(7.13)
Stagnation temperature:	$\dfrac{T_0}{T} = 1 + \dfrac{K-1}{2} M^2$	(7.14)
Stagnation density:	$\dfrac{\rho_0}{\rho} = \left[1 + \dfrac{K-1}{2} M^2\right]^{1/(K-1)}$	(7.15)

At critical conditions the stagnation properties become (K = 1.4)

$$\dfrac{p_0}{p^*} = \left[1 + \dfrac{K-1}{2}\right]^{K/(K-1)} = 1.893 \qquad (7.16)$$

$$\dfrac{T_0}{T^*} = 1 + \dfrac{K-1}{2} = 1.200 \qquad (7.17)$$

$$\dfrac{\rho_0}{\rho^*} = \left[1 + \dfrac{K-1}{2}\right]^{1/(K-1)} = 1.577 \qquad (7.18)$$

In addition,

$$V^* = c^* = \sqrt{\frac{2K}{K+1} RT_0}$$ (7.19)

$$\frac{A}{A^*} = \frac{1}{M}\left[\frac{1 + \frac{K-1}{2}M^2}{1 + \frac{K-1}{2}}\right]^{(K+1)/2(K-1)}$$ (7.20)

since, ρAV = constant

7.4 ISENTROPIC FLOW IN A CONVERGING NOZZLE

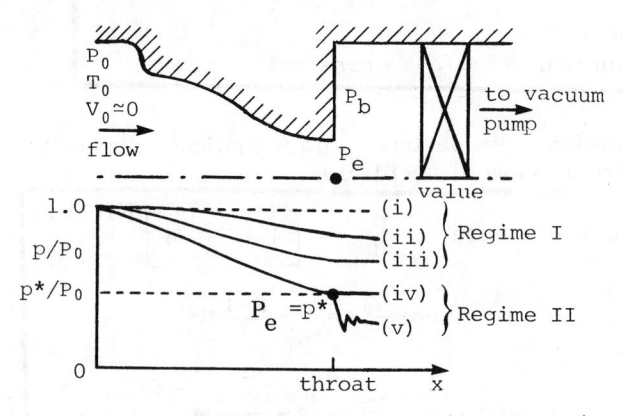

Fig. 7.3 Converging nozzle operating at various back pressures.

The effects of variations in back pressure, p_b, on the pressure distribution through a converging nozzle are graphically illustrated in Figure 7.3. The flow through a converging nozzle may be divided into two regimes:

REGIME I:

In this regime when the back pressure, P_b, decreases, the flowrate increases and the exit plane pressure, P_e, decreases (see curves b and c, Fig. 7.3). The flow at the exit plane will eventually reach a Mach number equal to unity (curve d) where the pressure is the critical pressure, P^* (M = 1, $P_b/P_0 = P^*/P_0$).

REGIME II :

In this regime, a reduction of the back pressure, P_b, below the critical pressure, P^*, has no effect on the flow conditions inside the nozzle. If $P_b \leq P^*$, the nozzle is said to be choked. If $P_b \leq P^*$, the flow leaving the nozzle expands to reach a lower back pressure (curve e). Figure 7.3 shows the T-S diagram for the flow processes in this regime.

7.5 ADIABATIC FLOW IN A CONSTANT AREA DUCT WITH FRICTION

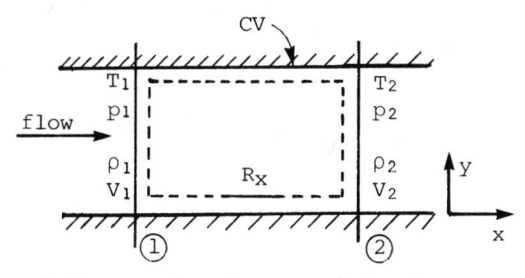

Fig. 7.4 Control volume used for integral analysis of frictional adiabatic flow.

The basic equations for the steady flow of an ideal gas, with constant specific heats applied to the control volume shown in Figure 7.4 are:

A) Continuity Equation

$$0 = \frac{\partial}{\partial t} \int_{cv} \rho \, d\forall + \int_{cs} \rho \vec{V} d\vec{A}$$

Because of a steady flow at each section, the continuity continuity equation becomes

C-52

$$\rho_1 V_1 = \rho_2 V_2 = \text{constant} \qquad (7.21)$$

B) Momentum Equation

$$F_{sx} + F_{Bx} = \cancelto{0}{\frac{\partial}{\partial t} \int_{cv} V_x \rho d\mathcal{V}} + \int_{cs} V_x \rho \vec{V} \cdot d\vec{A}$$

if $F_{Bx} = 0$ and $A_2 = A_1 = A$ then,

$$R_x + p_1 A - p_2 A = \dot{m} V_2 - \dot{m} V_1 \qquad (7.22)$$

C) First Law of Thermodynamics

$$\cancelto{0}{\dot{Q}} + \cancelto{0}{\dot{W}_s} + \cancelto{0}{\dot{W}_{shear}} + \cancelto{0}{\dot{W}_{other}}$$

$$= \cancelto{0}{\frac{\partial}{\partial t} \int_{cv} e \rho d\mathcal{V}} + \int_{cs} (e+pv)\rho \vec{V} \cdot d\vec{A}$$

where

$$e = u + \frac{V^2}{2} + \cancelto{0}{gz}$$

or

$$h_2 - h_1 + \frac{V_2^2 - V_1^2}{2} = 0$$

where

$$h_2 - h_1 = C_p(T_2 - T_1)$$

or

$$h_1 + \frac{V_1^2}{2} = h_2 + \frac{V_2^2}{2} = \text{constant}$$

or

$$h_{01} = h_{02} \qquad (7.23)$$

D) Second Law of Thermodynamics

$$\int_{cs} \frac{1}{T} \cancelto{0}{\frac{\dot{Q}}{A}} dA \leq \frac{\partial}{\partial t} \int_{cv} s \rho d\mathcal{V} + \int_{cs} s \rho \vec{V} \cdot d\vec{A}$$

Because the flow is frictional and irreversible,

$$\dot{m}(s_2 - s_1) > 0 \qquad (7.24)$$

C-53

where

$$s_2 - s_1 = C_p \ln \frac{T_2}{T_1} - R \ln \frac{p_2}{p_1}$$

E) Equation of State

$$p = \rho RT \qquad (7.25)$$

7.6 THE FANNO LINE

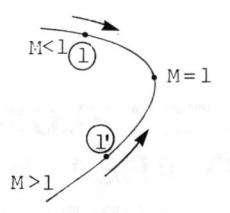

Fig. 7.5

Let us consider the governing equations developed in the last section and the relationship between h and T for an ideal gas (C_p = constant):

$$\boxed{\Delta h = C_p(T_2 - T_1)} \qquad (7.26)$$

Thus, we have six equations and seven unknowns. If the initial state is known, then there is an infinite number of final states (six equations and seven unknowns). The locus of the possible final states plotted on the T-S or h-S diagram is called the Fanno Line (Fig. 7.5). The Mach number is less than 1 (M<1) on the upper portion of the curve and it increases as we move to the right. The Mach number is greater than 1 (M>1) on the lower portion of the curve and it decreases as we move to the right. The Mach number is 1 (M=1) at the point of maximum entropy.

The effects of friction on flow properties in Fanno line flow are summarized in the following table:

Table 7.1

Property	M < 1	M > 1
Stagnation Temperature, T_0	Constant	Constant
Entropy, s	Increases	Increases
Stagnation Pressure, P_0	Decreases	Decreases
Temperature, T	Decreases	Increases
Velocity, V	Increases	Decreases
Mach number, M	Increases	Decreases
Density, ρ	Decreases	Increases
Pressure, p	Decreases	Increases
Stagnation Enthalpy	Constant	Constant

7.7 FRICTIONLESS FLOW IN A CONSTANT AREA DUCT WITH HEAT ADDITION

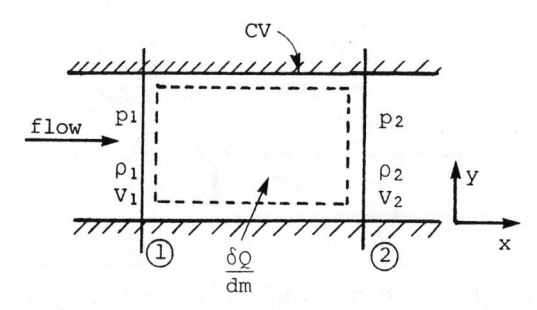

Fig. 7.6 Control volume used for integral analysis analysis of frictionless flow with heat transfer.

The basic equations for the steady, one-dimensional, frictionless flow of an ideal gas with constant specific heats applied to the control volume shown in (Fig. 7.6) are:

A) Continuity Equation

$$0 = \frac{\partial}{\partial t} \int_{cv} \rho d\forall + \int_{cs} \rho \vec{V} d\vec{A}$$

C-55

Because of a steady uniform flow at each section, the continuity equation becomes

$$\rho_1 V_1 = \rho_2 V_2 = \frac{\dot{m}}{A} \qquad (7.27)$$

B) Momentum Equation

$$F_{Sx} + \cancel{F_{Bx}}^{0} = \cancel{\frac{\partial}{\partial t} \int_{cv} V_x \rho\, d\Psi}^{0} + \int_{cs} V_x \rho \vec{V} d\vec{A} \quad \text{or,}$$

$$p_1 + \rho_1 v_1^2 = p_2 + \rho_2 v_2^2$$

or

$$p + \rho v^2 = \text{constant} \qquad (7.28)$$

C) First Law of Thermodynamics

$$\dot{Q} + \cancel{\dot{W}_s}^{0} + \cancel{\dot{W}_{shear}}^{0} + \cancel{\dot{W}_{other}}^{0} = \cancel{\frac{\partial}{\partial t} \int_{cv} e\rho\, d\Psi}^{0} + \int_{cs} (e+pv)\rho \vec{v} d\vec{A}$$

where

$$e = u + \frac{V^2}{2} + \cancel{gz}^{0}$$

or

$$\frac{\delta Q}{dm} = \left(h_2 + \frac{V_2^2}{2}\right) - \left(h_1 + \frac{V_1^2}{2}\right) = (h_2 - h_1) + \left(\frac{V_2^2 - V_1^2}{2}\right)$$

where

$$h_2 - h_1 = C_p(T_2 - T_1)$$

or

$$\boxed{\frac{\delta Q}{dm} = h_{02} - h_{01}} \qquad (7.29)$$

where

$$h_{01} = h_1 + \frac{V_1^2}{2}$$

$$h_{02} = h_2 + \frac{V_2^2}{2}$$

D) Second Law of Thermodynamics

C-56

$$\int_{cs} \frac{1}{T} \frac{\dot{Q}}{A} \, dA \leq \frac{\partial}{\partial t} \int_{cv} \cancelto{0}{S\rho \, d\mathcal{V}} + \int_{cs} s\rho \vec{V} d\vec{A}$$

or

$$\int_{cs} \frac{1}{T} \frac{\dot{Q}}{A} \, dA \leq \dot{m}(s_2 - s_1) \qquad (7.30)$$

where

$$\boxed{s_2 - s_1 = C_p \ln \frac{T_2}{T_1} - R \ln \frac{p_2}{p_1}}$$

E) Equation of State

$$p = \rho RT \qquad (7.31)$$

7.8 THE RAYLEIGH LINE

Let us consider the governing equations (7.27) to (7.31) developed in the last section and equation (7.26); thus, we have six equations and seven unknowns. If the initial state is known then there are an infinite number of final states. The locus of the possible final states is plotted on the T-S or h-S diagram is called the Rayleigh line (Fig. 7-7). The mach number is less than 1 (M<1) on the upper portion of the curve and it increases as we move to the right. The Mach number is greater than one (M>1) on the lower portion of the curve and it decreases as we move to the right. The Mach number is one (M=1) at the point of maximum entropy (point b).

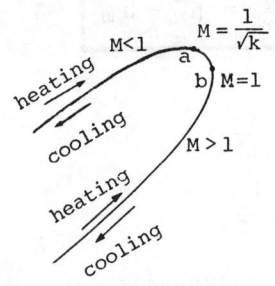

Fig. 7.7 Schematic Ts diagram for frictionless flow in a constant area duct with heat transfer (Rayleigh line flow).

The effects of heat transfer on properties in the steady, frictionless compressible flow of an ideal gas are summarized in the following table.

Table 7.2

Property	Heating M<1	Heating M>1	Cooling M<1	Cooling M>1
Entropy, S	Increases	Increases	Decreases	Decreases
Stagnation Temperature, T_0	Increases	Increases	Decreases	Decreases
Temperature, T	\multicolumn{4}{c}{$M < 1/\sqrt{K}$}
	Increases	Increases	Decreases	Decreases
	\multicolumn{4}{c}{$1/\sqrt{K} < M < 1$}
	Decreases	Increases	Increases	Decreases
Mach number, M	Increases	Decreases	Decreases	Increases
Pressure, p	Decreases	Increases	Increases	Decreases
Velocity, V	Increases	Decreases	Decreases	Increases
Density, ρ	Decreases	Increases	Increases	Decreases
Stagnation Pressure, p_0	Decreases	Decreases	Increases	Increases

7.9 NORMAL SHOCK

Normal shocks are discontinuities of flow, which are normal to the flow direction and can occur in any supersonic flow field. The basic equations applied to the control volume around a normal shock (Fig. 7.8) are:

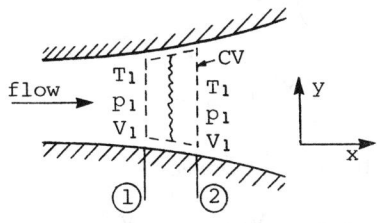

Fig. 7.8 Control volume used for analysis of normal shock.

C-58

A) Continuity Equation

$$0 = \frac{\partial}{\partial t} \int_{cv} \rho \, d\cancel{V}^{\,0} + \int_{cs} \rho \vec{V} d\vec{A}$$

Because of a steady uniform flow and, assuming that the area of change across the shock wave is negligible, (the shock is very thin, 0.2 microns), the continuity equation becomes

$$\rho_1 V_1 = \rho_2 V_2 = G \qquad (7.32)$$

B) Momentum Equation

$$F_{Sx} + \cancel{F_{Bx}}^{\,0} = \frac{\partial}{\partial t}\cancel{\int_{cv} V_x \rho \, dV}^{\,0} + \int_{cs} V_x \rho \vec{V} d\vec{A}$$

on

$$F_{Sx} = \dot{m}V_2 - \dot{m}V_1$$

where

$$F_{Sx} = p_1 A - p_2 A$$

Finally, the momentum equation becomes

$$\boxed{p_1 + \rho_1 V_1^2 = p_2 + \rho_2 V_2^2} \qquad (7.33)$$

C) First Law of Thermodynamics

$$\cancel{\dot{Q}}^{\,0} + \cancel{\dot{W}_s}^{\,0} + \cancel{\dot{W}_{shear}}^{\,0} + \cancel{\dot{W}_{other}}$$

$$= \frac{\partial}{\partial t} \int_{cv} e\rho \, dV + \int_{cs} (e+pv)\rho \vec{v} d\vec{A}$$

where

$$e = u + \frac{V^2}{2} + \cancel{gz}^{\,0}$$

Finally, the first law becomes

$$h_1 + \frac{V_1^2}{2} = h_2 + \frac{V_2^2}{2} \qquad (7.34a)$$

or, in terms of stagnation enthalpy,

$$h_{01} = h_{02} \tag{7.34b}$$

D) Second Law of Thermodynamics

$$\int_{cs} \frac{1}{T} \cancel{\frac{\dot{Q}}{A}}^{0} dA = \frac{\partial}{\partial t} \cancel{\int_{cv} s\rho d\mathcal{V}}^{0} + \int_{cs} s\rho \vec{V} d\vec{A}$$

or,

$$s_2 - s_1 > 0 \tag{7.35}$$

where

$$s_2 - s_1 = C_p \ln \frac{T_2}{T_1} - R \ln \frac{P_2}{P_1}$$

E) Equation of State

$$p = \rho RT \tag{7.36}$$

F) Relationship Between h and T for an Ideal Gas

$$\Delta h = C_p (T_2 - T_1) \tag{7.26}$$

If the initial state 1 is known, then there is a unique final state 2 for a given initial state.

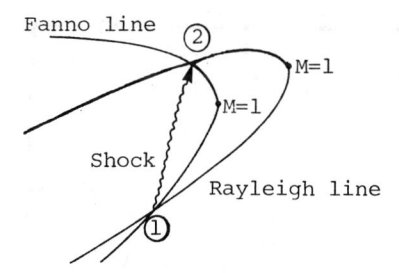

Fig. 7.9 Intersection of Fanno line and Rayleigh line as a solution of the normal shock equations.

The normal shock must satisfy the basic equations plus equation (7.26). A very informative procedure for finding the final state is to make use of the Fanno line and the Rayleigh line. The intersection of the two curves (Fig. 7.9, point 2) represents the conditions of the final state corresponding to that given initial state (point 1).

As a final remark, a normal shock occurs only in a supersonic flow and after the shock there must be a subsonic flow.

The summary of property changes across a normal shock are listed in the following table:

Table 7.3

Property	Effect
Stagnation temperature, T_0	Constant
Entropy, s	Increases
Stagnation pressure, p_0	Decreases
Temperature, T	Increases
Velocity, V	Decreases
Density, ρ	Increases
Pressure, p	Increases
Mach number, M	Decreases

The expressions for the property ratios across the normal shock are the following:

$$\frac{T_2}{T_1} = \frac{1 + \frac{K-1}{2} M_1^2}{1 + \frac{K-1}{2} M_2^2} \tag{7.37}$$

$$\frac{\rho_2}{\rho_1} = \frac{V_1}{V_2} = \frac{M_1}{M_2} \left[\frac{1 + \frac{K-1}{2} M_2^2}{1 + \frac{K-1}{2} M_1^2} \right]^{\frac{1}{2}} \tag{7.38}$$

$$\frac{p_2}{p_1} = \frac{1 + K M_1^2}{1 + K M_2^2} \tag{7.39}$$

$$\frac{p_{02}}{p_{01}} = \frac{p_2}{p_1} \left[\frac{1 + \frac{K-1}{2} M_2^2}{1 + \frac{K-1}{2} M_1^2} \right]^{K/(K-1)} \tag{7.40}$$

The above ratios can be supplied by the use of tables.

7.10 FLOW IN A CONVERGING-DIVERGING NOZZLE

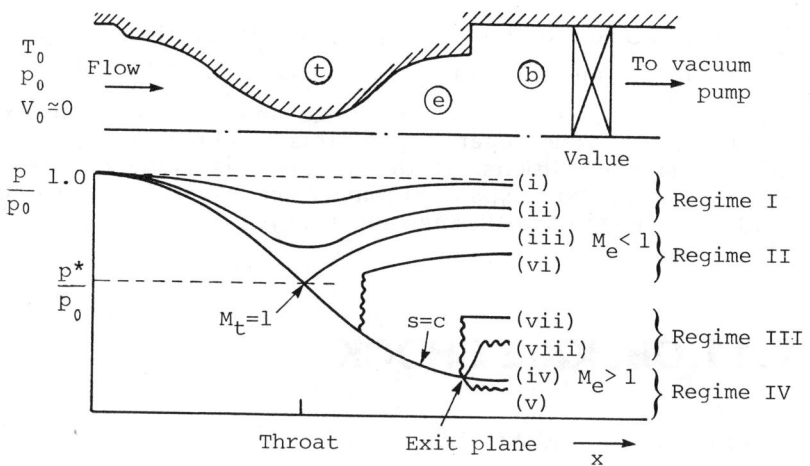

Fig. 7.10 Pressure distributions for flow in a converging-diverging nozzle as a function of back pressure.

The effect of variations in back pressure, P_b, on the pressure distribution through a converging-diverging nozzle is graphically illustrated in Figure 7.10. The flow through this type of nozzle may be divided into three regimes:

REGIME I : Subsonic Flow

The back pressure, P_b, is equal to the stagnation pressure, P_o; therefore there is no flow (curve a). As P_b continues to decrease the flowrate increases but the flow remains subsonic (curves b and c). Deceleration takes place in the diverging section.

REGIME II : Nozzle Over-Expanding

Here, the pressure at some point in the nozzle is less than the back pressure ($P_d > P_b > P_f$). The velocity of the

C-62

flow is critical, the pressure is minimum, and the flow is choked at the throat. The pressure distribution is represented by curve d. As the pressure is lowered, a shock occurs downstream from the throat and moves until it appears at the exit plane of the nozzle.

REGIME III : Nozzle Under-Expanding

A reduction in the back pressure lower than curve f has no effect on the flow in the nozzle. When $P_b = P_f$, the flow is isentropic through the nozzle and supersonic at the nozzle exit. Nozzles operating at this condition are said to be at design conditions. If the pressure is further lowered a series of oblique expansion waves occur and it cannot be studied using the one-dimensional theory.

7.11 OBLIQUE SHOCK

OBLIQUE SHOCK-RELATIONS

The normal shock is a special case of a more general family of oblique shocks that occur in a supersonic flow. An oblique shock occurs when a supersonic flow changes direction due to a boundary surface converging towards the flow.

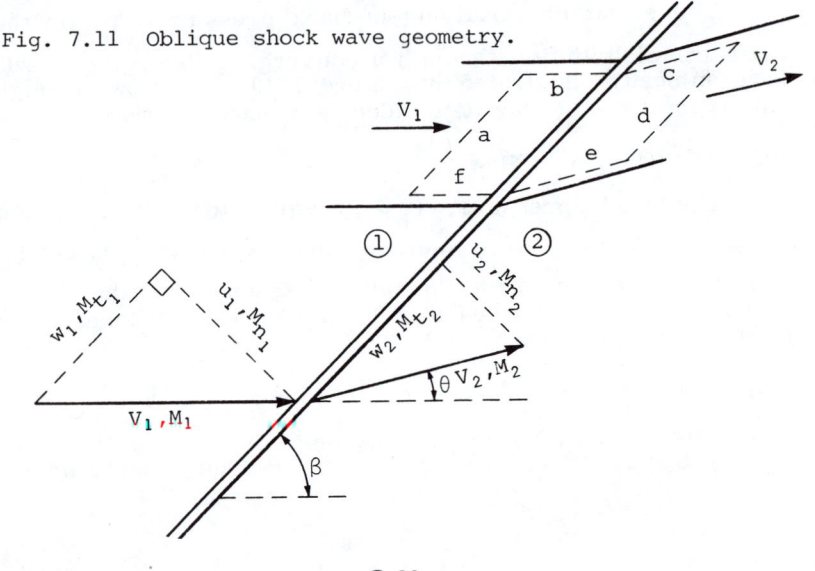

Fig. 7.11 Oblique shock wave geometry.

Figure 7.11 illustrates an oblique shock wave in two-dimensional flow.

The basic equations for a steady flow with no body forces applied to the control volume shown in Figure 7.11b are:

A) Continuity Equation

$$0 = \cancelto{0}{\frac{\partial}{\partial t} \int_{cv} \rho \, d\mathcal{V}} + \int_{cs} \rho \vec{V} \cdot d\vec{A}$$

or,

$$\rho_1 u_1 = \rho_2 u_2 \qquad (7.41)$$

where

ρ_1, ρ_2 = The densities at section 1 and 2, respectively

u_1 = The component of V_1 perpendicular to the shock wave

u_2 = The component of V_2 perpendicular to the shock wave

B) Momentum Equation

$$\cancelto{0}{F_{Sx}} + \cancelto{0}{F_{Bx}} = \cancelto{0}{\frac{\partial}{\partial t} \int_{cv} V_x \rho \, d\mathcal{V}} + \int_{cs} V_x \rho \vec{V} \cdot d\vec{A} \qquad (7.42)$$

If we decompose this equation into two components, we have:

a) Parallel to the shock

$$-(\rho_1 u_1) w_1 + (\rho_2 u_2) w_2 = 0$$

or,

$$w_1 = w_2 \qquad (7.43)$$

where,

w_1 = The component of V_1 parallel to the shock wave

w_2 = The component of V_2 parallel to the shock wave

b) Perpendicular to the shock

C-64

or,
$$-(-\rho_1 u_1)u_1 - (\rho_2 u_2)u_2 = -(p_1 - p_2) \quad (7.44)$$

$$P_1 + \rho_1 u_1^2 = p_2 + \rho_2 u_2^2$$

where

ρ_1, ρ_2 = The densities at section 1 and 2, respectively

C) First Law of Thermodynmaics

$$\dot{Q} + \dot{W}_s + \dot{W}_{shear} + \dot{W}_{other}$$

$$= \frac{\partial}{\partial t} \int_{cv} e\rho \, d\mathcal{V} + \int_{cv} (e+pv)\rho \vec{V} \cdot d\vec{A}$$

or,
$$-(-p_1 u_1 + p_2 u_2) = -\rho_1 \left[e_1 + \frac{V_1^2}{2}\right]u_1 + \rho_2 \left[e_2 + \frac{V_2^2}{2}\right]u_2$$

or,
$$h_1 + \frac{V_1^2}{2} = h_2 + \frac{V_2^2}{2} \quad (7.45)$$

or,
$$h_1 + \frac{u_1^2}{2} = h_2 + \frac{u_2^2}{2} \quad (7.46)$$

Looking at equations (7.41), (7.44), (7.46) we conclude that the changes across an oblique shock are governed by the normal components of the freestream velocity.

D) Perfect Gas

For an oblique shock wave with $M_{n_1} = M_1 \sin b$ and $M_{n_2} = M_2 \sin(b-\theta)$, we have for a perfect gas the following relations:

$$\frac{\rho_2}{\rho_1} = \frac{(K+1) M_{n_1}^2}{(K-1) M_{n_1}^2 + 2} \quad (7.47)$$

$$\frac{p_2}{p_1} = 1 + \frac{2K}{K+1}(M_{n_1}^2 - 1) \quad (7.48)$$

$$M_{n_2}^2 = \frac{M_{n_1}^2 + [2/(K-1)]}{[2K/(K-1)]M_{n_1}^2 - 1} \qquad (7.49)$$

$$\frac{T_2}{T_1} = \frac{p_2}{p_1} \frac{\rho_1}{\rho_2} \qquad (7.50)$$

where

M_{n_1} = Normal component of the upstream Mach number

M_{n_2} = Normal component of the downstream Mach number

K = Specific Heat Ratio

Combining the above equation we obtain the θ-b-M relation which specifies θ as a unique function of M_1 and b, or

$$\tan\theta = 2\cot b \left[\frac{M_1^2 \sin^2 b - 1}{M_1^2(K + \cos 2b) + 2}\right] \qquad (7.51)$$

C-66

CHAPTER 8

DIMENSIONAL ANALYSIS AND PHYSICAL SIMILARITY

8.1 DIMENSIONAL ANALYSIS, PHYSICAL SIMILARITY

Table 8.1

Quantity	Dimensions (M,L,T)	(F,L,T)
Acceleration	LT^{-2}	LT^{-2}
Area	L^2	L^2
Bulk modulus of elasticity	$ML^{-1}T^{-2}$	FL^{-2}
Density	ML^{-3}	FT^2L^{-4}
Discharge	L^3T^{-1}	L^3T^{-1}
Dynamic viscosity	$ML^{-1}T^{-1}$	$FT\,L^{-2}$
Force	MLT^{-2}	F
Gravity	LT^{-2}	LT^{-2}
Kinematic viscosity	L^2T^{-1}	L^2T^{-1}
Length	L	L
Mass	M	FT^2L^{-1}
Pressure	$ML^{-1}T^{-2}$	FL^{-2}
Specific weight	$ML^{-2}T^{-2}$	FL^{-3}
Tension	MT^{-2}	FL^{-1}
Time	T	T
Velocity	LT^{-1}	LT^{-1}

The study of fluid mechanics is based mainly on experimental results. One technique used to minimize the number of experiments required is dimensional analysis. Dimensional analysis does not provide a complete solution to a problem, but it reveals the mathematical relations among the variables involved. Table 8.1 shows a summary of the quantities, symbols, and dimensions used in fluid mechanics.

Many experiments in fluid mechanics are conducted on scale models rather than on prototypes. In all of these experiments, results taken from tests using the model are applied to the prototype. Physical similarity is a general proposition between the model and the prototype.

8.2 TYPES OF PHYSICAL SIMILARITY

For any comparison between prototype and model to be valid, they must satisfy the following conditions of physical similarity:

A) Geometric Similarity

This physical similarity requires, first, that the model and prototype have the same shape, and second, that their dimensions be related by a constant scale factor.

B) Kinematic Similarity

This physical similarity requires that the velocities be related in magnitude and direction by a constant scale factor.

C) Dynamic Similarity

This physical similarity requires that identical types of forces be related in magnitude and direction by a constant scale factor.

Some of the most important force ratios (dimensionless groups), in dynamic similarity are listed in table 8.2

Table 8.2   Force Ratios

Name of Ratio	Definition	Physical Meaning
Reynolds number, Re	$\dfrac{V \ell \rho}{\mu}$	$\dfrac{\text{Inertia force}}{\text{Viscous force}}$
Froude number, Fr	$\dfrac{V}{(\ell g)^{\frac{1}{2}}}$	$\dfrac{\text{Inertia force}}{\text{Gravity force}}$
Weber number, We	$V \left(\dfrac{\ell \rho}{\gamma} \right)^{\frac{1}{2}}$	$\dfrac{\text{Inertia force}}{\text{Surface tension force}}$
Mach number, M	$\dfrac{V}{a}$	$\dfrac{\text{Inertia force}}{\text{Elastic force}}$
Euler number, Eu	$\dfrac{\Delta p}{\frac{1}{2}\rho V^2}$	$\dfrac{\text{Pressure force}}{\text{Inertia force}}$

8.3 THE BUCKINGHAM'S Π THEOREM

This theorem states that the number of independent dimensionless groups used to describe a problem in which there are n variables and m dimensions is equal to n-m. Mathematically it can be represented as

$$f(\pi_1, \pi_2, \ldots, \pi_{n-m}) = 0$$

where $\pi \equiv$ The dimensionless group.

8.4 DIMENSIONAL ANALYSIS OF A PROBLEM

The dimensional analysis of a problem is performed in the following steps:

A) A list of parameters is selected.

B) The dimensionless π parameters are obtained by using the π theorem.

C-69

Six steps are used to find the dimensionless parameters:

1) List all the variables involved.

2) Select a set of primary dimensions.

3) List the dimensions of all variables in terms of primary dimensions.

4) Select from the list of variables, a number of repeating variables equal to the number of primary dimensions, m, including all the primary dimensions.

5) Write the π parameters in terms of the unknown exponents and write the equations so that the sum of exponents be zero. Solve the equation to obtain n-m dimensionless group.

6) Check if each group obtained is dimensionless.

C) The relation among π parameters is determined experimentally.

CHAPTER 9

INCOMPRESSIBLE VISCOUS FLOW

9.1 INTERNAL AND EXTERNAL FLOWS

Viscous flows may be divided into two categories:

A) Internal flows, which are enclosed by boundaries of interest such as pipes, ducts, and nozzles.

B) External flows, which flows around bodies such as airfoils, rockets, and surface vessels.

9.2 LAMINAR AND TURBULENT FLOWS

Laminar flow is one in which the fluid flows in a well-ordered pattern whereby fluid layers are assumed to slide over one another. If a transition is taking place from a previously well-ordered flow to an unstable flow, the flow is called a turbulent flow.

The difference between these two flows is shown in Figure 9.1. Figure 9.1 shows the traces of velocity at a fixed position for steady laminar, steady turbulent, and unsteady turbulent flows.

In turbulent flow the velocity, V', indicates fluctuations with respect to the mean velocity $\overline{V}$. The instantaneous velocity, V, may be written as

$$V = \overline{V} + V' \qquad (9.1)$$

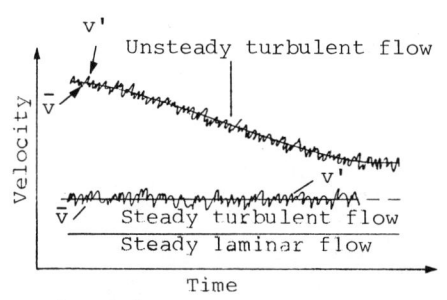

Fig. 9.1

9.3 FULLY DEVELOPED LAMINAR FLOW BETWEEN INFINITE PARALLEL PLATES

A) Both Plates Stationary

Boundary conditions

at $y = 0 \quad u = 0$
$\quad y = a \quad u = 0$
$\quad u = u(y) \quad v = w = 0$

where u, v, w are velocity components in the x, y, and z directions.

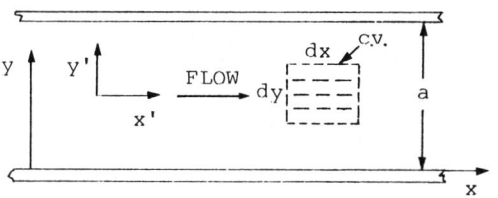

Fig. 9.2 Flow between stationary plates

C-72

The velocity profile is:

$$u = \frac{a^2}{2\mu}\left(\frac{\partial p}{\partial x}\right)\left[\left(\frac{y}{a}\right)^2 - \left(\frac{y}{a}\right)\right] \qquad (9.2)$$

or,

$$u = \frac{a^2}{2\mu}\left(\frac{\partial p}{\partial x}\right)\left[\left(\frac{y'}{a}\right)^2 - \frac{1}{4}\right] \qquad (9.3)$$

where

$$y' = \left(y - \frac{a}{2}\right)$$

The shear stress distribution is:

$$\tau_{yx} = a\left(\frac{\partial p}{\partial x}\right)\left[\frac{y}{a} - \frac{1}{2}\right] \qquad (9.4)$$

The volumetric flowrate per unit depth is:

$$\frac{Q}{\ell} = -\frac{1}{12\mu}\left(\frac{\partial p}{\partial x}\right)a^3 \qquad (9.5)$$

Flowrate as a function of pressure drop:

$$Q = \frac{a^3 \Delta p \ell}{12\mu L} \qquad (9.6)$$

The average velocity is:

$$\overline{V} = -\frac{1}{12\mu}\left(\frac{\partial p}{\partial x}\right)a^2 \qquad (9.7)$$

The point of maximum velocity is:

$$\text{at } y = \frac{a}{2} \quad u = u_{max} = \frac{3}{2}\overline{V} \qquad (9.8)$$

B) Upper Plate Moving with Constant Velocity, U

Boundary conditions:

at   y = 0   u = 0
at   y = a   u = U

C-73

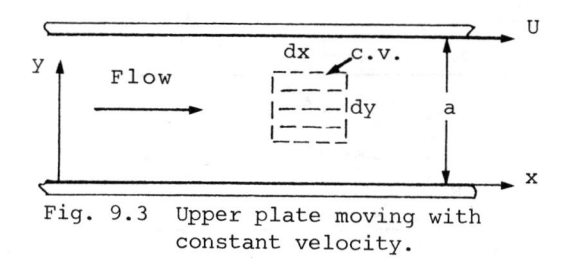

Fig. 9.3 Upper plate moving with constant velocity.

The velocity profile is:

$$u = \frac{Uy}{a} + \frac{a^2}{2\mu}\left(\frac{\partial p}{\partial x}\right)\left[\left(\frac{y}{a}\right)^2 - \left(\frac{y}{a}\right)\right] \quad (9.9)$$

The shear stress distribution is:

$$\tau_{yx} = \mu \frac{U}{a} + a\left(\frac{\partial p}{\partial x}\right)\left[\frac{y}{a} - \frac{1}{2}\right] \quad (9.10)$$

The volumetric flowrate per unit depth is:

$$\frac{Q}{\ell} = \frac{Ua}{2} - \frac{1}{12\mu}\left(\frac{\partial p}{\partial x}\right)a^3 \quad (9.11)$$

The average velocity is:

$$\overline{V} = \frac{U}{2} - \frac{1}{12\mu}\left(\frac{\partial p}{\partial x}\right)a^2 \quad (9.12)$$

The point of maximum velocity is at

$$\frac{du}{dy} = 0 \quad \text{at} \quad y = \frac{a}{2} - \frac{U/a}{(1/\mu)(\partial p/\partial x)} \quad (9.13)$$

9.4 FULLY-DEVELOPED LAMINAR FLOW IN A PIPE

The entrance length, L, which is the distance from the entrance to the position in the pipe, for fully-developed laminar flow has been determined to be

$$L = 0.058 \text{ Re } D \quad (9.14)$$

Fig. 9.4 Control volume for analysis of laminar flow in a pipe.

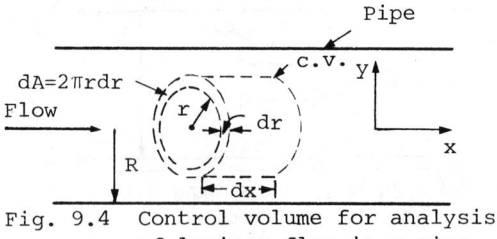

where D is the pipe diameter.

Boundary condition:

$$u = 0 \quad \text{at} \quad r = R$$

the velocity profile is

$$u = \frac{1}{4\mu}\left(\frac{\partial p}{\partial x}\right)(r^2 - R^2) = -\frac{R^2}{4\mu}\left(\frac{\partial p}{\partial x}\right)\left[1 - \left(\frac{r}{R}\right)^2\right] \qquad (9.15)$$

The shear distribution will be

$$\tau_{rx} = \mu \frac{du}{dr} = \frac{r}{2}\left(\frac{\partial p}{\partial x}\right) \qquad (9.16)$$

The volumetric flowrate is:

$$\boxed{Q = -\frac{\pi R^4}{8\mu}\left(\frac{\partial p}{\partial x}\right) = \frac{\pi \Delta p D^4}{128 \mu L}} \qquad (9.17)$$

The average velocity is:

$$V = -\frac{R^2}{8\mu}\left(\frac{\partial p}{\partial x}\right) \qquad (9.18)$$

The point of maximum velocity is at $r = 0$ when $\frac{du}{dr} = 0$.

Thus,

$$\boxed{\text{at } r = 0, \quad u = u_{max} = -\frac{R^2}{4\mu}\left(\frac{\partial p}{\partial x}\right) = 2\overline{V}} \qquad (9.19)$$

9.5 FLOW IN PIPES AND DUCTS

A) Velocity Profiles in Pipe Flow

For a fully-developed laminar pipe flow, the velocity profile is given by

$$\boxed{u = -\frac{R^2}{4\mu}\left(\frac{\partial p}{\partial x}\right)\left[1 - \left(\frac{r}{R}\right)^2\right]} \quad (9.20)$$

at $r = 0$, $u = u_{max} = U$, and

$$\frac{u}{U} = 1 - \left(\frac{r}{R}\right)^2 \quad (9.21)$$

For a turbulent flow the velocity profile can be represented by the empirical equation:

$$\frac{u}{U} = \left(1 - \frac{r}{R}\right)^{1/n} \quad (9.22)$$

where n varies with the Reynolds number.

This empirical equation is called the power-law equation, where n varies with the Reynolds number. Some values for n are:

$n = 6$ (Re = 4×10^3)
$n = 7$ (Re = 1.1×10^5)
$n = 10$ (Re = 3.2×10^6)

The ratio of the average velocity to the centerline velocity is given by

$$\boxed{\frac{\bar{V}}{U} = \frac{2n^2}{(n+1)(2n+1)}} \quad (9.23)$$

From this equation we can see that as n increases the ratio increases. Figure 9.5 shows the velocity profiles for laminar and turbulent flow at Reynolds number equal to 4×10^3.

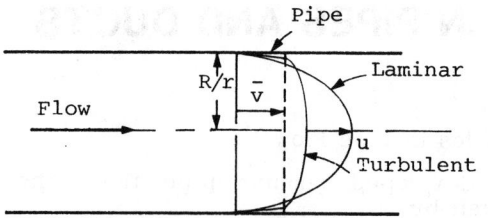

Fig. 9.5 Velocity profiles for laminar and turbulent flow in a smooth pipe.

B) Shear Stress Distribution in Fully-Developed Pipe Flow

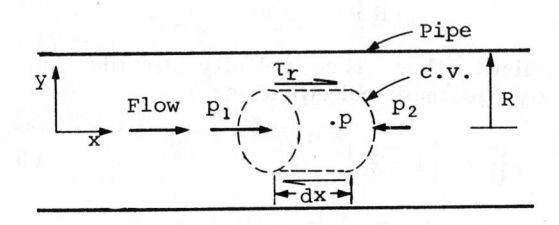

Fig. 9.6 Control volume for analysis of shear stress.

Assumptions:

a) Horizontal pipe, $F_{Bx} = 0$

b) Steady flow

c) Incompressible flow

d) Fully-developed flow

From the basic equation we have

$$F_{Sx} = 0 \tag{9.24}$$

From Figure 9.6 we get

$$F_{Sx} = p_1 - p_2 + \tau_r \tag{9.25}$$

$$F_{Sx} = \left(p - \frac{\partial p}{\partial x} \frac{\partial x}{2}\right) \pi r^2 - \left(p + \frac{\partial p}{\partial x} \frac{dx}{2}\right) \pi r^2$$
$$+ \tau_{rx} 2\pi r dx = 0 \quad (9.26)$$

Therefore,

$$\boxed{\tau_{rx} = \frac{r}{2} \frac{\partial p}{\partial x}} \quad (9.27)$$

at the surface of the pipe the shear stress is

$$\tau_w = -\frac{R}{2} \frac{\partial p}{\partial x} \quad (9.28)$$

C) Energy Considerations in Pipe Flow

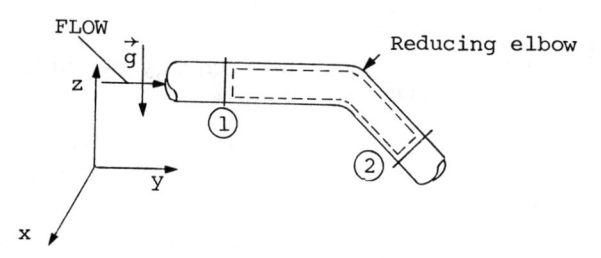

Fig. 9.7 Control volume and coordinates for energy analysis.

Applying the first law of thermodynamics for the control volume between sections 1 and 2 we have

$$\boxed{\begin{array}{l} \dot{Q} + \dot{W}_s + \dot{W}_{shear} + \dot{W}_{other} \\ = \frac{\partial}{\partial t} \int_{cv} e\rho\, d\mathcal{V} + \int_{cs} (e+pv)\, \rho \vec{V} \cdot d\vec{A} \end{array}} \quad (9.29)$$

where

$$e = u + \frac{V^2}{2} + gz \quad (9.30)$$

Under the following assumptions,

a) $\dot{W}_s = \dot{W}_{shear} = \dot{W}_{other} = 0$

b) Steady and incompressible flow

c) Pressure and internal energy uniform through the control volume,

the energy equation reduces to:

$$\left[\frac{p_1}{\rho} + \alpha_1 \frac{\bar{V}_1^2}{2} + gz_1\right] - \left[\frac{p_2}{\rho} + \alpha_2 \frac{\bar{V}_2^2}{2} + gz_2\right]$$

$$= (u_2 - u_1) - \frac{\delta Q}{\delta m} \qquad (9.31)$$

where

α = Kinetic energy flux coefficient

$\bar{V}$ = Average velocity

$\frac{p_1}{\rho} + \alpha_1 \frac{\bar{V}_1^2}{2} + gz_1$ = The mechanical energy per unit mass at section 1

$\frac{p_2}{\rho} + \alpha_2 \frac{\bar{V}_2^2}{2} + gz_2$ = The mechanical energy per unit mass at section 2

$(u_2 - u_1) - \frac{\delta Q}{\delta m}$ = The difference in mechanical energy per unit mass between sections 1 and 2

The term $(u_2 - u_1) - \frac{\delta Q}{\delta m}$ is identified as the total head loss, $h_{\ell T}$.

Then,

$$\left[\frac{p_1}{\rho} + \alpha_1 \frac{\bar{V}_1^2}{2} + gz_1\right] - \left[\frac{p_2}{\rho} + \alpha_2 \frac{\bar{V}_2^2}{2} + gz_2\right] = h_{\ell T} \qquad (9.32)$$

where

$$h_{\ell T} = h_\ell + h_{\ell m} \qquad (9.33)$$

h_ℓ = Head loss through the pipe

$h_{\ell m}$ = Head loss in fittings, valves, inlets, outlets, etc.

9.6 CALCULATION OF HEAD LOSS

Head loss is defined as the sum of major losses, h_ℓ, due to frictional effects, and minor losses, $h_{\ell m}$, due to entrances, fittings, valves, etc.

A) Major Losses; Friction Factor

   a) Laminar Flow

      For a fully developed flow in a horizontal pipe the head loss is given by

$$\boxed{h_\ell = \left(\frac{64}{Re}\right) \frac{L}{D} \frac{\bar{V}^2}{2}} \qquad (9.34)$$

where

    L = The length of the Pipe

    D = The diameter of the pipe

    $\bar{V}$ = The average velocity of the fluid

    Re = Reynolds number

   b) Turbulent Flow

      In a turbulent flow, we must use experimental results in order to calculate the head loss which is given by the formula

$$h_\ell = f \frac{L}{D} \frac{\bar{V}^2}{2} \qquad (9.35)$$

where

    f = Friction function, determined experimentally

    L = Length of the pipe

    D = Diameter of the pipe

    $\bar{V}$ = Average velocity of the fluid

The following steps should be taken to determine the head loss with known conditions:

1) Evaluate the Reynolds number.
2) Obtain the relative roughness, e/D, from Figure 9.8.
3) Obtain the friction coefficient, f, using the appropriate curve from Figure 9.8.

4) Find the head loss, introducing the friction coefficient value into equation (9.35).

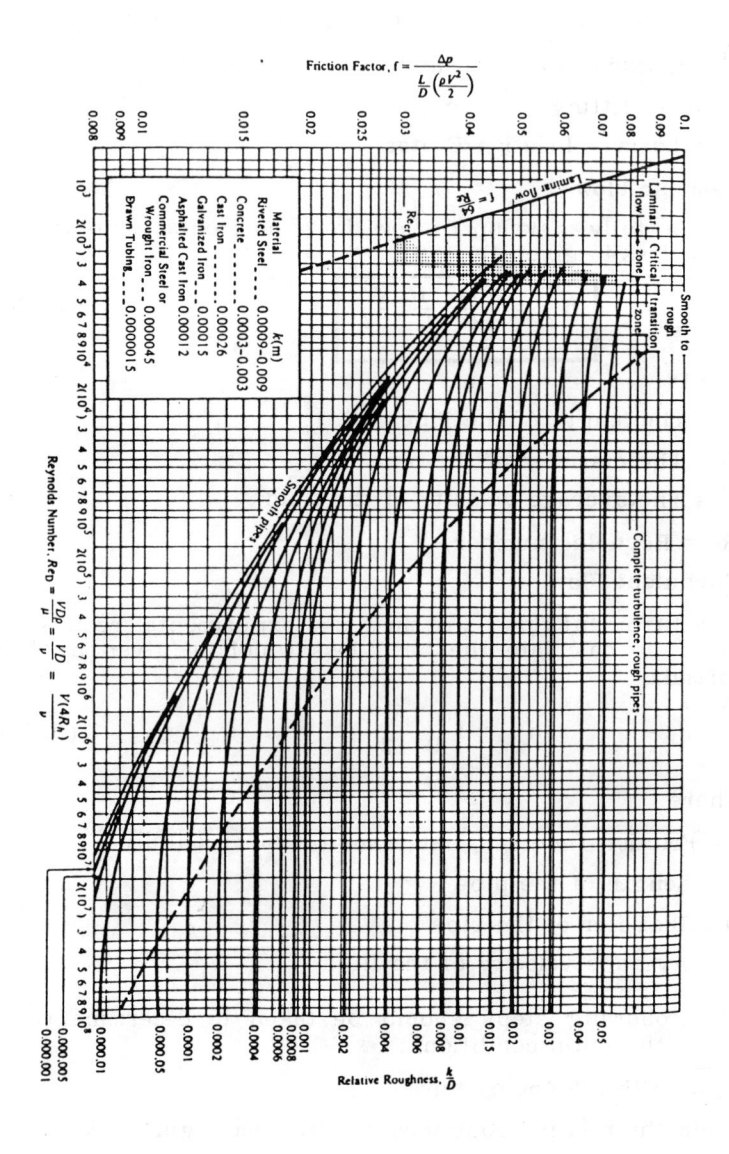

Fig. 9.8   Moody Diagram

C-81

MINOR LOSSES

The minor head loss may be expressed as

$$h_{\ell m} = K \frac{\bar{V}^2}{2} \tag{9.36}$$

where the loss coefficient, K, must be determined experimentally for each situation. (K for commercial fittings can be found in handbooks.)

Also, the minor head loss may be written as a function of the friction factor, f, as:

$$\boxed{(h_\ell)_m = f \frac{L_e}{D} \frac{\bar{V}^2}{2}} \tag{9.37}$$

where L_e is the equivalent length of a straight pipe.

A) Inlets and Exits

Inlets and exits in a pipe can cause pressure variations. The minor head loss can be calculated using equation (9.36). Some values of the minor loss coefficient, K, for three basic inlet and exit geometries are shown in table 9.1.

Table 9.1 Loss Coefficients for Pipe Inlets and Exits

Type	Coefficient K	
	Inlet	Exit
Square - edged	1	1
Rounded	0.5	1
Reentrant or Projecting Pipe	1	1

B) Expansions and Contractions

Minor loss coefficients for sudden expansion in circular ducts are given in Figure 9.9. The coefficients for gradual and sudden contractions are given in table 9.2 and Figure 9.10, respectively.

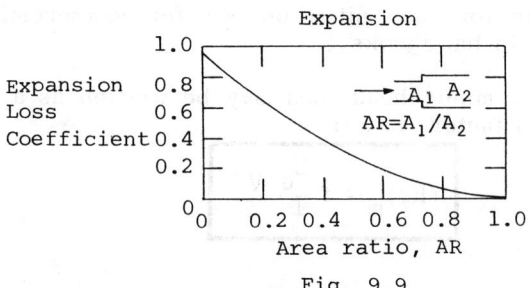

Expansion Loss Coefficient vs Area ratio, AR ($AR = A_1/A_2$)

Fig. 9.9

Table 9.2	Loss Coefficients for Gradual Contractions	
Contraction	Angle. θ. Degrees	Loss Coefficient, K
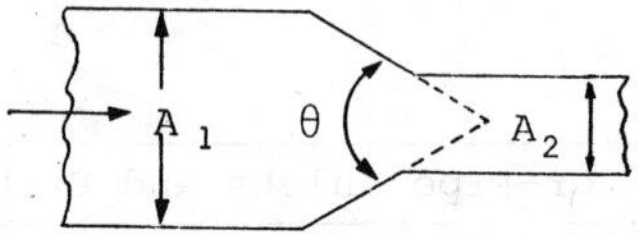	30	0.02
	45	0.04
	60	0.07

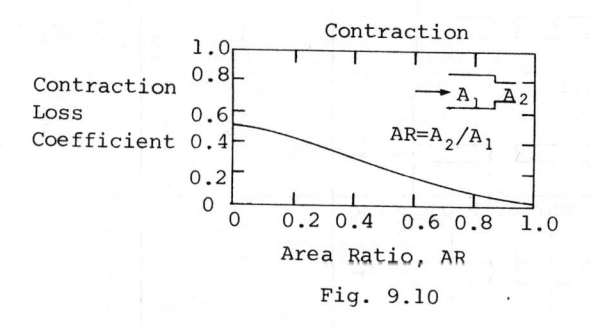

Contraction Loss Coefficient vs Area Ratio, AR ($AR = A_2/A_1$)

Fig. 9.10

Losses in diffusers depend on geometric and flow variables. These losses are expressed by the formula,

$$h_{\ell\,m} = \frac{V_1^2}{2}\left[\left(1 - \frac{1}{(AR)^2}\right) - C_p\right] \quad (9.38)$$

where

C_p = Pressure recovery coefficient

$$= \frac{p_2 - p_1}{\frac{1}{2}\rho \overline{V}_1^2} \quad (9.39)$$

$$AR = \frac{A_2}{A_1} \quad (9.40)$$

Pressure recovery data for conical diffuser, with a fully developed turbulent pipe flow at the inlet, are represented in Figure 9.11.

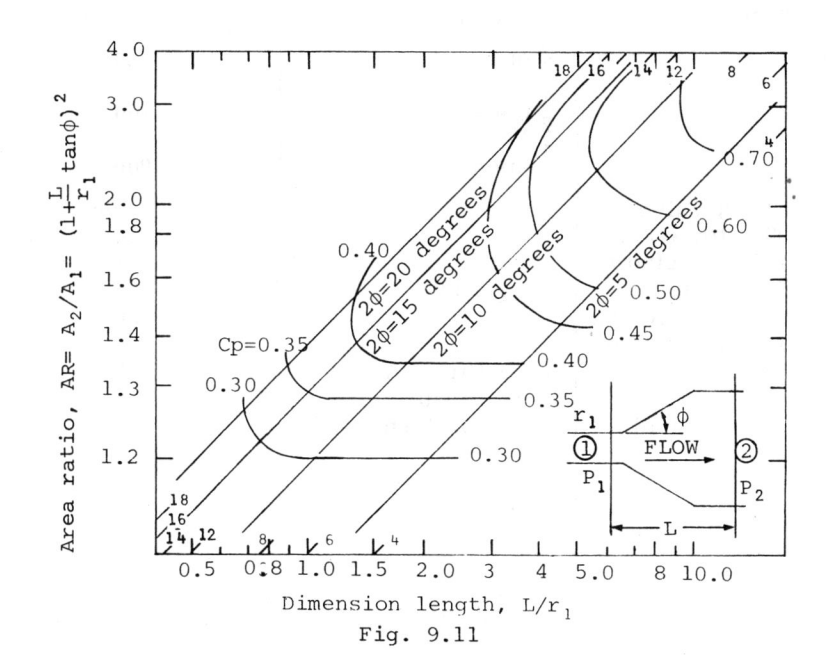

Fig. 9.11

C) Pipe Bends

The head loss in a bend pipe is larger when compared with the loss in a straight section of equal length. Calculations of these losses are based on experimental results. Equation (9.36) can be used for this purpose. The head loss can also be expressed by the equivalent length of a straight pipe. Data for some common situations are summarized in table 9.3.

D) Valves and Fittings

Losses through valves and fittings are also expressed in terms of the equivalent length of a straight pipe. Some values of interest are presented in table 9.3.

Table 9.3

Fitting Type	Description	Equivalent Length L_e/D*
Globe valve	Fully open	350
Gate valve	Fully open	13
	$\frac{3}{4}$ open	35
	$\frac{1}{2}$ open	160
	$\frac{1}{4}$ open	900
Check valve		50 - 100
90° std. elbow		30
45° std. elbow		16
90° elbow	Long radius	20
90° street elbow		50
45° street elbow		26
Tee	Flow through run	20
	Flow through branch	60
Return bend	Close pattern	50

Based on $h_{\ell m} = f \dfrac{L_e \bar{V}^2}{D^2}$

9.7 SOLUTION OF PIPE FLOW PROBLEMS

Pipe flow problems can be solved using the energy equation

$$\left[\frac{p_1}{\rho} + \alpha_1 \frac{\bar{V}_1^2}{2} + gz_1\right] - \left[\frac{p_2}{\rho} + \alpha_2 \frac{\bar{V}_2^2}{2} + gz_2\right] = h_{\ell T} \quad (9.41)$$

where
$$h_{\ell T} = h_\ell + h_{\ell m} \quad (9.42)$$

$$h_\ell = f \frac{L}{D} \frac{\bar{V}^2}{2} \quad (9.43)$$

$$h_{\ell m} = K \frac{\bar{V}^2}{2} \quad (9.44)$$

A) Single Path Systems

Equation (9.32) relates four variables Δp, L, Q, D. Any of these variables may be unknown. Thus, four general cases are possible.

a) Δp Unknown, L, Q, and D Known

1) Compute Re and e/D from the data and obtain a friction factor, f, from the Moody diagram.

2) Use equations (9.35) and (9.36) to calculate the total head loss.

3) Evaluate the pressure drop using equation (9.32).

b) L Unknown, Δp, Q and D Known

1) Use equation (9.32) to evaluate the total head loss, $h_{\ell T}$.

2) Compute Re and e/D from the data and obtain a friction factor, f, from the Moody diagram.

3) Determine the length L, using equation (9.35).

c) Q Unknown, Δp, L, and D Known

1) Combine all the above equations to obtain an expression for $\bar{V}$ or Q as a function of the friction factor f.

C-86

If f cannot be calculated (because of a large Reynolds number), a good first guess for f must be assumed in the fully-rough region of the Moody diagram.

2) Using the assumed value for f, a first approximation of $\bar{V}$ is computed.

3) The Reynolds number is calculated for this value of $\bar{V}$, and a new friction factor is obtained. Repeat step b) to obtain $\bar{V}$. More than two iterations are rarely required.

d) D Unknown, Δp, L, and Q Known

1) Assume a value for pipe diameter.

2) Using the assumed value, calculate the Reynolds number.

3) Obtain a friction factor value from the Moody diagram.

4) The head loss is computed from equations (9.35) and (9.36).

5) Compute the pressure drop from equation (9.32).

6) The resulting pressure drop is compared to the system requirement.

B) Multiple-Path Systems

The techniques developed in solving single path systems can be used to analyze flow in multiple-path pipe systems. The procedure is analogous to that used in solving direct current electric circuits, where the flowrate and the pressure drop are analogous to the current and voltage, respectively. However, the linear relation between voltage and current (Ohm's law) does not apply to the fluid flow system because the pressure drop varies proportionally to the square of the flow rate, making an iterative solution necessary.

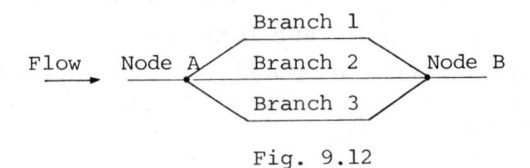

```
                    Branch 1
Flow   Node A ─────  Branch 2  ───── Node B
                    Branch 3
```

Fig. 9.12

C-87

C) Noncircular Ducts

The equations developed for pipe flow may be applied for ducts of square or rectangular geometry if the aspect ratio (ar = $\frac{h}{b}$) is less than 3 or 4. The equations for turbulent pipe flow may be applied but the hydraulic diameter D_h must be introduced in place of the diameter. That is,

$$D_h = \frac{4A}{P_w}$$

where P_w is the wetted perimeter. For a rectangular duct we have

$$D_h = \frac{4bh}{2(b+h)}$$

where b is the width and h is the height of the cross section.

For ducts of more irregular shapes such as triangular or flat ducts, these correlations are not applicable and experimental results must be used.

CHAPTER 10

BOUNDARY LAYER THEORY

10.1 INTRODUCTION

The boundary layer is a thin region adjacent to a solid boundary which is particularly sensitive to the effect of viscosity. A laminar region begins at the leading edge and grows in thickness as the flow moves along the surface. Then, a transition zone is reached where the flow becomes turbulent and the boundary layer is thicker. Figure 10.1 shows the boundary layer variation along a flat plate.

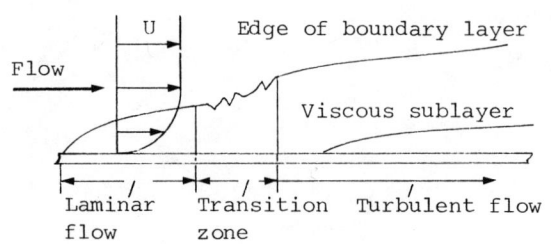

Fig. 10.1 Boundary layer details on a flat plate.

The thin layer near the solid boundary in the turbulent region is the viscous sublayer where the viscous effect is predominant.

10.2 BOUNDARY LAYER THICKNESS

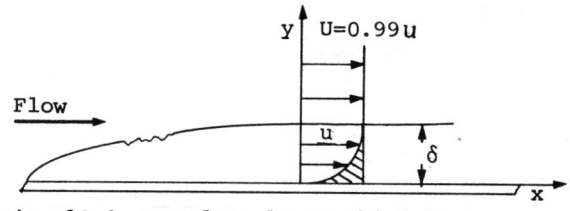

Fig. 10.2 Boundary layer thickness.

The boundary layer thickness (δ) is defined as the distance from the surface where the fluid velocity is ninety-nine percent (99%) of the freestream velocity (Fig. 10.2).

The boundary layer displacement thickness (δ^*) is defined as the distance by which the solid boundary would have to be displaced to keep the same mass flow in a hypothetical frictionless flow (see Fig. 10.3).

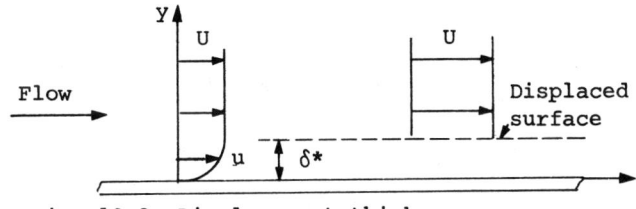

Fig. 10.3 Displacement thickness.

For incompressible flow we have:

$$\delta^* = \int_0^\infty \left(1 - \frac{u}{U}\right) dy \qquad (10.1)$$

10.3 INTEGRAL MOMENTUM EQUATION

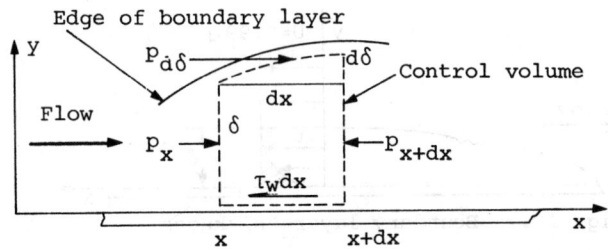

Fig. 10.4 Differential control volume in the boundary layer.

Assuming uniform pressure(p) at a section and neglecting hydrostatic pressure, from Figure 10.4 we obtain the force in the x-direction as:

$$dF_x = P_x - P_{x+dx} + P_{d\delta} - \tau_w dx$$

where as, considering the unit width into the plane of paper, the forces

$$P_x = p\delta$$

$$P_{x+dx} = \left(p + \frac{dp}{dx} dx \right)(\delta + d\delta)$$

and

$$P_{d\delta} = \left(p + \frac{1}{2} \frac{dp}{dx} dx \right) d\delta$$

$$dF_x = p\delta - \left(p + \frac{dp}{dx} dx \right)(\delta + d\delta) + \left(p + \frac{1}{2} \frac{dp}{dx} dx \right) d\delta - \tau_w dx$$

where τ_w is the shear stress at the wall.

Combining the linear momentum efflux in the x-direction and the continuity equation for the chosen control volume, we get the general form of the integral momentum equation

$$-\delta \frac{dp}{dx} - \tau_w = \frac{d}{dx} \int_0^\delta \rho u^2 dy - U \frac{\partial}{\partial x} \int_0^\delta \rho u\, dy \qquad (10.2)$$

In order to evaluate the boundary layer thickness for an incompressible laminar flow over a flat plate we must

A) Determine dp/dx from the irrotational flow analysis outside the boundary layer.
B) Assume a reasonable velocity profile inside the boundary layer.
C) Use $\mu(\partial u/\partial y)$ at the wall for τ_w.

10.4 FLOW OVER A FLAT PLATE

For this case:

U = Constant

p = Constant $\therefore \quad \frac{\partial p}{\partial x} = 0$

For an incompressible flow, we have from (10.2)

$$\tau_w = \rho U^2 \frac{\partial}{\partial x} \int_0^\delta \frac{u}{U}\left(1 - \frac{u}{U}\right) dy \qquad (10.3)$$

Changing the variable of integration from y to y/δ we get:

$$\eta = \frac{y}{\delta}$$

$$dy = \delta\, d\eta$$

Thus,

$$\boxed{\tau_w = \rho U^2 \frac{d\delta}{dx} \int_0^1 \frac{u}{U}\left(1 - \frac{u}{U}\right) d\eta} \qquad (10.4)$$

That is the momentum integral equation for zero pressure gradient. Equations (10.3) and (10.4) are restricted to:

A) Steady flow with no body forces
B) Incompressible and two-dimensional flow
C) Flat plate flow.

10.5 THE MOMENTUM INTEGRAL EQUATION FOR ZERO PRESSURE GRADIENT FLOW

A) Laminar Flow Over a Flat Plate

A reasonable velocity profile is assumed

$$u = A + By + Cy^2 \qquad (10.5)$$

The boundary conditions are:

$u = 0$ at $y = 0$

$u = U$ at $y = \delta$

$\frac{du}{dy} = 0$ at $y = \delta$

Solving equation (10.5), and combining it with Newton's law, at the wall, we find the wall shear stress as:

$$\tau_w = \frac{2\mu U}{\delta} \qquad (10.6)$$

The skin-friction coefficient is defined as:

$$\boxed{C_f = \frac{\tau_w}{\frac{1}{2}\rho U^2}} \qquad (10.7)$$

or,

$$C_f = \frac{0.730}{\sqrt{Re_x}} \qquad (10.8)$$

The laminar boundary layer thickness is given by

$$\boxed{\frac{\delta}{x} = \sqrt{\frac{30\mu}{\rho U x}} = \frac{5.48}{\sqrt{Re_x}}} \qquad (10.9)$$

The displacement thickness is given by

$$\frac{\delta^*}{x} = \frac{1.835}{\sqrt{Re_x}} \qquad (10.10)$$

B) Turbulent Flow Over a Flat Plate

A reasonable velocity profile for the flow is assumed:

$$\frac{u}{U} = \left(\frac{y}{\delta}\right)^{1/7} \qquad (10.11)$$

The shear stress experimentally determined is given by

$$\tau_w = 0.0225 \rho U^2 \left(\frac{\nu}{U\delta}\right)^{1/4} \qquad (10.12)$$

The boundary layer thickness is given by

$$\boxed{\frac{\delta}{x} = \frac{0.37}{Re_x^{1/5}}} \qquad (10.13)$$

The displacement thickness is given by

$$\frac{\delta^*}{x} = \frac{0.0463}{Re_x^{1/5}} \qquad (10.14)$$

The skin-friction coefficient is given by

$$C_f = \frac{0.0577}{Re_x^{1/5}} \quad \text{for} \quad 5 \times 10^5 < Re_x < 10^7 \qquad (10.15)$$

10.6 TURBULENT BOUNDARY-LAYER SKIN-FRICTION COEFFICIENT FOR SMOOTH PLATES

For higher Reynolds number values, the skin-friction coefficient is expressed as follows:

$$C_f = 0.074(Re_\ell)^{-1/5} \quad \text{for} \quad 5 \times 10^5 < Re_\ell < 10^7 \quad (10.16)$$

where $Re_\ell = U\ell/\nu$, the Reynolds number being based on the length of the plate. Also,

$$C_f = \frac{0.455}{(\log Re_\ell)^{2.58}} \quad \text{for} \quad Re_\ell > 10^7 \quad (10.17)$$

This is an empirical equation (H. Schlichting).

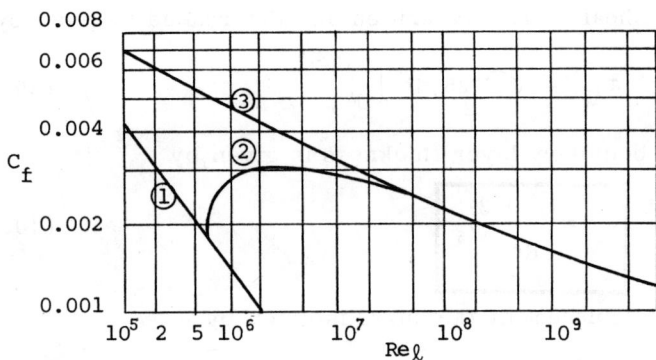

Fig. 10.5 Skin-friction coefficient for smooth plates. Curve ①, laminar. Curve ②, transition to turbulent. Curve ③, turbulent.

10.7 PRESSURE GRADIENT IN BOUNDARY-LAYER FLOW SEPARATION

Separation is an effect of the pressure gradient. It can be described as the breaking away of the main flow from the solid boundary and results from an adverse pressure gradient, $dp/dx > 0$ (see Fig. 10.5).

Fig. 10.6 Separation $\left(\dfrac{du}{dy}\right)_{y=0} = 0$

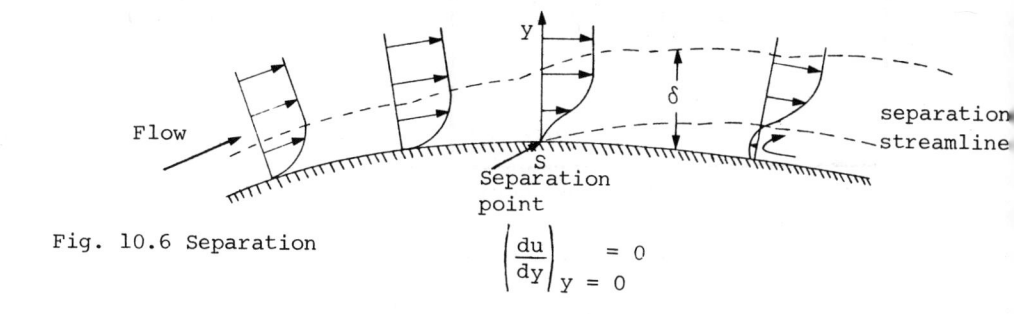

The point of separation is on the solid boundary where $du/dy\big|_{y=0} = 0$. The flow direction in the separated region (see Fig. 10.5) is opposite to the main flow direction.

Note: The adverse pressure gradient ($dp/dx > 0$) is a necessary (but still not a sufficient) condition for separation.

10.8 FLOW AROUND IMMERSED BODIES

The net force exerted on an immersed body by a moving fluid can be broken down into two components:

A) A parallel force which is called drag force (F_D) and

B) A perpendicular force called lift force (F_L). The drag force is written as:

$$F_D = C_D \dfrac{A \rho V^2}{2} \quad (10.18)$$

or

$$C_D = \dfrac{F_D}{\frac{1}{2} \rho V^2 A} \quad (10.19)$$

where C_D is the drag coefficient, A is the projected area in the direction of the flow, and V is the flow velocity. The lift force is expressed in the following form:

$$F_L = C_L \frac{A\rho V^2}{2} \qquad (10.20)$$

or,

$$C_L = \frac{F_L}{\frac{1}{2}\rho V^2 A} \qquad (10.21)$$

where C_L is the lift coefficient.

A) Flow Over a Flat Plate Parallel to the Flow

$$F_D = \int_{\text{plate surface}} \tau_w \, dA \qquad (10.22)$$

Thus,

$$C_D = \frac{\int_{PS} \tau_w \, dA}{\frac{1}{2}\rho V^2 A} \qquad (10.23)$$

C_D has to be determined from empirical data because the boundary layer varies from laminar to turbulent along the plate (see Fig. 10.7).

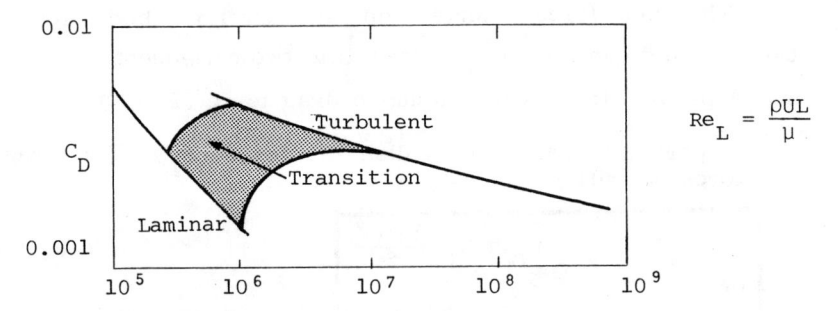

Fig. 10.7 Variation of the drag coefficient with the Reynolds number for a flow over a flat plate. $Re_L = \frac{\rho UL}{\mu}$

B) Flow Normal to the Plate

The drag coefficient for this case is determined using experimental results. For low Reynolds number values C_D depends on the aspect ratio and on the Reynolds number. For Reynolds number values greater than approximately 1,000, the drag coefficient can be considered dependent only on the aspect ratio. Figure 10.8 shows the variation of C_D with the aspect ratio.

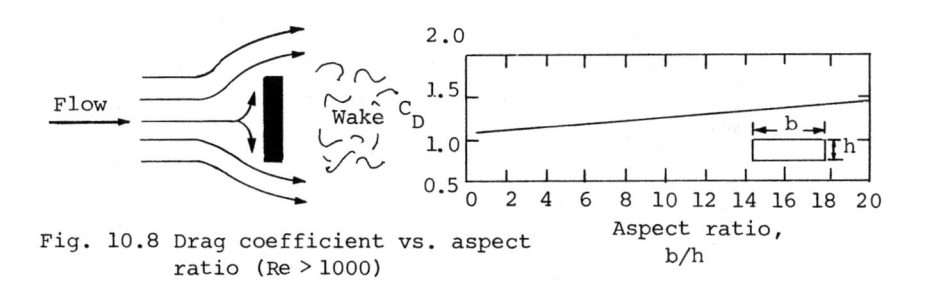

Fig. 10.8 Drag coefficient vs. aspect ratio (Re > 1000)

At any inclination of the plate with respect to the horizontal, lift and drag coefficients will be present in different amounts depending on α as is shown in Figure 10.9.

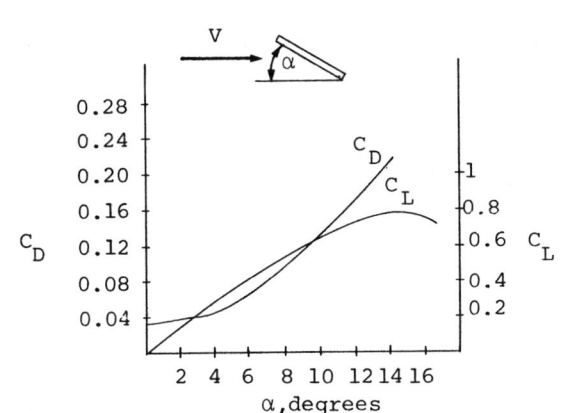

Fig. 10.9 Coefficients of lift and drag for a flat plate at varying inclination α.

C) Flow Over a Sphere

In this case, both lift and drag coefficients are present. Figure 10.10 shows the drag coefficient as a function of the Reynolds number.

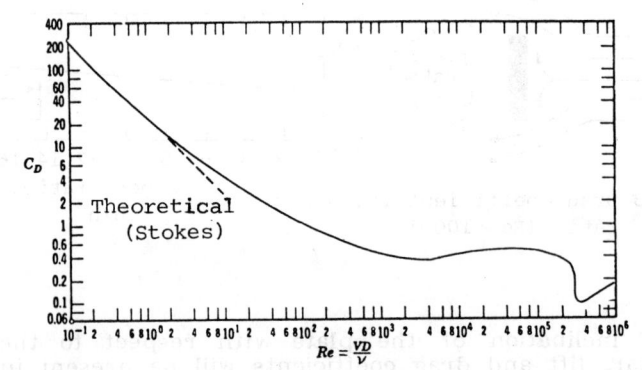

Fig. 10.10 Drag coefficient of a sphere as a function of Reynolds number.

For Re $\leq$ 1, the wake is laminar and the drag is mainly a friction drag. The drag force for a sphere of diameter d, moving at velocity V is given by

$$F_D = 3\pi \mu V d \qquad (10.24)$$

The drag coefficient is given by

$$C_D = \frac{24}{Re} \qquad (10.25)$$

Figure 10.11 shows the pressure distribution around a sphere for a laminar and a turbulent boundary layer flow. The separation process of the boundary layer occurs toward the top and bottom sections of the sphere. The pressure in the back drops and consequently the drag increases.

$$C_p = \frac{p-p^\infty}{\frac{1}{2}\rho v^2}$$

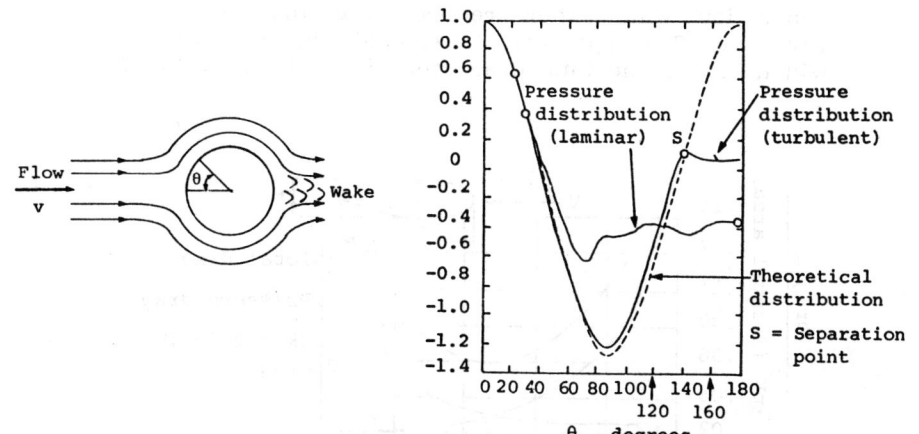

Fig. 10.11 Flow around a sphere and its pressure distribution.

D) Flow Around a Cylinder

The flow around a cylinder has the same characteristics as observed in the flow around the sphere, but the drag coefficient has larger values as is shown in Figure 10.12.

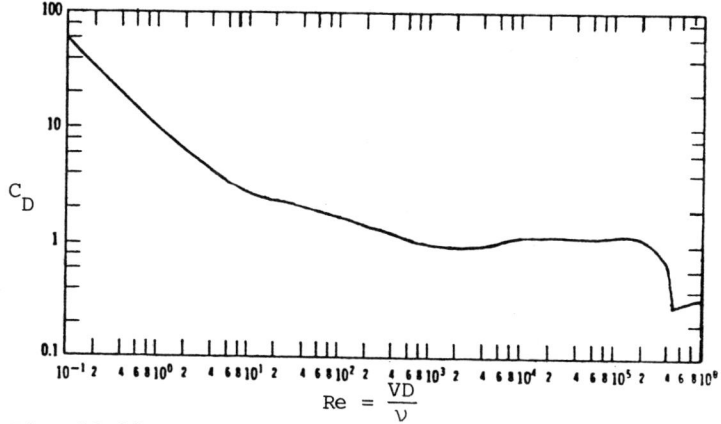

$$Re = \frac{VD}{\nu}$$

Fig. 10.12 Drag coefficient for cylinders as a function of Reynolds number.

C-100

E) Streamlining

Streamlining is a procedure used to reduce the pressure drag in bodies. The addition of material to the downwind part of the body delays the boundary layer separation which then reduces the adverse pressure gradient. The optimum streamlined shape of the body will decrease the total drag coefficient (see Fig. 10.13).

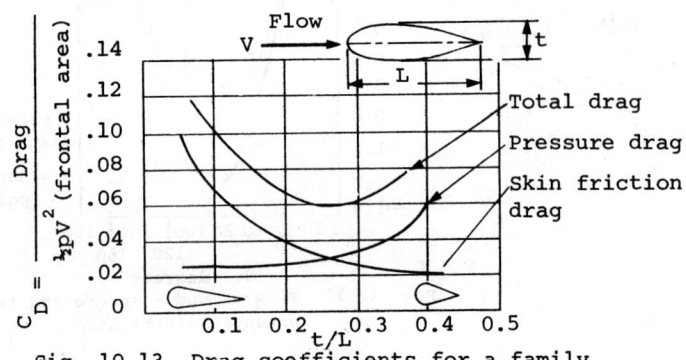

Fig. 10.13 Drag coefficients for a family of struts.

CHAPTER 11

IDEAL-FLUID FLOW

11.1 INTRODUCTION

For purposes of mathematical simplification and a better understanding of certin flows, the theory of ideal fluid flow is introduced. The ideal-fluid flow requires the following conditions: steady flow, incompressibility, and irrotationality ($\nabla \times \vec{V} = 0$).

11.2 MATHEMATICAL CONSIDERATIONS

A) Circulation (Γ):

Circulation is the integral performed around a fixed closed path of the tangent velocity component.

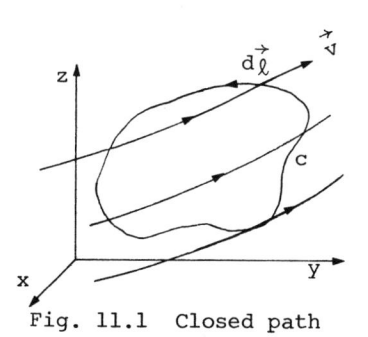

Fig. 11.1 Closed path

$$\Gamma = \oint_c \vec{V} \cdot d\vec{\ell} \qquad (11.1)$$

where c is the closed path and $\vec{V}$ the velocity.

B) Stokes' Theorem

This theorem relates an integral along a path to an equivalent surface integral. The surface S is arbitrary, but must be bounded by the closed path c (Fig. 11.2).

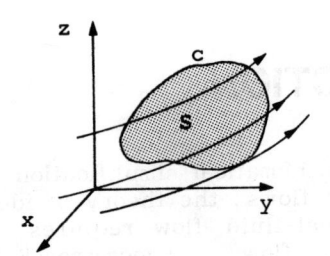

Fig. 11.2 Curved surface

$$\oint_c \vec{V} d\vec{\ell} = \int_S \hat{n} \cdot (\vec{V} \times \vec{A}) d\vec{S} \qquad (11.2)$$

The circulation in irrotational flows is zero for all paths.

C) The Velocity Potential

For an irrotational flow, the velocity field can be expressed in terms of the gradient of a scalar function ϕ. That is,

$$\vec{V} = -\vec{\nabla}\phi \qquad (11.3)$$

or,

$$\boxed{\begin{aligned} u &= -\frac{\partial \phi}{\partial x} \\ v &= -\frac{\partial \phi}{\partial y} \\ w &= -\frac{\partial \phi}{\partial z} \end{aligned}} \qquad \begin{aligned} &(11.4)\\ &(11.5)\\ &(11.6) \end{aligned}$$

D) The Stream Function

The stream function (ψ) is a mathematical tool that describes the characteristics of a flow.

11.3 THE STREAM FUNCTION AND IMPORTANT RELATIONS

A) The Stream Function and the Velocity Field

The stream function for a two-dimensional and incompressible flow is defined as follows:

$$u = -\frac{\partial \psi}{\partial y} \qquad (11.7)$$

$$v = \frac{\partial \psi}{\partial x} \qquad (11.8)$$

Fig. 11.3

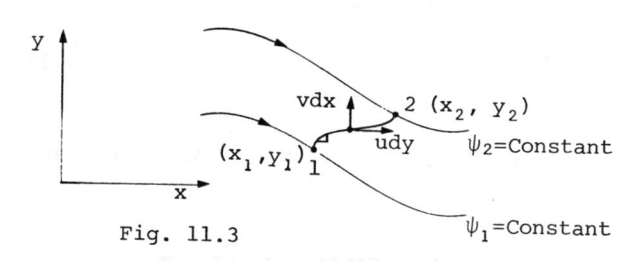

The stream function is a function of x and y, and the lines corresponding to ψ = constant are the streamlines. The volumetric flowrate between two streamlines is

$$\boxed{_1Q_2 = \psi_2 - \psi_1} \qquad (11.9)$$

B) The Stream Function and the Velocity Potential

Irrotational, two-dimensional, and incompressible flows satisfy the following relationship:

$$\boxed{\frac{\partial \psi}{\partial y} = \frac{\partial \phi}{\partial x}} \qquad (11.10)$$

$$\boxed{\frac{\partial \psi}{\partial x} = -\frac{\partial \phi}{\partial y}} \qquad (11.11)$$

C) The Streamlines and Potential Lines

The lines of constant stream function (streamlines) and the lines of constant vleocity potential (potential lines) are orthogonal to each other. The two sets of curves form a net known as a flow net (Fig. 11.4).

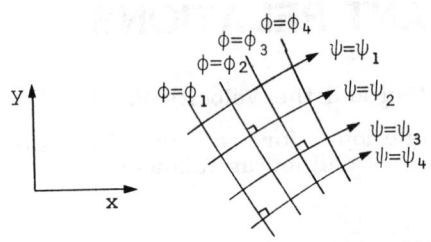

Fig. 11.4 Intersecting streamlines and potential lines at right angles (Flow net).

Mathematically, the flow net may be represented as

$$\left.\frac{dy}{dx}\right|_\phi = -\frac{1}{\left.\frac{dy}{dx}\right|_\psi}, \qquad (11.12)$$

where the subscripts ϕ and ψ stand for constant values.

11.4 BASIC EQUATIONS FOR IRROTATIONAL, INCOMPRESSIBLE, AND TWO-DIMENSIONAL FLOWS

A) Conservation of Mass

The continuity equation is given by:

$$\frac{\partial u}{\partial x} + \frac{\partial v}{\partial y} + \frac{\partial w}{\partial z} = 0 \qquad (11.13)$$

Introducing the velocity potential equations (11.6-11.16) into (11.13) we get

$$\frac{\partial^2 \phi}{\partial x^2} + \frac{\partial^2 \phi}{\partial y^2} = 0 \qquad (11.14)$$

or,

$$\boxed{\nabla^2 \phi = 0} \qquad (11.15)$$

for the stream function using equations (11.7) and (11.8) we get

$$\boxed{\nabla^2 \psi = 0} \qquad (11.16)$$

which means that under continuity considerations the velocity potential and the stream function must be harmonic functions. These equations are also known as Laplace equations.

B) Newton's Law

Newton's law is the integration of the Euler's equation, also known as Bernoulli's equation (see equation 5.36).

$$\frac{p}{\gamma} + \frac{V^2}{2g} + z = \text{constant} \qquad (5.36)$$

C) First Law of Thermodynamics

For a steady, incompressible and nonviscous flow the first law of thermodynamics is the same as Newton's law.

D) Second Law of Thermodynamics

The second law of thermodynamics is applied with no restrictions.

11.5 POLAR COORDINATES

$x = r\cos\theta$

$y = r\sin\theta$ \qquad (11.17)

$\theta = \tan^{-1} y/x$

$r = \sqrt{x^2 + y^2}$

The relationship between the velocity components and the velocity potential is given by

$$\boxed{V_r = -\frac{\partial \phi}{\partial r}} \qquad (11.18)$$

$$\boxed{V_\theta = -\frac{\partial \phi}{r \partial \theta}} \qquad (11.19)$$

C-106

where V_r and V_θ are the radial and transverse velocity components, respectively. For the stream function the relationship is:

$$V_r = -\frac{\partial \psi}{r \partial \theta} \qquad (11.20)$$

$$V_\theta = \frac{\partial \psi}{\partial r} \qquad (11.21)$$

The polar relations between the stream function and the velocity potential are:

$$\frac{\partial \phi}{\partial r} = \frac{\partial \psi}{r \partial \theta} \qquad (11.22)$$

$$\frac{\partial \phi}{r \partial \theta} = -\frac{\partial \psi}{\partial r} \qquad (11.23)$$

The condition of irrotationality is written as:

$$\frac{\partial (rV_\theta)}{\partial r} - \frac{\partial V_r}{\partial \theta} = 0 \qquad (11.24)$$

The Laplace equation is written as:

$$\nabla^2 \phi = \frac{1}{r} \frac{\partial \left(\frac{r \partial \phi}{\partial r}\right)}{\partial r} + \frac{1}{r^2} \frac{\partial^2 \phi}{\partial \theta^2} = 0 \qquad (11.25)$$

11.6 SIMPLE FLOWS

A) Uniform Flow

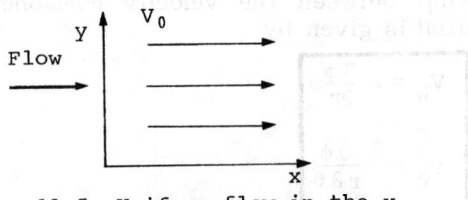

Fig. 11.5 Uniform flow in the x direction.

The uniform flow is represented by:

$$\phi = -V_0 \, x \qquad (11.26)$$

$$\psi = -V_0 \, y \qquad (11.27)$$

in polar coordinates

$$\psi = -V_0 \, r\sin\theta \qquad (11.28)$$

B) Sources and Sinks

The flow from a source is represented by:

$$\phi = -\frac{q}{2\pi} \ln r = \text{constant} \qquad (11.29)$$

$$\psi = -\frac{q\,\theta}{2\pi} = \text{constant} \qquad (11.30)$$

where q is the rate of flow and r is the modulus from the origin of a referece. Equation (11.30) represents a family of radial lines, for different constant values, emanating from the origin. Equation (11.29) represents the constant potential lines (see Fig. 11.6).

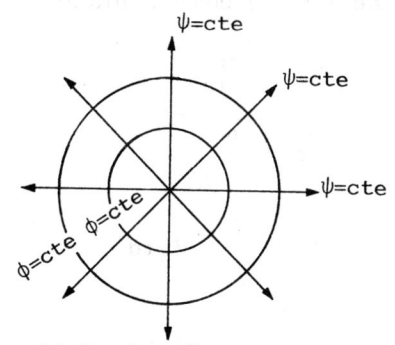

Fig. 11.6 Flow from a source.

The velocity components are

$$V_r = -\frac{\partial \phi}{\partial r} = \frac{q}{2\pi r} \qquad (11.31)$$

$$V_\theta = \frac{\partial \psi}{\partial r} = 0 \qquad (11.32)$$

A sink is represented by

$$\psi = \frac{q \ln r}{2\pi} = \text{constant} \qquad (11.33)$$

and the velocity potential is:

$$\phi = \frac{q}{2\pi} \ln r = \text{constant} \qquad (11.34)$$

In this case the direction of the flow is towards the origin. the origin of the sink is the only singular point (infinite velocity) of this flow. q is usually known as the strength of the sink or source.

C) The Simple Vortex

$$\boxed{\phi = -\frac{q\theta}{2\pi} = \text{constant} \qquad (11.35)}$$

$$\boxed{\psi = \frac{q}{2\pi} \ln r = \text{constant} \qquad (11.36)}$$

For various constant values of eq.(11.36) we will have a family of concentric circles as streamlines. Equation (11.35) represents the constant potential lines. The direction of the flow is given by the sign of equation (11.36). The velocity components are:

$$V_\theta = \frac{q}{2\pi r} \qquad (11.37)$$

$$V_r = 0 \qquad (11.38)$$

$$\Gamma = \int_0^{2\pi} V_\theta r d\theta = q \qquad (11.39)$$

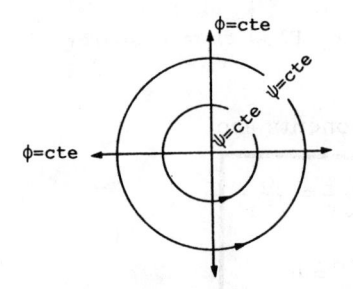

Fig. 11.7 Vortex flow

The only singular point is the origin.

D) The Doublet

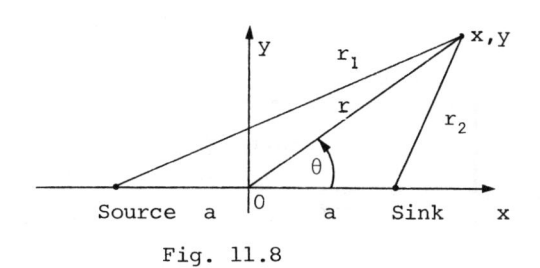

Fig. 11.8

The doublet is the result of a combination of a source and a sink. If we place a source and a sink of equal strength q at points which are a distance 'a' from the origin as is shown in Figure 11.9, the combined potential function can be written as

strengh q at equal distances a from the origin as is shown in Figure 11.9, the combined potential function can be written as

$$\phi = \frac{q}{2\pi}(\ln r_2 - \ln r_1) \qquad (11.40)$$

Using logarithmic properties and expanding logarithmic expressions into power series we get:

$$\phi = \frac{q}{4\pi}\left[\frac{-4r a \cos\theta}{r^2 + a^2} - \frac{2}{3}\left(\frac{2r a \cos\theta}{r^2 + a^2}\right)^3 + \ldots\right] \qquad (11.41)$$

(singular points must be avoided).

If the source and the sink are brought together so that

$$a \to 0 \quad \text{and} \quad q \to \infty \qquad (11.42)$$

such that

$$\frac{qa}{4\pi} \to K \qquad (11.43)$$

where K is a finite number, then in the limit the potential becomes

$$\phi = -\frac{K\cos\theta}{r} \qquad (11.44)$$

and the stream function,

$$\psi = \frac{K\sin\theta}{r} \qquad (11.45)$$

The streamlines are given by

$$\boxed{\frac{K\sin\theta}{r} = \text{constant}} \qquad (11.46)$$

or in cartesian coordinates,

$$x^2 + y^2 - \frac{K}{\text{cte}} y = 0 \qquad (11.47)$$

the lines of constant potential are given by

$$\boxed{\frac{K\cos\theta}{r} = \text{constant}} \qquad (11.48)$$

or in cartesian coordinates

$$x^2 + y^2 - \frac{K}{\text{cte}} x = 0 \qquad (11.49)$$

The family of circles for the doublet is shown in Figure 11.9. It can be seen that the streamlines are tangent to the x axis.

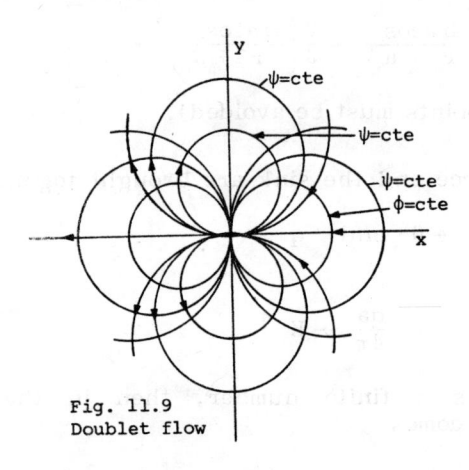

Fig. 11.9
Doublet flow

The velocity components are given by the equations

$$V_r = -\frac{K\cos\theta}{r^2} \quad (11.50)$$

$$V_\theta = -\frac{K\sin\theta}{r^2} \quad (11.51)$$

11.7 SUPERPOSITION OF SIMPLE FLOWS

Superposition of simple flows leads us to more general and practical solutions. Let us explore some of them.

A) Flow Around a Cylinder Without Circulation

This is the combination of a uniform flow and a doublet. The potential and stream functions for the combined flow become

$$\phi = -V_0 x - \frac{K\cos\theta}{r} \quad (11.52)$$

$$\psi = -V_0 y - \frac{K\sin\theta}{r} \quad (11.53)$$

For $\psi = 0$ we find that this streamline is a circle (Fig. 11.10). From equation (11.53) we get

$$\sin\theta \left(V_0 - \frac{K}{r} \right) = 0 \quad (11.54)$$

where

$$r = \sqrt{\frac{K}{V_0}} \quad (11.55)$$

satisfies equation (11.54). The velocity at the intersection of the circle (stagnation points 1, 2) can be found as follows:

$$V_r = -\frac{\partial \phi}{\partial r} = V_0 \cos\theta - \frac{K\cos\theta}{r^2} \quad (11.56)$$

$$V_\theta = -\frac{\partial \phi}{r \partial \theta} = -V_0 \sin\theta - \frac{K\sin\theta}{r^2} \quad (11.57)$$

for
$$\theta = 0, \quad V_\theta = 0$$
$$\theta = \pi, \quad V_r = 0$$

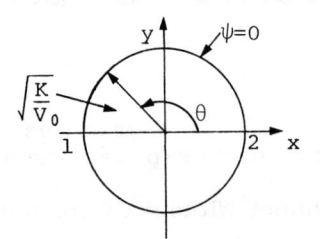

then,
$$\vec{V}_1 = \vec{V}_2 = 0 \qquad (11.58)$$

Fig. 11.10 Stream line $\psi=0$.

Thus, we can consider the circular region enclosed by the streamline $\psi = 0$ as a solid cylinder in a uniform flow (see Fig. 11.11).

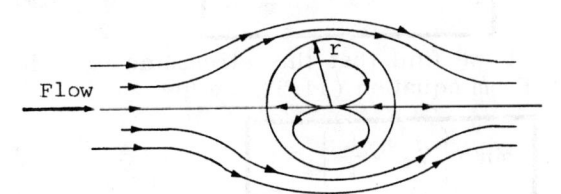

Fig. 11.11 Flow about a cylinder.

B) Lift and Drag for a Cylinder Without Circulation

The distribution of the pressure around a cylinder may be found using Bermoulli's equation:

$$\frac{V_0^2}{2g} + \frac{p_0}{\gamma} = \frac{V^{*2}}{2g} + \frac{p^*}{\gamma} \qquad (11.59)$$

where the asterisk (*) refers to the cylindrical boundary. The velocity at the surface V* is given by

$$\boxed{V^* = -\left.\frac{\partial \phi}{r\partial \theta}\right|_{r=(K/V_0)^{\frac{1}{2}}} = -2V_0 \sin\theta} \quad (11.60)$$

and from Bernoulli's equation,

$$P^* = \gamma\left[\frac{V_0^2}{2g} + \frac{P_0}{\gamma} - \frac{(2V_0\sin\theta)^2}{2g}\right] \quad (11.61)$$

The drag force will be given by

$$D = -\int_0^{2\pi} \gamma\left(\frac{K}{V_0}\right)^{\frac{1}{2}}\left[\frac{V_0^2}{2g} + \frac{P_0}{\gamma} - \frac{(2V_0\sin\theta)^2}{2g}\right]\cos\theta\, d\theta \quad (11.62)$$

Using the same procedure we find that the lift is equal to zero.

C) Circular Cylinder With Circulation

It is the combination of doublet, vortex, and uniform flow. The velocity potential and stream functions for the combined flow become

$$\boxed{\phi = -V_0 r\cos\theta - \frac{K\cos\theta}{r} + \frac{q}{2\pi}\theta} \quad (11.63)$$

$$\boxed{\psi = -V_0 r\sin\theta + \frac{K\sin\theta}{r} - \frac{q}{2\pi}\ln r} \quad (11.64)$$

The velocity components are

$$\boxed{V_r = V_0 \cos\theta - \frac{K\cos\theta}{r^2}} \quad (11.65)$$

$$\boxed{V_\theta = -V_0 \sin\theta - \frac{K\sin\theta}{r^2} - \frac{q}{2\pi r}} \quad (11.66)$$

The stagnation points occur at

$$r = \sqrt{\frac{K}{V_0}} \quad (11.67)$$

and

$$\theta = \sin^{-1}\left(\frac{-q}{4\pi\sqrt{KV_0}}\right) \quad (11.68)$$

There will be two stagnation points because there are two angles for a given sine (except for $\sin^{-1}(\pm 1)$). Thus,

$$\psi_{stag} = -\frac{q}{2\pi} \ln \sqrt{\frac{K}{V_0}} = cte \qquad (11.69)$$

If $\psi = \psi_{stag}$, we get

$$\sin\theta \left(-V_0 r + \frac{K}{r}\right) - \frac{q}{2\pi}\left(\ln r - \ln\sqrt{\frac{K}{V_0}}\right) = 0 \qquad (11.70)$$

All points along the circle $r = \sqrt{K/V_0}$ satisfy this equation. Thus we can consider the circular region as a cylinder. The flow around a cylinder with circulation is shown in Figure 11.13.

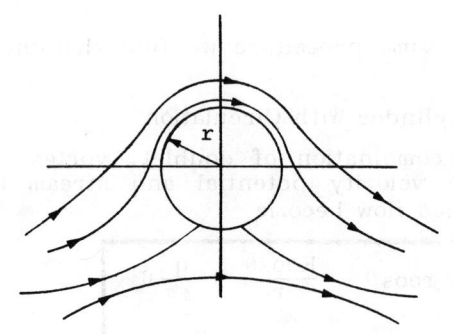

Fig. 11.12 Flow around a cylinder with circulation.

The circulation about the cylinder is equal to the strength of the vortex ($\Gamma = q$).

D) Lift and Drag for a Cylinder With Circulation

From Bernoulli's equation we have

$$P^* = \gamma\left(\frac{V_0^2}{2g} + \frac{P_0}{\gamma} - \frac{V^{*2}}{2g}\right) \qquad (11.71)$$

Then,

$$V^* = -\frac{\partial \phi}{r\,\partial\theta}\bigg|_{r=\sqrt{K/V_0}} = -2V_0 \sin\theta - \frac{q}{2\pi}\sqrt{\frac{V_0}{K}} \qquad (11.72)$$

The lift is given by (11.73)

$$\boxed{L = \rho V_0 q = \rho V_0 \Gamma}\;\text{(Magnus effect)}$$

The drag force is zero.

C-115

CHAPTER 12

FLOW IN OPEN CHANNELS

12.1 INTRODUCTION

An open channel is a conduit in which the liquid flows with a free surface subjected to certain prescribed conditions of pressure (usually atmospheric pressure). Streams, rivers, artificial channels, and irrigation ditches are some examples of open channel flows. However, a pipe not completely full of fluid will behave as an open channel as well.

A complete solution of open channel flow problems is more complicated because there are a large number of variables to be taken into account. Drastic variations of cross sections and boundary surfaces make the choice of an appropriate friction factor difficult. Besides that, the free surface allows unpredictable phenomena to occur which can change the behavior of the fluids.

12.2 STEADY UNIFORM FLOW

Steady uniform flow refers to the condition in which the depth flow, velocity and cross section remain constant over a given length of channel.

The shear stress at the walls, τ_w, is given by the equation

$$\tau_w = \rho g \, R_H \sin \alpha \qquad (12.1)$$

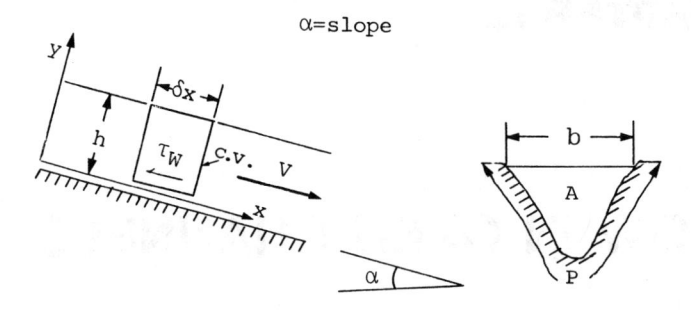

Fig. 12.1 Uniform flow in a prismatic channel.

where $R_H = \dfrac{A}{P}$ = Hydraulic radius

A = Area of the cross section of the liquid prism

P = Length of the wetted perimeter

g = Gravity constant

ρ = Density

A) Horizontal Velocity

The average horizontal velocity, V_H, of the fluid can be written as

$$\boxed{V_H = C\sqrt{R_H S}}\quad \text{(Chezy formula)} \qquad (12.2)$$

where
  S = Slope of water surface or the channel bottom ($\sin \alpha$)

  R_H = Hydraulic radius

  C = Chezy coefficient

The coefficient C can be obtained by using one of the following expressions:

$$C = \sqrt{\dfrac{8g}{f}} \qquad (12.3)$$

$$C = \dfrac{K}{n} R^{1/6} \quad \text{(Manning)} \qquad (12.4)$$

where K is 1.486 for the USCS system of units, and 1 for the SI system of units.

C-117

n is dependent on the relative roughness for various surfaces.

The slope S for uniform flow is given by the formula

$$S = \left(\frac{n}{K}\right)^2 \frac{1}{R_H^{4/3}} V^2 = \left(\frac{n}{K}\right)^2 \frac{1}{R_H^{4/3}} \frac{Q^2}{A^2} \quad (12.5)$$

where Q = Flowrate

In terms of the Manning formula, the flow rate is:

$$Q = AV = A\left(\frac{K}{n}\right) R_H^{2/3} S^{1/2} \quad (12.6)$$

To cover an hydraulically smooth zone flow, a transition zone flow, and a rough-zone flow by using the Moody's chart, the following equation is used:

$$V = \left(\frac{8gSR_H}{f}\right)^{1/2} \quad (12.7)$$

where f = Friction factor

For turbulent flow,

$$\frac{V}{y_t} = 2.5\sqrt{\tau_w/\rho}\ln(y/h) \quad (12.8)$$

To use equation (12.7), the following procedure is used:

a) Estimate a friction factor, f'

b) Determine v from the equation

c) Compute Reynolds number ($L=4R_H$)

d) With the computed Reynolds number and with a relative roughness ratio e/L, find f in the Moody chart

e) If the resulted f is not equal to f', choose another friction factor and repeat the above procedure until an agreeable value of f is reached.

B) Vertical Velocity

Vertical distribution of velocity in an open channel may be assumed as parabolic for laminar flow and logarithmic for turbulent flow.

For laminar flow,

$$\boxed{V_{y_L} = \frac{gS}{\nu}(yh_m - \frac{1}{2}y^2)} \qquad (12.9)$$

where
h_m = Mean depth of the channel = $\frac{A}{b}$

12.3 SURFACE WAVES

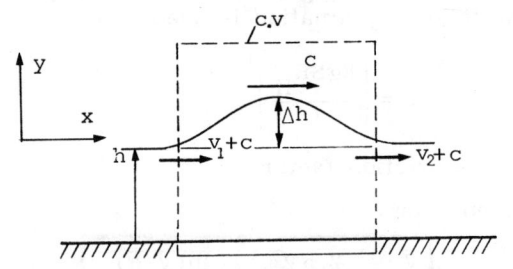

Fig. 12.2 Surface wave in open channel.

Surfave waves are products of temporary disturbances and travel radially outward from the source. The velocity of the wave relative to the undisturbed fluid is given by

$$\boxed{C = \sqrt{gh}} \qquad (12.10)$$

for waves of small height and large wavelength compared with the depth.

12.4 SPECIFIC ENERGY

Specific energy is defined as the energy per unit of weight relative to the bed of the channel:

C-119

$$E_s = \text{depth} + \text{velocity} = h + \frac{V^2}{2g} \qquad (12.11)$$

In terms of the flowrate it can be written as

$$\boxed{E_s = h + \frac{1}{2g}\left(\frac{q}{h}\right)^2,} \qquad (12.12)$$

where q is the volume of flow per unit of width.

For uniform flow, the specific energy remains constant from section to section. For non-uniform flow, the specific energy may increase or decrease along the channel. Figure 12.3 shows the specific energy diagram. The minimum specific energy occurs at the critical depth (h_c).

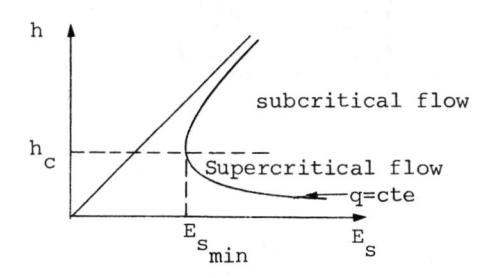

Fig. 12.3 Specific-energy diagram.

12.5 CRITICAL DEPTH

For each volume flow per unit width of the channel there is a point of minimum specific energy. The depth of this point is indicated as the critical depth. It occurs when the specific energy has a minimum value, or at $\frac{dE}{dh} = 0$. From equation (12.12) we get the critical depth (h_c) for a rectangular channel as

$$h_c = \left(\frac{q^2}{g}\right)^{1/3} \qquad (12.13)$$

The critical velocity at this point, v_c, is given by the equation

$$\boxed{v_c = \sqrt{gh_c} = C\sqrt{R_H S_c}} \qquad (12.14)$$

where $S_c = (g/c^2) = f/8$ (channel slope for steady, critical flow).

The quantity $V_c/\sqrt{gh_c}$ is called a Froude number (F_R).

When $F_R < 1$, $F_R = 1$, $F_R > 1$, subcritical, critical, and supercritical flows occur respectively.

The maximum flow occurs at the critical depth and is given by the formula

$$\boxed{q_{max} = \sqrt{gh_c^3}} \qquad (12.15)$$

For a channel with an arbitrary cross section (Fig. 12.1) we have

$$\frac{bQ^2}{gA_c} = 1 \qquad (12.16)$$

$$\left. E_s \right\}_{min} = \frac{A_c}{2b} + h_c \qquad (12.17)$$

$$S_c = \frac{A_c f}{8bR_H} = \frac{fP_c}{8b} \qquad (12.18)$$

where A_c = Critical area corresponding to $F_r = 1$ and
P_c = Wetted perimeter

12.6 THE HYDRAULIC JUMP

A hydraulic jump occurs when a supercritical flow changes to a subcritical flow.

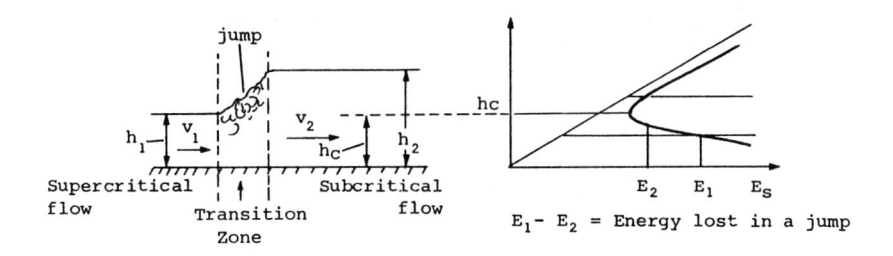

Fig. 12.4 Hydraulic jump.

After the hydraulic jump, the depth is changed to,

$$h_2 = \frac{-h_1 + \sqrt{h_1^2 + \frac{8q^2}{gh_1}}}{2} \qquad (12.19)$$

This equation applies under the following assumptions:
A) Horizontal bottom and uniform rectangular cross-section.
B) Uniform velocity through the cross-section.
C) Uniform depth across the width.
D) No friction at the boundaries.
E) No surface tension effects.

The energy lost in a jump is given by

$$h_L = \frac{(h_2 - h_1)^3}{4h_1 h_2} \qquad (12.20)$$

12.7 NON-UNIFORM FLOW

In a non-uniform flow, the bottom slope, the roughness, and the cross-sectional areas change along the channel (Fig. 12.5).

Fig. 12.5 Non uniform flow.

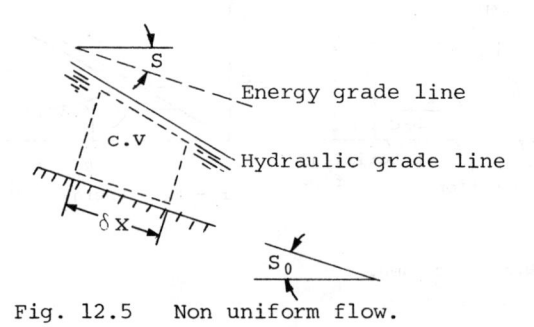

To estimate the depth h versus x along a gradually varied flow, the following formula is used:

$$\Delta L = \left\{ \frac{1 - Q^2 b/gA^3}{S_0 - (n/K)^2 [Q^2/R_h^{4/3} \; A^2]} \right\} \Delta h \qquad (12.21)$$

where

ΔL = Length taken along the channel bed

Δh = Change in elevation of the free surface corresponding to a change in position ΔL along the channel

The slope of the liquid surface in wide rectangular channels for non-uniform flow is given by the following formula:

$$\frac{dh}{dx} = \frac{S_0 [1 - (h_N/h)^{10/3}]}{1 - F_R^2} \qquad (12.22)$$

Table 12.1 shows a summary of the ten different types of varied flow.

Table 12.1 Types of Flow in Open Channels

Channel Slope	Depth Relation	Slope dy/dx	Symbol	Types of flow	Form of Profile
Mild $S_0 > 0$	$h > h_N > h_c$	+	M_1	Subcritical	
	$h_N > h > h_c$	−	M_2	Subcritical	
	$h_N > h_c > h$	+	M_3	Supercritical	
Horizontal $S_0 = 0$ $h_N = \infty$	$h > h_c$	−	H_1	Subcritical	
	$h_c > h$	+	H_2	Supercritical	
Steep $S_0 > S_c > 0$	$h > h_c > h_N$	+	S_1	Subcritical	
	$h_c > h > h_N$	−	S_2	Supercritical	
	$h_c > h_N > h$	+	S_3	Supercritical	
Adverse $S_0 < 0$	$h > h_c$	−	A_1	Subcritical	
	$h_c > h$	+	A_2	Supercritical	

h = Depth

h_N = Normal depth

h_c = Critical depth

12.8 THE OCCURENCE OF CRITICAL CONDITIONS

A) The Broad-Crested Weir:

The rate of the flow is

$$Q = g^{1/2} b \left(\frac{2E_s}{3} \right)^{3/2} \qquad (12.23)$$

or

$$Q = (1.705 \text{ mts}^{1/2}/\text{sec}) b E^{3/2} \qquad (12.24)$$

(SI System)

$$Q = (3.09 \text{ ft}^{1/2}/\text{sec}) bE^{3/2} \qquad (12.25)$$

(English System)

where E is over the crest of the weir.

B) Rapid Flow Approaching a Weir or Other Obstruction

The discharge per unit of depth is given by

$$\boxed{q = C_d h_1 \sqrt{2g(h_0 - h_1)}} \qquad (12.26)$$

where $(h_0 - h_1)$ is the difference of head across the sluice opening. C_d is close to unity for a well-rounded aperture.

C) Venturi Flume

The flow at the throat is critical and the flume is under free discharge. The discharge is given by

$$Q = b_2 h_2 \sqrt{g h_2} \qquad (12.27)$$

where b_2 is the width of the throat and h_2 the corresponding depth.

CHAPTER 13

TURBOMACHINERY

13.1 INTRODUCTION

A turbomachine is a device, consisting of a rotor with vanes or blades, in which there is a transfer of energy between the rotor and a fluid. The transference of energy may be to a shaft (turbine) or from a shaft (pump). Turbomachines can be classified according to the direction of the fluid in the rotor as axial flow turbomachine, radial flow turbomachine, or mixed-flow turbomachine.

13.2 SIMILARITY RELATIONS FOR TURBOMACHINES

A) Dimensionless Parameters

The use of the principles of dynamics similarity will enable us to predict the performance of a machine under conditions different from the tests done on a geometrically similar machine (geometric similarity is a requirement of dynamic similarity).

The following are the relevant variables in turbomachines:

D: Size of the machine (diameter)
N: Rotational speed
Q: Volume flowrate through the machine
ν: Kinematic viscosity
g: Gravity
ΔH: Difference in total head through the machine

p: Power transferred between fluid and rotor
ρ: Density

The following dimensionless parameters can be obtained:

$$\pi_1 = \frac{Q}{ND^3} \quad \text{(Flow coefficient)} \tag{13.1}$$

$$\pi_2 = \frac{g\,\Delta H}{N^2 D^2} \quad \text{(Head coefficient)} \tag{13.2}$$

$$\pi_3 = \frac{ND^2}{\nu} \quad \text{(Reynolds number)} \tag{13.3}$$

$$\pi_4 = \frac{P}{\rho N^3 D^5} \quad \text{(Power)} \tag{13.4}$$

The head coefficient can be written as a function of the flow coefficient as

$$\frac{g\,\Delta H}{N^2 D^2} = f\left(\frac{Q}{ND^3}\right) \tag{13.5}$$

The parameter π_3, is proportional to the Reynolds number and may be disregarded because the viscosity effects in these type of machines are very small. Combining parameter π_4 with π_1 and π_2, we obtain the hydraulic efficiency (η_h) for the turbomachine.

$$\eta_{h\ \text{turbine}} = \frac{\text{Power utilized by unit}}{\text{Power supplied by water}}$$

$$= \frac{P}{\rho g Q \Delta H} \tag{13.6}$$

$$\eta_{h\ \text{pump}} = \frac{\rho g Q \Delta H}{P} = \frac{\text{Power output}}{\text{Power input}} \tag{13.7}$$

where

ΔH = Change in total head through the machine, $H = \frac{P}{\gamma} + z + \frac{v^2}{2g}$

Q = Volume flow = $C_Q\, D^2 \sqrt{\Delta H}$

C_Q = Coefficient depends on the fluid viscosity

For a sufficiently large turbomachine, the efficiency will be a function of the flow coefficient or:

$$\eta_h = f\left(\frac{Q}{ND^3}\right) \qquad (13.8)$$

where N = Rotative speed.

Note: The unit rotative speed, N_u, is defined as the speed of a geometrically similar rotating element having a diameter of 1 in. operating under a head of 1 ft.

B) Specific Speed (N_S)

The specific speed, N_s, is defined as the speed of a rotating element of such diameter that it develops one horsepower for a head of one ft. It may be expressed as

$$\boxed{\begin{aligned} N_{s,\text{turbine}} &= \frac{N\sqrt{P}}{(g\Delta H)^{5/4}} \\ N_{s,\text{pump}} &= \frac{N\sqrt{Q}}{(g\Delta H)^{3/4}} \end{aligned}} \qquad \begin{aligned} (13.9) \\ \\ (13.10) \end{aligned}$$

Figure 13.1 shows curves of maximum efficiency, (η_h) versus specific speed (N_S) for various types of turbomachines. This graph allows us to pick the most efficient turbomachine for a given job.

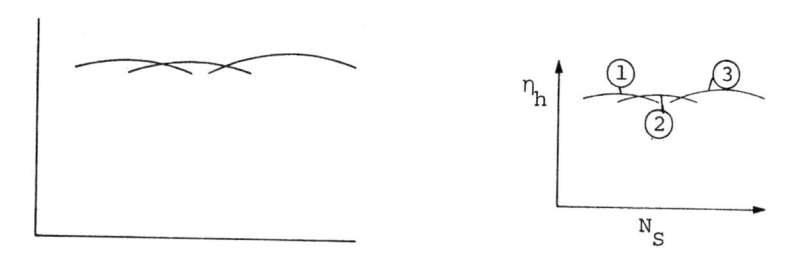

Fig. 13.1 Maximum efficiency vs. specific speed for turbomachines. (1) Radial flow, (2) Mixed flow and (3) Axial flow.

13.3 THE BASIC LAWS

For a steady, one-dimensional flow with no heat transfer and no chemical reactions in a turbomachine (Fig. 13.2) we have:

A) Continuity Equation

$$_1Q_2 = \rho_1 (V_1)_n A_1 = \rho_2 (V_2)_n A_2$$

For an incompressible flow:

$$\boxed{_1Q_2 = (V_1)_n A_1 = (V_2)_n A_2} \qquad (13.11)$$

where subscripts 1 and 2 refer to the inlet and outlet of the turbomachine.

B) Momentum Equation

$$M_x = [\bar{r}_2 (V_\theta)_2 - \bar{r}_1 (V_\theta)_1] \rho Q \qquad (13.12)$$

where M_x is the torque and $\bar{r}_1, \bar{r}_2$ are the radial distances of the inlet and outlet flow, respectively.

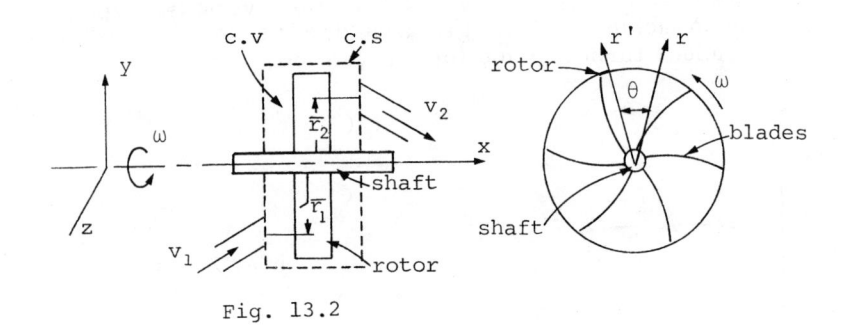

Fig. 13.2

C) First Law of Thermodynamics (13.13)

$$\boxed{\left(\frac{V_1^2}{2} + gy_1 + h_1\right) - \frac{dw_{shaft}}{dm} = \left(\frac{V_2^2}{2} + gy_2 + h_2\right)}$$

For steady incompressible flow we have

$$\boxed{\dfrac{dw_s}{dt} = \gamma Q(h_1 - h_2) = \gamma Q \Delta H}$$ (13.14)

D) Second Law of Thermodynamics

For incompressible flow the second law of thermodynamics is written as

$$g \Delta H = [(U_t V_t)_2 - (U_t V_t)_1],$$

or

$$\boxed{\dfrac{dw_s}{dm} = [(U_t)_1 (V_t)_1 - (U_t)_2 (V_t)_2]}$$ (13.15)

where $U_t = \bar{r}w$ (speed of the blade)

13.4 TURBINES

There are two classifications of turbines:

A) Impulse turbines, in which the fluid expansion from high to low pressure takes place in nozzles external to the blades. (The fluid pressure across the runner does not change.)

B) Reaction turbines, in which the fluid expansion from high to low pressure takes place externally and partially within the blades. (The fluid pressure is reduced as it flows through the runner.)

  a) Impulse Turbines

For an axial flow impulse turbine (Fig. 13.3) we have the following equation for the power:

$$\boxed{\dfrac{dw_s}{dt} = 2\rho Q U_t (V_1 \cos \alpha_1 - U_t)}$$ (13.16)

assuming that $V_{r_1} = V_{r_2}$,

C-130

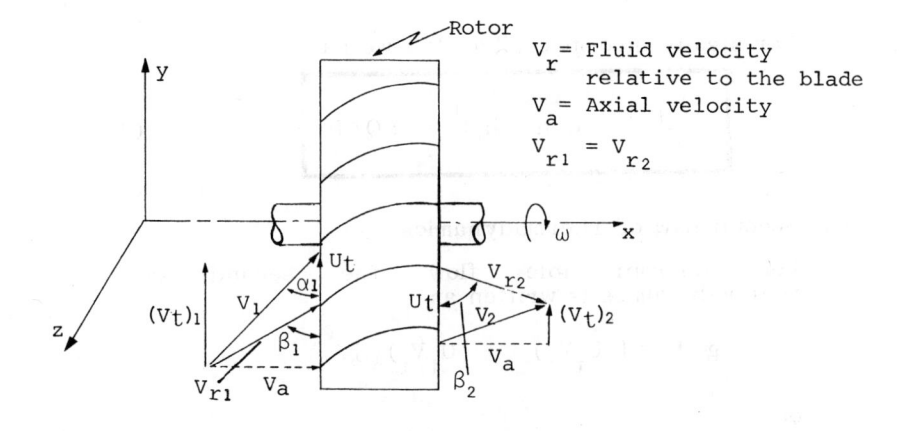

Fig.13.3 Axial-flow impulse turbine

The speed of the blade U_t for maximum power will be

$$U_t = \frac{1}{2} V_1 \cos \alpha_1 \qquad (13.17)$$

We can see that the smaller the angle α, the greater the power (for the Pelton wheel $U_t/V_1 = \frac{1}{2}$). From the momentum equation we find the torque as:

$$M_x = -2R \rho Q V_a \cot \beta \qquad (13.18)$$

where $\beta = \beta_1 = \beta_2$
$R = \bar{r}_1 = \bar{r}_2$
V_a = Axial velocity

b) Axial Flow Reaction Turbines

1) Kaplan Turbine (Fig. 13.4)

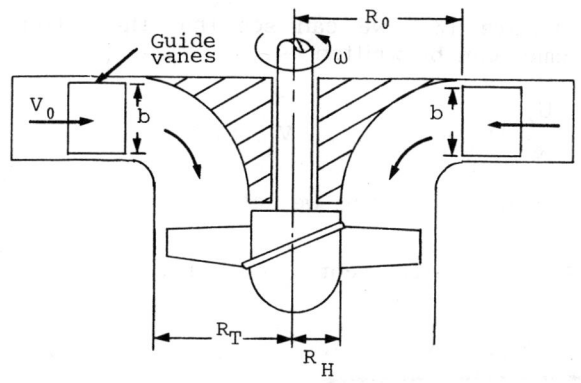

Fig. 13.4 Axial flow turbine (Kaplan).

The tangential velocity V_t is given by

$$V_t = \frac{V_0 R_0 \cos\alpha}{r} \qquad (13.19)$$

the axial velocity by:

$$V_a = \frac{2\pi R_0 b V_a \sin\alpha}{\pi(R_T^2 - R_H^2)} = \frac{Q}{\pi(R_T^2 - R_H^2)} \qquad (13.20)$$

where α is the guide vane angle.

2) Many-Bladed Reaction Turbines
   Under these types of turbines, compressors can also be considered.

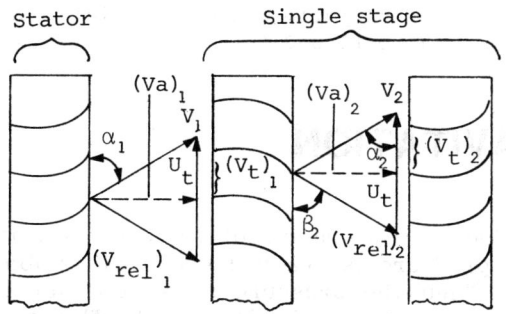

Fig. 13.5 Velocity diagrams for axial flow turbines.

From Figure 13.5 we can see that the second law of thermodynamics can be written as

$$\Delta H = \frac{U_t}{g} [U_t - V_a \cot \beta_2 - V_a \cot \alpha_1] \qquad (13.21)$$

Written in dimensionless form we get

$$\frac{\Delta Hg}{U_t^2} = 1 - \frac{V_a}{U_t} (\cot \alpha_1 + \cot \beta_2) \qquad (13.22)$$

if $K = (\cot \alpha_1 + \cot \beta_2)$ we have

$$\boxed{\frac{\Delta Hg}{U_t^2} = 1 - \frac{V_a}{U_t} K} \qquad (13.23)$$

The type of turbomachine under study can be deduced from this equation. Thus, for

$$\boxed{\begin{array}{ll} \Delta H > 0 & \text{(Compressor)} \\ \Delta H < 0 & \text{(Turbine)} \end{array}}$$

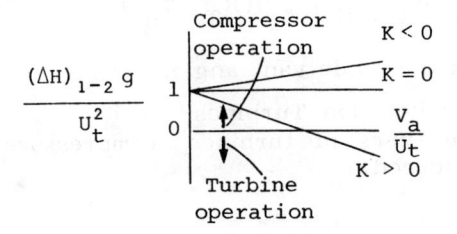

Fig. 13.6 Ideal performance for axial-flow many bladed turbomachines.

13.5 CAVITATION

Cavitation is the result of pressure variation in the flow; vapor pressure is reached and small bubbles of vapor are formed. When the pressure rises, the bubbles collapse and higher local pressure is developed. This causes severe damage to the surface of the rotor. Vibration, noise, and low efficiency are also associated with cavitation. The point of minimum pressure is at the outlet of the runner blade, on the leading side. For a reaction turbine we have

$$\frac{P_{min}}{\rho g} + \frac{V^2}{2g} + z - h_f = \frac{P_{atm}}{\rho g} \qquad (13.24)$$

where h_f is the head lost due to friction,

z is the height of the turbine runner above the tail water surface

P_{min}, P_{atm} is absolute pressures.

This equation shows that the outlet velocity should be small so as to avoid cavitation.

The critical cavitation parameter, σ_c, derived from the above equation is given by

$$\boxed{\sigma_c = \frac{(P_{atm}/\rho g) - (P_{min}/\rho g) - z}{H}} \qquad (13.25)$$

where H is the net head across the machine. The Thoma's cavitation parameter is defined as

$$\sigma = \frac{(P_{atm}/\rho g) - (P_v/\rho g) - z}{H} \qquad (13.26)$$

where p_v is the vapor pressure. Thus, when

$$P_{min} > P_v \quad \text{or} \quad \sigma > \sigma_c$$

cavitation is avoided.

figures 13.7 and 13.8 show caviation limits for reaction turbines and the effect of cavitation on the efficiency.

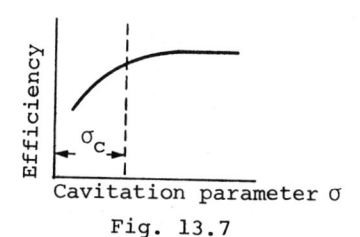

Fig. 13.7

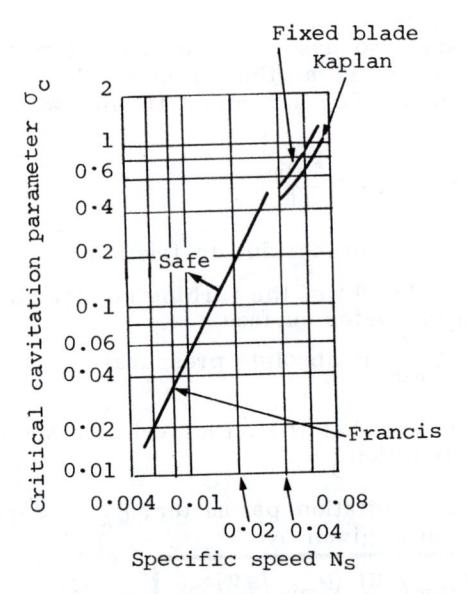

Fig. 13.8 Cavitation limits for for reaction turbines

13.6 RADIAL-FLOW PUMPS AND BLOWERS

In these types of machines the fluid enters axially in the center of the impeller, flows outward in the direction of the blades, and is discharged into the pump casing which converts kinetic energy to pressure head (Fig. 13.9). When the fluid used is liquid, the machine is called a centrifugal pump. When the fluid is a gas, it is called a compressor.

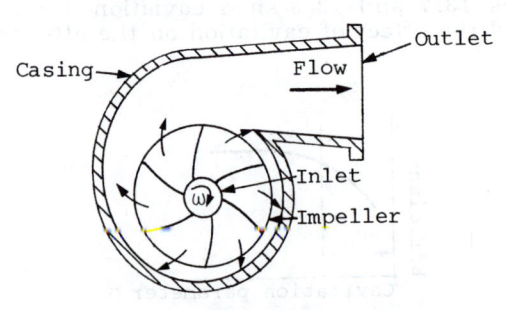

Fig. 13.9 Centrifugal Pump.

The shaft torque from the moment of momentum equation is given by the following expression:

$$T_s = -r_1(V_t)_1 [\rho_1(V_n)_1 A_1] + r_2(V_t)_2 [\rho_2(V_n)_2 A_2] \quad (13.27)$$

If we consider that the entire flow entering the impeller is in the radial direction, then $(V_t)_1 = 0$,

$$(V_n)_2 = (V_r)_2 \quad \text{(see Fig. 13.10)}.$$

Thus, the above equation can be written as

$$T_s = r_2 \, \rho_2 \, A_2 (V_r)_2 (V_t)_2 \quad (13.28)$$

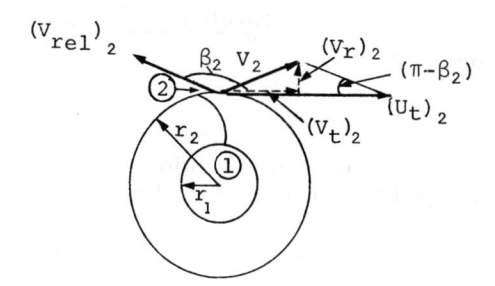

Fig. 13.10 Outlet velocity diagram

From the velocity diagram (Fig. 13.10), $(V_t)_2$ is

$$\boxed{(V_t)_2 = (U_t)_2 + (V_r)_2 \cot \beta_2} \quad (13.29)$$

Finally, T_s is written as:

$$\boxed{T_s = r_2 \dot{m}[(U_t)_2 + (V_r)_2 \cot \beta_2]} \quad (13.30)$$

where losses were not taken into account. the mechanical efficiency η_m is defined as

$$\boxed{\eta_m = \frac{T_s}{(T_s)_{actual}}} \quad (13.31)$$

The actual power to run the pump or compressor is:

$$\boxed{(P)_{input} = \frac{W}{\eta_m} r_2 \dot{m}[(U_t)_2 + (V_r)_2 \cot\beta_2]} \qquad (13.32)$$

The maximum shaft work per unit of mass when the fluid is liquid, will be

$$\left.\frac{dW_p}{dm}\right|_{theoretical} = \left[\frac{(V_2)^2_{theoretical}}{2} - \frac{V_1^2}{2}\right]$$
$$+ \frac{p_2 - p_1}{\rho} + g(z_2 - z_1) \qquad (13.33)$$

When the fluid is a perfect gas ($\rho \neq$ constant), the maximum shaft work will be

$$\left.\frac{dW_p}{dm}\right|_{theoretical} = \frac{(V_2)^2_{theoretical} - V_1^2}{2} + C_p T_1\left[\left(\frac{p_2}{p_1}\right)^{(K-1)/K} - 1\right] \qquad (13.34)$$

Ideally, for a frictionless fluid the total increase of total energy across the pump is given by:

$$H = (V_t)_2 \left[(V_t)_2 - \frac{Q}{A}\cot\beta_2\right] \qquad (13.35)$$

For a fixed speed, the variation of H with Q is linear. Figure 13.11 shows this fact:

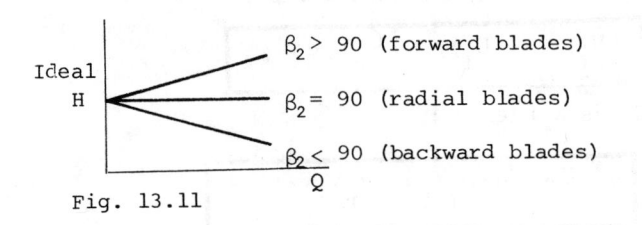

```
              β₂ > 90 (forward blades)
Ideal
  H           β₂ = 90 (radial blades)
              β₂ < 90 (backward blades)
                    Q
Fig. 13.11
```

13.7 CAVITATION IN CENTRIFUGAL PUMPS

The energy equation between the surface in the reservoir, and the entrance to the impeller where the pressure is a minimum can be written as:

C-137

$$\frac{p_0}{\rho g} - h_f = \frac{p_{min}}{\rho g} + \frac{V_1^2}{2g} + z_1 \qquad (13.36)$$

for steady conditions, where

V_1 = Velocity at p_{min}

p_0 = Pressure at surface reservoir

h_f = Head losses

The critical cavitation parameter is given by

$$\sigma_c = \frac{p_0/\rho g - p_{min}/\rho g - z_1 - h_f}{H} \qquad (13.37)$$

Cavitation does not occur when $P_{min} > P_v$ or when $\sigma > \sigma_c$, where

$$\sigma = \frac{p_0/\rho g - p_v/\rho g - z_1 - h_f}{H} \qquad (13.38)$$

and H is the total head. The numerator of equation (13.38) is known as the Net Positive Suction Head.

CHAPTER 14

FLOW MEASUREMENT

14.1 INTRODUCTION

A standard engineering practice is to use different classes of devices to measure the flowrate of fluids. There are simple and sophisticated flow metering methods and the use of any one in particular will depend upon the accuracy desired, cost, service life, etc.

14.2 DEFINITION OF COEFFICIENTS

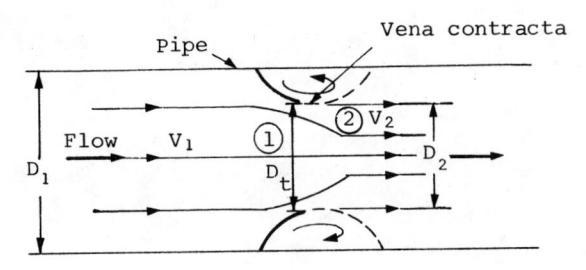

Fig. 14.1 Internal flow through a generalized nozzle.

A) Coefficient of Contraction

The coefficient of contraction, C_c, is defined as:

$$C_c = \frac{A_v}{A_t} \qquad (14.1)$$

where A_v = Area contracted

A_t = Meter throat area

B) Coefficient of Velocity

The coefficient of velocity is defined as:

$$C_v = \frac{V_{2\,actual}}{V_{2\,ideal}} \qquad (14.2)$$

where $V_{2\,actual}$ = Actual mean velocity of the flow

$V_{2\,ideal}$ = Ideal velocity (no friction)

C) Coefficient of Discharge

The coefficient of discharge is defined as:

where,

$$C_d = \frac{Q_{actual}}{Q_{ideal}} \qquad Q_{actual} = \text{Actual flowrate} \qquad (14.3)$$

also $C_d = C_v C_c \qquad Q_{ideal}$ = Ideal flowrate (14.4)

Note: The value of the coefficients are determined experimentally.

14.3 FLOWMETERS FOR INCOMPRESSIBLE INTERNAL FLOWS

A) The Flow Nozzle

Fig. 14.2 Nozzle flowmeter

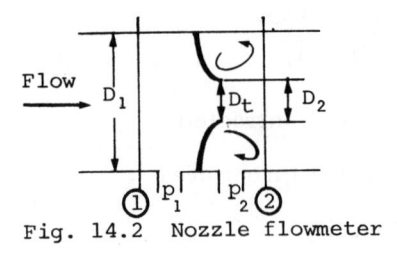

Combining Bernoulli's equation and the continuity equation between points 1 and 2 (Fig. 14.2) we get

$$V_2 = \sqrt{\frac{2(p_1-p_2)}{\rho[1-(A_2/A_1)^2]}} \qquad (14.5)$$

Thus, the flowrate is given by

$$Q_{ideal} = A_2 V_2 = A_2 \sqrt{\frac{2(p_1-p_2)}{\rho[1-(A_2/A_1)^2]}} \qquad (14.6)$$

Taking into account the frictional effects, the coefficient of discharge C_d is introduced. That is,

$$Q_{actual} = C_d A_2 \sqrt{\frac{2(p_1-p_2)}{\rho[1-(A_2/A_1)^2]}} \qquad (14.7)$$

B) The Orifice Plate

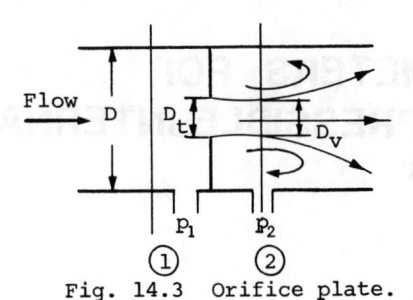

Fig. 14.3 Orifice plate.

The area of the vena contracta A_v is unknown. Using the coefficient of contraction formula, it may be found:

C-141

$$A_v = C_c A_t \qquad (14.8)$$

$$Q_{actual} = C_d^* A_t \sqrt{\frac{2(p_1 - p_2)}{\rho[1 - (A_2/A_1)^2]}} \qquad (14.9)$$

where C_d^* is a new coefficient of discharge for square-edge orifices.

C) Venturi Meter

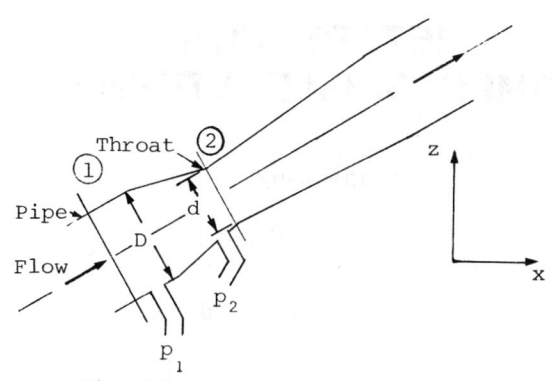

Fig. 14.4 Venturi.

Combining Bernoulli's equation with the continuity equation at 1 and 2 (Fig. 14.4) the flowrate is given by

$$Q_{ideal} = A_2 \sqrt{\frac{2[(p_1 - p_2)/\rho] + (z_1 - z_2)g}{1 - (A_2/A_1)^2}} \qquad (14.10)$$

$$Q_{actual} = (C_d)_{venturi} A_2 \sqrt{\frac{2[(p_1 - p_2)/\rho] + (z_1 - z_2)g}{1 - (A_2/A_1)^2}} \qquad (14.11)$$

where $(C_d)_{venturi}$ depends upon the ratio D/d and the Reynolds number (Re) is based on the outside diameter of the pipe.

D) The Pitot Tube

The pitot tube measures the velocity at a point. From

Figure 14.5 the velocity at B is given by the following formula:
$$V_B = C\sqrt{2\rho(p_1 - p_2)} \quad (14.12)$$

Fig. 14.5 Pitot tube.

14.4 FLOWMETERS FOR COMPRESSIBLE INTERNAL FLOWS

For an isentropic perfect gas, using a flow nozzle or a venturi, the mass flow is given by

$$\dot{m} = Y\rho_1 \, C_d \, A_2 \sqrt{\frac{2(p_1 - p_2)}{\rho[1 - (A_2/A_1)^2]}} \quad (14.13)$$

where Y is the compressibility factor defined as:

$$Y = \left\{ \frac{[K/(K-1)](p_2/p_1)^{2/K}[1 - (p_2/p_1)^{(K-1)/K}][1 - (A_2/A_1)^2]}{[1 - (A_2/A_1)^2 (p_2/p_1)^{2/K}][1 - p_2/p_1]} \right\}^{\frac{1}{2}} \quad (14.14)$$

and K is the isentropic (adiabatic) exponent ($K_{air} = 1.4$).

Equation (14.13) applies for a steady, subsonic, and isentropic perfect gas in which the potential energy of gravity has been neglected.

14.5 WEIRS

A) Rectangular Weirs

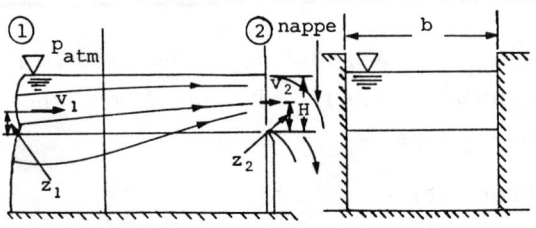

Fig. 14.6 Rectangular weir.

Assumptions:

a) $p = \rho gh$

b) The surface remains horizontal until the plane of the notch.

c) Pressure throughout the nappe is atmospheric.

d) No effects of viscosity and surface tension.

The ideal discharge will be

$$Q_{ideal} = \frac{2}{3} b \sqrt{2g} \left[\left(H + \frac{V_1^2}{2g} \right)^{3/2} - \left(\frac{V_1^2}{2g} \right)^{3/2} \right] \qquad (14.15)$$

Usually $V_1^2/2g \ll H$, thus

$$Q_{ideal} = \frac{2}{3} b \sqrt{2g} \, H^{3/2} \qquad (14.16)$$

Taking into account fricitonal losses, the actual discharge will be

$$Q_{actual} = \frac{2}{3} C_d \, b \sqrt{2g} \left[\left(H + \frac{V_1^2}{2g} \right)^{3/2} - \left(\frac{V_1^2}{2g} \right)^{3/2} \right] \qquad (14.17)$$

where C_d is the coefficient of discharge for a rectangular weir.

B) V-Shaped Weir

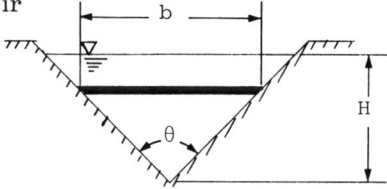

Fig. 14.7 V-shaped weir.

With the same assumptions used for the rectangular Weir the ideal discharge for a V-shaped weir is given by

$$Q_{ideal} = \frac{8}{15} \tan \frac{\theta}{2} \sqrt{2g} \, H^{5/2} \qquad (14.18)$$

The actual discharge will be

$$Q_{actual} = \frac{8}{15} C_d \tan \frac{\theta}{2} \sqrt{2g} \, H^{5/2} \qquad (14.19)$$

where C_d is the coefficient of discharge for a V-shaped weir.

Handbook of Mechanical Engineering

SECTION D
Thermodynamics

CHAPTER 1

BASIC CONCEPTS

1.1 THERMODYNAMIC SYSTEMS

The term system as used in thermodynamics refers to a definite quantity of matter bounded by some closed surface which is impervious to the flow of matter. This surface is called the boundary of the system. Everything outside the boundary of a system constitutes it surroundings.

Depending on the nature of the boundary involved, we can classify a thermodynamic system in one of the following three categories:

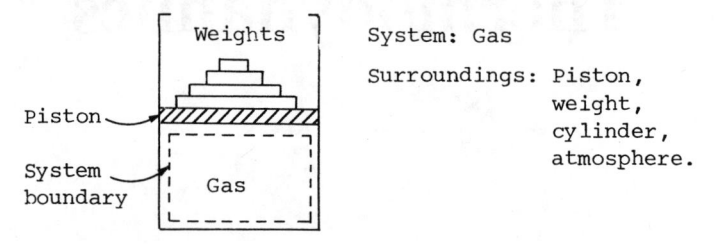

System: Gas

Surroundings: Piston, weight, cylinder, atmosphere.

Fig. 1.1   Example of a system

a) An Isolated System allows neither heat nor work transfer across the boundary.

b) An Open System allows exchange of both matter and energy.

c) A Closed System allows only exchange of energy.

1.2 PROPERTIES OF SYSTEMS

The state of a system is its condition as identified by coordinates which can usually be observed quantitatively, such as volume, density, temperature, etc. These coordinates are called properties.

All properties of a system can be divided into two types:

a) An Intensive Property is independent of the mass. (Pressure, density, temperature)

b) An extensive property has a value which varies directly with the mass. (Volume, energy, entropy)

1.3 PROCESSES

When a thermodynamic system changes from one state to another, it is said to execute a process, which is described in terms of the end states. A cycle is a process in which the end states are identical.

Processes are classified according to the following categories:

a) An Isothermal Process is a constant-temperature process.

b) An Isobaric Process is a constant-pressure process.

c) An Isometric Process is a constant-volume process.

d) An Adiabatic Process is a process in which heat does not cross the system boundary.

e) A Quasistatic Process consists of a succession of equilibrium states, such that at every instant the system involved departs only from the equilibrium state.

f) In a Reversible Process, the initial state of the system involved can be restored with no observable effects in the system and its surroundings. This is called an Ideal Process.

g) In an Irreversible Process, the initial state of the system involved cannot be restored without observable effects in the system and its surroundings.

1.4 THERMODYNAMIC EQUILIBRIUM

When a system is not subject to interactions and a change of state cannot occur, then the system is in a state of equilibrium. There are four kinds of equilibrium: stable, neutral, unstable and metastable. Of these, a stable equilibrium is the one most encountered in thermodynamics. A system is in a state of stable equilibrium if a finite change of state of the system cannot occur without leaving a corresponding, finite alteration in the state of the environment. Figure 1.2, of a marble at rest in a covered bowl is an example of such a state. To have thermodynamic equilibrium, the conditions of mechanical, chemical and thermal equilibrium must be satisfied.

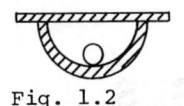

Fig. 1.2

A system is in mechanical equilibrium when it has no unbalanced force within it and when the force it exerts on its boundary is balanced by external forces.

A system is in thermal equilibrium when its temperature is uniform throughout, and equal to the temperature of the surroundings.

A system is in chemical equilibrium when the chemical composition of the system remains unchanged.

1.5 MUTUAL EQUILIBRIUM

Two systems are in mutual equilibrium if they are brought into communication and there is no change in either system.

1.6 ZEROTH LAW OF THERMODYNAMICS

When two systems are each in thermal equilibrium with a third system, they are also in thermal equilibrium with each other.

1.7 PRESSURE

Pressure is defined as the force per unit area on a surface whose dimensions are large, or

$$P = \frac{F}{A}$$

where

P = Pressure,   F = Normal force on the area element A.

Pressure as defined above is called the absolute pressure. Most pressure gauges read the difference between the absolute pressure in a system and the absolute pressure of the atmosphere. This difference is called the gauge pressure. Relations between these pressures are shown in Fig. 1.3.

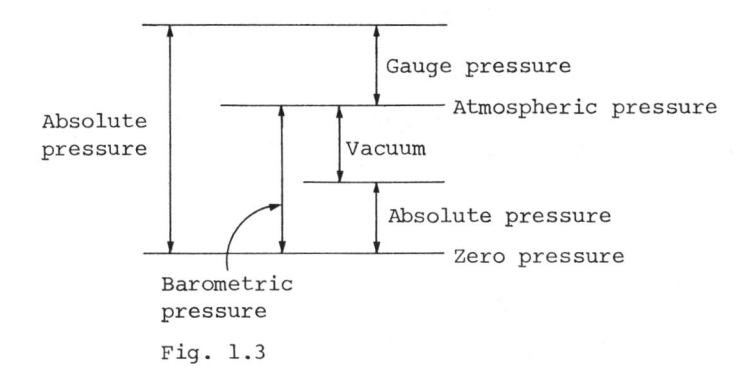

Fig. 1.3

1.7.1 UNITS OF PRESSURE

(I) Engineering system: One pound force per square foot (lbf/ft^2)

(II) International system: (a) Pascal (Pa) = $1 \, N/M^2$ (1.1)
                                (b) 1 bar = $10^5 Pa$ = 0.1 MPa   (1.2)

A pressure of one standard atmosphere is defined as the pressure produced by a column of mercury exactly 76 cm in length, of density 13.5951 gm/cm^3 at a point where g is 980.665 cm/sec^2.

$$1 \text{ standard atmosphere} = 1.01324 \times 10^6 \, \frac{dynes}{cm^2} = 14.6959 \, lb/in^2$$

1.8 OTHER PROPERTIES

a) Average specific volume: $\bar{v} = \dfrac{V}{m}$ (1.3)

where

    V = Total volume of the system
    m = Total mass of the system

b) The average density ρ of a system is the reciprocal of of the average specific volume, or

$$\boxed{\bar{\rho} = \frac{1}{\bar{v}} = \frac{m}{V}}$$

                                                               (1.4)

c) A specific value of a property can be defined at each point of a system as the average specific value of the property in a physically infinitesimal volume element that includes the point.

(I) Specific volume = $\dfrac{dV}{dm}$ (1.5)

(II) Density = $\dfrac{1}{v} = \dfrac{dm}{dV}$ (1.6)

1.9 DIMENSIONS AND UNITS

Dimensions are names given to physical quantities. Some examples of dimensions are length, time, mass, force, volume and velocity.

A unit is a definite standard or measure of a dimension. For example, foot, meters and angstroms are all different units of the common dimension of length.

There are four systems of international units:

a) International system
b) English Engineering System
c) Absolute engineering system
d) Absolute metric system

The dimensions and units of the four basic systems are summarized in table 1.

Dimensions and Units of Different Dimensional Systems

Name of System	Primary Quantities			Derived Quantities		g_c in $F = \frac{1}{g_c} ma$
	Mass	Length	Time	Force	Mass	Force	
SI	kg	m	s	—	—	N	$g_c = 1$
English engineering	lbm	ft	s	lbf	—	—	$g_c = 32.174 \frac{ft\text{-}lbm}{lbf\text{-}s^2}$
Absolute engineering	—	ft	s	lbf	slug	—	$g_c = 1$
Absolute metric (cgs)	gm	cm	s	—	—	dyne	$g_c = 1$

D-6

CHAPTER 2

PROPERTIES AND STATES OF A PURE SUBSTANCE

2.1 THE PURE SUBSTANCE

The term "pure substance" designates a substance which is homogeneous throughout and has the same chemical composition from one phase to another.

Water is a pure substance since its chemical composition is the same in all phases. A mixture of gases can be considered a pure substance. But if the mixture is cooled, the new phase would have a different composition from the old and the system would no longer be a pure substance.

2.2 P-V-T BEHAVIOR OF A PURE SUBSTANCE

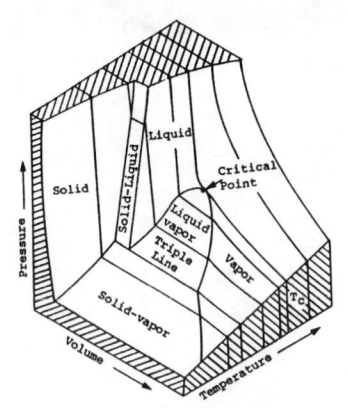

Fig. 2.1 p-v-T Surface for a Substance that contracts on Freezing

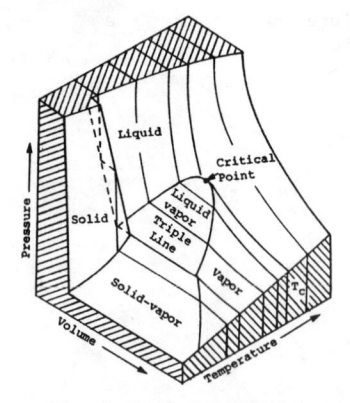

Fig. 2.2 p-v-T Surface for a Substance that expands on Freezing

Figures 2.1 and 2.2 are schematic diagrams of the P-V-T surface for a pure substance. They clearly show that a pure substance can exist only in the vapor, liquid or solid phase for certain ranges. To these diagrams, the following terms apply:

a) Critical point is the point beyond which there is no latent heat of vaporization and no other characteristics which normally works a change in phase.

b) The pressure, temperature and specific volume at the critical point are known as the critical pressure P_c, the critical temperature T_c and the critical specific volume V_c.

c) In the liquid-vapor region, the vapor in an equilibrium mixture is called a saturated vapor, and the liquid, a saturated liquid.

d) Saturation temperature is one at which vaporization takes place at a given pressure (called the saturation pressure).

2.3 P-T PROGRAM

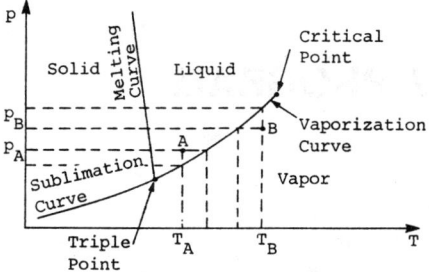

Fig. 2.4 p-T Diagram for a Substance That Expands on Freezing

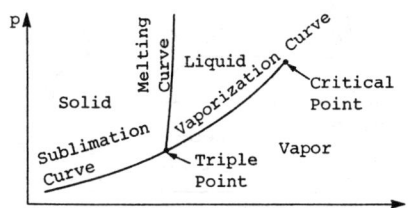

Fig. 2.3 p-T Diagram for a Substance That Contracts on Freezing

D-8

To these diagrams, the following terms apply:

a) Triple point
The point at which all three phases can coexist in equilibrium.

b) Sublimation Curve
The curve along which the solid phase may exist in equilibrium with the vapor phase.

c) Vaporization Curve
The curve along which the liquid phase may exist in equilibrium with the vapor phase.

d) Melting Curve or Fusion Curve
The curve along which the solid phase may exist in equilibrium with the liquid phase.

e) State A (shown in Fig. 2.4) is known as a subcooled liquid or a compressed liquid.
State B (also shown in Fig. 2.4) is known as a superheated vapor.

f) A transition from one solid phase to another is called an allotropic transformation.

2.4 T-V PROGRAM

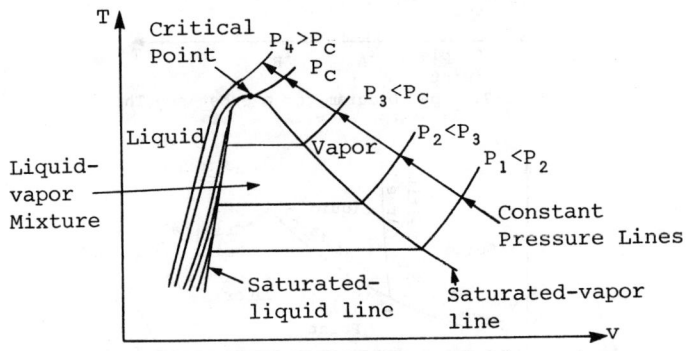

Fig. 2.5 Vapor Dome on a T-v Diagram

2.5 P-V DIAGRAM

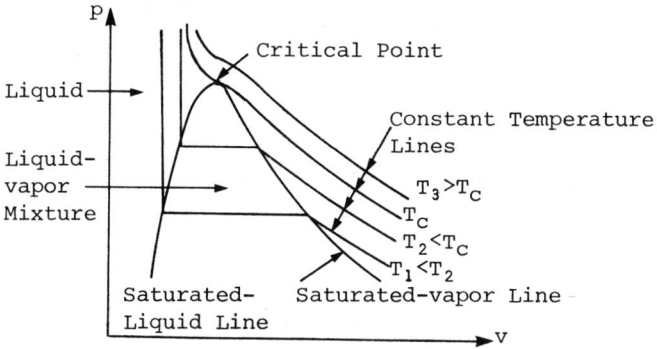

Fig 2.6 Vapor Dome on a p-v Digram

2.6 T-S DIAGRAM

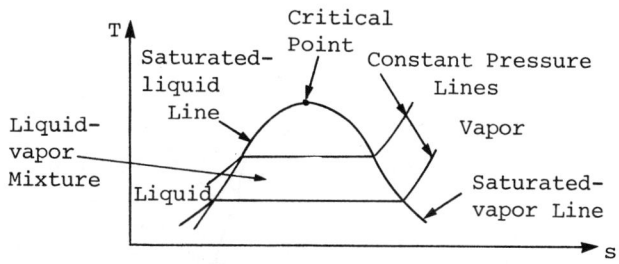

Fig 2.7 Vapor Dome on a T-s Diagram

2.6.1 LATENT HEAT OF VAPORIZATION

The change in enthalpy between the liquid phase and the vapor phase.

2.6.2 LATENT HEAT OF FUSION

The change in enthalpy between the solid phase and the liquid phase.

2.7 H-S DIAGRAM

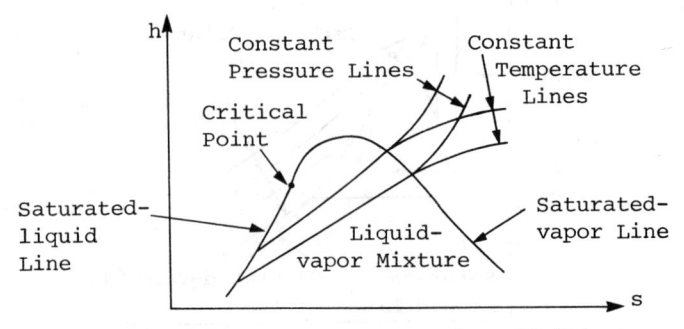

Fig. 2.8 Vapor Dome on an h-s (Mollier) Diagram

2.8 P-H DIAGRAM

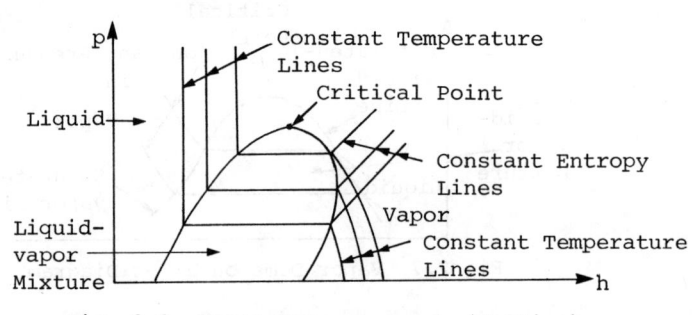

Fig. 2.9 Vapor Dome on a p-h (Mollier) Diagram

2.9 TABLES OF THERMODYNAMIC PROPERTIES

Tables of thermodynamic properties of many substances are available. They consist principally of tabulations of specific volume, enthalpy, and internal energy.

D-11

The notations commonly used for the properties listed in thermodynamics tables are:

s = Saturated

f = Saturated liquid

g = Saturated vapor or gas

sf = Fusion

fg = Vaporization

sg = Sublimation

From our study of thermodynamic diagrams, we've seen that the data we need may be given in the following tables:

I. Table for Saturated Liquids

II. Table for Saturated Vapors

III. Table for Superheated Vapors

IV. Table for Compressed Liquids

2.9.1 QUALITY

The quality of a substance is defined as:

$$x = \frac{m_g}{m_g + m_f} = \frac{m_{vapor}}{m_{total}}$$

where

m_g = Mass of vapor

m_f = Mass of liquid

2.9.2 MOISTURE

The moisture of a substance is defined as:

$$y = \frac{m_f}{m_f + m_g} = \frac{m_f}{m} = 1 - x$$

where

m_f = Mass of liquid

m_g = Mass of vapor

c) Each property for a substance having a given quality can be found by the equation:

$$P = P_f + x\, P_{fg} = xP_g + (1-x)P_f$$

where

P = Any property (v, u, h, s)

x = Quality

CHAPTER 3

WORK AND HEAT

3.1 DEFINITION OF WORK

The work W done by a force F, when the point of application of the force undergoes a displacement of dx, is defined as:

$$W = \int_1^2 F\,dx \qquad (3.1)$$

3.2 THERMODYNAMIC WORK

When looking at thermodynamics from a microscopic point of view, it is advantageous to relate the definition of work to the concepts of system, property and process. We therefore define work in the following manner: Work is an interaction between a system and its surroundings, and is said to be "done" by a system if the sole effect in the surrounds could be the raising of a weight.

Work done by a system is considered positive and work done on a system is considered negative. Let us illustrate the definition of work with an example.

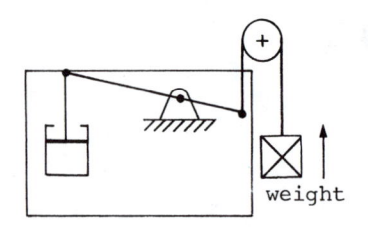

Fig. 3.1

A weight is raised by means of a cord wrapped around an external frictionless pulley, which is connected to a mechanism inside the box (system). Assume weightless links and cord.

3.3 WORK DONE ON A SIMPLE COMPRESSIBLE SYSTEM

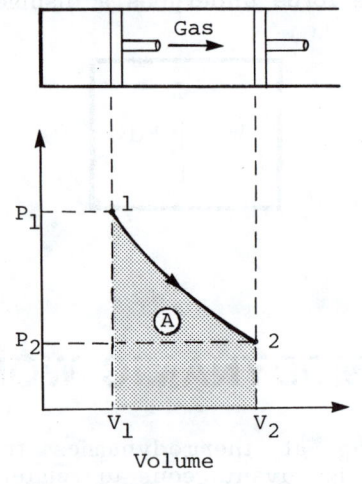

Fig. 3.2

Consider as a system the gas contained in a cylinder as shown in fig. 3.2.

For any small expansion in which the gas increases in volume by dv, the work done by the gas is

D-15

$$W = \int_1^2 p\,dv \qquad (3.2)$$

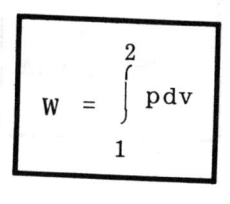

where p is the pressure exerted on the piston.

The interval $\int_1^2 p\,dv$ is the area under the curve on the P-V diagram. Since we can go from state 1 to state 2 along many different paths, it is evident that the amount of work involved in each case is a function not only of the end states of the process, but it is also dependent on the path that is followed in going from one state to another. For this reason, work is called a path function, or in mathematical language, work is an inexact differential.

3.4 OTHER MODES OF THERMODYNAMIC WORK

In the preceding section, we considered work in P-V-T systems. Work in different kinds of systems is also important. Several different kinds of work are summarized below:

I. Stretched wire work

$$_1W_2 = -\int_1^2 T\,dl \qquad (3.3)$$

where T = Tension
dl = change of length of the wire

II. Surface film work

D-16

$$_1W_2 = -\int_1^2 y\,dA \qquad (3.4)$$

where

y = Surface Tension
dA = Change of area

III. Magnetic work

$$_1W_2 = \int_1^2 \mu_0 H\,dlVM \qquad (3.5)$$

where

μ_0 = Permeability
V = Volume
H = Intensity of the magnetic field
M = Magnetization

IV. Electrical work

$$_1W_2 = \int_1^2 E\,dz \qquad (3.6)$$

where

E = Electric field
dz = Amount of electrical energy flow into the system

3.5 HEAT

Heat is defined as the form of energy that is transferred across the boundary of a system at a given temperature to another system at a lower temperature by virtue of the temperature difference between the two systems.

Positive heat transfer is heat addition to a system and negative heat transfer is heat rejection by the system.

3.6 COMPARISON OF HEAT AND WORK

There are many similarities between heat and work, and these are summarized here:

a) Heat and work are transient phenomena. Systems never possess heat or work. Heat and work, only cross system boundaries when systems undergo changes of state.

b) Both heat and work are boundary phenomena. Both are observed only at the boundaries of systems, and both represent energy crossing boundaries.

c) Both heat and work are path functions and inexact differentials.

3.7 UNITS OF WORK AND HEAT

	International System	English Engineering System
Work	Joule (J)*, 1(J) = 1(N)/ (m)	(FT-LBF) (BTU)*
Power	Watt (W), 1(W) = 1(J)/ (sec)	$\frac{(FT-LBF)}{(sec)}$ $\frac{(BTU)}{(sec)}$
Heat	The units for heat, and for any other forms of energy, are the same as the units for work.	

*Joule is defined as the amount of energy needed to raise a weight; that is, the product of a unit force (one Newton) acting through a unit distance (one meter).

*BTU is defined as the amount of energy required to raise 1 lbm of water from 59.5°F to 60.5°F.

CHAPTER 4

ENERGY AND THE FIRST LAW OF THERMODYNAMICS

4.1 ENERGY OF A SYSTEM

Since for a given closed system, the work done is the same in all adiabatic processes between the equilibrium states, it follows that a property of the system can be defined such that the change between any two equilibrium states is equal to the adiabatic work. We define this property as the energy E of the system or

$$\Delta E = E_2 - E_1 = W_{adiabatic} \quad (4.1)$$

For a system we would write,

E = Internal energy + Kinetic energy + Potential energy
where:

a) Internal energy (U) is an extensive property, since it depends upon the mass of the system it represents energy modes on the microscopic level, such as energy associated with nuclear spin, molecular binding, magnetic dipole moment and so on.

b) Kinetic energy (K.E.) is the kind of energy that a body has because of its motion. The kinetic energy of a system having a mass m with a velocity v is given by

$$KE = \tfrac{1}{2}mv^2 \quad (4.2)$$

D-19

c) Potential energy (P.E.) is the kind of energy that a body has because of its position in a potential field. The potential energy of a system having a mass m, and an elevation z, above a defined plane in a gravitational field with a constant g is given by

$$PE = mgz \qquad (4.3)$$

4.2 THE FIRST LAW OF THERMODYNAMICS

The principle of conservation of energy, the first law of thermodynamics, may simply be stated as:

change in stored energy = energy input - energy output

4.3 THE FIRST LAW FOR A SYSTEM UNDERGOING A CYCLE

Observations have led to the formulation of the first law of thermodynamics which in equation form is written:

$$J \oint dQ = \oint dW \qquad (4.4)$$

where

$\oint dQ$ = Cyclic integral of the heat transfer

$\oint dW$ = Cyclic integral of the work

J = Proportionality factor

4.4 THE FIRST LAW FOR A CHANGE IN THE STATE OF A SYSTEM

For a system undergoing a cycle, changing from state 1 to state 2, we have:

$$dQ - dW = dE \qquad (4.5)$$

where

Q = Heat transferred to the system during the process
W = Work transferred from the system during the process
E is the total energy of the system.

The integrated form of (4.5) with g constant becomes:

$$_1Q_2 - {_1W_2} = U_2 - U_1 + \frac{m(v_2^2 - v_1^2)}{2} + mg(z_2 - z_1) \qquad (4.6)$$

where

$_1Q_2$ = Change of heat from state 1 to state 2
$U_2 - U_1$ = change in internal energy
$m(v_2^2 - v_1^2)/2$ = Change in kinetic energy

$mg(z_2 - z_1)$ = Change in potential energy
$_1W_2$ = Work done by or in the system

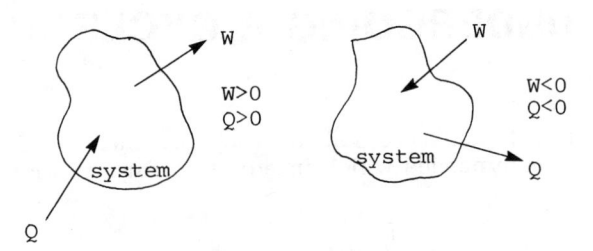

Fig. 4.1 Signs of work and heat related with a system*

Notes:

I. E depends only on the initial and final states and not on the path followed between the two states; therefore, E is a point function and is considered the differential of a property of the systems.

II. The net change of the energy of the system is always equal to the net transfer of energy across the system boundary as heat and work.

III. Equation (4.6) gives only changes in internal energy, both kinetic and potential. We cannot learn absolute value of these quantities from the equation.

IV. Equation (4.5) is a consequence of the first law, and not the first law itself. The first law includes the additional information that energy E is a property.

V. The first step for applying equation (4.6) to the solution of any problem must be the description of a closed system and its boundaries.

*This sign convention is not a universal convention.

4.5 THE THERMODYNAMIC PROPERTY, ENTHALPY

It is convenient to define a new extensive property, enthalpy, as

$$\boxed{H = U + PV}$$

(4.7)

or, per unit mass,

$$h = u + Pv$$

(4.8)

H, h = Enthalpy

U, u = Internal energy

P = Pressure

V, $\bar{v}$ = Volume, Specific Volume

4.6 THE CONSTANT-VOLUME AND CONSTANT-PRESSURE SPECIFIC HEATS

a) $Cv = \left(\dfrac{\delta u}{\delta T}\right)_v$ Constant-volume specific heat (4.9)

D-22

b) $Cp = \left(\frac{\delta h}{\delta T}\right)_p$ Constant-pressure specific heat   (4.10)

4.7 CONTROL VOLUME

A control volume is any specified volume in a fixed region in space through which fluid flow takes place. Its surface is called the control surface.

4.8 LAW OF CONSERVATION OF MASS

For any system the law of conservation of mass states that:

mass added - mass removed = change in mass storage.

4.9 CONSERVATION OF MASS FOR AN OPEN SYSTEM

$$\dot{m}_{out} - \dot{m}_{in} = \frac{dm}{dt} \qquad (4.11)$$

where

$\dot{m}_{out}$ = Mass flow out of the control volume

$\dot{m}_{in}$ = Mass flow into the control volume

$\frac{dm}{dt}$ = Rate of change in the mass within the control volume

4.10 THE FIRST LAW OF THERMODYNAMICS FOR A CONTROL VOLUME

The rate equation of the first law for a control volume is:

$$\dot{Q}_{cv} - \dot{W}_{cv} = \frac{dE_{cv}}{dt} + \Sigma m_e (h_e + \frac{v_e^2}{2} + gz_e) = \Sigma m_i (h_i + \frac{v_i^2}{2} + gz_i)$$
(4.12)

where

$\dot{Q}_{cv}$ = Rate of heat transfer into the control volume.

$\dot{W}_{cv}$ = Work that is associated with the displacement of the control surface and that crosses the surface.

$\Sigma m_i (h_i + \frac{v_i^2}{2} + gz_i)$ = Rate of energy flowing as a result of mass transfer

$\Sigma m_e (h_e + \frac{v_e^2}{2} + gz_e)$ = Rate of energy flowing out as a result of mass transfer

$\frac{dE_{cv}}{dt}$ = Rate of change of energy inside the CV.

4.11 THE STEADY-STATE STEADY-FLOW PROCESS (SSSF)

Let us consider the following assumptions:

I. The control volume does not move relative to the coordinate frame.

II. The state of the mass at each point in the CV does not vary with time.

III. The mass flux does not vary with time.

IV. The rates at which heat and work cross the control surface remain constant.

These assumptions lead to a reasonable model which is called the steady-state, steady-flow process (SSSF).

D-24

For this type of process we can write:

Continuity equation $\quad \dot{m}_i = \dot{m}_e = \dot{m}$ \hfill (4.13)

First Law

$$\dot{Q}_{cv} - \dot{W}_{cv} = \dot{m}(h_e + \frac{v_e^2}{2} + gz_e) - \dot{m}(h_i + \frac{v_i^2}{2} + gz_i) \quad (4.14)$$

4.12 THE JOULE-THOMSON COEFFICIENT AND THE TROTTLING PROCESS

a) The Joule-Thomas coefficient is defined as

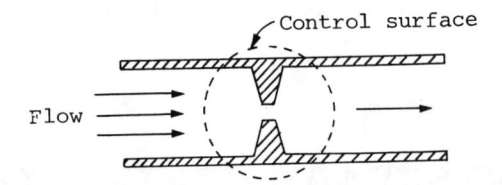

$$\mu_J = \left(\frac{\delta T}{\delta P}\right)_h \quad (4.15)$$

Fig. 4.2 The throttling process

Consider the throttling process in (Fig. 4.2), which is a SSSF process across a restriction, resulting in a drop in pressure. For such a process the μ_J is significant. A positive μ_J means that the temperature drops during throttling, and when μ_J is negative the temperature rises, during throttling.

b) The isothermal Joule-Thomson coefficient is defined as:

$$\mu_J = \left(\frac{\delta h}{\delta P}\right)_t \quad (4.16)$$

4.13 THE UNIFORM-STATE UNIFORM FLOW PROCESS (USUF)

Let us consider the following assumptions:

I. The cv remains constant relative to the coordinate frame.
II. The state of mass may change with time in the cv, but at any instant of time the state is uniform throughout the entire CV.
III. The state of the mass crossing all the areas of flow is constant in respect to the control surface but the mass flow rates may vary with time.

These assumptions lead to a model which is called a uniform-state, uniform-flow process (USUF).

For this type of process we can write:

Continuity equation (For a period of time t)

$$(m_2 - m_1)_{CV} + \Sigma m_e - \Sigma m_i = 0 \qquad (4.17)$$

First Law

$$Q_{cv} - W_{cv} = \Sigma m_e(h_e + \frac{v_e^2}{2} + gz_e) - \Sigma m_i(h_i + \frac{v_i^2}{2} + gz_i)$$
$$+ \left[m_2(u_2 + \frac{v_2^2}{2} + gz_2) - m_1(u_1 + \frac{v_1^2}{2} + gz_1) \right]_{cv}$$

$$(4.18)$$

CHAPTER 5

ENTROPY AND THE SECOND LAW OF THERMODYNAMICS

5.1 THE HEAT ENGINE

A heat engine is a system which operates in a cycle while only heat and work cross its boundaries.

A steam power plant is a heat engine which receives heat from a high-temperature system at the boiler, rejects heat to a lower-temperature system at the condenser, and delivers useful work.

5.2 HEAT ENGINE EFFICIENCY

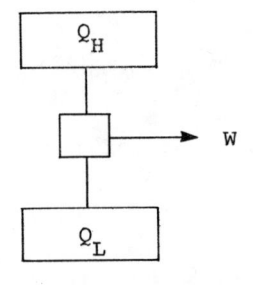

Fig. 5.1

a) We define the efficiency of a work producing heat engine as the ratio of the net work delivered to the environment, to the heat received from the source:

$$\boxed{\eta = \frac{W}{Q_H} = \frac{Q_H - Q_L}{Q_H} = 1 - \frac{Q_L}{Q_H}}$$

(5.1)

where Q_H = Amount of heat added to an engine

Q_L = Amount of heat rejected by an engine

W = Net amount of work produced by an engine

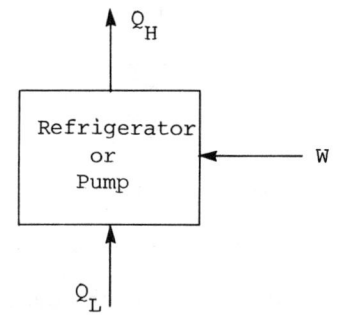

Fig. 5.2

b) Refrigerators and heat pumps are simply heat engines operating in reverse. The concept of coefficient of performance is used in these cases. The coefficient of performance of a refrigerator is given by:

$$\beta_R = \frac{Q_L}{W} = \frac{Q_L}{Q_H - Q_L} = \frac{1}{\frac{Q_H}{Q_L} - 1}$$

(5.2)

and the coefficient of performance of a heat pump is given by:

$$\beta_{HP} = \frac{Q_H}{W} = \frac{Q_H}{Q_H - Q_L} = \frac{1}{1 - \frac{Q_L}{Q_H}}$$

(5.3)

where Q_L = Amount of heat transferred from a low-temperature reservoir

Q_H = Amount of heat transferred to a warmer body

W = Net amount of work required

The expressions for the thermal efficiencies of heat engines, if they are reversible, are given by the following relations:

$$\eta_{th} = 1 - \frac{T_L}{T_H} \qquad (5.4)$$

$$\beta_R = \frac{1}{\frac{T_H}{T_L} - 1} \qquad (5.5)$$

$$\beta_{HP} = \frac{1}{1 - \frac{T_L}{T_H}} \qquad (5.6)$$

where T_L = Low temperature heat reservoir

T_H = High temperature heat reservoir

T_H, T_L = Absolute temperatures

Notes

a) The efficiency of a work-producing heat engine operating between two systems is always less than unity, or

$$\eta_{th} < 1.$$

b) The efficiency of any irreversible heat engine is less than that of any reversible engine, or

$$\eta_I \leq \eta_R.$$

c) For any reversible heat engine the net work and net heat in a cycle are zero:

$$\oint_{rev} dQ = \oint_{rev} dW = 0 \qquad (5.7)$$

D-29

5.3 THE SECOND LAW

There are two classical statements of the second law:

5.3.1 KELVIN-PLANK STATEMENT

It is impossible to construct a device that will operate in a cycle and produce no effect other than the raising of a weight and the exchange of heat with a single reservoir.

5.3.2 CLAUSIUS STATEMENT

It is impossible to construct a device that operates in a cycle and produces no effect other than the transfer of heat from a cooler body to a hotter body.

5.4 PERPETUAL MOTION MACHINES

5.4.1 PERPETUAL MOTION MACHINE OF THE FIRST KIND (PMM1)

A PMM1 is defined as any system which undergoes a cycle and has no external effect except the raising of a weight.

5.4.2 PERPETUAL MOTION MACHINE OF THE SECOND KIND (PMM2)

A PMM2 is defined as any cyclic device which has heat interactions with a single system, and delivers work.

Notes:

a) A corollary of the first law is that a PMM1 is impossible.

b) A corollary of the second law is that a PMM2 is impossible.

c) The efficiencies of PMM1 and PMM2 are equal to unity.

5.5 THE CARNOT CYCLE

By combining several reversible processes for a system, we may construct a reversible heat engine. The Carnot engine is an example of such an engine. When a large heat reservoir is brought into heat communication with a cylinder containing a certain amount of air confined by a piston, the following processes take place:

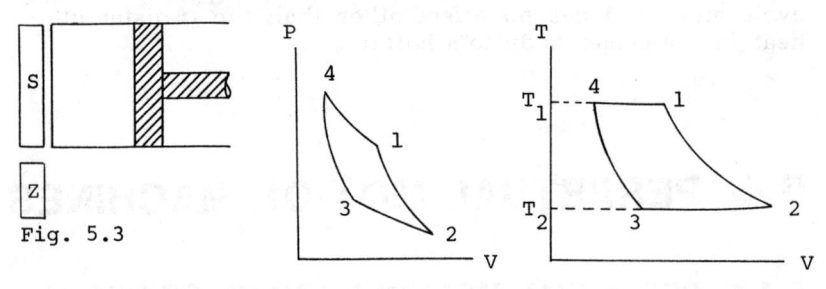

Fig. 5.3

Process 1-2

A reversible adiabatic process in which the temperature of the working fluid decreases from the high temperature to the low temperature

Process 2-3

A reversible isothermal process in which heat is transferred to or from the low-temperature reservoir

Process 3-4

A reversible adiabatic process in which the temperature of the working fluid increases from the low temperature to the high temperature

Process 4-1

A reversible isothermal process in which heat is transferred to or from the high-temperature reservoir

5.6 THERMODYNAMIC TEMPERATURE SCALES

The second law permits the definition of a temperature scale. Such a scale can be defined as follows:

Let us consider a system A (Fig. 5.4) in a state of equilibrium, and a reservoir, R, at constant temperature (for example, the melting point of ice). Operating a heat engine between A and R, the ratio of the heat quantities (Q_A/Q_R) that the engine exchanges with A and R can be evaluated. For a fixed reservoir R at constant temperature T_2, the quantity Q_A/Q_R depends only on the temperature T of system A. Thus

$$\frac{Q_A}{Q_R} = f(T).$$

This can also be expressed in the form,

$$T = F_R \left(\frac{Q_A}{Q_R} \right) \qquad (5.8)$$

The function F_R defines a temperature scale which is independent of the nature of any particular thermometric substance and can be determined experimentally.

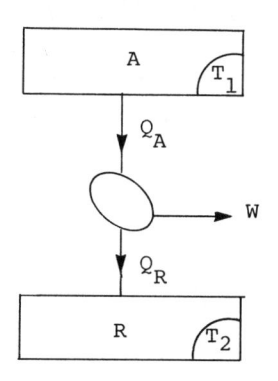

Fig. 5.4

a) The Kelvin scale defined by the relation:

$$T = F_R\left[\frac{Q_A}{Q_R}\right] = -T_R\frac{Q_A}{Q_R} \quad (5.9)$$

where T_R denotes the temperature in degrees kelvin of a heat reservoir at the triple state of water, $T_R = 273.16\ °K$.

b) Other thermodynamic temperature scales frequently used are the Rankine, Fahrenheit, and Celsius scales. The relations among them are:

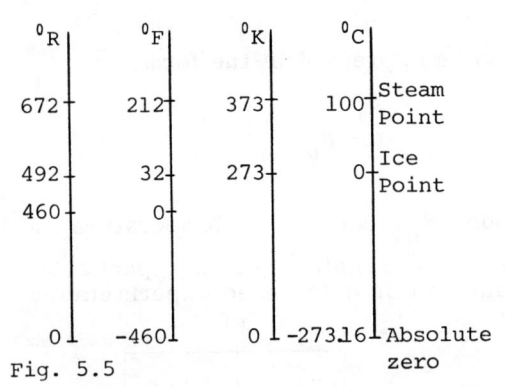

$$T(°K) = t(°C) + 273.15 \quad (5.10)$$

$$t(°F) = T(°R) - 459.67 \quad (5.11)$$

Fig. 5.5

Temperature scales compared at certain identifiable levels of temperature

5.7 ENTROPY

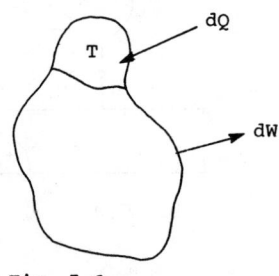

Fig. 5.6

Let us consider a system executing a cyclic reversible process (Fig. 5.6). For this process we define the quantity:

$$dS = \left(\frac{dQ}{T}\right)_{rev} \quad (5.12)$$

where dQ = Heat supplied to the system
T = Absolute temperature of the system

The quantity dS represents the change in the value of the property of the system, and it is called the entropy.

Notes

a) The equation (5.12) is valid for any reversible process.

b) Entropy is an extensive property, and it is a function of the end points only (a point function).

c) The change in the entropy of a system may be found by integrating (5.12)

$$S_2 - S_1 = \int_1^2 \frac{dQ}{T} \quad (5.13)$$

5.8 INEQUALITY OF CLAUSIUS

For any irreversible cycle the integral $\oint \frac{dQ}{T}$ is always less than zero,

$$\oint \frac{dQ}{T} < 0 \quad (5.14)$$

5.9 PRINCIPLE OF THE INCREASE OF ENTROPY

For any isolated system,

$$dS_{isol} \geq 0 \quad (5.15)$$

a) Principle of the increase of Entropy for a Control Value

$$\frac{dS_{cv}}{dt} + \frac{dS_{surr}}{dt} \geq 0 \qquad (5.16)$$

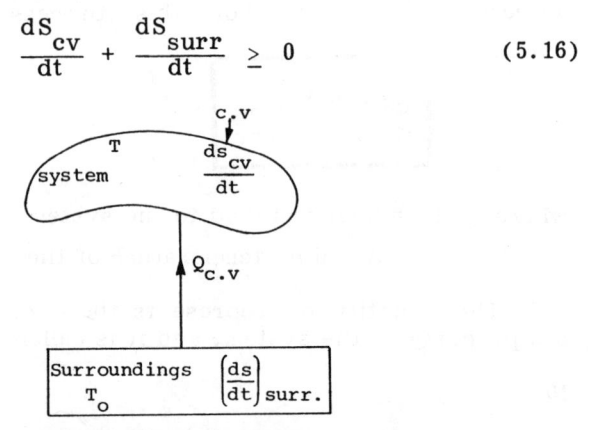

Fig. 5.7 Entropy change for a control volume plus surroundings.

5.10 ENTROPY CHANGE OF A SYSTEM DURING AN IRREVERSIBLE PROCESS

$$dS \geq \frac{dQ}{T} \quad \text{or} \quad S_2 - S_1 \geq \int \frac{dQ}{T} \qquad (5.17)$$

5.11 TWO THERMODYNAMIC RELATIONS

$$\boxed{\begin{array}{l} TdS = dU + PdV \\ TdS = dH - Vdp \end{array}}$$

(5.18)
(5.19)

Equations (5.18) and (5.19) hold for any process, reversible or irreversible, connecting equilibrium states of a simple system.

5.12 THE SECOND LAW OF THERMO-DYNAMICS FOR A CONTROL VOLUME

$$\frac{dS_{cv}}{dt} + \Sigma m_e s_e - \Sigma m_i s_i = \int_A \left[\frac{\dot{Q}_{cv}/A}{T}\right] dA + \int_V \left[\frac{L\dot{W}_{cv}/V}{T}\right] dV \quad (5.20)$$

This expression states that the rate of change of entropy inside the control volume plus the net rate of entropy flow out is equal to the sum of two terms, the integrated heat transfer term and the positive, internal reversibility term.

5.12.1 THE STEADY-STATE STEADY-FLOW PROCESS

For the SSSF, $\frac{dS_{cv}}{dt} = 0$.

$$\Sigma m_e s_e - \Sigma m_i s_i - \int_A \left[\frac{\dot{Q}_{cv}/A}{T}\right] dA \quad (5.21)$$

For an adiabatic process, $s_e \geq s_i$.

5.12.2 THE UNIFORM-STATE UNIFORM-FLOW PROCESS

$$\left[m_2 s_2 - m_1 s_1\right]_{cv} + \Sigma m_e s_e - \Sigma m_i s_i = \int_0^t \left[\frac{\dot{Q}_{cv} + L\dot{W}_{cv}}{T}\right] dt \quad (5.22)$$

5.12.3 THE REVERSIBLE STEADY-STATE, STEADY FLOW PROCESS

1. For a reversible adiabatic process,

$$W = -\int_i^e VdP + \frac{(v_i^2 - v_e^2)}{2} + g(z_i - z_e) \quad (5.23)$$

2. If we consider a reversible steady-state, steady-flow process in which the work is zero and the fluid is incompressible, equation (5.23) can be integrated to give,

$$V(P_e - P_i) + \frac{(v_e^2 - v_i^2)}{2} + g(z_e - z_i) = 0 \quad (5.24)$$

This is known as Bernoulli's equation.

3. For a reversible and isothermal process,

$$T(s_e - s_i) = \frac{\dot{Q}_{cv}}{\dot{m}} = q \quad (5.25)$$

CHAPTER 6

AVAILABILITY FUNCTIONS

6.1 REVERSIBLE WORK

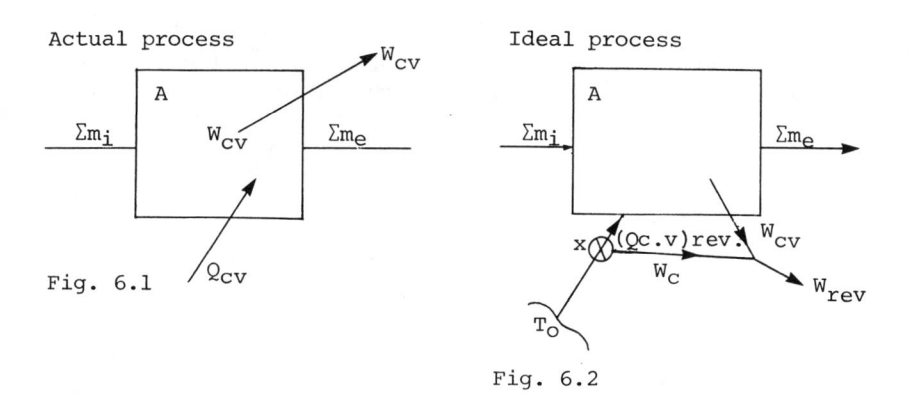

Actual process

Fig. 6.1

Ideal process

Fig. 6.2

In Fig. 6.1, we consider a system which shows a control volume undergoing a USUF process. There are irreversibilities present as in every actual process. In Fig. 6.2, we imagine an ideal process where all quantities and states are the same as for the actual process. All processes in this case are completely reversible.

The reversible work is determined by the relation:

$$W_{rev} = (W_{cv})_{rev} + W_c \qquad (6.1)$$

D-38

where W_{rev} ≡ Reversible work

$(W_{cv})_{rev}$ ≡ Work crossing the control surface for the reversible case.

W_c ≡ Work output of the reversible heat engine X

Q_{cv} ≡ Heat transfer with surroundings at temperature T_0

The difference between the reversible work and the work done in the first case (Fig. 6.1) is called the Irreversibility and is defined,

$$\boxed{I = W_{rev} - W_{cv}} \qquad (6.2)$$

a) The reversible work of a control volume that exchanges heat with the surroundings at temperature T_0 is:

$$W_{rev} = \Sigma m_i(h_i - T_0 s_i + \frac{v_i^2}{2} + gz_i) - \Sigma m_e(h_e - T_0 s_e + \frac{v_e^2}{2} + gz_e)$$
$$- \left[m_2(u_2 - T_0 s_2 + \frac{v_2^2}{2} + gz_2) - m_1(u_1 - T_0 s_1 + \frac{v_1^2}{2} + gz_1) \right]_{cv}$$
$$(6.3)$$

where:

$\Sigma m_i(h_i - T_0 s_i + \frac{v_i^2}{2} + gz_i)$ = Rate of energy flowing in as a result of mass transfer

$\Sigma m_e(h_e - T_0 s_e + \frac{v_e^2}{2} + gz_e)$ = Rate of energy flowing out as a result of mass transfer

$m_2(u_2 - T_0 s_2 + \frac{v_2^2}{2} + gz_2)$ = State within the control volume at state 2

$m_1(u_1 - T_0 s_1 + \frac{v_1^2}{2} + gz_1)$ = State within the control volume at state 1

1. For a system with fixed mass, the reversible work is:

$$\left(\frac{W_{rev}}{m}\right)_{1}^{2} = {}_1W_{rev\,2} = \left[(u_1 - T_0 s_1 + \frac{v_1^2}{2} + gz_1) - (u_2 - T_0 s_2 + \frac{v_2^2}{2} + gz_2)\right] \quad (6.4)$$

2. For a steady-state, steady-flow process the reversible work is:

$$W_{rev} = \Sigma m_i (h_i - T_0 s_i + \frac{v_i^2}{2} + gz_i) - \Sigma m_e (h_e - T_0 s_e + \frac{v_e^2}{2} + gz_e) \quad (6.5)$$

b) The irreversibility of a control volume that exchanges heat with the surroundings at temperature T_0 is:

$$I = \Sigma m_e T_0 s_e - \Sigma m_i T_0 s_i + m_2 T_0 s_1 - T_0 s_1 - Q_{cv} \quad (6.6)$$

1. For a system with fixed mass, we have from above:

$$_1I_2 = mT_0(s_2 - s_1) - {}_1Q_2 \quad (6.7)$$

2. For the steady-state, steady-flow process ($m_2 T_0 s_2 = m_1 T_0 s_1$):

$$I = \Sigma m_e T_0 s_e - \Sigma m_i T_0 s_i - Q_{cv} \quad (6.8)$$

6.2 AVAILABILITY

The maximum reversible work that can be done by a system in a given state is called availability, and is defined (per unit mass in the absence of kinetic and potential energy):

$$\phi = (W_{rev})_{max} - W_{surr} \quad (6.9)$$

where

$$(W_{rev})_{max} = (u - T_0 s) - (u_0 - T_0 s_0)$$

$$W_{surr} = P_0(V_0 - V) = -mP_0(v - v_0)$$

D-40

$$\therefore \phi = (u-u_0) + P_0(v-v_0) - T_0(s-s_0) \qquad (6.10)$$

where u = Internal energy of the system

v = Specific volume of the system

s = Entropy of the system

u_0, v_0, s_0 = Internal energy, specific volume, and entropy of the surroundings

CHAPTER 7

GASES

7.1 GAS CONSTANT

It has been found experimentally that for all homogeneous simple systems the ratio $\frac{PV}{T}$ approaches a finite limit as the pressure approaches zero. This limit is known as the gas constant R:

$$R = \lim_{P \to 0} \left(\frac{PV}{T} \right) \tag{7.1}$$

7.1.1 UNIVERSAL GAS CONSTANT

The universal gas constant $\bar{R}$ is defined by the relation:

$$\bar{R} = 32\, R_{O_2} \tag{7.2}$$

where R_{O_2} = gas constant of atmospheric oxygen.

The value of the universal gas constant, as determined experimentally, is as follows:

$\bar{R}$ = 1.9859 cal/°K g-mole, Btu/°R lb-mole

= 82.06 cm³ atm/°K g-mole

= 83.15 cm³ bar/°K g-mole

= 0.8478 kg m/°K g-mole

= 1545.3 ft lbf/°R lb-mole.

The gas constant R is related to the universal gas constant $\bar{R}$ through its molecular weight M in the following manner:

$$R = \frac{\bar{R}}{M} \qquad (7.3)$$

7.2 BOLZMANN'S CONSTANT

The universal gas constant per molecule, usually denoted by K, is called Boltzmann's constant:

$$K = \frac{\bar{R}}{N_0} \qquad (7.4)$$

$N_0 = 6.0232 \times 10^{23}$ (Avogadro's number)

7.3 COMPRESSIBILITY

The ratio PV/RT for any simple system is called the compressibility and is denoted z:

$$z = \frac{PV}{RT} \qquad (7.5)$$

where   P = Pressure
         V = Volume

D-43

R = Universal gas constant

T = Temperature

7.4 SEMIPERFECT GAS

A semiperfect gas is any homogeneous system obeying exactly the following relation for all ranges of pressure and temperature:

$$PV = RT \qquad (7.6)$$

7.4.1 PROPERTIES OF A SEMIPERFECT GAS

1. P-V-T Relation

$$PV = RT \qquad (7.7)$$

2. Internal Energy

$$\boxed{u = \int_{T_0}^{T} C_v dT} \qquad (7.8)$$

3. Enthalpy

$$h = RT_0 + \int_{T_0}^{T} C_p dT \qquad (7.9)$$

D-44

4. Entropy

$$s - s_0 = \int_{T_0}^{T} C_p \frac{dT}{T} - R \ln \frac{P}{P_0} \qquad (7.10)$$

where subscript ($_0$) refers to an arbitrarily selected state.

5. Specific heat of a semiperfect gas

$$C_v = \frac{1}{K-1} R \qquad C_p = \frac{K}{K-1} R$$

where $K = \dfrac{C_p}{C_v}$

7.5 PERFECT GAS (IDEAL GAS)

The perfect gas is a special case of a semiperfect gas obeying the equation,

$$\frac{PV}{RT} = 1 \qquad (7.11)$$

This equation may be written in the following alternative forms:

$$PV = n\bar{R}T \qquad (7.12)$$
$$PV = mRT \qquad (7.13)$$

where m is the mass, and n is the number of moles.

7.5.1 PROPERTIES OF A PERFECT GAS

1. P-V-T Relation

$$PV = RT \qquad (7.14)$$

2. Internal Energy

$$u = u(T)$$

$$u - u_0 = \int_{T_0}^{T} C_v \, dT = C_v(T-T_0) \qquad (7.15)$$

3. Enthalpy Change

$$h = h(T)$$

$$h - h_0 = \int_{T_0}^{T} C_p \, dT = C_p(T-T_0) \qquad (7.16)$$

4. Entropy Change

$$s - s_0 = \int_{T_0}^{T} \frac{C_v \, dT}{T} + R\ln\frac{v}{v_0} = C_v \ln\frac{T}{T_0} + R\ln\frac{v}{v_0} \qquad (7.17)$$

$$s - s_0 = \int_{T_0}^{T} \frac{C_p \, dT}{T} - R\ln\frac{p}{p_0} = C_p \ln\frac{P}{P_0} - R\ln\frac{P}{P_0} \qquad (7.18)$$

5. Specific Heats

$$K = K(T) \qquad C_v = C_v(T) \qquad C_p = C_p(T)$$

$$C_p - C_v = R$$

$$C_p = \frac{RK}{K-1}$$

$$C_v = \frac{R}{K-1} \qquad K = \frac{C_p}{C_v}$$

D-46

7.5.2 TWO DIAGRAMS FOR A PERFECT GAS

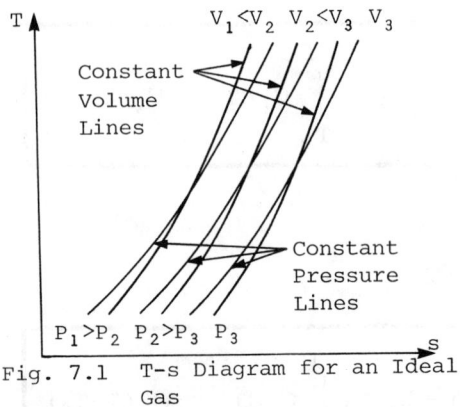

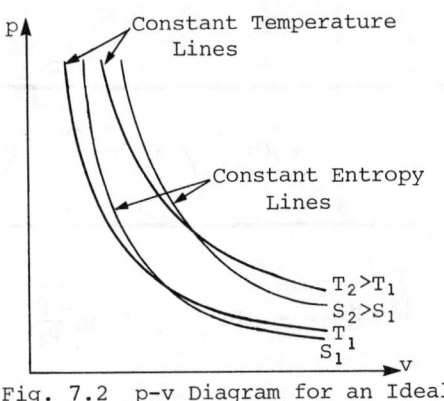

Fig. 7.1 T-s Diagram for an Ideal Gas

Fig. 7.2 p-v Diagram for an Ideal Gas

a) The reversible polytropic process for a perfect gas

A reversible polytropic process is one for which the pressure-volume relation is given by the relation,

$$PV^n = \text{constant} \qquad (7.19)$$

The polytropic processes for various values of n are shown on the P-r and T-s diagrams below:

D-47

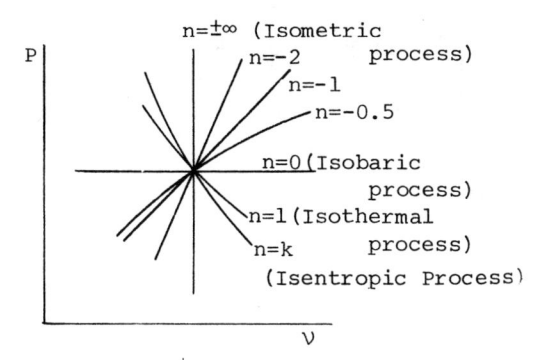

```
       n=±∞ (Isometric
P    n=-2      process)
       n=-1
       n=-0.5
       n=0 (Isobaric
              process)
       n=1 (Isothermal
       n=k     process)
       (Isentropic Process)
                                v
```

Fig. 7.3 Polytropic processes

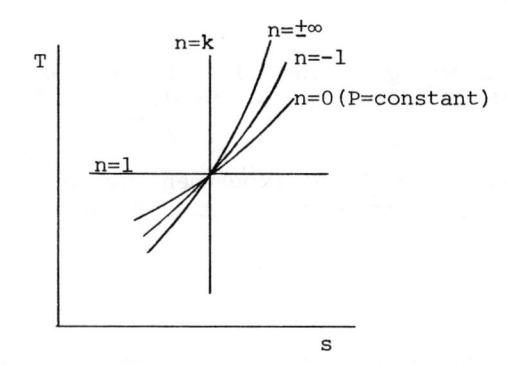

```
           n=±∞
      n=k  n=-1
T          n=0 (P=constant)
   n=1
                         s
```

1. For a polytropic process we have the following relations:

$$\frac{T_2}{T_1} = \left(\frac{P_2}{P_1}\right)^{(n-1)/n} = \left(\frac{V_1}{V_2}\right)^{n-1} \quad (7.20)$$

2. For a system consisting of a perfect gas, the work done at the moving boundary during a reversible polytropic process is:

$$_1W_2 = \frac{P_2V_2 - P_1V_1}{1-n} = \frac{mR(T_2-T_1)}{1-n} \quad (7.21)$$

3. Isentropic process for a perfect gas:

$$\frac{T_2}{T_1} = \left(\frac{P_2}{P_1}\right)^{(K-1)/K} = \left(\frac{V_1}{V_2}\right)^{K-1}$$

$$P_1 V_1^K = P_2 V_2^K \qquad (7.22)$$

where $K = C_p/C_v$

7.6 REAL GAS

For a real gas, the value of z may be greater or less than unity. Thus z gives the deviation from perfect gas behavior.

It has been found experimentally that for many gases the compressibility factor z may be approximated by the relation,

$$z = z(p_r, T_R) \qquad (7.23)$$

which is the basis of the generalized compressibility chart in which z versus p_R is given for various values of T_R. This is shown in (Fig. 7.4):

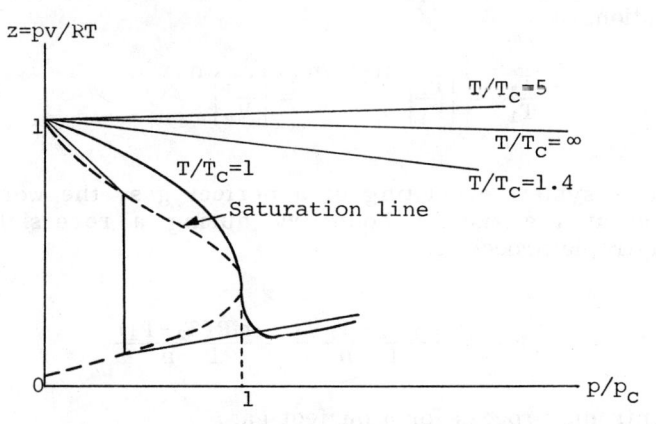

Fig. 7.4 Generalized compressibility chart

where $P_r = \dfrac{P}{P_c}$ = Reduced pressure,

$T_r = \dfrac{T}{T_c}$ = Reduced temperature

P_c = Critical pressure

T_c = Critical temperature

7.6.1 GENERALIZED ENTHALPY CHART FOR REAL GAS

To use the chart given in Fig. 7.5, all we need to know is the critical pressure p_c and the critical temperature T_c of the gas. Making use of this chart, we may give the change of enthalpy between two states as:

$$\bar{h}_2 - \bar{h}_1 = T_c\left[\left(\dfrac{\bar{h}^*-\bar{h}}{T_c}\right)_{T_1,P_1} - \left(\dfrac{\bar{h}^*-\bar{h}}{T_c}\right)_{T_2,P_2}\right]$$
$$+ (\bar{h}^*_{T_2,P_2} - \bar{h}^*_{T_1,P_1}) \qquad (7.24)$$

where

$\bar{h}^*_{T_2,P_2} - \bar{h}^*_{T_1,P_1} = \displaystyle\int_{T_1}^{T_2} \bar{C}_p^* dT$ is enthalpy change due to perfect gas behavior

$\bar{C}_p^*$ = Ideal gas constant-pressure specific heat as a function of temperature

$\bar{h}^*$ = Enthalpy of the gas at zero pressure and a given temperature

$\bar{h}$ = Enthalpy of the gas at the same temperature and any pressure p.

D-50

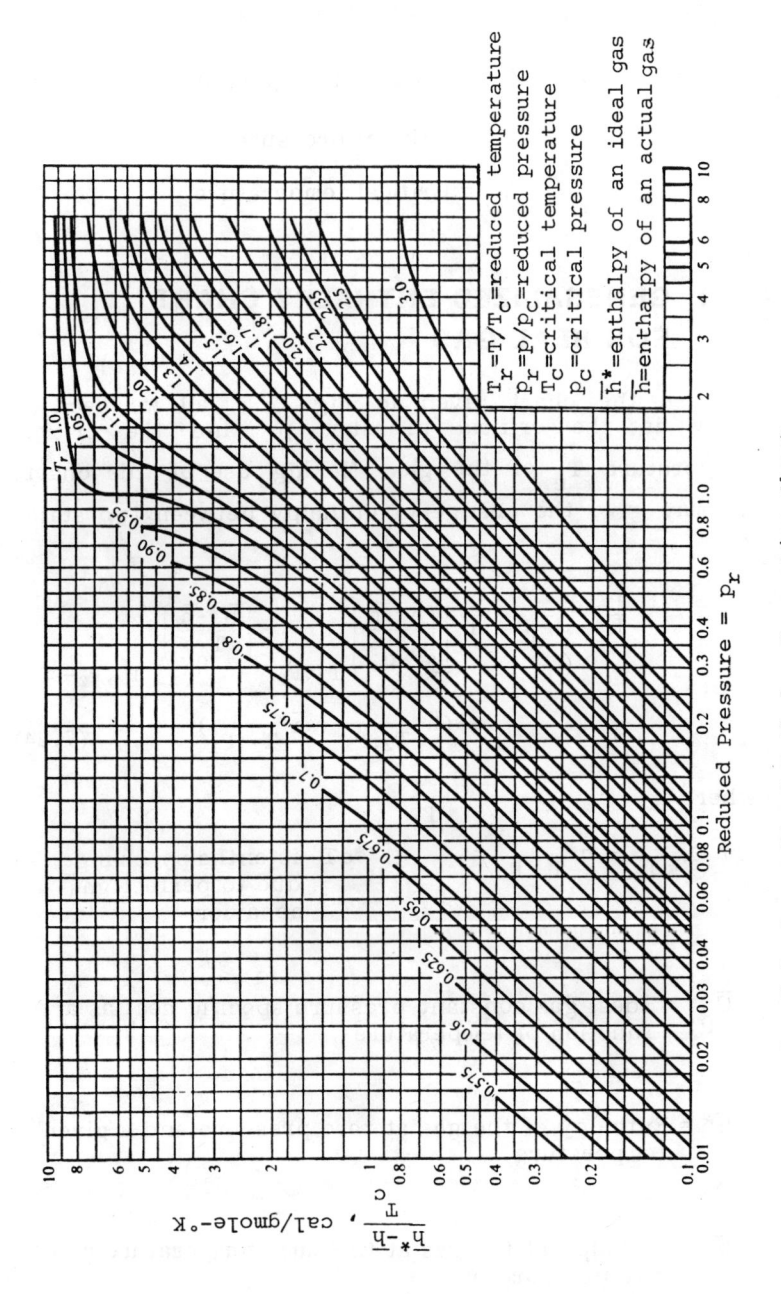

Fig. 7.5 Generalized Enthalpy Correction Chart

7.6.2 GENERALIZED ENTROPY CHART FOR REAL GASES

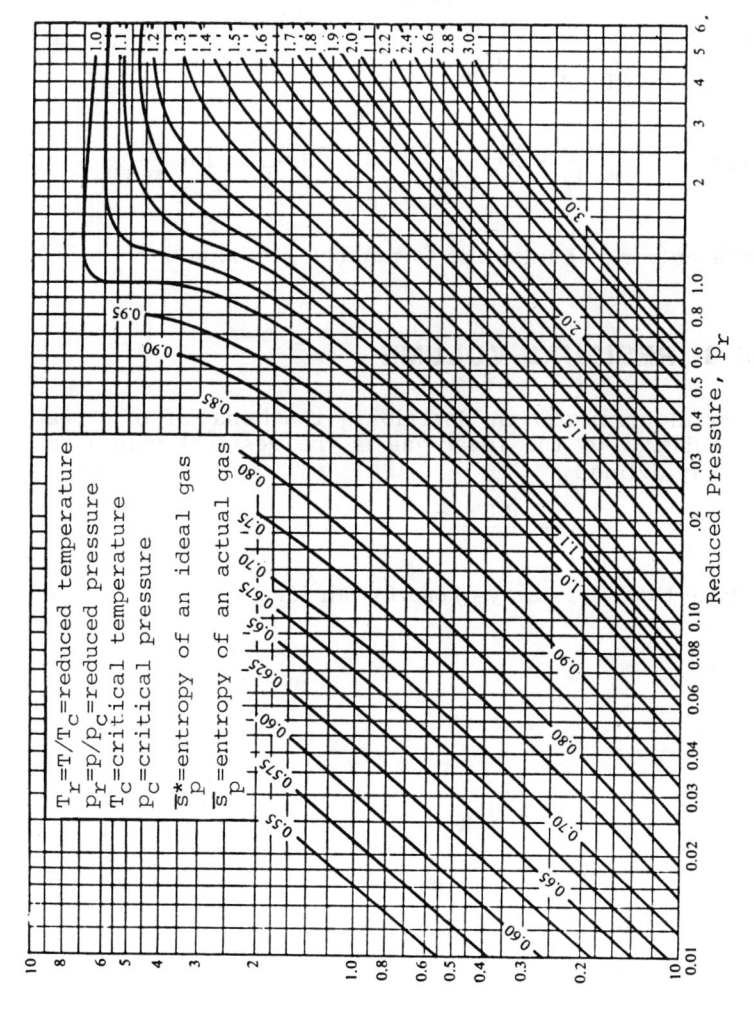

Fig. 7.6 Generalized Entropy Correction Chart

D-52

Using the generalized entropy correction chart (Fig. 7.6), we can give the change of entropy between two states as:

$$\overline{S}_2 - \overline{S}_1 = (\overline{S}_p^* - \overline{S}_p)_{T_1,P_1} - (\overline{S}_p^* - \overline{S}_p)_{T_2,P_2}$$
$$+ \overline{S}^*_{T_2,P_2} - \overline{S}^*_{T_1,P_1} \qquad (7.25)$$

where $\quad \overline{S}^*_{T_2,P_2} - \overline{S}^*_{T_1,P_1} = \int_{T_1}^{T_2} \overline{C}_p^* \frac{dT}{T} - \overline{R}\ln\frac{p_2}{p_1}$

is the entropy change due to perfect gas behavior.

7.6.3 RESIDUAL VOLUME

A useful parameter in describing the behavior of a real gas relative to the ideal gas is the residual volume α, which is defined,

$$\alpha = \frac{\overline{R}T}{P} - \overline{v} \qquad (7.26)$$

We note that α is always zero for an ideal gas.

CHAPTER 8

THERMODYNAMIC RELATIONS

8.1 THE MAXWELL RELATIONS

The Maxwell Relations are:

$$\left(\frac{\partial S}{\partial V}\right)_T = \left(\frac{\partial p}{\partial T}\right)_V \qquad (8.1)$$

$$\left(\frac{\partial T}{\partial V}\right)_S = -\left(\frac{\partial p}{\partial S}\right)_V \qquad (8.2)$$

$$\left(\frac{\partial S}{\partial p}\right)_T = -\left(\frac{\partial V}{\partial T}\right)_p \qquad (8.3)$$

$$\left(\frac{\partial T}{\partial p}\right)_S = \left(\frac{\partial V}{\partial S}\right)_p \qquad (8.4)$$

These relations are important because they relate the entropy to easily measured properties - pressure, volume and temperature.

8.2 CLAPEYRON EQUATION

The Clapeyron Equation is an important relation involving the saturation pressure and temperature, the change of enthalpy associated with a change of phase, and the specific volumes of the two phases.

The form of the Clapeyron equation is:

$$\frac{dp}{dT} = \frac{h_{fg}}{Tv_{fg}} \tag{8.5}$$

where h_{fg} is the enthalpy of vaporization.

8.3 OTHER THERMODYNAMIC RELATIONS

a) There are a number of other useful relations that can be easily derived:

$$\left(\frac{\partial U}{\partial S}\right)_V = T \tag{8.6}$$

$$\left(\frac{\partial U}{\partial V}\right)_S = -p \tag{8.7}$$

$$\left(\frac{\partial H}{\partial S}\right)_P = T \tag{8.8}$$

$$\left(\frac{\partial H}{\partial p}\right)_S = V \tag{8.9}$$

$$\left(\frac{\partial \psi}{\partial T}\right)_V = -S \tag{8.10}$$

$$\left(\frac{\partial \psi}{\partial V}\right)_T = -p \tag{8.11}$$

$$\left(\frac{\partial Z}{\partial T}\right)_p = -S \tag{8.12}$$

$$\left(\frac{\partial Z}{\partial p}\right)_T = V \tag{8.13}$$

D-55

b) Enthalpy

$$dh = C_p dT + \left[v - T \left(\frac{\partial v}{\partial T}\right)_p\right] dp \quad (8.14)$$

c) Internal Energy

$$du = C_v dT + \left[T \left(\frac{\partial p}{\partial T}\right)_v - p\right] dv \quad (8.15)$$

d) Entropy

$$ds = C_v \frac{dT}{T} + \left(\frac{\partial p}{\partial v}\right)_T dv \quad (8.16)$$

e) Specific heats

$$C_v = \left(\frac{\partial u}{\partial T}\right)_v = T \left(\frac{\partial S}{\partial T}\right)_v \quad (8.17)$$

$$C_p = \left(\frac{\partial h}{\partial T}\right)_p = T \left(\frac{\partial S}{\partial T}\right)_p \quad (8.18)$$

$$\left(\frac{\partial C_v}{\partial v}\right)_T = T \left(\frac{\partial^2 p}{\partial T^2}\right)_v \quad (8.19)$$

$$\left(\frac{\partial C_p}{\partial p}\right)_T = -T \left(\frac{\partial^2 v}{\partial T^2}\right)_p \quad (8.20)$$

8.4 EQUATIONS OF STATE

The P-V-T relation is often stated in the form of an equation which is called an equation of state:

$$PV = RT \quad (8.21)$$

a) Van der Waals Equation

$$p = \frac{RT}{v-b} - \frac{a}{v^2} \quad (8.22)$$

a, b are constants for any one substance.

D-56

b) Dieterici Equation

$$p = \frac{RT}{v-b} \exp\left(-\frac{a}{vRT}\right) \qquad (8.23)$$

c) Beattie-Bridgeman Equation

$$p = \frac{RT(1-\varepsilon)}{v^2}(v+B) - \frac{A}{v^2} \qquad (8.24)$$

$A = A_0(1-a/v)$, $B = B_0(1-B/v)$, $\varepsilon = c/UT^3$ and A_0, a, B_0, b and c are constants for different gases. Values for constants are given in table 1.

d) Benedict-Webb-Rubin Equation

$$p = \frac{RT}{v} + \left(B_0 RT - A_0 - \frac{C_0}{T^2}\right)\frac{1}{v^2} + (bRT-a)\frac{1}{v^3} + \frac{a\alpha}{v^6}$$
$$+ \frac{c\left(1+\frac{\gamma}{v^2}\right)}{T^2} \cdot \frac{1}{v^3} \exp\left(-\frac{\gamma}{v^2}\right) \qquad (8.25)$$

Values for the constants are given in table 2.

Constants of the Beattie-Bridgeman Equation of State

$$R = 0.08206 \frac{(atm)(liters)}{(g\text{-mole})(°K)}$$

Gas	A_0	a	B_0	b	$10^{-4}c$
Helium	0.0216	0.05984	0.01400	0.0	0.0040
Neon	0.2125	0.02196	0.02060	0.0	0.101
Argon	1.2907	0.02328	0.03931	0.0	5.99
Hydrogen	0.1975	-0.00506	0.02096	-0.04359	0.0504
Nitrogen	1.3445	0.02617	0.05046	-0.00691	4.20
Oxygen	1.4911	0.02562	0.04624	0.004208	4.80
Air	1.3012	0.01931	0.04611	-0.001101	4.34
CO_2	5.0065	0.07132	0.10476	0.07235	66.00
$(C_2H_5)_2O$	31.278	0.12426	0.45446	0.11954	33.33
C_2H_4	6.152	0.04964	0.12156	0.03597	22.68
Ammonia	2.3930	0.17031	0.03415	0.19112	476.87
CO	1.3445	0.02617	0.05046	-0.00691	4.20
N_2O	5.0065	0.07132	0.10476	0.07235	66.0
CH_4	2.2769	0.01855	0.05587	-0.01587	12.83
C_2H_6	5.8800	0.05861	0.09400	0.01915	90.00
C_3H_8	11.9200	0.07321	0.18100	0.04293	120
$n\text{-}C_4H_{10}$	17.794	0.12161	0.24620	0.09423	350
$n\text{-}C_7H_{16}$	54.520	0.20066	0.70816	0.19179	400

Table 1

Empirical Constants for Benedict-Webb-Rubin Equation

Units: Atmospheres, liters, moles, °K.   Gas Constants: $R = 0.08207$; $T = 273.13 + t(°C)$

Gas	A_0	B_0	$C_0 \times 10^{-6}$	a	b	$c \times 10^{-6}$	$\alpha \times 10^3$	$\gamma \times 10^2$
Nitrogen	1.19250	0.0458000	0.00588907	0.0149000	0.00198154	0.000548064	0.291545	0.750000
Methane	1.85500	0.0426000	0.0225700	0.494000	0.00338004	0.00254500	0.124359	0.60000
Ethylene	3.33958	0.0556833	0.131140	0.259000	0.0086000	0.021120	0.178000	0.923000
Ethane	4.15556	0.0627724	0.179592	0.345160	0.0111220	0.0327670	0.243389	1.18000
Propylene	6.11220	0.0850647	0.439182	0.774056	0.0187059	0.102611	0.455696	1.82900
Propane	6.87225	0.0973130	0.508256	0.947700	0.0225000	0.129000	0.607175	2.20000
i-Butane	10.23264	0.137544	0.849943	1.93763	0.0424352	0.286010	1.07408	3.40000
i-Butylene	8.95325	0.116025	0.927280	1.69270	0.0348156	0.274920	0.910889	2.95945
n-Butane	10.0847	0.124361	0.992830	1.88231	0.0399983	0.316400	1.10132	3.40000
i-Pentane	12.7959	0.160053	1.74632	3.75620	0.0668120	0.695000	1.70000	4.63000
n-Pentane	12.1794	0.156751	2.12121	4.07480	0.0668120	0.824170	1.81000	4.75000
n-Hexane	14.4373	0.177813	3.31935	7.11671	0.109131	1.51276	2.81086	6.66849
n-Heptane	17.5206	0.199005	4.74574	10.36475	0.151954	2.47000	4.35611	9.00000

Table 2

e) Serial Forms of the Equations of State

The P-V-T relation may be expressed in terms of an infinite series in powers of the density as follows:

$$z = \frac{Pv}{RT} = 1 + \frac{B(T)}{v} + \frac{C(T)}{v^2} + \frac{D(T)}{v^3} + \ldots \quad (8.26)$$

where $B(T)$, $C(T)$, $D(T)$ are called virial coefficients.

Any of the equations of state discussed above may be expanded into a series form similar to (8.26). For example, the Van der Waals equation yields simply:

$$B(T) = b - \frac{a}{RT} \quad (8.27)$$

$$B(T) = b^2$$

$$C(T) = b^3 \quad \text{etc.}$$

8.5 LAW OF CORRESPONDING STATES

The law of corresponding states declares that there is a single functional relationship of the form,

$$v_R = f(p_R, T_R),$$

which holds for all substances.

8.6 VOLUME EXPANSION AND ISOTHERMAL AND ADIABATIC COMPRESSIBILITY

a) Coefficient of linear expansion

$$\delta_P = \frac{1}{L}\left(\frac{\partial L}{\partial T}\right)_P \quad (8.28)$$

b) Volume

$$\alpha_P = \frac{1}{V}\left(\frac{\partial V}{\partial T}\right)_P = \frac{1}{v}\left(\frac{\partial V}{\partial T}\right)_P \tag{8.29}$$

c) Isothermal compressibility

$$\boxed{\beta_T = -\frac{1}{V}\left(\frac{\partial V}{\partial P}\right)_T = -\frac{1}{v}\left(\frac{\partial v}{\partial P}\right)_T} \tag{8.30}$$

d) Adiabatic compressibility

$$\boxed{\beta_S = -\frac{1}{v}\left(\frac{\partial v}{\partial P}\right)_S} \tag{8.31}$$

e) Adiabatic bulk modulus

$$\boxed{\beta_S = -v\left(\frac{\partial P}{\partial v}\right)_S} \tag{8.32}$$

CHAPTER 9

POWER AND REFRIGERATION CYCLES

9.1 VAPOR POWER CYCLES

9.1.1 SIMPLE RANKINE CYCLE

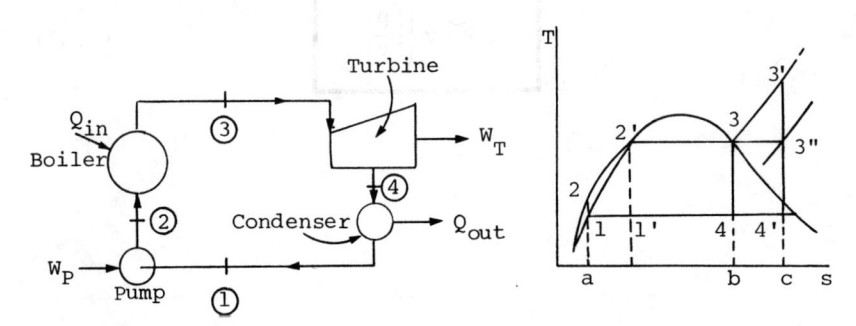

Fig. 9.1 Simple steam power plant which operates on the Rankine cycle.

A Rankine cycle consists of the following processes:

1-2 Reversible adiabatic pumping process in the pump

2-3 Constant-pressure transfer of heat in the boiler

3-4 Reversible adiabatic expansion in the turbine

4-1 Constant pressure transfer of heat in the condenser

D-61

Assuming steady-state, steady-flow processes throughout and neglecting changes in kinetic energy and potential energy across each piece of cycle component, we have, from the first law, the following results:

Boiler: $Q_{in} = h_3 - h_2$

Turbine: $W_T = h_3 - h_4$

Condenser: $Q_{out} = h_4 - h_1$

Pump: $W_p = h_2 - h_1 = v(P_2 - P_1)$

The thermal efficiency can be expressed in terms of properties at various points in the cycle:

$$\eta_{th} = \frac{W_{net}}{Q_{in}} = \frac{W_T - W_P}{Q_{in}} = \frac{(h_3 - h_2) - (h_4 - h_1)}{h_3 - h_2} = \frac{(h_3 - h_4) - (h_2 - h_1)}{h_3 - h_2}$$

(9.1)

The efficiency of the cycle can be increased by lowering the exhaust pressure, increasing the pressure during heat addition, or superheating the steam.

9.1.2 RANKINE CYCLE WITH SUPER HEATER

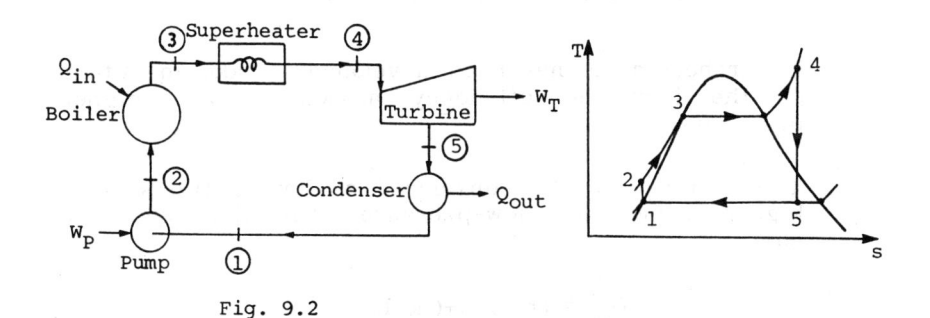

Fig. 9.2

We may operate a Rankine cycle with superheat as shown in Fig. 9.2. Using this method, we achieve a

D-62

higher mean temperature of heat addition without increasing the maximum pressure of the cycle.

For this cycle we have:

$$Q_{in} = h_4 - h_2$$
$$W_T = h_4 - h_5$$
$$Q_{out} = h_5 - h_1$$
$$W_P = h_2 - h_1 = v_1(P_2 - P_1) \quad \text{and}$$

$$\boxed{n_{th} = \frac{(h_4-h_5)-(h_2-h_1)}{h_4-h_2}} \quad (9.2)$$

9.1.3 THE REHEAT CYCLE

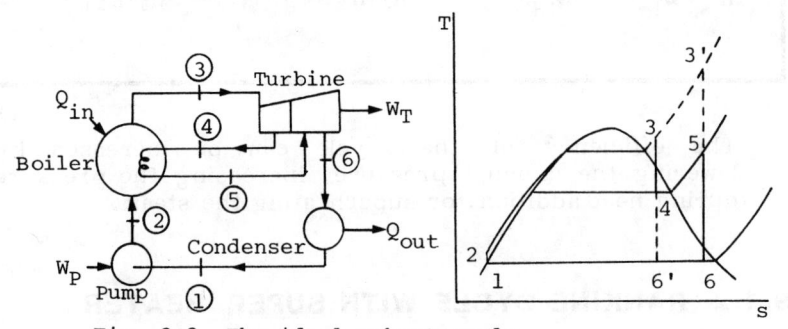

Fig. 9.3 The ideal reheat cycle

The reheat cycle has been developed to take advantage of the increased efficiency associated with higher pressures.

The turbine may be considered as having two stages: high-pressure and low-pressure. For this cycle we have:

$$Q_{in} = (h_3-h_2)+(h_5-h_4)$$
$$W_T = (h_3-h_4)+(h_5-h_6)$$

D-63

$$Q_{out} = (h_6-h_1)$$

$$W_P = (h_2-h_1) = v_1(P_2-P_1)$$

$$\boxed{n_{th} = \frac{(h_3-h_4)+(h_5-h_6)-(h_2-h_1)}{(h_3-h_2)+(h_5-h_4)}} \qquad (9.3)$$

9.1.4 THE REGENERATIVE CYCLE WITH OPEN FEEDWATER HEATER

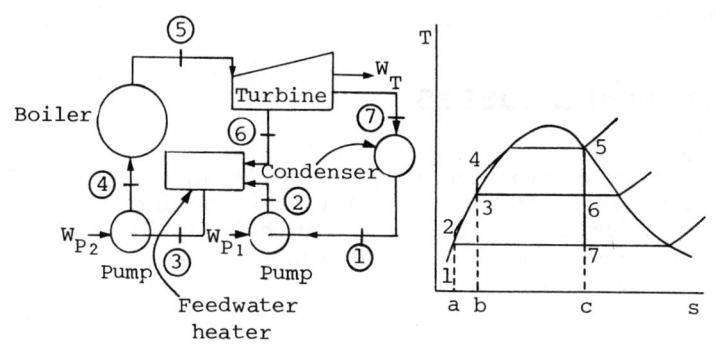

Fig. 9.4 Regenerative cycle with open feedwater heater.

This regenerative cycle involves the extraction of some of the vapor after it has partially expanded, and the use of feedwater heaters (Fig. 9.4). The number of stages of extraction is determined by economic considerations. For this cycle we have:

$$Q_{in} = h_5 - h_4$$

$$W_T = (h_5-h_6)+(1-m_1)(h_6-h_7)$$

$$Q_{out} = (h_1-h_7)$$

$$W_{P_2} = (h_4-h_3) = v(P_4-P_3)$$

$$W_{P_1} = (h_2-h_1) = v(P_2-P_1)$$

$$n_{th} = \frac{W_T - (1-m_1)W_{P_1} - W_{P_2}}{(h_5-h_4)} \qquad (9.4)$$

D-64

Around the feedwater heater we have:

$$m_1(h_6) + (1-m_1)h_2 = h_3$$

9.2 DEVIATION OF ACTUAL CYCLES FROM IDEAL CYCLES

The most important reasons for the deviation of actual cycles from ideal cycles are:

9.2.1 PIPING LOSSES

Observed pressure drops are due to frictional effects and heat transfer to the surroundings. The heat transfer causes a decrease in entropy. Also, both the pressure drops and heat transfer cause a decrease in the availability.

9.2.2 TURBINE LOSSES

The losses in the turbine are associated with the flow of the working fluid. The effects are the same as those outlined for piping losses.

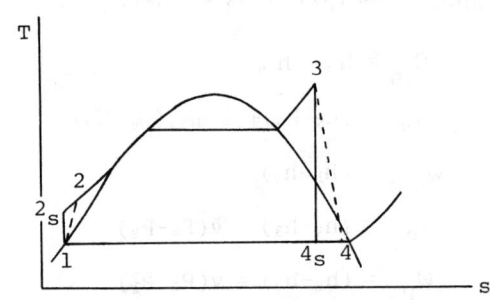

Fig. 9.5 Temperature-entropy diagram showing effect of turbine and pump inefficiencies on cycle performance.

The efficiency of the turbine has been defined,

$$\boxed{n_t = \frac{W_t}{h_3 - h_{4s}}} \qquad (9.5)$$

4 represents the actual state leaving the turbine and 4s represents the state after an isentropic expansion.

9.2.3 PUMP LOSSES

The losses in the pump are similar to those of the turbine. The pump efficiency is defined as

$$n_p = \frac{h_2 s - h}{W_p} \qquad (9.6)$$

9.2.4 CONDENSER

The losses in the condenser are relatively small. One of these is the cooling below saturation temperature of the liquid leaving the condenser.

9.3 VAPOR REFRIGERATION CYCLES

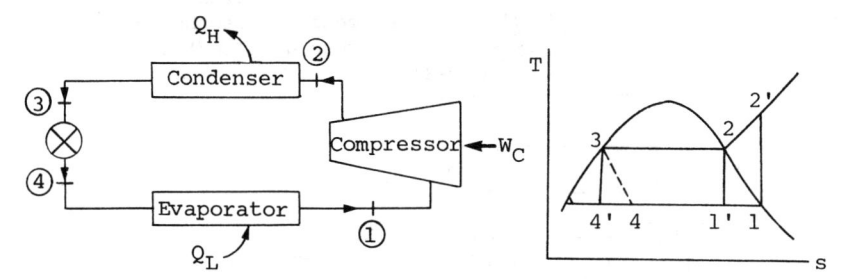

Fig. 9.6 The ideal vapor-compression refrigeration cycle.

The vapor refrigeration cycle is essentially the same as the Rankine cycle in reverse. The only difference is that the expansion valve replaces the pump. The performance of a regrigerator cycle is given as:

$$\boxed{\beta = \frac{Q_L}{W_c} = \frac{h_1 - h_4}{h_2 - h_1}}\qquad(9.7)$$

$$\text{capacity (tons)} = \frac{h_1 - h_4}{12,000 \text{ Btu/hr}}\qquad(9.8)$$

9.4 AIR-STANDARD POWER CYCLES

An Air-standard cycle operates under the following assumptions:

1. A fixed mass of air is the working fluid and the air is always an ideal gas.
2. The combustion process is replaced by a heat transfer process from an external source.
3. The cycle is completed by heat transfer to the surroundings.
4. All processes are internally reversible.
5. Air has a constant specific heat.

9.4.1 THE CARNOT CYCLE

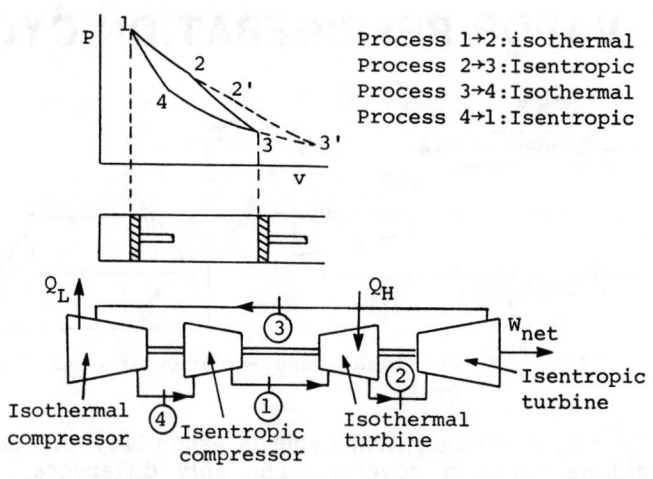

Process 1→2: isothermal
Process 2→3: Isentropic
Process 3→4: Isothermal
Process 4→1: Isentropic

Fig. 9.7 The air-standard Carnot cycle.

The thermal efficiency of the Carnot cycle is:

$$n_{th} = 1 - \frac{T_L}{T_H} = 1 - \frac{T_4}{T_1} = 1 - \frac{T_3}{T_2} \qquad (9.9)$$

The efficiency may also be expressed by the pressure ratio or compression ratio.

Isentropic pressure ratio: $r_{ps} = \frac{P_1}{P_4} = \frac{P_2}{P_3} = \left(\frac{T_3}{T_2}\right)^{K/(1-K)}$ (9.10)

Isentropic compression ratio:

$$r_{vs} = \frac{V_4}{V_1} = \frac{V_3}{V_2} = \left(\frac{T_3}{T_2}\right)^{1/(1-K)} \qquad (9.11)$$

$$\therefore \quad n_{th} = 1 - r_{ps}^{(1-K)/K} = 1 - r_{vs}^{1-K} \qquad (9.12)$$

9.4.2 THE OTTO CYCLE

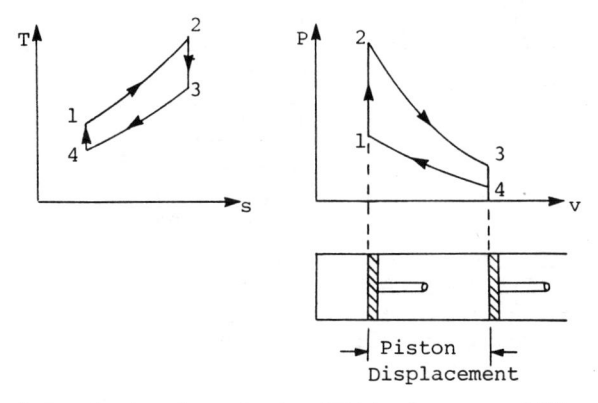

1. Constant-volume heat addition (process 1→2).
2. Isentropic expansion (process 2→3).
3. Constant-volume heat rejection (process 3→4).
4. Isentropic compression (process 4→1).

Fig. 9.8 Otto Cycle

On the basis of unit mass of gas, we have for the Otto cycle:

D-68

$$Q_{in} = Q_{12} = C_v(T_2-T_1)$$

$$Q_{out} = U_{34} = C_v(T_3-T_4)$$

$$n_{th} = \frac{W_{net}}{Q_{in}} = \frac{Q_{in}-Q_{out}}{Q_{in}}$$

$$\therefore \quad n_{th} = 1 - \frac{(T_3-T_4)}{(T_2-T_1)} = 1 - \frac{1}{r_v^{(K-1)}} \qquad (9.13)$$

where $r_v = v_3/v_2 = v_4/v_1$ is known as the compression rate of the cycle

Note: The Otto cycle efficiency increases with increased compression ratio.

9.4.3 THE DIESEL CYCLE

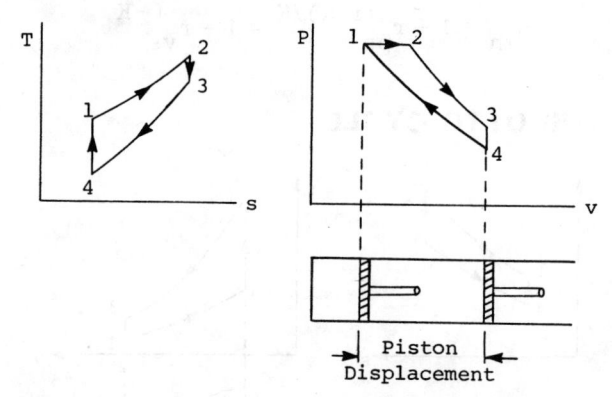

1. Constant-pressure heat addition (process 1→2).
2. Isentropic expansion (process 2→3).
3. Constant-volume heat rejection (process 3→4).
4. Isentropic compression (process 4→1).

Fig. 9.9 Diesel Cycle

On the basis of unit mass of gas, we have for the Diesel cycle:

$$Q_{in} = Q_{12} = C_p(T_2-T_1)$$

$$Q_{out} = Q_{34} = C_v(T_3-T_4)$$

$$n_{th} = \frac{Q_{in} - Q_{out}}{Q_{in}} = 1 - \frac{(T_3-T_4)}{K(T_2-T_1)}$$

∴

$$\boxed{n_{th} = 1 - \frac{1}{r_v^{K-1}} \cdot \frac{r_c^K - 1}{K(r_c - 1)}} \qquad (9.14)$$

$z_v = \frac{V_4}{V_1}$ is the compression ratio.

$z_c = \frac{V_2}{V_1}$ is the cutoff ratio.

Note: The efficiency of a Diesel cycle is always lower than that of an Otto cycle with the same compression ratio.

9.4.4 THE DUAL CYCLE

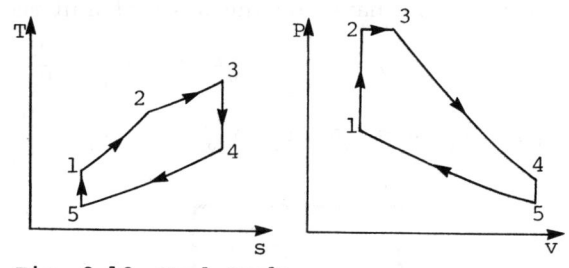

Fig. 9.10 Dual Cycle

The Dual cycle is a result of combining the processes of the Otto and Diesel cycles.

On the basis of unit mass, we have for the dual cycle:

$$Q_{in} = Q_{12} + Q_{23} = Cv(T_2-T_1) + Cp(T_3-T_2)$$

$$Q_{out} = Q_{45} = Cv(T_5-T_4)$$

$$n_{th} = 1 - \frac{T_4 - T_5}{(T_2-T_1)+K(T_3-T_2)}$$

∴

$$n_{th} = 1 - \frac{1}{r_v^{K-1}} \left[\frac{r_p r_c^K - 1}{Kr_p(r_c-1)+r_p-1} \right] \qquad (9.15)$$

D-70

When $r_p = 1$, equation 9.15 reduces to the Diesel efficiency equation:

$$r_v = \frac{V_5}{V_1}, \quad r_c = \frac{V_3}{V_2}, \quad r_p = \frac{P_2}{P_1}$$

9.4.5 STIRLING CYCLE

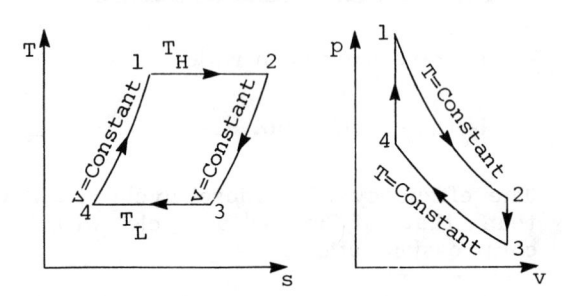

Fig. 9.11 Stirling Cycle

For a Stirling cycle, we have, on the basis of unit mass:

$$Q_{in} = Q_{12} = T_H(s_2 - s_1) = RT_H \ln \frac{v_2}{v_1} = -RT_H \ln \frac{P_2}{P_1}$$

$$Q_{out} = Q_{34} = T_L(s_3 - s_4) = RT_L \ln \frac{v_4}{v_3} = -RT_L \ln \frac{P_2}{P_1}$$

$$\boxed{n_{th} = \frac{T_H - T_L}{T_H}} \quad (9.16)$$

9.4.6 ERICSSON CYCLE

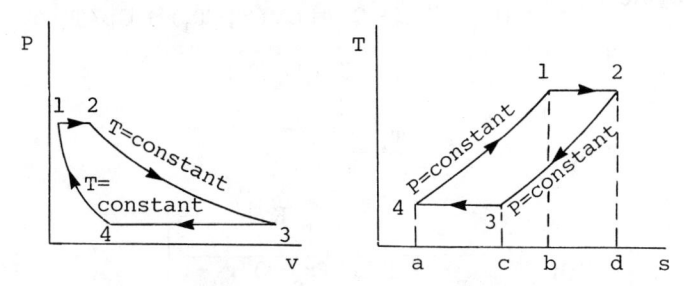

Fig. 9.12 The air-standard Ericsson cycle.

D-71

The constant-volume processes of the Stirling cycle are replaced by constant-pressure processes in the Ericsson cycle.

9.4.7 THE BRAYTON CYCLE

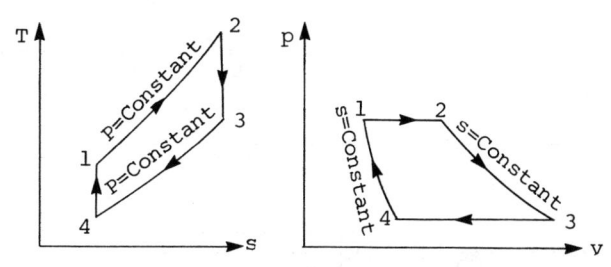

Fig. 9.13 Brayton Cycle

For the air-standard Brayton cycle, we have on the basis of unit mass:

$$Q_{in} = Q_{12} = Cp(T_2 - T_1)$$

$$Q_{out} = Q_{34} = Cp(T_4 - T_3)$$

$$\boxed{n_{th} = 1 - \frac{T_3 - T_4}{T_2 - T_1} = 1 - \frac{1}{r_p^{(K-1)/K}}} \quad (9.17)$$

$r_p = p_1/p_4 = p_2/p_3$ is known as the pressure ratio.

Two general types of gas-turbine power plants have been developed based on the Brayton cycle: closed-cycle and open-cycle.

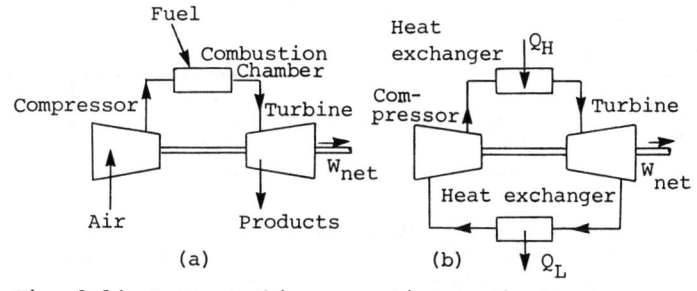

Fig. 9.14 A gas turbine operating on the Brayton cycle. (a) Open cycle. (b) Closed cycle.

9.4.8 BRAYTON CYCLE WITH REGENERATOR

Brayton Cycle with Regenerator

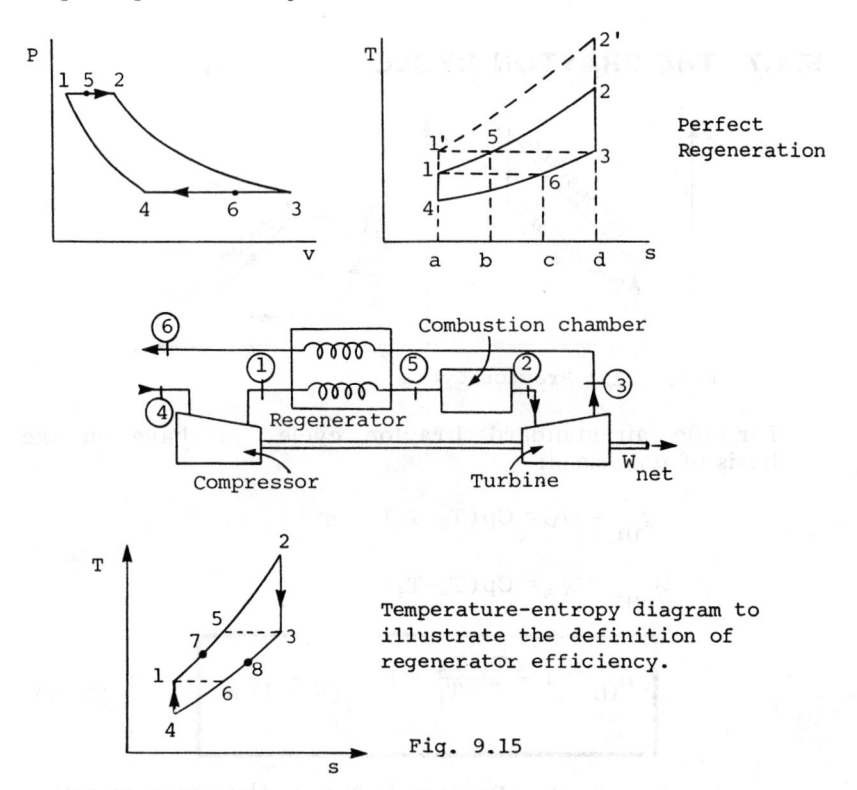

Perfect Regeneration

Temperature-entropy diagram to illustrate the definition of regenerator efficiency.

Fig. 9.15

For this cycle we have:

$$W_{net} = Cp(T_2-T_3) - Cp(T_2-T_4)$$

$$Q_{in} = Q_{52} = Cp(T_2-T_5) = Cp(T_2-T_3)$$

$$n_{th} = \frac{(T_2-T_3)-(T_1-T_4)}{(T_2-T_3)} = 1 - \frac{T_4}{T_2}\left(\frac{p_1}{p_4}\right)^{(K-1)/K} \qquad (9.18)$$

In this case, the heat absorbed by the air leaving the compressor is identical to the heat given up by the gas leaving the turbine.

D-73

The regenerator effectiveness is defined:

$$n_{reg} = \frac{T_7 - T_1}{T_3 - T_1} \tag{9.19}$$

9.4.9 GAS REFRIGERATION CYCLE

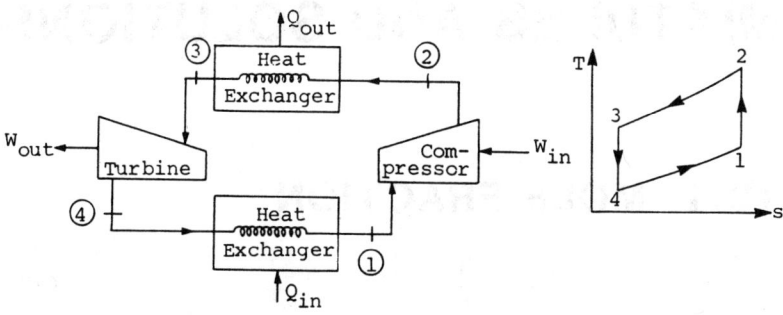

Fig. 9.16 Gas Refrigeration Cycle

1. Isentropic compression (process 1-2)
2. Cooling at constant pressure (process 2-3)
3. Isentropic expansion (process 3-4)
4. Refrigeration at constant pressure (process 4-1)

For this ideal cycle, assuming an ideal gas with constant specific heat:

$$W_{in} = h_2 - h_1 = Cp(T_2 - T_1)$$

$$W_{out} = h_3 - h_4 = Cp(T_3 - T_4)$$

$$Q_{in} = h_1 - h_4 = Cp(T_1 - T_4)$$

The coefficient of performance is given as:

$$\boxed{\beta_R = \frac{Q_{in}}{|W_{net}|} = \frac{1}{\left(\frac{T_2}{T_1}\right) - 1} = \frac{1}{\left(\frac{T_3}{T_4}\right) - 1} = \frac{1}{r_p^{(K-1)/K} - 1}} \tag{9.20}$$

where $r_p = P_2/P_1 = P_3/P_4$ is the pressure ratio.

D-74

CHAPTER 10

MIXTURES AND SOLUTIONS

10.1 MOLE FRACTION

$$x_i = \frac{n_i}{n} \qquad (10.1)$$

where $n = \sum_i n_i$ = Number of moles in the mixture

n_i = Number of moles of component i

10.2 MASS FRACTION

$$M_f = \frac{m_i}{m} \qquad (10.2)$$

where m_i = the mass of component i

$m = \sum_i m_i$ = Mass of mixture

10.3 DALTON'S RULE OF PARTIAL PRESSURE

The total pressure of a mixture of ideal gases is equal to the sum of the partial pressures when the partial

D-75

pressures are determined at the volume and temperature of the mixture, i.e.,

$$P = P_1 + P_2 + P_3 + \ldots + P_n = \sum_i p_i \qquad (10.3)$$

10.4 AMAGAT-LEDUC RULE OF PARTIAL VOLUME

The total volume of a mixture of ideal gases is equal to the sum of the partial volumes of the constituent gases when the partial volumes are determined at the pressure and temperature of the mixture.

$$V = \sum_i V_i \qquad (10.4)$$

V_i is the partial volume

V is the volume of the mixture

10.5 EXPRESSIONS FOR PERFECT GASES

The following expressions are valid for a mixture of perfect gases:

1. Entropy $\qquad S = \sum_i s_i = \sum_i n_i s_i \qquad (10.5)$

2. Internal energy $\qquad U = \sum_i m_i u_i, \quad u = \dfrac{\sum_i m_i u_i}{m} \qquad (10.6)$

3. Enthalpy $\qquad H = \sum_i m_i h_i, \quad h = \dfrac{\sum_i m_i h_i}{m} \qquad (10.7)$

4. Specific heats $\qquad C_v = \dfrac{\sum_i m_i C_{vi}}{m}, \quad C_p = \dfrac{\sum_i m_i C_{pi}}{m} \qquad (10.8)$

where
  m_i = Mass of component i
  m = Mass of the mixture

D-76

u_i = Internal energy of component i

u = Internal energy of the mixture

h_i = Enthalpy of component i

h = Enthalpy of the mixture

C_{pi} = Constant-pressure specific heat of component i

C_{vi} = Constant-volume specific heat of component i

P_i = Pressure of component i

T_i = Temperature of component i

10.6 MIXTURES INVOLVING GASES AND VAPOR

A gas-vapor mixture is an important type of gas mixture from which one or more of the constituent gases can be condensed out. The component that may condense out is called the vapor of the mixture.

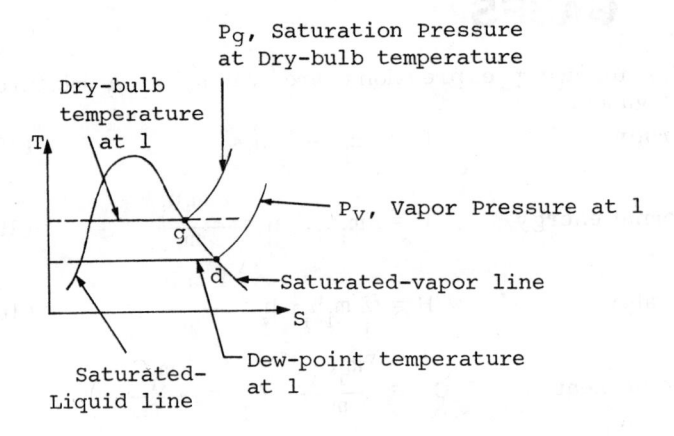

Fig. 10.1 Vapor Pressure, Saturation Pressure, Dew-Point Temperature, and Dry-Bulb Temperature

10.6.1 DEW POINT TEMPERATURE

The dew point of a gas vapor mixture is the temperature at which the vapor condenses or solidifies when it is cooled at constant pressure.

a) Saturated air is a mixture of dry air and saturated water vapor.

b) Unsaturated air is a mixture of dry air and superheated vapor.

c) Dry-bulb temperature is the equilibrium temperature of the mixture indicated by an ordinary thermometer.

d) Wet-bulb temperature is the temperature indicated by a wet-bulb thermometer which has been covered with a water-saturated cotton wick.

e) Specific humidity is defined as the ratio of the mass of water vapor to the mass of dry air in a given volume of mixture or

$$\omega = \frac{m_v}{m_a} \qquad (10.9)$$

Also:

$$\omega = 0.622 \frac{p_v}{p_a} = 0.622 \frac{p_v}{p-p_v} \qquad (10.10)$$

where

p_v = Partial pressure of vapor

p_a = Partial pressure of dry air in the same volume of mixture

p = Pressure of mixture

f) Relative humidity is defined as the ratio of the partial pressure of water vapor in a mixture to the saturation pressure of water at the dry-bulb temperature, or

$$\phi = \frac{p_v}{p_g} \qquad (10.11)$$

Also:

$$\phi = \frac{p_v}{p_g} = \frac{p_a \omega}{0.622 \, p_g} \qquad (10.12)$$

where p_v = Partial density of the water vapor

p_g = Density of water at the temperature of water

D-78

10.7 ENTHALPY AND ENTROPY OF A GAS-VAPOR MIXTURE

a) Enthalpy

$$h = h_a + \omega h_v \qquad (10.13)$$

b) Entropy

$$s = s_a + \omega s_v \qquad (10.14)$$

where

h_a, s_a = Specific enthalpy and entropy of the dry air in Btu/lbm of dry air

h_v, s_v = Specific enthalpy and entropy of water vapor

ω = Specific humidity

h, s = Specific enthalpy and entropy of the mixture

10.8 ADIABATIC SATURATION PROCESS

a) The adiabatic saturation process is an important process involving an air-water vapor mixture, in which an air-vapor mixture comes in contact with a body of water in a well-insulated duct (Fig. 10.2):

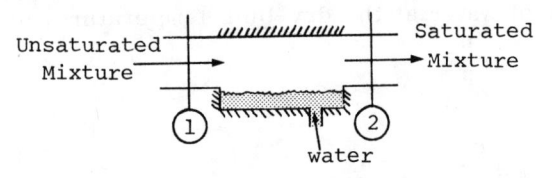

Fig. 10.2 The adiabatic saturation process

The relative humidity and the humidity ratio of the entering air-vapor mixture can be determined from the measurements of the pressure and temperature of the air-vapor mixture entering and leaving the adiabatic saturator. The adiabatic saturation process is one means of determining the humidity of an air-vapor mixture.

b) The humidity of an air-water vapor mixture is usually found from dry-bulb and wet-bulb data. These data are obtained by use of a phychrometer (Fig. 10.3):

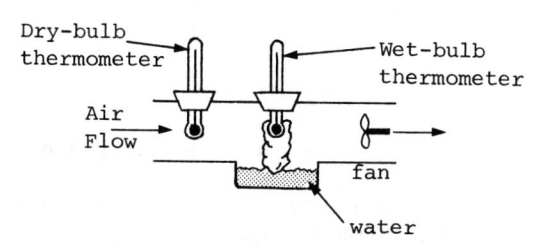

Fig. 10.3 Steady-flow phychrometer for measuring dry and wet temperatures.

c) Properties of air-water vapor mixtures are given in graphical form on phychrometric charts. The basic phychrometric chart consists of a plot of dry-bulb temperature (abscissa) and ratio (ordinate).

10.9 FUGACITY

The fugacity f_A of component A in a mixture is defined,

$$(d\bar{G}_A)_T = RTd(\ln \bar{f}_A)_T \qquad (10.15)$$

with the requirement that

$$\lim_{p \to 0}\left(\frac{\bar{f}_A}{Y_A P}\right) = 1$$

D-80

10.10 ACTIVITY

a) The activity a_A of a component A in a mixture at T,P is defined:

$$\boxed{a_A = \frac{\bar{f}_A}{f_A^0}} \qquad (10.16)$$

where f_A^0 = Fugacity of pure substance A at T,P⁰

P^0 = Standard state pressure (for gaseous mixtures P = 1 atm)

10.10.1 ACTIVITY COEFFICIENT

Another parameter commonly used in the description of mixtures is the activity coefficient γ, which for any component A, is defined in terms of its activity and mole fraction as

$$\boxed{\gamma_A = \frac{a_A}{Y_A}} \qquad (10.17)$$

CHAPTER 11

CHEMICAL REACTIONS

11.1 THE COMBUSTION PROCESS

Combustion is a process involving the reaction of a fuel and an oxidizer, in which the stored chemical energy in the fuel is released. A complete combustion reaction can be represented as:

$$C_xH_y + \underbrace{aO_2 + 3.76\, bN_2}_{\text{air}} \rightarrow cCO_2 + dH_2O + eO_2 + fN_2 \quad (11.1)$$

where x,y determine the type of fuel a,b,c,d,e,f are moles of each particular component.

a) Theoretical Air (TA)

The minimum amount of oxygen for the complete combustion of all the elements in the fuel.

b) Excess Air (EA)

The amount of air supplied over and above the theoretical air.

c) Air-Fuel Ratio (AF)

The ratio of the mass of theoretical air to the mass of the fuel.

d) The combustion efficiency η_{comb} is defined,

$$\boxed{\eta_{comb} = \frac{F.A.\ Ideal}{F.A.\ actual}} \quad (11.2)$$

where F.A. is the fuel-air ratio.

D-82

11.2 ENTHALPY OF REACTION

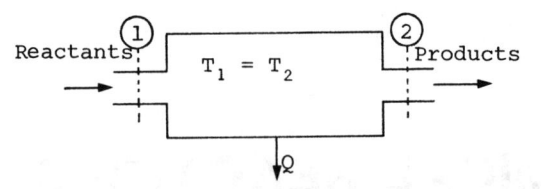

Fig. 11.1 Combustion Process

The concept of the enthalpy of reaction can be understood with the aid of Figure 11.1, where a chemical reaction takes place in a steady-flow process at constant pressure with no work transfer. Neglecting KE and PE and applying the first law of thermodynamics, we have

$$Q = H_2^P - H_1^R = \Delta H \qquad (11.3)$$

Q = Heat flow in
H_1^R = Enthalpy of the reactants at state 1
H_2^P = Enthalpy of the products at state 2

If the reactants and products are both at the same temperature, the quantity DH is called the enthalpy of reaction.

11.3 ENTHALPY OF FORMATION (h_f^o)

By definition, the standard enthalpy of formation of a compound is the enthalpy of reaction for the formation of the compound from its elements at stable state ($25\,^0C$ and 1 atm).

Notes
a) Exothermic reactions ($\Delta H < 0$) are those which liberate heat.
 Endothermic reactions ($\Delta H > 0$) are those which absorb heat.
b) The heating value of a fuel is numerically equal to its enthalpy of reaction but with opposite sign:

$$H_1^R - H_2^P = -\Delta H \qquad (11.4)$$

c) HHV = LHV + $m_{H_2O} h_{fg}$ $\qquad$ (11.5)

HHV = Higher heating value of fuel

LHV = Lower heating value of fuel

m_{H_2O} = Amount of H_2O formed

h_{fg} = Enthalpy of vaporization of H_2O

d) The enthalpy of formation of an element is zero.

e) The total molal enthalpy at any temperature and pressure $\bar{h}_{T,P}$ is:

$$\bar{h}_{T,P} = \bar{h}_f^0 + (\bar{h}_{T,P} - \bar{h}_{298(atm)}) \qquad (11.6)$$

where $\bar{h}_{T,P} - h_{298\cdot(atm)}$ = Difference in enthalpy between any given state and the enthalpy at the reference state of $298\,^0K$ and 1 atm

$\bar{h}_f^0$ = Enthalpy of formation of a substance

f) Standard enthalpy of reaction, $\Delta H^0 = H_2^0 - H_1^0$

g) The Gibbs function of formation, g_f^0, has been defined by a procedure similar to the enthalpy of formation.

11.4 FIRST LAW ANALYSIS OF REACTING SYSTEMS

Applying the first law to steady-state, steady-flow processes involving a chemical reaction and negligible changes in kinetic and potential energy we can write:

$$Q_{cv} - W_{cv} = \sum_P n_e(\bar{h}_f + \Delta\bar{h})_e - \sum_R n_i(\bar{h}_j + \Delta\bar{h})_i \qquad (11.7)$$

where

a) R,P refer to the reactants and products respectively

b) n_i = Number of reactants, n_e = Number of products

c) $\bar{h}_j$ = Molal enthalpy of formation

d) $\Delta\bar{h} = \bar{h}_f^0 - \bar{h}_{298}^0$ of the substance

Notes

1. $\Delta \bar{h}$ can be found directly from tables.

2. If the deviation from ideal gas behavior is significant but no tables are available, the value for $\Delta \bar{h}$ can be found from the generalized charts and the values for Cp_0 or $\Delta \bar{h}$ at 0.1 MPa pressure.

11.5 ADIABATIC FLAME TEMPERATURE

For any process that takes place adiabatically and with no work or changes in kinetic or potential energy, the temperature of the products is referred to as the adiabatic flame temperature or adiabatic combustion temperature.

11.6 THE THIRD LAW OF THERMODYNAMICS AND ABSOLUTE ENTROPY

The third law of thermodynamics states that the entropy of a pure crystalline substance is zero at the absolute zero of temperature ($0\,^0K$). The important result of this law is that it permits the determination of the absolute entropy for ideal-gas mixtures which is valid for most practical problems as follows:

$$\boxed{\bar{S}_{T,P} = \bar{S}_T^0 - \bar{R} \ln p} \qquad (11.8)$$

where

$\bar{S}_{T,P}$ = Absolute entropy at 0.1 MPa and temperature T

P = Pressure expressed in atmospheres
$\bar{R}$ = Universal gas constant

11.7 SECOND LAW ANALYSIS OF REACTING SYSTEMS

The second law for any reactive process may be written:

$$\Delta S = S_P - S_R - \Sigma \frac{Q_{cv}}{T} \geq 0 \qquad (11.9)$$

where S_P = Entropy of the products

S_R = Entropy of the reactants

$\Sigma \dfrac{Q_{cv}}{T}$ = Entropy transfer into control volume due to heat transfer

The concepts of reversible work, irreversibility and availability are applied to the reacting systems. The basic formulas are listed below (SSSF) (in the absence of kinetic and potential energy, changes):

a) Reversible work

$$W_{rev} = \underset{R}{\Sigma} n_i \left[h_j^0 + \Delta \bar{h} - T_0 \bar{s} \right]_i - \underset{P}{\Sigma} n_e \left[h_f^0 + \Delta \bar{h} - T_0 \bar{s} \right]_e \qquad (11.10)$$

b) Irreversibility

$$I = \underset{P}{\Sigma} n_e T_0 \bar{s}_e - \underset{R}{\Sigma} n_i T_0 \bar{s}_i - Q_{cv} \qquad (11.11)$$

c) Availability

$$\psi = (h - T_0 s) - (h_0 - T_0 s_0) \qquad (11.12)$$

11.8 GIBB'S FUNCTION

The Gibbs function is defined:

$$\boxed{G = H - TS} \quad (11.13)$$

where H = Enthalpy
S = Entropy
T = Temperature

For a chemical reaction carried out at constant temperature and pressure we have,

$$\boxed{\Delta G = \Delta H - T\Delta S \leq 0} \quad (11.14)$$

This means that a chemical reaction is possible only if the Gibbs function for the products is less than the Gibbs function for the reactants.

11.9 CHEMICAL POTENTIAL

Consider a homogeneous mixture. If there are i constituents, we may express the internal energy in functional form as

$$U = f(s, v, n_1, n_2, \ldots, n_i).$$

A differential change in the internal energy is given,

$$(11.15)$$

$$dU = \left(\frac{\partial U}{\partial s}\right)_{v, N_i} dS + \left(\frac{\partial U}{\partial V}\right)_{s, N_i} dV + \Sigma \left(\frac{\partial U}{\partial N_i}\right)_{s, v, N} dN_i$$

The partial derivatives in the summation of the equation are defined as the chemical potential u_i. Also we have the equivalent relations for the chemical potential:

$$u_i = \left(\frac{\partial H}{\partial N_i}\right)_{s,p,N_j} = \left(\frac{\partial A}{\partial N_i}\right)_{T,V,N} = \left(\frac{\partial G}{\partial N_i}\right)_{T,P,N_j} \quad j \neq i$$

(11.16)

where   H = Enthalpy = u + pv
           A = Helmholtz function = U − TS
           G = Gibbs function = H − TS

CHAPTER 12

CHEMICAL EQUILIBRIUM

12.1 REQUIREMENTS FOR CHEMICAL EQUILIBRIUM

Applying the Gibbs function to a reactive system, we find that a chemical reaction carried out at constant pressure and temperature can proceed only if the Gibbs function of the system will continually decrease. The reaction will stop where the Gibbs function of the system has reached a minimum (Fig. 12.1).

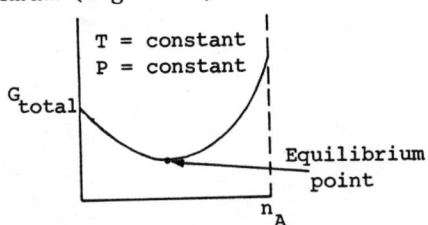

Fig. 12.1 Illustration of the requirement for the chemical equilibrium

We thus see that the equilibrium composition of any reactive system of known temperature and pressure is governed by

$$dG_{T,P} = 0 \qquad (12.1)$$

12.2 EQUILIBRIUM AND CHEMICAL POTENTIAL

To find the equilibrium composition of a mixture of gases undergoing a chemical reaction, we need an expression for dG in terms of the moles of reactants and products present at any given time:

$$dG = VdP - SdT + \sum_i u_i \, dN_i \qquad (12.2)$$

where

$$u_i = \left(\frac{\partial G}{\partial N_i} \right)_{P,T,N_i}$$

N_i: Number of moles of each chemical species within the system at some time.

12.3 EQUILIBRIUM CONSTANT OF A REACTIVE MIXTURE OF IDEAL GASES

We define the equilibrium constant k_p for ideal gas reactions by the relation:

$$K = \frac{(P_C)^{V_C} (P_D)^{V_D}}{(P_A)^{V_A} (P_B)^{V_B}} \qquad (12.3)$$

where

P_A, P_B, P_C, P_D are the partial pressures of the chemical constitutents

V_A, V_B, V_C, V_D are the stoichiometric coefficients.

Other forms of the equilibrium constant are:

a)
$$\ln K = - \frac{\Delta G^0}{RT} \qquad (12.4)$$

where

$$\Delta G^0 = V_C \bar{g}_C^0 + V_D \bar{g}_D^0 - V_A \bar{g}_A^0 - V_B \bar{g}_B^0$$

$\bar{g}_C^0$, $\bar{g}_D^0$, $\bar{g}_A^0$, $\bar{g}_B^0$ are the standard Gibbs functions.

T = Temperature of the environment
$\overline{R}$ = Universal gas constant

b)
$$K = \frac{a_C^{V_C} \cdot a_D^{V_D}}{a_A^{V_A} \cdot a_B^{V_B}} \qquad (12.5)$$

a_A, a_B, a_C, a_D are the activity coefficients.

c)
$$K = \frac{x_C^{V_C} \cdot x_D^{V_D}}{x_A^{V_A} \cdot x_B^{V_B}} (P)^{V_C + V_D - V_A - V_B} \qquad (12.6)$$

x_A, x_B, x_C, x_D are the mole fractions.

(P) is the mixture pressure.

12.4 EQUILIBRIUM BETWEEN TWO PHASES OF A PURE SUBSTANCE

Under equilibrium conditions the Gibbs function of each phase of a pure substance is equal.

12.5 EQUILIBRIUM OF A MULTICOM‑PONENT, MULTIPHASE SYSTEM

The requirement for equilibrium is that the chemical potential of each component is the same in all phases.

12.6 GIBB'S PHASE RULE

a) For an open system we have:

$$\boxed{F = n + 2 - r} \qquad (12.7)$$

where n = Number of components
r = Number of phases
F = Number of independent intensive properties

b) For a closed system we have:

$$\boxed{\begin{array}{l} F = n + 2 - r \\ n \leq r \end{array}} \qquad (12.8)$$

CHAPTER 13

FLOW THROUGH NOZZLES AND BLADE PASSAGES

13.1 CONSERVATION OF MASS FOR THE CONTROL VOLUME

$$\frac{\partial}{\partial t} \int_V \rho dV + \int_A \rho \vec{V} \cdot d\vec{A} = 0 \qquad (13.1)$$

where

$\frac{\partial}{\partial t} \int_V \rho dV$ = Rate of change of mass within the control volume

$\int_A \rho \vec{V} dA$ = Net rate of mass efflux through the control surface

13.1.1 SPECIAL CASES

For an incompressible flow (ρ=constant), equation (13.1) becomes

$$\int_A \rho \vec{V} \cdot d\vec{A} = 0 \qquad (13.2)$$

13.2 MOMENTUM EQUATION FOR THE CONTROL VOLUME

$$\Sigma F_j = \frac{1}{g_c} \left[\frac{d}{dt} \int_V V_j \rho dV + \int_A V_j \rho V_{rn} dA \right] \qquad (13.3)$$

D-93

For the steady-state, steady-flow process we have:

$$\Sigma F_j = \frac{1}{g_c}\left[\Sigma \dot{m}_e(v_e)_j - \Sigma \dot{m}_i(v_i)_j\right] \qquad (13.4)$$

where

$\dot{m}_i, \dot{m}_e$ = Rate of mass entering and leaving the C.V.

v_i, v_e = Velocity of the mass entering and leaving the C.V.

i = x,y,z (directions)

13.3 SPEED OF SOUND

13.3.1 MACH NUMBER

The Mach number is defined: $M = \frac{V}{C}$ \qquad (13.5)

where   V = Local flow speed
c = Local speed of sound

For an ideal gas $c = \sqrt{kRT}$, where $k = \frac{C_p}{C_v}$:

if M > 1, the flow is supersonic.

if M = 1, the flow is sonic.

if M < 1, the flow is subsonic.

13.4 LOCAL ISENTROPIC STAGNATION PROPERTIES

Local isentropic stagnation properties are those properties that would be obtained at any point in a flow field if the fluid at that point were decelerated from local conditions to zero velocity following a frictionless adiabatic - that is, isentropic - process.

For an ideal gas the isentropic stagnations (denoted by the subscript o) are:

$$\frac{P_0}{P} = \left[1 + \frac{k-1}{2} M^2\right]^{k/(k-1)} \quad (13.6)$$

$$\frac{T_0}{T} = 1 + \frac{k-1}{2} M^2 \quad (13.7)$$

$$\frac{\rho_0}{\rho} = \left[1 + \frac{k-1}{2} M^2\right]^{1/(k-1)} \quad (13.8)$$

13.5 CRITICAL CONSTANTS

The conditions at the throat of the nozzle can be found by noting that $M = 1$ at the throat. These properties (denoted by an asterisk *) are referred to as critical pressure, critical temperature and critical density and the ratios are given below:

$$\frac{P^*}{P_0} = \left(\frac{2}{k+1}\right)^{k/(k-1)} \quad (13.9)$$

$$\frac{T^*}{T_0} = \left(\frac{2}{k+1}\right) \quad (13.10)$$

$$\frac{\rho^*}{\rho_0} = \left(\frac{2}{k+1}\right)^{1/(k-1)} \quad (13.11)$$

13.6 EFFECTS OF AREA VARIATION ON FLOW PROPERTIES IN ISENTROPIC FLOW

In considering the effect of area variation on flow properties in isentropic flow, the following equation may be used:

$$\boxed{\frac{dA}{A} = \frac{dP}{\rho V^2}(1 - M^2)}\qquad (13.12)$$

where M = Mach number

dA = Change in the area

From this equation we can draw the following conclusions (Fig. 13.1) about the proper shape for nozzles and diffusers:

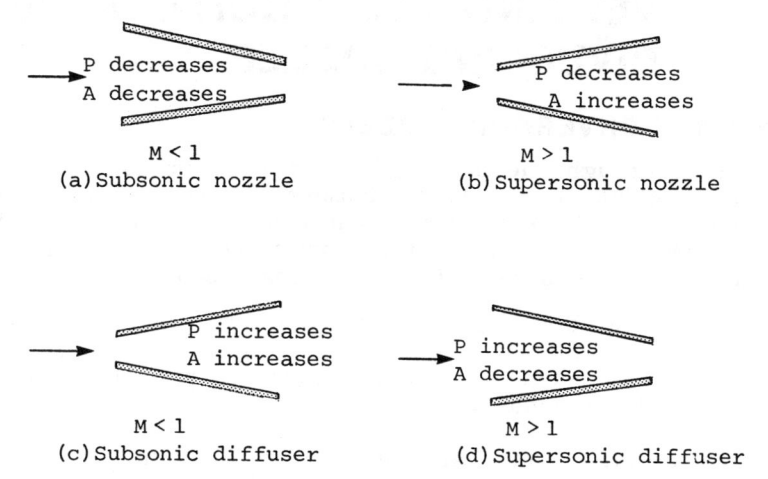

P decreases
A decreases

M < 1
(a) Subsonic nozzle

P decreases
A increases

M > 1
(b) Supersonic nozzle

P increases
A increases

M < 1
(c) Subsonic diffuser

P increases
A decreases

M > 1
(d) Supersonic diffuser

Fig. 13.1 Required area changes for nozzles and diffusers

13.7 ISENTROPIC FLOW OF AN IDEAL GAS

For the isentropic flow of an ideal gas we can summarize the basic equations as follows:

Continuity: $\quad \rho_1 V_1 A_1 = \rho_2 V_2 A_2 = \rho V A = \dot{m} \qquad (13.13)$

Momentum: $\quad R_x + p_1 A_1 - p_2 A_2 = \dot{m} V_2 - \dot{m} V_1 \qquad (13.14)$

First law: $\quad h_1 + \frac{V_1^2}{2} = h_2 + \frac{V_2^2}{2} = h + \frac{V^2}{2} \qquad (13.15)$

D-96

Second law: $\qquad s_1 = s_2 = s \qquad$ (13.16)

Equation of state: $\qquad p = \rho RT \qquad$ (13.17)

Process equation: $\qquad p/\rho^k = \text{constant} \qquad$ (13.18)

13.8 ISENTROPIC FLOW IN A CONVERGING AND A CONVERGING / DIVERGING NOZZLE

13.8.1 CONVERGING NOZZLE

Flow through the converging nozzle shown in Fig. 13.2 is supplied from a large chamber, where the conditions are assumed to be stagnation conditions. The flow is induced by a vacuum pump downstream and is controlled by the valve shown. The back pressure p_b to which the nozzle discharges is controlled by the valve. The upstream stagnation conditions (p_0, T_0, etc.) are maintained constant. The pressure in the exit plane of the nozzle is denoted p_e.

The effect of variations in p_b on the pressure distribution through the nozzle, on the mass flow rate, and on the exit plane pressure are illustrated graphically in Fig. 13.2:

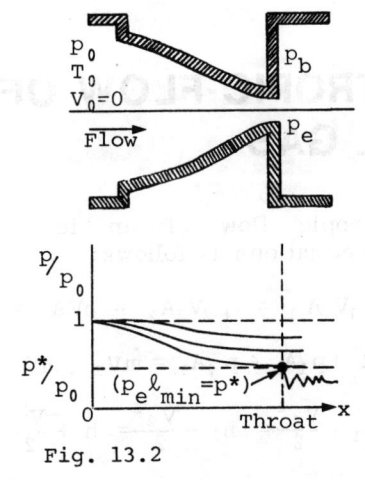

Fig. 13.2

D-97

13.8.2 CONVERGING-DIVERGING NOZZLE

As in the previous case, the flow through the converging-diverging nozzle is induced by a vacuum pump, and is controlled by the valve shown. The upstream stagnation conditions are assumed constant; the pressure in the exit plane of the nozzle is denoted p_e; the nozzle discharges to the back pressure p_b. The effects of variations in back pressure on the pressure distribution through the nozzle are illustrated graphically in Fig. 13.3:

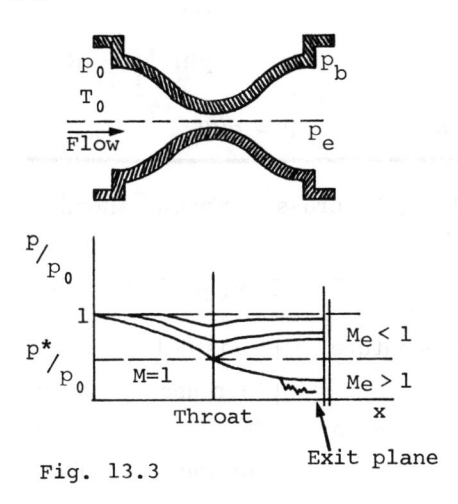

Fig. 13.3

13.9 NORMAL SHOCKS

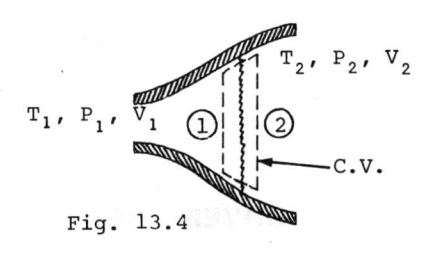

Fig. 13.4

a) A shock wave involves an extremely rapid and abrupt change of state. Figure 13.4 shows a control volume that can be used to analyze such a normal shock. The basic equations applied to the thin control volume shown in Fig. 13.4 are:

Continuity:	$P_1 V_1 = P_2 V_2 = G$	(13.19)
Momentum Equation:	$P_1 + P_1 V_1^2 = P_2 + P_2 V_2^2$	(13.20)
First Law:	$h_1 + \dfrac{V_1^2}{2} = h_2 + \dfrac{V_2^2}{2}$	(13.21)
Enthalpy:	$h_{01} = h_{02}$	(13.22)
Second Law:	$s_2 - s_1 = C_p \ln \dfrac{T_2}{T_1} - R \ln \dfrac{P_2}{P_1}$	(13.23)
Equation of State:	$P = \rho RT$	(13.24)

Property changes across a normal shock are summarized in table 13.1:

Property	Effect	Obtained from
Stagnation temperature, T_0	Constant	Energy equation
Entropy, s	Increase	Second Law
Stagnation pressure, p_0	Decrease	Ts diagram
Temperature, T	Increase	Ts diagram
Velocity, V	Decrease	Energy equation, and effect on T
Density, ρ	Increase	Continuity equation, and effect on V
Pressure, p	Increase	Momentum equation, and effect on V
Mach number, M	Decrease	$M = V/c$, and effect on V and T

Table 13.1

13.9.1 FLOW IN A CONVERGING-DIVERGING NOZZLE

Since we have considered normal shocks, we are now in a position to complete our discussion of flow in a converging-diverging nozzle. The pressure distribution

D-99

through the nozzle for different back pressures is shown in Fig. 13.5.

In Regime 1 the flow is subsonic throughout. At condition (iii), the flow at the throat is sonic, that is, $M_t = S$.

In Regime 2 the exit flow is subsonic, and consequently $p_e = p_b$.

In Regime 3 the back pressure is higher than the exit pressure but not sufficiently high to sustain a normal shock in the exit plane.

In Regime 4 the flow adjusts to the lower back pressure through a series of oblique expansion waves.

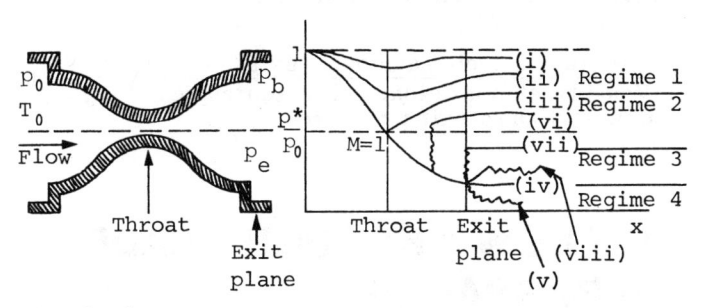

Fig. 13.5

13.10 NOZZLE AND DIFFUSER COEFFICIENTS

a) Nozzle efficiency is defined:

$$n_N = \frac{\text{Actual kinetic energy at nozzle exit}}{\text{Kinetic energy at nozzle exit with isentropic flow to same exit pressure}} \quad (13.25)$$

b) The coefficient of discharge C_p is defined by the relation:

$$C_p = \frac{\text{Actual mass rate of flow}}{\text{Mass rate of flow with isentropic flow}} \qquad (13.26)$$

c) The efficiency of a diffuser is defined:

$$n_p = \frac{(1 + \frac{k-1}{2} M_1^2)\left[\left(\frac{P_{02}}{P_{01}}\right)^{(k-1)/K} - 1\right]}{\frac{k-1}{2} M_1^2} \qquad (13.27)$$

where

1. states 1 and 01 are the actual and stagnation states of the fluid entering the diffuser
2. states 2 and 02 are the actual and stagnation states of the fluid leaving the diffuser

d) The velocity coefficient C_v is defined:

$$C_v = \frac{\text{Actual velocity at nozzle exit}}{\text{Velocity at nozzle exit with isentropic flow and same exit pressure}} \qquad (13.28)$$

13.11 FLOW THROUGH BLADE PASSAGES

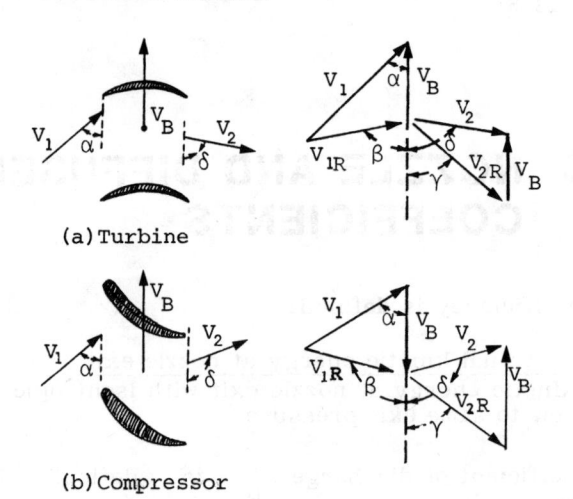

(a) Turbine

(b) Compressor

Fig. 13.6 Velocity vector diagrams

D-101

Figure 13.6 shows vector diagrams for both a turbine and a compressor.

In the diagrams:

a) V_1, V_{1R} represents the velocity and relative velocity of the fluid entering at angles α and β, respectively.

b) V_2, V_{2R} represents the velocity and relative velocity of the fluid leaving at angles δ and γ, respectively.

Applying the first law, assuming steady adiabatic flow, for a stationary and moving observer we have:

$$W = \frac{(V_1^2 - V_{1R}^2) - (V_2^2 - V_{2R}^2)}{2} \qquad (13.29)$$

where W is the work done on the blade.

Applying the second law to this process we conclude that

$$s_2 \geq s_1 \qquad (13.30)$$

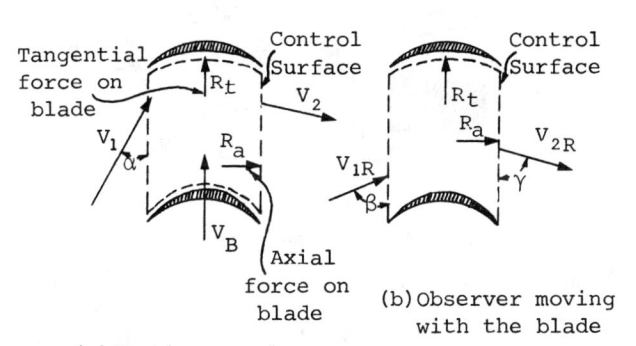

(a) Stationary observer

(b) Observer moving with the blade

Fig. 13.7 Analysis of forces on a turbine

If we apply the momentum equation in the tangential direction for both cases (see Fig. 13.6) we have:

a) For a stationary observer,

$$R_t = \dot{m}(V_1 \cos \alpha + V_2 \cos \delta) \qquad (13.31)$$

$$R_a = \dot{m}(V_2 \sin\delta - V_1 \sin\alpha) - (P_1A_1 - P_2A_2) \quad (13.32)$$

b) For a moving observer,

$$R_t = \dot{m}(V_{1R}\cos\beta + V_{2R}\cos\gamma) \quad (13.33)$$

$$R_a = \dot{m}(V_{2R}\sin\gamma - V_{1R}\sin\beta) - (P_1A_1 - P_2A_2) \quad (13.34)$$

where $\dot{m}$ represents the mass flow rate out and in of the control volume.

13.12 IMPULSE AND REACTION STAGES FOR TURBINES

13.12.1 IMPULSE STAGE

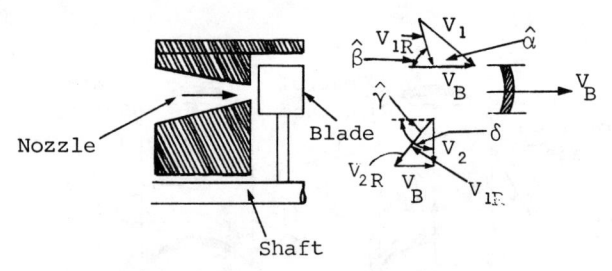

Fig. 13.8 Impulse stage

In an impulse stage, the entire pressure drop taken place in a stationary nozzle, and the pressure remains constant as the fluid flows through the blade passage. There is a decrease in the kinetic energy of the fluid as it flows through the blade passage, and the enthalpy will increase because of reversibilities associated with the fluid flow.

The blade efficiency η_B is defined:

$$\eta_B = \frac{W}{V_1^2/2} = 2\frac{V_B}{V_1}\left(\cos\alpha - \frac{V_B}{V_1}\right)\left[\frac{1 + \dfrac{K_B \cos\gamma}{\cos\alpha - V_B/V_1}}{\sqrt{(\cos - V_B/V_1)^2 + \sin^2\alpha}}\right]$$

$$(13.35)$$

where

$$K_B = \frac{V_{2R}}{V_{1R}}, \text{ blade velocity coefficient}$$

$$\eta_B = f(V_B/V_1, \alpha, \gamma, K_B)$$

13.12.2 REACTION STAGE

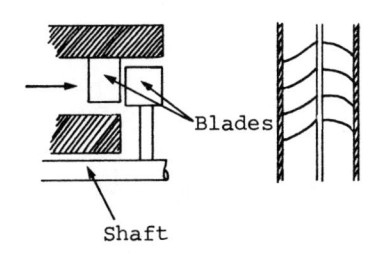

Fig. 13.9   Reaction stage

In the pure reaction stage, the entire pressure drop occurs as the fluid flows through the moving blades. Thus the moving blade acts as a nozzle, and the blade passage must have the proper contour for a nozzle. In the pure reaction stage the only purpose of the stationary blade is to direct the fluid into the moving blade at the proper angle and velocity.

A comparison of the impulse stage with the reaction stage regarding the blade speed ratio for maximum efficiency yields:

$$\frac{V_B}{V_1} = \frac{V_B}{V_0} = 0.5 \text{ for the impulse stage} \qquad (13.36)$$

$$\frac{V_B}{V_1} = \frac{V_B}{V_0} = \frac{1}{\sqrt{2}} \text{ for the reaction stage} \qquad (13.37)$$

where $V_0 = \sqrt{2g_c Dh_s}$ \qquad (13.38)

D-104

Handbook of Mechanical Engineering

SECTION E
Heat Transfer

CHAPTER 1

INTRODUCTION

1.1 HEAT TRANSFER

Heat transfer is the science which enables us to predict the rate of energy transferred and the temperature distribution between bodies which are at different temperatures.

1.2 MODES OF HEAT TRANSFER

There are three fundamental modes of heat transfer:

1.2.1 CONDUCTION

Conduction heat transfer takes place when a temperature gradient exists within a material and is governed by Fourier's Law of Conduction,

$$q = -kA \frac{\partial T}{\partial x} \qquad (1.1)$$

where q is the rate of heat transfer, $\partial T/\partial x$ is the temperature gradient normal to the surface, A is area and k is a material property independent of its shape called thermal conductivity. The minus sign is inserted in accordance with the second law of thermodynamics.

1.2.2 CONVECTION

Convection heat transfer takes place when a material is exposed to a moving fluid which is at different temperature. It is governed by the Newton's Law of Cooling

$$q = hA \Delta T \qquad (1.2)$$

where h is the convective heat transfer coefficient and ΔT is the temperature gradient between the surface and the fluid.

1.2.3 RADIATION

Radiation is the energy transferred between two separated bodies at different temperatures by means of electromagnetic waves. The fundamental law is the Stefan-Boltzman's Law of Thermal Radiation for black bodies.

$$E_b = \sigma T^4 \qquad (1.3)$$

the absolute temperature of the body and σ is the Stefan-Boltzman constant. All the actual surfaces (gray bodies) emit radiation at a lower rate than black bodies. Taking this fact into account equation (1.3) becomes

$$E_a = \varepsilon \sigma T^4 \qquad (1.4)$$

where ε the total emissivity of the surface.

1.3 THERMAL DIFFUSIVITY

The ratio of thermal conductivity to thermal capacity of the material is defined as thermal diffusivity α

$$\alpha = \frac{k}{\rho c_p} \qquad (1.5)$$

1.4 VISCOSITY

1.4.1 DYNAMIC VISCOSITY

Newton's Law of Viscosity states that in a fully developed, laminar, two-dimensional flow along a plate or duct the shear stress τ, produced by the relative motion of the fluid layers, is proportional to the normal velocity gradient. That is

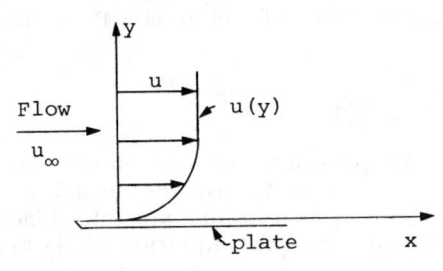

$$\boxed{\tau = \mu \frac{du}{dy}} \qquad (1.6)$$

Fig. 1.1

where u is the fluid velocity and μ is the coefficient of dynamic viscosity which is a molecular fluid property, independent of the fluid motion.

1.4.2 KINEMATIC VISCOSITY

The ratio of the dynamic viscosity μ to the fluid density ρ, is called kinematic viscosity, ν.

$$\boxed{\nu = \frac{\mu}{\rho}} \qquad (1.7)$$

1.5 UNITS

The systems of units are summarized in table 1.1.

Table of Units

Quantity	Unit S.I.	English
Force	1 newton = $1 \frac{kg\text{-}m}{s^2}$, N	1 pound force, lb_f
Length	1 meter, M	1 foot, ft
Mass	1 kilogram, kg	1 pound mass, lbm
Time	1 second, s	1 second, s
Temperature	degree Kelvin, °K	degree Rankine °R
Energy	1 Joule=1N-m, J	$ft\text{-}lb_f$
Heat flow rate, q	watt = 1J/s, W	Btu/s, Btu/h
Thermal conductivity, k	W/m-°C	Btu/h-ft-°F
Heat transfer coefficient, h	W/m^2-°C	$Btu/h\text{-}ft^2$°F
Thermal diffusivity, α	m^2/s	ft^2/h
Dynamic viscosity, μ	kg/m-s	lbm/ft-h
Kinematic viscosity, ν	m^2/s	ft^2/s, ft^2/h

CHAPTER 2

STEADY STATE HEAT CONDUCTION IN ONE DIMENSION

2.1 THE GENERAL HEAT CONDUCTION EQUATION

Let's consider the differential control volume shown in Figure 2.1. The general heat conduction equation for an isentropic and rigid solid is given by:

2.1.1 CARTESIAN COORDINATES

$$\rho C_p \frac{\partial T}{\partial t} = \frac{\partial}{\partial x}\left(k \frac{\partial T}{\partial x}\right) + \frac{\partial}{\partial y}\left(k \frac{\partial T}{\partial y}\right) + \frac{\partial}{\partial z}\left(k \frac{\partial T}{\partial z}\right) + q''' \quad (2.1)$$

where q''' = Internal heat generation or dissipation per unit of volume.

k = Variable thermal conductivity

T = Temperature

t = Time

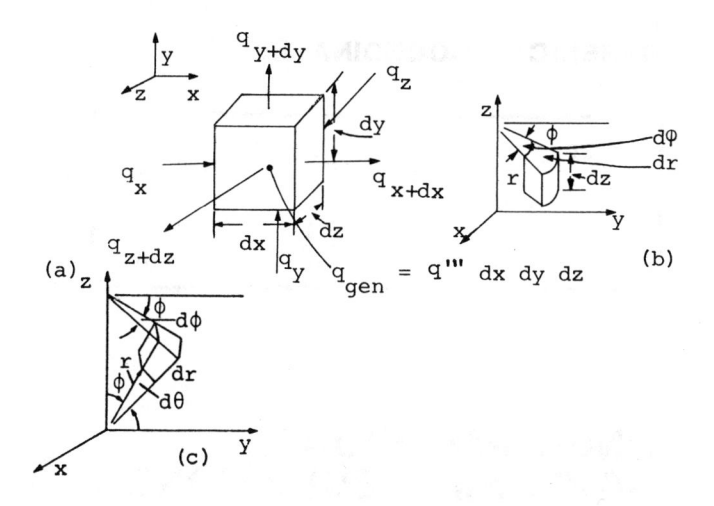

Fig. 2.1 Differential control volume for three-dimensional heat-conduction analysis. (a) Cartesian coordinates; (b) cylindrical coordinates; (c) spherical coordinates.

If the thermal conductivity k is constant, equation (2.1) can be written as

$$\frac{1}{\alpha}\frac{\partial T}{\partial t} = \left(\frac{\partial^2 T}{\partial x^2}\right) + \left(\frac{\partial^2 T}{\partial y^2}\right) + \left(\frac{\partial^2 T}{\partial z^2}\right) + \frac{q'''}{k} \quad (2.2)$$

where $\alpha = \dfrac{k}{\rho C_p}$

2.1.2 CYLINDRICAL COORDINATES

$$\frac{1}{\alpha}\frac{\partial T}{\partial t} = \left(\frac{\partial^2 T}{\partial r^2} + \frac{1}{r}\frac{\partial T}{\partial r} + \frac{1}{r^2}\frac{\partial^2 T}{\partial \theta^2} + \frac{\partial^2 T}{\partial z^2}\right) + \frac{q'''}{k} \quad (2.3)$$

2.1.3 SPHERICAL COORDINATES

$$\frac{1}{\alpha} \cdot \frac{\partial T}{\partial t} = \frac{1}{r} \frac{\partial^2}{\partial r^2}(rT) + \frac{1}{r^2 \sin\theta} \frac{\partial}{\partial \theta}\left(\sin\theta \frac{\partial T}{\partial \theta}\right) + \frac{1}{r^2 \sin^2\theta} \frac{\partial^2 T}{\partial \phi} + \frac{q'''}{k} \quad (2.4)$$

2.2 ONE-DIMENSIONAL GOVERNING EQUATIONS

For the steady-state, one-dimensional conduction with no heat generation, the general heat conduction equation (2.1) reduces to:

$$\frac{\partial^2 T}{\partial x^2} = 0 \quad (2.5)$$

In cylindrical coordinates equation (2.3) reduces to:

$$\frac{\partial^2 T}{\partial r^2} + \frac{1}{r} \frac{\partial T}{\partial r} = 0 \quad (2.6)$$

2.3 THE PLANE WALL WITH SPECIFIED SURFACE TEMPERATURES

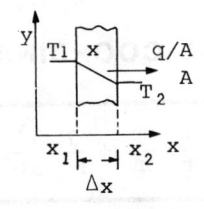

Fig. 2.2

E-7

Assumptions:

A) Steady state
B) One-dimensional conduction
C) Uniform temperature at each face
D) No internal heat generation ($q''' = 0$)
E) Constant conductivity (k=constant)

The equation to solve is:

$$\frac{\partial^2 T}{\partial x^2} = 0$$

Boundary conditions
at $x = x_1$, $T = T_1$
at $x = x_2$, $T = T_2$

Solution:

$$\boxed{\frac{q}{A} = -k \frac{(T_2 - T_1)}{\Delta x}} \qquad (2.7)$$

2.4 THE MULTILAYER WALL WITH SPECIFIED SURFACE TEMPERATURES

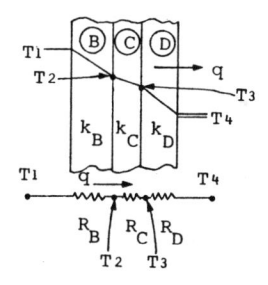

Fig. 2.3 Heat transfer through a multilayer wall and its electrical analog

E-8

Assumptions:
A through D same as in section 3, plus, C. Both conductivity and thickness vary depending on the layer.

General equation:

$$\frac{\partial^2 T}{\partial x^2} = 0$$

Solution:

$$q = \frac{T_1 - T_4}{\frac{\Delta x_b}{Ak_B} + \frac{\Delta x_C}{Ak_C} + \frac{\Delta x_D}{Ak_D}} \quad (2.8)$$

where the term $\Delta x/Ak$ is called the thermal resistance (R). The electrical analogy of the heat transfer is a very useful tool when solving more complex problems. Thus, the general solution for a multilayer wall may be written

$$q = \frac{\Delta T_{overall}}{\Sigma R} \quad (2.9)$$

where R represents the thermal resistance of the different wall layers.

2.5 THE SINGLE CYLINDER WITH SPECIFIED SURFACE TEMPERATURES

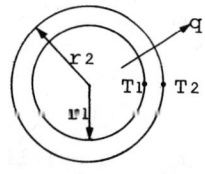

Fig. 2.4

E-9

Assumptions:

A-E same as in section 3.

The general equation (2.6)

$$\frac{d^2T}{dr^2} + \frac{1}{r}\frac{dT}{dr} = 0$$

Boundary conditions
- at $r = r_1$, $T = T_1$
- at $r = r_2$, $T = T_2$

Solution:

$$\boxed{q = \frac{(T_1 - T_2)}{\ln\left(\frac{r_2}{r_1}\right)/2\pi kL}} \quad (2.10)$$

where L is the length of the cylinder and $\ln(r_2/r_1)/2\pi kL$ is the thermal resistance (R).

2.6 THE MULTILAYER CYLINDER WITH SPECIFIED SURFACE TEMPERATURES

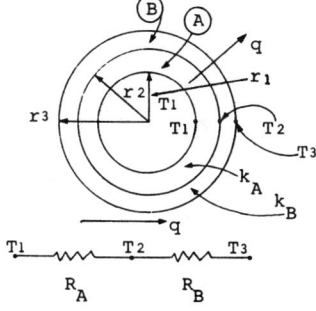

Fig. 2.5 Radial conduction across a multilayer cylinder and its respective electrical analog

E-10

Assumptions:

A-E same as in section 4.

Solution:

$$q = \frac{(T_1 - T_3)}{\frac{\ln(r_2/r_1)}{2\pi k_A L} + \frac{\ln(r_3/r_2)}{2\pi k_B L}} \qquad (2.11)$$

The general solution for a multilayer cylinder may be written:

$$q = \frac{\Delta T_{overall}}{\Sigma R} \qquad (2.12)$$

where R is the thermal resistance of all different layers.

2.7 THE PLANE WALL FLUIDS OF BOUNDED BY SPECIFIED TEMPERATURES

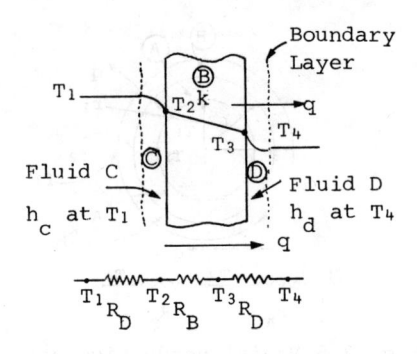

Fig 2.6

E-11

Assumptions:

A-E same as in section 4.

The general equation:

$$\frac{d^2 T}{dx^2} = 0$$

Solution:

$$\boxed{q = \frac{T_1 - T_4}{\dfrac{1}{Ahc} + \dfrac{\Delta X_B}{Ak_B} + \dfrac{1}{Ah_D}}} \qquad (2.13)$$

where $\Delta x / Ak$ is the thermal resistance and $1/Ah$ is called the convective thermal resistance. Equation (2.12) is the general equation for this case.

2.8 CYLINDRICAL SURFACES BY FLUIDS OF SPECIFIED TEMPERATURES

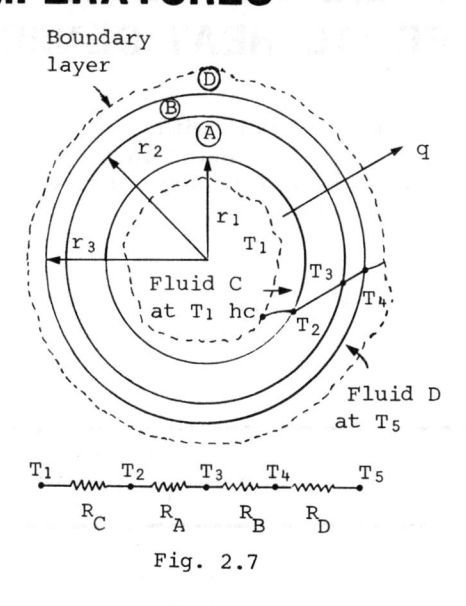

Fig. 2.7

Assumptions:

A-E same as in section 4.

The general equation is:

$$\frac{d^2T}{dr^2} + \frac{1}{r}\frac{dT}{dr} = 0$$

Solution:

$$q = \frac{T_1 - T_5}{\frac{1}{2\pi r_1 h_c L} + \frac{\ln(r_2/r_1)}{2\pi k_A L} + \frac{\ln(r_3/r_2)}{2\pi k_B L} + \frac{1}{2\pi r_3 hL}} \quad (2.14)$$

where $1/2\pi rhL$ is the convective thermal resistance. Equation (2.12) is also the general solution for this case.

2.9 THE PLANE WALL WITH INTERNAL HEAT GENERATION

Consider now a wall with constant internal heat generation, q''', uniformly distributed (Fig. 2.8). For steady-state conditions and constant thermal conductivity the general equation (2.2) becomes:

$$\frac{d^2T}{dx^2} + \frac{q'''}{k} = 0 \quad (2.15)$$

Solution:

$$T = \frac{q'''}{2k}(Lx - x^2) + \left(\frac{T_2 - T_1}{L}\right)x + T_1 \quad (2.16)$$

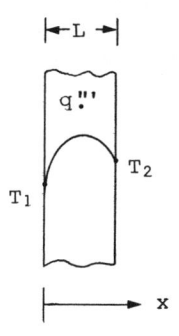

Fig. 2.8 Heat Conduction with internal heat generation.

2.10 CYLINDER WITH INTERNAL HEAT GENERATION

For a cylinder with constant internal heat generation, q''', uniformly distributed and radius R (Fig. 2.9) we have,

Assumptions:

A) Steady-state conditions
B) Temperature is only a function of radius
C) Constant thermal conductivity

The general equation (2.3) becomes:

$$\frac{d^2T}{dr^2} + \frac{1}{r}\frac{dT}{dr} + \frac{q'''}{k} = 0 \qquad (2.17)$$

Solution:

$$T - T_2 = \frac{q'''}{4k}(R^2 - r^2) \qquad (2.18)$$

or

$$\frac{T - T_2}{T_1 - T_2} = 1 - \left(\frac{r}{R}\right)^2 \qquad (2.19)$$

Equation (2.19) is in dimensionless form.

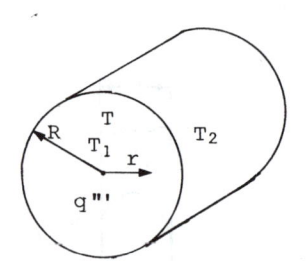

Fig. 2.9 Cylinder with internal heat generation

2.11 FINS

Extended surfaces or fins are used to increase the rate of heat exchange between a source and an ambient fluid by artificially increasing the effective surface area of heat transmission. Assuming that the temperature varies only in the x-direction, the generalized fin equation is given by:

$$\frac{d^2(T-T_\infty)}{dx^2} + \frac{1}{A}\frac{dA}{dx}\frac{d(T-T_\infty)}{dx} - \frac{h}{kA}\frac{ds}{dx}(T-T_\infty) = 0 \quad (2.20)$$

where $T = T(x)$ = Fin temperature distribution

T_∞ = Fluid temperature
$A = A(x)$ = Cross-sectional area for conduction
$S = S(x)$ = Surface area for convection
h = Convection heat transfer coefficient

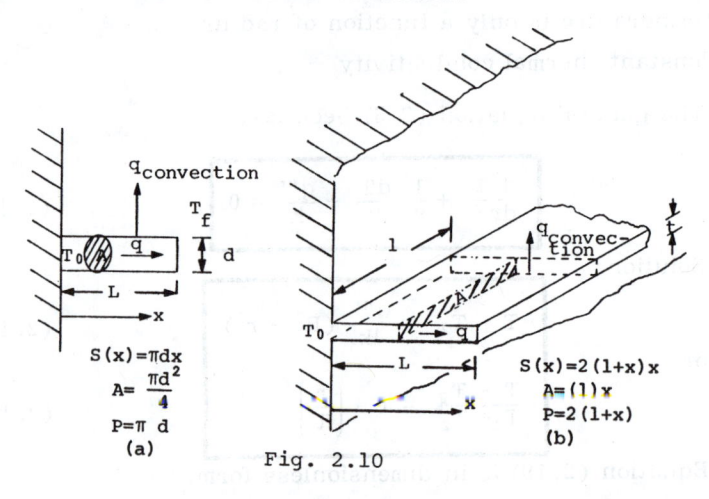

Fig. 2.10

E-15

2.11.1 UNIFORM CROSS-SECTION

The general equation (2.20) becomes:

$$\frac{d^2\theta}{dx^2} - m^2\theta = 0 \tag{2.21}$$

where $\theta = (T(x)-T_\infty)$

$m = \sqrt{\frac{hP}{kA}}$

P = Perimeter
A = Cross-sectional area

One boundary condition is

$$\theta(0) = \theta_0 = (T_0 - T_\infty)$$

Solution:

We will have three different solutions depending on the other boundary condition.

a) A very long fin

B.C. at $x = \infty$, $\theta(x) = 0$

thus,

$$\boxed{\frac{\theta}{\theta_0} = \frac{T-T_\infty}{T_0-T_\infty} = e^{-mx}} \tag{2.22}$$

$$\boxed{q = kAm\,\theta_0} \tag{2.23}$$

b) A fin of length L with insulated end

B.C. at $x = L$, $\frac{d\theta}{dx} = 0$

Thus,

$$\boxed{\frac{\theta}{\theta_0} = \frac{T-T_\infty}{T_0-T_\infty} = \frac{\cosh[m(L-x)]}{\cosh mL}} \tag{2.24}$$

E-16

$$\boxed{q = kAm\,\theta_0\,\text{Tanh}\,mL} \qquad (2.25)$$

c) A fin of length L losing heat by convection at the end

$$\boxed{\frac{\theta}{\theta_0} = \frac{T-T_\infty}{T_0-T_\infty} = \frac{\cosh m(L-x)+H\sinh m(L-x)}{\cosh mL + H\sinh mL}} \qquad (2.26)$$

where

$$H = \frac{h}{km}$$

$$\boxed{q = \sqrt{hPkA}\,(T_0-T_\infty)\frac{\sinh mL + H\cosh mL}{\cosh mL + H\sinh mL}} \qquad (2.27)$$

2.11.2 NONUNIFORM CROSS-SECTION

a) Straight triangular fin

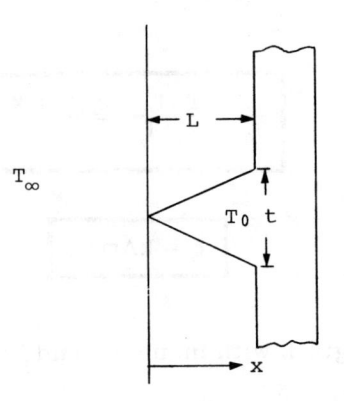

Fig. 2.11

$$S(x) = 2x\sqrt{1+(t/2L)^2}$$

$$A(x) = \frac{tx}{L}$$

E-17

The general equation (2.20) becomes:

$$\frac{d^2\theta}{dx^2} + \frac{1}{x}\frac{d\theta}{dx} - P^2\frac{\theta}{x} = 0 \qquad (2.28)$$

where
$$P = \sqrt{\frac{2fhL}{kt}} \quad \text{and} \quad f = \sqrt{1+(t/2L)^2}$$

Boundary conditions:
$$\theta(0) = \text{finite}$$
$$\theta(L) = \theta_0 = (T_0 - T)$$

Solution:

$$\boxed{\frac{\theta}{\theta_0} = \frac{T-T_\infty}{T_0-T_\infty} = \frac{I_0(2PX^{\frac{1}{2}})}{I_0(2PL^{\frac{1}{2}})}} \qquad (2.29)$$

where I_0 is the modified Bessel function of the first kind.

b) Annular Fin of Uniform Thickness

The general equation (2.20) becomes;

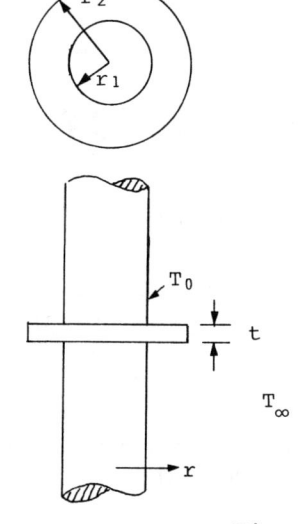

Fig 2.12

E-18

$$\frac{d^2\theta}{dr^2} + \frac{1}{r}\frac{d\theta}{dr} - \frac{2h}{kt}\theta = 0 \tag{2.30}$$

$$S(r) = \pi(r^2 - r_1^2)$$

$$A(r) = 2\pi r t$$

Boundary conditions:

at $r = r_1$, $\theta = \theta_0 = (T_0 - T_\infty)$
at $r = r_2$, $\frac{d\theta}{dr} = 0$

Solution:

$$\boxed{\frac{\theta}{\theta_0} = \frac{I_0(mr)K_1(mr_2) + K_0(mr)I_1(mr_2)}{I_0(mr_1)K_1(mr_2) + K_0(mr_1)I_1(mr_2)}} \tag{2.31}$$

where I_0, K_0, K_1, I_1 are the modified Bessel functions.

The heat transfer rate is

$$\boxed{q = 2\pi kt\, \theta_0(mr_1)\frac{K_1(mr_1)I_1(mr_2) - I_1(mr_1)k_1(mr_2)}{I_0(mr_1)K_1(mr_2) + K_0(mr_1)I_1(mr_2)}}$$
$$\tag{2.32}$$

2.12 FIN EFFICIENCY

The fin efficiency η_f is defined as follows:

$$\boxed{\eta_f = \frac{\text{Actual heat transfer}}{\text{Ideal heat transfer}}}$$

where the ideal heat transfer is the heat transferred if the entire fin surface were at the wall temperature (T_0).

Figure 2.13 shows the efficiencies for fins of different geometries.

Corrected Length (Lc)	Profile Area (A*)	Fin Geometry
L + t/2	tLc	Rectangular
L	t/2 L	Triangular
L + t/2	$t(r_{2c} - r_1)$	Circumferential and Rectangular Profile

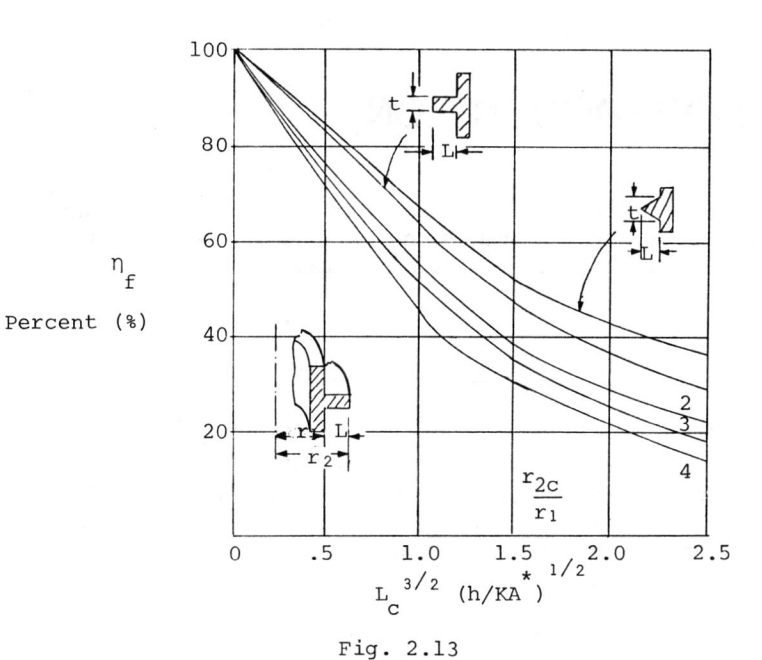

η_f Percent (%)

$L_c^{3/2} (h/KA^*)^{1/2}$

Fig. 2.13

CHAPTER 3

STEADY STATE HEAT CONDUCTION IN TWO DIMENSIONS

3.1 INTRODUCTION

The equation that governs the two-dimensional steady-state heat conduction (assuming constant thermal conductivity, k) is the Laplace equation

$$\frac{\partial^2 T}{\partial x^2} + \frac{\partial^2 T}{\partial y^2} = 0 \qquad (3.1)$$

from which the temperature distribution may be obtained using analytical, graphical, or numerical methods. The heat transfer is then calculated from the Fourier equation (equation 1.1) in each direction.

$$q_x = -kA_x \frac{\partial T}{\partial x} \qquad (3.2)$$

$$\boxed{q_y = -kA_x \frac{\partial T}{\partial y}} \qquad (3.3)$$

3.2 ANALYTICAL METHOD

3.2.1 STEADY STATE CONDUCTION IN A RECTANGULAR PLATE

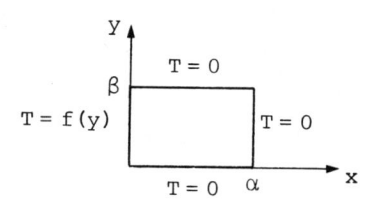

Fig. 3.1

The solution of the Laplace equation for a rectangular plate depends upon the boundary conditions as well as the initial conditions of the plate. To show the analytical method of solution, consider the following example:

Find the temperature distribution for the rectangular plate shown in Fig. 3.1.

The differential equation controlling the two-dimensional steady-state heat conduction is:

$$\frac{\partial^2 T}{\partial x^2} + \frac{\partial^2 T}{\partial y^2} = 0 \qquad (3.1)$$

Boundary Conditions:

a) $T(x,0) = 0$
b) $T(x,\beta) = 0$
c) $T(\alpha,y) = 0$
d) $T(0,y) = f(y)$

Using the separation of variables method to solve equation (3.1) we assume:

$$T(x,y) = X\,Y \qquad (3.4)$$

Substituting eq. (3.4) into (3.1) we get

$$Y \frac{d^2X}{dx^2} + X \frac{d^2Y}{dy^2} = 0$$

$$\frac{1}{X} \frac{d^2X}{dx^2} = -\frac{1}{Y} \frac{d^2Y}{dy^2} = \pm \lambda^2 \qquad (3.5)$$

where λ is a constant and in this case $\lambda^2 > 0$ to get a sine and cosine series in y. Thus,

$$\frac{d^2X}{dx^2} - \lambda^2 X = 0 \qquad (3.6)$$

$$\frac{d^2Y}{dy^2} + \lambda^2 Y = 0 \qquad (3.7)$$

Solving these two differential equations we find

$$X = (A\cosh \lambda x + B\sinh \lambda x) \qquad (3.8)$$

$$Y = (C\cos \lambda y + D\sin \lambda y) \qquad (3.9)$$

Equation 3.4 can now be written,

$$\boxed{T(x,y) = (A\cosh \lambda x + B\sinh \lambda x)(C\cos \lambda y + D\sin \lambda y)}$$
$$(3.10)$$

Now, using the four boundary conditions we find the constants A, B, C, D. Let's consider

$T(x,0) = 0 = (A\cosh \lambda x + B\sinh \lambda x)(C + 0)$

then $C = 0$

$$\therefore \boxed{T(x,y) = D\sin \lambda y(A\cosh \lambda x + B\sinh \lambda x)} \qquad (3.11)$$

B.C. #2

$T(x, \beta) = 0 = D\sin \lambda \beta (A\cosh \lambda x + B\sinh \lambda x)$

as $D \neq 0$ then $\sin \lambda \beta = 0$

and $\lambda \beta = 0 = n\pi$

E-23

$$\therefore \quad \lambda = \frac{n\pi}{\beta} \text{ for } n = 1, 2, 3, \ldots$$

thus the solution is

$$T(x,y) = D\sin\frac{n\pi}{\beta}y \left[A\cosh\frac{n\pi}{\beta}x + B\sinh\frac{n\pi}{\beta}x\right] \quad (3.12)$$

Now, B.C. #3 $T(\alpha,y) = 0$ we find that

$$A = -B\tanh\frac{n\pi}{\beta}\alpha$$

so
$$T(x,y) = D\sin\frac{n\pi}{\beta}y \left[B\left(-\tanh\frac{n\pi}{\beta}\alpha\cosh\frac{n\pi}{\beta}x + \sinh\frac{n\pi}{\beta}x\right)\right]$$

or
$$T(x,y) = \sum_{n=1}^{\infty} B_n \sin\frac{n\pi}{\beta}y \left[\sinh\frac{n\pi}{\beta}x - \tanh\frac{n\pi}{\beta}\alpha\cos\frac{n\pi}{\beta}x\right]$$

B.C. #4
$$T(0,y) = f(y) = \sum_{n=1}^{\infty} B_n \sin\frac{n\pi}{\beta}y \left(-\tanh\frac{n\pi}{\beta}\alpha\right)$$

from where

$$B_n = \frac{-2}{\beta\tanh\frac{n\pi\alpha}{\beta}} \int_0^\beta f(y)\sin\frac{n\pi}{\beta}y \, dy \quad (3.13)$$

Finally the temperature distribution is

$$T(x,y) = \sum_{n=1}^{\infty} B_n \sin\left(\frac{n\pi y}{\beta}\right)\left[\sinh\left(\frac{n\pi}{\beta}\right)x \right.$$
$$\left. - \tanh\left(\frac{n\pi}{\beta}\right)\alpha\cos\left(\frac{n\pi}{\beta}\right)x\right] \quad (3.14)$$

where B_n is given by equation (3.13)

Note: This procedure should be followed by the student in order to solve rectangular plate problems with different boundary conditions.

3.2.2. STEADY STATE CONDUCTION IN A CIRCULAR CYLINDER

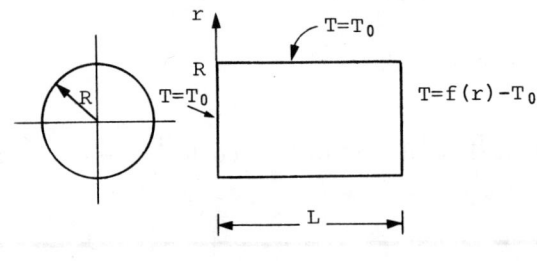

Fig. 3.2

The general equation is given by

$$\frac{\partial^2 \theta}{\partial r^2} + \frac{1}{r} \cdot \frac{\partial \theta}{\partial r} + \frac{\partial^2 \theta}{\partial z^2} = 0 \tag{3.15}$$

where $\theta = T-T_0$. The method of solution for this type of problems follows the same procedure mentioned in section 3.2. Let us consider the following example:

Find the temperature distribution for the cylinder shown in Figure 3.2. The general equation is

$$\frac{\partial^2 \theta}{\partial r^2} + \frac{1}{r} \frac{\partial \theta}{\partial r} + \frac{\partial^2 \theta}{\partial z^2} = 0$$

where $\theta = T-T_0$.

Boundary conditions:

a) $\theta(R,Z) = 0$

b) $\dfrac{\partial \theta(0,Z)}{\partial Z} = 0$

E-25

c) $\theta(r,0) = 0$
d) $\theta(r,L) = f(r) - T_0$

Using the separation of variables method we assume the following solution:

$$\theta(r,z) = R(r) \cdot Z(z)$$

Thus, the temperature distribution is given by

$$\theta = \frac{Z}{R^2} \sum_{n=1}^{\infty} \frac{\sinh \lambda_n Z}{\sinh \lambda_n L} \cdot \frac{J_0(\lambda_n r)}{J_1^2(\lambda_n R)} \cdot \int_0^R r[f(r)-T_0] \cdot J_0(\lambda_n r) dr$$

where λ_n is obtained from $J_0(\lambda_n R) = 0$.

3.2.3 TWO-DIMENSIONAL HEAT CONDUCTION IN POLAR COORDINATES

The general equation is given by

$$\frac{\partial^2 T}{\partial r^2} + \frac{1}{r}\frac{\partial T}{\partial r} + \frac{1}{r^2}\frac{\partial^2 T}{\partial \theta^2} = 0 \qquad (3.16)$$

Using the separation of variables method we assume the following solution

$$T = R(r) \Theta(\theta)$$

following the same procedure as in section 2a, we find that the temperature distribution is given by

$$T(r,\theta) = (A\cos\lambda\theta + B\sin\lambda\theta)(Cr^\lambda + Dr^{-\lambda})$$

where A,B,C,D are constants to be determined using the boundary conditions.

3.3 GRAPHICAL METHOD

The graphical technique is an approximated method to solve more complex problems. The method itself consists of plotting isotherms and heat flow lines (lines orthogonal to each other) and calculating the heat flow across each curvilinear section (see Fig. 3.3).

If $\Delta x = \Delta y$, the total heat transfer is given by

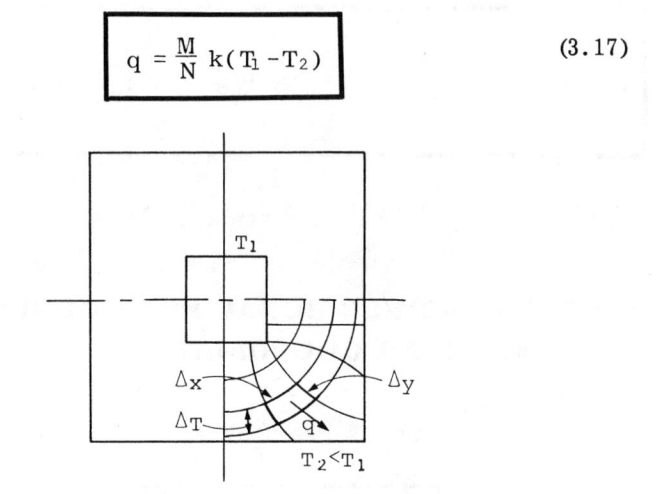

$$q = \frac{M}{N} k(T_1 - T_2) \qquad (3.17)$$

Fig. 3.3

where N is the number of temperature gradients and M is the number of heat flow lines. The ratio M/N is also known as the conduction shape factor (s)

$$S = \frac{M}{N} \qquad (3.18)$$

then equation (3.17) can be written

$$q = kS(T_1 - T_2) \qquad (3.19)$$

Many tables exist where the conduction shape factor (s) is tabulated for different geometries.

3.4 NUMERICAL METHOD

The numerical technique uses the relaxation method which is a successive approximation method and its accuracy depends upon the number of finite increments in the temperature and space coordinates. Let us consider Figure (3.4) where $\Delta x = \Delta y$.

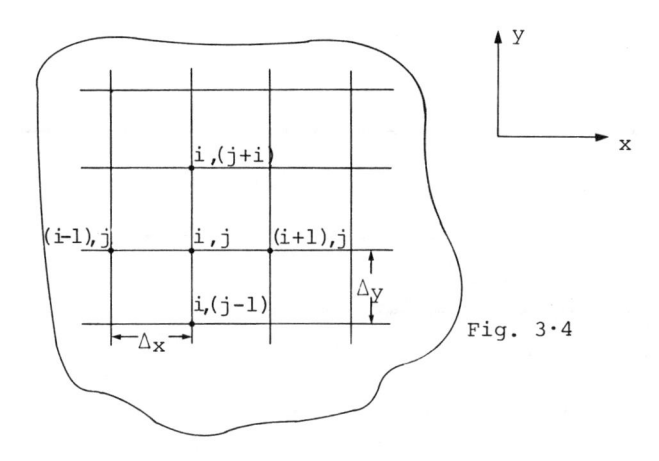

Fig. 3·4

The introduction of this temperature gradient into the Laplace equation (3.1) leads us to the general nodal equation:

$$T_{(i+1),j} + T_{(i-1),j} + T_{i,(j+1)} + T_{i,(j-1)} - 4T_{i,j} + \frac{q'''(\Delta x)^2}{k} = 0 \quad (3.20)$$

A nodal equation must be written for each node inside the body and then solved for the temperature at each node. For a large number of nodes a computer solution should be used. For a small number of nodes the relaxation method may be used. The steps to follow in relaxation analysis are:

A) Assume reasonable temperature values at each node.

B) Using the assumed temperatures, evaluate the residuals at each node.

C) Relax the largest residual (absolute value) to zero. In order to do this, change the corresponding estimated temperature.

D) Recompute all the residuals using the new temperature value.

E) Continue to relax residuals until all of them are very close to zero.

In order to evaluate the exterior nodes temperatures convection heat transfer must be considered. It can be shown that the nodal temperature equation for a plane surface is

$$T_{i,J}\left(\frac{h\Delta x}{k} + z\right) - \frac{h\Delta x}{k} T_\infty - \frac{1}{2}(2T_{i-1,J} + T_{i,J+1} + T_{i,J-1}) = 0$$

(3.21)

and for a corner is

$$2T_{i,J}\left(\frac{h\Delta x}{k} + 1\right) - \frac{2h\Delta x}{k} T_\infty - (T_{i-1,J} + T_{i,J-1}) = 0$$

(3.22)

Equations like (3.21) and (3.22) must be written for each external node.

E-29

CHAPTER 4

UNSTEADY STATE HEAT CONDUCTION

4.1 INTRODUCTION

The governing equation for unsteady-state heat conduction is given by the Laplace equation (2.1). In one dimension we have

$$\frac{\partial^2 T}{\partial x^2} = \frac{1}{\alpha} \frac{\partial T}{\partial t} \qquad (4.1)$$

where the temperature distribution T is also a function of time. The analytical method discussed in chapter 3 may be used to solve the equation.

4.2 LUMPED SYSTEM

To study the transient problem by this method the following assumptions must be made:

A) The internal resistance of the body is negligible.

B) Uniform temperature throughout the body.

Fig. 4.1

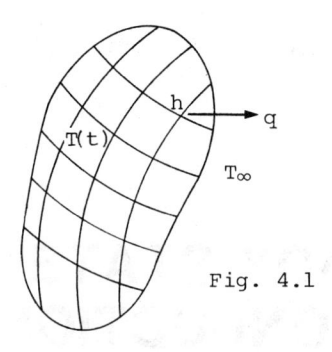

Let us consider the body shown in Figure 4.1. The body is losing heat from the surface. Thus, the energy balance will be

$$\text{Change of internal energy with time} = \text{Heat lost by convection}$$

or

$$\frac{dE}{dt} = hA(T-T_\infty) \qquad (4.2)$$

where $E = C_p \rho V T$, is the heat energy. Inserting into equation (4.2) and arranging terms we get:

$$\frac{dT}{dt} + \frac{hA}{\rho C_p V}(T-T_\infty) = 0 \qquad (4.3)$$

For a n-lump system, n simultaneous differential equations must be written. The solution to equation (4.3) is

$$\frac{T-T_\infty}{T_0-T_\infty} = e^{-\left(\frac{hA}{\rho C_p V}\right)t} \qquad (4.4)$$

where T_0 is the temperature of the body at $t = 0$.

Note: This method is applicable when the ratio $\frac{h(V/A)}{k}$ is less than one tenth (0.1).

E-31

4.3 ONE DIMENSIONAL TRANSIENT HEAT CONDUCTION

4.3.1 SUDDEN TEMPERATURE CHANGE AT THE SURFACE OF A SEMI-INFINITE BODY

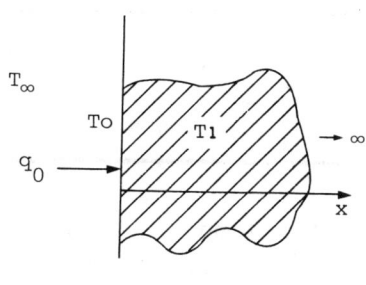

Fig. 4.2

Let us consider the body in figure 4.2.
The general equation is

$$\frac{\partial^2 T}{\partial x^2} = \frac{1}{\alpha} \frac{\partial T}{\partial t} \qquad (4.5)$$

Boundary conditions:

$T(0,t) = T_0$
$T(\infty,t) = T_1$

Initial Condition: $T(x,0) = T_1$

condition $T \to T_1$ as $x \to \infty$ for $t > 0$

Solution:

$$\boxed{\frac{\theta}{\theta_i} = \operatorname{erf} \frac{x}{2\sqrt{\alpha t}}} \qquad (4.6)$$

where $\theta = T(x,t) - T_0$
$\theta_1 = T_1 - T_0$ T_1: initial temperature

and the gauss error function is defined by

$$\text{erf}\frac{x}{2\sqrt{\alpha t}} = \frac{2}{\sqrt{\pi}} \int_0^{x/2\sqrt{\alpha t}} e^{-\eta^2} d\eta \tag{4.7}$$

Substituting eq.(4.7) in (4.6) we get the temperature distribution as

$$T(x,t) = \frac{2(T_1-T_0)}{\sqrt{\pi}} \int_0^{x/2\sqrt{\alpha t}} e^{-\eta^2} d\eta + T_0 \tag{4.8}$$

Figure 4.3 shows the temperature distribution in a semi-infinite body.

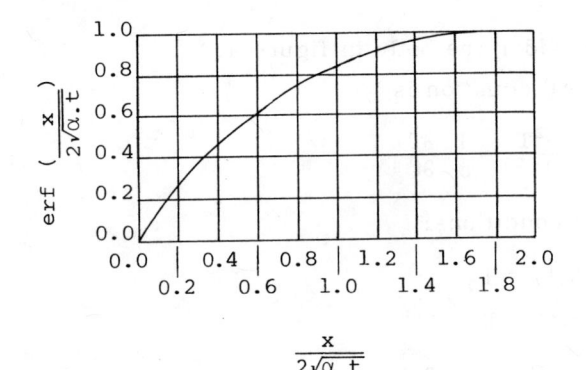

[Graph: erf$\left(\frac{x}{2\sqrt{\alpha.t}}\right)$ vs $\frac{x}{2\sqrt{\alpha.t}}$]

Gauss's error integral

Fig. 4.3

The heat flow at any position in the x-direction is given by

$$q = \frac{kA(T_0-T_1)}{\sqrt{\pi \alpha t}} \exp(-x^2/4\alpha t) \tag{4.9}$$

4.3.2 SEMI-INFINITE BODY WITH CONVECTION BOUNDARY CONDITIONS

Let us consider the semi-infinite body but this time with a convection boundary condition at x = 0 (surface), so that

$$hA(T_\infty - T) = -KA \frac{\partial T}{\partial x} \quad \text{at } x = 0 \text{ (surface)} \quad (4.10)$$

where h is the convective heat transfer coefficient at the surface. The solution of eq.(4.10) is given by

$$\frac{T-T_1}{T_\infty - T_1} = 1 - \mathrm{erf}\left(\frac{x}{\sqrt{4\alpha t}}\right) - \left[\exp\left(\frac{hx}{k} + \frac{h^2 \alpha t}{k^2}\right)\right]\left[1 - \mathrm{erf}\left(\frac{x}{\sqrt{4\alpha t}} + \frac{h\sqrt{\alpha t}}{k}\right)\right] \quad (4.11)$$

where T_∞ is the fluid temperature. Figure 4.4 shows the temperature distribution for this case.

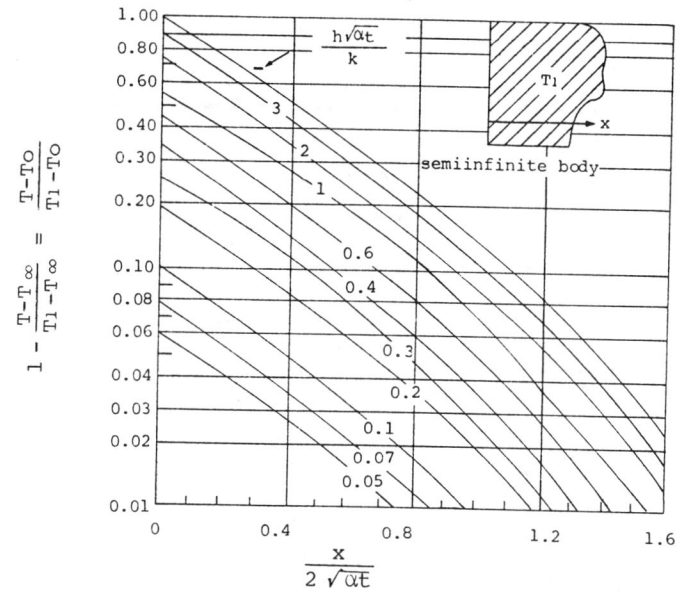

Fig. 4.4

4.3.3 CHARTS FOR SOLVING PROBLEMS WITH CONVECTION BOUNDARY CONDITIONS

Two important dimensionless parameters have to be considered:

$$B_i = \frac{hs}{k} \quad \text{(Biot modulus)} \quad (4.12)$$

$$F_o = \frac{\alpha t}{s^2} \quad \text{(Fourier modulus)} \quad (4.13)$$

where s is a characteristic dimension of the body.

$$s = \frac{V}{A} = \frac{\text{Volume}}{\text{Area}} \quad (4.14)$$

a) The finite slab

General equation is

$$\boxed{\frac{\partial \theta}{\partial t} = \alpha \frac{\partial^2 \theta}{\partial x^2}} \quad (4.15)$$

where $\theta = T - T_\infty$

Initial condition: $\theta(x,0) = T_1 - T_\infty$

Boundary Conditions
$$\begin{cases} \frac{\partial \theta}{\partial x}(0,t) = 0 \\ \frac{\partial \theta(L,t)}{\partial x} = -\frac{h}{k}\theta(L,t) \end{cases}$$

Solution:

$$\boxed{\frac{T_{\mathcal{C_L}} - T_\infty}{T_1 - T_\infty} = z \sum_{n=1}^{\infty} e^{-\delta_n^2 F_0} \frac{\sin \delta_n}{\delta_n + (\sin \delta_n)(\cos \delta_n)}}$$

(4.16)

where $T_{\mathcal{C_L}}$ is the centerline temperature.

The heat flow is

$$\boxed{Q = 2Q_1 \sum \frac{1}{\delta_n} \frac{\sin^2 \delta_n}{\delta_n + (\sin \delta_n)(\cos \delta_n)} \left[1 - e^{(1-\delta_n^2 F_0)} \right]}$$

(4.17)

where $Q_1 = \rho CV(T_1 - T_\infty) = \rho CV\theta_1$ is the initial internal energy. The Heisler Charts of the centerline temperature are given in Figures 4.5 and 4.6. The heat flow chart is in Figure 4.7.

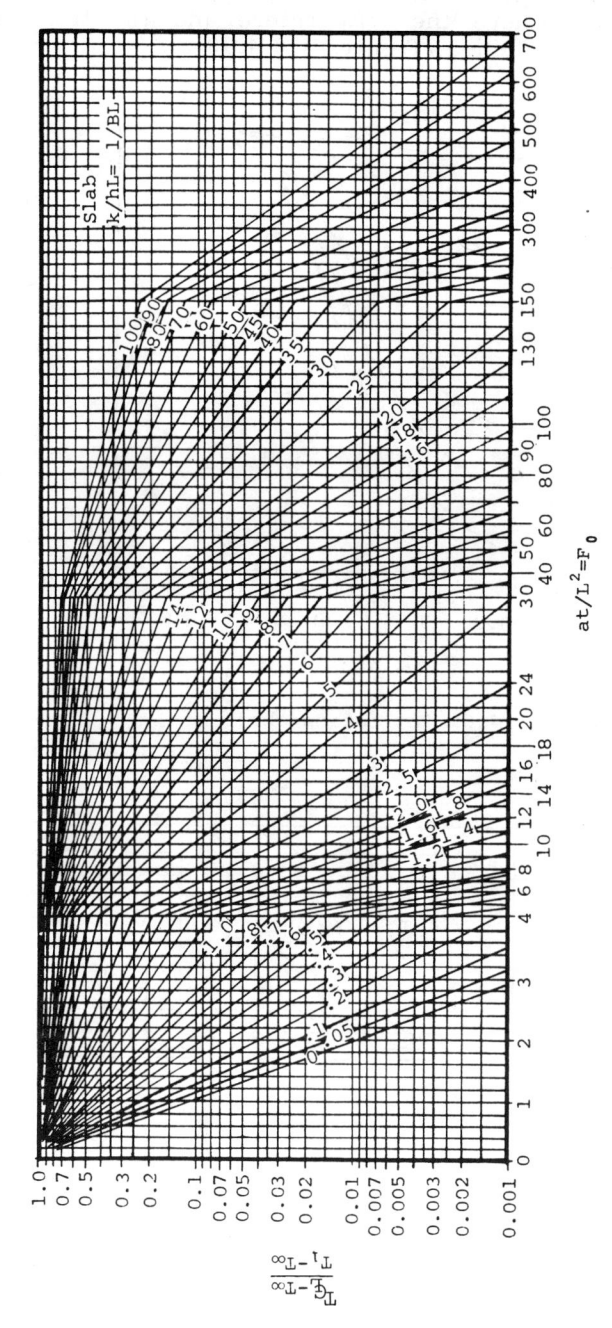

Fig. 4.5 Centerline temperature for a finite slab

In the same fashion, there are charts for cylinders and spheres from where the temperature and the transfer of energy can be obtained (Figures 4.8-4.13).

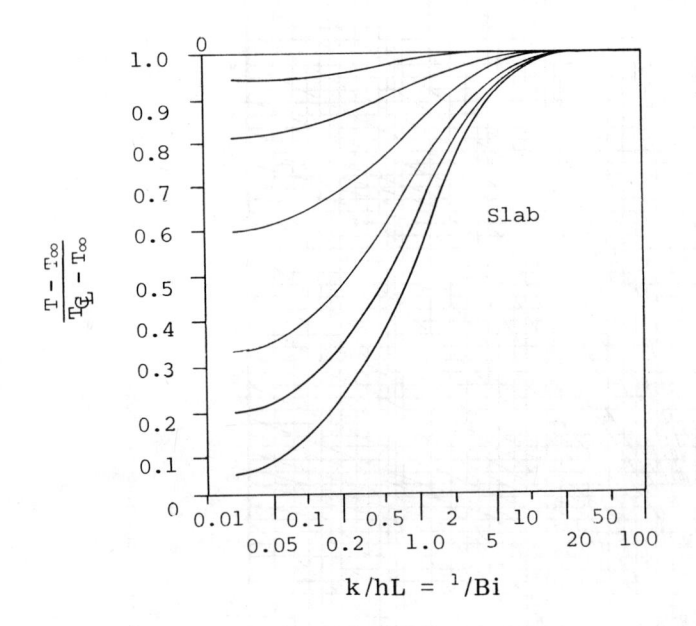

Fig. 4.6 Correction chart

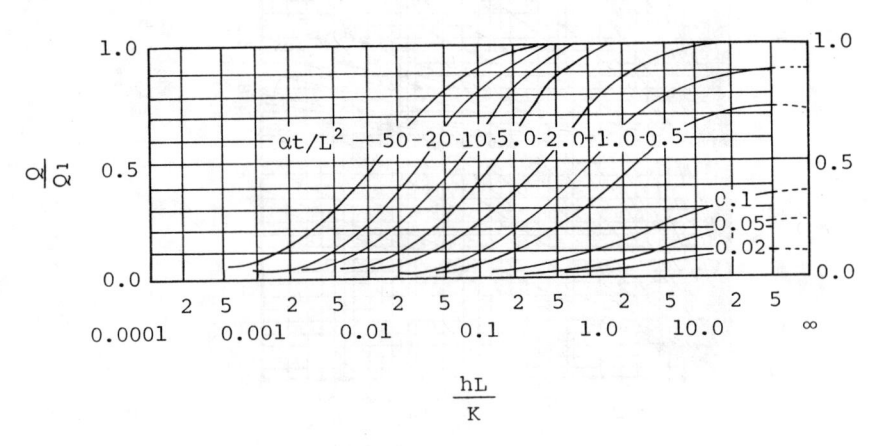

Fig. 4.7 Dimensionless heat flow for a slab

E-37

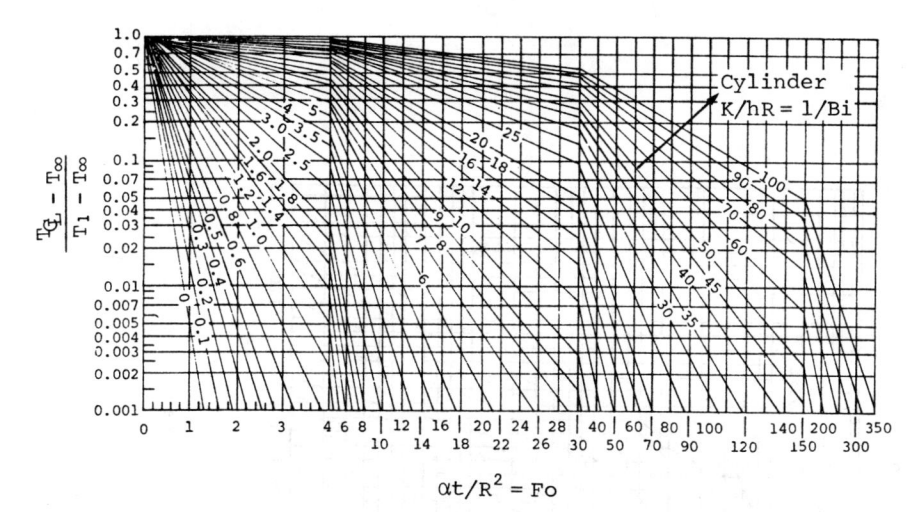

Fig. 4.8 Centerline temperature for a cylinder of radius R

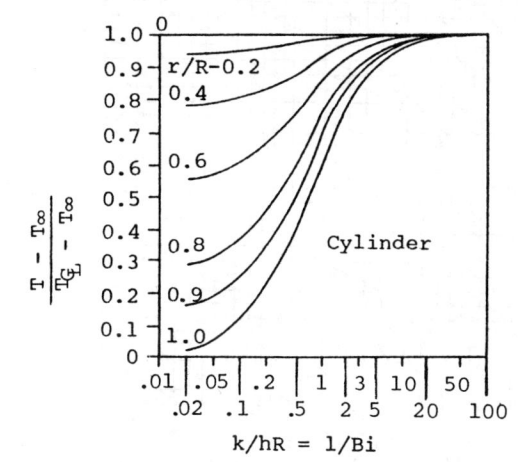

Fig. 4.9 Correction factor chart

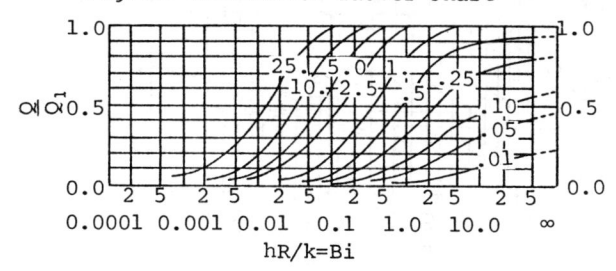

Fig. 4.10 Dimensionless heat flow for a cylinder of radius R

E-38

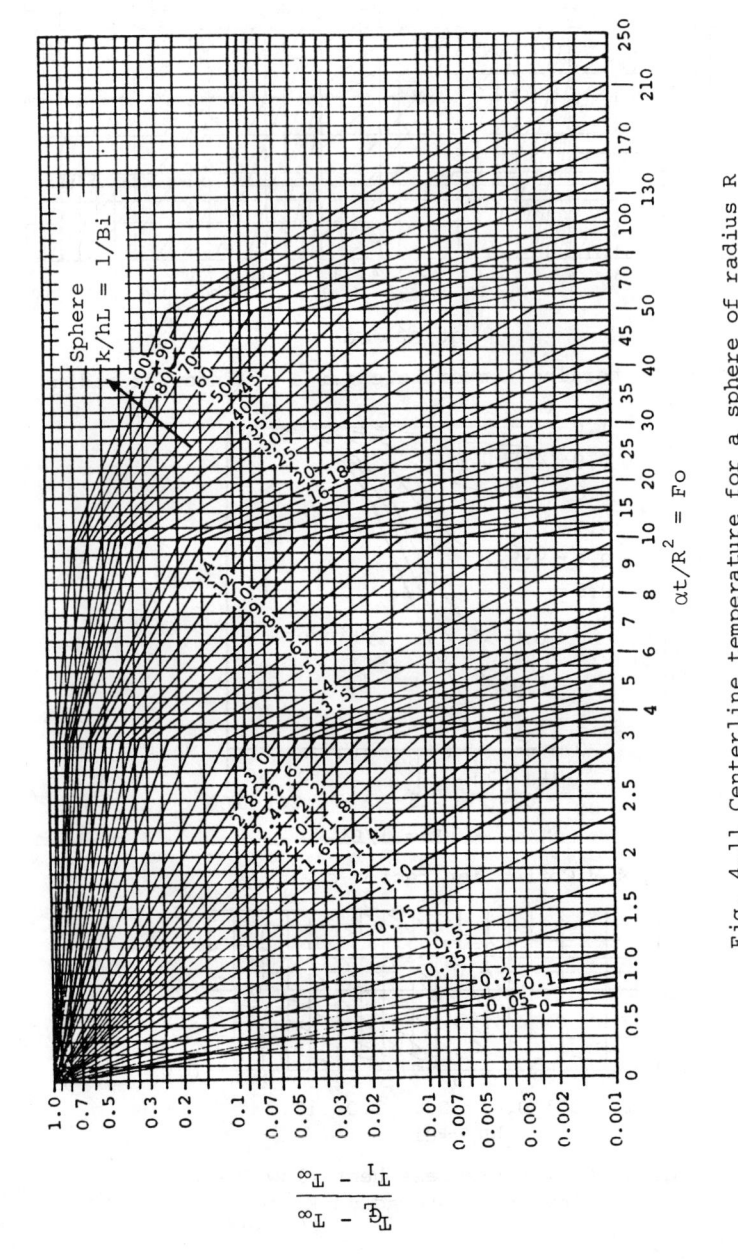

Fig. 4.11 Centerline temperature for a sphere of radius R

E-39

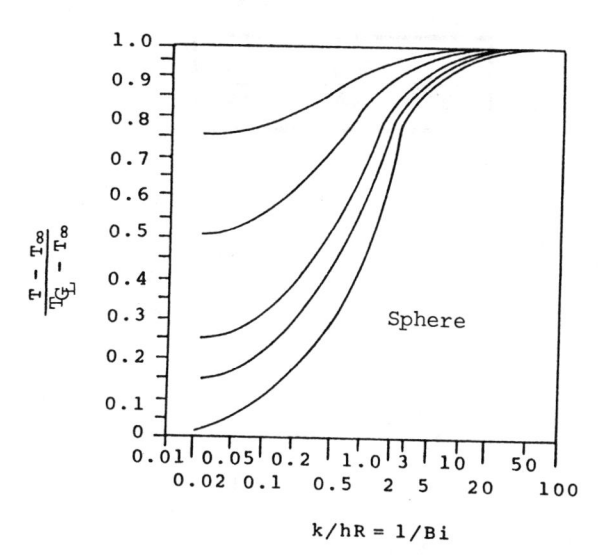

Fig. 4.12 Correction factor chart

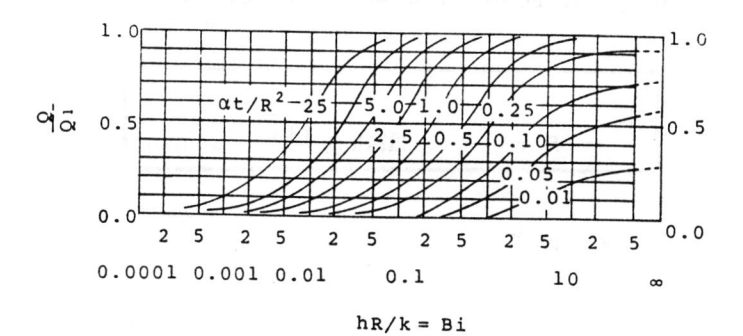

Fig. 4.13 Dimensionless heat flow of a sphere of radius R

4.4 MULTIPLE-DIMENSIONS SYSTEMS

The combination of solutions of one-dimensional systems will result in the solution of a multiple-dimension system. Let us take for example the bar shown in Figure 4.14. The governing equation is given by

$$\frac{\partial^2 T}{\partial y^2} + \frac{\partial^2 T}{\partial z^2} = \frac{1}{\alpha}\frac{\partial T}{\partial t}$$

(4.18)

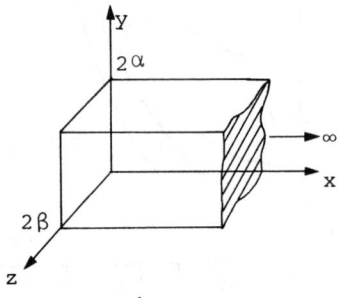

Fig. 4.14

Using the separation of variables method, it can be shown that the temperature distribution can be written as follows:

$$\left(\frac{\theta}{\theta_1}\right)_{Bar} = \left(\frac{\theta}{\theta_1}\right)_{2\alpha,\,slab} \left(\frac{\theta}{\theta_1}\right)_{2\beta,\,slab} \quad (4.19)$$

where $\theta = (T - T_\infty)$
$\theta_1 = (T_1 - T_\infty)$

For a cylinder of height 2h and radius R we get

$$\left(\frac{\theta}{\theta_1}\right)_{short\ cylinder} = \left(\frac{\theta}{\theta_1}\right)_{2R,\,cylinder} \left(\frac{\theta}{\theta_1}\right)_{2h,\,slab} \quad (4.20)$$

In general, the product of the solutions of one-dimensional systems is the solution of a multi-dimensional system.

Note: The one-dimensional solutions can be easily obtained from the charts discussed in section 4.3.

E-41

4.5 NUMERICAL SOLUTION

Let us consider Figure 4.15 a two-dimensional conduction problem where $\Delta x = \Delta y$. The governing equation is:

$$\frac{\partial^2 T}{\partial x^2} + \frac{\partial^2 T}{\partial y^2} = \frac{1}{\alpha}\frac{\partial T}{\partial t} \qquad (4.21)$$

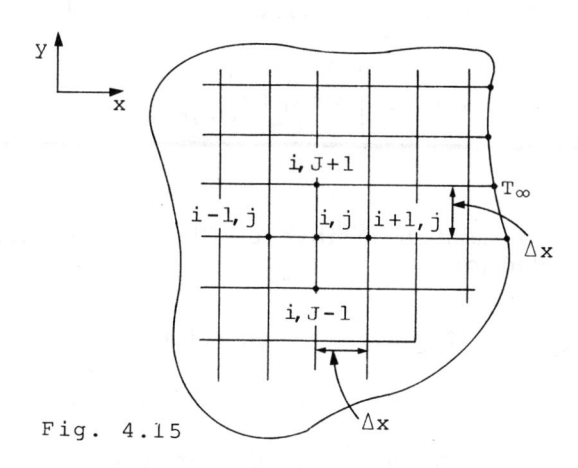

Fig. 4.15

Introducing the temperature gradients in eq.(4.21) we get the general nodal equation for the two-dimensional system:

$$T_{i,j}^{t+1} = \frac{1}{M}(T_{i+1,J}^{t} + T_{i-1,J}^{t} + T_{i,J+1}^{t} + T_{i,J-1}^{t}) + \left[1 - \frac{4}{M}\right]T_{i,J}^{t} \qquad (4.22)$$

where $M = \Delta x^2 / \alpha \Delta t$

i, J = Position of the node

T^{t+1} = Temperature after a time increment

In the same way we can conclude that for a one-dimensional system the nodal equation is given by

E-42

$$T_i^{t+1} = \frac{1}{M}(T_{i+1}^t + T_{i-1}^t) + \left[1 - \frac{2}{M}\right]T_i^t \qquad (4.23)$$

and for three-dimensional systems

$$T_{i,J,k}^{t+1} = \frac{1}{M}(T_{i+1,J,k}^t + T_{i-1,J,k}^t + T_{i,J+1,k}^t + T_{i,J-1,k}^t + T_{i,J,k+1}^t + T_{i,J,k-1}^t)$$
$$+ \left[1 - \frac{6}{M}\right]T_{i,J,k}^t$$
$$(4.24)$$

In order to maintain positive coefficients, the values of M are restricted to:

$M \geq 2$ For one-dimensional systems

$M \geq 4$ For two-dimensional systems

$M \geq 6$ For three-dimensional systems

Exterior nodes are exposed to convection heat transfer so the temperatures have to be calculated differently. It can be shown that for one-dimensional systems the nodal equation is

$$T_i^{t+1} = \frac{2}{M}\left[T_i^t + \frac{h \Delta x}{k} T_\infty\right] + \left[1 - \frac{2}{M}\left(\frac{h \Delta x}{k} + 1\right)\right]T_i^t$$
$$(4.25)$$

where $M \geq 2\left(\frac{h \Delta x}{k} + 1\right)$ to ensure convergence.

E-43

For two-dimensional systems

$$T_{i,J}^{t+1} = \frac{1}{M}\left(T_{i+1,J}^t + 2T_{i-1,J}^t + T_{i,J+1}^t + 2\frac{h\Delta x}{k}T_\infty\right)$$
$$+ \left[1 - \frac{2}{M}\left(\frac{h\Delta x}{k} + 2\right)\right]T_{i,J}^t$$

(4.26)

where $M \geq 2\left(\frac{h\Delta x}{k} + 2\right)$ to ensure convergence.

CHAPTER 5

FLUID MECHANICS REVIEW

5.1 INTRODUCTION

The dynamics of the fluid flow will help us to understand the mechanisms of the convection heat transfer. As it was stated in chapter one, the convection energy transfer is calculated using the Newton's cooling law:

$$q = hA \Delta T \qquad (5.1)$$

where the convection heat transfer coefficient, h, is a function of different parameters such as the velocity of the flow, the thermal properties of the fluid, and the geometry of the body under consideration among others.

This chapter will review parts of the fluid mechanics that are going to be useful in the analysis of the forced convection heat transfer.

5.2 VISCOUS AND INVICID FLOWS

A) Invicid flow is an imaginary flow wherein the fluid viscosity is assumed to be zero.

B) Viscous flows are all real fluid flows which experience viscous forces proportional to the normal velocity gradient. For one-dimensional viscous flow it may be expressed as

$$T = \mu \frac{du}{dy} \qquad (5.2)$$

where u is the velocity component in the x-direction and μ is the dynamic viscosity (see chapter 1, section 1.4).

5.3 LAMINAR AND TURBULENT FLOWS

A) Laminar flow - It is a well-ordered pattern flow in which the fluid particles slide parallel to each other.

B) Turbulent flow - It is an unstable flow in which the fluid particles move disorderly in all directions.

The criterion for the transition from laminar to turbulent flow is given by the Reynolds number:

$3.2 \times 10^5 < Re < 10^6$ transition for a flat plate

$Re < 2300$ the flow is laminar in a pipe.

5.4 BASIC EQUATIONS

The basic equations apply under the following assumptions:

    A) Continuum

    B) Constant Properties

    C) Steady State

    E) Newtonian Fluids

    F) Two-Dimensional

5.4.1 CONTINUITY EQUATION

$$\frac{\partial u}{\partial x} + \frac{\partial v}{\partial y} = 0 \qquad (5.3)$$

where u and v are the velocity components in the x-direction.

5.4.2 MOMENTUM EQUATION

x-direction

$$\boxed{u \frac{\partial u}{\partial x} + v \frac{\partial u}{\partial y} = -\frac{1}{\rho} \frac{\partial P}{\partial x} + \nu \left(\frac{\partial^2 u}{\partial x^2} + \frac{\partial^2 u}{\partial y^2} \right)}$$

y-direction

$$u \frac{\partial v}{\partial x} + v \frac{\partial v}{\partial y} = -\frac{1}{\rho} \frac{\partial P}{\partial y} + \nu \left(\frac{\partial^2 v}{\partial x^2} + \frac{\partial^2 v}{\partial y^2} \right)$$

5.4.3 ENERGY EQUATION

$$K \left(\frac{\partial^2 T}{\partial x^2} + \frac{\partial^2 T}{\partial y^2} \right) + \mu \left[2 \left(\frac{\partial u}{\partial y} \right)^2 + 2 \left(\frac{\partial v}{\partial y} \right)^2 + \left(\frac{\partial u}{\partial y} + \frac{\partial v}{\partial x} \right)^2 \right]$$

$$= \rho C_p \left(u \frac{\partial T}{\partial x} + v \frac{\partial T}{\partial y} \right)$$

5.5 NAVIER-STOKES EQUATIONS

Assumptions:

A) Laminar flow
B) Unsteady-state flow
C) Incompressible fluid

D) Newtonian Fluid
E) Constant properties
F) Two-dimensional flow

$$\frac{\partial u}{\partial t} + u\frac{\partial u}{\partial x} + v\frac{\partial u}{\partial y} = B_x - \frac{1}{\rho}\frac{\partial P}{\partial x} + \nu\left(\frac{\partial^2 u}{\partial x^2} + \frac{\partial^2 u}{\partial y^2}\right)$$

$$\frac{\partial v}{\partial t} + u\frac{\partial v}{\partial x} + v\frac{\partial v}{\partial y} = B_y - \frac{1}{\rho}\frac{\partial P}{\partial y} + \nu\left(\frac{\partial^2 v}{\partial x^2} + \frac{\partial^2 v}{\partial y^2}\right)$$

where B_x and B_y are the body forces in the x and y directions, respectively.

5.6 EULER EQUATIONS

This is a special case of the Navier-Stokes equations, when $\mu = 0$

$$\frac{\partial u}{\partial t} + u\frac{\partial u}{\partial x} + v\frac{\partial u}{\partial y} = B_x - \frac{1}{\rho}\frac{\partial P}{\partial x}$$

$$\frac{\partial v}{\partial t} + u\frac{\partial v}{\partial x} + v\frac{\partial v}{\partial y} = B_y - \frac{1}{\rho}\frac{\partial P}{\partial y}$$

5.7 IMPORTANT PHYSICAL PARAMETERS

A) Eckert number: $Ec = \dfrac{U^2}{C_p \cdot \Delta T}$: Due to viscous dissipation

B) Grashof number: $Gr = \dfrac{gl^3 B \Delta T}{\nu^2}$: Due to buoyancy forces

C) Nusselt number: $Nu_L = \dfrac{hL}{k}$: Dimensionless heat transfer coefficient

D) Prandtl number: $Pr = \dfrac{\mu \cdot C_p}{k}$: A measure of momentum transferred and heat conducted in the fluid
$ = \dfrac{\nu}{\alpha}$

E) Rayleigh number: $Ra = Gr \cdot Pr$

F) Reynolds number: $Re = \dfrac{VL}{\nu}$: A measure of the magnitude of the inertia forces in the fluid to the viscous forces

CHAPTER 6

FORCED CONVECTION HEAT TRANSFER

6.1 LAMINAR BOUNDARY LAYER ON A FLAT PLATE

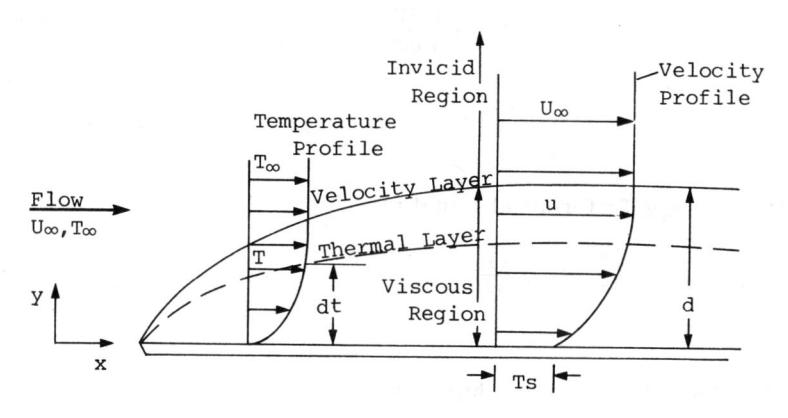

Fig. 6.1 Laminar Boundary Layer

δ = Velocity boundary layer thickness
U_∞ = Free-stream velocity
δ_t = Thermal boundary layer
T_∞ = Fluid temperature far away from plate
T_s = Plate surface temperature

E-50

Governing Equations

The governing equation for two-dimensional boundary layer apply under the following assumptions:

1) Newtonian fluid
2) Steady state
3) Constant properties

A) Continuity Equation

$$\frac{\partial u}{\partial x} + \frac{\partial v}{\partial y} = 0 \qquad (6.1)$$

B) Momentum Differential Equation

$$u\frac{\partial u}{\partial x} + v\frac{\partial u}{\partial y} = -\frac{1}{\rho}\frac{dP}{dx} + \nu\frac{\partial^2 u}{\partial y^2} \qquad (6.2)$$

where $P = P(x)$.

C) Energy Differential Equation

$$u\frac{\partial T}{\partial x} + v\frac{\partial T}{\partial y} = \alpha\frac{\partial^2 T}{\partial y^2} \qquad (6.3)$$

D) Momentum Integral Equation

$$\frac{d}{dx}\int_0^\delta u(U_\infty - u)dy = \nu\left(\frac{\partial u}{\partial y}\right)_{y=0} \qquad (6.4)$$

E) Integral Energy Equation

$$\frac{d}{dx}\int_0^{\delta_t} (T_\infty - T)u\,dy = \alpha\left(\frac{\partial T}{\partial y}\right)_{y=0} \qquad (6.5)$$

6.2 SOLUTION TO THE VELOCITY BOUNDARY LAYER

6.2.1 BLASIUS SOLUTION

This is an exact solution derived from the differential equations of the boundary layer (6.1) and (6.2).

$$u(0) = 0$$
$$u(\infty) = U_\infty$$
$$u = U_\infty \quad \text{at } x = 0$$

Using the similarity transformation method it can be shown that

$$f'''(\eta) + \frac{1}{2} f(\eta) f''(\eta) = 0 \quad \text{(Blasius equation)} \tag{6.6}$$

where $\eta = y \sqrt{\dfrac{U_\infty}{\nu x}} = \eta(x,y)$

$$u(x) = U_\infty f'(\eta)$$

Thus, the boundary conditions become

$$f(0) = 0 \qquad f''(0) = 0$$
$$f'(0) = 0 \qquad f'''(0) = 0$$
$$f'(\infty) = 1$$

The solution is shown in Fig. 6.2 and table 6.1.

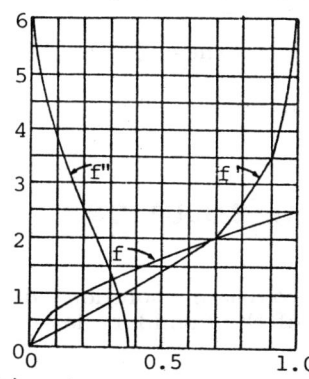

Fig. 6.2 Blasius Solution

η	f	f'	f''
0	0	0	0.3321
0.4	0.0266	0.133	0.3315
0.8	0.1061	0.265	
1.2	0.238	0.394	
1.6	0.420	0.517	
2.0	0.650	0.630	
2.8	1.230	0.812	
3.6	1.930	0.923	
4.4	2.69	0.976	0.039

Table 6.1

The local skin coefficient is defined by

$$C_f = \frac{\tau_0}{\rho U_\infty^2 /2} = 0.664 \, Re_x^{-\frac{1}{2}} \tag{6.7}$$

where τ_0 is the shear stress at the surface.

The average skin-friction coefficient is given by

$$\overline{C}_f = \frac{\overline{\tau_0}}{\rho U_\infty^2 /2} = 1.328 \, Re_x^{-\frac{1}{2}} \tag{6.8}$$

The boundary layer thickness (δ):

$$\delta = 5.0 \times Re^{-\frac{1}{2}} \tag{6.9}$$

6.2.2 INTEGRAL SOLUTION

This is an approximate solution derived from the integral equations of the boundary layer eq.(6.4).

E-53

Boundary conditions:

$$u(0) = 0$$
$$u(\delta) = U_\infty, \quad \left.\frac{\partial u}{\partial y}\right)_{y=\delta} = 0$$

$$\left.\frac{\partial^2 u}{\partial y^2}\right)_{y=0} = 0 \quad \text{(for constant pressure, } dP/dx = 0\text{)}$$

Now, we have to assume a velocity profile in such a way that it will satisfy the boundary conditions. Let us take a cubic polynomial equation:

$$u(x,y) = A + By + Cy^2 + Dy^3 \qquad (6.10)$$

where A, B, C, D are functions of x to be determined using the boundary conditions. The solution is

$$\boxed{u = U_\infty \left[\frac{3}{2} \frac{y}{\delta} - \frac{1}{2}\left(\frac{y}{\delta}\right)^3\right]} \qquad (6.11)$$

Inserting equation (6.11) in the momentum integral equation (6.4) yields,

$$\frac{d}{dx} \int_0^\delta U_\infty \left[\frac{3}{2}\frac{y}{\delta} - \frac{1}{2}\left(\frac{y}{\delta}\right)^3\right] \left[U_\infty - U_\infty\left(\frac{3}{2}\frac{y}{\delta} - \frac{1}{2}\left(\frac{y}{\delta}\right)^3\right)\right] dy$$
$$= \frac{3}{2} \nu \frac{U_\infty}{\delta} \qquad (6.12)$$

Integrating the above equation we get the approximate solution for the boundary layer thickness

$$\boxed{\delta = 4.64 \times Re_x^{-\frac{1}{2}}} \qquad (6.13)$$

The skin-friction coefficient

$$C_f = 0.646 \, Re^{-\frac{1}{2}} \qquad (6.14)$$

E-54

6.3 SOLUTION TO THE THERMAL BOUNDARY LAYER

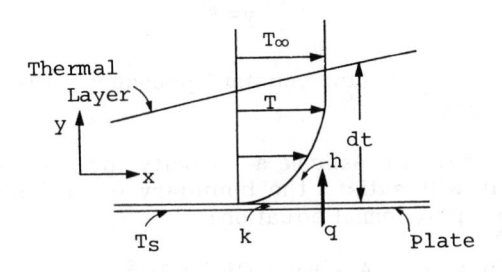

Fig. 6.3

From Fig. 6.3 we can see that the heat conducted through the plate must be equal to the heat convected to the fluid close to the plate. Then, the energy balance at the plate surface is:

$$q = -kA \left(\frac{\partial T}{\partial y}\right)_{surface} = hA(T_s - T_\infty) \quad (6.15)$$

from where the convection heat transfer coefficient is obtained:

$$\boxed{h = \frac{-k(\partial T/\partial y)_{surface}}{T_s - T_\infty}} \quad (6.16)$$

In order to evaluate the heat transfer coefficient, we must find the temperature distribution in the thermal boundary layer. Solving this problem in a similar fashion used to solve the velocity boundary layer, we have:

6.3.1 POHLHAUSEN SOLUTION

This is an exact solution derived from the energy differential equation (eq. 6.3).

E-55

Boundary conditions

$$T(0) = T_o \quad \left.\frac{\partial^2 T}{\partial y^2}\right|_{y=0} = 0$$

$$T(\delta_t) = 0 \quad \left.\frac{\partial T}{\partial y}\right|_{y=\delta_t} = 0$$

Using the similarity transformation method it can be shown that

$$\boxed{\theta'' + \frac{1}{2} \Pr f(\eta)\theta' = 0} \qquad (6.17)$$

where

$$\eta = y\sqrt{\frac{u_\infty}{\nu x}}$$

$$\theta = \frac{T(\eta) - T_o}{T_\infty - T_o} = \theta(\eta)$$

Boundary conditions become

$$\theta(0) = 0$$
$$\theta(\infty) = 1$$

with the function $f(\eta)$ given, the solution is

$$\boxed{\theta(\eta) = \frac{\int_0^\eta \exp\left[-\frac{\Pr}{2}\int_0^\eta f(\eta)d\eta\right] d\eta}{\int_0^\infty \exp\left[-\frac{\Pr}{2}\int_0^\eta f(\eta)d\eta\right] d\eta}} \qquad (6.18)$$

Figure 6.4 shows a graphical result of the temperature profile for various Prandtl numbers.

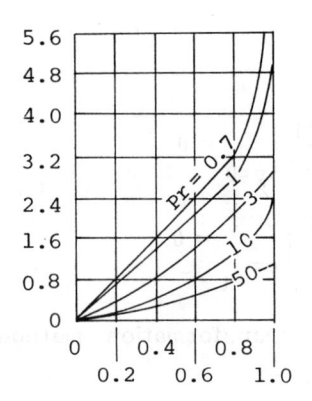

Fig. 6.4 $\theta = \dfrac{T-T_0}{T_\infty-T_0}$

At the surface, the temperature gradient is approximately

$$\left.\frac{\partial T}{\partial y}\right|_{\text{surface}} = (T_\infty - T_s)\sqrt{\frac{u_\infty}{\nu x}}(0.332)\text{Pr}^{1/3} \qquad (6.19)$$

and the dimensionless heat transfer coefficient is

$$\text{Nu}_x = \frac{hx}{k} = 0.332\,\text{Pr}^{1/3}\,\text{Re}_x^{1/2} \qquad (6.20)$$

the average Nusselt number is

$$\overline{\text{Nu}}_L = \frac{hL}{k} = 0.664\,\text{Pr}^{1/3}\,\text{Re}_L^{1/2} \qquad (6.21)$$

6.3.2 INTEGRAL SOLUTION

This is an approximate solution derived from the integral energy equation (eq. 6.5).

Boundary conditions

$$T(0) = T_s \qquad \left.\frac{\partial T}{\partial y}\right|_{y=\delta_t} = 0$$

$$T(\delta_t) = T_\infty \quad \left.\frac{\partial^2 T}{\partial y^2}\right|_{y=0} = 0 \quad (\text{when } v = 0)$$

Now, as in the velocity boundary layer, we have to assume a temperature profile that will satisfy the boundary conditions. Let us take a cubic polynomial equation

$$T(x,y) = A + By + Cy^2 + Dy^3 \tag{6.22}$$

again, A,B,C,D are functions of x to be determined using the boundary conditions. Finally, the solution is

$$\boxed{\frac{\theta}{\theta_0} = \left[\frac{3}{2} y/\delta_t - \frac{1}{2}(y/\delta_t)^3\right]} \tag{6.23}$$

where
$$\theta = T - T_s$$
$$\theta_0 = T_\infty - T_s$$

Now, an expression for the thermal boundary layer thickness δ_t, must be found. This can be done by an analysis of the energy equation of the boundary layer. It can be shown that the thermal boundary layer thickness is given by

$$\delta_t = \frac{\delta}{1.026} Pr^{-1/3} \quad \text{for } Pr > 0.7 \tag{6.24}$$

where δ is the velocity boundary layer thickness. Thus, the dimensionless convective heat transfer coefficient is

$$\boxed{Nu_x = \frac{hx}{k} = 0.332\, Pr^{1/3}\, Re_x^{1/2} \left[1 - \left(\frac{x_0}{x}\right)^{3/4}\right]^{-1/3}}$$
$$\tag{6.25}$$

where x_0 is the unheated part of the plate. If $x_0 = 0$, the plate is totally heated and

$$Nu_x = 0.332\, Pr^{1/3}\, Re_x^{1/2} \tag{6.26}$$

E-58

The average Nusselt number is

$$\overline{Nu}_L = 2Nu \Big|_{x=L} \qquad (6.27)$$

Note: The fluid properties should be evaluated at the film temperature, T_f.

$$T_f = \frac{T_s + T_\infty}{2} \qquad (6.28)$$

6.4 LAMINAR FLOW INSIDE PIPES

6.4.1 CIRCULAR PIPES

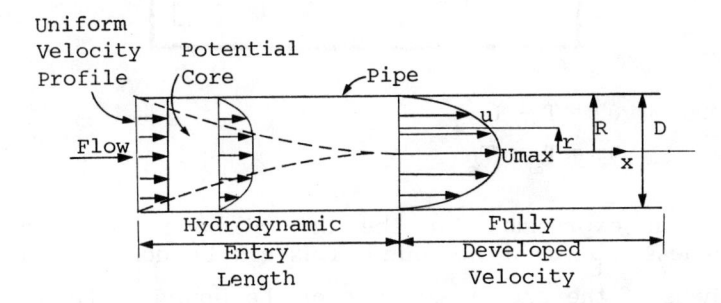

Fig. 6.5 Development of Velocity Profile

Governing Equation

The governing equation for laminar flow inside tubes or pipes applies under the following conditions

a) Newtonian fluids

b) Steady flow

c) Constant properties

$$\frac{\mu}{r} \frac{d}{dr}\left(r \frac{\partial u}{\partial r}\right) = \frac{dP}{dx} \qquad (6.29)$$

Boundary conditions

a) $\left. \dfrac{\partial u}{\partial r} \right|_{r=0} = 0$

b) $u(R) = 0$

Solving equation (6.29) for the velocity distribution, u, we get

$$u(r) = 2U_m \left[1 - \left(\frac{r}{R}\right)^2\right] \quad (6.30)$$

where U_m is the mean velocity.

The friction factor is given by

$$f = \frac{64}{Re_D} \quad (6.31)$$

Entry length (Le) - It is the length required to establish a constant velocity profile along the axis of the pipe.

$$Le = 0.0575 \, D \, Re_D \quad (6.32)$$

(Langhaar's formula)

Reynolds number

$$Re_D = \frac{U_m D}{\nu} \quad (6.33)$$

6.4.2 NONCIRCULAR PIPES

The governing equation is given by

$$\mu \left(\frac{\partial^2 u}{\partial x^2} + \frac{\partial^2 u}{\partial y^2} + \frac{\partial^2 u}{\partial z^2}\right) = \frac{dP}{dx} \quad (6.34)$$

The solution of this equation is given in Figure 6.7 for a family of rectangular tubes and concentric annuli. The Reynold number is defined as

$$\boxed{Re = \frac{4r_h \bar{V}}{\mu}} \qquad (6.35)$$

where $r_h = \frac{A_c L}{P}$ (Hydraulic radius)

A_c = Cross-sectional area

P = Perimeter

L = Tube length

$\bar{V} = \frac{\dot{m}}{A_c}$ (mean mass velocity)

$R^* = \frac{r_1}{r_2}$

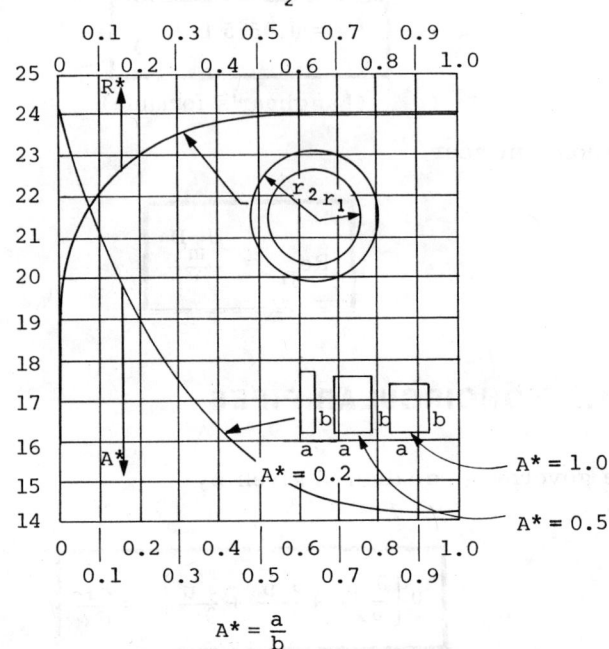

$\frac{1}{4} fRe$

$A^* = \frac{a}{b}$

Fig. 6.6

6.5 LAMINAR FLOW HEAT TRANSFER IN CIRCULAR PIPES

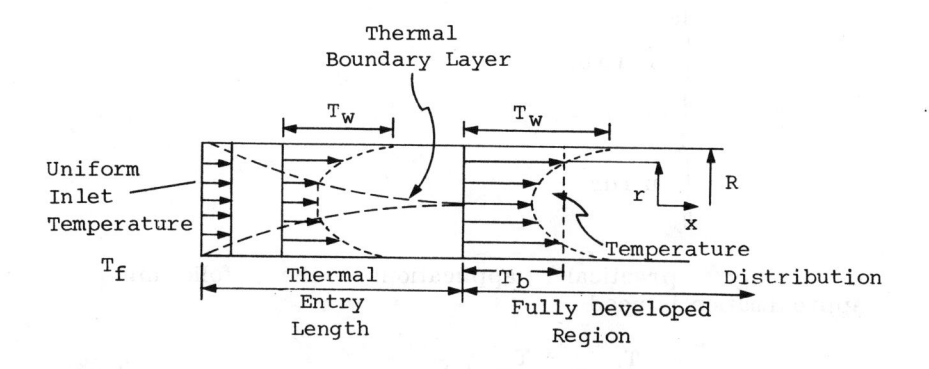

Fig. 6.7 Development of temperature profile when Tw = constant and $T_w > T_f$

6.5.1 GOVERNING EQUATIONS

The governing equation for heat transfer inside tubes or pipes applies under the following conditions:

a) No axial conduction

b) Incompressible fluid

c) Steady laminar flow

d) Constant properties

$$\frac{1}{r} \frac{\partial}{\partial r}\left(r \frac{\partial T}{\partial r}\right) = \frac{u}{\alpha} \frac{dT}{dx} \qquad (6.36)$$

6.5.2 DEFINITIONS

Bulk fluid temperature or mean fluid temperature, T_b, is defined as

$$T_b = \frac{\int_0^R T u \, r dr}{\int_0^R u \, r dr} \qquad (6.37)$$

in many practical applications the following approximation is used

$$T_b = \frac{T_{inlet} + T_{outlet}}{2} \qquad (6.38)$$

The fully developed thermal profile is defined as

$$\frac{\partial \theta}{\partial x} = 0 \qquad (6.39)$$

where $\theta = \dfrac{T_s - T}{T_s - T_b}$

6.5.3 DIMENSIONLESS HEAT TRANSFER COEFFICIENT

a) Constant-Wall Temperature (Fig. 6.8)

$$Nu_D = \frac{hD}{k} = 3.658 \qquad (6.40)$$

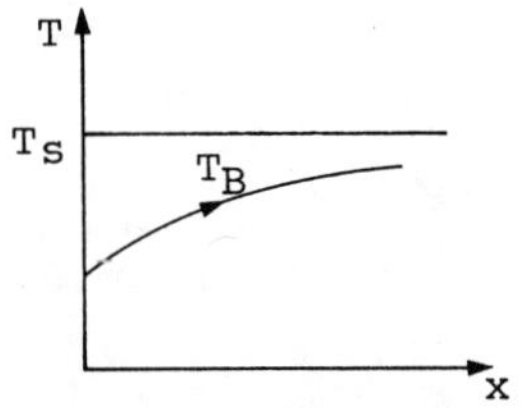

Fig. 6.8

E-63

b) Constant Heat Rate (Fig. 6.9)

$$Nu_D = \frac{hD}{k} = 4.364 \qquad (6.41)$$

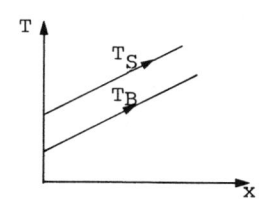

Fig. 6.9

6.6 LAMINAR FLOW HEAT TRANSFER IN NON CIRCULAR PIPES

The governing equation is given by

$$\frac{\partial^2 T}{\partial y^2} + \frac{\partial^2 T}{\partial z^2} = \frac{u}{\alpha} \frac{\partial T}{\partial x} \qquad (6.42)$$

The dimensionless heat transfer coefficients (Nu) for some non-circular pipes are shown in table 6.2 where the temperature has been assumed constant around the periphery and the conductance is an average with respect to the cross-sectional area. The Nusselt number is defined as

$$\boxed{Nu = \frac{hD_h}{k}} \qquad (6.43)$$

where D_h is the hydraulic diameter.

Table 6.2

Cross-section	b/a	Nu (Tw = Constant)	Nu (Heat rate constant)
□ a	1.0	2.98	3.61
b ▯ b / a	2.0	3.39	4.12
	3.0	3.96	4.79
	4.0	4.44	5.33
	8.0	5.60	6.49
══	∞	7.54	8.235
△		2.35	3.00

6.7 TURBULENT BOUNDARY LAYER ON A FLAT PLATE

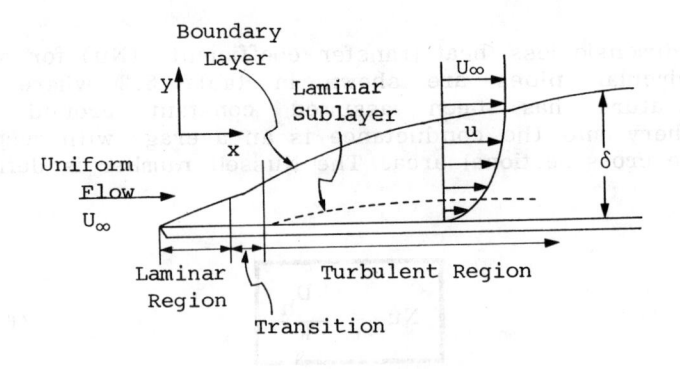

Fig. 6.10 Turbulent Boundary Layer

E-65

Governing Equations

The governing equations for two-dimensional boundary apply under the following assumptions:
A) Steady flow
B) Constant properties
C) Newtonian fluid

A) Continuity Equation

$$\frac{\partial \bar{u}}{\partial x} + \frac{\partial \bar{v}}{\partial y} = 0 \qquad (6.44)$$

where $\bar{u} = u - u'$
$\bar{v} = v - v'$

and u', v' are the velocity deviations in the x and y directions respectively.

B) Momentum Differential Equation

$$\boxed{\bar{u}\frac{\partial \bar{u}}{\partial x} + \bar{v}\frac{\partial \bar{v}}{\partial y} + \frac{1}{\rho}\frac{d\bar{P}}{dx} = \frac{\partial}{\partial y}\left[\nu\frac{\partial \bar{u}}{\partial y} - \overline{u'v'}\right]} \qquad (6.45)$$

where $\overline{u'v'} = -\varepsilon_M \frac{\partial \bar{u}}{\partial y}$ (Apparent turbulent shear stress)

ε_M = Eddy diffusivity for momentum

$\bar{P} = P - P'$ (Pressure deviation)

C) Energy Differential Equation

$$\boxed{\bar{u}\frac{\partial \bar{T}}{\partial x} + \bar{v}\frac{\partial \bar{T}}{\partial y} - \frac{\partial}{\partial y}\left[\alpha\frac{\partial \bar{T}}{\partial y} - \overline{T'v'}\right] = 0} \qquad (6.46)$$

where $\overline{T'v'} = -\varepsilon_H \frac{\partial \bar{T}}{\partial y}$ (Apparent turbulent heat flux)

ε_H = Eddy diffusivity for heat transfer

D) Integral Momentum Equation

$$\boxed{\frac{d}{dx}\int_0^\delta u(u_\infty - u)dy = \frac{\tau_s}{\rho}} \qquad (6.47)$$

where τ_s = Shear stress at the surface

6.8 SOLUTION TO THE VELOCITY BOUNDARY LAYER (TURBULENT)

In order to calculate the turbulent boundary layer thickness let us assume an approximate velocity profile (Von-Karman).

$$\frac{u}{u_\infty} = \left(\frac{y}{\delta}\right)^{1/7} \qquad \text{(one-seventh power law)} \qquad (6.48)$$

also, let us assume a value for the shear stress (from experimental results),

$$\tau_s = 0.0225\, \rho u_\infty^2 \left(\frac{\nu}{u_\infty \delta}\right)^{1/4} \qquad (6.49)$$

Introducing equations (6.48) and (6.49) in the integral momentum equation (6.47) and solving for δ, we get,

$$\boxed{\frac{\delta}{x} = 0.376\, Re^{-1/5}} \qquad (6.50)$$

E-67

The skin-friction coefficient is given by

$$\boxed{C_f = 0.0576 \, Re_x^{-1/5}} \qquad (6.51)$$

The average skin-friction coefficient is

$$\boxed{\overline{C}_f = 0.072 \, Re_L^{-1/5}} \qquad (6.52)$$

6.9 TURBULENT BOUNDARY LAYER WITH HEAT TRANSFER (FLAT PLATE)

From the energy differential equation (6.46) we have,

$$\boxed{\overline{u} \, \frac{\partial \overline{T}}{\partial x} + v \, \frac{\partial \overline{T}}{\partial y} - \frac{\partial}{\partial y} \left[(\alpha + \varepsilon_H) \, \frac{\partial \overline{T}}{\partial y} \right] = 0} \qquad (6.53)$$

Reynolds Analogy - It is a relation between the fluid friction and the heat transfer. By definition for turbulent boundary layer we have,

$$\boxed{\frac{\tau}{\rho} = \varepsilon_M \, \frac{d\overline{u}}{dy}} \qquad (6.54)$$

$$\boxed{\frac{q''}{\rho C_p} = \varepsilon_H \, \frac{d\overline{T}}{dy}} \qquad (6.55)$$

by comparison, it may be seen that

$$\varepsilon_H = \varepsilon_M \tag{6.56}$$

which corresponds to Pr = 1. This is a reasonable approximation for real fluids (Pr > 0.5). From this analogy, it can be shown that the Stanton number is,

$$\boxed{St_x = \frac{Nu_x}{Re_x Pr} = \frac{h_x}{\rho C_p u_\infty} = \frac{C_f}{2}} \tag{6.57}$$

A modified Reynolds analogy useful for 0.6 < Pr < 50 is the Colburn's Analogy,

$$\boxed{St_x \, Pr^{2/3} = \frac{C_{fx}}{2}} \tag{6.58}$$

The average heat transfer coefficient for combined laminar and turbulent flow is given by

$$\boxed{\overline{Nu} = \frac{\overline{h}L}{k} = (0.036 \, Re_L^{0.8} - 836) Pr^{1/3} \text{ for } Re = 5 \times 10^5} \tag{6.59}$$

The Von Karman's Analogy is an equation for Pr ≠ 1

$$\boxed{St_x = \frac{C_f/2}{1 + \sqrt{C_f/2} \, \{Pr-1+\ln[1 + 5/6(Pr-1)]\}}} \tag{6.60}$$

Note that for Pr = 1 the above equation reduces to the Reynolds analogy eq.(6.57).

Film Temperature - It is the temperature at which all the temperature dependent parameters should be evaluated. It is defined as,

$$T_f = \frac{T_s + T_\infty}{2} \qquad (6.61)$$

6.10 TURBULENT FLOW INSIDE PIPES

6.10.1 CIRCULAR PIPES

The governing equation for turbulent flow inside smooth pipes applies under the following conditions:

a) Fully developed flow

b) $\bar{v}_r = 0$, $\bar{u} = \bar{u}(r)$

c) Steady flow

d) Constant properties

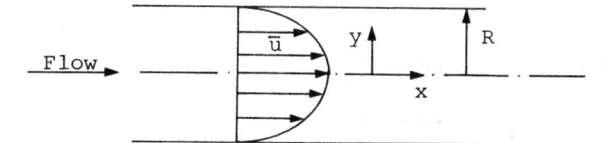

$$\frac{1}{r}\frac{d}{dr}\left[(\nu+\varepsilon_M)r\frac{d\bar{u}}{dr}\right] = \frac{1}{\rho}\frac{d\bar{P}}{dx} \qquad (6.62)$$

Fig. 6.11 Fully Developed Turbulent Velocity Profile

The one-seventh law profile is a good approximation for the velocity profile ($10^4 < Re_b < 10^6$).

$$\boxed{\frac{\bar{u}}{u_c} = \left(\frac{y}{R}\right)^{1/7}} \qquad (6.63)$$

where u_c is the centerline velocity.

E-70

Law of the Wall

$$u^* = \sqrt{\frac{\tau_s}{\rho}} \qquad (6.64)$$

Von Karman's law of the wall

$$u^+ = 2.5 \ln y^+ + 5.5 \quad \text{for } y^+ > 30 \qquad (6.65)$$
$$u^+ = 5.0 \ln y^+ - 3.05 \quad \text{for } 5 < y^+ < 30 \qquad (6.66)$$
$$u^+ = y^+ \quad \text{for } 0 < y^+ < 5 \qquad (6.67)$$

where
$$u^+ = \frac{\bar{u}}{u^*} \qquad (6.68)$$

$$y^+ = \frac{y u^*}{\nu} \qquad (6.69)$$

Mean velocity

$$u_m = 0.817 \, U_c \qquad (6.70)$$

The skin-friction coefficient is

$$\frac{C_f}{2} = 0.039 \, Re^{-0.25} \qquad \text{for } 10^4 < Re < 5 \times 10^4$$
(6.71)

6.10.2 NONCIRCULAR PIPES

The cross-section shape of a pipe has little effect on the shear stress at the surface because the velocity changes in turbulent flow occur near the pipe surface. The formulas mentioned above apply to non-circular pipes introducing the concept of hydraulic radius,

E-71

$$\boxed{r_h = \frac{A_c L}{P}}$$ (6.72)

where $r_h = D_h/4$

Note: The equations do not apply for triangular ducts with very small angles.

6.11 FLOW INSIDE PIPES WITH HEAT TRANSFER

The energy equation for fully-developed flow inside pipes is given by

$$\boxed{\frac{1}{r}\frac{\partial}{\partial r}\left[r(\alpha+\varepsilon_H)\frac{\partial \bar{T}}{\partial r}\right] = \bar{u}\frac{\partial \bar{T}}{\partial x}}$$ (6.73)

The Reynolds analogy for pipe flow is given by

$$\boxed{St_x = \frac{h}{\rho C_p U_m} = \frac{Nu_D}{Re_D Pr} = \frac{C_f}{2}}$$ (6.74)

A modified Reynolds analogy useful for $0.5 < Pr < 100$ is the Colburn's analogy

$$\boxed{St_x Pr^{2/3} = \frac{C_{fx}}{2}}$$ (6.75)

Bulk or mean temperature

$$T_b = \frac{T_i + T_o}{2}$$ (6.76)

where T_i and T_o are the inside and outside fluid temperatures, respectively.

E-72

CHAPTER 7

CORRELATIONS FOR FORCED CONVECTION

7.1 DEFINITIONS

A) Bulk Temperature

$$T_b = \frac{\int_0^R urTdr}{\int_0^R urdr} \qquad (7.1)$$

B) Film Temperature

$$T_f = \frac{T_s + T_\infty}{2} \qquad (7.2)$$

C) Local Nusselt Number

$$Nu_x = \frac{hx}{k} \qquad (7.3)$$

E-73

D) Average Nusselt Number

$$\boxed{Nu_L = \frac{hL}{k}} \qquad (7.4)$$

E) Critical Reynolds Numbers

$$\boxed{Re_c = \frac{UX_c}{\nu}} \qquad (7.5)$$

where X_c is the distance at which transition from laminar to turbulent flow takes place.

F) Stanton Number

$$\boxed{St_x = \frac{Nu_x}{Re_x\, Pr} = \frac{h_x}{\rho\, C_p\, U_\infty}} \qquad (7.6)$$

G) Peclet Number

$$\boxed{Pe_x = Re_x\, Pr = \frac{U_\infty X}{\alpha}} \qquad (7.7)$$

7.2 FLOW OVER A FLAT PLATE

7.2.1 LAMINAR FLOW

$$Nu_x = 0.332\, Re_x^{1/2} \cdot Pr^{1/3} \qquad (7.8)$$

$$Nu_L = 0.664\, Re_L^{1/2} \cdot Pr^{1/3} \qquad (7.9)$$

Conditions

a) $0.6 \leq Pr \leq 50$
b) $100 < Re < Re_{x,c} \simeq 5 \times 10^5$
c) Evaluate properties at T_f
d) Dissipation is neglected

$$Nu_x = 0.565(Re_x \cdot Pr)^{\frac{1}{2}} \qquad (7.10)$$
$$Nu_L = 1.130(Re_L \cdot Pr)^{\frac{1}{2}} \qquad (7.11)$$

Conditions

a) $Pr < 0.05$
b) $Re < Re_{x,c} \simeq 5 \times 10^5$
c) Evaluate properties at T_f
d) Dissipation is neglected.

7.2.2 TURBULENT FLOW

$$\boxed{Nu_x = 0.0296\, Re_x^{0.8} \cdot Pr^{1/3}, \quad 5 \times 10^5 < Re_x < 10^7}$$

(7.12)

$$\boxed{Nu_x = 0.185\, Re_x (\log_{10} Re_x)^{-2.584} \cdot Pr^{1/3}, \quad Re_x > 10^7}$$

(7.13)

Conditions

a) $0.6 \leq Pr \leq 60$
b) Properties at T_f
c) Dissipation is neglected

$$\boxed{Nu_L = [0.037 Re_L^{0.8} - \frac{A}{2}]Pr^{1/3},}$$ (7.14)

$5 \times 10^5 < Re_L < 10^7$

$$\text{Nu}_L = [0.228 \text{Re}_L (\log_{10} \text{Re}_L)^{-2.584} - \frac{A}{2}] \text{Pr}^{1/3}, \quad (7.15)$$

$10^7 < \text{Re}_L < 10^9$

Conditions

a) $0.6 \leq \text{Pr} \leq 60$
b) $\text{Re}_{x,c} = 5 \times 10^5$
c) Evaluate properties at T_f
d) Dissipation is neglected
e) Obtain A from table 7.1

Table 7.1

Transition Reynolds Number $\text{Re}_{x,c}$	A
3×10^5	1055
5×10^5	1743
1×10^6	3341
3×10^6	8944
Laminar portion is ignored	0

7.3 FLOW INSIDE CYLINDRICAL PIPES AND TUBES

7.3.1 LAMINAR FLOW

$$\text{Nu}_d = \frac{h \cdot d}{k} = 3.66 \quad (7.16)$$

Conditions

a) Fully developed flow

E-76

b) $Re_d < 2300$

c) Constant surface temperature, T_s

d) Evaluate properties at T_b

e) $(d/L)Re_d \cdot Pr < 10$

f) Smooth tubes

$$Nu_d = 1.86 \left[\left(\frac{d}{L} \right) Re_D \cdot Pr \right]^{1/3} \left(\frac{\mu}{\mu_3} \right)^{0.14} \quad (7.17)$$

Conditions

a) $0.48 < Pr < 16,700$

b) Evaluate properties, except μ_3, at T_b; μ_s at T_s

c) $(D/L)Re_D \cdot Pr > 10$

7.3.2 TURBULENT FLOW

$$Nu_d = 0.023 Re_d^{0.8} \cdot Pr^n \quad (7.18)$$

Conditions

a) $n = 0.4$ for $T_s > T_b$
   $n = 0.3$ for $T_s < T_b$

b) $0.7 < Pr < 160$
   $10^4 < Re_D < 10^6$

c) $|T_s - T_b| < 6°C$ for liquids
   $|T_s - T_b| < 60°C$ for gases

d) Evaluate properties at T_b

e) Fully developed flow;
Effect of starting length is negligible;
$L/d \geq 60$

f) $Re_d > 2300$

$$Nu_d = 0.027 Re_d^{0.8} \cdot Pr^{1/3} \left(\frac{\mu}{\mu_s}\right)^{0.14} \quad (7.19)$$

Conditions

a) $0.7 < Pr < 160$
b) $10^4 < Re_d < 10^6$
c) Evaluate properties, except μ_s, at T_b; μ_s at T_s
d) Fully developed flow;
Effect of starting length is negligible;
$L/d \geq 60$

7.4 FLOW NORMAL TO A CIRCULAR CYLINDER

$$Nu_d = C \cdot Re_d^m \cdot Pr^n \left(\frac{Pr_\infty}{Pr_s}\right)^{\frac{1}{4}} \quad (7.20)$$

Conditions

a) $0.7 < Pr < 500$
b) $1 < Re_d < 10^6$

c) $n = 0.37$, $Pr \leq 10$
   $n = 0.36$, $Pr > 10$

d) Evaluate properties at $T_f = \dfrac{T_\infty + T_s}{2}$
   except Pr_∞, Pr_s

   Pr_∞ at T_∞

   Pr_s at T_s

e) Single tube or cylinder

f) Obtain C and m from table 7.2

Table 7.2

Range of Re_d	C	m
1 - 40	0.75	0.4
$4 \times 10 - 1 \times 10^3$	0.51	0.5
$1 \times 10^3 - 2 \times 10^5$	0.26	0.6
$2 \times 10^5 - 1 \times 10^6$	0.076	0.7

7.5 FLOW NORMAL TO NONCIRCULAR CYLINDERS

$$\boxed{Nu_d = C\, Re_d^m\, Pr^{1/3}} \qquad (7.21)$$

Conditions

a) Correlation applicable to gases

b) Evaluate properties at T_f

c) Obtain C and m from table 7.3

Table 7.3

Geometric Configuration	Re_d	C	m
■	5×10^3-1×10^5	0.102	0.657
◆	5×10^3-1×10^5	0.246	0.588
⬢	5×10^3-1.95×10^4	0.160	0.638
	1.95×10^4-1×10^5	0.0385	0.782
⬡	5×10^3-1×10^5	0.153	0.638
▮	4×10^3-1.5×10^4	0.228	0.731

7.6 FLOW OVER BANKS OF TUBES

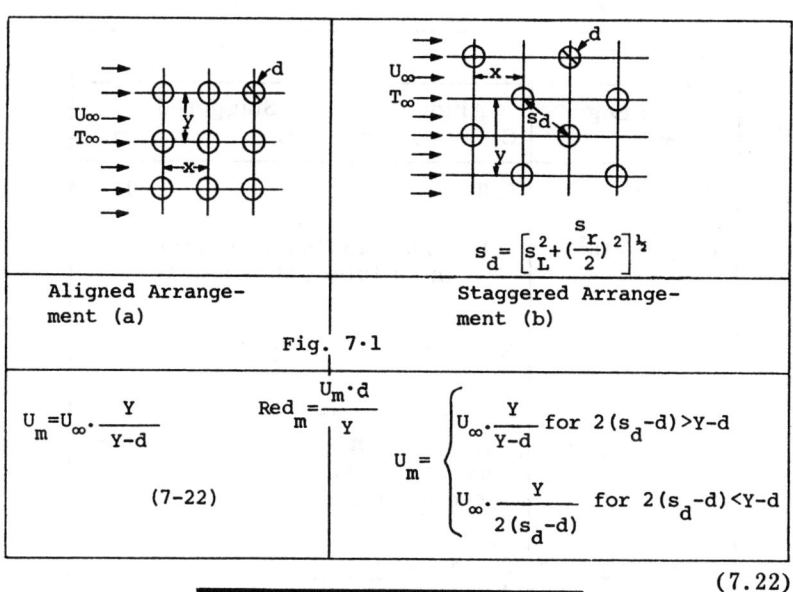

Aligned Arrangement (a)

Staggered Arrangement (b)

$$s_d = \left[s_L^2 + \left(\frac{s_T}{2}\right)^2\right]^{1/2}$$

Fig. 7·1

$$U_m = U_\infty \cdot \frac{Y}{Y-d} \qquad (7\text{-}22)$$

$$Re_{d_m} = \frac{U_m \cdot d}{\nu}$$

$$U_m = \begin{cases} U_\infty \cdot \dfrac{Y}{Y-d} & \text{for } 2(s_d-d) > Y-d \\[2mm] U_\infty \cdot \dfrac{Y}{2(s_d-d)} & \text{for } 2(s_d-d) < Y-d \end{cases}$$

(7.22)

$$\boxed{Nu_d = C Re_{dm}^n Pr^{0.36}\left(\frac{Pr_\infty}{Pr_s}\right)^{\frac{1}{4}}}$$

(7.23)

E-80

Conditions

a) $0.7 < Pr < 500$

b) $10 < Re_{dm} < 10^6$

c) Evaluate properties at T_f, except Pr_∞, Pr_s
   Pr_∞ at T_∞
   Pr_s at T_s

d) Obtain C and n from table 7.4

e) Correlation is valid for 20 or more rows. For less than 20 rows a different figure is used in which a correction factor is included:

$$\frac{(Nu_d)_{n\ rows}}{(Nu_d)_{20\ rows}}$$

Table 7.4				
Re_{D_m} Range	Aligned C	n	Staggered C	n
$10 - 10^2$	0.8	0.4	0.9	0.4
$10^2 - 1 \times 10^3$	Use Zhukauskas correlation for a single tube (section 7.4)			
	$\frac{Y}{X} < 0.7$:		$\frac{Y}{X} < 2$:	
	heat transfer is very poor and this design isn't recommended		$0.35\left(\frac{Y}{X}\right)^{1/5}$	0.6
$1 \times 10^3 - 2 \times 10^5$	$\frac{Y}{X} > 0.7$:		$\frac{Y}{X} > 2$:	
	0.27	0.63	0.40	0.6
$2 \times 10^5 - 1 \times 10^6$	0.021	0.84	0.022	0.84

E-81

7.7 LIQUID-METAL HEAT TRANSFER

7.7.1 TURBULENT FLOW ON A FLAT PLATE

$$\boxed{Nu = 0.53(Re_x Pr)^{\frac{1}{2}} = 0.53\, Pe^{\frac{1}{2}}}\qquad(7.24)$$

7.7.2 TURBULENT FLOW INSIDE TUBES

a)
$$\boxed{Nu_d = 0.625(Re_d Pr)^{0.4}}\qquad(7.25)$$

Conditions
1) Fully developed turbulent flow
2) Smooth tubes
3) Uniform heat flux at the surface
4) Evaluate properties at bulk temperature
5) Valid for $10^2 < Pe < 10^4$ and $\frac{L}{d} > 60$

b)
$$\boxed{Nu_d = 5 + 0.025(Re_d Pr)^{0.8}}\qquad(7.26)$$

Conditions
1) Constant surface temperature
2) Valid for $Pe > 10^2$ and $\frac{L}{d} > 60$
3) Evaluate properties at bulk temperature

c)
$$\boxed{Nu = 6.3 + 0.0167\, Re^{0.85}\, Pr^{0.93}}\qquad(7.27)$$

Conditions
1) Fully developed velocity and temperature profiles
2) Constant heat rate

E-82

7.8 HEAT TRANSFER IN HIGH SPEED FLOW

A) Laminar Boundary Layer

$$(St_x)_{ref}(Pr)_{ref}^{2/3} = 0.332(Re_x)_{ref}^{-\frac{1}{2}} \qquad (7.28)$$

for $Re < 5 \times 10^5$

B) Turbulent Boundary Layer

$$(St_x)_{ref}(Pr)_{ref}^{2/3} = 0.0288(Re_x)_{ref}^{-1/5} \qquad (7.29)$$

for $5 \times 10^5 < Re_x < 10^7$

$$(St_x)_{ref}(Pr)_{ref}^{2/3} = \frac{0.185}{[\log(Re_x)_{ref}]^{2.584}} \qquad (7.30)$$

for $10^7 < Re_x$

Note: The subscript ref, means that the properties must be evaluated at the reference temperature defined as

$$T_{ref} = T_\infty + 0.5(T_s - T_\infty) + 0.22(T_{as} - T_\infty) \qquad (7.31)$$

where T_s = Surface temperature
T_{as} = Actual adiabatic surface temperature
T_∞ = Flow temperature

CHAPTER 8

FREE CONVECTION

8.1 INTRODUCTION

Free convection heat transfer takes place when the fluid flows due to density changes arising from temperature gradients within the fluid in contact with a hot or cold surface. Heating units, nuclear reactor cooling systems, pipes carrying refrigerants are some examples of heat transfer by free convection.

8.2 BOUNDARY LAYER GOVERNING EQUATIONS

The governing equations for the boundary layer for free convection apply under the following assumptions:

A) Newtonian fluid

B) Steady flow

C) Two-dimensional flow

D) No viscous dissipation

E) No pressure gradients

a) Continuity Equation

$$\frac{\partial u}{\partial x} + \frac{\partial v}{\partial y} = 0 \qquad (8.1)$$

b) Momentum Equation

$$u\frac{\partial u}{\partial x} + v\frac{\partial u}{\partial y} = g\beta(T-T_\infty) + \nu\frac{\partial^2 u}{\partial y^2} \qquad (8.2)$$

where β is the volumetric coefficient of thermal expansion:

$$\beta = \frac{1}{T} \quad \text{(ideal gases)}$$

where T is absolute temperature

$\beta \approx$ Constant   (incompressible liquids)

c) Energy Equation

$$u\frac{\partial T}{\partial x} + v\frac{\partial T}{\partial y} = \alpha\frac{\partial^2 T}{\partial y^2} \qquad (8.3)$$

The boundary conditions for the above equations are:

at $y = 0$: $T = T_s$, $u = v = 0$
at $y = \infty$: $T = T_\infty$, $u = 0$
at $x = 0$: $T = T_\infty$, $u = 0$

8.3 SOLUTION OF THE PROBLEM OF LAMINAR FREE CONVECTION PAST A VERTICAL SURFACE

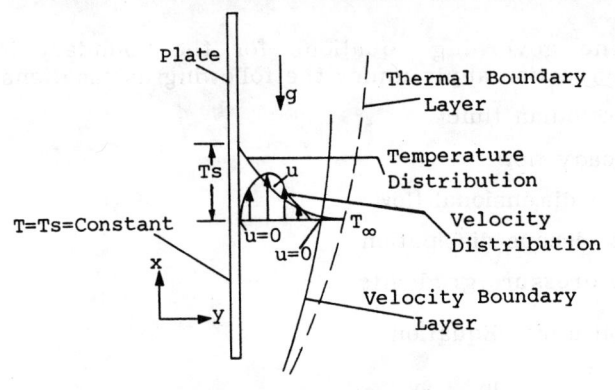

Fig. 8·1 Free-convection Velocity and Thermal Boundary Layer

E-85

8.3.1 THE SIMILARITY SOLUTION

Similarity Parameter

$$\eta = \frac{y}{x}\left[\frac{\beta g(T_s - T_\infty)}{4\nu^2}\right]^{\frac{1}{4}} = \frac{y}{x}\left(\frac{Gr_x}{4}\right)^{\frac{1}{4}} \quad (8.4)$$

where Gr_x is a dimensionless parameter called Grashof number which describes the ratio of buoyancy to viscous forces,

$$Gr_x = \frac{g\beta x^3(T_s - T_\infty)}{\nu^2} \quad (8.5)$$

The differential equation for T_s = constant are:

$$F''' + 3FF'' - 2(F')^2 + \phi = 0 \quad (8.6)$$

$$\phi'' + 3Pr\, F\phi = 0 \quad (8.7)$$

where

$$\phi = \frac{T - T_\infty}{T_s - T_\infty} = \phi(\eta) \quad \text{and} \quad F = F(\eta)$$

The boundary conditions become

$F(0) = F'(0) = 0, \quad \phi(0) = 1$

$F'(\infty) = 0, \quad \phi(\infty) = 0$

Results:

Figures 8.2 and 8.3 show the solution of the differential equations for different Prandtl numbers.

The local heat transfer coefficient is given by

$$Nu_x = -\phi'(0)\left(\frac{Gr_x}{4}\right)^{\frac{1}{4}} \quad (8.8)$$

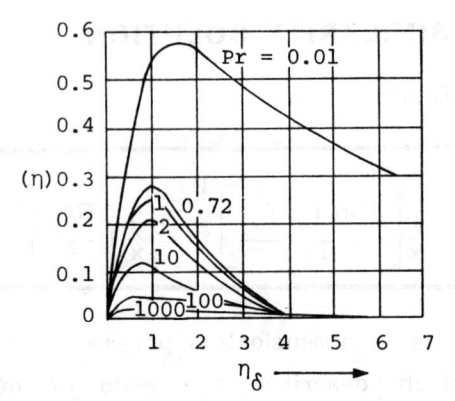

Fig. 8·2 Velocity Profile

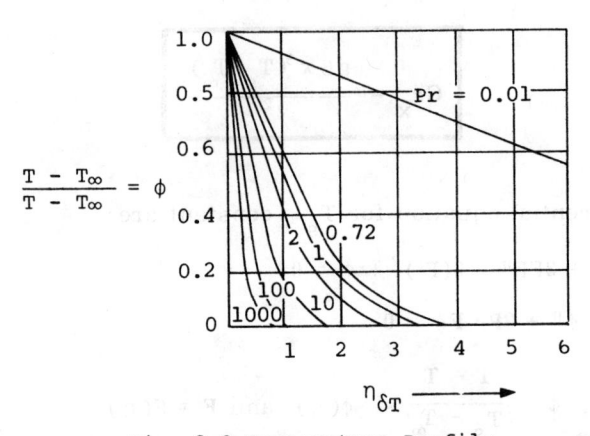

Fig. 8·3 Temperature Profile

Then, the local heat transfer coefficient can be found by either using the following formula

$$Nu_x = \frac{3}{4}\left[\frac{2}{5} \cdot \frac{Pr}{(1+2Pr^{\frac{1}{2}}+2Pr)}\right]^{\frac{1}{4}} (Gr_x Pr)^{\frac{1}{4}} \qquad (8.9)$$

or the following table:

Pr	0.01	0.1	0.72	1.0	10	100	1000
$Nu_x Gr_x^{-\frac{1}{4}}$	0.0570	0.164	0.357	0.401	0.827	1.55	2.80

E-87

The average heat transfer coefficient is

$$\boxed{Nu_L = \frac{hL}{k} = -\frac{4}{3}\phi'(0)\left(\frac{Gr_L}{4}\right)^{\frac{1}{4}}}$$ (8.10)

8.3.2 THE INTEGRAL SOLUTION

The governing equations are the momentum integral equation,

$$\boxed{\frac{d}{dx}\int_0^\delta u^2 dy - \int_0^\delta g\beta(T-T_\infty)dy = -\nu\left.\frac{\partial u}{\partial y}\right|_{surface}}$$ (8.11)

and the energy integral equation,

$$\boxed{\frac{d}{dx}\int_0^\delta u(T-T_\infty)dy = \alpha\left.\frac{\partial T}{\partial y}\right|_{surface}}$$ (8.12)

To solve the above equations we have to assume velocity and temperature distributions.

$$\frac{u}{U} = \frac{y}{\delta}\left(1-\frac{y}{\delta}\right)^2$$ (8.13)

$$\frac{T-T_\infty}{T_s-T_\infty} = \left(1-\frac{y}{\delta}\right)^2$$ (8.14)

where δ is the thermal or momentum boundary layer thickness ($\delta = \delta_t$ has been assumed), and U is a fictitious velocity. The results are as follows:

The boundary layer thickness is,

$$\boxed{\frac{\delta}{x} = 3.93\left(\frac{0.952 + Pr}{Pr^2}\right)^{\frac{1}{4}} Gr_x^{-\frac{1}{4}}}$$ (8.15)

E-88

The local heat transfer coefficient is,

$$\text{Nu}_x = 0.508 \left(\frac{\text{Pr}^2}{0.952+\text{Pr}}\right)^{\frac{1}{4}} \text{Gr}_x^{\frac{1}{4}} \quad (8.16)$$

The average heat transfer coefficient is,

$$\text{Nu}_L = 0.678 \left(\frac{\text{Pr}^2}{0.952+\text{Pr}}\right)^{\frac{1}{4}} \text{Gr}_x^{\frac{1}{4}} \quad (8.17)$$

8.4 TURBULENT FLOW ON A VERTICAL FLAT SURFACE

For the vertical flat surface, transition from laminar to turbulent flow occurs when the Rayleigh (Ra) number equals approximately 10^9; this is then the critical Rayleigh number

$$\text{Ra}_L = \text{Gr}_L \, \text{Pr} = \frac{L^3 g \beta \Delta T \text{Pr}}{\nu^2} \quad (8.18)$$

$$\text{Ra}_{L_c} = 10^9 \quad \text{(critical Rayleigh number)} \quad (8.19)$$

where

$$L_c = \left[\frac{10^9 \nu^2}{\beta g (T_s - T_\infty) \text{Pr}}\right]^{1/3} \quad (8.20)$$

$x < L_c$: Laminar flow
$x > L_c$: Turbulent flow

Introducing the one-seventh power law for the velocity and temperature distribution into the integral equations (8.11) and (8.12), we obtain the following expression for the local heat transfer coefficient,

$$\mathrm{Nu}_x = 0.0295 \left[\frac{\mathrm{Pr}^7}{(1+0.494\mathrm{Pr}^{2/3})^6} \right]^{1/15} \mathrm{Gr}_x^{2/5} \qquad (8.21)$$

Note: All properties are evaluated at the film temperature.

8.5 CORRELATIONS FOR FREE CONVECTIONS

8.5.1 VERTICAL FLAT SURFACES AND VERTICAL CYLINDERS

$$\overline{\mathrm{Nu}} = 0.56(\mathrm{GrPr})^{\frac{1}{4}} \qquad (8.22)$$

for $10^4 < \mathrm{GrPr} < 10^9$

$$\overline{\mathrm{Nu}} = 0.13(\mathrm{GrPr})^{1/3} \qquad (8.23)$$

for $10^9 < \mathrm{GrPr} < 10^{12}$

$$\overline{\mathrm{Nu}} = 0.0246 \mathrm{Gr}^{0.4} \left[\frac{\mathrm{Pr}^{7/15}}{(1+0.494\mathrm{Pr}^{2/3})^{0.4}} \right] \qquad (8.24)$$

for $\mathrm{Gr} > 10^{10}$

The above correlations are valid for vertical cylinder, as long as

$$\frac{D}{L} > \frac{35}{\mathrm{Gr}^{\frac{1}{4}}} \qquad (8.25)$$

E-90

8.5.2 HORIZONTAL PLATES

a) Heated Surface Facing Up

$$\boxed{\overline{Nu}_{L_c} = 0.54 \, Ra_{L_c}^{\frac{1}{4}}} \qquad (8.26)$$

for $2.6 \times 10^4 < Ra_{L_c} < 10^7$

$$\boxed{\overline{Nu}_{L_c} = 0.15 \, Ra_{L_c}^{1/3}} \qquad (8.27)$$

for $10^7 < Ra_{L_c} < 3 \times 10^{10}$

b) Heated Surface Facing Down

$$\boxed{\overline{Nu}_{L_c} = 0.27 \, Ra_{L_c}^{\frac{1}{4}}} \qquad (8.28)$$

for $3 \times 10^5 < Ra_{L_c} < 3 \times 10^{10}$

where L_c is the characteristic length,

$$L_c = \frac{\text{Plate Area}}{\text{Plate Perimeter}}$$

8.5.3 HORIZONTAL CYLINDERS

$$\boxed{\overline{Nu} = 0.53 \, Ra^{\frac{1}{4}}} \qquad (8.29)$$

for $10^4 < Ra < 10^9$

$$\boxed{\overline{Nu} = 0.13 \, Ra^{\frac{1}{4}}} \qquad (8.30)$$

for $10^9 < Ra < 10^{12}$

8.5.4 SPHERES

$$\boxed{\overline{Nu} = 2 + 0.426\, Ra^{\frac{1}{4}}}$$ (8.31)

for $1 < Gr < 10^5$

8.5.5 RECTANGULAR BLOCKS

Equations (8.30) and (8.31) can be used for rectangular blocks introducing the characteristic length concept which is defined as follows:

$$\boxed{\frac{1}{L_c} = \frac{1}{L_v} + \frac{1}{L_h}}$$ (8.32)

where L_v and L_h are the vertical and horizontal dimensions respecitvely.

Note: All the properties, except β, are evaluated at the film temperature, T_f.

$$\beta \text{ at } \begin{cases} T_f & \text{for liquids} \\ T_\infty & \text{for gases} \end{cases}$$

8.6 FREE CONVECTION IN CONFINED FLUIDS

8.6.1 BETWEEN HORIZONTAL ISOTHERMAL PLATES

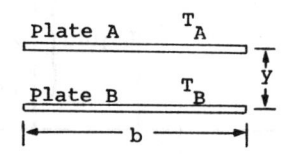

Plate A T_A

Plate B T_B

$\longleftarrow b \longrightarrow$

$$Gr_y = \frac{g\beta(T_A-T_B)y^3}{\nu^2}$$

Table 8.1

Surface Characteristic	$\overline{Nu}_y = \frac{\overline{h}y}{k}$	Fluid	Conditions
$T_A > T_B$	1.0	Air, liquid	
	$0.195\ Gr_y^{\frac{1}{4}}$	Air	$10^4 < Gr_y < 3.7\times10^5$
$T_A < T_B$	$0.068\ Gr_y^{1/3}$	Air	$3.7\times10^5 < Gr_y < 10^7$
	$0.069(Gr_y Pr)^{1/3} Pr^{0.074}$	Liquid	$1.5\times10^5 < Gr_y Pr < 10^9$

Note: All properties are evaluated at $\frac{1}{2}(T_A+T_B)$.

8.6.2 BETWEEN VERTICAL PLATES

Table 8.2

Surface Characteristic	$Nu_y = \frac{\overline{h}y}{k}$	Fluid	Conditions
Constant Heat Flux	1.0	Air	$Gr_y < 2,000$
	$0.18\ Gr_y^{\frac{1}{4}}\left(\frac{b}{y}\right)^{-1/9}$	Air	$2\times10^4 < Gr_y < 2\times10^5,\ \frac{b}{y} > 3$
	$0.065\ Gr_y^{1/3}\left(\frac{b}{y}\right)$	Air	$2\times10^5 < Gr_y < 11\times10^6,\ \frac{b}{y} > 3$
Isothermal plates	$0.42(Gr_y Pr)^{\frac{1}{4}} Pr^{0.012}\left(\frac{b}{y}\right)^{-0.30}$	Liquid	$10^4 < Gr_y Pr < 10^7$ $1 < Pr < 2\times10^4$ $10 < \frac{b}{y} < 40$
	$0.046(Gr_y Pr)^{1/3}$	Liquid	$10^6 < Gr_y Pr < 10^9$ $1 < Pr < 20$ $1 < \frac{b}{y} < 40$

Note: All properties are evaluated at $\frac{1}{2}(T_A+T_B)$.

8.7 MIXED, FREE AND FORCED CONVECTION

Free and forced convection become important when the ratio of buoyancy forces to inertial forces is approximatley one. That is,

$$Gr/Re^2 \simeq 1.0 \qquad (8.33)$$

Figures 8.4 and 8.5 show the regimes for mixed free and forced convection in vertical and horizontal tubes respectively. These figures apply only for constant surface temperature and when

$$10^{-2} < Pr\left(\frac{d}{L}\right) < 1 \qquad (8.34)$$

where d and L are the pipe diameter and length, respectively.

The following correlations are used in Figure 8.5 for the different regimes:

A) Forced convection, turbulent flow

$$Nu = 0.116\left[1 + \left(\frac{D}{L}\right)^{2/3}\right](Re^{2/3} - 125)Pr^{1/3}\left(\frac{\mu_b}{\mu_s}\right) \qquad (8.35)$$

where μ_b and μ_s are evaluated at the bulk and surface temperature, respectively.

B) Forced convection laminar flow

$$Nu = 1.86 Pe^{1/3}\left(\frac{\mu_b}{\mu_s}\right)^{0.14} \qquad (8.36)$$

where the Peclet number, Pe = Re Pr

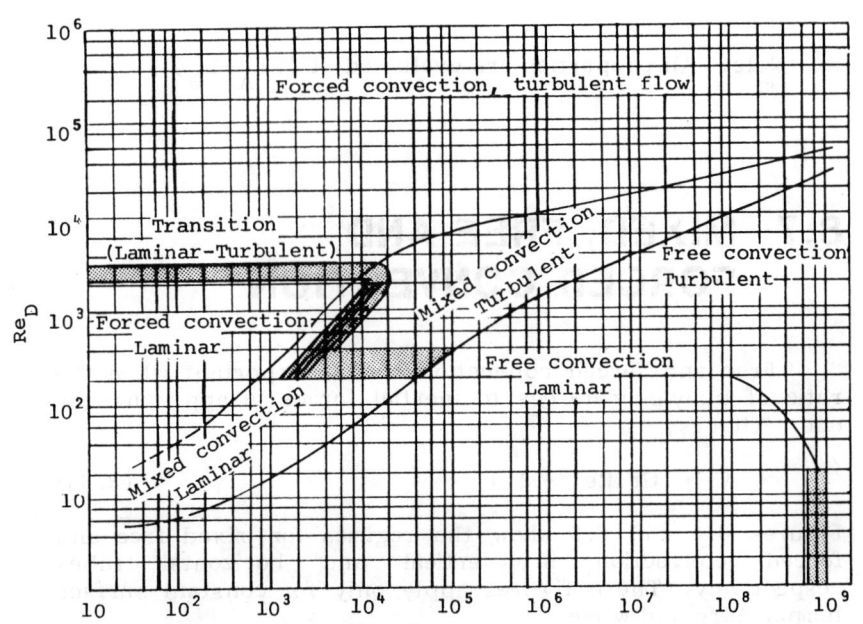

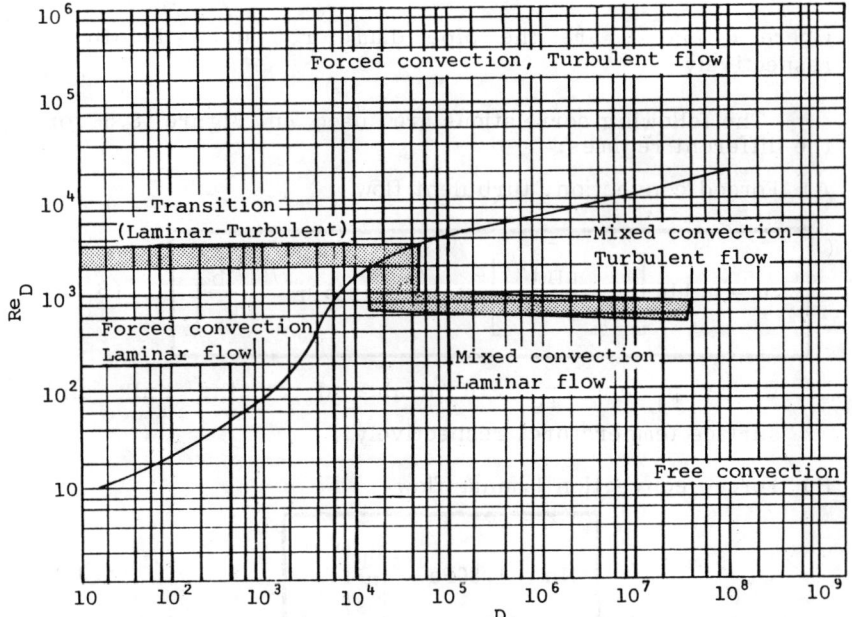

Fig. 8·4 Regimes of flow through vertical tubes $Gr_D(\frac{D}{L})$

Fig. 8·5 Regimes of flow through horizontal tubes $Gr_D(\frac{D}{L})$

C) Mixed convection, turbulent flow

$$\mathrm{Nu} = 4.69\,\mathrm{Re}^{0.27}\,\mathrm{Pr}^{0.21}\,\mathrm{Gr}^{0.07}\left(\frac{D}{L}\right)^{0.36} \qquad (8.37)$$

D) Mixed convection, laminar flow

$$\mathrm{Nu} = 1.75\left(\frac{\mu_b}{\mu_s}\right)^{0.14}[G_z + 0.0083(\mathrm{GrPr})^{0.75}]^{1/3} \qquad (8.38)$$

where the Graetz number, Gz, is defined as

$$Gz = \mathrm{Re}_D\,\mathrm{Pr}\left(\frac{D}{L}\right) \qquad (8.39)$$

Note: Properties are evaluated at film temperature.

CHAPTER 9

HEAT EXCHANGERS

9.1 TYPES OF HEAT EXCHANGERS

A heat exchanger is a device designed to accomplish the transfer of thermal energy between fluids at different temperatures.

9.1.1 DOUBLE-PIPE ARRANGEMENT

In this arrangement of either parallel flow or counterflow, the hot or cold fluid flows through the annular space while the other flows inside the inner tube, as shown in figure 9.1.

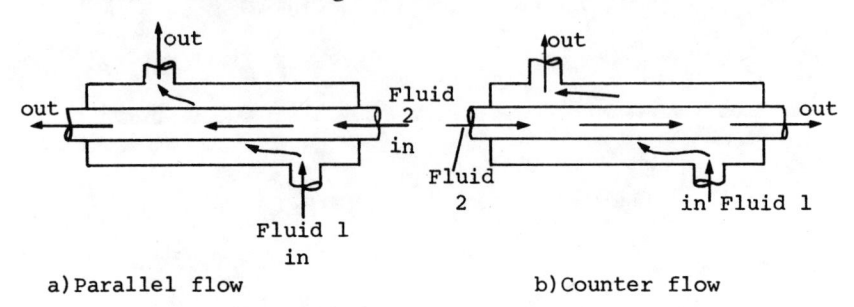

a) Parallel flow   b) Counter flow

Fig. 9.1 Double-pipe Arrangement

9.1.2 SHELL AND TUBE ARRANGEMENT

In this arrangement, one fluid flows inside the tubes and the other flows over the tubes. Figure 9.2 shows the one-shell pass, one-tube pass with baffles. Baffles are used to ensure uniformity of flow across the tubes which increases the heat tansfer.

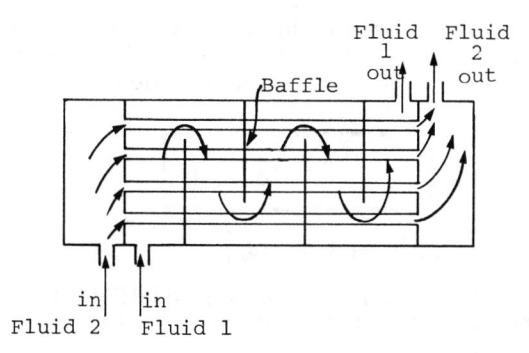

Fig. 9·2 One-shell pass, one-tube pass heat exchange

9.1.3 CROSS-FLOW HEAT EXCHANGERS

Two representative arrangements of this type of heat exchanger are shown in Figure 9.3. In both the cooling or heating fluid is confined to flow in the tubes while the other fluid flows normal over the tubes. In case A, the fluid flows unrestricted (mixed) over the tubes, while in case B, the fluid flows between the fins as it crosses the tubes (unmixed).

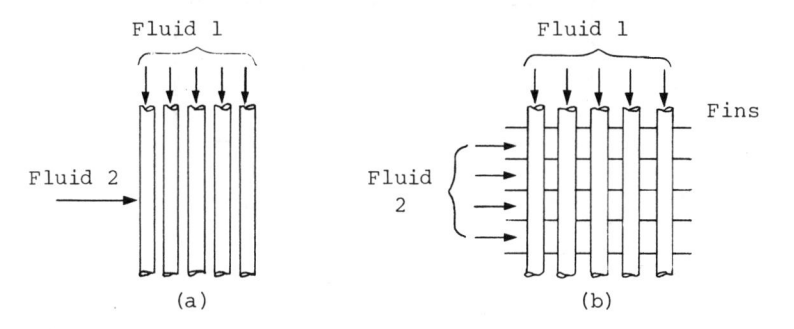

Fig. 9·3 Cross-flow heat exchangers

(a) mixed flow, (b) unmixed flow

E-98

9.2 THE OVERALL HEAT TRANSFER COEFFICIENT

The equations derived in chapter two for the heat transfer through walls and cylinders, may be expressed in terms of the overall heat transfer coefficient. The general form is given by

$$q = UA \Delta T_{overall} \qquad (9.1)$$

where q = The overall heat transfer

U = The overall heat transfer coefficient, which is proportional to the inverse of the total thermal resistance

A = Appropriate surface area for the heat transfer.

For the case of the plane wall bounded by fluids (eq. 2.13), the overall heat transfer coefficient is defined as

$$U = \frac{1}{\frac{1}{h_C} + \frac{\Delta x_B}{k_B} + \frac{1}{h_D}} \qquad (9.2)$$

For a double-pipe heat exchanger (Fig. 9.1), the overall heat transfer is given by

$$q = \frac{T_2 - T_1}{\frac{1}{h_i A_i} + \frac{\ln(r_o/r_i)}{2\pi k L} + \frac{1}{h_o A_o}} \qquad (9.3)$$

where the subscripts i and o correspond to the inside and outside of the inner tube, respectively (Fig. 9.4).

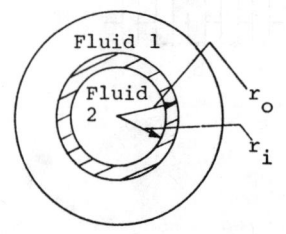

Fig. 9.4 Double-pipe heat exchanger cross-section

E-99

The overall heat transfer coefficient may be based on either the inside or the outside area of the tube (see Fig. 9.4)

$$U_i = \cfrac{1}{\cfrac{1}{h_i} + \cfrac{A_i \ln(r_o/r_i)}{2\pi kL} + \cfrac{r_i}{r_o}\cfrac{1}{h_o}} \qquad (9.4)$$

$$U_o = \cfrac{1}{\cfrac{r_o}{r_i}\cfrac{1}{h_i} + \cfrac{A_o \ln(r_o/r_i)}{2\pi kL} + \cfrac{1}{h_o}} \qquad (9.5)$$

9.3 FOULING FACTOR

The factor that takes into account the effects of dirt accumulated on the surface of a heat exchanger is called fouling factor (R_f) and is defined as

$$R_f = \frac{1}{U_d} - \frac{1}{U_c} \qquad (9.6)$$

where U_d = Overall heat transfer coefficient on the dirty surface

U_c = Overall heat transfer coefficient on the clean surface

9.4 THE LOG-MEAN TEMPERATURE DIFFERENCE (LMTD)

The log-mean temperature difference is an appropriate mean temperature difference throughout a heat exchanger.

The LMTD for a parallel flow double-pipe heat exchanger is expressed as

$$LMTD = \frac{(T_{h_1}-T_{c_1})-(T_{h_2}-T_{c_2})}{\ln[(T_{h_1}-T_{c_1})/(T_{h_2}-T_{c_2})]} \qquad (9.7)$$

where T_{h_1} and T_{h_2} are the inlet and outlet temperatures of the hot fluid; T_{c_1} and T_{c_2} are the inlet and outlet temperatures of the cold fluid (see Fig. 9.5).

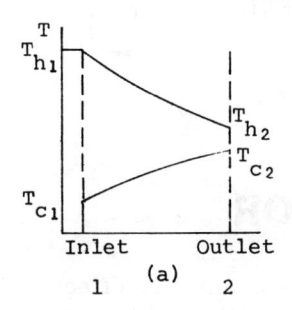

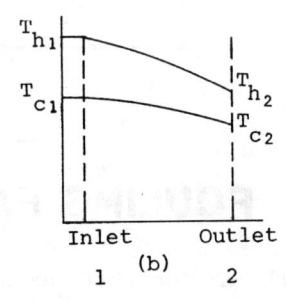

Fig. 9·5 Temperature plots for (a) parallel flow, (b) counter flow

The above equation is useful when the inlet and outlet temperatures are known or are easily determined. This equation applies under the following assumptions:

A) The fluid specific heats do not change with temperature.

B) The overall heat transfer coefficient is constant across the heat exchanger.

C) No external heat losses.

For heat exchangers other than the above mentioned, a correction factor (F) applied to the LMTD must be used to evaluate the heat tansfer. Then equation (9.1) becomes,

$$q = UAF \Delta T_m \qquad (9.8)$$

Values of the correction factor (F) are given in Figures 9.6-9.10 where $T_{s\ in}$ and $T_{s\ out}$ are the inlet and outlet

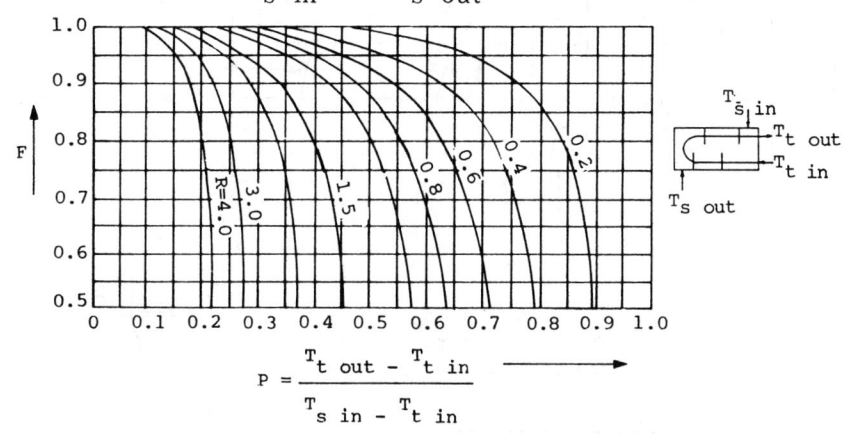

$$P = \frac{T_{t\ out} - T_{t\ in}}{T_{s\ in} - T_{t\ in}}$$

Fig. 9·6 One shell pass and two, four or any multiple of tube passes

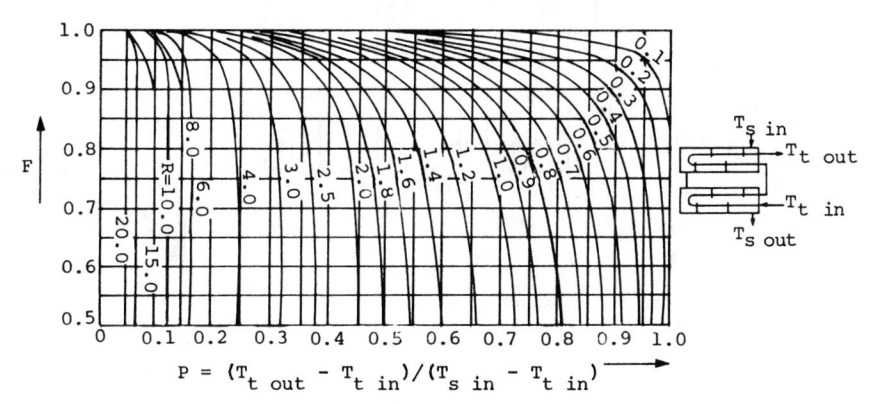

$$P = (T_{t\ out} - T_{t\ in})/(T_{s\ in} - T_{t\ in})$$

Fig. 9·7 Correction factor plot for two shell passes and four, eight, or any multiple of tube passes

temperatures of the fluid that flows through the shell $T_{t\ in}$ and $T_{t\ out}$ are the inlet and outlet temperatures of the fluid that flows through the tubes. Also, R is defined as

$$R = \frac{T_{s\ in} - T_{s\ out}}{T_{t\ out} - T_{t\ in}} \qquad (9.9)$$

Fig. 9·8 Corrections factor plot cross flow with one fluid mixed

$$P = \frac{T_{t\,out} - T_{t\,in}}{T_{s\,in} - T_{t\,in}}$$

Fig. 9·9 Correction factor plot for crossflow with both fluids unmixed

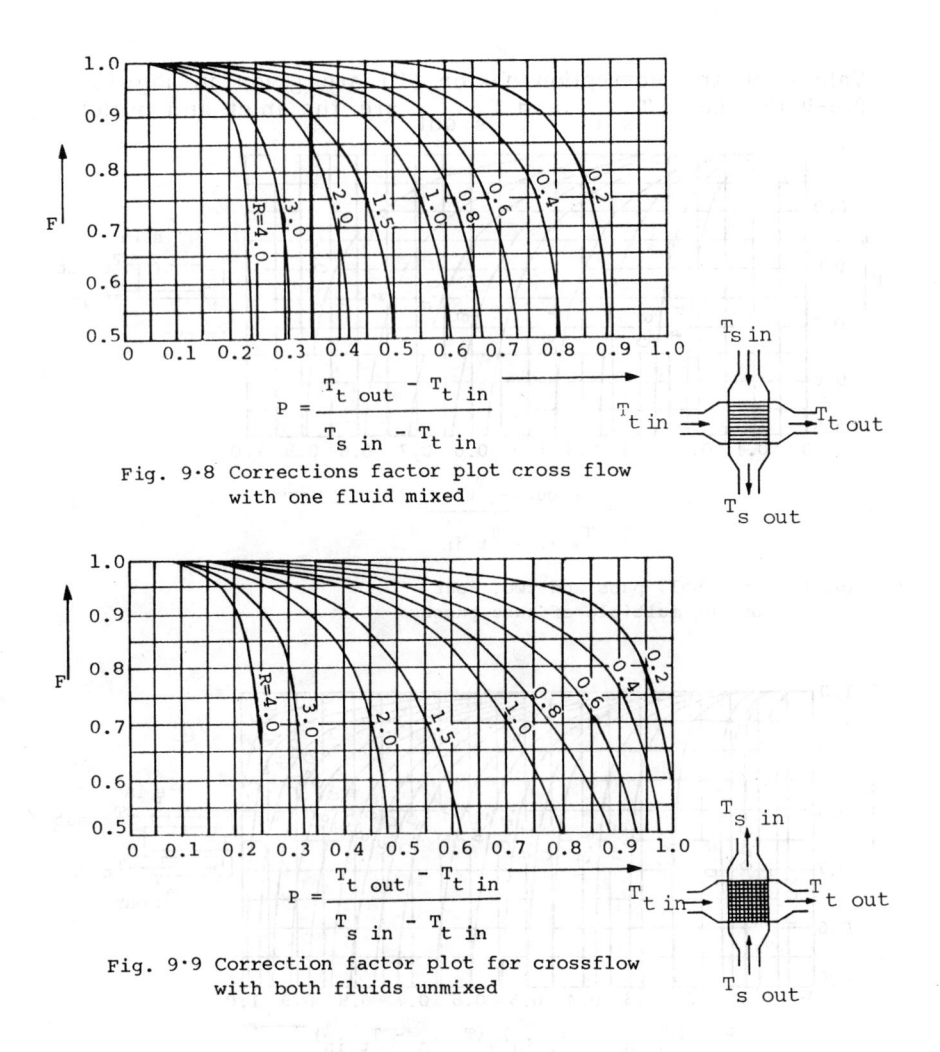

$$P = \frac{T_{t\,out} - T_{t\,in}}{T_{s\,in} - T_{t\,in}}$$

9.5 EFFECTIVENESS-NTU METHOD

This method is very useful when the inlet and outlet temperatures in the heat exchanger are not known. The effectiveness (ε) is defined as

$$\varepsilon = \frac{\text{Actual heat transfer}}{\text{Maximum possible heat transfer}} \quad (9.10)$$

For parallel and counterflow heat exchangers, the maximum heat transfer is

$$q_{max} = (mc)_{min}(T_{hi}-T_{ci}) = C_{min}(T_{hi}-T_{ci}) \quad (9.11)$$

where the subscript i denotes conditions at inlet and $C = \dot{m}c$ is the capacity rate. For the parallel flow exchanger the effectiveness is given by

$$\varepsilon_h = \frac{T_{h_1}-T_{h_2}}{T_{h_1}-T_{c_1}} \quad (9.12)$$

$$\varepsilon_c = \frac{T_{c_2}-T_{c_1}}{T_{h_1}-T_{c_1}} \quad (9.13)$$

where the subscripts h and c on ε represent the fluid which has the minimum value of capacity rate (c).

For the counterflow exchanger we have,

$$\varepsilon_h = \frac{T_{h_1}-T_{h_2}}{T_{h_1}-T_{c_2}} \quad (9.14)$$

$$\varepsilon_c = \frac{T_{c_1}-T_{c_2}}{T_{h_1}-T_{c_2}} \quad (9.15)$$

Another expression for the effectiveness in parallel flow is derived from the LMTD equation and is given by

$$\varepsilon = \frac{1 - \exp[(-UA/C_h)(1+C_h/C_c)]}{1 + C_c/C_h} \quad (9.16)$$

where UA/C_{min} = N° of transfer units (NTU) and is indicative of the size of the heat exchanger.

The above expression applies for both cases (either the cold or the hot fluid is the minimum fluid), except that

C_h and C_c are interchanged. Figures 9.10, 9.11, and 9.12 show effectiveness ratios for some heat exchangers.

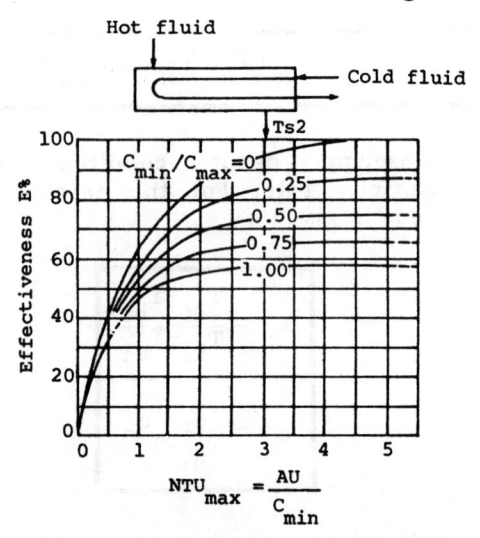

Fig. 9·10 Parallel counterflow exchanger

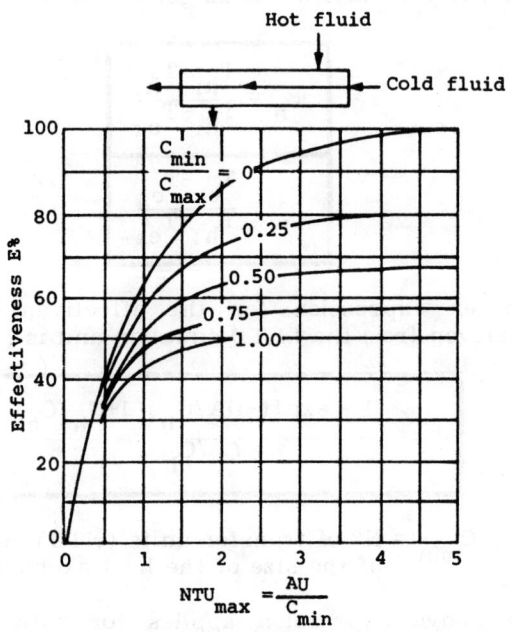

Fig. 9·11 Parallel-flow exchanger

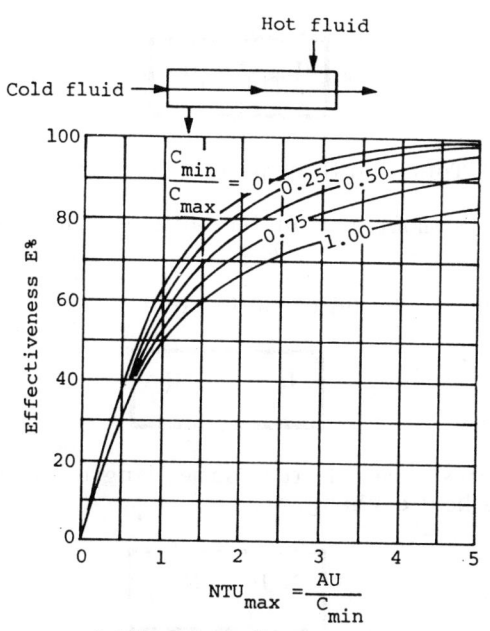

9·12 Counter-flow exchanger

For parallel-flow heat exchanger the following expression is widely used,

$$\varepsilon = \frac{1 - \exp[(-NTU)(1+C_{min}/C_{max})]}{1 + C_{min}/C_{max}} \quad (9.17)$$

and for counterflow heat exchanger we have

$$\varepsilon = \frac{1-\exp[(-NTU)(1-C_{min}/C_{max})]}{1-(C_{min}/C_{max})\exp[(-NTU)(1-C_{min}/C_{max})]}$$

(9.18)

9.6 TRANSFER UNITS IN HEAT EXCHANGERS

If calculations are based on inner fluid,

E-106

$$\boxed{L_{Ti} = N_i H_i} \qquad (9.19)$$

where L_{Ti} = Total length of path of inner fluid

N_i = Number of transfer units of inner fluid.

if calculations are based on outer fluid,

$$\boxed{L_{To} = N_o H_o}$$

if both paths are of the same length (e.g., as in a double-pipe heat exchanger),

$$\boxed{N_o H_o = N_i H_i}$$

The number of transfer unit $N = \int_{T_a}^{T_b} \dfrac{dT}{(T_h - T_c)}$

Assuming inner fluid as basis:

$$dA_i = \pi D_i \, dL_c$$

$$dq = \dot{m}_i \, Cp_i \, dT_i$$

where $\dot{m}_i$ = rate of flow of inner fluid

Cp_c = specific heat of inner fluid

Substituting above equations into the equation below,

$$\dfrac{dq}{dA} = U \Delta T = U(T_h - T_c) \qquad \text{gives}$$

$$dL_i = \dfrac{\dot{m}_i \, Cp_i}{\pi D_i U_i} \, \dfrac{dT_i}{(T_o - T_i)}$$

integrating over the entire exchanger we get:

$$L_{Ti} = \frac{\dot{m}_i Cp_i}{\pi D_i U_i} \int_{T_{ib}}^{T_{ia}} \frac{dT_i}{(T_o - T_i)}$$

In the case where U_i and Cp_i are constant

$$\boxed{H_{ti} = \frac{\dot{m}_i Cp_i}{\pi D_i U_i}} \qquad (9.20)$$

and if outer fluid were chosen as the basis

$$\boxed{H_{to} = \frac{\dot{m}_o Cp_o}{\pi D_o U_o}} \qquad (9.21)$$

CHAPTER 10

RADIATION HEAT TRANSFER

Radiation is the transmission of energy in the form of electromagnetic waves with the speed of light through space or regions devoid of any matter.

10.1 SPECTRUM OF ELECTROMAGNETIC RADIATION

Various kinds of radiations are identified by the range of wavelengths they include. The wavelength λ, and frequency ν, are related by the equation below:

$$\boxed{\lambda = c/\nu} \qquad (10.1)$$

where $\quad c = 2.9979 \times 10^{10} \text{cm sec}^{-1}$

The electromagnetic radiation consists of particles called photons which have zero charge and zero mass. Their energy, E, is given by:

$$\boxed{E = h\nu} \qquad (10.2)$$

where $\quad h = $ Planck's constant $= 6.624 \times 10^{-27}$ erg-sec.

Thermal radiation is one kind of electromagnetic radiation that takes place in the following range,

$$10^{-7} < \lambda < 10^{-4}$$

where λ is in meters (Fig. 10.1).

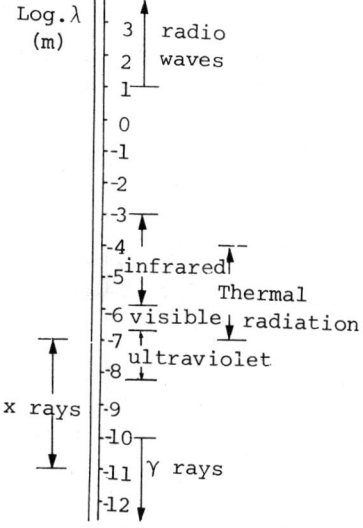

```
Log.λ
 (m)    3 | radio
        2 | waves
        1
        0
       -1
       -2
       -3
       -4
       -5  infrared
       -6  visible    Thermal
       -7           radiation
       -8  ultraviolet
x rays -9
       -10
       -11  γ rays
       -12
```

Fig. 10.1 Spectrum of Electromagnetic Radiation

10.2 RADIATION PROPERTIES

Reflectivity ρ : Fraction reflected
Absorptivity α : Fraction absorbed
Transmissivity τ : Fraction transmitted

A part of the radiation incident upon a medium is reflected, a part transmitted and the rest absorbed (Fig. 10.2)

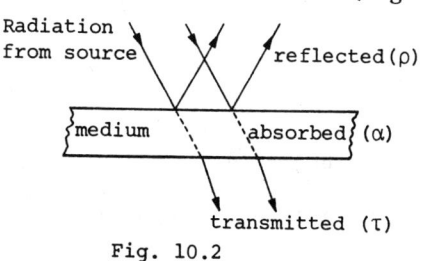

Fig. 10.2

E-110

$$\boxed{\rho + \alpha + \tau = 1} \qquad (10.3)$$

The reflection may be a) specular or b) diffuse. In the first case, the angle of incidence equals to the angle of reflection while in the second case, the reflection is distributed uniformly in different directions (Fig. 10.3).

(a) Specular (b) Diffuse

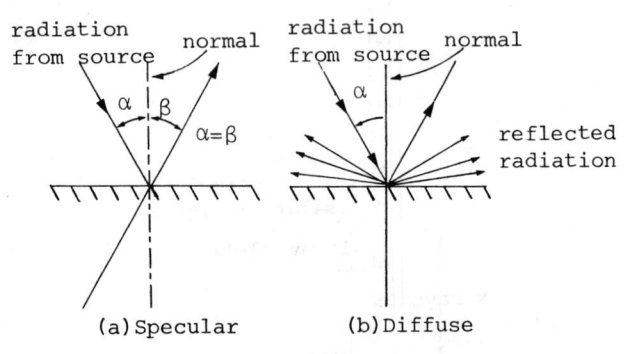

Fig. 10.3

10.3 CONCEPTS OF THERMAL RADIATION

10.3.1 BLACK SURFACE

It is an ideal surface that absorbs all energy incident upon it, reflects none, and emits all the radiant energy it absorbs. The expression for radiation is given by the Stefan-Boltzman's law:

$$\boxed{E_b = \sigma T^4} \qquad (10.4)$$

where E_b = Emissive power of a black surface

σ = Stefan-Boltzmann's constant

= 5.669×10^{-8} w/m² · °K⁴

= 0.1714×10^{-8} Btu/hr-ft² · °R⁴

T = Absolute temperature

E-111

10.3.2 TOTAL HEMISPHERIC EMISSIVITY (ε)

It is the ratio of the emissive power of a body (E) to the emissive power of a black body (E_b) at the same temperature,

$$\varepsilon = \frac{E}{E_b} \tag{10.5}$$

For black bodies in equilibrium, the total emissivity is equal to the absorptivity. That is,

$$\varepsilon = \alpha \quad \text{(Kirchoff's identity)} \tag{10.6}$$

Equation 10.5 can also be expressed in terms of the monochromatic emissivity (ε_λ)

$$\boxed{\varepsilon = \frac{E}{E_b} = \frac{\int_0^\infty \varepsilon_\lambda E_{b\lambda} \, d\lambda}{\sigma T^4}} \tag{10.7}$$

where

$$\varepsilon_\lambda = \frac{E_\lambda}{E_{b\lambda}}$$

$E_{b\lambda}$ = Emissive power of a black body per unit of wavelength, given by the Plank's law:

$$\boxed{E_{b\lambda} = \frac{C_1 \lambda^{-5}}{\exp(C_2/\lambda T) - 1}} \tag{10.8}$$

where

$C_1 = 1.187 \times 10^8$ Btu-μ^4/hr-ft^3
$\quad\quad (3.743 \times 10^8$ W-μ^4/m$^2)$

$C_2 = 2.5896 \times 10^4$ μ-°R
$\quad\quad (1.4387 \times 10^4 \mu$ °K)

λ = Wavelength, μ

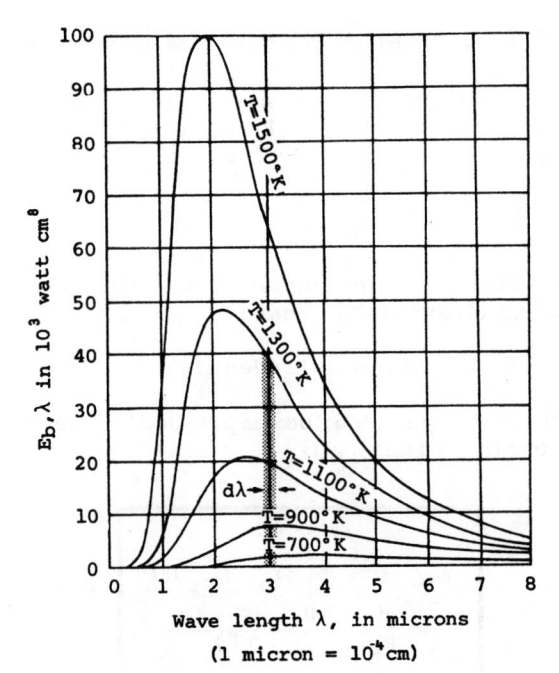

Fig. 10.4
Black body emissive power distribution

Figure 10.4 shows emissive power of the black body for various absolute temperatures as the function of the wavelength. (The wavelength, at which the maximum radiation takes place, becomes smaller as the temperature of the emitting body increases.) This situation is expressed by the Wien's displacement law as,

$$\lambda_{max} T = 2897.6 \text{ } \mu \text{ °K} = 5215.6 \text{ } \mu \text{ °R} \tag{10.9}$$

The total energy radiated from a black body between zero and any given wavelength (λ) is given by,

$$\frac{E_{b(0-\lambda)}}{E_{b(0-\infty)}} = \frac{\int_0^\lambda E_{b\lambda} \, d\lambda}{\int_0^\infty E_{b\lambda} \cdot d\lambda} \tag{10.10}$$

E-113

The results are tabulated in table 10.1.

λτ μm·R	λτ μm·K	Eb(0-λτ) aT⁴	λτ μm·R	λτ μm·K	Eb(0-λτ) aT⁴	λτ μm·R	λτ μm·K	Eb(0-λτ) aT⁴
1000	556	0	7200	4000	.4809	13400	7445	.8317
1200	667	0	7400	4111	.5007	13600	7556	.8370
1400	778	0	7600	4222	.5199	13800	7667	.8421
1600	889	.0001	7800	4333	.5381	14000	7778	.8470
1800	1000	.0003	8000	4445	.5558	14200	7889	.8517
2000	1111	.0009	8200	4556	.5727	14400	8000	.8563
2200	1222	.0025	8400	4667	.5890	14600	8111	.8606
2400	1333	.0053	8600	4778	.6045	14800	8222	.8648
2600	1445	.0098	8800	4889	.6194	15000	8333	.8688
2800	1556	.0164	9000	5000	.6337	16000	8889	.8868
3000	1667	.0254	9200	5111	.6474	17000	9445	.9017
3200	1778	.0368	9400	5222	.6606	18000	10000	.9142
3400	1889	.0506	9600	5333	.6731	19000	10556	.9247
3600	2000	.0667	9800	5444	.6851	20000	11111	.9335
3800	2111	.0850	10000	5556	.6966	21000	11667	.9411
4000	2222	.1051	10200	5667	.7076	22000	12222	.9475
4200	2333	.1267	10400	5778	.7181	23000	12778	.9531
4400	2445	.1496	10600	5889	.7282	24000	13333	.9589
4600	2556	.1734	10800	6000	.7378	25000	13889	.9621
4800	2667	.1979	11000	6111	.7474	26000	14445	.9657
5000	2778	.2229	11200	6222	.7559	27000	15000	.9689
5200	2889	.2481	11400	6333	.7643	28000	15556	.9718
5400	3000	.2733	11600	6445	.7724	29000	16111	.9742
5600	3111	.2983	11800	6556	.7802	30000	16667	.9765
5800	3222	.3230	12000	6667	.7876	40000	22222	.9881
6000	3333	.3474	12200	6778	.7947	50000	27778	.9941
6200	3445	.3712	12400	6889	.8015	60000	33333	.9963
6400	3556	.3945	12600	7000	.8081	70000	38887	.9981
6600	3667	.4171	12800	7111	.8144	80000	44445	.9987
6800	3778	.4391	13000	7222	.8204	90000	50000	.9990
7000	3889	.4604	13200	7333	.8262	100000	55556	.9992
								1.0000

Also, the radiant energy emitted between two wavelengths λ_1 and λ_2 is given by

$$E_{b(\lambda_1-\lambda_2)} = E_{b(0-\infty)} \left[\frac{E_{b(0-\lambda_2)}}{E_{b(0-\infty)}} - \frac{E_{b(0-\lambda_1)}}{E_{b(0-\infty)}} \right] \quad (10.11)$$

where $E_{b(0-\infty)} = \sigma T^4$ (Radiation over all wavelengths)

10.3.3 GRAY SURFACE

It is a surface which behaves in such a way that ε_λ can be considered independent of the wavelength (λ) and equal to the total hemispheric emissivity (ε). Figure 10.5 shows the difference of emissive power among black surfaces, gray surfaces and real surfaces.

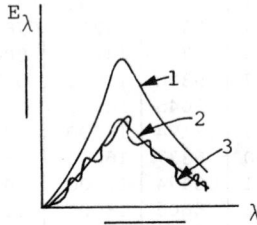

Fig. 10·5 Emissive 1) black surface,
2) gray surface, 3) real surface

10.4 INTENSITY OF RADIATION

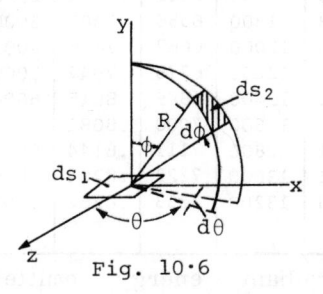

Fig. 10·6

It is defined as the rate of radiation emitted from an elemental surface, ds_1, per unit solid angle and per unit projected area on a surface, ds_2, normal to the direction of the radiation (Fig. 10.6). Mathematically, it is expressed as

$$I_b = \frac{dE_b}{(ds_1 \cos\phi)\frac{ds_2}{R^2}} \qquad (10.12)$$

E-115

where $\frac{ds_2}{R^2}$ is the solid angle, dw, defined as

$$dw = \frac{ds_2}{R^2} = \sin\phi\, d\phi\, d\theta$$

From the above equation, the total radiation over the enclosing hemisphere is given by

$$ds_1 E_b = I_b ds_1 \int_0^{2\pi} \int_0^{2\pi} \sin\phi \cos\phi\, d\phi\, d\phi \qquad (10.13)$$

from where

$$E_b = \pi I_b \qquad \text{(for } I_b = \text{constant)} \qquad (10.14)$$

Equation (10.14) is the total emissive power of a black surface (diffuse emitter).

10.5 RADIATION SHAPE FACTOR

When radiation between two black surfaces at different temperatures takes place, the portion of radiation emitted from one surface that reaches the other surface is called radiation shape factor or angle factor (Fig. 10.7).

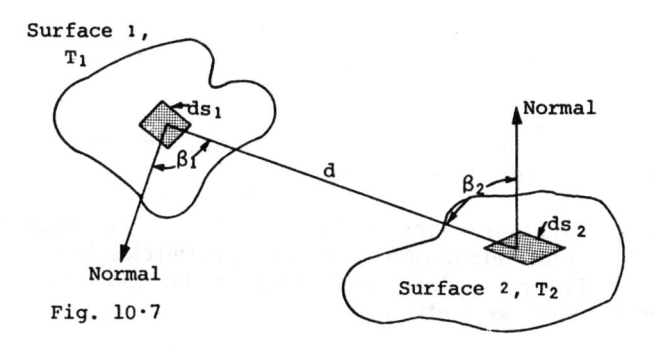

Fig. 10·7

E-116

General notation:

F_{mn} - Portion of radiation from surface m that strikes surface n.

The radiation shape factor for the surfaces shown in Figure 10.7 is given by

$$F_{12} = \frac{1}{s_1} \int_{s_1} \int_{s_2} \frac{\cos\beta_1 \cos\beta_2}{\pi d^2} ds_1 ds_2 \qquad (10.15)$$

in general,

$$\boxed{F_{mn} = \frac{1}{s_m} \int_{s_m} \int_{s_n} \frac{\cos\beta_m \cos\beta_n}{\pi d^2} ds_m ds_n} \qquad (10.16)$$

and

$$A_m F_{mn} = A_n F_{nm} \qquad (10.17)$$

Eq.(10.17) is the reciprocity theorem. For convex and flat surfaces,

$$F_{mn} = 0 \qquad (10.18)$$

For black enclosures where a surface m interacts with another surface n,

$$F_{mn} = 1 \qquad (10.19)$$

in general

$$\sum_{n=1}^{k} F_{mn} = 1 \qquad (10.20)$$

where m is the radiant surface that interacts with k different surfaces, including m.

The radiation shape factor is a function of geometrical parameters and calculations for some geometries become very difficult. Figures 10.8, 10.9 and 10.10 show the shape factors for some geometries.

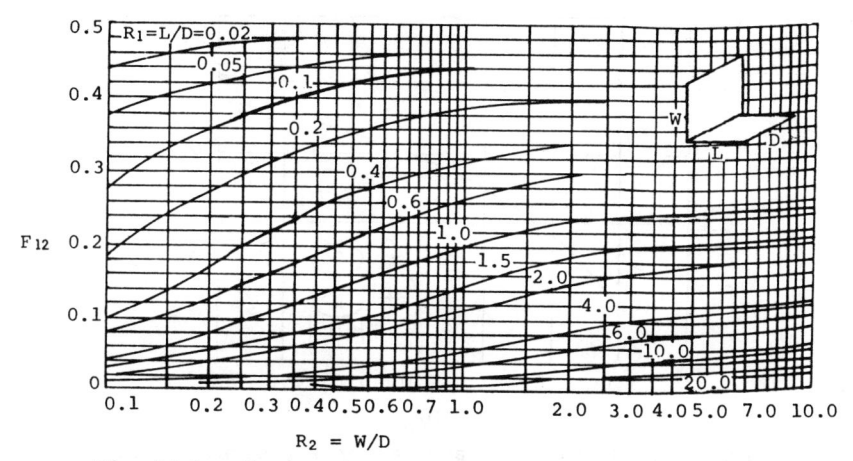

Fig. 10·8 Radiation shape factor for perpendicular rectangles with a common edge

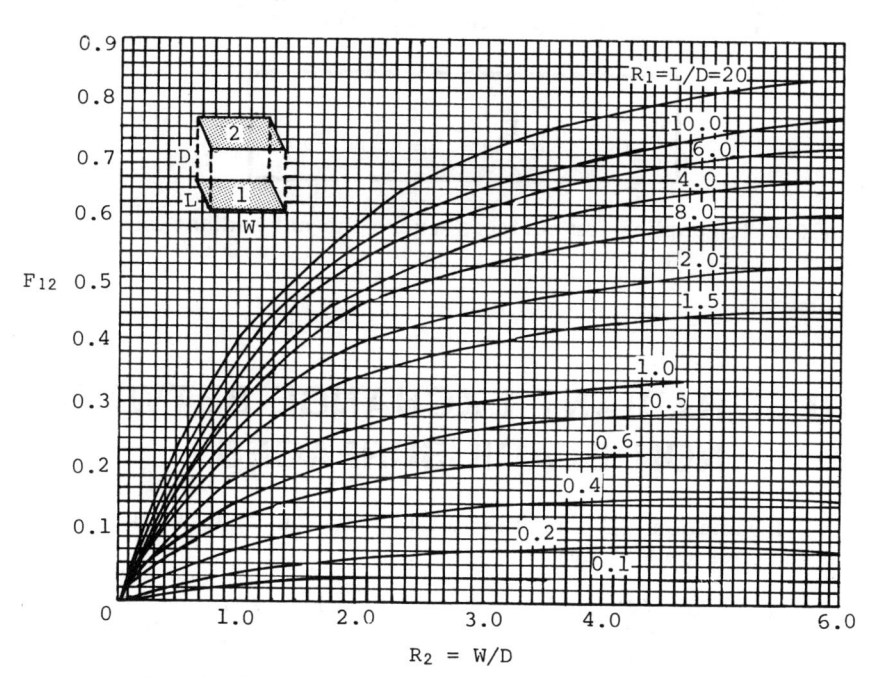

Fig. 10·9 Radiation shape factor for parallel rectangle

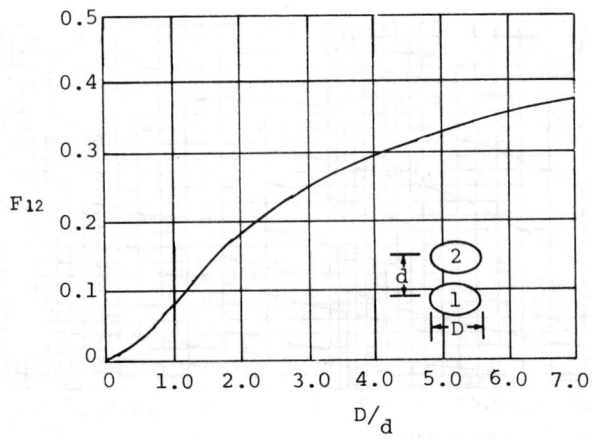

Fig. 10·10 Radiation shape factor for parallel discs

For black enclosures, the net heat transfer rate, q_m, from the mth surface of a system consisting of k black surfaces at different temperatures is given by

$$q_m = A_m(E_{bm} - \sum_{n=1}^{k} F_{mn}E_{bn}) \qquad (10.21)$$

Eq.(10.21) applies for surfaces where the net radiant loss is zero.

10.6 RADIANT EXCHANGE BETWEEN NON-BLACK SURFACES

Irradiation (G) is the total radiation that strikes a surface per unit of time.

Radiosity (J) is the radiation that leaves a surface per unit of time and unit of area.

In order to apply this method, the following assumptions must be considered:

A) All surfaces are at uniform temperature.

E-119

B) ε and ρ are constant along the surface.
C) G and J are uniform on each surface.

When no radiation is transmitted, radiosity, J, is defined as

or

$$J = \varepsilon E_b + \rho G \qquad (10.22)$$

$$J = \varepsilon E_b + (1-\varepsilon)G \qquad (10.23)$$

Let us consider now an enclosure of k different gray surfaces which can see each other and themselves. The heat loss, q_m, from surface m, is given by

$$q_m = \frac{E_{bm} - J_m}{(1-\varepsilon_m)/(A_m \varepsilon_m)} \qquad (10.24)$$

The above equation is analogous to an electric network with a resistance of $(1-\varepsilon_m)/(A_m \varepsilon_m)$, called the surface resistance, q_m is the current, and the numerator $(E_{bm} - J_m)$ is the potential difference (see Fig. 10.11).

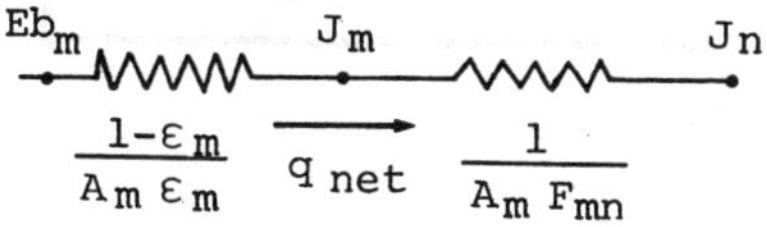

Fig 10·11 Surface and space resistances in the network method

The net rate of radiant exchange between surface m and any other surface n, is given by

$$q_{mn} = F_{mn} A_m (J_m - J_n) \qquad (10.25)$$

or

$$q_{mn} = \frac{J_m - J_n}{\frac{1}{A_m F_{mn}}} \qquad (10.26)$$

where the denominator $1/F_{mn} A_m$ is called the space resistance (Fig. 10.11). The final network is solved for any particular radiosity (J) using standard methods for electrical networks.

Special cases:

10.6.1 RADIATION BETWEEN TWO GRAY SURFACES

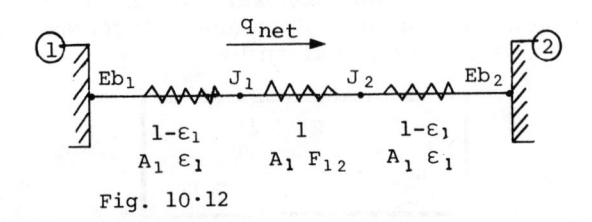

Fig. 10·12

The net heat transfer is given by

$$q_{net} = \frac{E_{b1} - E_{b2}}{\frac{(1-\varepsilon_1)}{A_1 \varepsilon_1} + \frac{1}{A_1 F_{12}} + \frac{1-\varepsilon_2}{A_2 \varepsilon_2}} \qquad (10.27)$$

where $E_b = \sigma T^4$

10.6.2 INFINITE PARALLEL PLANES

In this case, $A_1 = A_2$ and $F_{12} = 1$. Introducing these values in equation (10.27), the net heat transfer per unit area is,

$$\frac{q}{A} = \frac{E_{b1} - E_{b2}}{\frac{1}{\varepsilon_1} + \frac{1}{\varepsilon_2} - 1} \qquad (10.28)$$

10.6.3 TWO CONCENTRIC CYLINDERS

From equation (10.27) and using $F_{12} = 1$, the net heat transfer is given by

$$q = \frac{A_1(E_{b_1} - E_{b_2})}{\frac{1}{\varepsilon_1} + \left(\frac{A_1}{A_2}\right)\left(\frac{1}{\varepsilon_2 - 1}\right)} \quad (10.29)$$

10.7 RADIATION SHIELDS

Radiation shields are used to reduce radiant heat transfer between two particular surfaces (Fig. 10.13). The heat transfer betwen surfaces S_1 and S_2 with no shield, is given by

$$\frac{q}{A} = \frac{\sigma(T_1^4 - T_2^4)}{\frac{1}{\varepsilon_1} + \frac{1}{\varepsilon_2} - 1} \quad (10.30)$$

while the heat transfer between the surfaces with the radiation shield is

$$\frac{q}{A} = \frac{\sigma(T_1^4 - T_3^4)}{\frac{1}{\varepsilon_1} + \frac{1}{\varepsilon_3} - 1} = \frac{\sigma(T_3^4 - T_2^4)}{\frac{1}{\varepsilon_3} + \frac{1}{\varepsilon_2} - 1} \quad (10.31)$$

If we assume the same emissivity for the three surfaces, the heat transfer is one-half of the heat transfer without shield. That is,

$$\frac{q}{A} = \frac{\frac{1}{2}\sigma(T_1^4 - T_2^4)}{\frac{1}{\varepsilon_1} + \frac{1}{\varepsilon_2} - 1} \quad (10.32)$$

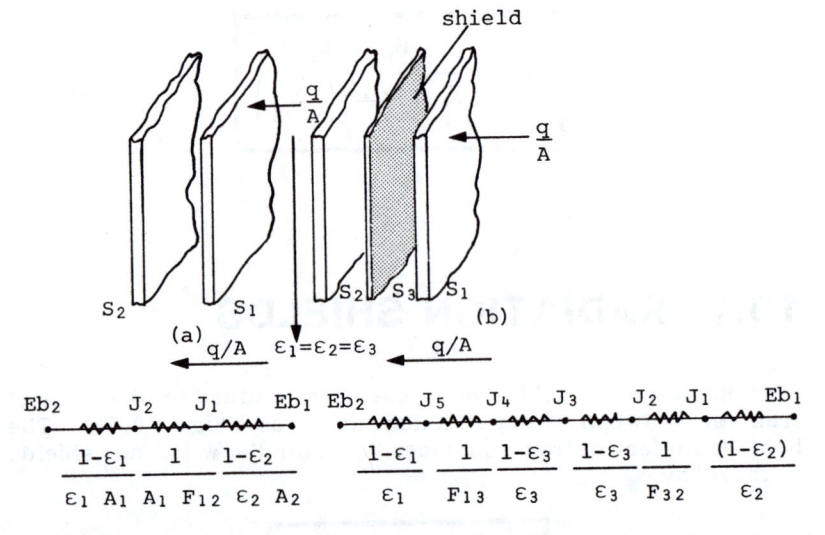

Fig. 10·13 Radiation shield and electrical analog

When more than one shield is present, and the emissivities of all surfaces are equal, the heat transfer is given by

$$\left\{\frac{q}{A}\right\}_{\text{with } k \text{ shields}} = \frac{1}{k+1} \left\{\frac{q}{A}\right\}_{\text{with no shields}} \quad (10.33)$$

where k is the number of shields.

10.8 GAS RADIATION

Gases such as nitrogen, oxygen, and in general, gases with non-polar symmetrical molecules, are basically transparent while gases with non-symmetrical molecules such as carbon dioxide, water vapor, and some other hydrocarbon gases radiate and absorb appreciably. The ranges of wavelength where the absorption or emission is observed are called bands.

The decrease in intensity of a monochromatic radiation upon a layer of gas is given by

E-123

$$dI_\lambda = -C_\lambda I_\lambda dx \qquad (10.34)$$

where C_λ is the monochromatic absorption coefficient and dx is the gas layer thickness (Fig. 10.14).

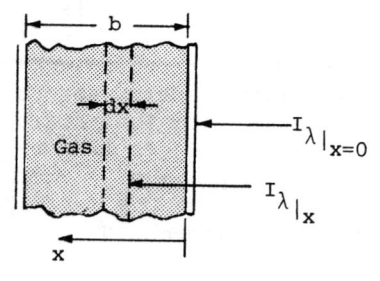

Fig. 10·14

Integrating eq.(10.27), we have

$$\boxed{\frac{I_{\lambda x}}{I_{\lambda o}} = e^{-C_\lambda X}} \quad \text{(Beer's Law)} \quad (10.35)$$

If $\rho = 0$, then the monochromatic absorptivity is

$$\boxed{\alpha \lambda = 1 - \tau_\lambda = 1 - e^{-C_\lambda X}} \qquad (10.36)$$

because

$$\tau_\lambda = e^{-C_\lambda X}$$

10.9 HEAT EXCHANGE BETWEEN GAS VOLUME AND BLACK ENCLOSURE

Let us consider a gas volume at uniform temperature

T_g enclosed by a black surface which is at temperature T_s (steam pipes, engine exhaust pipes, etc.).

The net heat transfer from gas to enclosure is given as,

$$\boxed{\frac{q}{A} = \underbrace{\varepsilon_{g,Tg}\, \sigma T_g^4}_{} - \underbrace{\alpha_{g,Ts}\, \sigma T_s^4}_{}} \quad (10.37)$$

 energy emitted energy absorbed
 by gas by the gas from
 enclosure

where $\varepsilon_{g,Tg}$ = Gas emittance at T_g

 $\alpha_{g,Ts}$ = Gas absorptance at T_s

10.10 HEAT EXCHANGE BETWEEN GAS VOLUME AND GRAY ENCLOSURE

The net heat transfer to a gray enclosure with an emissivity, ε_s, is obtained by multiplying equation (10.37) by a factor such that,

$$\boxed{\left(\frac{q_{gray}}{q_{black}}\right) = \frac{\varepsilon_s + 1}{2}} \quad \text{for } \varepsilon_s > 0.8$$

$$(10.38)$$

10.11 COMBINED HEAT TRANSFER COEFFICIENT : RADIATION AND CONVECTION

In many practical situations, radiation and convection heat transfer are combined to solve problems. The total

E-125

heat transfer is given by adding the convection and radiation,

$$q = (h_{conv} + h_{rad}) A_1 (T_s - T_\infty) \qquad (10.39)$$

where h_{conv} and h_{rad} are the convective and radiation heat transfer coefficients. The radiation heat transfer coefficient is a function of the temperature. Eq.(10.30) applies when the second surface is at the fluid temperature ($T_2 = T_\infty$) and enclosed in the first surface.

CHAPTER 11

BOILING AND CONDENSATION

11.1 BOILING PHENOMENON

When a hot surface is brought in contact with a liquid and the surface temperature is greater than the saturation temperature of the liquid, bubbles will appear on the surface and eventually will move upwards to the free surface where vaporization takes place. This phenomenon is termed boiling or ebullition and depends on the temperature difference ($T_s - T_{sat}$). Boiling may occur in two forms:

11.1.1 POOL BOILING

The boiling occurs when a hot surface is submerged into a liquid.

11.1.2 FORCED CONVECTION BOILING

The boiling occurs when a liquid is forced to pass through a hot surface.

Many experiments have been performed to analyze the boiling phenomena. Figure 11.1 shows the results of an electrically heated chromel wire submerged in water in which four specific regions can be distinguished:

A) Heat transfer takes place by natural or free convection.

B) Vapor bubbles appear on the heat surface, grow and move upwards as the surface temperature increases. The heat transfer increases (Fig. 11.1, curve A from point 1 to 2) and the bubbles form columns that reach the free surface.

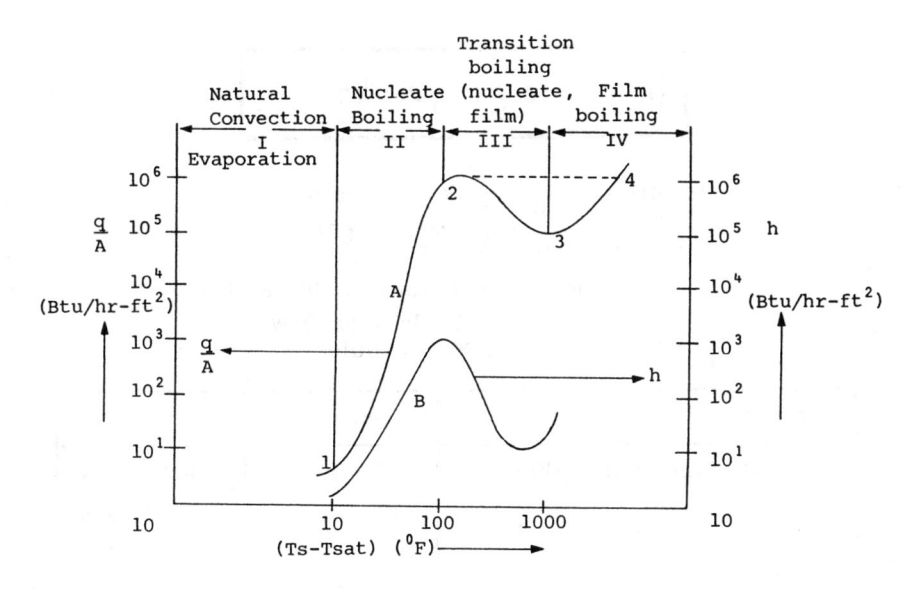

Fig. 11.1 Heat flux and the heat transfer coefficient through the boiling regions.

C) Vapor bubbles grow so rapidly that they do not separate from the surface forming in this way a vapor film that acts as an insulator. The heat transfer decreases because of the increase in thermal resistance due to the film (curve A, point 2 to 3).

D) As the difference in temperature increases, the heat transfer takes place mainly by radiation and the heat flux increases again (point 3 to 4). The point 4 usually is not reached because the temperature is above the melting point of the material. This point is known as the "burnout" point.

11.2 HEAT TRANSFER CORRELATIONS

11.2.1 REGIME I: FREE CONVECTION

General correlation

$$(q/A) = C \frac{k}{L} (Gr_L Pr)^m (T_s - T_b) \qquad (11.1)$$

where T_b = Bulk temperature
 $Gr_L = g\beta(T_s - T_b)L^3/\nu^2$
 m,c = constants from table 11.1
 $m,c = \begin{array}{ll} 1/4 & \text{laminar flow} \\ 1/3 & \text{turbulent flow} \end{array}$

Table 11.1

Geometry	GrPr	c	m
Vertical planes and Cylinders	$10^4 - 10^9$	0.59	1/4
	$10^9 - 10^{13}$	0.021	2/5
	$10^4 - 10^{13}$	0.1	1/3
Horizontal Cylinders	$0 - 10^{-5}$	0.4	0
	$10^4 - 10^9$	0.53	1/4
	$10^9 - 10^{12}$	0.13	1/3
Horizontal Plates Upper surface heated or lower surface cooled	$10^5 - 2\times10^7$	0.54	1/4
Lower surface heated or upper surface cooled	$3\times10^5 - 3\times10^{10}$	0.27	1/4
Upper surface heated or lower surface cooled	$2\times10^7 - 3\times10^{10}$	0.14	1/3

11.2.2 REGIME II: NUCLEATE BOILING

General Correlation

$$\frac{C_p(T_s - T_{sat})}{h_{fg} Pr^7} = C_{sf} \left[\frac{q/A}{\mu h_{fg}} \sqrt{\frac{g_c \sigma}{g(\rho - \rho_v)}} \right]^{1/3} \quad (11.2)$$

where the subscript v on the properties refers to the saturated vapor otherwise are the saturated liquid properties. Also,

C_p = Specific heat (Btu/lbm °F)
h_{fg} = Enthalpy of vaporization (Btu/lbm)
Pr = Prandtl number
q/A = Heat transfer per unit area (Btu/hr-ft^2)
μ = Viscosity (lbm/hr-ft)
C_{sf} = Constant (Table 11.2)
ρ = Density (lbm/ft^3)
σ = Surface tension (lb/ft)

Table 11.2

Surface-Fluid	C_{sf}
Water-Brass	0.006
Water-Copper	0.013
Water-Platinum	0.013
n-Butyl Alcohol-copper	0.00305
n-Pentane-Chromium	0.015
Isopropyl Alcohol-Copper	0.00225
Benzene-Chromium	0.010
Ethyl Alcohol-Chromium	0.027

The peak heat flux (point 2, Fig. 11.1) can be found using the following suggested correlation:

$$\left(\frac{q}{A} \right)_{max} = \frac{\pi}{24} h_{fg} \rho_v \left[\frac{\sigma g}{\rho_v} (\rho - \rho_v) \right]^{\frac{1}{4}} \left[\frac{\rho + \rho_v}{\rho} \right]^{\frac{1}{2}} \quad (11.3)$$

E-130

where σ is again the surface tension. Notice that the peak flux does not depend on the surface material.

11.2.3 REGIME III AND IV: FILM BOILING

A recommended empirical correlation for the total heat transfer coefficient on horizontal tubes is

$$h = h_{conv}\left(\frac{h_{conv}}{h}\right)^{1/3} + h_{rad} \quad (11.4)$$

$$h_{conv} = 0.62\left[\frac{k_v^3 \rho_v g(\rho - \rho_v)(h_{fg} + 0.4 Cp_v \Delta T)}{d\mu_v \Delta T}\right]^{1/4} \quad (11.5)$$

$$h_{rad} = \frac{\sigma \varepsilon (T_s^4 - T_{sat}^2)}{(T_s - T_{sat})} \quad (11.6)$$

where σ = Stefan-Boltzmann constant

ε = Emissivity of the surface (ε =1, for the liquid)

In eq. (11.5), d is the tube diameter, and properties of the vapor are to be evaluated at film temperature,

$$T_f = 1/2(T_s + T_{sat}).$$

For vertical tubes, a suggested correlation for the heat transfer coefficient is,

$$h = (0.0020)Re^{0.6}\left[\frac{k_v^3 g \rho_v(\rho - \rho_v)}{\mu_v^2}\right]^{1/3} \quad (11.7)$$

where,

$$Re = \frac{4\dot{m}}{\pi d \mu_v}$$

E-131

$\dot{m}$ = Vapor mass flow rate

The minimum heat flux (point 3) can be found using the following suggested correlation

$$\left(\frac{q}{A}\right)_{min} = (0.09)\rho_v h_{fg} \left[\frac{g(\rho - \rho_v)}{\rho + \rho_v}\right]^{1/2} \left[\frac{g_c \sigma}{g(\rho - \rho_v)}\right]^{1/4}$$

(11.8)

A simplified correlation for water boiling on the outside of a submerged surface at atmospheric pressure, P_{atm}, is given by the following relationship:

$$h_p = h_{atm} \left(\frac{P}{P_{atm}}\right)^{0.4}$$

(11.9)

where h_p = Heat transfer coefficient at some pressure P.

h_{atm} = Heat transfer coefficient determined from Table 11.3.

Table 11.3

Surface	q/A(Btu/hr-ft²)	h(Btu/hr-ft²-°F)
Horizontal	$q/A < 5 \times 10^3$	$151(T_s - T_{sat})^{1/3}$
	$5 \times 10^3 < q/A < 7.5 \times 10^4$	$0.168(T_s - T_{sat})^3$
Vertical	$q/A < 10^3$	$87(T_s - T_{sat})^{1/7}$
	$10^3 < q/A < 2 \times 10^4$	$0.24(T_s - T_{sat})^3$

D) Heat Transfer in Forced-Convection Boiling

The total heat flux for forced-convection in smooth tubes is given by

$$\left(\frac{q}{A}\right)_{Total} = \left(\frac{q}{A}\right)_{boiling} + \left(\frac{q}{A}\right)_{convection}$$

(11.10)

where

$$\left(\frac{q}{A}\right)_{convection} = h^*(T_s - T_{sat}) \qquad (11.11)$$

where h^* is given by

$$\frac{h^* \cdot d}{k} = 0.019(Re_d)^{0.8} Pr^{0.4} \qquad (11.12)$$

Eq.(11.10) applies for local forced-convection boiling.

11.3 CONDENSATION

Condensation of vapor on a surface takes place when the surface temperature is lower than the saturation temperature of the vapor. Two types of condensation can be distinguished: a) dropwise, in which droplets of liquid are formed on the surface, and b) filmwise, in which a film of liquid is formed along the surface. In practice, dropwise condensation is difficult to achieve, so, design calculations are based on the assumption of film type condensation.

11.4 FILM CONDENSATION IN VERTICAL PLATES

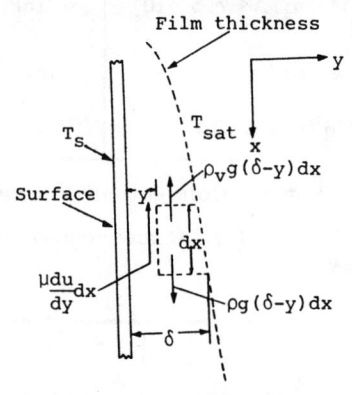

Fig. 11-2 Analysis of film condensation on a vertical surface.

Assumptions:

A) Shear stress at the edge of the film ($y=\delta$) is zero.
B) Linear temperature distribution through the film.
C) Unit depth.

From Figure 11.2 the force balance gives us the following equation

$$\mu \frac{du}{dy} dx = \rho g(\delta - y) dx - \rho_v g(\delta - y) dx \qquad (11.13)$$

Boundary conditions: $u = 0$ at $y = 0$

Integrating,

$$u = \frac{g}{\mu}(\rho - \rho_v)(\delta y - \tfrac{1}{2} y^2) \qquad (11.14)$$

The mass flow of condensate at any point x is given by

$$\dot{m} = \frac{\rho(\rho - \rho_v) g \delta^3}{3\mu} \qquad (11.15)$$

The convection heat transfer coefficient is given by

$$h_x = \left[\frac{\rho(\rho - \rho_v) g h_{fg} X^3}{4\mu k (T_{sat} - T_s)} \right]^{1/4} \qquad (11.16)$$

where
h_{fg} = Enthalpy.

In dimensionless form the heat transfer coefficient is

$$\boxed{Nu_x = \frac{hx}{k} = \left[\frac{\rho(\rho - \rho_v) g h_{fg} x^3}{4\mu k (T_{sat} - T_s)} \right]^{1/4}} \qquad (11.17)$$

The average heat-transfer coefficient along the surface is

$$\boxed{\bar{h} = \frac{1}{L} \int_0^L h_x dx = \frac{4}{3} h_{x=L}} \qquad (11.18)$$

E-134

$$\boxed{\bar{h} = 0.943 \left[\frac{\rho g(\rho - \rho_v) h_{fg} k}{\mu L (T_{sat} - T_v)} \right]^{1/4}} \qquad (11.19)$$

For laminar condensation on horizontal tubes, we have

$$\bar{h} = 0.725 \left[\frac{\rho(\rho - \rho_v) g h_{fg} k^3}{\mu d (T_{sat} - T_v)} \right]^{1/4} \qquad (11.20)$$

The criterion of laminar or turbulent flow in the condensation system is given by the Reynolds number, defined as:

$$Re = \frac{D_H \rho V}{\mu} \qquad (11.21)$$

where D_H is the hydraulic diameter and V is the average velocity of the flow,

or

$$Re = \frac{4\dot{m}}{P\mu} \qquad (11.22)$$

where $\dot{m} = \rho V A$

$$P = \begin{cases} 1 & \text{Vertical plate (unit depth)} \\ \pi d & \text{Vertical tube} \end{cases}$$

or

$$Re = \frac{4\Gamma}{\mu} \qquad (11.23)$$

where $\Gamma = \frac{\dot{m}}{L}$

The flow is turbulent for $Re > 1800$.

For vertical surfaces we have

$$\bar{h} = 1.47 \text{Re}^{-1/3} \left[\frac{gk^3 \rho_\ell (\rho_\ell - \rho_v)}{\mu^2} \right]^{1/3} \quad \text{Re} < 1800$$

(11.24)

$$\bar{h} = 0.0077 \text{Re}^{0.4} \left[\frac{gk^3 \rho_\ell (\rho_\ell - \rho_v)}{\mu^2} \right] \quad \text{Re} > 1800$$

(11.25)

Handbook of Mechanical Engineering

SECTION F
Laplace Transforms

CHAPTER 1

THE LAPLACE TRANSFORM

1.1 INTEGRAL TRANSFORMS

A class of transformations, which are called **Integral transforms**, are defined by

$$T\{f(t)\} = \int_{-\infty}^{\infty} K(s,t) f(t) \, dt = F(s) \qquad (1)$$

Given a function $K(s, t)$, called the **kernel** of the transformation, equation (1) associates with each function $f(t)$, of the class of functions for which the above exists, a function $F(s)$ defined by (1).

Various particular choices of the kernel function $K(s, t)$ in (1) have led to a special transformation, each with its own properties to make it useful in specific circumstances. The transform defined choosing the kernel

$$K(s,t) = \begin{cases} 0 & \text{for } t < 0 \\ e^{-st} & \text{for } t \geq 0. \end{cases} \qquad (2)$$

is called **Laplace transform**, which is the one to which this book is devoted.

1.2 DEFINITION OF LAPLACE TRANSFORM

Let $f(t)$ be a function for $t > 0$. The Laplace transform of $f(t)$, denoted by $L\{f(t)\}$, is defined by

$$L\{f(t)\} = \int_0^\infty e^{-st} f(t)\, dt \qquad (3)$$

1.3 NOTATION

The integral in (3) is a function of the parameter s that is called $F(s)$. It is customary to denote the functions of t by the lower case letters f, g, h, k, y, etc., and their Laplace transforms by the corresponding capital letters. Also some texts use capital letter L and show the Laplace transform of $f(t)$ by $L\{f(t)\}$. Therefore we may write

$$L\{f(t)\} = L\{f(t)\} = \int_0^\infty e^{-st} f(t)\, dt = F(s) \qquad (4)$$

If $f(x)$ is a function of $x \geq 0$, then its Laplace transform is denoted by $L\{f(x)\}$.

Example

Find $L\{\sin at\}$.

By definition

$$L\{\sin at\} = \int_0^\infty e^{-st} \sin at\, dt.$$

By employing integration by parts

$$e^{bx} \sin mx\, dx = \frac{e^{bx}(b\sin mx - m\cos mx)}{b^2 + m^2} + c.$$

Therefore

$$L\{\sin at\} = \left[\frac{e^{-st}(-s\sin at - a\cos at)}{(-s)^2 + a^2}\right]_{t=0}^{t=\infty}$$

F-2

Since for positive s, $e^{-st} \to 0$ at $t \to \infty$, and $\sin at$ and $\cos at$ are bounded functions, therefore the above yields

$$L\{\sin at\} = 0 - \frac{-a}{s^2 + a^2} = \frac{a}{s^2 + a^2}$$

1.4 LAPLACE TRANSFORMATION OF ELEMENTARY FUNCTIONS

The following table shows Laplace transforms of some elementary functions.

Note: Factorial n. For every integer $n > 0$,

$$\text{factorial } n = n! = 1 \cdot 2 \cdot \ldots \cdot n$$

and by definition $0! = 1$.

Table 1

$f(t)$	$L\{f(t)\} = F(s)$	
1	$\frac{1}{s}$	$s > 0$
t	$\frac{1}{s^2}$	$s > 0$
t^n for $n = 0, 1, 2, \ldots$	$\frac{n!}{s^{n+1}}$	$s > 0$
e^{at}	$\frac{1}{s-a}$	$s > a$
$\sin at$	$\frac{a}{s^2+a^2}$	$s > 0$
$\cos at$	$\frac{s}{s^2+a^2}$	$s > 0$
$\sinh at$	$\frac{a}{s^2-a^2}$	$s > |a|$
$\cosh at$	$\frac{s}{s^2-a^2}$	$s > |a|$
$\sqrt{t}$	$\frac{1}{2s}\sqrt{\frac{\pi}{s}}$	$s > 0$

1.5 SECTIONALLY OR PIECEWISE CONTINUOUS FUNCTIONS

The function $f(t)$ is said to be **piecewise continuous** or **sectionally continuous** over an interval $a < t < b$ if that interval can be divided into a finite number of intervals $c < t < d$ such that

1. $f(t)$ is continuous in the open interval $c < t < d$,

2. $f(t)$ approaches a finite limit as t approaches each end point within the interval $c < t < d$; that is, the limits

$$\lim_{\varepsilon \to 0} f(d-\varepsilon) \quad \text{and} \quad \lim_{\varepsilon \to 0} f(c+\varepsilon)$$

exist and are finite, where $\varepsilon > 0$.

An example of a sectionally continuous function is shown in the following Figure 1.1.

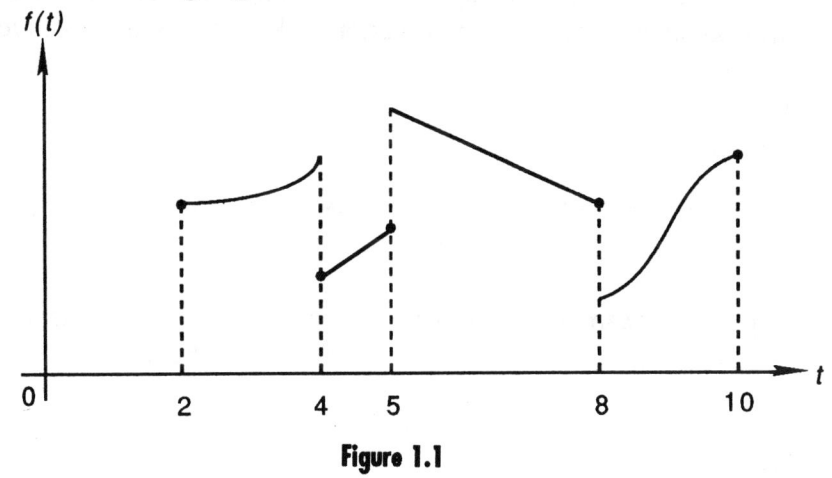

Figure 1.1

1.6 FUNCTIONS OF EXPONENTIAL ORDER

The function $f(t)$ is said to be of **exponential order** as $t \to \infty$ if there exists two constants $M > 0$ and b and a fixed t_0

F- 4

such that

$$|f(t)| < M e^{bt} \text{ for all } t \geq t_0 \quad (5)$$

We also say $f(t)$ is of **exponential order** b as $t \to \infty$, and we write

$$f(t) = 0(e^{bt}), \quad t \to \infty. \quad (6)$$

Example 1

Most elementary functions such as all power functions

$$f(t) = t^n \text{ for every } n \geq 0,$$

$\sin at$, $\cos at$, e^{at}, $\sinh at$, and $\cosh at$

are all of exponential order. Because for example $t^n \leq e^{nt}$, for all $t > 0$, therefore $f(t)$ is of exponential order n (one could show that the exponential order of the function t^n is as low as any positive number).

Example 2

The functions

$$e^{t^2} \text{ and } e^{t^3}$$

are some examples of functions that are not of exponential order. Because for any given number $b > 1$ we have

$$e^{t^3} > e^{t^2} > e^{bt}$$

for all $t > b$.

1.6.1 SUFFICIENT CONDITION FOR EXPONENTIAL ORDER FUNCTIONS

Let $f(t)$ be a function on $t \geq 0$, if a constant b exists such that the limit

$$\lim_{t \to \infty} [e^{-bt} f(t)]$$

exists, then the function $f(t)$ is of exponential order.

Example

The function $f(t) = t^5 + e^{2t}$ is of exponential order 2, since

$$\left| e^{-2t} f(t) \right| = \frac{t^5 + e^{2t}}{e^{2t}} \to 1 \quad \text{as} \quad t \to \infty.$$

1.7 EXISTENCE OF LAPLACE TRANSFORM

Theorem 1.1

If for all finite $N > 0$ the integral

$$\int_0^N e^{-st} f(t)\, dt$$

exists and $f(t)$ is of exponential order b as $t \to \infty$, then the Laplace transform

$$\mathcal{L}\{f(t)\} = \int_0^\infty e^{-st} f(t)\, dt = F(s)$$

exists for $s > b$.

1.7.1 BEHAVIOR OF LAPLACE TRANSFORMS AT INFINITY

Theorem 1.2

If $f(t)$ is sectionally continuous over every finite interval in the range $t \geq 0$ and if $f(t)$ is of exponential order,

$$f(t) = 0(e^{bt}) \quad \text{as} \quad t \to \infty,$$

then the Laplace transform $\mathcal{L}\{f(t)\}$ exists for $s > b$.

Theorem 1.3

If $f(t)$ is sectionally continuous over every finite interval in the range $t \geq 0$, and is of exponential order as $t \to \infty$, and if

$$\mathcal{L}\{f(t)\} = F(s)$$

then

$$\lim_{s \to \infty} F(s) = 0$$

Remark: It is important to realize that the conditions stated in Theorem 1.2 are sufficient to guarantee the existence of the Laplace transform. If the conditions are not satisfied, however, the Laplace transform may or may not exist. For example the function $t^{-1/2}$ is not sectionally continuous in every interval in the range $t \geq 0$. But

$$\mathcal{L}\{t^{-1/2}\} = (\pi/s)^{1/2}, \quad s > 0$$

exists.

1.8 SOME IMPORTANT PROPERTIES OF LAPLACE TRANSFORMS

In the following theorems we assume that all functions satisfy the sufficient conditions in Theorem 1.2 so that their Laplace transforms exist.

1.8.1 LINEARITY PROPERTY

Theorem 1.4

Let $f_1(t)$ and $f_2(t)$ be two functions with Laplace transforms $F_1(s)$ and $F_2(s)$ correspondingly. Then for all constants c_1 and c_2 we have

$$\mathcal{L}\{c_1 f_1(t) + c_2 f_2(t)\} = c_1 \mathcal{L}\{f_1(t)\} + c_2 \mathcal{L}\{f_2(t)\} \quad (7)$$
$$= c_1 F_1(s) + c_2 F_2(s)$$

Example

$$\mathcal{L}\{\sin 5t - 3e^{-2t} + 6t^5\} = \mathcal{L}\{\sin 5t\} - 3\mathcal{L}\{e^{-2t}\} + 6\mathcal{L}\{t^5\}$$

$$= \left(\frac{5}{s^2+25}\right) - 3\left(\frac{1}{s+2}\right) + 6\left(\frac{5!}{s^6}\right)$$

$$= \frac{5}{s^2+25} - \frac{3}{s+2} + \frac{720}{s^6}$$

1.8.2 CHANGES OF SCALE PROPERTY
Theorem 1.5

Let $f(t)$ be a function with the Laplace transform $F(s)$. Then for every $a > 0$, we have

$$\mathcal{L}\{f(at)\} = \frac{1}{a} F\left(\frac{s}{a}\right) \qquad (8)$$

Example

Since $\mathcal{L}\{\cos t\} = \dfrac{s}{s^2+1}$, then

$$\mathcal{L}\{\cos 5t\} = \frac{1}{5} \frac{(s/5)}{(s/5)^2+1} = \frac{s}{s^2+25}$$

1.8.3 FIRST SHIFT PROPERTY
Theorem 1.6

Let $\mathcal{L}\{f(t)\} = F(s)$, then

$$\mathcal{L}\{e^{at}f(t)\} = F(s-a) \qquad (9)$$

for all real numbers a, Figure 1.2, following.

Example

Find $\mathcal{L}\{t^3 e^{5t}\}$. Apply the first shift property for the function

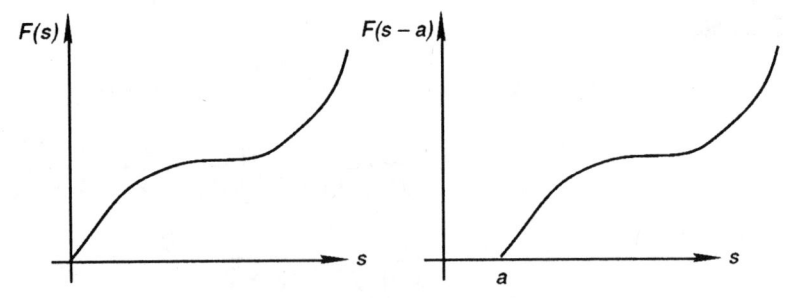

Figure 1.2

$f(t) = t^3$ and $a = 5$. Since

$$L\{t^3\} = \frac{3!}{s^{3+1}} = \frac{6}{s^4} \quad \text{for} \quad s > 0$$

then

$$L\{t^3 e^{5t}\} = \frac{6}{(s-5)^4} \quad \text{for} \quad s > 5.$$

1.8.4 SECOND SHIFT PROPERTY

Theorem 1.7

If

$$L\{f(t)\} = F(s) \quad \text{and} \quad g(t) = \begin{cases} f(t-a) & \text{if } t \geq a \\ 0 & \text{if } t < a, \end{cases}$$

then

$$L\{g(t)\} = e^{-as} F(s) \qquad (10)$$

Figure 1.3, following.

Example

Find the Laplace transform of the function

$$g(t) = \begin{cases} \sin(t-5) & t \geq 5 \\ 0 & t < 5 \end{cases}$$

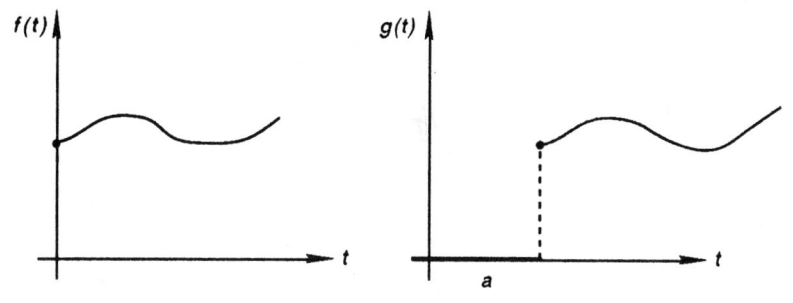

Figure 1.3

Since $\mathcal{L}\{\sin t\} = \dfrac{1}{s^2 + 1}$, then

$$\mathcal{L}\{\sin t - 5\} = e^{-5s}\mathcal{L}\{\sin t\}$$

$$= \dfrac{1}{e^{5s}(s^2 + 1)}.$$

1.8.5 LAPLACE TRANSFORM OF DERIVATIVE

Theorem 1.8

If $f(t)$ is continuous for $t \geq 0$ and of exponential order as $t \to \infty$, and if $f'(t)$ is sectionally continuous, then

$$\mathcal{L}\{f(t)\} = F(s)$$

implies that

$$\boxed{\mathcal{L}\{f'(t)\} = sF(s) - f(0) \qquad (11)}$$

Example

Since $\mathcal{L}\{\sqrt{t}\} = \dfrac{1}{2s}\sqrt{\dfrac{\pi}{s}}$ (Table 1) and $(\sqrt{t})' = \dfrac{1}{2\sqrt{t}}$, then we have

$$\mathcal{L}\left\{\dfrac{1}{\sqrt{t}}\right\} = 2\mathcal{L}\left\{\dfrac{1}{2\sqrt{t}}\right\},$$

by linearity

$$= 2(s\mathcal{L}\{\sqrt{t}\} - 0)$$

$$= \sqrt{\frac{\pi}{s}}.$$

Theorem 1.9

If in Theorem 1.8 $f(t)$ fails to be continuous at a point $a > 0$, then

$$\mathcal{L}\{f'(t)\} = sF(s) - f(0) = e^{-as}\{f(a+) - f(a-)\} \qquad (12)$$

Figure 1.4.

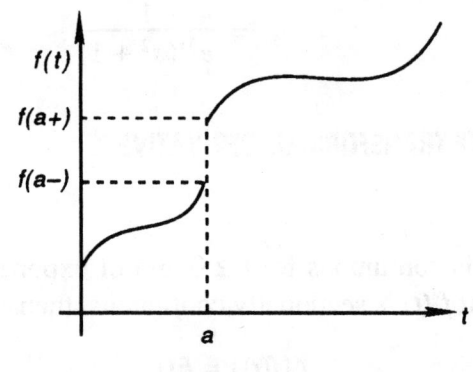

Figure 1.4

Theorem 1.10

If in Theorem 1.9 the function is discontinuous at $t = 0$, then

$$\mathcal{L}\{f'(t)\} = sF(s) - f(0+) \qquad (13)$$

Theorem 1.11

If $f(t), f'(t), f''(t), \ldots, f^{(n-1)}(t)$ are continuous and of exponential order as $t \to \infty$, and $f(n)(t)$ is sectionally continuous, then

$$\mathcal{L}\{f(t)\} = F(s)$$

implies that

$$\mathcal{L}\{f^{(n)}(t)\} = s^n F(s) - s^{n-1} f(0) - s^{n-2} f^{(1)}(0) - s^{n-3} f^{(2)}(0) - \ldots - f^{(n-1)}(0) \quad (14)$$

Example

Specially

$$\mathcal{L}\{f''(t)\} = s^2 F(s) - sf(0) - f'(0),$$

and

$$\mathcal{L}\{f^{(3)}(t)\} = s^3 F(s) - s^2 f(0) - sf'(0) - f''(0).$$

1.8.6 DERIVATIVE OF LAPLACE TRANSFORMS

Theorem 1.12

If $f(t)$ is sectionally continuous and is of exponential order as $t \to \infty$, then

$$\mathcal{L}\{f(t)\} = F(s)$$

implies that

$$F'(s) = \mathcal{L}\{-tf(t)\} \quad (15)$$

and more general

$$F^{(n)}(s) = \mathcal{L}\{(-t)^n f(t)\} \quad (16)$$

Example

Since

$$\mathcal{L}\{e^{5t}\} = \frac{1}{s-5} \text{ and } \left(\frac{1}{s-5}\right)' = \frac{-1}{(s-5)^2},$$

we have

$$\mathcal{L}\{te^{5t}\} = -\mathcal{L}\{-te^{5t}\}$$
$$= -\left(\frac{1}{s-5}\right)'$$
$$= \frac{1}{(s-5)^2}.$$

1.8.7 PERIODIC FUNCTIONS

The function $F(t)$ is said to be **periodic with period** T if we have

$$f(t + T) = F(t)$$

The functions

$$\sin at, \quad \cos at, \quad \sec at, \quad \text{and} \quad \csc at$$

are periodic with period $T = 2\pi/a$, and the functions

$$\tan at \quad \text{and} \quad \cot at$$

are periodic with period $T = \pi/a$.

Theorem 1.13

If $f(t)$ has Laplace transform and if $f(t) = f(t + w)$, then

$$\mathcal{L}\{f(t)\} = \frac{\int_0^w e^{-st}f(t)\,dt}{1 - e^{-sw}} \qquad (17)$$

CHAPTER 2

INVERSE LAPLACE TRANSFORM

2.1 DEFINITION OF INVERSE LAPLACE TRANSFORM

If

$$\mathcal{L}\{f(t)\} = F(s),$$

we say that $f(t)$ is an **Inverse Laplace transform**, or **Inverse transform**, of $F(s)$ and we write

$$f(t) = \mathcal{L}^{-1}\{F(s)\} \qquad (1)$$

Example

Since

$$\mathcal{L}\{\sin 5t\} = \frac{5}{s^2 + 25},$$

we can write

$$\mathcal{L}^{-1}\left\{\frac{5}{s^2 + 25}\right\} = \sin 5t.$$

2.2 UNIQUENESS OF INVERSE LAPLACE TRANSFORM
2.2.1 NULL FUNCTIONS

A function $n(t)$ is called a **null function** if for every positive number a we have

$$\int_0^a n(t)\, dt = 0 \qquad (2)$$

In general, a function that is zero for all but a countable set of real numbers is a null function. Notice that a set is called countable if one can put its elements in a one to one correspondence with the set of positive integers.

Example

The function

$$n(t) = \begin{cases} 1 & \text{if } t \text{ is integer} \\ 0 & \text{otherwise} \end{cases}$$

is a null function.

Theorem 2.1 — Lerch's Theorem

If

$$\mathcal{L}\{f_1(t)\} = \mathcal{L}\{f_2(t)\},$$

then for some null function $n(t)$ we have

$$f_1(t) - f_2(t) = n(t). \qquad (3)$$

That is, an inverse Laplace transform is unique except for the addition of an arbitrary null function.

Example

Let

$$f_1(t) = e^t \text{ and } f_2(t) = \begin{cases} e^t & \text{if } t \neq 1 \\ 5 & \text{if } t = 1 \end{cases}$$

The difference of these two functions is a null function:

$$N(t) = f_1(t) - f_2(t) = \begin{cases} 0 & \text{if } t \neq 1 \\ -5 & \text{if } t = 1 \end{cases}$$

while their Laplace transforms are the same

$$L\{f_1(t)\} = L\{f_2(t)\} = \frac{1}{t-1}.$$

2.3 SOME INVERSE LAPLACE TRANSFORMS

The following table shows some inverse Laplace transforms.

Table 2

$F(s)$	$\mathcal{L}^{-1}\{F(s)\} = f(t)$
$\frac{1}{s}$	1
$\frac{1}{s^2}$	t
$\frac{1}{s^{n+1}}\ n = 0, 1, 2, \ldots$	$\frac{t^n}{n!}$
$\frac{1}{s-a}$	e^{at}
$\frac{1}{s^2 + a^2}$	$\frac{\sin at}{a}$
$\frac{s}{s^2 + a^2}$	$\cos at$
$\frac{1}{s^2 - a^2}$	$\frac{\sinh at}{a}$
$\frac{s}{s^2 - a^2}$	$\cosh at$

2.4 SOME PROPERTIES OF INVERSE LAPLACE TRANSFORMS

2.4.1 LINEARITY PROPERTY

Theorem 2.2

If c_1 and c_2 are constant numbers then

$$\mathcal{L}^{-1}\{c_1 F_1(s) + c_2 F_2(s)\} = c_1 \mathcal{L}^{-1}\{F_1(s)\} + c_2 \mathcal{L}^{-1}\{F_2(s)\} \quad (4)$$

Example

Using linearity property and Table 2, we can write

$$\mathcal{L}^{-1}\left\{\frac{5}{s^2} + \frac{6s}{s^2 - 25} - \frac{12}{s - 6}\right\}$$

$$= 5\mathcal{L}^{-1}\left\{\frac{1}{s^2}\right\} + 6\mathcal{L}^{-1}\left\{\frac{1}{s^2 - 25}\right\} - 12\mathcal{L}^{-1}\left\{\frac{1}{s - 6}\right\}$$

$$= \frac{5}{t} + 6\cosh 5t - 12e^{6t}.$$

2.4.2 FIRST TRANSLATION OR SHIFTING PROPERTY

Theorem 2.3

$$\mathcal{L}^{-1}\{F(s)\} = e^{-at}\mathcal{L}^{-1}\{F(s-a)\} \quad (5)$$

Or

$$\mathcal{L}^{-1}\{F(s-a)\} = e^{at}\mathcal{L}^{-1}\{F(s)\} \quad (6)$$

Example

Since we have

$$\mathcal{L}^{-1}\left\{\frac{1}{s^2 + 1}\right\} = \sin t,$$

F-18

we have

$$\mathcal{L}^{-1}\left\{\frac{1}{s^2-6s+10}\right\} = \mathcal{L}^{-1}\left\{\frac{1}{(s-3)^2-9+10}\right\}$$

$$= \mathcal{L}^{-1}\left\{\frac{1}{(s-3)^2+1}\right\}$$

$$= e^{3t}\mathcal{L}^{-1}\left\{\frac{1}{s^2+1}\right\}$$

$$= e^{3t}\sin t.$$

2.4.3 SECOND TRANSLATION OR SHIFTING PROPERTY
Theorem 2.4

If

$$\mathcal{L}^{-1}\{F(s)\} = f(t),$$

then

$$\mathcal{L}^{-1}\{e^{-as}F(s)\} = \begin{cases} f(t-a) & \text{for } t \geq a \\ 0 & \text{for } t < a. \end{cases} \quad (7)$$

Example

Since

$$\mathcal{L}^{-1}\left\{\frac{1}{s^7}\right\} = \frac{t^6}{6!},$$

we have

$$\mathcal{L}^{-1}\left\{\frac{e^{-5s}}{s^7}\right\} = \begin{cases} \dfrac{(t-5)^6}{6!} & \text{for } t \geq 5 \\ 0 & \text{for } t < 5. \end{cases}$$

2.4.4 CHANGE OF SCALE PROPERTY

Theorem 2.5

If
$$\mathcal{L}^{-1}\{F(s)\} = f(t),$$

then for every constant number $k > 0$ we have

$$\mathcal{L}^{-1}\{F(ks)\} = \frac{1}{k} f\left(\frac{t}{k}\right) \qquad (8)$$

Example

Since
$$\mathcal{L}^{-1}\left\{\frac{1}{s^2 - 1}\right\} = \sinh t,$$

we have

$$\mathcal{L}^{-1}\left\{\frac{1}{s^2 - 25}\right\} = \mathcal{L}^{-1}\left\{\frac{1}{25\left(\frac{s}{5}\right)^2 - 25}\right\}$$

$$= \mathcal{L}^{-1}\left\{\frac{1}{25} \frac{1}{\left(\frac{s}{5}\right)^2 - 1}\right\}$$

$$= \frac{1}{25} \mathcal{L}^{-1}\left\{\frac{1}{\left(\frac{s}{5}\right)^2 - 1}\right\}$$

$$= \frac{1}{5} \sinh 5t.$$

2.4.5 INVERSE LAPLACE TRANSFORM OF DERIVATIVES

Theorem 2.6

If
$$\mathcal{L}^{-1}\{F(s)\} = f(t)$$

then

$$\mathcal{L}^{-1}\{F'(s)\} = -t f(t) \qquad (9)$$

and more general

$$\mathcal{L}^{-1}\{F^{(n)}(s)\} = (-t)^n f(t) \qquad (10)$$

Example

Since

$$\mathcal{L}^{-1}\left\{\frac{s}{s^2+1}\right\} = \cos t \quad \text{and} \quad \left(\frac{s}{s^2+1}\right)' = \frac{1-s^2}{(s^2+1)^2},$$

we have

$$\mathcal{L}^{-1}\left\{\frac{1-s^2}{(s^2+1)^2}\right\} = (-t)\cos t = -t\cos t.$$

2.4.6 INVERSE LAPLACE TRANSFORM OF INTEGRALS

Theorem 2.7

If

$$\mathcal{L}^{-1}\{F(s)\} = f(t),$$

then

$$\mathcal{L}^{-1}\left\{\int_s^\infty F(x)\,dx\right\} = \frac{f(t)}{t} \qquad (11)$$

Example

Since

$$\mathcal{L}^{-1}\left\{\frac{1}{s(s-1)}\right\} = \mathcal{L}^{-1}\left\{-\frac{1}{s} + \frac{1}{s-1}\right\}$$

$$= -\mathcal{L}^{-1}\left\{\frac{1}{s}\right\} + \mathcal{L}^{-1}\left\{\frac{1}{s-1}\right\}$$
$$= -1 + e^t,$$

and

$$\int_s^\infty \frac{1}{x(x-1)} dx = \ln\left(\frac{s}{s-1}\right),$$

we have

$$\mathcal{L}^{-1}\left\{\ln\left(\frac{s}{s-1}\right)\right\} = \frac{-1+e^t}{t}.$$

2.5 THE CONVOLUTION PROPERTY

2.5.1 DEFINITION

We call the integral

$$\int_0^t f(u)g(t-u)\,du = f*g(t) \qquad (12)$$

the **convolution integral**, or **convolution**, of f and g.

Example

Find the convolution $f*g(t)$ if

$$f(t) = t \quad \text{and} \quad g(t) = t^2.$$

Since $f(u) = u$ and $g(t-u) = (t-u)^2 = t^2 - 2tu + u^2$, we have

$$f*g(t) = \int_0^t u(t^2 - 2tu + u^2)\,du$$
$$= \int_0^t (t^2 u = 2tu^2 + u^3)\,du$$
$$= \left[\frac{1}{2}t^2 u^2 - \frac{2}{3}tu^3 + \frac{1}{4}u^4\right]_{u=0}^{u=t}$$

F-22

$$= \frac{1}{2}t^4 - \frac{2}{3}t^4 + \frac{1}{4}t^4 - 0$$
$$= \frac{1}{12}t^4$$

2.5.2 PROPERTIES OF CONVOLUTION
Theorem 2.8

Convolution of the functions f and g obeys the following laws.

(a) Commutative law

$$f*g = g*f \qquad (13)$$

(b) Associative law

$$f*\{g*h\} = \{f*g\}*h \qquad (14)$$

(c) Distributive law

$$f*\{g + h\} = f*g + f*h \qquad (15)$$

(d) Derivative property

$$(f*g)' = f'*g = f*g' \qquad (16)$$

2.5.3 INVERSE LAPLACE TRANSFORM OF THE CONVOLUTION
Theorem 2.9

If

$$\mathcal{L}^{-1}\{F(s)\} = f(t) \quad \text{and} \quad \mathcal{L}^{-1}\{G(s)\} = g(t),$$

then

$$\mathcal{L}^{-1}\{F(s)G(s)\} = \int_0^t f(u)\,g(t-u)\,du = f*g(t) \quad (17)$$

Example

To find
$$\mathcal{L}^{-1}\left\{\frac{1}{s^2(s+1)^2}\right\},$$
one can use convolution. We have
$$\mathcal{L}^{-1}\left\{\frac{1}{s^2}\right\} = t \quad \text{and} \quad \mathcal{L}^{-1}\left\{\frac{1}{(s+1)^2}\right\} = te^{-t}.$$

Then using integration by part we have

$$\mathcal{L}^{-1}\left\{\frac{1}{s^2(s+1)^2}\right\} = \int_0^t (ue^{-u})(t-u)\,du$$

$$= \left[(ut-u^2)-(t-2u)+(-2)(-e^{-u})\right]_{u=0}^{u=t}$$

$$= te^{-t} + 2e^{-t} + t - 2.$$

CHAPTER 3

SOME SPECIAL FUNCTIONS

3.1 THE GAMMA FUNCTION

We use the Gamma function to calculate the Laplace transform of non-integral powers of t.

If $r > 0$, we define the **Gamma function** by

$$\Gamma(r) = \int_0^\infty t^{r-1} e^{-t}\, dt. \qquad (1)$$

Example

Find $\Gamma(1)$.

$$\Gamma(1) = \int_0^\infty t^{1-1} e^{-t}\, dt]$$

$$= \int_0^\infty e^{-t}\, dt$$

$$= [-e^{-t}]_{t=0}^{t=\infty}$$

$$= 1.$$

Therefore

$$\Gamma(1) = 1 \qquad (2)$$

3.1.1 PROPERTIES OF THE GAMMA FUNCTION

Theorem 3.1

For $r > 0$, we have

$$\Gamma(r + 1) = r\, \Gamma(r). \qquad (3)$$

Example

$$\Gamma\left(\frac{3}{2}\right) = \Gamma\left(\frac{1}{2} + 1\right)$$
$$= \frac{1}{2}\Gamma\left(\frac{1}{2}\right).$$

$\Gamma(1/2)$ is calculated in example of Theorem 3.3.

Theorem 3.2

For any positive integer n we have

$$\Gamma(n + 1) = n! \qquad (4)$$

Example

$$\Gamma(6) = 5! = 120.$$

Theorem 3.3

For every number $r > -1$ we have

$$\mathcal{L}\{t^r\} = \frac{\Gamma(r+1)}{s^{r+1}}; \quad s > 0. \qquad (5)$$

Example

Find $\Gamma(1/2)$. Since

$$\mathcal{L}\left\{\sqrt{\frac{1}{t}}\right\} = \sqrt{\frac{\pi}{s}},$$

if in (4) we put $r = -1/2$, we have

$$\mathcal{L}\left\{\sqrt{\frac{1}{t}}\right\} = = \frac{\Gamma\left(\frac{1}{2}\right)}{s^{(1/2)}}.$$

Or

$$\Gamma\left(\frac{1}{2}\right) = s^{1/2} \mathcal{L}\left\{\sqrt{\frac{1}{t}}\right\}$$

$$= s^{1/2}\sqrt{\frac{\pi}{2}}$$

$$= \sqrt{\pi}.$$

Hence

$$\Gamma\left(\frac{1}{2}\right) = \sqrt{\pi} \qquad (6)$$

3.2 BESSEL FUNCTIONS

The **Bessel function** of **order** n is defined by

$$J_n(t) = \sum_{k=0}^{\infty} \frac{(-1)^k \left(\frac{1}{2}t\right)^{2k}}{k!\,\Gamma(k+n+1)} \qquad (7)$$

This function is also called the **Bessel function of first kind** and of **index** n. We will use $J_n(t)$ in an application of the differential equations method of chapter 5.

Example

For $n = 0$

$$J_0(t) = \sum_{k=0}^{\infty} \frac{(-1)^k \left(\frac{1}{2}t\right)^{2k}}{k!\,\Gamma(k+1)}$$

$$= \sum_{k=0}^{\infty} \frac{(-1)^k \left(\frac{1}{2}t\right)^{2k}}{k!\,k!}.$$

And

$$J_0(2) = \sum_{k=0}^{\infty} \frac{(-1)^k}{(k!)^2}.$$

3.2.1 MODIFIED BESSEL FUNCTION

For every number n the **modified Bessel function** of order n is defined by

$$I_n(t) = (-i)^{-n} J_n(it) \qquad (8)$$

where i is the imaginary unit, i.e.

$$i = \sqrt{-1}.$$

3.2.2 SOME PROPERTIES OF BESSEL FUNCTIONS

Theorem 3.4

Bessel functions satisfy the following properties

(i) For all positive n

$$J_{-n}(t) = (-1)^n J_n(t) \qquad (9)$$

(ii) For all n

$$J_{n+1}(t) = {}^{2n}/{}_t\, J_n(t) - J_{n-1}(t) \qquad (10)$$

(iii) For all $n \neq 0$

$$\frac{d}{dt}[t^n J_n(t)] = t^n J_{n-1}(t) \tag{11}$$

(iv)
$$J'_0(t) = -J_1(t) \tag{12}$$

(v) The following formula is called the **Generating function** for the Bessel function.

$$e^{(t/2)(u-1/u)} = \sum_{n=-\infty}^{\infty} J_n(t) u^n \tag{13}$$

(vi) $J_n(t)$ satisfies the **Bessel's differential equation**

$$t^2 y'' + t y' + (t^2 - n^2) y = 0 \tag{14}$$

3.3 THE ERROR FUNCTION

The **Error function**, abbreviated **erf**, is defined by

$$\text{erf}(t) = \frac{2}{\sqrt{\pi}} \int_0^t e^{-u^2} du \tag{15}$$

We use erf(t) to find the inverse Laplace transform of some certain simple functions.

Example

Since
$$\mathcal{L}^{-1}\left\{\sqrt{\frac{1}{s}}\right\} = \frac{1}{\sqrt{\pi t}},$$

and therefore by the first shifting property, formula (5) of chapter 2, we have

$$\mathcal{L}^{-1}\left\{\frac{1}{\sqrt{s+1}}\right\} = \frac{e^{-1}}{\sqrt{\pi t}}.$$

Then the convolution theorem yields

$$\mathcal{L}^{-1}\left\{\frac{1}{s}\frac{1}{s+1}\right\} = \int_0^t 1 \cdot \frac{e^{-u}}{\sqrt{\pi u}}\,du$$

$$= \frac{2}{\sqrt{\pi}}\int_0^{\sqrt{t}} e^{-z^2}\,dz$$

$$= \mathrm{erf}(\sqrt{t}).$$

3.3.1 SOME PROPERTIES OF erf(t)
Theorem 3.5

$\mathrm{erf}(t)$ satisfies the following properties

(i)
$$\frac{d}{dt}\mathrm{erf}(t) = \frac{2}{\sqrt{\pi}} e^{-t^2} \qquad (16)$$

(ii)
$$\mathrm{erf}(t) = \frac{2}{\sqrt{\pi}} \sum_{n=0}^{\infty} \frac{(-1)^n t^{2n+1}}{(2n+1)\,n!} \qquad (17)$$

(iii)
$$\lim_{t\to\infty} \mathrm{erf}(t) = 1 \qquad (18)$$

3.4 THE COMPLEMENTARY ERROR FUNCTION

The **Complementary Error function**, abbreviated **erfc**, is defined by

$$\mathrm{erfc}(t) = 1 - \mathrm{erf}(t) = 1 - \frac{2}{\sqrt{\pi}}\int_0^t e^{-u^2}\,du$$
$$= \frac{2}{\sqrt{\pi}}\int_t^{\infty} e^{-u^2}\,du \qquad (19)$$

3.4.1 A PROPERTY OF erfc(t)
Theorem 3.6

For every constant m

$$\lim_{t \to \infty} t^m \,\text{erfc}(t) = 0 \qquad (20)$$

3.5 THE SINE AND COSINE INTEGRALS

The **Sine Integral function** is defined by

$$si(t) = \int_0^t \frac{\sin u}{u}\,du = \int_0^1 \frac{\sin(tv)}{v}\,dv. \qquad (21)$$

And the **Cosine Integral function** is defined by

$$ci(t) = \int_t^\infty \frac{\cos u}{u}\,du = \int_1^\infty \frac{\cos(tv)}{v}\,dv. \qquad (22)$$

3.6 THE EXPONENTIAL INTEGRAL

The **Exponential Integral function** is defined as

$$ei(t) = \int_t^\infty \frac{e^{-u}}{u}\,du = \int_1^\infty \frac{e^{-tv}}{u}\,dv. \qquad (23)$$

3.7 THE UNIT STEP FUNCTION OR THE HEAVISIDE FUNCTION

The **unit step function** or the **Heaviside function** is defined as

$$h_1(t-t) = \begin{cases} 0 & \text{if } t < t_0 \\ 1 & \text{if } t \geq t_0 \end{cases} \qquad (24)$$

where we assume that $t_0 \geq 0$. A generalization for the unit step function is

$$h_c(t-t) = \begin{cases} 0 & \text{if } t < t_0 \\ c & \text{if } t \geq t_0 \end{cases} \quad (25)$$

Example

One can write the **square-wave function,** Figure 3.1, in terms of Heaviside functions.

$$f(t) = h_c(t) - h_c(t-1) + h_c(t-2) \ldots$$

$$= \sum_{k=0}^{\infty} (-1)^k h_c(t-k). \quad (26)$$

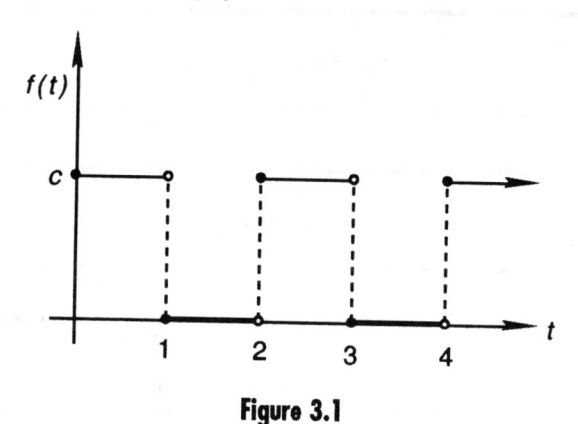

Figure 3.1

3.7.1 SOME PROPERTIES OF THE UNIT STEP FUNCTION
Theorem 3.7

The unit step function satisfies the following properties

(i)

$$h_c(t-t_0) = c\, h_1(t-t_0) \quad (27)$$

(ii) The graph of $h_1(t-t)$ is given in Figure 3.2.

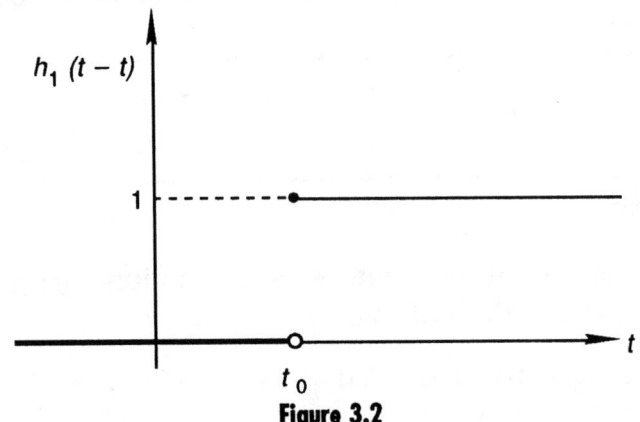

Figure 3.2

(iii)
$$\mathcal{L}\{h_1(t-t_0)\} = \frac{e^{-st_0}}{s} \tag{28}$$

(iv)
$$\mathcal{L}\{h_1(t)\} = \frac{1}{s} \tag{29}$$

(v)
$$\{h_c(t-t_0)\} = \frac{c\,e^{-st_0}}{s} \tag{30}$$

(vi) If $\mathcal{L}\{f(t)\} = F(s)$, then
$$\{f(t-t_0)h_1(t-t_0)\} = e^{-st_0}F(s) \tag{31}$$

Example

Express the function

$$f(t) = \begin{cases} 6 & \text{if } 0 \leq t < 4 \\ 2t+1 & \text{if } t \geq 4 \end{cases}$$

in terms of the Heaviside function, h, and find $\mathcal{L}\{f(t)\}$.

Since $h_1(t) = 0$ for $t < 0$ and $h_1(t) = 1$ for $t \geq 0$, we build $f(t)$ in the following way. We write

$$f_1(t) = 6\, h_1(t)$$

that works for $0 < t < 4$, to knock off 6 for $t > 4$ we write

$$f_2(t) = 6\, h_1(t) - 6\, h_1(t-4).$$

Then we add the term $(2t + 1)\, h_1(t-4)$ and finally have

$$f(t) = 6\, h_1(t) - 6\, h_1(t-4) + (2t+1)\, h_1(t-4)$$

or

$$f(t) = h_6(t) - h_5(t-4) + t h_2(t-4).$$

Now we calculate the Laplace transform

$$\mathcal{L}\{f(t)\} = \mathcal{L}\{h_6(t) - h_5(t-4) + t h_2(t-4)\}$$

$$= \mathcal{L}\{h_6(t)\} - \mathcal{L}\{h_5(t-4)\} + \mathcal{L}\{t h_2(t-4)\}.$$

Hence since

$$\mathcal{L}\{h_6(t)\} = \frac{6}{s},\ \mathcal{L}\{h_5(t-4)\} = \frac{5e^{-4s}}{s},$$

and

$$\mathcal{L}\{t h_2(t-4)\} = 2e^{-4s}\, \mathcal{L}\{t\} = 2e^{-4s}\, \frac{1}{s^2},$$

we have

$$\mathcal{L}\{f(t)\} = \frac{6}{s} - \frac{5e^{-4s}}{s} + \frac{2e^{-4s}}{s^2}$$

$$= \frac{6s - 5s\, e^{-4s} + 2e^{-4s}}{s^2}.$$

F-34

3.8 THE UNIT IMPULSE FUNCTION

Consider the function

$$f_\varepsilon(t) = \begin{cases} \dfrac{1}{\varepsilon} & \text{if } 0 \leq t \leq \varepsilon \\ 0 & \text{if } t > \varepsilon \end{cases}, \qquad (32)$$

that is shown in the figure 3.3.

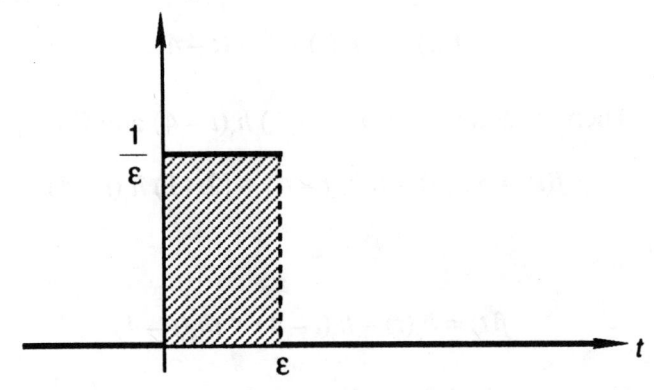

Figure 3.3

It is true that for all value of $\varepsilon > 0$, the area of the shaded region is always equal to 1, i.e.

$$\int_0^\infty f_\varepsilon(t)\, dt = 1. \qquad (33)$$

The **unit impulse function** or **Dirac delta function**, denoted by $\delta(t)$ is defined as the following limiting function

$$\delta(t) = \lim f_\varepsilon(t) \text{ as } \delta \to 0. \qquad (34)$$

3.8.1 SOME PROPERTIES OF THE UNIT IMPULSE FUNCTION

Theorem 3.8

$\delta(t)$ satisfies the following properties.

F-35

(i)
$$\int_0^\infty \delta(t)\, dt = 1 \qquad (35)$$

(ii) For any continuous function g(t)

$$\int_0^\infty \delta(t) g(t)\, dt = g(0) \qquad (36)$$

(iii) For any continuous function g(t)

$$\int_0^\infty \delta(t - t_0) g(t)\, dt = g(t_0) \qquad (37)$$

(iv) $\delta(t)$ is the identity element for the convolution operator, i.e. for any function $g(t)$ we have

$$\delta * g = g \qquad (38)$$

(v)
$$\mathcal{L}\{\delta(t)\} = 1 \qquad (39)$$

(vi)
$$\mathcal{L}\{\delta(t - t_0)\} = e^{-st_0} \qquad (40)$$

3.9 THE BETA FUNCTION

If $m > 0$ and $n > 0$, the **Beta function** is defined by

$$B(m,n) = \int_0^1 u^{m-1}(1-u)^{n-1}\, du. \qquad (41)$$

Theorem 3.9

The Beta function satisfies the following properties.

(i)
$$B(m,n) = \frac{\Gamma(m)\,\Gamma(n)}{\Gamma(m+n)} \qquad (42)$$

(ii)
$$B(m,n) = 2\int_0^{\pi/2} \sin^{2m-1} u \cos^{2n-1} u \, du \qquad (43)$$

Example

$$\int_0^{\pi/2} \sin^7 u \cos^5 u \, du = \frac{1}{2} B(4,3)$$

$$= \frac{1}{2} \frac{\Gamma(4)\,\Gamma(3)}{\Gamma(4+3)}$$

$$= \frac{1}{2} \frac{(3!)\,(2!)}{6!} = \frac{1}{120}.$$

3.10 MORE PROPERTIES OF INVERSE LAPLACE TRANSFORMS

3.10.1 MULTIPLICATION BY s^n

Theorem 3.10

In Laplace transform, multiplication by s effects differentiation on $f(t)$. If $\mathcal{L}^{-1}\{F(s)\} = f(t)$, then

$$\mathcal{L}^{-1}\{s\,F(s)\} = f'(t) \quad \text{when} \quad f(0) = 0, \qquad (44)$$

and

$$\mathcal{L}^{-1}\{s\,F(s)\} - f(0)\} = f'(t) \quad \text{when} \quad f(0) \neq 0, \qquad (45)$$

or

$$\mathcal{L}^{-1}\{s\,F(s)\} = f'(t) + f(0)\,\delta(t) \quad \text{when} \quad f(0) \neq 0. \qquad (46)$$

Example

Since
$$\mathcal{L}^{-1}\left\{\frac{1}{s^2+1}\right\} = \sin t \quad \text{and} \quad \sin 0 = 0,$$

we have
$$\mathcal{L}^{-1}\left\{\frac{1}{s^2+1}\right\} = (\sin t)' = \cos t.$$

Example

Since
$$\mathcal{L}^{-1}\left\{\frac{1}{s^2+1}\right\} = \cos t \quad \text{and} \quad \cos 0 = 1,$$

we have
$$\mathcal{L}^{-1}\left\{\frac{s^2}{s^2+1}\right\} = (\cos t)' + (\cos 0)\,\delta(t)$$
$$= -\sin t + \delta(t).$$

3.10.2 DIVISION BY S

Theorem 3.11

In Laplace transform, division by s effects integration on $f(t)$. If $\mathcal{L}^{-1}\{F(s)\} = f(t)$, then

$$\mathcal{L}^{-1}\left\{\frac{F(s)}{s}\right\} = \int_0^t f(u)\,du. \qquad (47)$$

Or

$$\mathcal{L}^{-1}\left\{\frac{F(s)}{s}\right\} = \int_0^t \{F(s)\}(u)\,du. \qquad (48)$$

From (48) we obtain the following more general formula

$$\mathcal{L}^{-1}\left\{\frac{F(s)}{s}\right\} = \int_0^t \mathcal{L}^{-1}\left\{\frac{F(s)}{s^{n-1}}\right\}(u)\,du. \qquad (49)$$

for $n = 1, 2, 3, \ldots$

Example 1

Using (47) since

$$\mathcal{L}^{-1}\left\{\frac{1}{s-1}\right\} = e^t,$$

we have

$$\mathcal{L}^{-1}\left\{\frac{1}{s(s-1)}\right\} = \int_0^t e^u\,du$$

$$= [e^u]_{u=0}^{u=t}$$

$$= e^t - 1.$$

Example 2

Using (49) three times and example 1, we have

$$\mathcal{L}^{-1}\left\{\frac{1}{s^3(s-1)}\right\} = \int_0^t \mathcal{L}^{-1}\left\{\frac{1}{s^2(s-1)}\right\}(u)\,du$$

$$= \int_0^t \left[\int_0^u \mathcal{L}^{-1}\left\{\frac{1}{s(s-1)}\right\}(v)\,dv\right]du$$

$$= \int_0^t \left[\int_0^u \left(\int_0^v \mathcal{L}^{-1}\left\{\frac{1}{(s-1)}\right\}(w)\,dw\right)dv\right]du$$

$$= \int_0^t \int_0^u \int_0^v (e^w - 1)\, dw\, dv\, du$$

$$= e^t - \frac{t^3}{6} - \frac{t^2}{2} - t - 1$$

Theorem 3.12

$$\mathcal{L}^{-1}\left\{\frac{F(s)}{s^n}\right\} = \int_0^t \mathcal{L}^{-1}\left\{\frac{F(s)}{s^{n-1}}\right\}(u_1)\, du_1$$

$$= \int_0^t \int_0^{u_1} \mathcal{L}^{-1}\left\{\frac{F(s)}{s^{n-2}}\right\}(u_2)\, du_2\, du_1$$

or

$$= \ldots$$

$$\boxed{\begin{array}{c} \mathcal{L}^{-1}\left\{\dfrac{F(s)}{s^n}\right\} \\ = \int_0^t \int_0^{u_1} \int_0^{u_2} \ldots \int_0^{u_{n-1}} f(u_n)\, du_n\, du_{n-1}\, \ldots\, du_1 \end{array}} \quad (50)$$

Example

Since

$$\mathcal{L}^{-1}\left\{\frac{1}{s^2 - 1}\right\} = \sinh t,$$

we have

$$\mathcal{L}^{-1}\left\{\frac{1}{s^3(s^2 - 1)}\right\}$$

$$= \int_0^t \mathcal{L}^{-1}\left\{\frac{1}{s^2(s^2 - 1)}\right\}(u)\, du$$

$$= \int_0^t \int_0^u \mathcal{L}^{-1}\left\{\frac{1}{s(s^2 - 1)}\right\}(v)\, dv\, du$$

$$= \int_0^t \int_0^u \int_0^v \mathcal{L}^{-1}\left\{\frac{1}{(s^2 - 1)}\right\}(w)\, dw\, dv\, du$$

$$= \int_0^t \int_0^u \int_0^v \sinh w \, dw \, dv \, du$$

$$= \int_0^t \int_0^u (\cosh v - 1) \, dv \, du$$

$$= \int_0^t (\sinh u - u) \, du$$

$$= \cosh t - \frac{1}{2} t^2 - 1.$$

CHAPTER 4

APPLICATION OF ORDINARY LINEAR DIFFERENTIAL EQUATIONS

4.1 ORDINARY DIFFERENTIAL EQUATION WITH CONSTANT COEFFICIENTS

The Laplace transformation transforms a linear differential equation with constant coefficients into an algebraic equation.

To solve an initial value problem (IVP):

$$a_0 \frac{d^n y}{dt^n} + a_1 \frac{d^{n-1} y}{dt^{n-1}} + \ldots + a_{n-1} \frac{dy}{dt} + a_n y = f(t) \quad (1)$$

$$y(0) = c_0, \, y'(0) = c_1, \, \ldots, \, y^{(n-1)}(0) = c_{n-1} \quad (2)$$

by means of the Laplace transform, we perform the following procedure.

a. Let $Y(s)$ be the Laplace transform of the unknown function $y(t)$, and $F(s)$ be the Laplace transform of the known function $f(t)$, i.e.

$$Y(s) = \mathcal{L}\{y(t)\} \text{ and } F(s) = \mathcal{L}\{f(t)\}.$$

b. Take the Laplace transform of both sides of (1).

c. Use the linearity property, Theorem 1.4, to separate the terms and use the derivative property, Theorem 1.11,

$$\mathcal{L}\{y^{(n)}\} = s^n \mathcal{L}\{y(t)\} - s^{n-1}y(0) - s^{n-2}y'(0) - \ldots - y^{(n-1)}(0)$$

$$= s^n Y(s) - c_0 s^{n-1} - c_1 s^{n-2} - \ldots - c_{n-1}.$$

d. Form the algebraic equation

$$[a_0 s^n + a_1 s^{n-1} + \ldots + a_{n-1} s + a_n] Y(s)$$

$$- c_0 [a_0 s^{n-1} + a_1 s^{n-2} + \ldots + a_{n-1}]$$

$$- c_1 [a_0 s^{n-2} + a_1 s^{n-3} + \ldots + a_{n-2}]$$

$$\vdots$$

$$- c_{n-2}[a_0 s + a_1] - c_{n-1} a_0 = F(s).$$

e. Solve the resulting equation of step d to determine $Y(s)$.

f. Use the Laplace inverse transform to determine the solution

$$y(t) = \mathcal{L}^{-1}\{Y(s)\}.$$

Example 1

Solve the initial value problem

$$y'' - 2y' + y = 0, \qquad (3)$$

$$y(0) = 1, y'(0) = 0. \qquad (4)$$

Taking the Laplace transform of (3), we have

F-43

$$\mathcal{L}\{y'' - 2y' + y\} = \mathcal{L}\{0\}$$

or

$$\mathcal{L}\{y''\} - 2\mathcal{L}\{y'\} + \mathcal{L}\{y\} = 0.$$

Now we substitute the initial values (4), to obtain

$$s^2 Y(s) - s\, y(0)\, y'(0) - 2(s\, Y(s) - 1) + Y(s) = 0$$

or

$$(s^2 - 2s + 1)\, Y(s) - s + 2 = 0.$$

Therefore

$$Y(s) = \frac{s-2}{s^2 - 2s + 1} = \frac{s-2}{(s-1)^2}.$$

Hence

$$y(t) = \mathcal{L}^{-1}\left\{\frac{s-2}{(s-1)^2}\right\}$$

$$= \mathcal{L}^{-1}\left\{\frac{1}{s-1}\right\} - \mathcal{L}^{-1}\left\{\frac{1}{(s-1)^2}\right\}$$

$$= e^t - t e^t.$$

Example 2

Solve the IVP

$$y'' - 6y' + 8y = t \sin t + \cos t \qquad (5)$$

$$y(0) = 0,\ y'(0) = 1. \qquad (6)$$

Taking the Laplace transform of the equation (5), we have

$$\mathcal{L}\{y'' - 6y' + 8y\} = \mathcal{L}\{t \sin t + \cos t\}.$$

Separating the terms, we obtain

$$\mathcal{L}\{y''\} - 6\mathcal{L}\{y'\} + 8\mathcal{L}\{y\} = \mathcal{L}\{t\sin t\} + \mathcal{L}\{\cos t\}.$$

Using the derivative property and the table of transforms, we find

$$s^2 Y(s) - s y(0) - y'(0) - 6 [s Y(s) - y(0)] + 8 Y(s)$$

$$= \frac{2s}{(s^2+1)^2} + \frac{s}{s^2+1}.$$

Use the initial values (6) and simplify to find

$$(s^2 - 6s + 8)Y = \frac{2s}{(s^2+1)^2} + \frac{s}{s^2+1} + 1.$$

Therefore

$$Y = \frac{2s}{(s^2+1)^2(s-2)(s-4)} + \frac{s}{(s^2)(s-2)(s-4)}$$

$$+ \frac{1}{(s-2)(s-4)}.$$

Now to prepare $Y(s)$ for the inverse Laplace transform, we use the partial fractions method of chapter 5 to write

$$Y(s) = \frac{85963}{173400} \frac{1}{s^2+1} - \frac{8712}{14450} \frac{1}{(s^2+1)^2} + \frac{365}{578} \frac{1}{s-4}$$

$$- \frac{39}{50} \frac{1}{s-2}.$$

Therefore

$$y(t) = \mathcal{L}^{-1}\{Y(s)\}$$

$$= \frac{85963}{173400} \mathcal{L}^{-1}\left\{\frac{1}{s^2+1}\right\} - \frac{8712}{14450} \mathcal{L}^{-1}\left\{\frac{1}{(s^2+1)^2}\right\}$$

$$+ \frac{365}{578} \mathcal{L}^{-1}\left\{\frac{1}{s-4}\right\} - \frac{39}{50} \mathcal{L}^{-1}\left\{\frac{1}{s-2}\right\}.$$

Thus, using the general table of Laplace transforms, we have

$$y(t) = \frac{85963}{173400} \sin t - \frac{8712}{14450} \frac{\sin t - t\cos t}{2} + \frac{365}{578} e^{4t} - \frac{39}{50} e^{2t}.$$

Or

$$y(t) = \frac{9736699}{50112600} \sin t + \frac{4356}{14450} t\cos t + \frac{365}{578} e^{4t} - \frac{39}{50} e^{2t}.$$

4.2 ORDINARY DIFFERENTIAL EQUATIONS WITH VARIABLE COEFFICIENTS

The method of Laplace transform also can be used to solve some differential equations with variable coefficients. By Theorem 1.12, we have

$$\mathcal{L}\{t^n y(t)\} = (-1)^n Y^{(n)}(s), \qquad (7)$$

i.e., multiplying the function by powers of t effects the transform in derivatives, the method of Laplace transform is more effective for linear differential equations with terms of the following form:

$$t^m \frac{d^n}{dt^n} y(t). \qquad (8)$$

Example

We use the Laplace transform to solve the **Laguerre equation**

$$t \frac{d^2 y}{dt^2} + (1-t) \frac{dy}{dt} + 2y = 0. \qquad (9)$$

We take the Laplace transform of the equation (9)

$$L\left\{t\frac{d^2y}{dt^2}+(1-t)\frac{dy}{dt}+2y\right\}=0.$$

We separate the terms to obtain

$$L\left\{t\frac{d^2y}{dt^2}\right\}+L\left\{\frac{dy}{dt}\right\}-L\left\{t\frac{dy}{dt}\right\}+2L\{y\}=0.$$

Using the derivative property of Laplace transforms, formula (15) of chapter 1, the above equation will be written as

$$-\frac{d}{ds}L\left\{\frac{d^2y}{dt^2}\right\}+L\left\{\frac{dy}{dt}\right\}+\frac{d}{ds}L\left\{\frac{dy}{dt}\right\}+2L\{y\}=0.$$

Using formula (14) of chapter 1, we will have

$$-\frac{d}{ds}\left[s^2Y(s)-s\,y(0)-y'(0)\right]+\left[sY(s)-y(0)\right]$$
$$+\frac{d}{ds}\left[sY(s)-y(0)\right]+2Y(s)=0.$$

This equation is simplified in the following form

$$s(1-s)\frac{d}{ds}Y(s)+(3-s)Y(s)=0.$$

Or

$$\frac{\frac{d}{ds}Y(s)}{Y(s)}=\frac{3-s}{s(s-1)}=-\frac{3}{s}+\frac{2}{s-1}$$

Solving this differential equation, we obtain the general solution

$$Y(s)=\frac{(s-1)^2}{s^3}=C\left[\frac{1}{s}-\frac{2}{s^2}+\frac{1}{s^3}\right].$$

Therefore

$$y(t) = C\left[\mathcal{L}^{-1}\left\{\frac{1}{s}\right\} - 2\mathcal{L}^{-1}\left\{\frac{1}{s^2}\right\} + \mathcal{L}^{-1}\left\{\frac{1}{s^3}\right\}\right].$$

Thus by the Laplace inverse formulas we have

$$y(t) = C\left[1 - 2t + \frac{1}{2}t^2\right]$$

where C is a constant number that is determined by using an initial value.

4.3 SYSTEMS OF LINEAR DIFFERENTIAL EQUATIONS

A system of linear differential equations involves two or more linear differential equations of two or more unknown functions of a single independent variable. A simple example of a system of linear differential equations is

$$\begin{cases} x' = x + 5y & , \ x(0) = 0 \\ y' = -x - y & , \ y(0) = 1 \end{cases} \tag{10}$$

where x and y are unknown functions of the variable t. Another example of a system of linear differential equations is

$$\begin{cases} 2x' + y - y = t & , \ x(0) = 1 \\ x' + y' = t^2 & , \ y(0) = 0. \end{cases} \tag{11}$$

When the initial conditions are specified, the Laplace transform will reduce a system of linear differential equations with constant coefficients to a set of simultaneous algebraic equations, in the transformed functions.

Example 1

Solve the system of linear equations (10) by means of the Laplace transforms.

We denote

$$X(s) = \mathcal{L}\{x(t)\} \quad \text{and} \quad Y(s) = \mathcal{L}\{y(t)\}.$$

Taking Laplace transform of each equation of the system (10), we obtain

$$\begin{cases} (sX(s) - x(0) = X(s) + 5Y(s) \\ sY(s) - y(0) = -X(s) - Y(s). \end{cases}$$

Or

$$\begin{cases} (s-1)X - 5Y = 0 \\ X + (s+1)Y = 1. \end{cases} \quad (12)$$

There are several ways to solve an algebraic system of equations. Using the substitution method and solving the first equation of (12) for Y in terms of X, we obtain

$$Y = \frac{s-1}{5} X \quad (13)$$

Substituting (13) into the second equation of (12), we obtain an equation for X

$$X + \frac{(s+1)(s-1)}{5} X = 1. \quad (14)$$

Simplifying (14) and solving for X, we get

$$X(s) = \frac{5}{s^2 + 4}. \quad (15)$$

Therefore

$$x(t) = \mathcal{L}^{-1}\{X(s)\}$$
$$= \mathcal{L}^{-1}\left\{\frac{5}{s^2+4}\right\}$$
$$= \frac{5}{2} \mathcal{L}^{-1}\left\{\frac{2}{s^2+2^2}\right\} = \frac{5}{2} \sin 2t.$$

To find $y(t)$, one may substitute $X(s)$ in one of the equations of (12) to find $Y(s)$, then using $Y(s)$ we can find $y(t)$ by the inverse Laplace transform. However, it is easier to substitute $x(t)$ in the first equation of (10), then by solving the resulting equation

$$(5/2 \sin 2t)' = 5/2 \sin 2t + 5 y(t),$$

Or

$$5 \cos 2t = 5/2 \sin 2t + 5 y(t),$$

we find

$$y(t) = \cos 2t - 1/2 \sin 2t.$$

Example 2

To solve the system of linear equations (11), we take the Laplace transform of its equations to get

$$\begin{cases} 2[sX(s) - x(0)] + [sY(s) - y(0)] - Y(s) = \dfrac{1}{s^2} \\ [sX(s) - x(0)] + [sY(s) - y(0)] = \dfrac{2}{s^3} \end{cases} \quad (16)$$

where $X(s) = \mathcal{L}\{x(t)\}$ and $Y(s) = \mathcal{L}\{y(t)\}$. Use the initial conditions $x(0) = 1$ and $y(0) = 0$ to find

$$\begin{cases} 2sX(s) + (s-1)Y(s) = 2 + \dfrac{1}{s^2} \\ sX(s) + sY(s) = 1 + \dfrac{2}{s^3} \end{cases} \quad (17)$$

Solving the algebraic system (17) for X and Y, we obtain

$$Y(s) = \frac{4-s}{s^3(s+1)}.$$

To prepare $Y(s)$ for the Laplace inverse transform we use the partial fractions method of chapter 5 to write

$$Y(s) = \frac{5}{s} - \frac{5}{s^2} + \frac{4}{s^3} - \frac{5}{s+1}$$

Thus

$$y(t) = 5\mathcal{L}^{-1}\left\{\frac{1}{s}\right\} - 5\mathcal{L}^{-1}\left\{\frac{1}{s^2}\right\} + 2\mathcal{L}^{-1}\left\{\frac{2!}{s^3}\right\}$$

$$- 5\mathcal{L}^{-1}\left\{\frac{1}{s+1}\right\}$$

or

$$y(t) = 5 - 5t + 2t^2 - 5e^{-t}.$$

By substituting $Y(s)$ in the second equation of (17), we have

$$X(s) = -Y(s) + \frac{1}{s} + \frac{2}{s^4},$$

therefore

$$x(t) = -\mathcal{L}^{-1}\{Y(s)\} + \mathcal{L}^{-1}\left\{\frac{1}{s}\right\} + \frac{1}{3}\mathcal{L}^{-1}\left\{\frac{3!}{s^4}\right\}$$

$$= -y(t) + t + \frac{1}{3}t^3.$$

or

$$x(t) = -4 + 5t - 2t^2 + \frac{1}{3}t^3 + 5e^{-t},$$

$$y(t) = 5 - 5t + 2t^2 - 5e^{-t}.$$

4.4 THE VIBRATION OF SPRING

Suppose a mass m is attached to the lower end of a light coil spring, suspended from a point on a support (figure 4-1(a)), and brought to the point of equilibrium, E, where it remains at rest (Figure 4-1(b)). Once the mass m is moved from the point of equilibrium E (figure 4-1(c)), the motion of the mass m will be determined by a differential equation with some initial conditions.

Figure 4.1

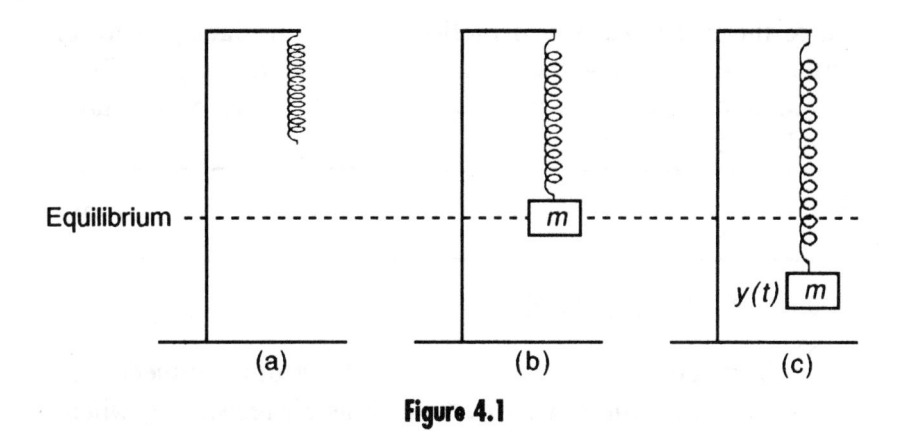

If $y(t)$ denotes the instantaneous displacement of m at time t from the equilibrium or rest position, then by the Hooke's law there will be a **restoring force acting on m**

$$F = -k\, y(t) \qquad (18)$$

where k is a constant depending on the spring that is called the **spring constant**. On the other hand by the Newton's second law of motion **the net force acting on m is equal to the mass times acceleration**, i.e.

$$F = m\, \frac{d^2 y(t)}{dt^2} \qquad (19)$$

Therefore by equating (18) and (19) **the differential equation of the motion of the mass** is

$$m\, y'' = -k\, y \qquad (20)$$

Or since usually the time depending derivatives are shown by $\dot{y}$ we have

$$\ddot{y} + \frac{k}{m}\, y = 0 \qquad (21)$$

since the mass was set to motion by first displacing it to an initial position y_0, and then giving it an initial velocity v_0. Thus, together with the differential equation (21), we have two initial conditions

$$y(0) = y_0, \quad \dot{y}(0) = v_0. \tag{22}$$

4.4.1 DAMPED VIBRATIONS

In practice, a vibrating spring is most often subjected to frictional and some other forces, such as air resistance, which act to retard (dampen) the motion to come to rest. Experiments have shown that

the magnitude of the damping force is approximately proportional to the velocity of the mass,

provided that the velocity of the mass is small. Since the damping force acts in opposite direction of the motion, we can express the **damping force** as

$$-b \frac{dy}{dt} \tag{23}$$

where b is called the **damping constant**. Thus the **differential equation** of the **damped vibration motion** is

$$m\ddot{y} + b\dot{y} + ky = 0 \tag{24}$$

Furthermore, when some time varying **external force** also acts on m, the equation of the **damped vibration motion with external force** is

$$m\ddot{y} + b\dot{y} + ky = f(t) \tag{25}$$

Thus, in general, a mathematical model for the **linear mass-spring system** formulated in an initial value problem of the form

$$m y'' + b y' + k y = f(t) \ (t > 0)$$
$$y(0) = y_0, \ y'(0) = y'_0$$
(26)

where $m > 0$ is the mass, $b \geq 0$ is the damped constant, $k > 0$ is the spring constant, and $f(t)$ is a time-varying external force. One can solve the IVP (26) by the means of Laplace transformation.

Damped Vibration is the linear mass system (26) when $b > 0$.

4.4.2 UNDAMPED VIBRATION

Undamped Vibration is the linear mass system (26) when $b = 0$.

4.4.3 FREE VIBRATION

If in a linear spring system there is no external force, i.e. in equation (26) $f(t) = 0$, then the system is called a **free vibration system**.

Solution of the free vibration system.

$$my'' + by' + ky = 0.$$
$$y(0) = 0, \ y'(0) = y'_0$$
(27)

Case 1 (undamped)

If in (27) $b = 0$.

$$my'' + ky = 0.$$
$$y(0) = 0, \ y'(0) = y'_0$$
(28)

Taking the Laplace transform of (28) and using the initial conditions, we will have

$$y(t) = c\cos(\omega_0 t - \delta)$$
$$\omega_0 = \sqrt{\frac{k}{m}}, \quad c = \sqrt{y_0^2 + (y'_0/\omega_0)^2} \qquad (29)$$
$$\cos\delta = y_0/c, \quad \sin\delta = y'_0/(w_0 c)$$

It represents **simple harmonic motion** with amplitude c, frequency ω_0, and period

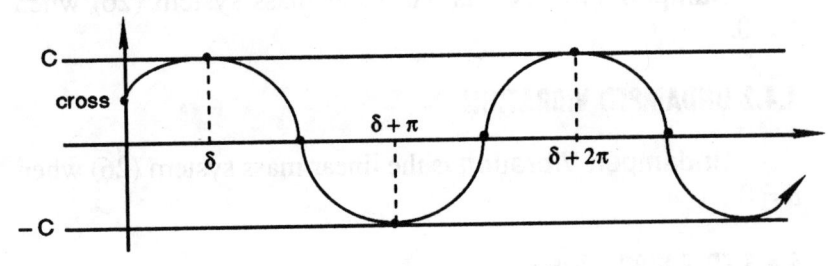

Figure 4.2

Case 2 (damped)

If in (27) $b > 0$. The general solution of the **damped free vibration model**:

$$my'' + by' + ky = 0$$

depends on the sign of the quantity $b^2 - 4km$, that is given in three cases.

(a) If $b^2 - 4km > 0$ (overdamping) (Figure 4-3(a)):

$$y(t) = c_1 e^{z_1 t} + c_2 e^{z_2 t}, \quad z_1 < 0, \; z_2 < 0$$
$$z_1 = \frac{-b + \sqrt{b^2 - 4km}}{2m}, \quad z_2 = \frac{-b - \sqrt{b^2 - 4km}}{2m} \qquad (30)$$

F-55

(b) If $b^2 - 4km = 0$ (critical damping) (Figure 4-3(b)):

$$y(t) = c_1 e^{z_1 t} + c_2 t e^{z_1 t}, \quad z_1 = -\frac{b}{2m} < 0. \qquad (31)$$

(c) If $b^2 - 4km < 0$ (underdamping) (Figure 4-3(c)):

$$y(t) = c\, e^{-bt/2m} \cos(\lambda_0 t - \delta)$$

$$\lambda_0 = \sqrt{\frac{4km - b^2}{4m^2}}, \quad c^2 = c_1^2 + c_2^2 \qquad (32)$$

$$\cos \delta = c_1/c, \quad \sin \delta = c_2/c$$

In this case of damping each solution is an **oscillation** with **amplitude** $ce^{-bt/2m}$ and period $2\pi / \lambda_0$; it is decreasing to zero.

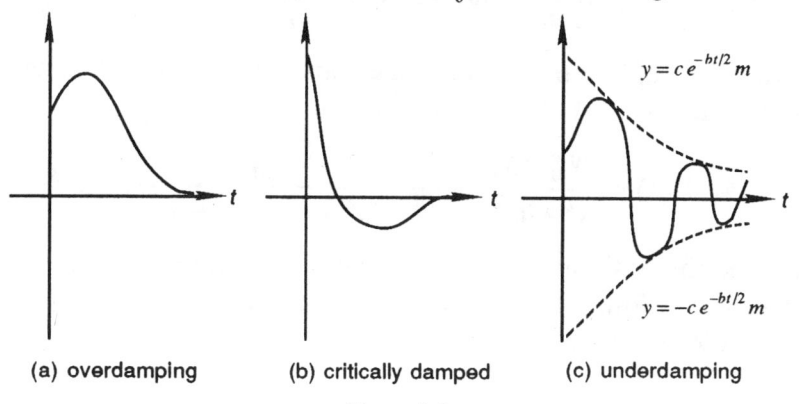

(a) overdamping   (b) critically damped   (c) underdamping

Figure 4.3

4.5 RESONANCE

An important example in the problem of undamped vibration of a spring is

$$my''(t) + ky(t) = F_0 \cos \omega t$$
$$y(0) = y_0, \quad y'(0) = \dot{y}_0 \qquad (33)$$

F-56

If we let
$$w_0 = \sqrt{\frac{k}{m}} \qquad (34)$$
then the problem will turn into the following standard form

$$y''(t) + \omega_0^2 y(t) = \frac{F_0}{m} \cos \omega t \qquad (35)$$
$$y(0) = y_0,\ y'(0) = \dot{y}_0$$

where ω_0^2 is called the **natural frequency** and ω is called the **applied frequency** of the undamped mass-spring system.

By putting
$$Y(s) = \mathcal{L}\{y(t)\},$$

taking the Laplace transform of the equation of (35), and using the initial values of (35), we will obtain
$$Y(s) = \frac{sy_0 + \dot{y}_0}{s^2 + \omega_0^2} + \frac{F_0}{m} \cdot \frac{\omega}{(s^2 + \omega_0^2)(s^2 + \omega^2)}. \qquad (36)$$

To find the position function $y(t)$, we have to find the Laplace inverse transform of (36), that differs according to whether $\omega = \omega_0$ or $\omega \ne \omega_0$.

Case 1

If $\omega \ne \omega_0 = \sqrt{k/m}$, the solution of the undamped vibration model (33) is

$$y(t) = y_0 \cos \omega_0 t + \frac{\dot{y}_0}{\omega_0} \sin \omega_0 t$$
$$= \frac{F_0}{m(\omega_0^2 - \omega^2)} \quad [\cos \omega t = \cos \omega_0 t] \qquad (37)$$

F-57

Case 2

If $\omega = \omega_0 = \sqrt{k/m}$, the solution of (33) is

$$y(t) = y_0 \cos \omega_0 t + \frac{\dot{y}_0}{\omega_0} \sin \omega_0 t + \frac{F_0}{2m\omega_0} t \sin \omega_0 t. \quad (38)$$

In this case the solution $y(t)$ obtained by (38) may be regarded as an oscillation with frequency ω_0 and amplitude

$$\frac{F_0 t}{(2m\omega_0)}$$

which increases with t (Figure 4.4). This phenomenon is called **resonance.** It occurs when the applied frequency is equal to the natural frequency of an undamped spring system.

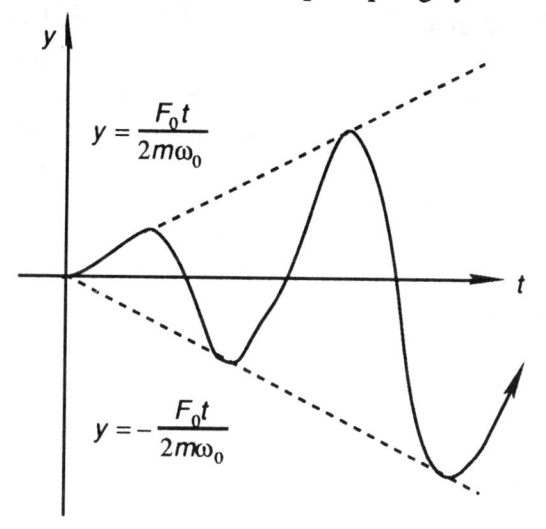

Figure 4.4

4.6 THE SIMPLE PENDULUM

A pendulum is made of a rod of length L feet that is suspended by one end so that it can swing freely in a vertical plane. Let a weight of m pounds be attached to the free end of the rod, and let the weight of the rod, compared to the weight m, be negligible.

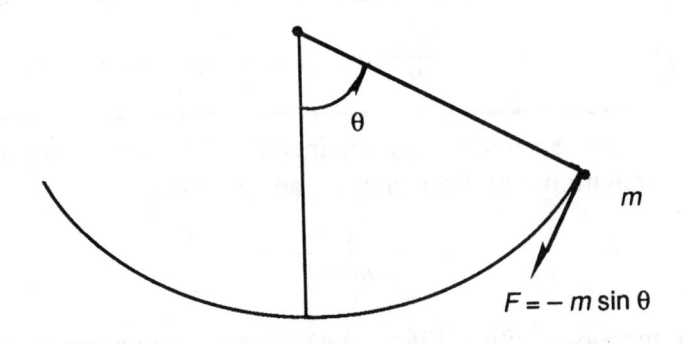

Figure 4.5

Let θ (radians) be the angular displacement from the vertical position (Figure 4.5), at time t (second). The tangential component of the force m is $m \sin \theta$ tending to decrease θ. Therefore, along the circular trajectory curve, the mass m is under the force

$$F = -m \sin \theta \qquad (39)$$

On the other hand by the Newton's second law of motion

$$F = \frac{m}{g} \frac{d^2 s}{dt^2} \qquad (40)$$

where $s = L\theta$ is the arc length from the vertical position. Thus equating (39) and (40) we conclude that

$$\frac{m}{g} \frac{d^2 s}{dt^2} = -m \sin \theta.$$

Since $s = L\theta$, from this equation it follows that

$$\frac{d^2 \theta}{dt^2} + \frac{g}{L} \sin \theta = 0 \qquad (41)$$

The second order nonlinear differential equation (41) governs the motion of the pendulum; its solution involves an elliptic integral which is not easy. If θ is small, however, sin θ and θ are approximately equal and (41) closely approximated by the second order linear differential equation

$$\frac{d^2\theta}{dt^2} + \frac{g}{L}\theta = 0. \tag{42}$$

Requiring $-.3 < \theta < .3$ (radian), the solution of (42) with pertinent initial values is easily obtained by the means of Laplace transformation.

4.7 ELECTRIC CIRCUITS

The notations used in simple electrical circuits for circuit elements are:

1. t (second) for time.

2. q (coulombs) quantity of electricity; could be charge on a capacitor.

3. i (amperes) current, time rate of flow of electricity.

4. E (volts) electromotive force, could be supplied by a generator or battery.

5. R (ohms) resistance by a resistor.

6. L (henrys) inductance by an inductor.

7. (farads) capacitance by a capacitor.

8. K for key or switch.

A circuit is treated as a network containing only one closed path.

Example

An "RLC" circuit is shown in Figure 4-6.

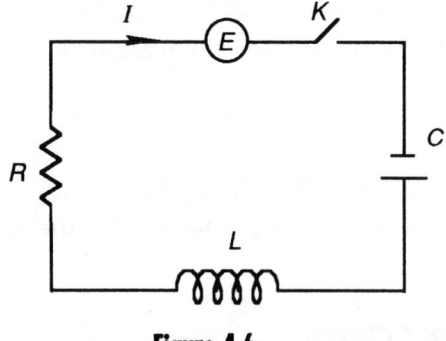

Figure 4.6

When the switch K is closed a charge q will flow to the capacitor plates, then the following relations between the elements will be set:

(a) By definition current is

$$i = \frac{dq}{dt} \tag{43}$$

(b) Voltage drop across a resistor is

$$R\,i = R\,\frac{dq}{dt} \tag{44}$$

(c) Voltage drop across an inductor is

$$L\,\frac{di}{dt} = L\,\frac{d^2 q}{dt^2} \tag{45}$$

(d) Voltage drop across a capacitor is

$$\frac{q}{C} \tag{46}$$

(e) Voltage drop across a generator is

$$-\text{voltage rise} = -E \quad (47)$$

Problem

An important problem in an electrical circuit is to find the charge on the capacitor and the currents as functions of time. For this purpose we find a differential equation by using the Kirchhoff's laws:

Kirchhoff's Laws:

(i) In an electrical circuit the algebraic sum of the currents flowing toward any junction point is equal to zero.

(ii) The algebraic sum of the potential drops, or voltage drops, around any closed loop is equal to zero.

Example

From the Kirchoff's laws for the RLS circuit of Figure 4.6, we get the following equation

$$L\,i(t) + R\,i(t) + \tfrac{1}{c}\,q(t) = E(t) \quad (t \geq 0). \quad (48)$$

And if $i(0) = 0$, then

$$i'(0) = \frac{1}{L[E(0) - 1/C\,q(0)]}.$$

Differentiating this equation and using $q'(t) = i(t)$ we obtain the IVP

$$L\,i''(t) + R\,i'(t) + \tfrac{1}{c}\,i(t) = E'(t) \quad (49)$$
$$i(0) = 0,\, i'(0) = \tfrac{1}{L}(E(0) - \tfrac{1}{c}\,q(0)).$$

Notice that if L, R, and C are positive constants, the IVP (44) is the same as the IVP (27) for the linear damped-spring system.

If in (9) we use $i(t) = q'(t)$ and $i'(t) = q''(t)$, we obtain the IVP

$$L q''(t) + R q'(t) + \tfrac{1}{c} q(t) = E(t), \qquad (50)$$
$$q(0) = q_0, \, q'(0) = i(0) = 0.$$

4.8 BEAMS

Consider a beam of length l, (Figure 4.5). Denote distance from one end of the beam by x, and the deflection of the beam by y. If the beam is subjected to a vertical load $w(x)$, the deflection y must satisfy the equation

$$E I \frac{d^4 y}{dx^4} = w(x) \quad \text{for} \quad 0 < x < 2C. \qquad (51)$$

where the quantity $(E\ I)$ is called the **flexural rigidity** of the beam.

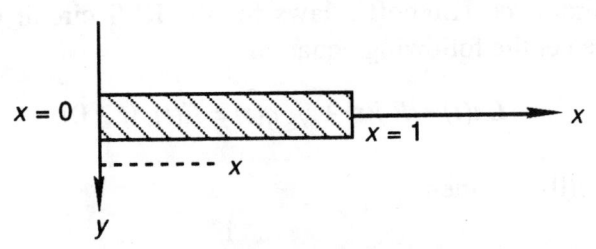

Figure 4.7

The boundary conditions associated with the differential equation (46) depend on the way that beam is supported. The common boundary conditions are of the following types:

(i) Beam imbedded in a support:

$$y = 0,\ y' = 0 \text{ at the point.}$$

(ii) Beam simply supported:

$$y = 0, y'' = 0 \text{ at the point.}$$

(iii) Beam free:

$$y'' = 0, y''' = 0 \text{ at the point.}$$

CHAPTER 5

METHODS OF FINDING LAPLACE TRANSFORMS AND INVERSE TRANSFORMS

5.1 INITIAL VALUE THEOREM

Theorem 5.1

If the limits exist, then

$$\lim_{t \to 0+} f(t) = \lim_{s \to \infty} s F(s) \qquad (1)$$

5.2 FINAL VALUE THEOREM

Theorem 5.2

If limits exist, then

$$\lim_{t \to \infty} f(t) = \lim_{s \to 0} s F(s) \qquad (2)$$

5.3 METHODS OF FINDING LAPLACE TRANSFORMS

In the following we list some common methods for determining Laplace transforms.

5.3.1 DIRECT METHOD

This involves direct use of the definition of the Laplace transforms, formula (3) of chapter 1:

$$\mathcal{L}\{f(t)\} = \int_0^\infty e^{-st} f(t)\, dt. \qquad (3)$$

Example.

Find $\mathcal{L}\{f(t)\}$ if

$$f(t) = \begin{cases} 7 & \text{if } 0 < t < 10 \\ 0 & \text{if } t \geq 10 \end{cases}$$

Using (3), we have

$$\mathcal{L}\{f(t)0\} = \int_0^\infty e^{-st} f(t)\, dt$$

$$= \int_0^{10} e^{-st} \cdot 7\, dt + \int_{10}^\infty e^{-st} \cdot 0\, dt$$

$$= 7\int_0^{10} e^{-st}\, dt$$

$$= 7\frac{e^{-st}}{-s}\Big|_0^{10} = 7\frac{(1 - e^{-10s})}{s}.$$

5.3.2 POWER SERIES METHOD

If $f(t)$ has a power series expansion (Taylor series) given by

$$f(t) = a_0 + a_1 t + a_2 t^2 + \ldots = \sum_{n=0}^\infty a_n t^n \qquad (4)$$

where

$$a_n = \frac{f^{(n)}(0)}{n!} \qquad (5)$$

then

$$\mathcal{L}\{f(t)\} = \frac{a_0}{s} + \frac{a_1}{s^2} + \frac{2!a_2}{s^3} + \ldots = \sum_{n=0}^{\infty} \frac{n!\, a_n}{s^{n+1}} \qquad (6)$$

provided that (6) is convergent for $s > b$, for some $b > 0$, or

$$\mathcal{L}\{f(t)\} = \sum_{n=0}^{\infty} \frac{f^{(n)}(0)}{s^{n+1}}. \qquad (7)$$

Example 1

Find $\mathcal{L}\{e^{at}\}$.

We use the power series method. The Taylor series for e^t is

$$e^t = 1 + \frac{t}{1!} + \frac{t^2}{2!} + \frac{t^3}{3!} + \ldots = \sum_{n=0}^{\infty} \frac{t^n}{n!}.$$

Therefore by substituting (at) for t in this, we will have

$$e^{at} = 1 + at + a^2 \frac{t^2}{2!} + a^3 \frac{t^3}{3!} + \ldots = \sum_{n=0}^{\infty} a^n \frac{t^n}{n!}.$$

Thus

$$\mathcal{L}\{e^{at}\} = \frac{1}{s} + a\frac{1}{s^2} + a^2 \frac{2!}{2!\, s^3} + a^3 \frac{3!}{3!\, s^4} + \ldots$$

$$= \frac{1}{s} \sum_{n=0}^{\infty} \left(\frac{a}{s}\right)^n = \frac{1}{s} \frac{1}{1 - \frac{a}{s}}$$

$$= \frac{1}{s - a}.$$

Example 2

Find $\mathcal{L}\{\sin \sqrt{t}\}$.

Since

$$\sin t = t - \frac{t^3}{3!} + \frac{t^5}{5!} - \frac{t^7}{7!} + \ldots = \sum_{n=0}^{\infty} (-1)^n \frac{t^{2n+1}}{(2n+1)!}$$

Then by replacing t by $\sqrt{t}$ in this, we will have

$$\sin\sqrt{t} = \sqrt{t} - \frac{\sqrt{t}^3}{3!} + \frac{\sqrt{t}^5}{5!} - \ldots = \sum_{n=0}^{\infty} (-1)^n \frac{t^{(2n+1)/2}}{n!}.$$

Since $\mathcal{L}\{t^p\} = \dfrac{\Gamma(p+1)}{s^{p+1}}$, the Laplace transform is

$$\mathcal{L}\{\sin\sqrt{t}\} = \sum_{n=0}^{\infty} (-1)^n \frac{\Gamma((2n+1)/2 + 1)}{n!\, s^{(2n+1)/2 + 1}} \qquad (8)$$

Now using (3) of chapter 3, we have

$$\Gamma\left(\frac{2n+3}{2}\right) = \frac{2n+1}{2}\,\Gamma\left(\frac{2n+1}{2}\right)$$
$$= \frac{2n+1}{2}\,\frac{2n-1}{2}\,\Gamma\left(\frac{2n-1}{2}\right) = \ldots$$
$$= \frac{(2n+1)!}{2^{2n+1}\,n!}\,\Gamma\left(\frac{1}{2}\right) = \frac{(2n+1)!}{2 \cdot 4^n\, n!}\sqrt{\pi}$$

using this in (8), we will obtain

$$\mathcal{L}\{\sin\sqrt{t}\} = \frac{\sqrt{\pi}}{2\, s^{3/2}} \sum_{n=0}^{\infty} (-1)^n \frac{1}{n!}\left(\frac{1}{4s}\right)^n.$$

This is the Taylor series for

$$\mathcal{L}\{\sin\sqrt{t}\} = \frac{\sqrt{\pi}}{2s^{3/2}} = \frac{\sqrt{\pi}}{2s^{3/2}}\, e^{=1/4s}.$$

5.3.3 METHOD OF DIFFERENTIAL EQUATIONS

This method involves finding a suitable differential equation satisfying the given function $f(t)$, and using the properties of Laplace transforms, such as derivative and linearity properties of chapter 1.

Example

To find $\mathcal{L}\{J_0(t)\}$ where $J_0(t)$ is the Bessel function of order zero, we use the method of differential equations. Using the properties of the Bessel function given in chapter 4, the function

$$y = J_0(t)$$

satisfies the initial value problem

$$t\,y''(t) + y(t) + t\,y(t) = 0, \tag{9}$$
$$y(0) = 1,\ y'(0) = 0.$$

Let

$$Y(s) = \mathcal{L}\{y(t)\}.$$

Taking the Laplace transform of the equation of (9), we have

$$-\frac{d}{ds}[s^2Y - s\,y(0) - y'(0)] + [sY - y(0)] - \frac{dY}{ds} = 0$$

which, using the initial conditions of (7), reduces to

$$-\frac{d}{ds}[s^2Y - s] + [sY - 1] - \frac{dY}{ds} = 0$$

or

$$\frac{dY}{Y} = -\frac{s\,ds}{s^2 + 1}$$

Integrating this equation, we obtain

$$Y = \frac{c}{\sqrt{s^2 + 1}}$$

To evaluate c, we use the following facts

$$\lim_{s \to \infty} sY(s) = \frac{c\,s}{\sqrt{s^2 + 1}} = c \quad \text{and} \quad \lim_{t \to 0} J_0(t) = 1$$

F-69

and the initial value theorem, Theorem 5.1 formula (1),

$$\lim_{s \to \infty} sY(s) = \lim_{t \to 0} J_0(t),$$

to find $c = 1$. Therefore

$$\mathcal{L}\{J_0(t)\} = \frac{1}{\sqrt{s^2+1}}.$$

5.3.4 METHOD OF DIFFERENTIATION WITH RESPECT TO A PARAMETER

Another method of finding the Laplace transform of a function is to differentiate with respect to a parameter. The idea is through the following fact. If

$$\mathcal{L}\{f(at)\} = \int_0^\infty e^{-st} f(at)\, dt$$

Then

$$\frac{d}{da}\mathcal{L}\{f(at)\} = \frac{d}{da}\int_0^\infty e^{-st} f(at)\, dt$$

$$= \int_0^\infty e^{-st} t\, f'(at)\, dt$$

$$= \mathcal{L}\{t f'(at)\}.$$

Therefore

$$\mathcal{L}\{t f'(at)\} = \frac{d}{da} \mathcal{L}\{f(at)\} \qquad (10)$$

Example

Since

$$\mathcal{L}\{\sin at\} = \frac{a}{s^2+a^2} \quad \text{and} \quad \frac{d}{da}\sin at = t\cos at,$$

therefore we have

$$\mathcal{L}\{t\cos at\} = \frac{d}{da}\mathcal{L}\{\sin at\}$$

$$= \frac{d}{da}\frac{a}{s^2+a^2}$$

$$= \frac{s^2+a^2-2a^2}{(s^2+a^2)^2}$$

$$= \frac{s^2-a^2}{(s^2+a^2)^2}.$$

5.3.5 MISCELLANEOUS METHODS – MULTIPLICATION BY T^N PROPERTY AND DIVISION BY T PROPERTY

This method uses the properties like convolution (formula (17) of Chapter 1) and **multiplication by t^n** property, formula (16) of chapter 1, that is:

$$\mathcal{L}\{t^n f(t)\} = (-1)^n F^{(n)}(s) \qquad (11)$$

Also, it uses the **division by t property**, formula (11) of chapter 1, that is:

$$\mathcal{L}\left\{\frac{f(t)}{t}\right\} = \int_s^\infty F(x)\,dx \qquad (12)$$

Example 1

To find $\mathcal{L}\{t\cos at\}$, we may use the formula (11) and

$$\mathcal{L}\{\cos at\} = \frac{s}{s^2+a^2}$$

to find

$$\mathcal{L}\{t\cos at\} = (-1)\frac{d}{ds}\mathcal{L}\{\cos at\}$$

$$= -\frac{d}{ds}\frac{s}{s^2+a^2}$$

$$= -\frac{s^2 + a^2 - 2s^2}{(s^2 + a^2)^2}$$

$$= \frac{s^2 - a^2}{(s^2 + a^2)^2}.$$

Example 2

To find $\mathcal{L}\{e^{-t^2}\}$ we can use the division by t property, formula (12), to write

$$\mathcal{L}\{e^{-t^2}\} = \mathcal{L}\left\{\frac{te^{-t^2}}{t}\right\} = \int_s^\infty ue^{-u^2} du$$

$$= \frac{1}{2}e^{-u^2}\Big|_{u=s}^{u=\infty} = \frac{1}{2}e^{-s^2}.$$

5.4 METHODS OF FINDING INVERSE TRANSFORMS

5.4.1 PARTIAL FRACTION METHOD

Rational functions. A quotient or ratio of two polynomial functions is called a **rational function**. Therefore a rational function $R(s)$ looks like

$$R(s) = \frac{P(s)}{Q(s)} = \frac{a_n s^n + a_{n-1} s^{n-1} + \ldots + a_0}{b_m s^m + b_{m-1} s^{m-1} + \ldots + b_0} \quad (13)$$

where a_i and b_j for all $i = 0, 1, \ldots, n$ and $j = 0, 1, \ldots, m$, are real numbers.

Proper rational function. We say that the rational function (13) is **proper** if the degree of its numerator is lower than the degree of its denominator, i.e. $n < m$.

The idea of partial fraction method is based on a procedure and some rules such as the following:

Partial fractions procedure.

(i) By eliminating the common factors of numerator and denominator, reduce the fraction to its lowest form.

(ii) If the result is not a proper fraction, then we divide $Q(s)$ into $P(s)$ so as to obtain a quotient plus a remainder which is a proper fraction.

(iii) Factor the denominator of the remainder fraction, and write the fraction as a sum of fractions of the forms:

$$\frac{A}{(s-a)^k} \quad \text{and} \quad \frac{Bs+C}{(s^2+bs+c)^l} \qquad (14)$$

Partial fractions rules.

1. If a factor $(s-a)$ appears k times in the factorization of $Q(s)$, then assume that the partial fractions of $\frac{P(s)}{Q(s)}$ contains the terms

$$\frac{A_1}{(s-a)} + \frac{A_2}{(s-a)^2} + \ldots + \frac{A_k}{(s-a)^k}. \qquad (15)$$

2. If an irreducible factor s^2+bs+c appears 1 time in factorization of $Q(s)$, then assume the partial fraction of $\frac{P(s)}{Q(s)}$ contains the terms

$$\frac{B_1 s+c_1}{(s^2+bs+c)} + \frac{B_2 s+c_2}{(s^2+bs+c)^2} + \ldots + \frac{B_l s+c_l}{(s^2+bs+c)^l}. \qquad (16)$$

Example 1

Find

$$\mathcal{L}^{-1}\left\{\frac{2s^2-4}{(s+1)(s-2)(s-3)}\right\}.$$

F-73

We write

$$\frac{2s^2-4}{(s+1)(s-2)(s-3)} = \frac{A}{s+1} + \frac{B}{s-2} + \frac{C}{s-3}. \quad (17)$$

To determine the coefficients A, B, and C we may use one of the following two methods:

Method 1

Multiply (17) by $(s+1)(s-2)(s-3)$, to obtain

$$2s^2 - 4 = A(s-2)(s-3) + B(s+1)(s-3) \\ + C(s+1)(s-2). \quad (18)$$

Letting $s = -1$, this will give

$$2 - 4 = A(-3)(-4) + 0 + 0,$$

i.e., $A = -\frac{1}{6}$.

Similarly, by letting $s = 2$ and $s = 3$, (18) will give $B = -\frac{4}{3}$ and $C = \frac{7}{2}$ respectively.

Method 2

Multiplying both sides of (17) by $(s+1)(s-2)(s-3)$, we obtain (18). Then simplifying (18) we have

$$2s^2 - 4 = (B+C)s^2 + (A - 2B - C)s + (-2A - 3B - 2C).$$

Equating the coefficients, we will have the system of linear equations

$$\begin{cases} B + C = 2 \\ A - 2B - C = 0 \\ -2A - 3B - 2C = -4 \end{cases}$$

Solving this system, we obtain the coefficients A, B, and C.

Thus, using the values of A, B, and C from the method 1 in the formula (17), we have

$$\mathcal{L}^{-1}\left\{\frac{2s^2-4}{(s+1)(s-2)(s-3)}\right\}$$

$$= \mathcal{L}^{-1}\left\{\frac{-1/6}{s+1} + \frac{-4/3}{s-2} + \frac{7/2}{s-3}\right\}$$

$$= -\frac{1}{3}e^{-1} - \frac{7}{2}t^2e^{2t} + 4te^{2t} + \frac{1}{3}e^{2t}.$$

Example 2

Find

$$\mathcal{L}^{-1}\left\{\frac{3s+1}{(s-1)(s^2+1)}\right\}.$$

We write

$$\frac{3s+1}{(s-1)(s^2+1)} = \frac{A}{s-1} + \frac{Bs+c}{s^2+1}.$$

Multiplying this by $s-1$ and letting $s=1$, then we will have A = 2. Now we have

$$\frac{3s+1}{(s-1)(s^2+1)} = \frac{2}{s-1} + \frac{Bs+c}{s^2+1}$$

To determine B and C, we substitute for s any number other than 1, for example $s = 0$ and 2; then

$$-1 = -2 + C \quad \text{and} \quad \frac{7}{5} = 2 + \frac{2B+C}{5}$$

from which we obtain $C = 1$ and $B = -2$. Thus

$$\mathcal{L}^{-1}\left\{\frac{3s+1}{(s-1)(s^2+1)}\right\}$$

$$= \mathcal{L}^{-1}\left\{\frac{2}{s-1} + \frac{-2s+1}{s^2+1}\right\}$$

$$= 2\mathcal{L}^{-1}\left\{\frac{1}{s-1}\right\} - 2\mathcal{L}^{-1}\left\{\frac{2}{s^2+1}\right\} + \mathcal{L}^{-1}\left\{\frac{1}{s^2+1}\right\}$$
$$= 2e^t - 2\cos t + \sin t.$$

5.4.2 THE HEAVISIDE EXPANSION FORMULA

Let $R(s)$ be a proper rational function

$$R(s) = \frac{P(s)}{Q(s)}$$

where $Q(s)$ has m distinct real zeros a_k, $k = 1, 2, \ldots, m$, i.e.

$$Q(s) = (s - a_1)(s - a_2) \ldots (s - s_m).$$

Then we have the **Heaviside expansion formula:**

$$\mathcal{L}^{-1}\left\{\frac{P(s)}{Q(s)}\right\} = \sum_{k=1}^{m} \frac{P(a_k)}{Q'(a_k)} e^{a_k t}. \qquad (19)$$

Example

Since

$$s^3 - 6s^2 + 11s - 6 = (s-1)(s-2)(s-3), \text{ and}$$
$$(s^3 - 6s^2 + 11s\ 6)' = 3s^2 - 12s + 11,$$

We have

$$\mathcal{L}^{-1}\left\{\frac{s^2+1}{s^3 - 6s^2 + 11s - 6}\right\} = \frac{2}{2}e^t + \frac{5}{-1}e^{2t} + \frac{10}{2}e^{3t}$$
$$= e^t - 5e^{2t} + 5e^{3t}.$$

CHAPTER 6

FOURIER TRANSFORMS

6.1 FOURIER SERIES

Let $f(x)$, $-L < x < L$, be a real-valued function. The **Fourier series** of f is the **trigonometric series**

$$a_0 + \sum_{n=1}^{\infty} \left(a_n \cos \frac{n\pi x}{L} + b_n \sin \frac{n\pi x}{L} \right) \quad (1)$$

where (a_n, b_n) are called the **Fourier coefficients** and are defined by

$$a_0 = \frac{1}{2L} \int_{-L}^{L} f(x)\, dx$$
$$a_n = \frac{1}{L} \int_{-L}^{L} f(x) \cos \frac{n\pi x}{L}\, dx \quad (2)$$
$$b_n = \frac{1}{L} \int_{-L}^{L} f(x) \cos \frac{n\pi x}{L}\, dx$$

for $n = 1, 2, \ldots$

Theorem 6.1 EVEN FUNCTION

If f is an **even function**, i.e.

$$f(x) = f(-x),$$

then

$$b_n = 0 \text{ for } n = 1, 2, \ldots \qquad (3)$$

ODD FUNCTION

If f is an **odd function**, i.e.

$$f(-x) = -f(x)$$

then

$$A_n = 0 \text{ for } n = 0, 1, 2, \ldots \qquad (4)$$

Example 1

Let $f(x) = x, -L < x < L$. Then $f(x)$ is odd and we have

$$A_n = 0 \quad \text{for all } n$$

$$B_n = \frac{1}{L} \int_{-L}^{L} x \sin \frac{n\pi x}{L} \, dx$$

$$= \frac{2}{L} \int_{0}^{L} x \sin \frac{n\pi x}{L} \, dx$$

Using integration by parts, it follows that

$$B_n = \left(\frac{2L}{n\pi}\right)(-1)^{n+1},$$

therefore the Fourier series of $f(x) = x, -L < x < L$, is

$$f(x) = \frac{2L}{\pi} \sum_{n=1}^{\infty} \frac{(-1)^{n+1}}{n} \sin \frac{n\pi x}{L}. \qquad (5)$$

PARTIAL SUM OF ORDER N

The **partial sum of order N** of the trigonometric series (1) is the function

$$\boxed{f_N = A_0 + \sum_{n=1}^{N} \left(A_n \cos \frac{n\pi x}{L} + B_n \sin \frac{n\pi x}{L} \right) \qquad (6)}$$

Example 2

The partial sums f_N for $N = 1, 2, 3$ of the Fourier series of $f(x) = x$, $\pi < x < \pi$, are shown in the Figure 6.1. These are found from Example 1 using $L = \pi$.

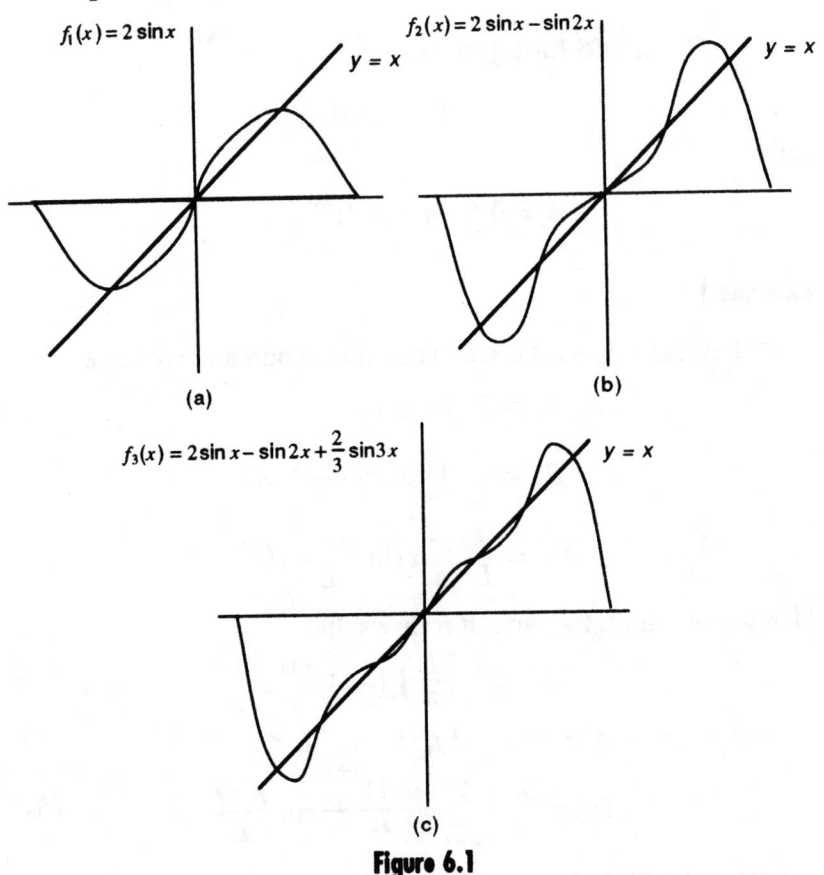

$f_1(x) = 2\sin x$

$f_2(x) = 2\sin x - \sin 2x$

(a)   (b)

$f_3(x) = 2\sin x - \sin 2x + \frac{2}{3}\sin 3x$

(c)

Figure 6.1

6.2 FOURIER SINE AND COSINE SERIES

Suppose a function $f(x)$ is defined for $0 < x < L$ and we want a Fourier series representation of it. To get this, we can extend f to the interval $(-L, L)$ in two ways which follow.

6.2.1 ODD EXTENSION

Odd extension is denoted by $f_o(x)$ and is defined by

$$f_o(x) = \begin{cases} f(x) & \text{for } 0 < x < L \\ -f(-x) & \text{for } -L < x < 0 \\ 0 & \text{for } x = 0 \end{cases} \quad (7)$$

which is an odd function, thus we have a **Fourier sine series** representation for $f(x)$:

$$\sum_{n=1}^{\infty} B_n \sin\frac{n\pi x}{L}; \quad 0 \le x \le L$$

$$B_n = \frac{2}{L}\int_0^L f(x)\sin\frac{n\pi x}{L}dx. \quad (8)$$

6.2.2 EVEN EXTENSION

Even extension of f from the interval $(0, L)$ to the interval $(-L, L)$ is denoted by f_E and defined by

$$f_E(x) = \begin{cases} f(x) & \text{for } 0 < x < L \\ -f(-x) & \text{for } -L < x < 0 \\ 0 & \text{for } x = 0 \end{cases} \quad (9)$$

which is an even function, thus we have the **Fourier cosine series** representation for $f(x)$:

$$a_0 + \sum_{n=1}^{\infty} a_n \cos\frac{n\pi x}{L}; \quad 0 \le x \le L$$

$$a_0 = \frac{1}{L}\int_0^L f(x)dx,$$

$$a_n = \frac{2}{L}\int_0^L f(x)\cos\frac{n\pi x}{L}dx \text{ for } n = 1, 2, \ldots \quad (10)$$

6.3 PIECEWISE SMOOTH FUNCTIONS

A function $f(x)$, $0 , x < b$, is said to be **piecewise smooth** if f and all its derivatives are piecewise continuous.

Example

The function
$$f(x) = |x|; -\pi < x < \pi$$
is continuous on the entire interval, while its derivative
$$f'(x) = \begin{cases} 1 & \text{if } x > 0 \\ -1 & \text{if } x < 0 \end{cases}$$
is piecewise continuous, with
$$f'(0+0) = 1 \text{ and } f'(0-0) = -1.$$
All higher derivatives, for $n = 2, 3, \ldots$, are
$$f^{(n)}(x) = 0 \text{ for } x \neq 0$$
which are piecewise continuous on the interval. Thus f is piecewise smooth on $(-\pi, \pi)$.

6.4 PERIODIC EXTENSIONS

The **periodic extension** of a function $f(x)$, defined on $a < x < b$, Figure 6.2(a), is denoted by $\overline{f}$ and is defined by setting

$$\boxed{\overline{f}[x + n(b-a)] = f(x) \qquad (11)}$$

for $a < x < b$ and $n = \ldots, -2, -1, 0, 1, 2, \ldots$, Figure 6.2(b)

6.5 THEOREM 2 (CONVERGENCE THEOREM)

Let f be a piecewise smooth function on $(-L, L)$, then the

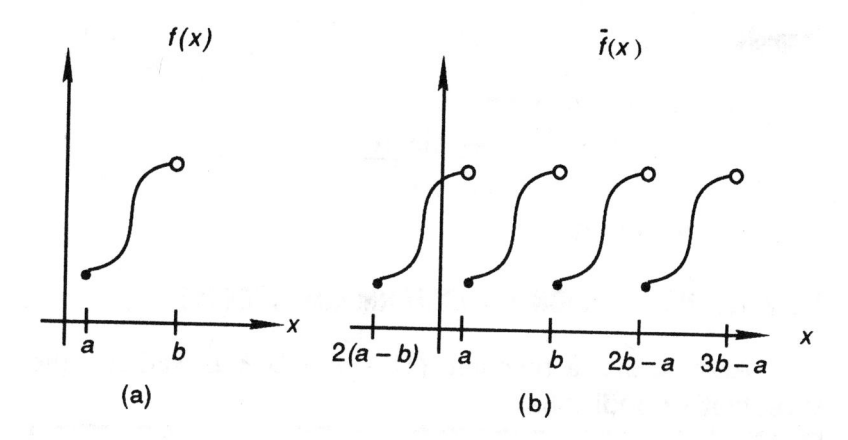

Figure 6.2

(a)

(b) with markers at $2(a-b)$, a, b, $2b-a$, $3b-a$

Fourier series of f converges for all x to the average value

$$\tfrac{1}{2}[\overline{f}(x+0)+\overline{f}(x-0)] \qquad (12)$$

where $\overline{f}$ is the periodic extension of f and

$$\overline{f}(x+0) = \lim_{\varepsilon \to 0+} f(x+\varepsilon) \quad \text{and} \quad \overline{f}(x-0) = \lim_{\varepsilon \to 0-} f(x-\varepsilon) \qquad (13)$$

Thus

$$\frac{1}{2}[\overline{f}(x+0)+\overline{f}(x-0)] \qquad (14)$$

$$= a_0 + \sum_{n=1}^{\infty}\left(a_n \cos\frac{n\pi x}{L} + b_n \sin\frac{n\pi x}{L}\right)$$

6.5.1 FIRST CRITERION FOR UNIFORM CONVERGENCE

Let $f(x)$, $-L < x < L$, be a piecewise smooth function. If the Fourier coefficients $\{a_n\}$, $\{b_n\}$ satisfy

$$\sum_{n=1}^{\infty}\left(|a_n|+|b_n|\right) < \infty, \qquad (15)$$

then the Fourier series converges uniformly.

Example

The Fourier sine series
$$\sum_{n=1}^{\infty} \frac{\sin nx}{n^2}$$
converges uniformly.

6.5.2 SECOND CRITERION FOR UNIFORM CONVERGENCE

Assume that a function $f(x)$, $-L < x < L$, satisfies the **Direchlet's conditions:**

> (i) f is piecewise smooth on $(-L, L)$,
> (ii) f is continuous and $(-L, L)$, and (16)
> (iii) $f(-L+0) = f(L-0)$

Then the Fourier series converges uniformly.

6.6 GENERAL CRITERION FOR DIFFERENTIATION

Theorem 6.3

If $f(x)$ satisfies the Direchlet's conditions (16), then

$$\frac{1}{2}[f'(x+0)+f'(x-0)] \qquad (17)$$
$$= \sum_{n=1}^{\infty} \frac{n\pi}{L}\left(b_n \cos \frac{n\pi x}{L} - a_n \sin \frac{n\pi x}{L}\right)$$

Example

We know that the Fourier series of the even function $f(x) = x^2$, $-\pi < x < \pi$, is of the following form
$$a_0 + \sum_{n=1}^{\infty} a_n \cos nx .$$

To determine the coefficients $\{a_n\}$, we use (17) to find

$$2x = -\sum_{n=1}^{\infty} n a_n \sin nx.$$

Now we use the Fourier series of $f(x) = x$, formula (5) for $L = \pi$, to obtain

$$2x = 4\sum_{n=1}^{\infty} (-1)^{n+1} \frac{\sin nx}{n}$$

Therefore $a_n = 4(-1)^n/n^2$ for $n = 1, 2, \ldots$. To compute a_0 we must return to the definition

$$a_0 = (1/2\pi)\int_{-\pi}^{\pi} x^2 dx = \pi^2/3.$$

Therefore we have

$$x^2 = \frac{\pi^2}{3} + \sum_{n=1}^{\infty} \frac{(-1)^n}{n^2} \cos nx \quad \text{for } -\pi < x < \pi.$$

6.7 PARSEVAL'S THEOREM AND MEAN SQUARE ERROR

Let $f(x)$, $-L < x < L$, be a piecewise smooth function with Fourier series

$$a_0 + \sum_{n=1}^{\infty} \left(a_n \cos \frac{n\pi x}{L} + b_n \sin \frac{n\pi x}{L} \right).$$

Then:

Parseval's theorem states that

$$\frac{1}{2L}\int_{-L}^{L} [f(x)]^2 dx = a_0^2 + \frac{1}{2}\sum_{n=1}^{\infty} (a_n^2 + b_n^2) \quad (18)$$

Mean square error σ_N^2 is defined by

$$\sigma_N^2 = \frac{1}{2L}\int_{-L}^{L} [f(x) - f_N(x)]^2 dx \quad (19)$$

where $f_N(x)$ is the N-th partial sum, defined by (6). We have

$$\sigma_N^2 = \frac{1}{2}\sum_{n=N+1}^{\infty} (a_n + b_n) \quad (20)$$

Example

Let $f(x) = x$, then we have

$$a_n = 0, \ b_n (-1)^{n-1}(2/n)$$

therefore the mean square error is

$$\sigma_N^2 = \frac{1}{2} \sum_{n=N+1}^{\infty} \frac{4}{m^2} = 2 \sum_{n=N+1}^{\infty} \frac{1}{m^2}.$$

6.8 COMPLEX FORM OF FOURIER SERIES

With the convention of the **De Moivre's formula**

$$e^{i\theta} = \cos\theta + i\sin\theta \ ; \ i = \sqrt{-1} \qquad (21)$$

the Fourier series of $f(x)$, $-L < x < L$, in complex notation, assumes the form

$$f(x) = \sum_{-\infty}^{\infty} \alpha_n e^{in\pi x/L} \qquad (22)$$

where the coefficients $\{\alpha_n\}$ for $n = 0, +1, -1, +2, -2, \ldots$, are defined by

$$\alpha_n = \frac{1}{2L} \int_{-L}^{L} f(x) e^{-(in\pi x/L)} dx \qquad (23)$$

6.9 PARSEVAL'S IDENTITY

Parseval's Identity will take the form:

$$\frac{1}{2L} \int_{-L}^{L} |f(x)|^2 dx = \sum_{-\infty}^{\infty} |\alpha_n|^2. \qquad (24)$$

Example

Compute the complex Fourier series of

$$f(x) = e^{2x}, \quad -\pi < x < \pi$$

$$\alpha_n = \frac{1}{2\pi}\int_{-\pi}^{\pi} e^{2x}e^{-inx}dx = \frac{1}{2\pi}\int_{-\infty}^{\infty} e^{(2-in)x}dx$$

$$= \frac{(-1)^n}{2\pi(2-in)}(e^{2\pi} - e^{-2\pi}) = \frac{1}{\pi}\sinh 2\pi \frac{(-1)^n(2+in)}{4+n^2}.$$

Therefore the complex Fourier series of $f(x) = e^{2x}, -\pi < x < \pi$, is

$$f(x) = \frac{1}{\pi}\sinh 2\pi \sum_{-\infty}^{\infty} \frac{(-1)^n(2+in)}{4+n^2}e^{inx}$$

6.10 FOURIER INTEGRAL TRANSFORMS

We assume that $f(x)$ is a real-valued piecewise smooth function on every bounded interval (a,b) and that all the relevant integrals are absolutely convergent. Then we define the **Fourier Integral transform**, or simply **Fourier transform**, of f by

$$\mathcal{F}(\mu) = \frac{1}{2\pi}\int_{-\infty}^{\infty} f(x)e^{-i\mu x}dx \qquad (25)$$

then the function $f(x)$ is recovered by the **Fourier Inversion formula**

$$f(x)\int_{-\infty}^{\infty} \mathcal{F}(\mu)e^{i\mu x}d\mu \qquad (26)$$

The functions $f(x)$ and $\mathcal{F}(\mu)$, (25) and (26) respectively, are called **Fourier transform pairs**.

6.10.1 PARSEVAL'S THEOREM

Parseval's theorem for Fourier transforms is

$$\int_{-\infty}^{\infty} |f(x)|^2 dx = 2\pi \int_{-\infty}^{\infty} |\mathcal{F}(\mu)|^2 d\mu \qquad (27)$$

6.10.2 INVERSION THEOREM FOR FOURIER TRANSFORMS

Also called **convergence theorem for Fourier transforms**.

Let $f(x)$, $-\infty < x < \infty$, be a piecewise smooth function on each interval with

$$\int_{-\infty}^{\infty} |f(x)| \, dx$$

finite. Define the Fourier transform $\mathcal{F}(\mu)$ by (25). Then for each x we have

$$\lim_{L \to \infty} \int_{-L}^{L} \mathcal{F}(\mu) e^{i\mu x} dx = \frac{1}{2}[f(x+0) + f(x-0)] \quad (28)$$

Example

(Square wave function). We find the Fourier transform of the square wave function

$$f(x) = \begin{cases} 0 & \text{for} \quad x < a \\ 1 & \text{for} \quad a \leq x \leq b \\ 0 & \text{for} \quad x > b \end{cases} \quad (29)$$

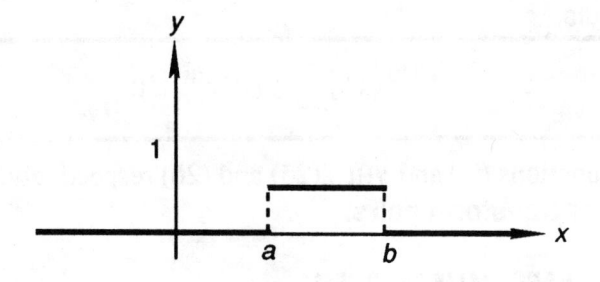

Figure 6.3

We have

$$\mathcal{F}(\mu) = \frac{1}{2\pi} \int_{-\infty}^{\infty} f(x) e^{-i\mu x} dx = \frac{1}{2\pi} \int_{a}^{b} e^{-i\mu x} dx$$

F-87

$$= \begin{cases} \dfrac{e^{-i\mu b} - e^{-i\mu a}}{-2\pi i\mu} & \text{for } \mu \neq 0 \\ \dfrac{b-a}{2} & \text{for } \mu = 0 \end{cases}$$

By the Fourier inversion theorem we have

$$\lim_{L\to\infty}\int_{-L}^{L} \mathcal{F}(\mu)e^{i\mu x}dx = \lim_{L\to\infty}\int_{-L}^{L} \frac{\left(e^{-i\mu b} - e^{-i\mu a}\right)e^{-i\mu x}}{-2\pi i\mu}d\mu$$

$$= \begin{cases} 0 & \text{for } x < a \\ 1/2 & \text{for } x = a \\ 1 & \text{for } a < x < b \\ 1/2 & \text{for } x = b \\ 0 & \text{for } x > b \end{cases}$$

And the Parseval's theorem states that

$$\frac{1}{2\pi}\int_{-\infty}^{\infty} \frac{\left|e^{-i\mu b} - e^{-i\mu a}\right|^2}{\mu^2} d\mu = b - a.$$

6.11 FOURIER COSINE FORMULAS

Let $f(x)$ be defined for all $x > 0$. We extend $f(x)$ to an even function (even extension) by

$$f(x) = f(-x) \text{ for } x < 0,$$

then we have the **Fourier cosine formulas**

$$\boxed{f(x)=\int_0^\infty \mathcal{F}_c(\mu)\cos\mu x\, dx,\ \mathcal{F}_c(\mu)=\frac{2}{\pi}\int_0^\infty f(x)\cos\mu x\, dx} \quad (30)$$

6.12 FOURIER SINE FORMULAS

Let $f(x)$ be defined for all $x > 0$. We extend f to a an odd function (odd extension) by

$$f(x) = -f(-x) \text{ for all } x < 0,$$

then we have the **Fourier sine formulas**

$$f(x) = \int_0^\infty \mathcal{F}_s(\mu)\sin\mu x\, dx, \mathcal{F}_s(\mu) = \frac{2}{\pi}\int_0^\infty f(x)\sin\mu x\, dx \quad (31)$$

6.13 THE CONVOLUTION THEOREM

If $h(x)$ is the convolution of $f(x)$ and $g(x)$, i.e.

$$h(x) = (f*g)(x) = \int_{-\infty}^\infty f(u)\, g(x-u)\, du \quad (32)$$

Then

$$\int_{-\infty}^\infty h(x)e^{-i\mu x}dx = \left\{\int_{-\infty}^\infty f(x)e^{-i\mu x}dx\right\}\left\{\int_{-\infty}^\infty g(x)e^{-i\mu x}dx\right\} \quad (33)$$

or

$$H(\mu) = \mathcal{F}(\mu)\, G(\mu) \quad (34)$$

where H, $\mathcal{F}$, and G are the Fourier transforms of h, f, and g respectively.

6.14 RELATIONSHIP OF FOURIER AND LAPLACE TRANSFORMS

Let the function $f(t)$ be defined for all $t > 0$. We consider the function

$$g(t) = \begin{cases} e^{-xt}f(t) & \text{for } t > 0 \\ 0 & \text{for } t < 0 \end{cases} \quad (35)$$

then from (25), with μ replace by y, we see that the Fourier transform of $g(t)$ is

$$\mathcal{F}\{g(t)\} = \int_0^\infty e^{-(x+iy)t} f(t)\, dt = \int_0^\infty e^{-st} f(t)\, dt \quad (36)$$

where

$$s = x + iy \quad (37)$$

The right side of (36) is the Laplace transform of $f(t)$ in complex variable. This result gives a relationship between Fourier and Laplace transforms

$$\mathcal{F}\{e^{-xt} f(t)\} = \mathcal{L}\{f(t)\} \quad (38)$$

or

$$\mathcal{F}\{f(t)\} = \mathcal{L}\{e^{xt} f(t)\} \quad (39)$$

CHAPTER 7

APPLICATIONS OF LAPLACE TRANSFORMS TO INTEGRAL AND DIFFERENCE EQUATIONS

7.1 INTEGRAL EQUATIONS

An **integral equation,** is an equation of the form

$$y(t) = f(t) + \int_a^b k(x,t)\, y(x)\, dx \qquad (1)$$

where $f(t)$ and $k(x,t)$ are known functions, and a and b are either constant numbers or some known functions of t. The function $y(t)$ is an unknown which has to be determined.

Kernel

The function $k(x,t)$ in (1) is called the kernel of the integral equation.

7.1.1 FREDHOLM INTEGRAL EQUATION

This equation is an integral equation of type (1) where a and b are constant numbers.

7.1.2 VOLTERRA INTEGRAL EQUATION

This equation is an integral equation of type (1) where a

is a constant while $b = t$, i.e., it is of the form

$$y(t) = f(t) + \int_0^t k(x,t)\, y(x)\, dx \qquad (2)$$

7.1.3 INTEGRAL EQUATION OF CONVOLUTION TYPE

If in the integral equation (1) we put

$$a = 0,\ b = t \text{ and } k(x,t) = k(t-x)$$

then we have an **integral equation of convolution type**, i.e. it is of the form

$$y(t) = f(t) + \int_0^t k(t-x)\, y(x)\, dx \qquad (3)$$

This integral equation of convolution type is also written as

$$y(t) = f(t) + k * y(t) \qquad (4)$$

SOLUTION OF AN INTEGRAL EQUATION OF CONVOLUTION TYPE

To solve the integral equation (3) or (4), we assume

$$Y(s) = \mathcal{L}\{y(t)\},\ F(s) = \mathcal{L}\{f(t)\},\ \text{and}\ K(s) = \mathcal{L}\{k(t)\}$$

then we take the Laplace transform of both sides of (4) to obtain

$$Y(s) = F(s) + K(s)\, Y(s) \qquad (5)$$

solving (5) for $Y(s)$, we find

$$Y(s) = \frac{F(s)}{1 - K(s)} \qquad (6)$$

Thus by Laplace inversion

$$y(t) = \mathcal{L}^{-1}\left\{\frac{F(s)}{1-K(s)}\right\} \quad (7)$$

Example

To solve

$$y(t) = t^2 + \int_0^t \sin(t-x)\,y(x)\,dx,$$

we write

$$y(t) = t^2 + (y * \sin)(t)$$

Thus

$$\mathcal{L}\{y(t)\} = \mathcal{L}\{t^2 + (y * \sin)(t)\}$$

or

$$Y(s) = \frac{2}{s^3} + \frac{Y(s)}{s^2+1}$$

or

$$Y(s) = \frac{2}{s^3} + \frac{Y(s)}{s^5}.$$

Hence

$$y(t) = \mathcal{L}^{-1}\left\{\frac{2}{s^3}\right\} + \mathcal{L}^{-1}\left\{\frac{2}{s^5}\right\}$$

$$= t^2 + \frac{1}{12}t^4.$$

7.2 INTEGRO-DIFFERENTIAL EQUATIONS

An equation involving an integral and some derivatives of an unknown function $y(t)$ is called an **integro-differential equation**.

Example

The following is an integro-differential equation

$$y'(t) - y(t) = t + \int_0^t \sin(t-x)\, y(x)\, dx \qquad (8)$$

The solution of an integro-differential equation like (8) is easily obtained by taking the Laplace transform of both sides of the equation. It is then solvable subject to some given initial conditions.

Example

To solve (8), we write it in the following convolution form

$$y'(t) + y(t) = t + (\sin * y)(t)$$

with initial condition

$$y(0) = 0.$$

We take the Laplace transform of the equation, to obtain

$$\mathcal{L}\{y'(t) + y(t)\} = \mathcal{L}\{t + (\sin * y)(t)\}$$

or

$$sY(s) - y(0) + Y(s) = \frac{1}{s^2} + \frac{Y(s)}{s^2+1}.$$

Therefore using the initial condition $y(0) = 0$, we have

$$(s+1)Y - \frac{Y}{s^2+1} = \frac{1}{s^2}$$

Solving this equation, we will obtain

$$Y = \frac{s^2+1}{s^3(s^2+s+1)}$$

$$= \frac{1}{s^3} - \frac{1}{s^2} + \frac{1}{s} - \frac{s}{s^2+s+1}.$$

Thus

$$y(t) = \mathcal{L}^{-1}\left\{ \frac{1}{s^3} - \frac{1}{s^2} + \frac{1}{s} - \frac{s}{s^2+s+1} \right\}$$

$$= \mathcal{L}^{-1}\left\{\frac{1}{s^3}\right\} - \mathcal{L}^{-1}\left\{\frac{1}{s^2}\right\} + \mathcal{L}^{-1}\left\{\frac{1}{s}\right\} - \mathcal{L}^{-1}\left\{\frac{s}{s^2+s+1}\right\}$$

$$= \frac{t^2}{2} - t + e^{-1/2t}\left(\cos\frac{\sqrt{3}}{t}t - \frac{\sqrt{3}}{3}\sin\frac{\sqrt{3}}{2}t\right).$$

7.3 DIFFERENCE EQUATIONS

An equation involving an unknown function $y(t)$ with one or more shifted forms of it like $y(t-a)$, where a is a constant, is called a difference equation.

Example

Solve the difference equation

$$y(t) - y(t-1) = t.$$

We take the Laplace transform of the equation, to have

$$\mathcal{L}\{y(t)\} - \mathcal{L}\{y(t-1)\} = \mathcal{L}\{t\}$$

or

$$Y(s) - \mathcal{L}\{y(t-1)\} = \frac{1}{s^2}.$$

By the second shifting theorem, property 4 of chapter 1 formula (10), we have

$$\mathcal{L}\{y(t-1)\} = e^{-s}Y(s).$$

Therefore

$$Y(s) - e^{-s}Y(s) = \frac{1}{s^2}.$$

Thus

$$Y(s) = \frac{1}{s^2(1-e^{-s})}.$$

Hence we have

$$y(t) = \mathcal{L}^{-1}\left\{\frac{1}{s^2(1-e^{-s})}\right\}$$

$$= \mathcal{L}^{-1}\left\{\frac{1}{s^2} + e^{-s}\frac{1}{s^2} + e^{-2s}\frac{1}{s^2} + \ldots\right\}.$$

Now we use the second translation property, property 3 formula (7) of chapter 2, to obtain

$$y(t) = \begin{cases} t & \text{for} \quad t < 1 \\ t + (t-1) & \text{for} \quad 1 \leq t < 2 \\ t + (t-1) + (t-2) & \text{for} \quad 2 \leq t < 3 \\ \vdots \end{cases}$$

$$= \begin{cases} t & \text{for} \quad t < 1 \\ 2t - 1 & \text{for} \quad 1 \leq t < 2 \\ 3t - 3 & \text{for} \quad 2 \leq t < 3 \\ 4t - 7 & \text{for} \quad 3 \leq t < 4 \\ \vdots \\ nt - (n-1)n/2 & \text{for} \quad n \leq t < (n+1) \end{cases}$$

7.4 DIFFERENTIAL-DIFFERENCE EQUATIONS

A difference equation involving some derivatives of the unknown function y(t) is called a **differential-difference equation**.

Example

$$y''(t) = y(t-2) + 5\sin t$$

is a differential-difference equation where by taking Laplace transform of its both sides, using formula (10) of chapter 1, and using derivative property formula (14) of chapter 1, we will obtain

$$s^2 Y(s) - s y(0) - y'(0) = e^{-2s} Y(s) + \frac{5}{s^2+1},$$

where by assuming some initial values, e.g.

$$y(0) = 0 \text{ and } y'(0) = 1,$$

it will reduce to

$$s^2 Y(s) - 1 = e^{-2s} Y(s) + \frac{5}{s^2+1}$$

or

$$Y(s) = \frac{1}{s^2 - e^{-2s}} + \frac{5}{(s^2+1)(s^2 - e^{-2s})}.$$

Therefore

$$y(t) = \mathcal{L}^{-1}\left\{\frac{1}{s^2 - e^{-2s}}\right\} + \mathcal{L}^{-1}\left\{\frac{5}{(s^2+1)(s^2 - e^{-2s})}\right\}.$$

CHAPTER 8

APPLICATIONS TO BOUNDARY-VALUE PROBLEMS

8.1 FUNCTIONS OF TWO VARIABLES

A function of two variables is denoted by

$$u = u(x,y)$$

and its **partial derivatives** by

$$u_x = \frac{\partial u}{\partial x},\ u_y = \frac{\partial u}{\partial y},\ u_{xx} = \frac{\partial^2 u}{\partial x^2},\ u_{xy} = \frac{\partial^2 u}{\partial x \partial y},\ u_{yy} = \frac{\partial^2 u}{\partial y^2}$$

Example

If $u(x,y) = \sin x + \cos y + x y$, then

$$u_x = \cos x + y,\ u_y = -\sin y + x,$$

$$u_{xx} = -\sin x,\ u_{xy} = u_{yx} = 1,\ \text{and } y_{yy} = -\cos y.$$

8.2 PARTIAL DIFFERENTIAL EQUATION

A **linear second order partial differential equation** is an equation of the form

$$A u_{xx} + 2 B u_{xy} + C u_x + D u_y + E u = G \qquad (1)$$

where A, B, C, D, E, and G are all functions of x and y.

If in (1) $G = 0$, then the equation is called **homogeneous**.

8.3 SOME IMPORTANT PARTIAL DIFFERENTIAL EQUATIONS

Laplace's equation.

$$u_{xx} + u_{yy} = 0 \qquad (2)$$

The wave equation.

$$u_{tt} - C^2 u_{xx} = 0 \qquad (3)$$

The heat equation.

$$u_t - K u_{xx} = 0 \qquad (4)$$

The telegraph equation.

$$u_{tt} - C^2 u_{xx} + 2\beta u_t + \alpha u = 0 \qquad (5)$$

Poisson's equation.

$$u_{xx} + u_{yy} = G \qquad (6)$$

8.4 CLASSIFICATION OF SECOND-ORDER PARTIAL DIFFERENTIAL EQUATIONS

Depending on the second-order coefficients A, B, and C, the equation (1) is classified in the following three types:

$$\begin{array}{l}\text{Elliptic if } A\,C - B^2 > 0 \\ \text{Hyperbolic if } A\,C - B^2 < 0 \\ \text{Parabolic if } A\,C - B^2 = 0\end{array} \quad (7)$$

Example

Poisson's and Laplace's equations are elliptic. Wave and telegraph equations are hyperbolic. The heat equation is parabolic.

8.5 BOUNDARY CONDITIONS

Unlike ordinary differential equations, it is hard to formulate a general method of solving for all second-order partial differential equations. Instead, we specify a solution in terms of some certain boundary conditions. Natural types of boundary conditions are:

8.5.1 DIRICHLET PROBLEM

It is to specify the values of $u(x,y)$ on the boundary of a bounded plane region with smooth boundary. It is used for elliptic equations.

Example

In a physical problem of determining the electrostatic potential function u in a cylinder where the charge density is specified in the interior and the boundary is required to be an equipotential surface, the Dirichlet's problem is

$$\begin{cases} u_{xx} + u_{yy} = -\rho & \text{for } x^2 + y^2 < R^2, \\ u(x,y) = C & \text{for } x^2 + y^2 = R^2. \end{cases}$$

8.5.2 CAUCHY PROBLEM

It is naturally used for a hyperbolic equation. It is to specify the solution $u(x,t)$ and its time derivative $u_t(x,t)$ on the line $t = 0$. Also, we may need to specify boundary conditions at the end of a finite interval or, in case of infinite intervals, to restrict the growth of $u(x,t)$ when x becomes very large.

Example

The following Cauchy problem is treated for the **vibrating string** with fixed ends.

$$u_{tt} - C^2 u_{xx} = 0 \quad \text{for} \quad t > 0, 0 < x < L$$

$$u(x,0) = f_1(x) \quad \text{for} \quad 0 < x < L$$

$$u_t(x,0) = f_2(x) \quad \text{for} \quad 0 < x < L$$

$$u(0,t) = 0 \quad \text{for} \quad t > 0$$

$$u(L,t) = 0 \quad \text{for} \quad t > 0.$$

Where the functions f_1 and f_2 correspond respectively to the initial position and velocity of the vibrating string.

8.6 SOLUTION OF BOUNDARY-VALUE PROBLEMS BY LAPLACE TRANSFORMS

8.6.1 ONE-DIMENSIONAL BOUNDARY VALUE PROBLEMS

In this case our unknown function depends only on two variables, either x and t or x and y. We may use the Laplace transformation with respect to only one of the variables (t or x), to reduce the partial differential equation to an ordinary differential equation. Then we solve the resulting differential equation to find the Laplace transform of the original equation. By inverting the solution of the ordinary differential equation, using the Laplace inverse formulas and methods, we will obtain the solution.

Example

To solve the heat equation

$$u_t = u_{xx} \text{ for } x > 0, t > 0 \tag{8}$$

with the boundary-value

$$u(x,0) = 1, u_x(0,1) = -u(0,1), |u(x,t)| < M \tag{9}$$

we take the Laplace transform of (8) with respect to t

$$\mathcal{L}\{u_t\} = \mathcal{L}\{u_{xx}\}, \tag{10}$$

to obtain the initial value problem

$$sU - 1 = \frac{d^2U}{dx^2}, \tag{11}$$

$$U_x(0,s) = -U(0,s), \text{ and } U(x,s) \text{ is bounded.} \tag{12}$$

The equation (11) is a second order differential equation with the general solution

$$U(x,s) = c_1 e^{\sqrt{s}x} + c_2 e^{-\sqrt{s}x} + \frac{1}{s}. \tag{13}$$

From the boundedness condition of (12), we have $c_1 = 0$. Therefore the solution (13) is simplified to

$$U(x,s) = c_2 e^{-\sqrt{s}x} + \frac{1}{s}. \tag{14}$$

From the first condition of (12) we find $c_2 = \dfrac{1}{s(\sqrt{s} - 1)}$,

$$U(x,s) = \frac{1}{s(\sqrt{s}-1)} e^{-\sqrt{s}x} + \frac{1}{s} \qquad (15)$$

Therefore by the inverse Laplace transformation we have

$$u(x,t) = \mathcal{L}^{-1}\left\{\frac{1}{s(\sqrt{s}-1)} e^{-\sqrt{s}x} + \frac{1}{s}\right\}$$

$$= 1\mathcal{L}^{-1}\left\{\frac{1}{s(\sqrt{s}-1)} e^{-\sqrt{s}x}\right\} \qquad (16)$$

8.6.2 TWO-DIMENSIONAL BOUNDARY VALUE PROBLEM

In this case our unknown function depends on three variables x, y, and t. We may use Laplace transformation to solve our boundary value problem by taking Laplace transformation of our equation twice. For example, apply Laplace transformation with respect to x variable then to t variable. Then our equation will reduce to an ordinary differential equation. Solving the resulting ordinary differential equation we obtain a solution, such that, by applying Laplace inverse transformation to it twice, we will obtain the solution of our boundary value problem. This method is called **Iterated Laplace transformation.**

CHAPTER 9

TABLES

9.1 TABLE OF GENERAL PROPERTIES

Original Function $f(t)$ $f(t) = \mathcal{L}^{-1}\{F(s)\}$	Transformed Function $F(s)$ $F(s) = \mathcal{L}\{f(t)\}$ $= \int_0^\infty e^{-st} f(t)\, dt$
Inversion formula $\dfrac{1}{2\pi i} \int_{c-i}^{c+i} e^{is} F(s)\, ds$	$F(s)$
Linearity property $a\, f(t) + b\, g(t)$	$a\, F(s) + b\, G(s)$
Differentiation $f'(t)$	$s\, F(s) - f(0+)$
General Differentiation $f^{(n)}(t)\, ;\quad n = 1, 2, \ldots$	$s^n F(s) - s^{n-1} f(0+)$ $- s^{n-2} f'(0+) - \ldots - f^{(n-1)}(0+)$

$f(t)$	$F(s)$
Integration $$\int_0^t f(u)\, du$$	Division by s $$\frac{1}{s} F(s)$$
Integration $$\int_0^t \int_0^u f(v)\, dv\, du$$	Division by s^2 $$\frac{1}{s^2} F(s)$$
Convolution $$\int_0^t f(t-u)\, g(u)\, du = f * g(t)$$	Product $$F(s) \cdot G(s)$$
Multiplying by t^n $$(-1)^n t^n f(t);\ n = 1, 2, \ldots$$	Differentiation $$F^{(n)}(s)$$
Dividing by t $$\frac{1}{t} f(t)$$	Integration $$\int_s^\infty F(x)\, dx$$
$e^{at} f(t)$	Shifting $F(s-a)$
$\frac{1}{c} f\left(\frac{t}{c}\right);\ c > 0$	$F(cs)$
$\frac{1}{c} e^{(b/c)t} f\left(\frac{t}{c}\right);\ c > 0$	$F(cs - b)$

$f(t)$	$F(s)$
Shifting $$f(t-b)\,h_1(t-b);\quad b>0$$	$e^{-bs}F(s)$
Periodic Functions $$f(t+a)=f(t)$$	$\dfrac{\int_0^a e^{-st}f(t)\,dt}{1-e^{-as}}$
$f(t+a)=-f(t)$	$\dfrac{\int_0^a e^{-st}f(t)\,dt}{1+e^{-as}}$
Half-Wave Rectification of $f(t)$ $$f(t)\sum_{n=0}^{\infty}(-1)^n h_1(t-na)$$	$\dfrac{F(s)}{1-e^{-as}}$
Full-Wave Rectification of $f(t)$ $$\lvert f(t)\rvert$$	$F(s)\coth\dfrac{as}{2}$
Heaviside Expansion $$\sum_{n=1}^{m}\dfrac{P(a_n)}{Q'(a_n)}e^{a_n t}$$	$\dfrac{P(s)}{Q(s)}$ where $P(s)$ is of degree $<m$ $Q(s)=(s-a_1)(s-a_2)\ldots(s-a_m)$
$$e^{at}\sum_{n=1}^{r}\dfrac{P^{(r-n)}(a)}{(r-n)!}\dfrac{t^{n-1}}{(n-1)!}$$	$\dfrac{P(s)}{(s-a)^r}$ P is a polynomial of degree $<r$

9.2 TABLE OF MORE COMMON LAPLACE TRANSFORMS

$f(t) = \mathcal{L}^{-1}\{F(s)\}$	$F(s) = \mathcal{L}\{f(t)\}$
1	$\dfrac{1}{s}$
t	$\dfrac{1}{s^2}$
$\dfrac{t^{n-1}}{(n-1)!}$; $n = 1, 2, \ldots$	$\dfrac{1}{s^n}$
$\dfrac{1}{\sqrt{\pi t}}$	$\dfrac{1}{\sqrt{s}}$
$2\sqrt{\dfrac{t}{\pi}}$	$s^{-3/2}$
$\dfrac{2^n t^{n-1/2}}{1\,3\,4\,5\ldots(2n-1)\sqrt{\pi}}$	$s^{-(n+1/2)}$; $n = 1, 2, 3, \ldots$
t^r ; $r > -1$	$\dfrac{\Gamma(r+1)}{s^{r+1}}$
e^{at}	$\dfrac{1}{s-a}$

$f(t) = \mathcal{L}^{-1}\{F(s)\}$	$F(s) = \mathcal{L}\{f(t)\}$
$t\,e^{at}$	$\dfrac{1}{(s-a)^2}$
$\dfrac{t^{n-1}e^{-at}}{(n-1)!}$	$\dfrac{1}{(s+a)^n}$; $n = 1, 2, \ldots$
$t^r e^{-at}$; $r > -1$	$\dfrac{\Gamma(r+1)}{(s+a)^{r+1}}$
$\dfrac{e^{-at} - e^{-bt}}{b-a}$; $a \neq b$	$\dfrac{1}{(s+a)(s+b)}$
$\dfrac{a\,e^{-at} - b\,e^{-bt}}{a-b}$; $a \neq b$	$\dfrac{s}{(s+a)(s+b)}$
$\sin st$	$\dfrac{a}{s^2 + a^2}$
$\cos at$	$\dfrac{s}{s^2 + a^2}$
$\sinh at$	$\dfrac{a}{s^2 - a^2}$
$\cosh at$	$\dfrac{s}{s^2 - a^2}$

$f(t) = \mathcal{L}^{-1}\{F(s)\}$	$F(s) = \mathcal{L}\{f(t)\}$
$\dfrac{1}{a^2}(1 - \cos at)$	$\dfrac{1}{s(s^2 + a^2)}$
$\dfrac{1}{a^3}(at - \sin at)$	$\dfrac{1}{s(s^2 + a^2)}$
$\dfrac{1}{2a^3}(\sin at - at \cos at)$	$\dfrac{1}{(s^2 + a^2)^2}$
$\dfrac{t}{2a}\sin at$	$\dfrac{s}{(s^2 + a^2)^2}$
$\dfrac{1}{sa}(\sin at + at \cos at)$	$\dfrac{s^2}{(s^2 + a^2)^2}$
$\dfrac{1}{b}e^{-at}\sin bt$	$\dfrac{1}{(s+a)^2 + b^2}$
$e^{-at}\cos bt$	$\dfrac{s+a}{(s+a)^2 + b^2}$
$e^{-at} - e^{at/2}\left(\cos\dfrac{at\sqrt{3}}{2} - \sqrt{3}\sin\dfrac{at\sqrt{3}}{2}\right)$	$\dfrac{3a^2}{s^3 + a^3}$

$f(t) = \mathcal{L}^{-1}\{F(s)\}$	$F(s) = \mathcal{L}\{f(t)\}$
$\sin at \cosh at - \cos at \sinh at$	$\dfrac{4a^3}{s^4 + 4a^4}$
$\dfrac{1}{2a^2} \sin at \sinh at$	$\dfrac{s}{s^4 + 4a^4}$
$(1 + a^2 t^2)\sin at - at \cos at$	$\dfrac{8a^3 s^2}{(s^2 + a^2)^3}$
$\dfrac{1}{2\sqrt{\pi t^3}}(e^{-bt} - e^{-at})$	$\sqrt{s+a} - \sqrt{s+b}$
$\dfrac{1}{\sqrt{\pi t}} - ae^{a^2 t} \operatorname{erfc} a\sqrt{t}$	$\dfrac{1}{\sqrt{s} + a}$
$\dfrac{1}{\sqrt{\pi t}} + ae^{a^2 t} \operatorname{erfc} a\sqrt{t}$	$\dfrac{\sqrt{s}}{s - a^2}$
$\dfrac{1}{a} e^{a^2 t} \operatorname{erf} a\sqrt{t}$	$\dfrac{1}{\sqrt{s}(s - a^2)}$
$e^{a^2 t} \operatorname{erfc} a\sqrt{t}$	$\dfrac{1}{\sqrt{s}(\sqrt{s} + a)}$
$\dfrac{1}{\sqrt{b-a}} e^{-at} \operatorname{erf}\left[\sqrt{(b-a)t}\right]$	$\dfrac{1}{(s+a)\sqrt{s+b}}$

$f(t) = \mathcal{L}^{-1}\{F(s)\}$	$F(s) = \mathcal{L}\{f(t)\}$
$a\,e^{-at}[I_1(at) + I_0(at)]$	$\dfrac{\sqrt{s+2a}}{\sqrt{s}} - 1$
$e^{-(a+b)t/2} I_0\!\left(\dfrac{a-b}{2}t\right)$	$\dfrac{1}{\sqrt{(s+a)(s+b)}}$
$\sqrt{\pi}\left(\dfrac{t}{a-b}\right)^{k-\frac{1}{2}} e^{-\frac{1}{2}(a+b)t} I_{k-\frac{1}{2}}\!\left(\dfrac{a-b}{2}t\right)$	$\dfrac{\Gamma(k)}{(s+a)^k (s+b)^k}\;;\;k>0$
$t\,e^{-\frac{1}{2}(a+b)t}\left[I_0\!\left(\dfrac{a-b}{2}t\right) + I_1\!\left(\dfrac{a-b}{2}t\right)\right]$	$\dfrac{1}{(s+a)^{\frac{1}{2}}(s+b)^{\frac{1}{2}}}$
$\dfrac{1}{t}e^{-at} I_1(at)$	$\dfrac{\sqrt{s+2a}-\sqrt{s}}{\sqrt{s+2a}+\sqrt{s}}$
$\dfrac{k}{t}e^{-\frac{1}{2}(a+b)t} I_k\!\left(\dfrac{a-b}{2}t\right)$	$\dfrac{(a-b)^k}{\left(\sqrt{s+a}+\sqrt{s+b}\right)^{2k}}\;\;k>0$
$\dfrac{1}{a^k}e^{-\frac{1}{2}at} I_k\!\left(\dfrac{1}{2}at\right)$	$\dfrac{\left(\sqrt{s+a}+\sqrt{s}\right)^{-2k}}{\sqrt{s}\sqrt{s+a}}\;\;v>-1$

$f(t) = \mathcal{L}^{-1}\{F(s)\}$	$F(s) = \mathcal{L}\{f(t)\}$
$J_0(at)$	$\dfrac{1}{\sqrt{s^2 + a^2}}$
$a^n J_n(at);\quad n > -1$	$\dfrac{\left(\sqrt{s^2+a^2} - s\right)^n}{\sqrt{s^2+a^2}}$
$\dfrac{\sqrt{\pi}}{\Gamma(k)}\left(\dfrac{t}{2a}\right)^{k-1/2} J_{k-1/2}(at)$	$\dfrac{1}{(s^2 + a^2)^k};\quad k > 0$
$\dfrac{ka^k}{t} J_k(at);\ k > 0$	$\left(\sqrt{s^2 + a^2} - s\right)^k$
$a^n I_n(at);\quad n > -1$	$\dfrac{\left(s - \sqrt{s^2 - a^2}\right)^n}{\sqrt{s^2 - a^2}}$
$\dfrac{\sqrt{\pi}}{\Gamma(k)}\left(\dfrac{t}{2a}\right)^{k-1/2} I_{k-1/2}(at)$	$\dfrac{1}{(s^2 - a^2)k};\quad k > 0$
$h_1(t - a)$	$\dfrac{1}{s} e^{-as}$

$f(t) = \mathcal{L}^{-1}\{F(s)\}$	$F(s) = \mathcal{L}\{f(t)\}$
$(t-a)\,h_1(t-a)$	$\dfrac{1}{s^2}e^{-as}$
$\dfrac{(t-a)^{r-1}}{\Gamma(r)}h_1(t-a)\;;\;r>0$	$\dfrac{1}{s^r}e^{-as}$
$h_1(t) - h_1(t-a)$	$\dfrac{1-e^{-as}}{s}$
Step Function $\sum\limits_{n=0}^{\infty} h_1(t-na)$	$\dfrac{1}{s(1-e^{-as})}$
$\sum\limits_{n=1}^{\infty} c^{n-1} h_1(t-na)$	$\dfrac{1}{s(e^{as}-c)}$

$f(t) = \mathcal{L}^{-1}\{F(s)\}$	$F(s) = \mathcal{L}\{f(t)\}$
Square Wave Function $h_1(t) + 2\sum_{n=1}^{\infty}(-1)^n h_1(t-2na)$ [graph: square wave with value 1, transitions at 2a, 4a, 6a]	$\dfrac{1}{2}\tanh as$
Triangular Wave Function $h_1(t) + 2\sum_{n=1}^{\infty}(-1)^n (t-2na)h_1(t-2na)$ [graph: triangular wave with peaks at 2a, 4a, 6a, 8a]	$\dfrac{1}{s^2}\tanh as$
$2\sum_{n=0}^{\infty} h_1[t-(2n+1)a]$ [graph: staircase function with steps of height 2 at a, 3a, 5a]	$\dfrac{1}{s\sinh as}$

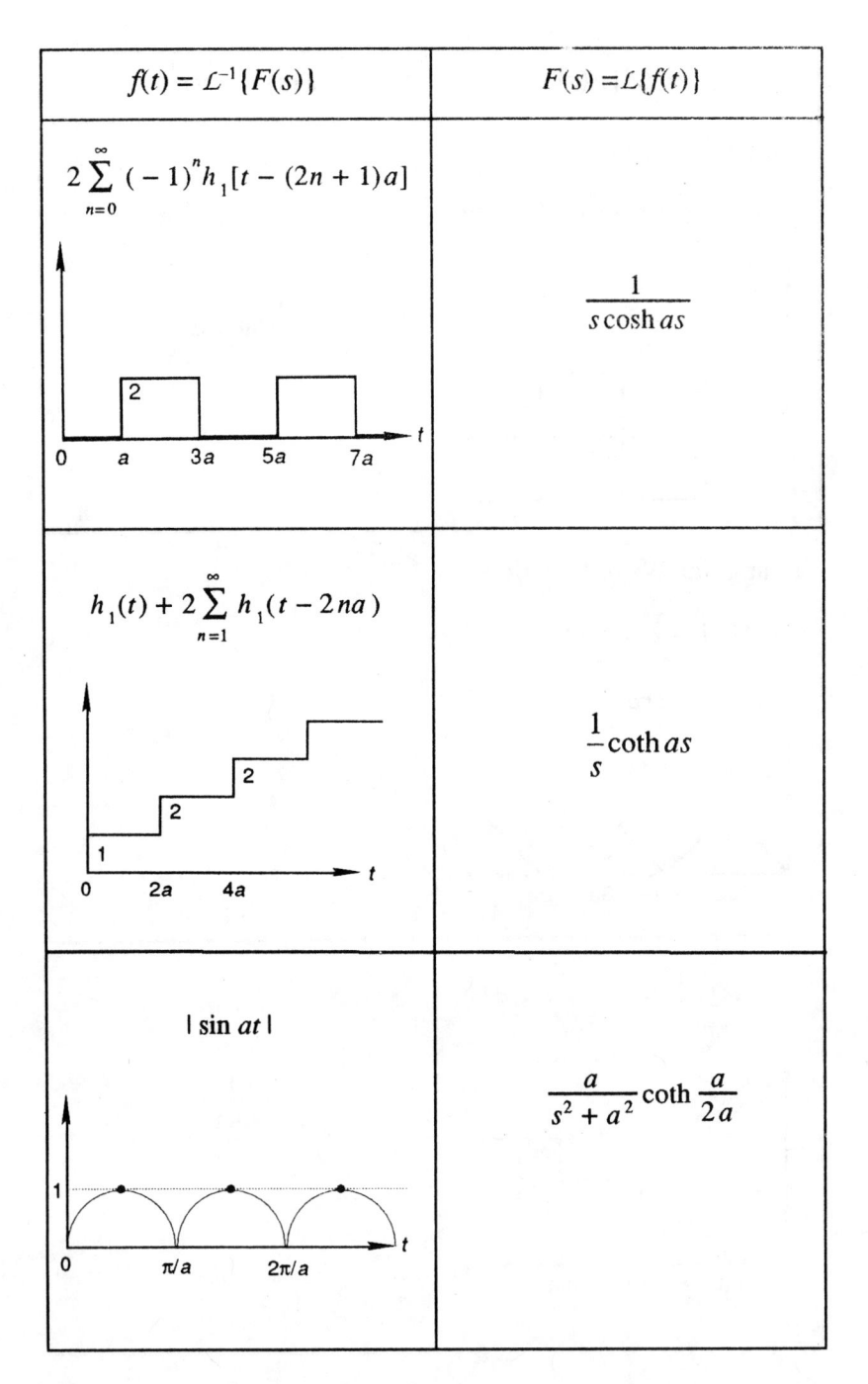

$f(t) = \mathcal{L}^{-1}\{F(s)\}$	$F(s) = \mathcal{L}\{f(t)\}$
$2\sum_{n=0}^{\infty}(-1)^n h_1[t-(2n+1)a]$	$\dfrac{1}{s\cosh as}$
$h_1(t) + 2\sum_{n=1}^{\infty} h_1(t-2na)$	$\dfrac{1}{s}\coth as$
$\lvert \sin at \rvert$	$\dfrac{a}{s^2+a^2}\coth\dfrac{a}{2a}$

$f(t) = \mathcal{L}^{-1}\{F(s)\}$	$F(s) = \mathcal{L}\{f(t)\}$
$\sum_{n=0}^{\infty} (-1)^n h_1(t - n\pi) \sin t$	$\dfrac{1}{(s^2 + 1)(1 - e^{-\pi s})}$
$J_0(2\sqrt{rt})$	$\dfrac{1}{s} e^{-\frac{r}{s}}$
$\dfrac{1}{\sqrt{\pi t}} \cos 2\sqrt{at}$	$\dfrac{1}{\sqrt{s}} e^{-\frac{a}{s}}$
$\dfrac{1}{\sqrt{\pi t}} \cosh 2\sqrt{at}$	$\dfrac{1}{\sqrt{s}} e^{\frac{a}{s}}$
$\dfrac{1}{\sqrt{\pi a}} \sin 2\sqrt{at}$	$\dfrac{1}{s^{3/2}} e^{-\frac{a}{s}}$
$\dfrac{1}{\sqrt{\pi a}} \sinh 2\sqrt{at}$	$\dfrac{1}{s^{3/2}} e^{\frac{a}{s}}$
$\left(\dfrac{t}{k}\right)^{\frac{r-1}{2}} J_{r-1}(2\sqrt{kt})$	$\dfrac{1}{s^r} e^{-\frac{k}{s}} \; ; \; r > 0$
$\left(\dfrac{t}{k}\right)^{\frac{r-1}{2}} I_{r-1}(2\sqrt{kt})$	$\dfrac{1}{s^r} e^{\frac{k}{s}} \; ; \; r > 0$

$f(t) = \mathcal{L}^{-1}\{F(s)\}$	$F(s) = \mathcal{L}\{f(t)\}$
$\dfrac{a}{2\sqrt{\pi t^3}} \exp\left(-\dfrac{k^2}{4t}\right)$	$\dfrac{1}{s} e^{-k\sqrt{s}}$; $k > 0$
$\operatorname{erfc} \dfrac{k}{2\sqrt{t}}$	$\dfrac{1}{s} e^{-k\sqrt{s}}$; $k \geq 0$
$\dfrac{1}{\sqrt{\pi t}} \exp\left(-\dfrac{k^2}{4t}\right)$	$\dfrac{1}{\sqrt{s}} e^{-k\sqrt{s}}$; $k \geq 0$
$2\sqrt{t}\, i\operatorname{erfc} \dfrac{k}{2\sqrt{t}}$	$\dfrac{1}{s^{3/2}} e^{-k\sqrt{s}}$; $k \geq 0$
$-\gamma - \ln t$ γ = Euler's constant = .577215…	$\dfrac{1}{s} \ln s$
$\ln t$	$-\dfrac{(\gamma + \ln s)}{s}$ γ = Euler's constant = .577215…
$\cos t\, si(t) - \sin t\, ci(t)$	$\dfrac{\ln s}{s^2 + 1}$
$-\sin t\, si(t) - \cos t\, ci(t)$	$\dfrac{s \ln s}{s^2 + 1}$
$\dfrac{1}{t}(e^{-bt} - e^{-at})$	$\ln \dfrac{s+a}{s+b}$

$f(t) = \mathcal{L}^{-1}\{F(s)\}$	$F(s) = \mathcal{L}\{f(t)\}$
$-2\,ci\left(\dfrac{t}{k}\right)$	$\dfrac{1}{s}\ln(1 + k^2 s^2)\,;\quad k > 0$
$\dfrac{2}{t}(1 - \cos at)$	$\ln\dfrac{s^2 + a^2}{s^2}$
$\dfrac{2}{t}(1 - \cosh at)$	$\ln\dfrac{s^2 - a^2}{s^2}$
$\dfrac{1}{t}\sin kt$	$\arctan\dfrac{k}{s}$
$\dfrac{1}{k\sqrt{\pi}}\exp\left(-\dfrac{t^2}{4k^2}\right)$	$e^{k^2 s^2}\,\mathrm{erfc}\,ks\,;\quad k > 0$

9.3 TABLE OF SPECIAL FUNCTIONS

Gamma Function $$\Gamma(r) = \int_0^\infty x^{r-1} e^{-x} dx \, ; \ r > 0$$
Bessel Function $$J_n(t) = \sum_{k=0}^\infty \frac{(-1)^k \left(\frac{1}{2} t\right)^{2k+n}}{k! \, \Gamma(k+n+1)}$$
Modified Bessel Function $$I_n(t) = (-i)^{-n} J_n(it)$$
The Error Function $$\text{erf}(t) = \frac{2}{\sqrt{\pi}} \int_0^t e^{-u^2} du$$
The Complementary Error Function $$\text{erfc}(t) = 1 - \text{erf}(t)$$
Sine Integral Function $$si(t) = \int_0^t \frac{\sin u}{u} du$$
Cosine Integral Function $$ci(t) = \int_0^\infty \frac{\cos u}{u} du$$

Exponential Integral Function

$$ei(t) = \int_{t}^{\infty} \frac{e^{-u}}{u} \, du$$

The Unit Step Function or Heaviside Function

$$h_1(t - t_0) = \begin{cases} 0 & \text{if } t < t_0 \\ 1 & \text{if } t \geq t_0 \end{cases}$$

The Unit Impulse Function

$$\delta(t) = \lim_{\varepsilon \to 0} f(t) \qquad f(t) = \begin{cases} \dfrac{1}{\varepsilon} & \text{if } 0 \leq t \leq \varepsilon \\ 0 & \text{if } t > \varepsilon \end{cases}$$

Beta Function

$$B(m,n) = \int_0^1 u^{m-1}(1-u)^{n-1} \, du = \frac{\Gamma(m)\,\Gamma(n)}{\Gamma(m+n)}$$

Handbook of Mechanical Engineering

SECTION G
Automatic Control Systems/Robotics

CHAPTER 1

SYSTEM MODELING: MATHEMATICAL APPROACH

1.1 ELECTRIC CIRCUITS AND COMPONENTS

The equations for an electric circuit obey Kirchoff's laws which can be stated as follows:

A) $\sum$ potential differences around a closed circuit = 0

B) $\sum$ currents at a junction or node = 0

Voltage drops across three basic electrical elements:

A) $v_R = Ri$

B) $v_L = L \frac{di}{dt} = LDi$

C) $v_C = \frac{q}{c} = \frac{1}{c} \int_0^t i\, d\tau + \frac{Q_o}{C} = \frac{1}{c} \cdot \frac{i}{D}$

where operator D is defined as the following:

$$DX = \frac{dx}{dt}$$

and $\quad \dfrac{X}{D} \equiv \displaystyle\int_{-\infty}^{t} x \, d\tau$

SERIES R-L CIRCUIT

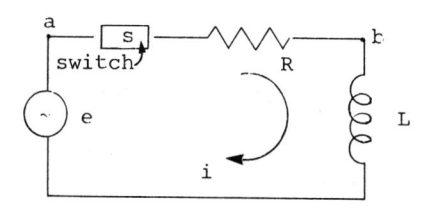

Fig. Simple R-L circuit

$$Ri + \dfrac{L di}{dt} = e = \dfrac{R}{L} \cdot \dfrac{v_L}{D} + v_L$$

(because $i_{inductor} = \dfrac{1}{LD} v_L$)

$$\dfrac{v_b - v_a}{R} + \dfrac{1}{L} \cdot \dfrac{v_b}{D} = 0 \quad \ldots \text{ Kirchoff's second law}$$

$$\left(\dfrac{1}{R} + \dfrac{1}{LD} \right) v_b - \dfrac{v_a}{R} = 0$$

SERIES R-L-C CIRCUIT

$$v_R + \dfrac{L}{R} Dv_R + \dfrac{1}{RC} \cdot \dfrac{v_R}{D} = e$$

MULTILOOP ELECTRIC CIRCUITS

Loop method:

    A loop current is drawn in each closed loop; then Kirchoff's voltage equation is written for each loop. These equations are solved simultaneously to obtain output (voltage) in terms of input (voltage) and the circuit parameters.

Node Method:

    The rules for writing the node equations:

A) The number of equations required is equal to the number of unknown node voltages.

B) An equation is written for each node.

C) The equation includes the following terms: the node voltage multiplied by the sum of all the admittances that are connected to this node, and the node voltage of the other end of each brand multiplied by the admittance connected between the two nodes.

1.2 MECHANICAL TRANSLATION SYSTEMS

The mechanical translation system is characterized by mass, elastance and damping.

Representation of the basic elements:

A) Mass

    a) It is the inertial element.

    b) Reaction force $f_M = M \times \text{acceleration} = Ma$

$$= M \frac{dv}{dt} = M \frac{d^2s}{dt^2}.$$

    c) Network representation:

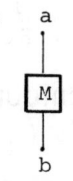

        a has the motion of the mass

        b has the motion of the reference

        f_M is a function of time

B) Elastance (or stiffness, k):

    a) Representation:

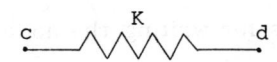

b) Reaction force $f_k = k(x_c - x_d)$; x_c is the position of c and x_d is the position of d.

C) Damping (viscous friction, B):

a) Representation:

— dashpot

e ▬ f
B

b) The reaction damping force $f_B = B(v_e - v_f)$

$$= B(Dx_e - Dx_f)$$

SIMPLE TRANSLATION SYSTEM

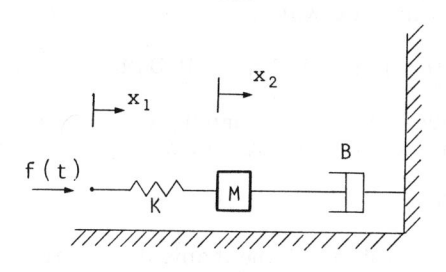

$$f = f_k = k(x_a - x_b)$$

$$f_k = f_M + f_B = MD^2 x_2 + BDx_2$$

These can be solved for displacements x_1 and x_2 and Dx_1 and Dx_2.

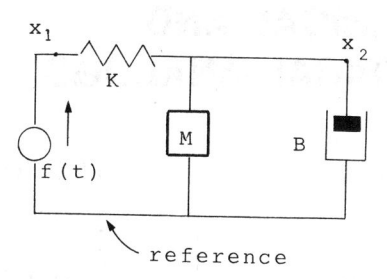

Fig: Mechanical Network

The system is initially at rest. To draw the mechanical method, x_1 and x_2 and the reference are located.

MULTI-ELEMENT SYSTEM

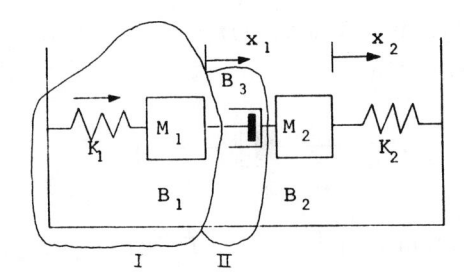

The system equations must be written in terms of two displacements x_1 and x_2. These are the nodes in the equivalent mechanical network:

$$(M_1 D^2 + B_1 D + B_3 D + K_1)x_1 - (B_3 D)x_2 = f \ldots$$

because the forces at node x_1 must add to zero. At node x_2, the equations are written by observing the above pattern:

$$(M_2 D^2 + B_2 D + K_2)x_2 - (B_3 D)x_1 = 0$$

Using these, an equivalent mechanical network can be drawn using the following table of electrical and mechanical analogies.

1.3 MECHANICAL AND ELECTRICAL ANALOGS

Mechanical Element

M - mass
f - force
$v = \dfrac{dv}{dt}$ - velocity
B - damping coefficient
K - stiffness coefficient

Electrical Element

C - capacitance
i - current
e or v - voltage
$G = \dfrac{1}{R}$ - conductance
$\dfrac{1}{L}$ = reciprocal inductance

G-5

An equivalent electrical network is drawn using the table of electrical and mechanical analogs.

These two networks have the same mathematical forms.

1.4 MECHANICAL ROTATIONAL SYSTEMS

Network elements of mechanical rotational systems:

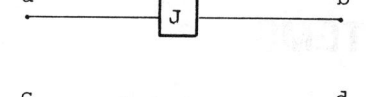

a ———[J]——— b   J: Moment of inertia

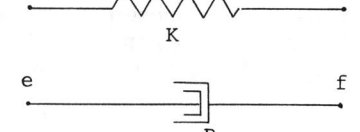

c ——/\/\/\——— d
         K

e ———⊐[ ]——— f
         B

A) The reaction torque $= T_J = J\alpha = JD\omega = JD^2\theta$

   where θ is the angular displacement.

B) Reaction spring torque $= T_k = K(\theta_c - \theta_d)$

   where θ_c and θ_d are the positions of the two ends of a spring ($\theta_c - \theta_d$) = the angle of twist.

C) Damping torque $T_B = B(\omega_e - \omega_f)\ldots$

   where $\omega_e - \omega_f$ = relative angular velocity of the ends of the dashpot.

SIMPLE ROTATIONAL SYSTEM

The governing equation is

$$JD^2\theta + BD\theta + K\theta = T(t)\ldots$$

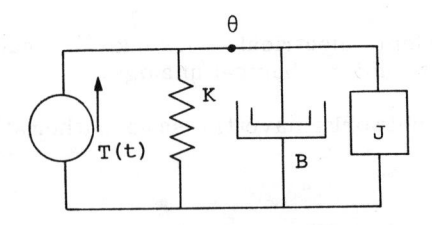

Only one equation is necessary because the system has only one node. The actual system consists of a shaft with fins of a moment of inertia, J, which is immersed in oil. The fluid has a damping factor of B.

1.5 THERMAL SYSTEMS

Only a few thermal systems are represented by linear differential equations. The basic requirement is that the temperature of the body should be assumed to be uniform.

NETWORK ELEMENTS

C: Thermal capacitance
R: Thermal resistance

A) $h = \text{Heat stored} = \frac{q}{D} C(T_2 - T_1)$ due to change in temperature $(T_2 - T_1)$

B) $q = CD(T_2 - T_1)$ in terms of rate of heat flow.

C) $q = \text{Rate of heat flow} = \frac{T_3 - T_4}{R}$ where T_3 and T_4 are two boundary temperatures.

Temperature is analogous to potential.

SIMPLE THERMAL SYSTEM: MERCURY THERMOMETER

A) Network representation

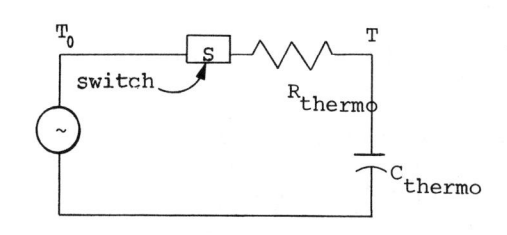

Fig: Simple Network

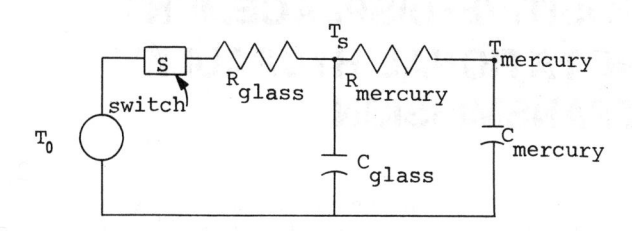

Fig: Exact Network

B) q = flow of heat = $(T_0 - T)/R$ where T_0 is the temperature of the bath and T is the temperature before immersing into the bath.

C) h = Heat entering the thermometer = $C(T - T_1)$

D) $RCDT + T = T_0$ is the governing equation.

E) More exact analysis:

T_s = The temperature at the inner surface between the glass and the mercury

For node s: $\left[\dfrac{1}{R_g} + \dfrac{1}{R_m} + C_g D \right] T_s - \dfrac{T_m}{R_m} = \dfrac{T_o}{R_g}$

For node m: $\dfrac{-T_s}{R_m} + \left[\dfrac{1}{R_m} + DC_m \right] T_m = 0$

G-8

$$\left[C_g C_m D^2 + \left(\frac{C_g}{R_m} + \frac{C_m}{R_g} + \frac{C_m}{R_m} \right) D + \frac{1}{R_g R_m} \right] T_m$$

$$= \frac{T_o}{R_g R_m} \quad \text{the governing equation.}$$

The latter equation is from the form

$$(A_2 D^2 + A_1 D + A_o) T_m = k T_o .$$

1.6 POSITIVE-DISPLACEMENT ROTATIONAL HYDRAULIC TRANSMISSION

The hydraulic transmission is used when a large torque is required. It contains a variable displacement pump driven at a constant speed. It is assumed that the transmission is linear over a limited range.

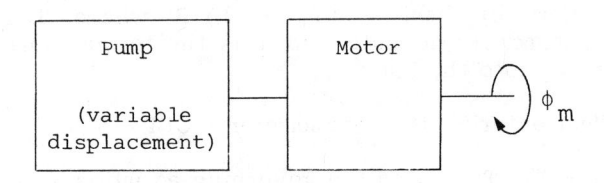

Pump (variable displacement) — Motor — ϕ_m

A) $q_p = q_m + q_l + q_c$ Since the fluid flow rate from the pump must equal the sum of the flow rates.

$q_p = x d_p \cdot \omega_p$

$q_m = d_m \cdot \omega_m$, $\quad q_l = LP_L \quad$ and $\quad q_c = Dv = \dfrac{vDP_L}{K_B}$

B) Torque at the motor shaft $= T = n_T \cdot d_m \cdot P_L$

$$= CP_L$$

G-9

where P_L is the load-induced pressure-drop across motor.

DIFFERENT CASES

A) Inertia load

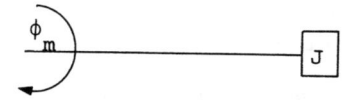

$$T = J \cdot D^2\phi_m = CP_L$$

B) Inertia load coupled through a spring

$$T = K(\phi_m - \phi_L) = JD^2\phi_L = CP_L$$

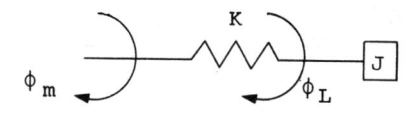

C) In case of inertia load, value of P_L obtained from equation $q_p = q_m + q_1 + q_c$ is substituted into $T = CP_L$. The resulting equation can be solved for ϕ_m in terms of x. The same procedure is adopted in case of a spring-coupled inertia load.

1.7 D-C AND A-C SERVOMOTOR

D-C SERVOMOTOR

$T(t)$ = The torque = $Ka\phi\, i_m \ldots i_m$ = armature current,

ϕ = the flux, K_a = constant of proportionality

G-10

Modes of Operation

A) An adjustable voltage is applied to the armature while the field current is held constant.

B) An adjustable voltage is applied to the field while the armature current is kept constant.

Armature Control

A constant field current is obtained by exciting the field from a fixed Dc source.

$$T(t) = k_T i_m \quad \text{where } \phi \text{ is a constant}$$
(since field current is a constant)

Back emf $= e_m = k_i \phi \omega_m = k_b \omega_m = kb \cdot D\theta_m \ldots \omega_m =$ speed

Armature-controlled d-c motor:

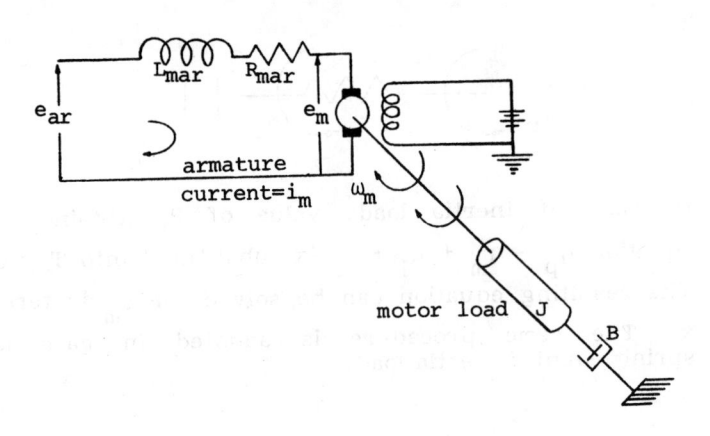

$$e_{ar} = L_m(L_{mar} + R_{mar}) + e_m$$

$$JD\omega_m + B\omega_m = T(t) \ldots \text{Torque equation with motor load.}$$

$$\boxed{\frac{L_{mar} J}{k_T} D^3 \theta_m + \frac{L_{mar} B + R_{mar} J}{k_T} D^2 \theta_m + \frac{R_{mar} B + K_b K_T}{k_T} D\theta_m = e_{ar}}$$

The system equation.

Field Control

In this case the armature current i_{mar} is constant so that T(t) is proportional only to the flux ϕ:

$$T(t) = K_3 \cdot \phi \cdot i_{mar} = K_3 K_2 \cdot i_{mar} \cdot i_{field} = k_f \cdot i_{field}$$

where i_{mar} is constant.

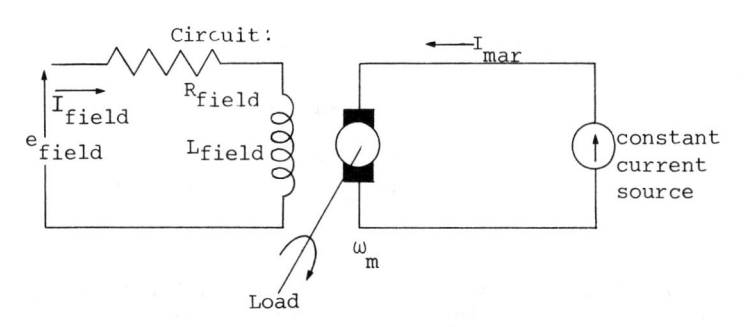

Circuit:

[Circuit diagram showing e_{field}, I_{field}, R_{field}, L_{field}, connected to motor with I_{mar} from constant current source, driving load at ω_m.]

$$e_{field} = L_{field} D_{field} + R_{field} \cdot i_{field}$$

$$(K_f \cdot D + R_{field})(JD + B)\omega_m = k_f \cdot e_{field}$$

The system equation.

A-C SERVOMOTOR

This may be considered to be a two-phase induction motor having two field coils positioned 90 electrical degrees apart.

In a two-phase induction motor, the speed which is a little below the synchronous speed is constant; however when the unit is used as a servomotor, the speed is proportional to the input voltage.

The two-phase induction motor shown below can be used as a servomotor by applying an ac voltage e to one of the windings. Thus e is fixed and when e_{c_0} is varied, the torque and speed are a function of this voltage.

Because these curves are non-linear, we must approximate them by straight lines in order to obtain linear differential equations.

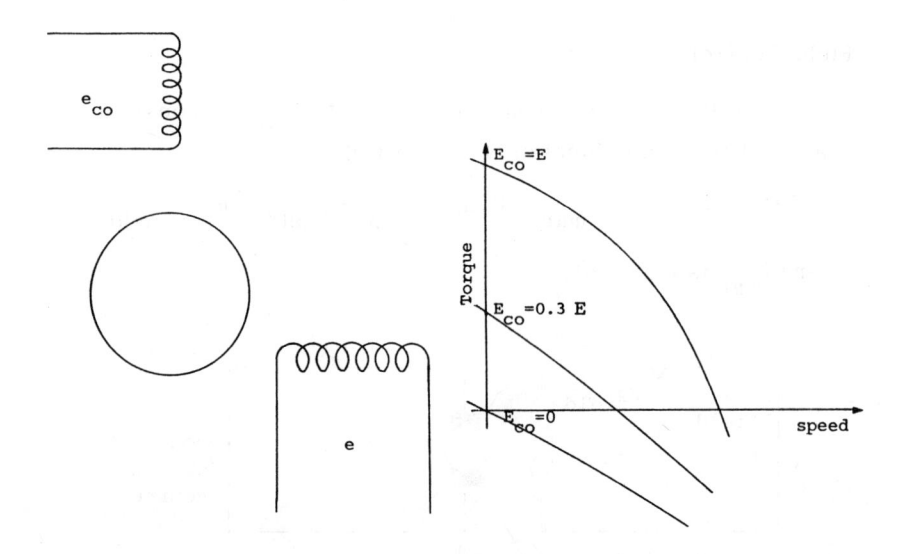

From these curves, it is noted that the generated torque is a function of the speed ω and voltage e_{c_0}.

The torque equation is given by

$$\boxed{e_{c_0} \frac{\partial T}{\partial e_{c_0}} + \frac{\partial T}{\partial \omega} \omega = T(e_{c_0}, \omega)} \qquad (a)$$

Because of the straight line approximation we are using, it is justifiable to let

$$\frac{\partial T}{\partial e_{c_0}} = K_{c_0} \quad \text{and} \quad \frac{\partial T}{\partial \omega} = K_\omega$$

Suppose we assume a load consisting of inertia and damping, then

$$T_L = JD\omega + B\omega \qquad (b)$$

However, the generated torque must be equal to the load torque. Therefore, equating equations (a) and (b) we get

$$K_{c_0} e_{c_0} + K\omega = JD\omega + B\omega$$

$$\boxed{JD\omega + (B - K)\omega = K_{c_0} e_{c_0}}$$

G-13

1.8 LAGRANGE'S EQUATION

$$\frac{d}{dt}\left(\frac{\partial T}{\partial q_n}\right) - \frac{\partial T}{\partial q_n} + \frac{\partial D}{\partial q_n} + \frac{\partial V}{\partial q_n} = Q_n$$

where n = 1,2,3,... are the independent coordinates or degrees of freedom which exist in the system and

T = total kinetic energy of system
D = dissipation function of system
V = total potential energy of system
Q_n = generalized applied force at the coordinate n
q_n = generalized coordinate
$\dot{q}_n = dq_n/dt$ (generalized velocity)

Energy functions for translational mechanical elements
Table 1.1

Element	Kinetic energy T	Potential energy V	Dissipation factor D	Forcing function Q
Force, f	—	—	—	f (force)
Mass, M	$\tfrac{1}{2}Mv^2 = \tfrac{1}{2}M\dot{x}^2$	—	—	—
Spring K, x_1, x_2	—	$\tfrac{K}{2}\left[\int(v_1 - v_2)dt\right]^2$ $= \tfrac{1}{2}K(x_1 - x_2)^2$ where x is the displacement	—	—
Damping, B, v_1, v_2	—	—	$\tfrac{1}{2}B(v_1 - v_2)^2$ $= \tfrac{1}{2}K(\dot{x}_1 - \dot{x}_2)^2$	—

Energy functions for electric circuits based on the loop or mesh analysis

Table 1.2

Element	Kinetic energy T	Potential energy V	Dissipation function D	Forcing function Q
Voltage source, e	—	—	—	e
Inductance, L	$\frac{1}{2}Li^2 = \frac{1}{2}L\dot{q}^2$ where q is the charge	—	—	—
Capacitance, C	—	$\dfrac{(\int i\, dt)^2}{2C} = \dfrac{q^2}{2C}$	—	—

CHAPTER 2

SOLUTIONS OF DIFFERENTIAL EQUATIONS: SYSTEM'S RESPONSE

2.1 STANDARDIZED INPUTS

Sinusoidal function: $r(t) = \cos\omega t$
Power-series function: $r = a_0 + a_1 t + a_2 t^2 + \ldots$
Unit step function: $r = u(t)$
Unit ramp function: $r = tu(t)$
Unit parabolic function: $r = t^2 u(t)$
Unit impulse function: $r = \delta(t)$

2.2 STEADY STATE RESPONSE

SINUSOIDAL INPUT

Input: $r(t) = A\cos(\omega t + \alpha)$
This is generally the form of input.
$= \text{real part of } (Ae^{j(\omega t + \alpha)}) = \text{Re}(Ae^{j(\omega t + \alpha)})$
$= \text{Re}(Ae^{j\alpha} e^{j\omega t}) = \text{Re}(Ae^{j\omega t})$

G-16

using Euler's identity.

Form of the equation to be solved:

$$A_m D^m C + A_{m-1} D^{m-1} C \ldots + A_{-n} D^{-n} C = r$$

$\overline{R} = Re^{j\alpha}$ phasor representation of the input.

$\overline{C}$ = Phasor representation of the output

$$= \frac{\overline{R}}{A_m (j\omega)^m + A_{m-1}(j\omega)^{m-1} \ldots + A_{-n}(j\omega)^{-n}}$$

Time response = $c(t) = |\overline{C}| \cos(\omega t + \phi)$

Steady-State Sinusoidal Response of Series RLC Circuit:

$$LDi + Ri + \frac{1}{CD} i = e$$

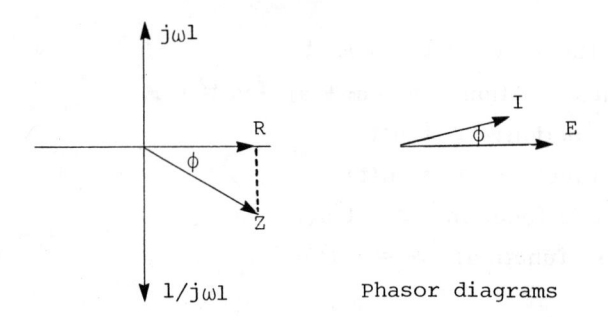

Phasor diagrams

The equation obtained after application of Kirchoff's voltage law.

$$e = Re[\sqrt{2}\, \overline{E}(j\omega)e^{j\omega t}]$$

The input to the RLC circuit.

$$\overline{I}(j\omega) = \text{Phasor current} = \frac{\overline{E}(j\omega)}{j\omega L + R + \dfrac{1}{j\omega c}}$$

$$\overline{Z}(j\omega) = j\omega L + R + \frac{1}{j\omega c}$$

Frequency transfer function:

$$\overline{G}(j\omega) = \frac{\overline{V}_R(j\omega)}{\overline{E}(j\omega)} = \frac{R}{j\omega L + R + \frac{1}{j\omega C}}$$

$\overline{V}_R(j\omega)$ is the voltage (output) across R in phasor form.

POWER SERIES INPUT

Consider the general differential equation:

$$A_m D^m C + \ldots + A_0 C + A_{-1} D^{-1} C + \ldots + A_n D^{-n} C = r$$

where $r(t) = a_0 + a_1 t + a_2 t^2 + \ldots + a_k t^k$.

To find the particular (steady state) solution of the response $c(t)$, we assume a solution of the form

$$(H) = b_0 + b_1 t + b_2 t^2 + \ldots + b_v t^v$$

Substituting in the differential equation and equating coefficients on both sides will enable us to find the constants $b_0, b_1, \ldots$.

Note: 1) k is the highest power of t on the right side. Therefore t^k must also appear on the left side of the equation.

2) The highest power of t on the left side will result from the lowest order derivative. Let x be the order of the lowest derivative.

$$v = k + x \qquad v \geq 0$$

Special Cases of Power Series Input

(i) Step-Function Input:

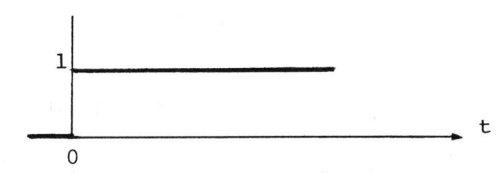

G-18

A) Step-function input:

Voltage equation: $A_2 D^2 \omega_m + A_1 D \omega_m + A_0 \omega_m = u(t)$
where $u(t)$ is the input.

The response is of the form $\omega_m = b_o$ because, for step f_n, the highest exponent of t is k = 0 for u(t).

$$D \omega_m = 0 = D^2 \omega_m$$

$$\boxed{b_o = \frac{1}{A_o}}$$

B) Ramp-function:

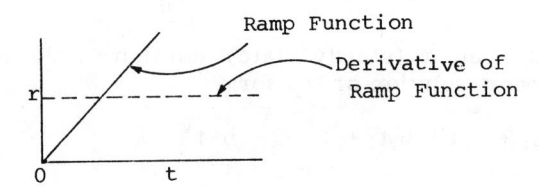

Assume we have a system whose equation is

$$C \cdot D\theta + A = m$$

By integrating, the response is

$$C \cdot \theta + \left(ns + \frac{1}{R}\right) D^{-1}\theta = q$$

One form of the response is

$$\boxed{\theta = b_o}$$

because the highest exponent of t in the input n = t is k = 1.

After integrating,

$$D^{-1}\theta = b_o t + C_o$$

$$b_o = \frac{1}{A}, \quad \theta = \frac{1}{A} \text{ at a steady state.}$$

C) Parabolic-function input:

G-19

```
                    Dr
                   ╱
                  ╱  Parabolic
           ┌─────╱─  Function
        r  │   ╱ ╱
           │  ╱ ╱ ──── D²r
           │ ╱╱
           │╱
           └──────────
                t
```

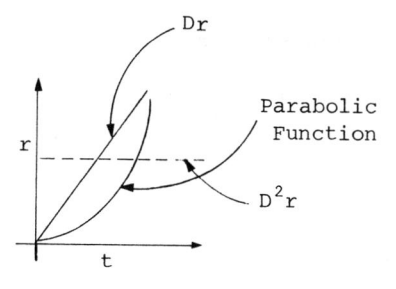

$$r = t^2 u(t) \quad D^2 r = 2u(t)$$

$$Dr = 2t \cdot u(t)$$

Example: Given a system whose equation is

$$AD^2 x_2 + BDx_2 + Kx_2 = Kx_1$$

k = 2 because the highest exponent of t in the input is 2.
Order of the lowest derivation in the system equation

$$x = 0$$

The form of the steady-state response is

$$\boxed{x_2(t) = b_0 + b_1 t + b_2 t^2}$$

$$Dx_2 = b_1 + 2b_2 t \quad \text{and} \quad D^2 x_2 = 2b_2$$

Steady-state solution:

$$x_2(t) = \frac{-2A}{K} + \frac{2B^2}{K^2} - \frac{2Bt}{K} + t^2.$$

2.3 TRANSIENT RESPONSE

Transient response of a differential equation:

REAL ROOTS

Steps:

1) Write a homogeneous equation by equating the given

differential equation to zero.

Differential equation:

$$b_m D^m c + b_{m-1} D^{m-1} c + \ldots + b_o D^o c + \ldots + b_{-n} D^{-n} c = r$$

Homogeneous equation:

$$b_m D^m c_t + b_{m-1} D^{m-1} c_t + \ldots + b_{-n} D^{-n} c_t = 0$$

2) Solution of the homogeneous equation gives the general expression for the transient response.
Assume a solution $C_t = e^{kt}$.

3) Substituting $C_t = e^{kt}$ in the equation results in the characteristic equation:

$$b_m k^m + b_{m-1} k^{m-1} + \ldots + b_o + \ldots + b_{-n} k^{-n} = 0$$

Its roots are $k_1, k_2, \ldots$

4) So if there are no multiple roots the transient response is $C_t = A_1 e^{k_1 t} + A_2 e^{k_2 t} \ldots$

Short-cut method:

We can obtain the characteristic equation by substituting into the homogeneous equation: k for DC_t, k^2 for $D^2 C_t$ etc. For the general equation we are using, our characteristic equation will consist of m + n constants. We must consequently have m + n initial conditions in order to set up m + n equations which will enable us to determine the transient response.

COMPLEX ROOTS

When the roots of the characteristic equation are complex, the above method cannot be used; instead, the response takes the form

$$A e^{\sigma t} \sin(\omega_d t + \phi) \qquad \text{(For 2 complex roots)}$$

[Figure: Exponentially damped sinusoid showing envelope $Ae^{\sigma t}$ and curve $Ae^{\sigma t}\sin(\omega_d t + \phi)$ with phase ϕ marked on t-axis]

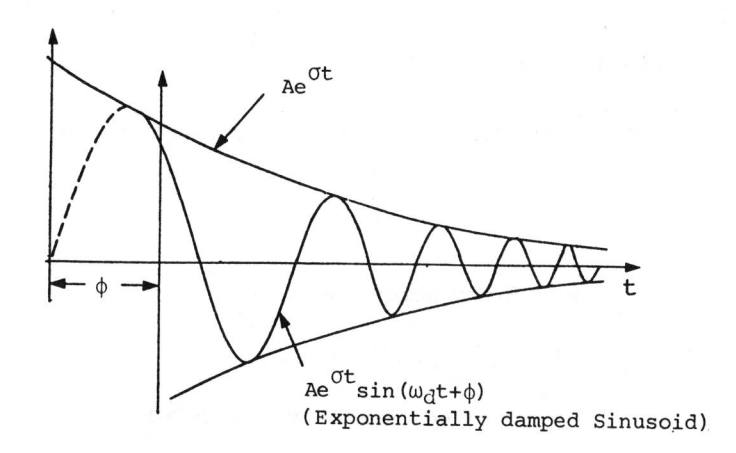

Damping ratio and undamped national frequency:

$$\text{Damping ratio } \zeta = \frac{\text{Actual damping}}{\text{Critical damping}}$$

If the characteristic equation has a pair of complex conjugate roots, the form of the quadratic factor is $b_2 m^2 + b_1 m + b_0$.

b_1 represents the effective damping constant of the system.

$$\zeta = \frac{\text{Actual damping}}{\text{Critical damping}} = \frac{b_1}{b_1{}^*} = \frac{b_1}{2\sqrt{b_2 b_0}}$$

If $\zeta < 1$, the response is said to be underdamped.
If $\zeta > 1$, the response is said to be overdamped.
If $\zeta = 0$,

A) $\omega_m = \sqrt{\dfrac{b_0}{b_2}}$ = undamped natural frequency

B) the zero damping constant means that the transient response is a sine wave of constant amplitude.

We can rewrite the above equation $b_2 m^2 + b_1 m + b_0$ as follows:

$$\frac{b_2}{b_0} m^2 + \frac{b_1}{b_0} m + 1 = \frac{1}{\omega_n^2} m^2 + \frac{2\zeta m}{\omega_n} + 1$$

where

$$\omega_m = \sqrt{\frac{b_0}{b_2}}$$

$$\to m^2 + 2\zeta\omega_n m + \omega_n^2$$

This is the general form of the characteristic equation whose roots are:

$$m_{1,2} = -\zeta\omega_n \pm j\omega_n\sqrt{1-\zeta^2}$$

Transient response for underdamped case:

$$Ae^{-\zeta\omega_n t}\sin(\omega_n\sqrt{1-\zeta^2}\,t + \phi)$$

ω_d - the damped frequency of oscillation

$$\omega_d = \omega_n\sqrt{1-\zeta^2}$$

Time constant:

The transient terms have the form Ae^{kt}. The value of time that makes the exponent of e equal to -1 is called the time constant T.

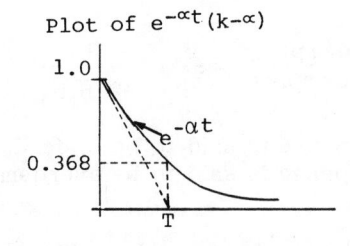

Plot of $e^{-\alpha t}$ $(k = -\alpha)$

1.0

0.368

$e^{-\alpha t}$

T

In case of the damped sinusoid

$$T = \frac{1}{|\sigma|}, \quad \text{for } k = \sigma + \omega_d$$

$$T = \frac{1}{\zeta}\omega_n$$

where the larger the $\zeta\omega_n$, the greater the rate of decay of the transient.

G-23

2.4 FIRST AND SECOND-ORDER SYSTEM

FIRST-ORDER SYSTEM

An example:

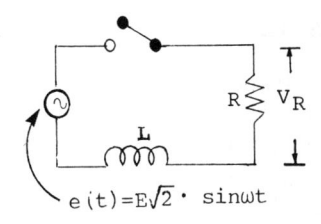

[Circuit diagram: AC source $e(t) = E\sqrt{2} \cdot \sin\omega t$ with switch, resistor R with voltage v_R, and inductor L in series]

Applying KVL to series RL circuit,

$$e = \frac{L}{R} D v_R + v_R$$

The characteristic equation is therefore

$$\frac{L}{R} x + 1 = 0 \quad \text{or,} \quad x = -\frac{R}{L}$$

So the transient solution is

$$v_{R,t} = A e^{-(R/L)t}$$

The steady-state solution in phasor form is,

$$\overline{V}_{R,SS}(j\omega) = \frac{E(j\omega)}{1 + \left(\frac{L}{R}\right) j\omega} = \frac{E(j\omega)}{\left[i + \left(\frac{\omega L}{R}\right)^2 \right]} \underline{/-\tan^{-1} \frac{\omega L}{R}}$$

Therefore, the steady-state voltage in time domain is

$$V_{R,SS} = \frac{E\sqrt{2}}{\left[1 + \left(\frac{\omega L}{R}\right)^2 \right]^{\frac{1}{2}}} \sin\left(\omega t - \tan^{-1} \frac{\omega L}{R} \right)$$

The complete solution is:

G-24

$$\boxed{V_R = \frac{E\sqrt{2}}{1 + \left(\frac{\omega}{L}\right)^2} \sin\left[\omega t - \tan^{-1}\frac{\omega L}{R}\right] + Ae^{-\frac{R}{L}t}}$$

|← steady-state solution → | |← transient → |
                                    solution

Evaluation of A using initial conditions :

initial conditions: $V_R = 0$ at $t = 0$

$$A = \frac{\omega \cdot R \cdot L \cdot E\sqrt{2}}{R^2 + (\omega L)^2}$$

obtained after putting $V_R = 0$ and $t = 0$ in the complete solution.

SECOND-ORDER SYSTEM

An example

$R_1 (10\Omega)$ $L(1H)$ $C(0.01F)$ $E(10V)$ $R_3 (10\Omega)$ $R_2 (10\Omega)$

$$E = i_2 \left(R_1 + \frac{R_1 R_2}{R_3} + \frac{R_1}{R_3} LD + \frac{R_1}{CDR_3} + R_2 + LD + \frac{1}{CD} \right)$$

$$10 = \left(2D + 40 + \frac{200}{D} \right) i_2 \quad \text{after application of KVL.}$$

The steady-state output is: $i_{2,ss} = 0$...since the branch contains a capacitor.

The characteristic equation: $m + 20 + \frac{100}{m} = 0$.

Its roots are $m_{1,2} = -10$...so the circuit is critically damped.

The output current:

$$i_2(t) = A_1 e^{-10t} + A_2 t e^{-10t}$$

$$Di_2(t) = -10A_1 e^{-10t} + A_2(1 - 10t)e^{-10t}$$

Initial conditions: $i_1(0^-) = i_2(0^-) = 0$

$$i_2(0^+) = i_2(0^-) = 0$$

because i_2 can't change instantly in the inductor.

A) $Di_2(t) = 10i_1(t) - 25i_2(t) - vi(t)$

B) $i_1(0^+) = 0.5$ because $i_2(0^+) = 0$

C) $v_c(0^-) = v_c(0^+)$ = steady-state value of = 10 capacitor voltage for $t < 0$

Hence, $Di_2(t) = -5$

$A_1 = 0$, $A_2 = -5$, so $i_2(t) = -5t\, e^{-10t}$

Second-order transients:

Simple second-order equation:

$$\boxed{\frac{D^2 c}{\omega_n^2} + \frac{2\zeta}{\omega_n} Dc + c = r}$$

because there are no derivatives of r on the right-hand side.

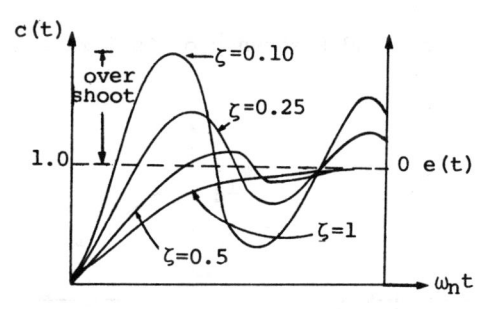

Representation of the Transient Response of a Second Order System

$$c(t) = 1 - \frac{e^{-\zeta \omega_n t}}{\sqrt{1-\zeta^2}} \sin(\omega_n \sqrt{1-\zeta^2}\, t + \cos^{-1}\zeta)$$

This is the response to a unit step with initial conditions set to zero.

G-26

e(t) - error in the system

e = r - c, where r = u(t)

Representation of the transient response of a second-order system.

Conclusion: The amount of overshoot (i.e. beyond 1.0) depends on the damping ratio ζ.

For the overdamped ($\zeta > 1$) and critically damped case ($\zeta = 1$), there is no overshoot.

For underdamped ($\zeta < 1$), the system oscillates around the steady state value before it settles down at the steady state.

$$t_p = \pi/\omega_n \sqrt{1-\zeta^2} = \text{The time at which the peak overshoot occurs.}$$

$$c_p = 1 + e^{-\zeta\pi/\sqrt{1-\zeta^2}} \quad \text{The value of the peak overshoot.}$$

$$M_o = \frac{c_p - c_{ss}}{c_{ss}} \quad \text{per unit overshoot.}$$

Peak overshoot v/s ζ and Frequency of Oscillation v/s Damping ratio

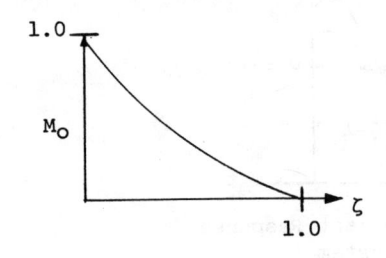

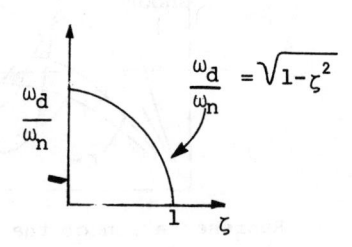

Error equation:

$$e = r - c = \frac{e^{-\zeta\omega_n t}}{\sqrt{1-\zeta^2}} \sin(\omega_n \sqrt{1-\zeta^2} \cdot t + \cos^{-1}\zeta)$$

RESPONSE CHARACTERISTICS

One of the characteristics of the system's response is the setting time. Setting time is the time required for the envelope of the transient to die out or to reduce to an insignificantly small value. This time is a function of a number of time constants involved in the transfer function of the system, and can be presented mathematically as follows:

$$T_s = \frac{\text{no. of time constants in the system's transfer function}}{(\text{undamped natural frequency})(\text{the damping constant})}$$

2.5 TIME-RESPONSE SPECIFICATIONS

The following terms are used as specifications to evaluate the performance of a system. These are:

A) Maximum overshoot (Cp): the magnitude of the first overshoot as shown.

B) t_p (Time to reach the maximum overshoot): the time required to reach the first overshoot.

C) Duplicating time (t_o): the time when the response of the system has the final value for the first time.

D) Settling time (t_s): the time taken by the control system to reach the final specified value and thereafter remain within a specified tolerance.

E) Frequency of oscillation (ω_d): the frequency of oscillation of the tansient response.

CHAPTER 3

APPLICATIONS OF LAPLACE TRANSFORM

3.1 DEFINITION OF LAPLACE TRANSFORM

$$L[f(t)] = \int_0^\infty f(t)e^{-st}dt = F(s)$$

provided f(t) is piecewise continuous over every finite interval and is of exponential order.

Laplace Transform of Some Simple Functions:

$L[\text{step function } u(t)] = \frac{1}{s} \ldots$ if $\sigma > 0$

$L[e^{-\alpha t}] = \frac{1}{s+\alpha} \ldots\ldots\ldots$ if $\sigma > -\alpha$

$L[\cos \omega t] = \frac{s}{s^2 + \omega^2} \ldots\ldots$ if $\sigma > 0$

$L[\text{Ramp function } f(\dot{t}) = t] = \frac{1}{s^2} \ldots$ if $\sigma > 0$

where s is a complex variable and has the form $\sigma + j\omega$.

G-29

Table 3.1 Table of Laplace-transform pairs

	F(s)	f(t) $0 \leq t$
1.	1	$u_1(t)$ unit impulse at $t = 0$
2.	$\dfrac{1}{s}$	1 or $u(t)$ unit step at $t = 0$
3.	$\dfrac{1}{s^2}$	$tu(t)$ ramp function
4.	$\dfrac{1}{s^n}$	$\dfrac{1}{(n-1)!} t^{n-1}$ n is a positive integer
5.	$\dfrac{1}{s} e^{-as}$	$u(t - a)$ unit step starting at $t = a$
6.	$\dfrac{1}{s}(1 - e^{-as})$	$u(t) - u(t - a)$ rectangular pulse
7.	$\dfrac{1}{s + a}$	e^{-at} exponential decay
8.	$\dfrac{1}{(s + a)^n}$	$\dfrac{1}{(n-1)!} t^{n-1} e^{-at}$ n is a positive integer
9.	$\dfrac{1}{s(s + a)}$	$\dfrac{1}{a}(1 - e^{-at})$
10.	$\dfrac{1}{s(s + a)(s + b)}$	$\dfrac{1}{ab}\left[1 - \dfrac{b}{b-a} e^{-at} + \dfrac{a}{b-a} e^{-bt}\right]$
11.	$\dfrac{s + \alpha}{s(s + a)(s + b)}$	$\dfrac{1}{ab}\left[\alpha - \dfrac{b(\alpha - a)}{b-a} e^{-at} + \dfrac{a(\alpha - b)}{b-a} e^{-bt}\right]$
12.	$\dfrac{1}{(s + a)(s + b)}$	$\dfrac{1}{b-a}(e^{-at} - e^{-bt})$
13.	$\dfrac{s}{(s + a)(s + b)}$	$\dfrac{1}{a-b}(ae^{-at} - be^{-bt})$
14.	$\dfrac{s + \alpha}{(s + a)(s + b)}$	$\dfrac{1}{b-a}[(\alpha - a)e^{-at} - (\alpha - b)e^{-bt}]$
15.	$\dfrac{1}{(s + a)(s + b)(s + c)}$	$\dfrac{e^{-at}}{(b-a)(c-a)} + \dfrac{e^{-bt}}{(c-b)(a-b)} + \dfrac{e^{-ct}}{(a-c)(b-c)}$
16.	$\dfrac{s + \alpha}{(s + a)(s + b)(s + c)}$	$\dfrac{(\alpha - a)e^{-at}}{(b-a)(c-a)} + \dfrac{(\alpha - b)e^{-bt}}{(c-b)(a-b)} + \dfrac{(\alpha - c)e^{-ct}}{(a-c)(b-c)}$
17.	$\dfrac{\omega}{s^2 + \omega^2}$	$\sin \omega t$
18.	$\dfrac{s}{s^2 + \omega^2}$	$\cos \omega t$
19.	$\dfrac{s + \alpha}{s^2 + \omega^2}$	$\dfrac{\sqrt{\alpha^2 + \omega^2}}{\omega} \sin(\omega t + \phi)$ $\phi = \tan^{-1}\dfrac{\omega}{\alpha}$
20.	$\dfrac{1}{s(s^2 + \omega^2)}$	$\dfrac{1}{\omega^2}(1 - \cos \omega t)$
21.	$\dfrac{s + \alpha}{s(s^2 + \omega^2)}$	$\dfrac{\alpha}{\omega^2} - \dfrac{\sqrt{\alpha^2 + \omega^2}}{\omega^2} \cos(\omega t + \phi)$ $\phi = \tan^{-1}\dfrac{\omega}{\alpha}$
22.	$\dfrac{1}{(s + a)(s^2 + \omega^2)}$	$\dfrac{e^{-at}}{a^2 + \omega^2} + \dfrac{1}{\omega\sqrt{a^2 + \omega^2}} \sin(\omega t - \phi)$ $\phi = \tan^{-1}\dfrac{\omega}{\alpha}$
23.	$\dfrac{1}{(s + a)^2 + b^2}$	$\dfrac{1}{b} e^{-at} \sin bt$
24.	$\dfrac{1}{s^2 + 2\zeta\omega_n s + \omega_n^2}$	$\dfrac{1}{\omega_n\sqrt{1 - \zeta^2}} e^{-\zeta\omega_n t} \sin \omega_n\sqrt{1 - \zeta^2}\, t$
25.	$\dfrac{s + a}{(s + a)^2 + b^2}$	$e^{-at} \cos bt$
26.	$\dfrac{1}{s^2(s + a)}$	$\dfrac{1}{a^2}(at - 1 + e^{-at})$
27.	$\dfrac{1}{s(s + a)^2}$	$\dfrac{1}{a^2}(1 - e^{-at} - ate^{-at})$
28.	$\dfrac{s + \alpha}{s(s + a)^2}$	$\dfrac{1}{a^2}[\alpha - \alpha e^{-at} + a(a - \alpha)te^{-at}]$

LAPLACE TRANSFORM THEOREMS

A) Linearity

$$L[a \cdot f(t)] = a \cdot L[f(t)] = a \cdot F(s) \quad \ldots \text{ if a is a constant}$$

B) Superposition

$$L[f_1(t) \pm f_2(t)] = F_1(s) \pm F_2(s)$$

C) Translation in time

$$L[f(t - a)] = e^{-as} F(s) \quad \ldots \text{ if a is a positive real, and } f(t - a) = 0 \text{ for } 0 < t < a.$$

D) Complex differentiation

$$L[t \cdot f(t)] = \frac{-d}{ds} F(s)$$

E) Translation in tne s domain

$$L[e^{at} \cdot f(t)] = F(s - a)$$

F) Real differentiation

$$L[Df(t)] = sF(s) - f(o^+)$$

$f(o^+)$ is the value of the limit of $f(t)$ as the origin, $t = 0$ is approached from the right-hand side.

$$L[D^2 f(t)] = s^2 F(s) - sf(o) - Df(o)$$

$$L[D^n f(t)] = s^n F(s) - s^{n-1} f(o) \ldots - D^{n-1} f(o)$$

G) Real integration

$$L[D^{-n} f(t)] = \frac{F(s)}{s^n} + \frac{D^{-1} f(o)}{s^n} + \ldots + \frac{D^{-n} f(o)}{s}$$

H) Final value F

if $f(t)$ and $DF(t)$ are Laplace transformable, then

$$\lim_{s \to 0} sf(s) = \lim_{t \to \infty} f(t)$$

According to the final value theorem we can find the

G-31

final value of the function f(t) by working in the s domain which saves us the work involved in taking the inverse Laplace transform.

Limitations of this theorem:

A) Whenever SF(s) has poles on the imaginary axis or in the right half s plane, there is no finite final value of f(t) since SF(s) becomes infinite and the theorem cannot be used.

B) Suppose the driving function is sinusoidal and is equal to $\sin\omega t$. The $L[\sin\omega t]$ has poles at $s = \pm j\omega$; furthermore, $\lim_{t \to \infty} \sin\omega t$ does not exist. Thus this theorem is invalid whenever the driving function is sinusoidal.

INITIAL VALUE THEOREM

$$\lim_{s \to \infty} S F(S) = \lim_{t \to 0} f(t) \ldots \text{ if } \lim S F(S) \text{ exists.}$$

COMPLEX INTEGRATION THEOREM

$$L\left[\frac{f(t)}{t}\right] = \int_s^\infty F(s)ds$$

providing the $\lim_{t \to 0^+} f(t)/t$ exists

In words this means that division by the variable in the real time domain is equivalent to integration with respect to s in the s domain.

3.2 APPLICATION OF LAPLACE TRANSFORM TO DIFFERENTIAL EQUATIONS

An example

Step 1) $AD^2x_2 + MDx_2 + kx_2 = kx_1$ ....

the differential equation where $x_1(t)$ is the input and $x_2(t)$ is called the response function. First of all take the Laplace transform of both sides:

$$L[AD^2x_2 + MDx_2 + kx_2] = L[kx_1].$$

2) Then take the Laplace transform of each term; after substituting in the original equation, rearrange the equation:

$$kx_1(s) = (As^2 + Ms + k)x_2(s)$$
$$- [Asx_2(o) + ADx_2(o) + Mx_2(o)]$$

characteristic function $x_2(s)$ - the transform equation

$$x_2(s) = \frac{kx_1(s) + Asx_2(o) + Mx_2(o) + ADx_2(o)}{As^2 + Ms + k}$$

3) $x_2(t) = L^{-1}[x_2(s)]$ ... the response function

3.3 INVERSE TRANSFORM

$$F(s) = P(s)/Q(s) = [a_n s^n + a_{n-1} s^{n-1} + \ldots]/[s^m + b_{m-1} s^{m-1} + \ldots]$$

$$= P(s)/[(s - s_1)(s - s_2)\ldots(s - s_m)]$$

after breaking $Q(s)$ into linear and quadratic factors.

Application of partial expansion in finding inverse Laplace transform.

A) $F(s)$ has first order real poles:

$$F(s) = \frac{P(s)}{Q(s)} = \frac{P(s)}{s(s-s_1)(s-s_2)} = \frac{A_0}{s} + \frac{A_1}{s-s_1} + \frac{A_2}{s-s_2}$$

$$\boxed{f(t) = L^{-1}[F(s)] = A_0 + A_1 e^{s_1 t} + A_2 e^{s_2 t}}$$

Two inverse transforms.

Evaluation of coefficients A_k:

$$A_k = \left[(s - s_k) \frac{P(s)}{Q(s)}\right]_{s=s_k}$$

A_k is called the residue of $F(s)$.

B) Multiple-order real poles:

$$F(s) = \frac{P(s)}{Q(s)} = \frac{P(s)}{(s-s_1)^2(s-s_2)} = \frac{A_{12}}{(s-s_1)^2} + \frac{A_{11}}{s-s_1} + \frac{A_2}{s-s_2}$$

$$f(t) = A_{12} \cdot t \cdot e^{s_1 t} + A_{11} \cdot e^{s_1 t} + A_2 e^{s_2 t}$$

A general formula for finding coefficients associated with the repeated real pole of order n,

$$A_{rn} \left[(s - s_r)^n \frac{P(s)}{Q(s)}\right]_{s=s_r}$$

$$A_{r(n-k)} = \frac{1}{k!} \frac{d^k}{ds^k} \left[(s - s_r)^n \frac{P(s)}{Q(s)}\right]_{s=s_r}$$

C) Complex conjugate poles:

$$f(s) = \frac{P(s)}{Q(s)} = \frac{P(s)}{(s^2 + 2\zeta\omega_n s + \omega_n^2)(s-s_3)}$$

$$= \frac{A_1}{s - s_1} + \frac{A_2}{s - s_2} + \frac{A_3}{s - s_3}$$

$$= \frac{A_1}{s + \zeta\omega_n - j\omega_n\sqrt{1-\zeta^2}} + \frac{A_2}{s + \zeta\omega_n + j\omega_n\sqrt{1-\zeta^2}}$$

$$+ \frac{A_3}{s - s_3}$$

Hence, the inverse function of $F(s)$ is,

$$f(t) = A_1 e^{(-\zeta\omega_n + j\omega_n\sqrt{1-\zeta^2})t} + A_2 e^{(-\zeta\omega_n - j\omega_n\sqrt{1-\zeta^2})t} + A_3 e^{s_3 t}$$

$$= 2|A_1| e^{6t} \sin(\omega_d t + \phi) + A_3 e^{s_3 t}$$

where ϕ = angle of A_1 + 90°

Now, $A_1 = [(s - s_1)F(s)]_{s=s_1}$ and $A_3 = [(s - s_3)F(s)]_{s=s_3}$

HAZONY AND RILEY RULE

For a normalized ratio of polynomials:

A) If the denominator is one degree higher than the numerator, the sum of the residues is one.

B) If the denominator is two or more degrees higher than the numerator, the sum of the residues is zero.

GRAPHICAL METHOD

$$F(s) = \frac{P(s)}{Q(s)} = k \frac{\prod_{m=1}^{\omega}(s - z_m)}{\prod_{k=1}^{\nu}(s - p_k)} = \frac{A_1}{s - p_1} + \frac{A_2}{s - p_2}$$

Pole-zero plot of the function F(s):

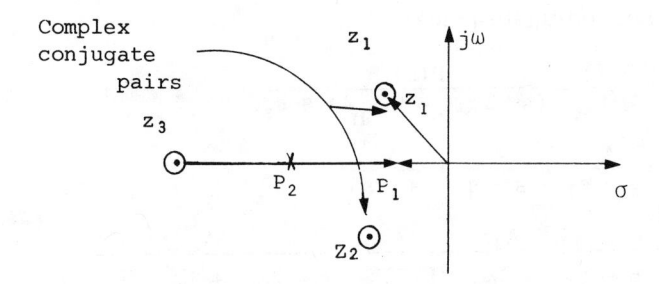

$$A_k = k \frac{\text{Product of directed distances from each zero to the pole } p_k}{\text{Product of directed distances from all other poles to the pole } p_k}$$

... except f ∿ k = 0.

G-35

An example:

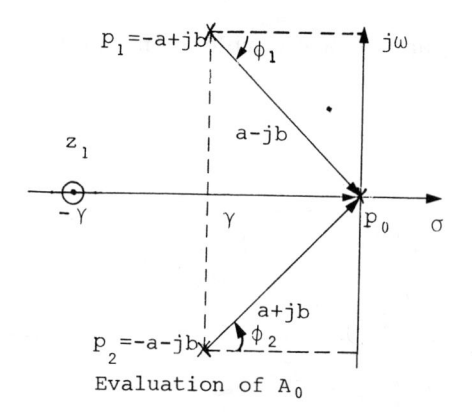

Evaluation of A_0

$$-F(s) = \frac{k(s+\gamma)}{s[(s+a)^2+b^2]} = \frac{k(s+\gamma)}{s(s+a-jb)(s+a+jb)} = \frac{k(s-z_1)}{s(s-p_1)(s-p_2)}$$

$$= \frac{A_0}{s} + \frac{A_1}{s+a-jb} + \frac{A_2}{s+a+jb}$$

$$A_0 = \frac{k\gamma}{(a-jb)(a+jb)} = \frac{k\gamma}{a^2+b^2}$$

$$A_1 = \frac{k[(\gamma-a)+jb]}{(-a+jb)(j2b)}$$

$$= \frac{k}{2b}\sqrt{\frac{(\gamma-a)^2+b^2}{a^2+b^2}}\; e^{j(\phi-\theta-\pi/2)}$$

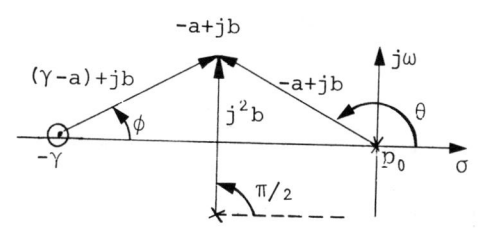

Fig. Evaluation of A_1.

where $\phi = \tan^{-1}\dfrac{b}{\gamma-a}$

$\theta = \tan^{-1}\dfrac{b}{-a}$

G-36

A_1 and A_2 are complex conjugate

$$\sigma - f(t) = A_0 + 2|A_1| e^{-\gamma t} \cdot \cos(bt + \phi - \theta - \pi/2)$$

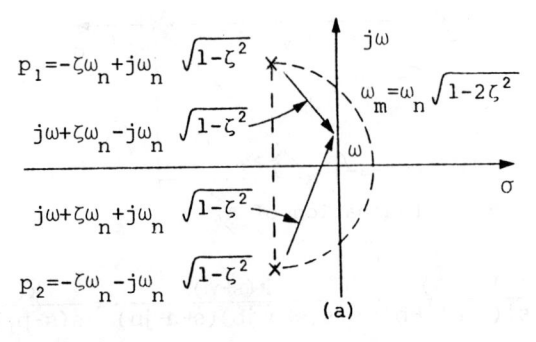

$p_1 = -\zeta\omega_n + j\omega_n \sqrt{1-\zeta^2}$

$j\omega + \zeta\omega_n - j\omega_n \sqrt{1-\zeta^2}$

$j\omega + \zeta\omega_n + j\omega_n \sqrt{1-\zeta^2}$

$p_2 = -\zeta\omega_n - j\omega_n \sqrt{1-\zeta^2}$

$\omega_m = \omega_n \sqrt{1-2\zeta^2}$

(a)

Pole-zero diagram.

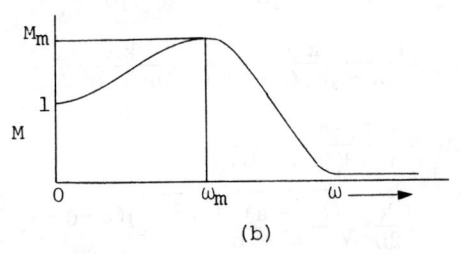

(b)

Magnitude of frequency response.

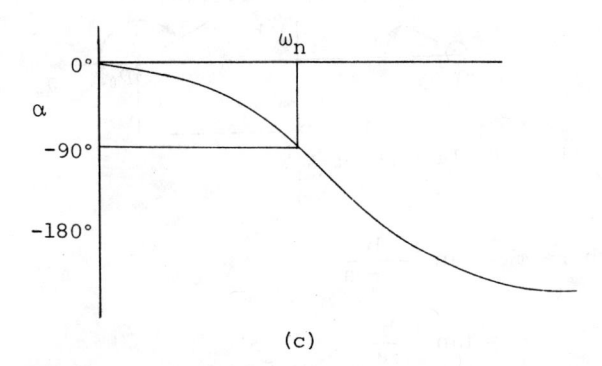

(c)

Angle of frequency response.

3.4 FREQUENCY RESPONSE FROM THE POLE-ZERO DIAGRAM

Frequency response: It is the steady-state response with a sine-wave forcing function for all values of frequency.

It is given by two curves:

A) M curve: the ratio of output amplitude to input amplitude as a function of frequency.

B) α curve: phase angle of the output α as a function of frequency.

Frequency-response characteristics:

A) at $\omega = 0$, the magnitude is a finite value and the angle is $0°$.

B) If the number of poles is more than the number of zeros, as $\omega \to \infty$, the magnitude approaches zero and the angle is $-90°$ times the difference between the number of poles and zeros.

C) In order to have a peak value M_m, there must be present complex poles near the imaginary axis. More precisely, the damping ratio must be less than 0.707.

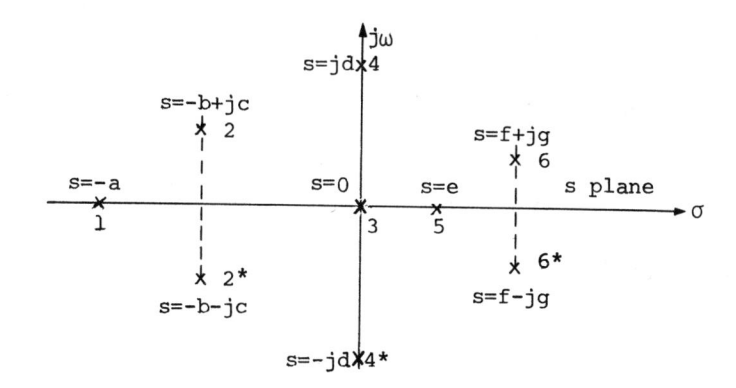

Location of poles in the s plane.

Table 3.1

Pole	Response	Characteristics
1	Ae^{-at}	Damped exponential
2-2*	$Ae^{-bt}\sin(ct + \phi)$	Exponentially damped sinusoid
3	A	Constant
4-4*	$A\sin(dt + \phi)$	Constant sinusoid
5	Ae^{et}	Increasing exponential (unstable)
6-6*	$Ae^{ft}\sin(gt + \phi)$	Exponentially increasing sinusoid (unstable)

3.5 ROUTH'S STABILITY CRITERION

Consider the characteristic equation

$$Q(s) = b_n s^n + b_{n-1} s^{n-1} + b_{n-2} s^{n-2} + \ldots + b_1 s + b_0 = 0$$

where all coefficients are real.

A) All powers of s from s^n to s^o must be present in the characteristic equation. The system is unstable if any coefficients other than b_o are zero or if any of the coefficients are negative. In this case, the roots are either imaginary or complex with positive real parts.

B) To determine the number of roots in the right half s plane:

The coefficients of the characteristic are arranged in the following Routhian array.

s^n	b_n	b_{n-2}	b_{n-4}	b_{n-6}	$\cdots$
s^{n-1}	b_{n-1}	b_{n-3}	b_{n-5}	b_{n-7}	$\cdots$
s^{n-2}	c_1	c_2	c_3		$\cdots$
s^{n-3}	d_1	d_2			$\cdots$
$\cdot$					
$\cdot$					
$\cdot$					
s^1	j_1				
s^0	k_1				

G-39

c_1, c_2, c_3, etc., are evaluated as follows:

$$c_1 = \frac{b_{n-1}b_{n-2} - b_n b_{n-3}}{b_{n-1}}$$

$$c_2 = \frac{b_{n-1}b_{n-4} - b_n b_{n-5}}{b_{n-1}}$$

$$c_3 = \frac{b_{n-1}b_{n-6} - b_n b_{n-7}}{b_{n-1}}$$

This pattern is continued until the rest of the c's are all equal to zero. The d row is formed by using the s^{m-1} and s^{m-2} row. The constants are

$$d_1 = \frac{c_1 b_{n-3} - b_{n-1} c_2}{c_1}$$

$$d_2 = \frac{c_1 b_{n-5} - b_{n-1} c_3}{c_1}$$

$$d_3 = \frac{c_1 b_{n-7} - b_{n-1} c_4}{c_1}$$

When no more d terms are present, the other rows are formed in a similar way.

Routh's criterion: The number of roots of the characteristic equation with positive real parts is equal to the number of changes of sign of the coefficients in the first column of the Routhian array. Thus a system will be stable if all terms in the first column have identical signs.

An example

The Routh's array

s^4	1	2	9
s^3	10	3	
s^2	$\frac{17}{10}$	9	
s^1	-50		
s^0	9		

There are two changes of sign, so there are two roots in the right-half of the s plane.

C) Theorems: (For special cases)

Division of a row: The coefficients of any row may be multiplied or divided by the number without changing the signs of the first column.

A zero coefficient in the first column: The procedure below can be used whenever the first term in a row is zero, but the other terms are not equal to zero.

Method A: A small positive number δ is substituted for the zero. The rest of the terms in the array are evaluated as usual.

Method B: Substitute in the original equation $s = \frac{1}{x}$; then solve for the roots of x with positive real parts. The number of roots x with positive real parts will be the same as the number of s roots with positive real parts.

A ZERO ROW

A) An auxiliary equation is formed from the preceding row.

B) The array is completed by replacing the all-zero row by the coefficients obtained by differentiating the auxiliary equation.

C) The roots of the auxiliary equation are also roots of the original equation. These roots occur in pairs and are the negatives of each other.

3.6 IMPULSE FUNCTION: LAPLACE-TRANSFORM AND ITS RESPONSE

$$f(t) = \frac{u(t) - u(t-a)}{a}$$ Analytical expression

$$f(s) = \frac{1 - e^{-as}}{as}$$ its Laplace-transform

Fig: Rectangular pulse

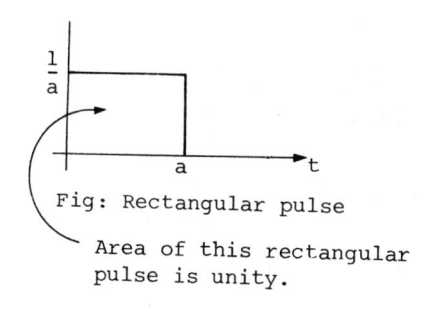

Area of this rectangular pulse is unity.

A unit impulse: The limit of f(t) as a → 0 is termed a unit impulse and is designated $b\delta(t)$.

$$\delta(t) = \lim_{a \to 0} f(t) = \lim_{a \to 0} \frac{u(t) - u(t-a)}{a}$$

$\Delta(s)$ = The Laplace-transform of the unit impulse = 1.

SECOND-ORDER SYSTEM WITH IMPULSE EXCITATION

$$F(s) = \frac{\omega_n^2}{s^2 + 2\zeta\omega_n s + \omega_2^2}$$

Laplace-transform with an impulse input

The impulse function is the derivative of the step function, so the response to an impulse is the derivative of the response to a step function.

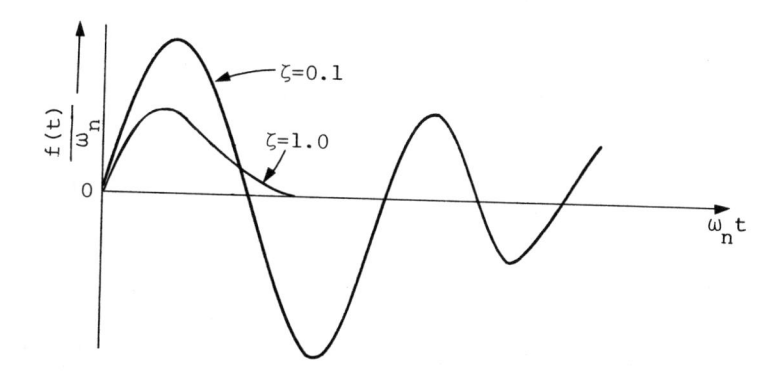

$\zeta = 0.1$

$\zeta = 1.0$

G-42

Fig: Maximum overshoot

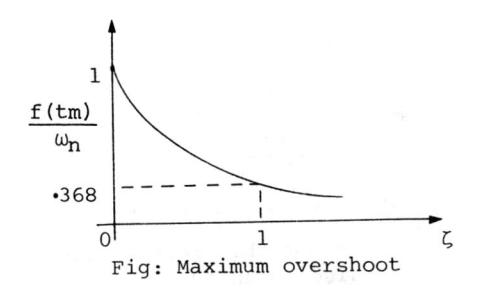

$$t_m = \frac{\cos^{-1}\zeta}{\omega_n \sqrt{1-\zeta^2}}$$

where the maximum overshoot occurs.

$$f(t_m) = \omega_n \, e^{-(\zeta \cos^{-1}\zeta)/\sqrt{1-\zeta^2}}$$

CHAPTER 4

MATRIX ALGEBRA AND Z-TRANSFORM

4.1 FUNDAMENTALS OF MATRIX ALGEBRA

OPERATIONS WITH MATRICES

A) Matrix equality:

$\overline{A} = \overline{B}$ if and only if $a_{jk} = b_{jk}$ for $1 \leq j \leq p$ and $1 \leq k \leq q$ where the size of both matrices is $p \times q$.

B) Addition and subtraction:

If $\overline{A} = [a_{jk}]$ and $\overline{B} = [b_{jk}]$ are both $p \times q$ matrices, then $\overline{A} + \overline{B} = \overline{C}$ and $\overline{A} - \overline{B} = \overline{D}$ implies that $\overline{C} = [c_{jk}]$ and $\overline{D} = [d_{jk}]$ are also of the size $p \times q$ and $c_{jk} = a_{jk} + b_{jk}$ and $d_{jk} = a_{jk} - b_{jk}$ where $j = 1, 2, \ldots, p$ and $k = 1, 2, 3, \ldots, q$.

C) Matrix multiplication:

Scalar multiplication:

$$a\overline{B} = \overline{B}a = [ab_{jk}]$$

where a is a scalar.

Matrix multiplication:

$\bar{A} = [a_{jk}]$ of size $p \times q$ and $\bar{B} = [b_{jk}]$ of size $1 \times m$. The product $\bar{A}\bar{B} = \bar{D}$ is defined only when $q = 1$, then $\bar{A}$ and $\bar{B}$ are conformable.

$$d_{jk} = \sum_{k'=1}^{a} a_{jk'} b_{k'k}, \quad \text{size of } \bar{D} \text{ is } p \times m.$$

D) Transpose, conjugate and the associate matrix:

If $\bar{B} = [b_{jk}]$ then the transpose of $\bar{B}$ is $\bar{B}^T = [b_{kj}]$. The matrix $\bar{B}$ is symmetric if $\bar{B} = \bar{B}^T$. If $\bar{B} = -\bar{B}^T$, then $\bar{B}$ is skew-symmetric. $(\bar{A}\bar{B})^T = \bar{B}^T \bar{A}^T$.

$A^* = $ conjugate of $A = [\overline{a_{jk}}]$

Associate matrix of $\bar{A}$ = conjugate transpose of $\bar{A}$.

Hermitian matrices: These are the matrices for which $\bar{B} = \bar{B}^T$, these are called skew-hermitian if $\bar{A} = -\bar{A}^T$.

E) Matrix inversion:

$\bar{B}\bar{A} = \bar{A}\bar{B} = \bar{I}$ then $\bar{A} = \bar{B}^{-1}$...A should be a square matrix.

$A^{-1} = \dfrac{\bar{C}^T}{|\bar{A}|}$ where $\bar{C}^T$ is the adjoint matrix denoted by Adj($\bar{A}$).

$A^{-1} = \text{Adj}(\bar{A})/|\bar{A}|$ Inverse of a non-singular matrix.

F) Special relationships:

$(\bar{A}\bar{B}\bar{C}\ldots\bar{F})^{-1} = \bar{F}^{-1}\ldots\bar{C}^{-1}\bar{B}^{-1}\bar{A}^{-1}$

If $\bar{A}^{-1} = A$, then $\bar{A}$ is called involutory.

If $\bar{B}^{-1} = \bar{B}^T$ then $\bar{B}$ is called orthogonal.

If $\bar{C}^{-1} = \bar{C}^{T*}$, then $\bar{C}$ is called unitary.

G) Cofactors and minors:

Minors: $\bar{B}$ is a $m \times m$ matrix, then minor M_{jk} is the determinant of $m-1 \times m-1$ matrix formed from $\bar{B}$ by eliminating the $j + n$ row and the kth column.

G-45

Cofactors:

The cofactors are given by $C_{jk} = (-1)^{j+k} M_{jk}$ where every element b_{jk} of $\bar{B}$ has a cofactor C_{jk}.

Determinants:

$\bar{B}$ is a m × n matrix, then $|\bar{B}| = \sum_{j=1}^{m} b_{ij} C_{ij}$ where i is any arbitrary row.

The Laplace expansion can be done with respect to any column j' to get $|\bar{B}| = \sum_{i=1} b_{ij'} c_{ij'}$

H) Integration and differentiation of matrices:

If $\bar{C}(t) = [c_{jk}(t)]$, then $\dfrac{d\vec{c}}{dt} = \left[\dfrac{d}{dt} c_{jk}(t)\right]$ and

$\int \bar{D}\, dt = \left[\int d_{ij}(t)dt\right]$

Useful role for differentiating a determinant:

$$\dfrac{\partial |\bar{B}|}{\partial b_{ij}} = \bar{C}_{ij}$$

PARTITIONED MATRICES

$\bar{C}\,\bar{D} = \bar{E}$ can be divided as follows:

$$\begin{bmatrix} \bar{C}_1 \\ \hline \bar{C}_2 \end{bmatrix} [D_1 \mid D_2] = \begin{bmatrix} \bar{C}_1\bar{D}_1 & \bar{C}_1\bar{D}_2 \\ \hline \bar{C}_2\bar{D}_1 & \bar{C}_2\bar{D}_2 \end{bmatrix} = \begin{bmatrix} \bar{F}_1 & \bar{F}_2 \\ \hline \bar{F}_3 & \bar{F}_4 \end{bmatrix}$$

$$\begin{bmatrix} \bar{C}_1 & \bar{C}_2 \\ \hline \bar{C}_3 & \bar{C}_4 \end{bmatrix} \begin{bmatrix} D_1 \\ \hline D_2 \end{bmatrix} = \begin{bmatrix} \bar{C}_1\bar{D}_1 + \bar{C}_2\bar{D}_2 \\ \hline \bar{C}_3\bar{D}_1 + \bar{C}_4\bar{D}_2 \end{bmatrix} = \begin{bmatrix} E_1 \\ \hline E_2 \end{bmatrix}$$

A position can be used to find the inverse of a matrix $\bar{C}$.

$$\bar{C}\,\bar{D} = \bar{I} \rightarrow \begin{bmatrix} \bar{C}_1 & \bar{C}_2 \\ \hline \bar{C}_3 & \bar{C}_4 \end{bmatrix} \begin{bmatrix} \bar{D}_1 & \bar{D}_2 \\ \hline \bar{D}_3 & \bar{D}_4 \end{bmatrix} = \begin{bmatrix} I & 0 \\ \hline 0 & I \end{bmatrix}$$

This position results in two simultaneous equations $\bar{C}_1\bar{D}_1 + \bar{C}_2\bar{D}_3 = I$ and $\bar{C}_3\bar{D}_1 + \bar{C}_4\bar{D}_3 = 0$ from which $\bar{D}_1$ and $\bar{D}_3$ can be obtained. Two other equations give $\bar{D}_2$ and $\bar{D}_4$ and ultimately result into $\bar{C}^{-1}$.

Diagonal, block diagonal and triangular matrices:

Diagonal matrix: If all the non-zero elements of a square matrix are on the main diagonal, then it is called a diagonal matrix.

Triangular matrix: A square matrix, all of whose elements below (or above) the main diagonal are equal to zero, is called an upper (or lower) triangular matrix.

4.2 Z-TRANSFORMS

This is especially useful for linear systems with sampled or discrete input signals.

Block diagram:

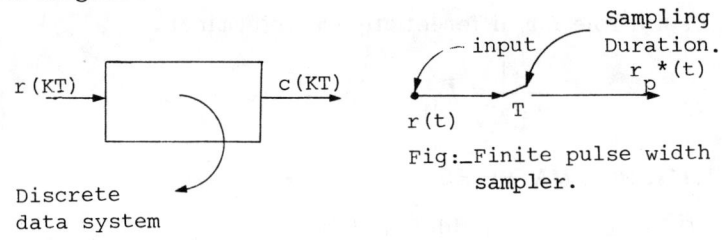

r(KT) → [ ] → c(KT)

Discrete data system

Fig: Finite pulse width sampler.

input, Sampling Duration. $r_p^*(t)$

The input for a discrete data system is given by

$$r^*(t) = \sum_{k=0}^{\infty} r(kT)\, \delta(t - kT)$$

where k varies from 0 to ∞ and these numbers are T seconds apart.

The output of the sampler is given by:

$$r_p^*(t) = r(t) \sum_{k=0}^{\infty} [u(t - kT) - u(t - kT - p)]$$

where u is the unit step function, T is the sampling period, and P is the sampling ration ($P \ll T$).

DEFINITION

The z-transform is defined as $z = e^{Ts}$, where s is the Laplace transform variable.

$$s = \frac{1}{T} \ln z$$

Table of z-transforms:

Table 4.1

Laplace Transform	Time Function	z-Transform
1	Unit impulse $\delta(t)$	1
$\frac{1}{s}$	Unit step $u(t)$	$\frac{z}{z-1}$
$\frac{1}{1-e^{-Ts}}$	$\delta_T(t) = \sum_{n=0}^{\infty} \delta(t-nT)$	$\frac{z}{z-1}$
$\frac{1}{s^2}$	t	$\frac{Tz}{(z-1)^2}$
$\frac{1}{s^3}$	$\frac{t^2}{2}$	$\frac{T^2 z(z+1)}{2(z-1)^3}$
$\frac{1}{s^{n+1}}$	$\frac{t^n}{n!}$	$\lim_{a \to 0} \frac{(-1)^n}{n!} \frac{\partial^n}{\partial a^n} \left(\frac{z}{z-e^{-aT}} \right)$
$\frac{1}{s+a}$	e^{-at}	$\frac{z}{z-e^{-aT}}$
$\frac{1}{(s+a)^2}$	te^{-at}	$\frac{Tze^{-aT}}{(z-e^{-aT})^2}$
$\frac{a}{s(s+a)}$	$1 - e^{-at}$	$\frac{(1-e^{-aT})z}{(z-1)(z-e^{-aT})}$
$\frac{\omega}{s^2+\omega^2}$	$\sin \omega t$	$\frac{z \sin \omega T}{z^2 - 2z\cos \omega T + 1}$
$\frac{\omega}{(s+a)^2+\omega^2}$	$e^{-at}\sin \omega t$	$\frac{ze^{-aT}\sin \omega T}{z^2 e^{2aT} - 2ze^{aT}\cos \omega T + 1}$
$\frac{s}{s^2+\omega^2}$	$\cos \omega t$	$\frac{z(z-\cos \omega T)}{z^2 - 2z\cos \omega T + 1}$
$\frac{s+a}{(s+a)^2+\omega^2}$	$e^{-at}\cos \omega t$	$\frac{z^2 - ze^{-aT}\cos \omega T}{z^2 - 2ze^{-aT}\cos \omega T + e^{-2aT}}$

INVERSE Z

Methods:

A) Partial-fraction expansion

G-48

B) Power-series

C) The inversion formula

POWER-SERIES METHOD

A) The z-transform is expanded into a power series in powers of z^{-1}.

B) The coefficient of z^{-k} is the value of $r(t)$ at $t = kT$.

INVERSION FORMULA

Inversion formula: $r(kT) = \dfrac{1}{2\pi j} \oint_L R(z) z^{k-1} dz$

where L is a circle of radius $|z| = t^{cT}$. The center of this is at the origin and c is such that all the poles of $R(z)$ are inside the circle.

THEOREMS

A) Addition and Subtraction

If $r_1(kT)$ and $r_2(kT)$ have z-transforms $R_1(z)$ and $R_2(z)$, respectively, then

$$\zeta[r_1(kT) \pm r_2(kT)] = R_1(z) \pm R_2(z)$$

B) Multiplication by a Constant

$$\zeta[ar(kT)] = a\zeta[r(kT)] = aR(z)$$

where a is a constant.

C) Real Translation

$$\zeta[r(kT - nT)] = z^{-n} R(z)$$

and

$$\zeta[r(kT + nT)] = z^n \left[R(z) - \sum_{k=0}^{n-1} r(kT) z^{-k} \right]$$

D) Complex Translation

$$\zeta[e^{\mp akT} r(kT)] = R(ze^{\pm aT})$$

G-49

E) Initial-Value Theorem

$$\lim_{k \to 0} r(kT) = \lim_{z \to \infty} R(z)$$

F) Final-value Theorem

$$\lim_{k \to \infty} r(kT) = \lim_{z \to 1} (1 - z^{-1})R(z)$$

if the function, $(1 - z^{-1})R(Z)$, has no poles on or outside the circle centered at the origin in the z-plane, $|z| = 1$.

CHAPTER 5

SYSTEM'S REPRESENTATION: BLOCK DIAGRAM, TRANSFER FUNCTIONS, AND SIGNAL FLOW GRAPHS

5.1 BLOCK DIAGRAM AND TRANSFER FUNCTION

Fundamentals:

A block diagram is a symbolic representation of:

A) the flow of information in a system, and

B) the functions performed by each component in the system.

The transfer function expresses the relationship between output and input of a system and is frequently given in terms of the Laplace variable s.

In finding the transfer function of a system, all initial conditions must be set to zero.

The transfer function may be expressed in terms of:

A) the operator D . . . . (D)

B) the Laplace-transform variable s . . . . C(j)

C) in phasor form . . . . C(jω)

This latter case is used in the sinusoidal steady-state analysis.

The denominator of the transfer function is the characteristic equation when it is set equal to zero.

Summation points in the block diagrams:

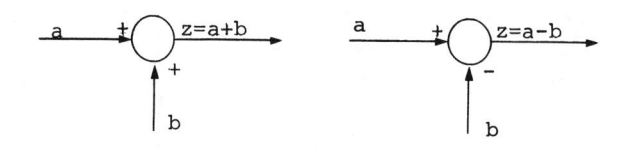

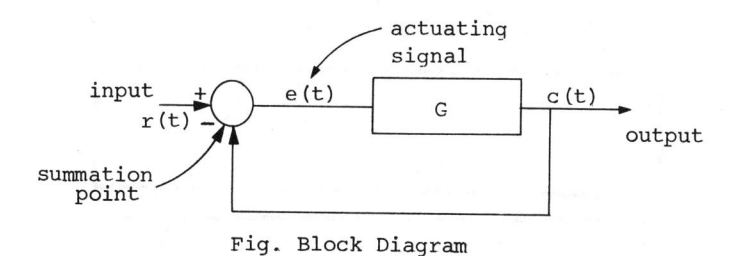

```
   a    +⊕  z=a+b         a    +⊕  z=a-b
         ↑+                     ↑-
         b                      b
```

```
                        actuating
                         signal
  input  +⊕   e(t)    ┌─────┐   c(t)
  r(t) -  ↑           │  G  │  ─────►
                      └─────┘        output
  summation ↑_____|
    point
```

Fig. Block Diagram

Transfer function:

$$G(D) = \text{Transfer function} = \frac{V_{out}}{V_{in}}$$

G(S) = Laplace-transform of the transfer function

$$= \frac{E_2(s)}{E_1(s)} = RCs/1 + RCs$$

$\overline{G}(j\omega)$ = Frequency transfer function = $\dfrac{\overline{E_2}(j\omega)}{\overline{E_1}(j\omega)}$

Blocks in Cascade:

If the operation of an element or a component can be described independently, then a block can be used to represent that component.

G-52

Combination of cascade blocks:

$$E_1(s) \rightarrow \boxed{G_1(s)} \xrightarrow{E_2(s)} \boxed{G_2(s)} \xrightarrow{E_3(s)} \equiv \xrightarrow{E_1(s)} \boxed{G_1(s)G_2(s)} \xrightarrow{E_3(s)}$$

Fig. Equivalent Block diagram after cascading.

Determination of control ratio:

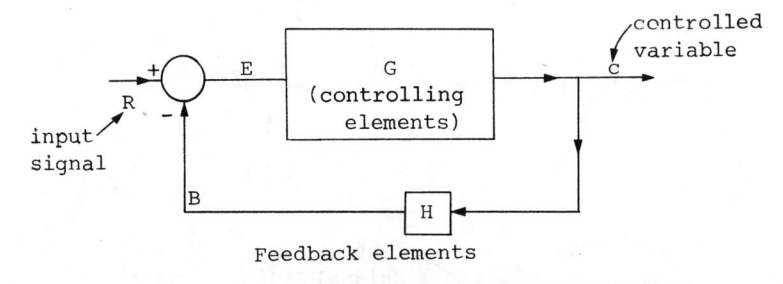

$$\boxed{\text{Control ratio} = \frac{C(s)}{R(s)} = \frac{G(s)}{1 + G(s)H(s)}}$$

$$= \text{Close-loop transfer function}$$

$$\boxed{\text{Characteristic equation: } 1 + G(s)H(s) = 0}$$

$$\boxed{\begin{array}{l}\text{Open-loop} \\ \text{transfer function}\end{array} = \frac{B(s)}{E(s)} = G(s)H(s)}$$

$$\boxed{\begin{array}{l}\text{Forward transfer} \\ \text{function}\end{array} = \frac{C(s)}{E(s)} = G(s)}$$

5.2 TRANSFER FUNCTIONS OF THE COMPENSATING NETWORKS

Lag compensator:
$$\frac{E_1(s)}{E_2(s)} = \frac{1 + (R_1 + R_2)Cs}{1 + R_2Cs}$$

[Circuit diagram: Lag compensator with R_1 in series, C and R_2 shunt]

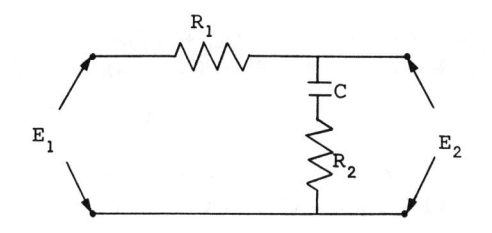

$$\frac{E_2(s)}{E_1(s)} = \frac{1 + R_2 C s}{1 + (R_1 + R_2)Cs} = G(s)$$

$$G(s) = \frac{1}{\alpha} \frac{s + 1/T}{s + 1/\alpha T}$$

$$\ldots \quad \alpha = \frac{R_1 + R_2}{R_2} \quad \text{and} \quad T = R_2 C$$

Lead compensator:

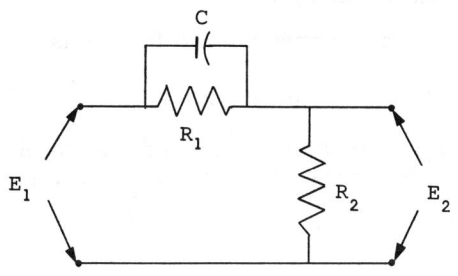

[Circuit diagram: R_1 parallel with C, in series; R_2 shunt]

$$G(s) = \frac{s + 1/T}{s + 1/\alpha T} \qquad \alpha = \frac{R_2}{R_1 + R_2}$$

$$\text{and} \quad T = R_1 C$$

Lag-lead compensator:

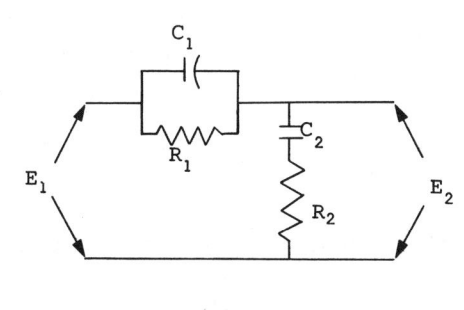

[Circuit diagram: R_1 parallel with C_1, in series; C_2 and R_2 shunt]

G-54

$$G(s) = \frac{E_2(s)}{E_1(s)} = \frac{1 + (T_1 + T_2)s + T_1 T_2 s^2}{1 + (T_1 + T_2 + T_{12})s + T_1 T_2 s^2}$$

$$\ldots T_1 = R_1 C_1, \quad T_2 = R_2 C_2 \quad T_{12} = R_1 C_2$$

$$= \frac{(1 + R_1 C_1 s)(1 + R_2 C_2 s)}{(1 + R_1 C_1 s)(1 + R_2 C_2 s) + R_1 C_2 s}$$

5.3 SIGNAL FLOW GRAPHS

This graph is the pictorial representation of a set of simultaneous equations. The nodes represent the system variable and a branch acts as a signal multiplier.

DEFINITION AND ALGEBRA

Node: A node performs two functions:

A) Addition of all the incoming signals.

B) Transmission or distribution of the total incoming signals of a node to all the outgoing branches.

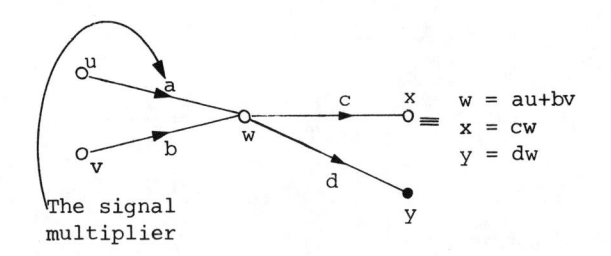

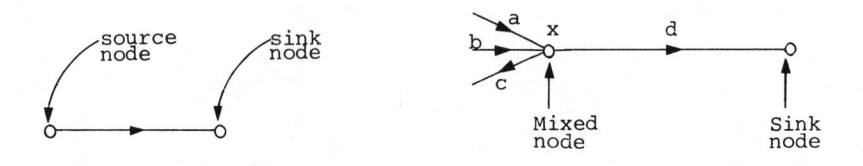

$w = au + bv$
$x = cw$
$y = dw$

The signal multiplier

source node   sink node

Mixed node   Sink node

These are the equations represented by the adjacent flow graph.

G-55

FLOW-GRAPH ALGEBRA
Rules

A) Series paths (cascade nodes): Series paths are combined into a single path by multiplying the transmittances of the individual paths.

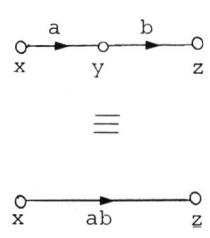

$$x \xrightarrow{a} y \xrightarrow{b} z \equiv x \xrightarrow{ab} z$$

B) Parallel paths: Parallel paths are combined by adding the transmittance.

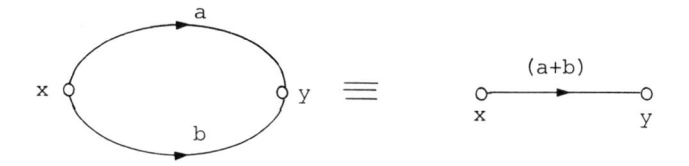

$$x \underset{b}{\overset{a}{\rightrightarrows}} y \equiv x \xrightarrow{(a+b)} y$$

C) Elimination of a node: A node representing a variable can be eliminated as follows:

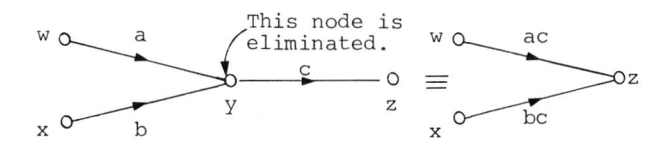

(This node is eliminated.)

$$\begin{matrix} w \xrightarrow{a} \\ x \xrightarrow{b} \end{matrix} y \xrightarrow{c} z \equiv \begin{matrix} w \xrightarrow{ac} \\ x \xrightarrow{bc} \end{matrix} z$$

MASON'S RULE gives the overall transmittance of a system and is defined as follows:

$$T = \text{overall transmittance} = \frac{\Sigma T \Delta}{\Delta} \text{ where}$$

A) Δ represents the determinant of the graph given by

$$\Delta = 1 - \Sigma L_1 + \Sigma L_2 - \Sigma L_3 + \ldots \text{ and}$$

a) L_1 is defined as the transmittance of each loop, thus ΣL_1 would be the sum of the individual transmittances of all loops.

b) ΣL_2 = sum of L_2 where L_2 is the product of the transmittances of 2 non-touching loops. Two loops are non-touching if they do not share a common node.

c) ΣL_3 = sum of all possible combinations of the product of transmittances of non-touching loops taken three at a time.

B) T denotes the transmittance of each forward path between a source and a sink node.

C) If we remove the path which has transmittance T, the determinant of the subgraph produced is denoted by Δ.

CHAPTER 6

SERVO CHARACTERISTICS: TIME-DOMAIN ANALYSIS

6.1 TIME-DOMAIN ANALYSIS USING TYPICAL TEST SIGNALS

Following signals are used for analysis:

A) Step function: This function is defined as follows.

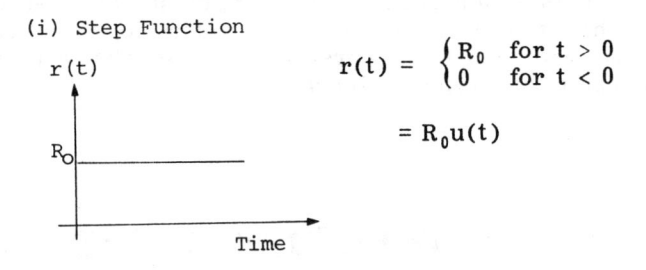

(i) Step Function

$$r(t) = \begin{cases} R_0 & \text{for } t > 0 \\ 0 & \text{for } t < 0 \end{cases}$$

$$= R_0 u(t)$$

where $u(t)$ is the unit step function.

B) Ramp function:

$$r(t) = R_0 t \text{ for } t \geq 0 \text{ and zero elsewhere}$$

$$= R_0 t u(t)$$

G-58

ii) Ramp Function

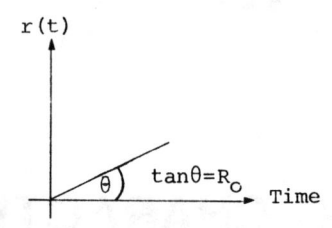

$r(t)$

$\tan\theta = R_o$

Time

C) Parabolic function:

(iii) Parabolic Function

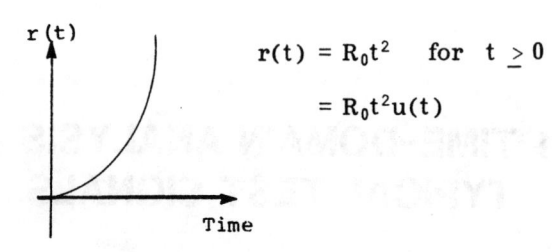

$r(t) = R_0 t^2$ for $t \geq 0$

$= R_0 t^2 u(t)$

Time

Any random input signal can be considered as being composed of these signals.

CONCLUSION

If any linear system is analyzed mathematically or experimentally for each of these input signals then it can be said that the response of this system to these basic test signals will be the representation of the actual system response to any random signal.

6.2 TYPES OF FEEDBACK SYSTEMS

The standard form of the transfer function is

$$G(s) = \frac{k_n(1 + a_1 s + a_2 s^2 + \ldots + a_m s^m)}{s^n(1 + b_1 s + b_2 s^2 + \ldots + b_p s^p)}$$

where $a_1, a_2, \ldots, b_1, b_2, \ldots$ are constant coefficients.

k_n = overall gain of transfer function G(s)

n = 0,1,2 denotes the type of the transfer function.

Characteristics of different types of systems:

A) Type 0: A constant error signal E(s) results in a constant value of the output signal C(s).

B) Type 1: A constant E(s) signal results in a constant rate of change of the output signal.

C) Type 2: A constant E(s) signal will produce a constant D^2C of the output variable.

Note: A type of the given system can be readily known from $G(j\omega)$ and log $G(j\omega)$ versus ω plots.

6.3 GENERAL APPROACH TO EVALUTION OF ERROR

IMPORTANT FACTS

A) Final-value theorem:

$$\lim_{t \to \infty} f(t) = \lim_{s \to 0} sF(s)$$

B) Differentiation theorem:

$$L[D^n C(t)] = s^n C(s)$$

when all the initial conditions are zero.

C) If the input signal to a unit feedback, stable system is a power series, then the steady-state output will have the same form as the input.

Steady-state error function: General Form

$$G(s) = \frac{C(s)}{E(s)} = \frac{k_n(1 + T_1 s)(1 + T_2 s)\dots}{s^n(1 + T_a s)(1 + T_b s)\dots}$$

$$E(s) = \frac{(1+T_a s)(1+T_b s)\ldots s^n C(s)}{s^n(1+T_a s)(1+T_b s)(1+T_c s)}$$

This is obvious after rearranging the top equation.

$$e(t)_{ss} = \lim_{s \to 0}[SE(s)] = \lim_{s \to 0}\left[\frac{s(1+T_a s)(1+T_b s)\ldots\ldots s^n C(s)}{k_n(1+T_1 s)(1+T_2 s)}\right]$$

$$= \lim_{s \to 0} \frac{s[s^n C(s)]}{K_n}$$

This is the steady-state error.

This can be written as follows:

$$\boxed{e(t)_{ss} = \frac{D^n C(t)_{ss}}{k_n}}$$

This equation is useful when $D^n C(t)_{ss}$ = constant. Then $e(t)_{ss}$ will be constant and is equal to E_o. So $k_n E_o = D^n C(t)_{ss}$ = constant = C_n.

$$E(s) = \frac{s^n(1+T_a s)(1+T_b s)\ldots R(s)}{s^n(1+T_a s)(1+T_b s) +\ldots + k_n(1+T_1 s)(1+T_2 s)}$$

When $H(s) = 1$, i.e. unity feedback.

$$\boxed{e(t)_{ss} = \lim_{s \to 0} s\left[\frac{s^n(1+T_a s)(1+T_b s)\ldots R(s)}{s^n(1+T_a s)(1+T_b s)+\ldots+k_n(1+T_1 s)(1+T_2 s)}\right]}$$

The general expression.

ERROR SERIES: CONCEPTS

This error series is useful when the input to a feedback system is an arbitrary function of time.

Mathematical representation:

$$E(s) = R(s)/1 + G(s)H(s)$$

$$e(t) = \int_{-\infty}^{t} A(\tau)r(t - \tau)d\tau$$

where $A(q)$ is the inverse Laplace transform of $1/(1 + G(s)H(s))$.

Error series:

A) $$e(t) = r(t)\int_{0}^{t} A(\tau)d\tau - r(t)\int_{0}^{t} \tau A(\tau)d\tau$$

$$+ r(t)\int_{0}^{t} \frac{\tau^2}{2!} A(\tau)d\tau$$

where $r(t - \tau) = r(t) - \tau r(t) + \frac{\tau^2}{2!} r(t)$ Taylor expansion

B) $$e_s(t) = C_0 r_s(t) + C_1 r_s(t) + \frac{C_2}{2!} r_s(t)$$

Where C_0, C_1 are called the error coefficients and $e_s(t)$ is known as the error series.

$$C_n = (-1)^n \int_{0}^{0} \tau^n A(\tau)d\tau$$

Evaluation of the error coefficients from $A(s)$:

A) $$C_0 = \lim_{s \to 0} A(s) = \lim_{s \to 0} \int_{0}^{\infty} A(s)e^{-\tau s}d\tau$$

B) $$C_1 = \lim_{s \to 0} \frac{d}{ds} A(s)$$

C) $$C_n = \lim_{s \to 0} \frac{d^n}{ds^n} A(s)$$

G-62

6.4 ANALYSIS OF SYSTEMS: UNITY FEEDBACK

TYPE 0 SYSTEM (n = 0)

Step input $(r(t) = R_0 u(t))$:

$$\boxed{\begin{aligned} e(t)_{ss} &= \frac{R_0}{1 + k_0} = \text{constant} \\ &= E_0 \end{aligned}}$$

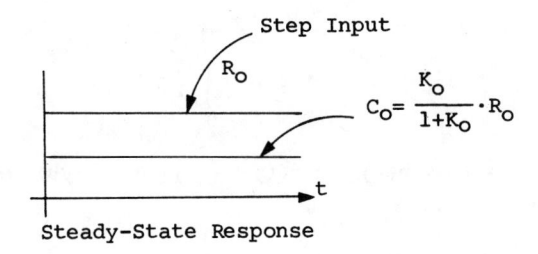

```
                    Step Input
         R₀  ↙
    ─────────────
                      ↖  C₀ = K₀/(1+K₀) · R₀
         ─────────────
                          → t
    Steady-State Response
```

This is obtained from the general expression for $e(t)_{ss}$. Since

$$R(s) = \frac{R_0}{s}$$

Ramp input:

$$\boxed{e(t)_{ss} = \infty}$$

since $r(t) = R_1 t\, u(t)$

$$R(s) = R_1/s^2$$

Parabolic input:

$$\boxed{\begin{aligned} e(t)_{ss} &= r(t)_{ss} - C(t)_{ss} \quad \text{which approaches a value of infinity} \\ &= \infty \quad\quad\quad\quad\quad\quad \text{since } r(t) = R_2 t^2 u(t) \end{aligned}}$$

G-63

Conclusions:

A) A constant input (i.e. step input) produces a constant value of the output with a constant error signal.

B) When a ramp-function input produces a ramp output with a smaller slope, there is an error which approaches a value of ∞ with increasing time.

C) Type 0 system cannot follow a parabolic input.

TYPE 1 SYSTEM (n = 1)

Step input:

$$\boxed{e(t)_{ss} = 0}$$

This is obtained from the general equation by putting n = 1 and

$$R(s) = \frac{R_0}{s}$$

Ramp input:

$$\boxed{e(t)_{ss} = \infty} \qquad \text{since } r(t) = R_2 t^2 u(t)$$
$$R(s) = 2R_2/s^3$$

Parabolic input:

$$\boxed{\begin{aligned} e(t)_{ss} &= \frac{R_1}{k_1} = \text{constant} \\ &= E_0 \neq 0 \end{aligned}} \qquad \text{since } R(s) = R_1/s^2$$

Conclusions:

A) Type 1 system with a constant input produces a

steady-state constant output of value equal to the input, i.e. zero steady-state error.

B) Type 1 system with a ramp input produces a ramp output with a constant error signal.

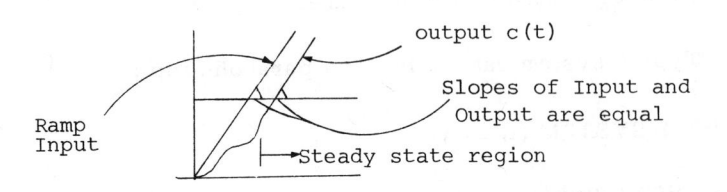

```
                                    output c(t)
                                    Slopes of Input and
                                    Output are equal
        Ramp
        Input
                     Steady state region
```

C) With a parabolic input, a parabolic output is produced with an error which increases with time $e(\infty) = \infty$.

TYPE 2 SYSTEM (n = 2)

Step input:

$$\boxed{e(t)_{ss} = 0}$$

This is obtained by putting $n = 2$ and $R(s) = R_0/s$.

Ramp input:

$$e(t)_{ss} = 0 \qquad \text{since } R(s) = R_1/s^2$$

Parabolic input:

$$e(t)_{ss} = \frac{2 \cdot R_2}{k_2} = \text{constant} \quad \text{since} \quad R(s) = \frac{2R_2}{s^3}$$

$$= E_0 \neq 0$$

Summary:

Table 6.1

Steady-state characteristics: Unity feedback systems

System type n	$r(t)_{ss}$ Steady-state (input)	(Steady-state error function)	(Steady-state output)	Value of $e(t)$ at $t = \infty$
0	Step	$\dfrac{R_0}{1+K_0}$	$\dfrac{K_0}{1+K_0} R_0$	$\dfrac{R_0}{1+K_0}$
	Ramp	$\dfrac{R_1}{1+K_0} t - C_0$	$\dfrac{K_0 R_1}{1+K_0} t + C_0$	∞
	Parabolic	$\dfrac{R_2}{1+K_0} t^2 - C_1 t - C_0$	$\dfrac{K_0 R_2}{1+K_0} t^2 + C_1 t + C_0$	∞
1	Step	0	R_0	0
	Ramp	$\dfrac{R_1}{K_1}$	$R_1 t - \dfrac{R_1}{K_1}$	$\dfrac{R_1}{K_1}$
	Parabolic	$-C_1 t - C_0$	$R_2 t^2 + C_1 t + C_0$	∞
2	Step	0	R_0	0
	Ramp	0	$R_1 t$	0
	Parabolic	$2\dfrac{R_2}{K_2}$	$R_2 t^2 - \dfrac{2 R_2}{K_2}$	$\dfrac{2 R_2}{K_2}$

CHAPTER 7

ROOT LOCUS

7.1 ROOTS OF CHARACTERISTIC EQUATIONS

Basic approach:

$$G(s) = \frac{C(s)}{E(s)} = \frac{k}{s(s+a)}$$

forward transfer function

STATIC LOOP SENSITIVITY

This is the value of k, when the transfer function is expressed in such form that the coefficients of s in both numerator and denominator are equal to unity.

$$\frac{C(s)}{R(s)} = \frac{k}{s^2 + 2s + k} = \frac{k}{s^2 + 2\zeta\omega_n s + \omega_n^2} \quad \text{for } a = 2$$

Roots of the characteristic equation:

$$s = -\zeta\omega_n \pm \omega_n \sqrt{\zeta^2 - 1}$$

where $\omega_n = \sqrt{k}$ and $\zeta = \frac{1}{\sqrt{k}}$

Plot of the roots of the characteristic equation:

G-67

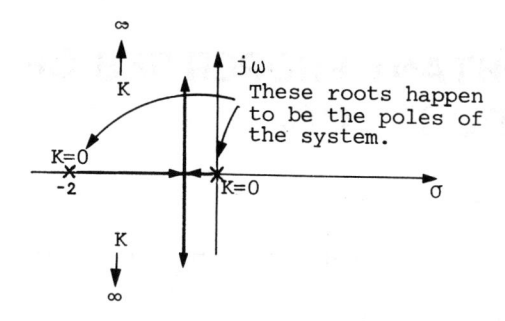

Roots are obtained for different values of k and these are plotted as shown.

A smooth curve is drawn (as shown by thick lines) through those points.

System performace v/s sensitivity k: salient points from the root locus.

An increase in the gain of the system results in:

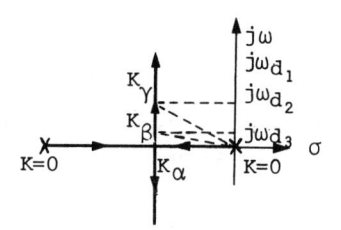

Fig. A sample plot.

A) a decrease in the damping ratio ζ, so the overshoot of the time response increases.

B) increase in ω_n (i.e., undamped natural frequency).

C) increase in ω_d (i.e., damped natural frequency)... ω_d is the imaginary component of the complex root.

D) The rate of decay, σ, is unchanged.

E) Root locus is a vertical line for $k \geq k_\alpha$ and

$$\zeta \omega_n = \sigma = \text{const.}$$

7.2 IMPORTANT PROPERTIES OF THE ROOT LOCI

EFFECT OF ADDITION OF POLES

When a pole is added to the function $G(s)H(s)$ in the left half of the s-plane, the net effect is of pushing the original locus towards the right half plane.

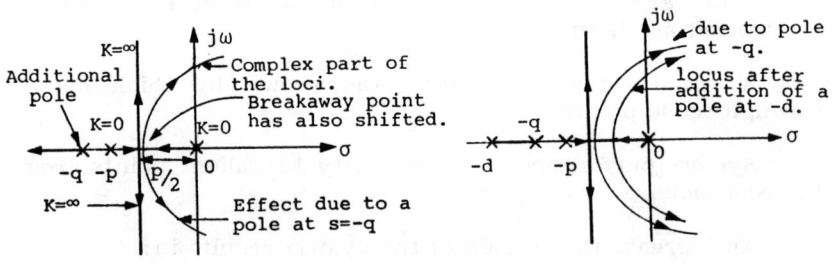

Fig. Original plot

Fig. Addition of pole at -d

ADDITION OF ZERO

Adding zeros has the effect of moving the root locus towards the left half of the s-plane.

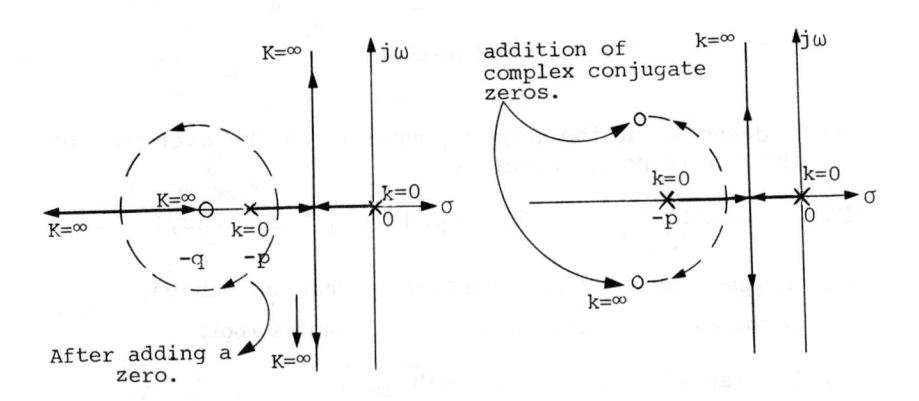

STEPS IN PLOTTING THE ROOT-LOCUS

A) The open loop transfer function $G(s)H(s)$ of the system is first determined.

G-69

B) The numerator and the denominator of the transfer function are then factorized into factors of the form (s + p).

C) Locate zeros and poles of the open-loop transfer function in the s-plane.

D) Determine the roots of the characteristic of the closed loop system, i.e. [1 = G(s)H(s) = 0]. The locus of the roots of the characteristic equation is then found from the plotted zeros and poles of the open loop function.

E) The locus is calibrated in terms of k. If the gain of the open-loop system is predetermined, the location of the exact roots of HG(s)H(s) is known. If the location of the roots is specified, k can be known.

F) Now that the roots are known, the system's response is the inverse Laplace transform.

G) If the specifications are not satisfied, then the shape that meets the desired specifications is determined and compensating networks are introduced in the system to meet these requirements.

Poles of the control ratio $\frac{C(s)}{R(s)}$:

$$\boxed{\frac{C(s)}{R(s)} = M(s) = \frac{N_1 D_2}{D_1 D_2 + N_1 N_2} = \frac{P(s)}{Q(s)}}$$

where $G(s) = \frac{N_1(s)}{D_1(s)}$, $H(s) = \frac{N_2(s)}{D_2(s)}$

and $B(s) = 1 + G(s)H(s) = D_1 D_2 + N_1 N_2 / D_1 D_2$

i.e. the zeros of B(s) are the poles of $\frac{C(s)}{R(s)}$ and these zeros determine the transient response.

Factors of Q(s) fall into the following categories:

Pole of C(s)	Corresponding inverse of the form
S	$u(t)$ - unit step
$S + \frac{1}{T}$	$e^{-t/T}$
$s^2 + 2\zeta\omega_n s + \omega_n^2$	$e^{-\zeta\omega_n t} \sin(\omega_n \sqrt{1-\zeta^2}\, t + \phi)$, $\zeta < 1$

G-70

P(s) only modifies the constant multiplier of these transients.

Conditions to plot the root-locus for k > 0:

$$|G(s)H(s)| = 1 \quad \text{Magnitude condition}$$

$$\underline{/G(s)H(s)} = (1+2m)180° \quad \text{for } m = 0, \pm 1, \pm 2 \ldots \text{for } k > 0.$$

Angle condition

Conditions for negative values of k:

$$|G(s)H(s)| = 1 \quad \text{Magnitude condition}$$

$$\underline{/G(s)H(s)} = (m)360° \ldots \text{for } m = 0, \pm 1, \pm 2 \ldots \text{for } k < 0$$

Angle condition

The magnitude and angle conditions:

APPLICATION

$$G(s)H(s) = \frac{k(s-z_1)}{s^n(s-p_1)(s-p_2)} \quad \text{General form of the open-loop transfer function.}$$

Note: $s - p_1, s - p_2, \ldots, s - z_1$ etc., are the complex numbers representing the directed line segments.

$$1 + G(s)H(s) = 0 \quad \text{The characteristic equation.}$$

Application of the magnitude and angle conditions results in

$$\frac{|k||s-z_1|\ldots|s-z_n|}{|s^n||s-p_1||s-p_2|\ldots} = 1$$

$$-\zeta = \underline{/s-z_1} \ldots + \underline{/s-z_w} - n\underline{/s} - \underline{/s-p_1}$$

$$= \begin{cases} (1+2m)180° & \ldots \text{if } k > 0 \\ (m)360° & \ldots \ldots \text{if } k < 0 \end{cases}$$

So modification of these equations results in:

A) $|k|$ = static loop sensitivity = $\dfrac{|s^n||s-p_1|\ldots|s-p_q|}{|s-z_1|\ldots|s-z_m|}$

G-71

B) $+\zeta = -$(angles of numerator terms) $+$ (angles of denominator terms)

$$= \begin{cases} -(1+2m)180° & \text{...if } k > 0 \\ -(m)360° & \text{.......if } k < 0 \end{cases}$$

Note: These two conditions are used in the graphical construction of the root locus.

Construction rules for plotting the root locus:

A) Total number of branches in the complete root loci:

The number of branches in a root loci is equal to the number of poles of the open-loop transfer function.

If the equation is expressed in the following form

$$s^q + b_1 \ldots s^{q-1} \ldots b_n + k(s^p + a_1 s^{p-1} \ldots) = 0$$

then the number of branches of the root loci is greater or equal to q and p.

B) Locus on the real axis:

If the total number of poles and zeros to the right of the search point on the real axis is odd, then this point is on the root locus.

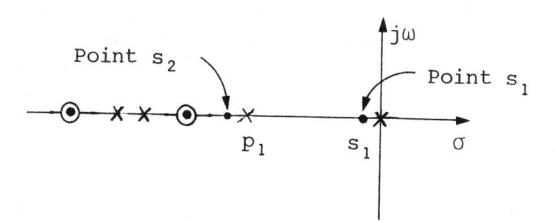

Note: Point s_1 lies on the real axis root locus but not s_2.

The following is the required angle condition so that the selected search point will be on the root locus.

$$(R_p - R_z)180° = (1+2m)180°$$

Where R_p = number of poles to the right of the search point on the real axis and R_z = number of zeros.

C) End points of the locus:

The starting points of the root-locus (i.e. for k = 0) are the open-loop poles and the end points (for k = ∞) are the open-loop zeros (∞ is an equivalent zero).

Asymptotes of the root loci (behavior as $s \to \infty$):

$$\theta = \frac{(1 + 2m)180°}{[\text{number of poles of } G(s)H(s)] - [\text{number of zeros of } G(s)H(s)]}$$

= This is the angle of straight lines or asymptotes.

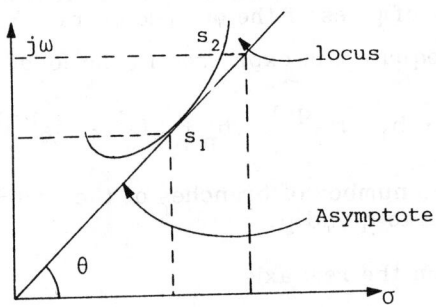

Fig. Asymptotic condition

The behavior of the root-locus near ∞ are important because when number of poles of $G(s)H(s) \neq$ number of zeros of $G(s)H(s)$, then 2|number of poles $G(s)H(s)$ - number of zeros of $G(s)H(s)$| will tend to infinity in the s-plane.

Intersection of the asymptotes (centroid):

$$\sigma_0 = \text{Real axis interception of the asymptotes} = \frac{\sum_{c=1}^{v} \text{Re}(P_c) - \sum_{m=1}^{w} \text{Re}(Z_m)}{v - w}$$

since the proportions of the locus away from the asymptotes and near the axes are important. σ_0 is also the centroid of the pole zero plot where

$$A(s) = \frac{\prod_{c=1}^{v}(s - p_c)}{\prod_{m=1}^{w}(s - z_m)} = -k$$

G-73

Note that there are 2 |number of poles of $G(s)H(s) = v - $ number of zeros ϕ of $G(s)H(s) = w$ |.

Asymptotes whose intersection lies on the real axis.

Also note that the point of intersection is always on the real axis.

Real axis break-away points (saddle points):

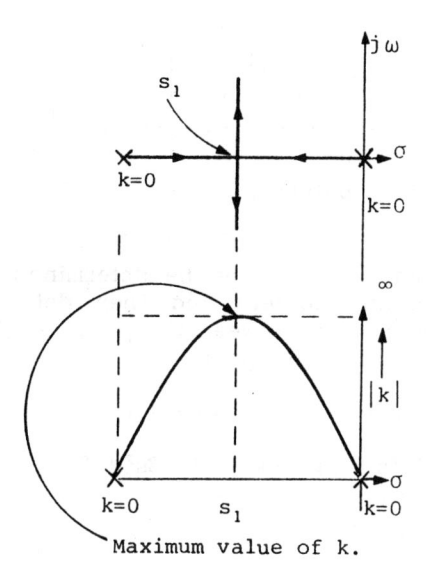

[Diagram showing root locus with poles marked k=0, break-away point s_1, and |k| axis. Caption: Maximum value of k.]

The break-away point: Since k starts with a value of zero at the poles and increases in value, as the locus moves away from the poles, there is a point somewhere in between where the k's for the two branches simultaneously reach a maximum value. This is the break-away point.

Break-away points on the root-loci of an equation correspond to multiple-order roots of the equation.

A root locus can have more than one break-away point and these need not always be on the real axis; however the break-away points may be real or complex conjugate pairs.

The break-in point: This is the value of σ for which $|k|$ is a minimum between 2 zeros. This is shown in the diagram.

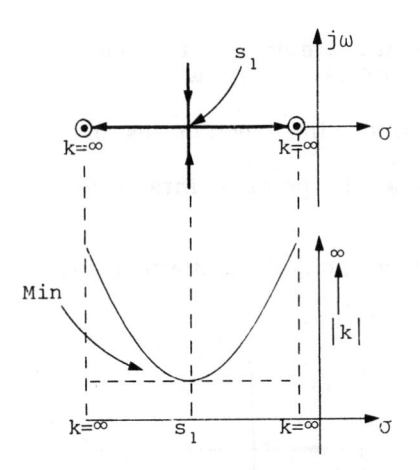

Break-in points

Break-away points: These can be determined by taking the derivative equating to zero and then determining its roots. The root that occurs between the poles (or the zeros) is the break-away (or break-in) point.

Example: $G(s)H(s) = k/s(s+1)(s+2)$

so $W(s) = s(s+1)(s+2) = -k = s^3 + 3s^2 + 2s$

$$\frac{dW(s)}{ds} = 3s^2 + 6s + 2 = 0$$

Hence, the roots are:

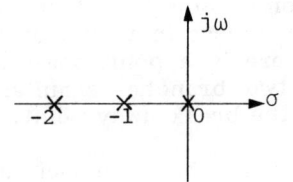

Fig. Pole-zero plot.

$s_{1,2} = -1 \pm 0.6$ i.e. -0.4 and -1.57

The break-away point for $k > 0$ lie between $s = 0$ and $s = -1$ in order to satisfy the angle condition, $s_1 = -0.4$. The other point, $s_2 = -1.57$ is the break-in point for $k < 0$.

$$-k = (-0.4)^3 + 3(-0.4)^2 + 2(-0.4)$$

so, $k = 0.38$

(This is the value of k at the break-away point for $k > 0$.)

The necessary angular condition near the break-away point: A break-away point is determined as follows:

Step 1: A check point is selected between two poles on the real axis at a distance d; then the angular contribution due to all poles and zeros which are on the real axis is $\frac{d}{L_i}$, where L_i is the distance between the check point and the pole or zero.

Step 2: The angular contribution towards this check point due to complex poles or zeros is

$$\Delta\phi = 2b\delta/b^2 + a^2 \quad \text{as shown}$$

Step 3: All the angular contributions with proper signs are summed to zero. The point which satisfies this zero condition is the break-away point.

Complex pole (or zero): Angle of departure

For $k > 0$, the angle of departure from a complex pole is equal to $180°(1 + 2m)$ minus the sum of the angles from the other poles plus the sum of the angles from the zeros. Any of these angles may be tve or -ve. For $k < 0$, the departure angle is $180°$ from that obtained for $k > 0$.

Imaginary-axis intercepting point:

The points where the complete root locus intersects the imaginary axis of the s-plane, and the corresponding values of k can be determined by means of the Routh-Hurwitz criterion:

Example: $s^3 + as^2 + bs + kd = 0$

The closed-loop characteristic equation.

A Routhian array is formed

s^3	1	b
s^2	a	
s^1	(ab-kd)a	kd
s^0	kd	

The Routhian array formed from the closed-loop characteristic equation.

Undamped oscillation occurs if the s^1 row is zero. Thus, the auxiliary equation obtained from the s^2 row is:

$$as^2 + kd = 0 \text{ and its roots } s_1 = +j\sqrt{\frac{kd}{a}} \text{ and } s_2 = -j\sqrt{\frac{kd}{a}}$$

i.e. $k = \frac{ab}{d}$

Root locus branches (non-intersection and intersection):

Properties

A) Any point on the root locus satisfies the angle condition. There are no root locus intersections at a point on the root locus if $\frac{dW(s)}{ds} \neq 0$ at this point. In this case there is only one branch of the root locus through the point.

B) A point on the root locus will have branches through it (i.e. it is an intersection point) if the derivatives of W(s) vanish at this point. Thus if the first y-1 derivatives of W(s) vanish at a given point on the root locus, there will be y branches approaching and y branches leaving this point. The angle between 2 approaching branches is

$$\lambda_y = \pm \frac{360°}{y}$$

while the angle between 2 branches (one leaving and the other approaching the same point) is

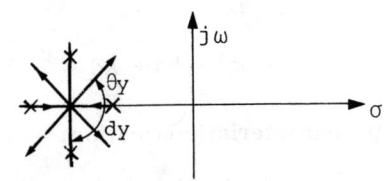

$$\phi_y = \pm \frac{180°}{y}$$

Fig. Root locus.

Conservation of the sum of the system of roots:

Grant's rule states that the sum of the closed-loop roots is equal to the sum of the open-loop poles. This is applicable when the open-loop transfer function is such that $v - z \geq W$.

Determination of roots on the root locus:

Application of Grant's rule

Grant's rule which is stated as follows, is used to find one real or two complex roots of the system provided the dominant roots of the characteristic equations is known.

$$\sum_{k=1}^{v} p_k = \sum_{l=1}^{v} r_l$$

7.3 FREQUENCY RESPONSE

DEFINITION

The frequency response gives the ratio of the phasor output to the phasor input for any inputs over a range of frequencies. The combined plots of the magnitude (M) and the angle (α) of $\frac{C(j\omega)}{R(j\omega)}$ versus the angular frequency is the frequency response of a control system. How to obtain the frequency response:

a) To determine the frequency response, the closed-loop control ratio should be known.

b) Then the control ratio is expressed as a function of frequency by substituting $s = j\omega$ as follows.

$$\frac{C(j\omega)}{R(j\omega)} = \frac{k(j\omega + a)}{(j\omega + b - cj)(j\omega + b + c)(j\omega + d - ej)(j\omega + d + ej)}$$

Please note that this is the most generalized form.

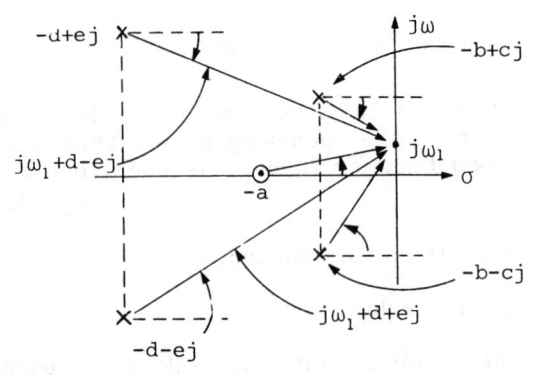

Fig. Frequency response from the plot.

c) For any frequency ω_1, a point ω_1 is chosen on the $j\omega$ axis, and the directed lines are drawn from all poles and zeros to this point. Then the lengths of the directed lines and angles these lines make with horizontal lines are determined as shown in the figure.

d) When the magnitudes and angles for each term (i.e. for each of the directed lines) of the $\frac{C(j\omega)}{R(j\omega)}$ equation are obtained, the value of $C(j\omega)/R(j\omega)$ is obtained for that particular frequency ω_1. Note that the clockwise angles are -ve and the anticlockwise angles are +ve.

e) This procedure is repeated for sufficient numbers of angular frequencies and a smooth curve is drawn as shown below.

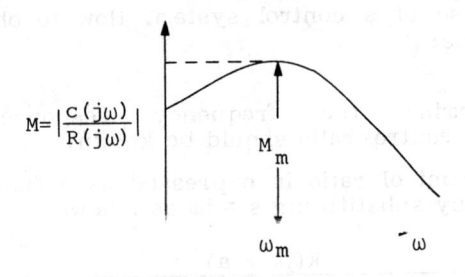

G-79

CHAPTER 8

SPECIAL POLE-ZERO TOPICS: DOMINANT POLES AND THE PARTITION METHOD

8.1 TRANSIENT RESPONSE: DOMINANT COMPLEX POLES

The conditions which are necessary for the time response to be nominated by only one pair of complex poles, require a pole-zero pattern with the following characteristics:

A) All other poles in the pole-zero diagram must be far to the left of the dominant poles, so the transient response due to these poles are small in amplitude.

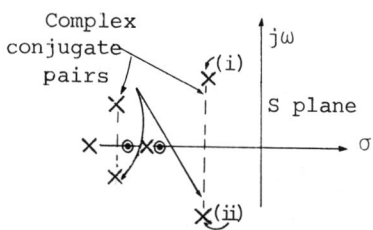

Pattern whose time response is dominated by complex poles.

G-80

B) In the pole-zero diagram, any pole which is not far away from the dominant complex poles must be near a zero, so that the magnitude of the transient response is small; since its effect will be modified by the zero. Note: The dominant poles are drawn darker.

The Time Response

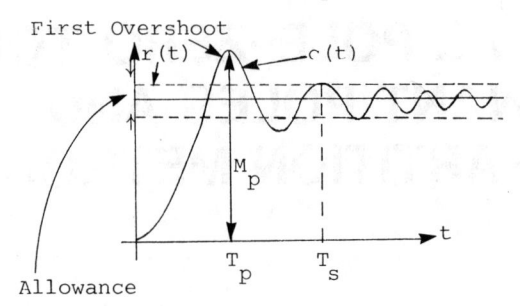

FIGURES OF MERIT

A) M_p (peak overshoot): The amplitude of the first overshoot.

B) t_p (peak time): The time to reach the peak overshoot from the initial starting time.

C) t_s (setting time): The time for the response to first reach and thereafter remain within the allowable limit (usually 2% of the final value).

D) n: The number of oscillations in the response up to the setting time. Not that there are two oscillations in a complete cycle.

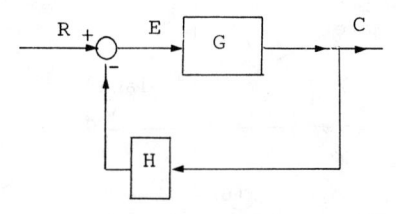

Non-Unity Feedback System

G-81

Figures of merit for the non-unity feedback system:

$$G(s)H(s) = \frac{k_G k_H \prod_{m=1}^{\omega} (s - z_m)}{\prod_{i=1}^{v} (s - p_i)}$$

where product $k_G k_H = k$ is the static loop sensitivity.

$$\frac{C(s)}{R(s)} = \frac{k_G \prod_{m=1}^{n} (s - z_n)}{\prod_{i=1}^{v} (s - p_i)}$$

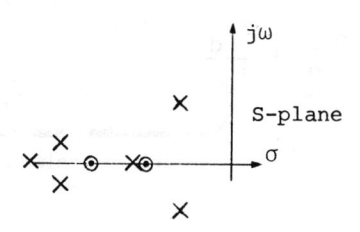

The desired Pole-Zero pattern

Here k_G is not k unless the system has unity feedback.

$$= \frac{A_0}{s} + \frac{A_1}{s - p_1} + \ldots$$

$$c(t) = \text{Time solution if the system has a dominant pole } p_1 = \sigma + j\omega_d = \frac{P(o)}{Q(o)} + 2 \left| \frac{k_G \prod_{m=1}^{n} (p_1 - z_m)}{p_1 \prod_{i=2}^{v} (p_1 - p_i)} \right|$$

$$e^{\sigma t} \cdot \cos[\omega_d t + \underline{/P(p_1)} - \underline{/p_1} - \underline{/Q'(p_1)}]$$

$$+ \sum_{k=3}^{v} \frac{P(p_k) e^{p_k t}}{p_k Q'(p_k)}$$

where
$$Q'(p_k) = \left.\frac{Q(s)}{s-p_k}\right]_{s=p_k}$$

$$T_p = \frac{1}{\omega_d}\left\{\frac{\pi}{2} - \begin{bmatrix}\text{sum of angles}\\ \text{from zeros of } \frac{C(s)}{R(s)}\\ \text{to } P_1,\text{ the dominant}\\ \text{pole}\end{bmatrix} + \begin{bmatrix}\text{sum of angles from}\\ \text{all other poles of } \frac{C(s)}{R(s)}\\ \text{to dominant pole } p_1\\ \text{(including conjugate}\\ \text{pole)}\end{bmatrix}\right\}$$

$$M_p = \text{The peak overshoot} = \underbrace{\frac{P(o)}{Q(o)}}_{\substack{\text{The}\\ \text{final}\\ \text{value}}} + \underbrace{\frac{2\omega_d}{\omega_n^2}\left|\frac{k_G \prod_{m=1}^{\omega^1}(p_1 - z_m)}{\prod_{i=2}^{v}(p_1 - p_i)}\right|e^{\sigma T_p}}_{\text{The overshoot } M_0}$$

$$\boxed{\begin{array}{l}t_s = \dfrac{m}{|\sigma|} = \dfrac{m}{\zeta\omega_n} = \begin{array}{l}\text{Four time constants}\\ \text{for 2\% error, i.e., } m = 4\end{array}\\[2ex] n = \dfrac{\text{Settling time}}{\text{Period}} = \dfrac{t_s}{2\pi/\omega_d} = \dfrac{2\omega_d}{\pi|\sigma|} = \dfrac{2}{\pi}\dfrac{\sqrt{1-\zeta^2}}{\zeta}\end{array}}$$

Additional Significant Poles:

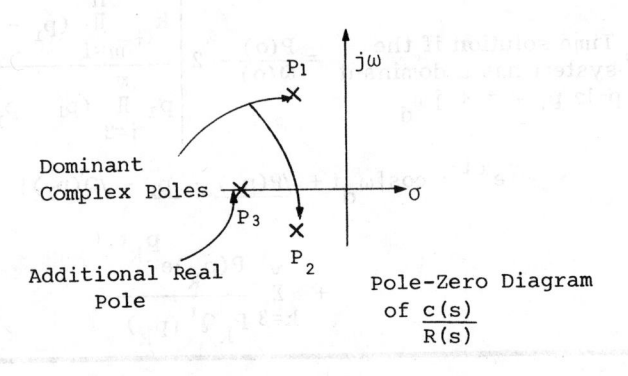

Pole-Zero Diagram of $\frac{c(s)}{R(s)}$

(Labels: P_1, P_2, P_3; Dominant Complex Poles; Additional Real Pole)

$$c(t) = 1 + 2|A_1|e^{-\zeta\omega_n t} \sin(\omega_n \sqrt{1-\zeta^2}\, t + \phi) + A_0 e^{P_0 t}$$

The time response to a unit step input for a system,

$$\frac{C(s)}{R(s)} = \frac{k}{(s + 2\zeta\omega_n s + \omega_n^2)/(s + P_0)}$$

The effects of an additional real pole can be given as follows:

Time Responses As A Function of Real-Pole Location

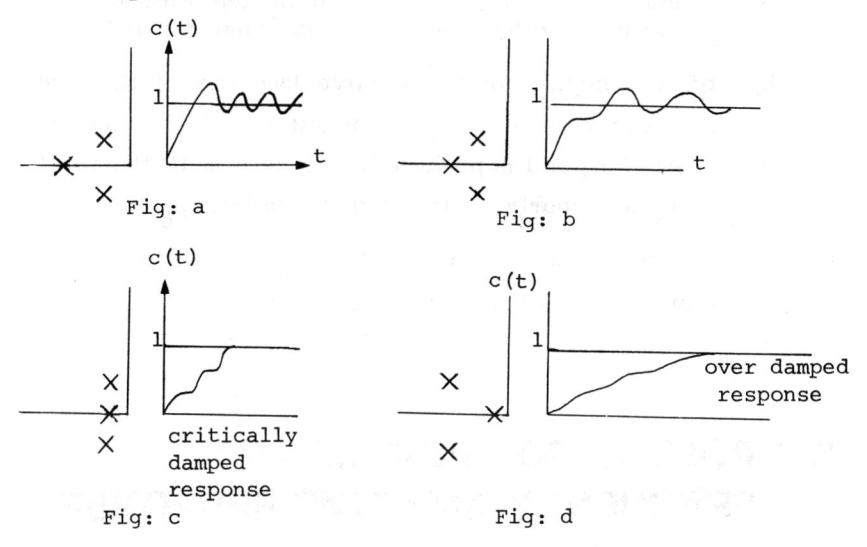

Fig: a

Fig: b

Fig: c critically damped response

Fig: d over damped response

A) Peak overshoot M_p is reduced.

B) The setting time t_s is increased because $A_0 e^{P_0 t}$ is the transient term due to P_0 and A_0 is negative.

C) $|A_0|$ depends on the relative location of P_0 with respect to dominant pole of the distance between P_0 and complex poles is large, then A_0 is small.

Effects of additional real pole and zero:

G-84

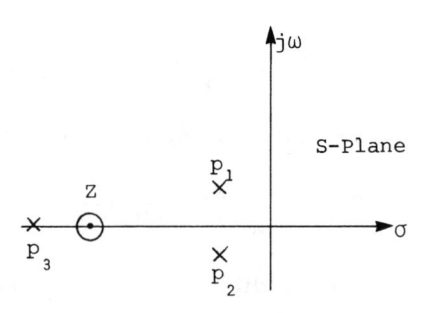

Fig. Pole-zero plot.

A) The complete time response to a unit-step function has the same form as in the case of an additional real pole.

B) Sign of A_o depends on the relative locations of the real pole and the real zero. A_o is negative if the zero is to the left of p_o and is positive if the zero is to the right of p_o. A_o is proportional to the distance from p_o to z.

C) If the zero is close to the pole, A_o is small, then the contribution of this transient term is small.

8.2 POLE-ZERO DIAGRAM AND FREQUENCY AND TIME RESPONSE

CHARACTERISTICS

A) Fig. a:
    a) Frequency response curve has a single peak M_m and $1.0 < M < M_m$ in the frequency range $0 < \omega < 1.0$.
    b) Time response: The first maxima of c(t) due to the oscillatory term is greater than $c(t)_{ss}$ and the c(t) response after this maxima oscillates around the value $c(t)_{ss}$.

Relation: An illustration.

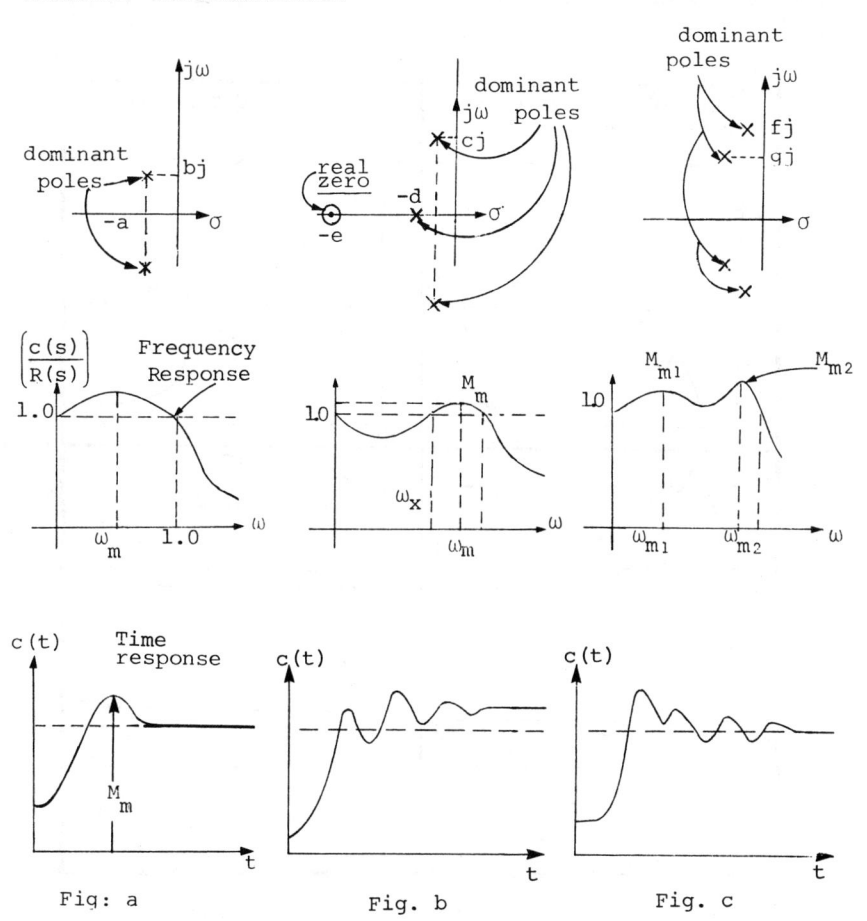

Fig:-Comparison of frequency and time responses

B) Fig. b: Time response: The first maximum of c(t) due to oscillatory term is less than $c(t)_{ss}$.

C) Fig. c: Time response: The first maxima of c(t) in the oscillation is greater than $c(t)_{ss}$, and the oscillatory portion of c(t) does not oscillate about a value of $c(t)_{ss}$.

The system's time-response can be predicted to a

greater extend from the shape of the frequency-response plot.

Table 8.1

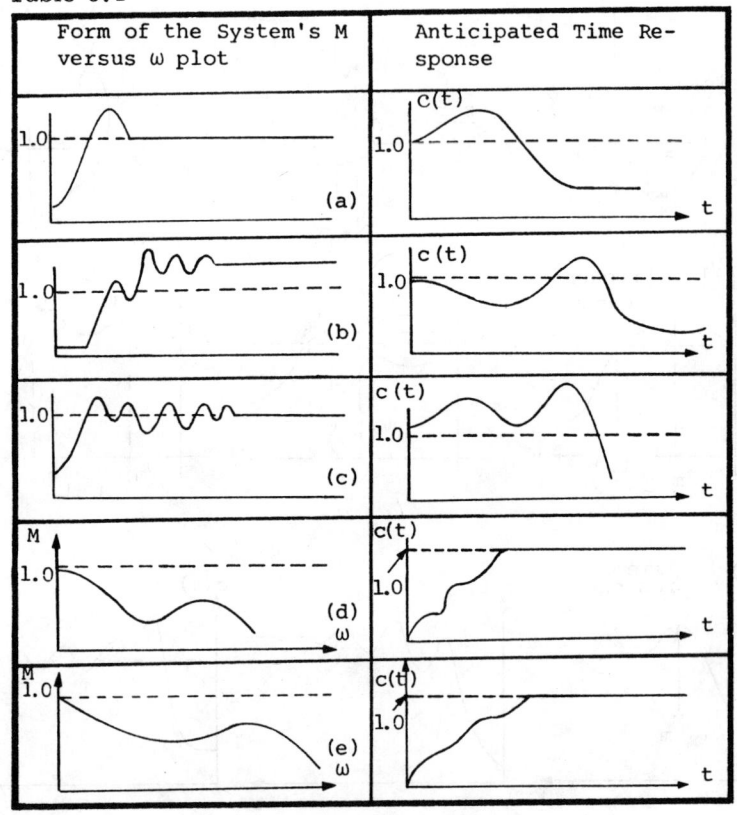

Form of the System's M versus ω plot	Anticipated Time Response
(a)	(a)
(b)	(b)
(c)	(c)
(d)	(d)
(e)	(e)

Correlation between Frequency and Time Response

8.3 FACTORING OF POLYNOMIALS USING ROOT-LOCUS

Partition method: The application of the root-locus method for factoring polynomials is called the partition method.

An example:

$$s^3 + es^2 + fs + g = 0$$

where e, f and g are constants.

A) This can be partitioned at s^3, s^2 and s.

B) Partition at s^3:

$$s^3 = -(es^2 + fs + g)$$

$$-1 = \frac{e(s^2 + \frac{f}{e}s + \frac{g}{e})}{s^3} = I_1(s)$$

$$= \frac{e(s + \beta)(s + \gamma)}{s^3}$$

C) Partition at s^2:

$$s^3 + es^2 = -(fs + g)$$

$$-1 = \frac{f(s + \frac{g}{f})}{s^2(s + e)} = I_2(s)$$

D) $s^3 + es^2 + fs = -g$

$$-1 = \frac{g}{s(s^2 + es + f)} = \frac{g}{s(s + \alpha)(s + \sigma)} = I_3(s)$$

E) Each resulting equation after partitioning has a form I(s) = -1. It should satisfy the angle and magnitude conditions. Since the resulting equation looks like G(s)H(s) = -1, the root-locus method can be utilized to determine the roots of a polynomial.

F) For a third-degree polynomial, partition at s^2 is preferred.

CHAPTER 9

SYSTEM ANALYSIS IN THE FREQUENCY-DOMAIN: BODE AND POLAR PLOTS

9.1 THE FREQUENCY RESPONSE AND THE TIME RESPONSE: RELATIONSHIP

If the frequency response of a system is known, its time-response can be known using the fourier integral.

FOURIER INTEGRAL

Definition

If f(t) is any time-variant input signal, its Fourier integral is given by

$$\overline{F}(j\omega) = \int_{-\infty}^{\infty} f(t)e^{-j\omega t} dt.$$

For a control-system shown below, the frequency response is given by taking Fourier integral of C(t) where

$$\overline{C}(j\omega) = \frac{\overline{G}(j\omega)}{1 + \overline{G}(j\omega)\overline{H}(j\omega)} \overline{R}(j\omega)$$

where R(jω) is the Fourier integral of the input signal (t).

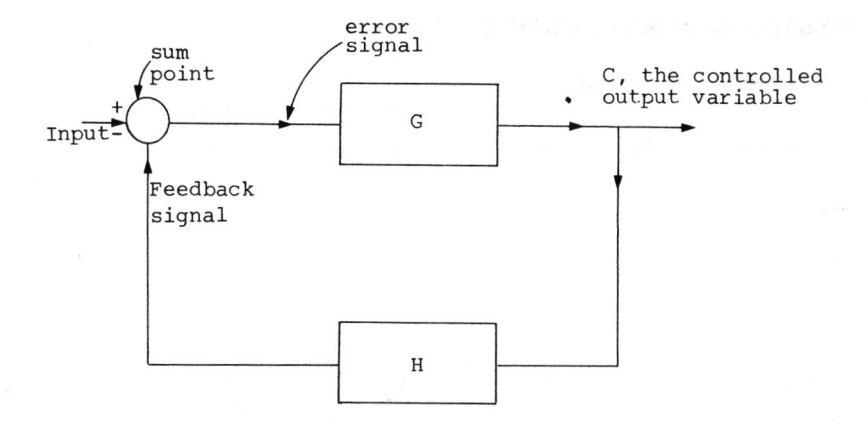

```
                    error
      sum          /signal
     /point       /
      +  ___     /    _____                 C, the controlled
Input─→(   )─────┼───→|  G   |─────┬────→ • output variable
         ‾‾‾     │    |_____|     │
                 Feedback           │
                 signal             │
                                    │
                   _____           │
             ┌────|  H   |←─────────┘
             │    |_____|
```

So now taking the inverse Fourier integral of $\overline{C}(j\omega)$ results in

$$c(t) = \text{Time-response of the system} = \int_{-\infty}^{\infty} \frac{\overline{C}(j\omega)}{2\pi} e^{j\omega t} d\omega.$$

9.2 FREQUENCY RESPONSE PLOTS

For a non-unity feedback control system, the control ratio is given by

$$\frac{\overline{C}(j\omega)}{\overline{R}(j\omega)} = \frac{\overline{G}(j\omega)}{1 + \overline{G}(j\omega)\overline{H}(j\omega)}.$$

For each value of $\omega = \omega_1$ these are two important quantities which are frequently used in the frequency response plot; these are

A) $\left|\dfrac{\overline{C}(j\omega_1)}{\overline{R}(j\omega_1)}\right|$ ... at $\omega = \omega_1$.

B) α = The angle between $\overline{C}(j\omega)$ and $\overline{R}(j\omega)$, where $\overline{C}(j\omega)$ and $\overline{R}(j\omega)$ are phasors.

SYSTEM CHARACTERISTICS

A) Ideal characteristics: $\alpha = 0$
   $R = C$ i.e. input = output.
   Note: α is the angle between phasors R and C.

B) Actual characteristics:

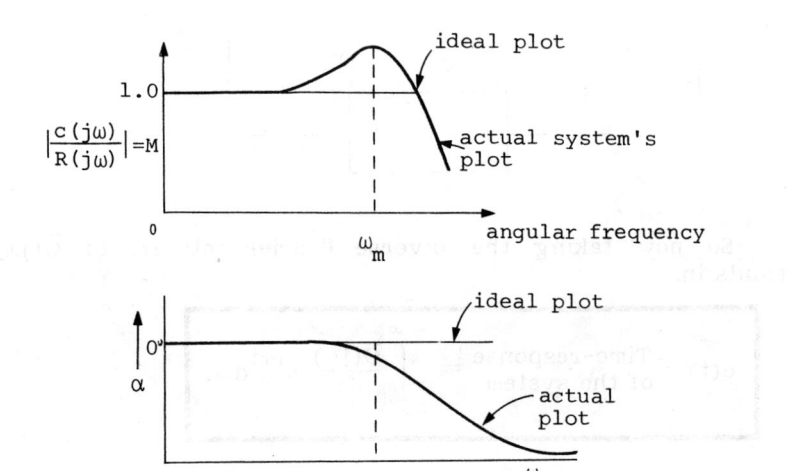

[Plot of $\left|\frac{C(j\omega)}{R(j\omega)}\right| = M$ versus angular frequency, showing value 1.0 at low frequency, a peak near ω_m (actual system's plot), and the ideal plot as a horizontal line dropping off. The actual plot differs from the ideal plot.]

[Plot of α versus ω, with ideal plot at $0°$ and actual plot decreasing past ω_m.]

Note: The actual plot is different from the ideal plot due to presence of energy-storage devices and energy dissipation.

C) The bandwidth: It is the frequency range from 0 to ω_{band}, where

$$\left|\frac{C(j\omega)}{R(j\omega)}\right| = 0.707 \text{ times the value at } \omega = 0.$$

For $\omega = \omega_{band}$,

$$\left|\frac{C(j\omega)}{R(j\omega)}\right| = 0.707 \left|\frac{C(o)}{R(o)}\right|.$$

D) $M \rightarrow 1$ and $\alpha \rightarrow 0°$ as ω approaches 0.

E) As $\omega \rightarrow \infty$ (or very large angular frequencies), $M \rightarrow 0$ and $\alpha \rightarrow -k\pi/2$ radians where $-k$ is equal to the degree of M.

G-91

9.3 POLAR PLOT

A) This method is used to represent the open-loop steady-state sinusoidal response. Using this polar plot, the stability and frequency response of the closed-loop system can be estimated.

B) The forward transfer function $= \overline{G}(j\omega) = \dfrac{\overline{C}(j\omega)}{\overline{E}(j\omega)}$.

Note that the bar represents the phasor.

For $\omega = \omega_1$, the phasor $G(j\omega)$ has the magnitude

$$\left| \dfrac{C(j\omega_1)}{E(j\omega_1)} \right|$$

and the angle $\theta(\omega_1)_j$. In this way many phasors are plotted for different values of the angular frequency and then a smooth curve is drawn passing through the tips of these phasors.

C) A sample polar plot:

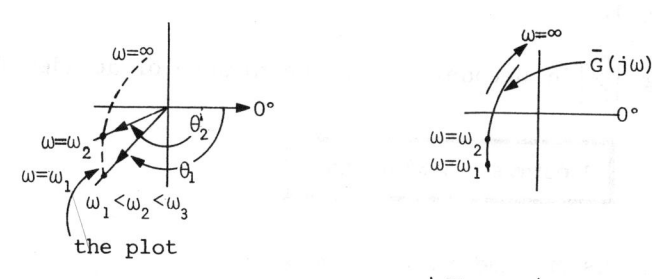

the plot

Note: The length of the arrow is $\left| \dfrac{\overline{C}(j\omega_1)}{\overline{E}(j\omega_1)} \right|$ for $\omega = \omega_1$.

9.4 LOGARITHMIC PLOTS

DEFINITIONS OF VARIOUS TERMS

Decibel: The log (always to the base 10) of the magnitude of a transfer function $\overline{G}(j\omega)$ is expressed in decibels as follows:

$20 \log |\overline{G}(j\omega)|$ (db)

$\ln |\overline{G}(j\omega)| e^{j\phi(\omega)} = \ln |\overline{G}(j\omega)| + j\phi(\omega)$

$\log |\overline{G}(j\omega)| e^{j\phi(\omega)} = \log |\overline{G}(j\omega)| + j 0.434 \phi(\omega)$

(since $\log_{10}^{e} = 0.434$.).

OCTAVE AND DECADE

These are the units used to express the frequency bands or the frequency ratios.

An octave is not a fixed frequency band but it can be defined as a band of frequencies with lower limit defined as f' and upper limit defined as f" such that f" = 2f'. Mathematically it is

$$\frac{\log(f''/f')}{\log 2} = 3.32 \log \frac{f''}{f'} \text{ octaves}$$

(i.e. any arbitrarily frequency can be substituted in the above formula to determine the length of the frequency band in octaves).

$\text{Log} \left(\frac{f''}{f'}\right)$ in decades, gives the number of decades from f' to f".

So $\boxed{1 \text{ octave} = 3.32 \text{ decade.}}$

Decibels of some common numbers:

Number	Decibels
0.1	-20
0.5	-6
1.0	0
2.0	6
10.0	20
100.0	40

The following are deductions drawn from the characteristics of logarithms:

A) When a number is doubled, its decimal value increases by 6 db.

B) When a number increases by a factor of 10, its decibal value increases by 20.

9.5 LOG-MAGNITUDE AND PHASE DIAGRAM: BASIC APPROACH

The transfer function is usually written in the following generalized form:

$$\overline{G}(j\omega) = \frac{K(1+j\omega T')(1+j\omega T'')^n}{(\omega j)^n \left[1 + j\omega T_1)(1 + j\frac{2\zeta}{\omega_n}\omega + \frac{-\omega^2}{\omega_n^2}\right]} = KG(j\omega)$$

where k represents the gain of the system.

The basic approach in drawing curves of log magnitude and angle versus log frequency is the following: curves are drawn for each factor and a complete curve is obtained with the help of asymptotic approximations.

Curves for the basic factors found in the numerator and denominator of the transfer function.

A) Constants:

The curve of $Lmk = 20\log k$ (db) is a horizontal straight line.

The angle is 0 if k is positive; it is 180° if the value of k is negative.

This constant has the effect of raising or lowering the log-magnitude curve of the transfer function along the vertical axis.

B) $j\omega$:

The curve of $Lm \frac{1}{j\omega} = 20\log \left|\frac{1}{j\omega}\right| = -20\log\omega$ is a straight line with a slope of -20db/decade or -6db/octave.

The value of the angle is (always) $-90°$.

When $j\omega$ appears in the numerator, the curve is a straight line with a slope of $+6db$/octave or $+20db$/decade.

Note: a) Both curves pass through the 0 db point at $\omega = 1$.

b) The curves for $j\omega$ and $1/j\omega$ differs very little; there is a change in the sign of the slope of the log-magnitude curve and a change in the sign of the angle of the phase curve.

For $(j\omega)^m$ term, the curve is a straight line of slope $\pm 6m$ (depending on whether it is in the numerator or the denominator) and angle is $\pm n 90°$.

C) $1 + j\omega T$:

$$Lm(1 + j\omega T) = 20\log |1 + j\omega T|$$
$$= 20\log \sqrt{1 + \omega^2 T^2}.$$

Since magnitude of $1 + j\omega T$ is $1 + \omega^2 T^2$.

Plot of $1 + j\omega T$:

A) When the frequency of the operation is very large, i.e. $\omega \gg 1/T$,

$$Lm(1 + j\omega T) = 20\log \omega T$$

where f is the frequency.

B) At very small values of frequencies, i.e. $\omega \ll 1/T$,

$$Lm(1 + j\omega T) = 0$$

because $\log 1 = 0$

Plot of $1/(1 + j\omega T)$;

The plot for low frequencies is a 0 db horizontal line and for very high frequencies the plot is a straight line with a slope of -6db per octave. This plot is shown below.

G-95

ASYMPTOTES For the log-magnitude plot shown below, i.e. Lm[1/1+jωT], there are two asymptotes.

A) One is of slope = 0 below $\omega = 2\pi f = \frac{1}{T}$.

B) The other of slope -6db/octave above $\omega = \frac{1}{T}$.

Corner frequency: It is the point of intersection of the asymptotes of the log magnitude plot.

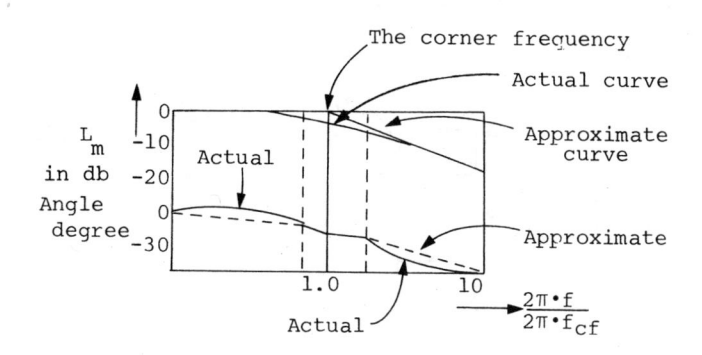

For $\frac{1}{1+j\omega T}$ function the angular corner frequency is

$$\omega_g = \frac{1}{T}$$

as shown in the figure.

C) $\omega = \omega_n$ is called the corner frequency; it is the point where the asymptotes intersect. At this frequency resonance occurs. ω_n is also called the undamped natural frequency.

D) The phase curve is also a function of ζ. The following are noted:
   a) the angle is 0° at zero frequency;
   b) the angle is -90° at the corner frequency;
   c) the angle is -180° at infinite frequency.

9.6 RELATION BETWEEN SYSTEM TYPE, GAIN AND LOG-MAGNITUDE CURVES

TYPE 0 SYSTEM

A type 0 system is characterized by the transfer function

Log-magnitude plot of the transfer function

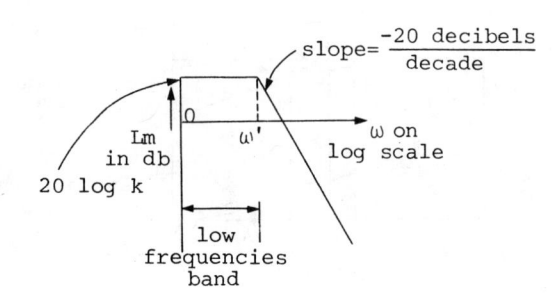

$$G = \frac{K}{1 + ST}$$

i.e. $G(j\omega) = \dfrac{K}{1 + j\omega T}$ since $s = j\omega$.

Facts:

A) For $2\pi f = \omega < \dfrac{1}{T}$, $Lm G(j\omega) = 20 \log k = $ constant.

B) Corner frequency $= \omega^1 = \dfrac{1}{T}$.

C) K = static error coefficient for the step input

TYPE 1 SYSTEM

This system is characterized by $G = \dfrac{k}{s(1 + sT)}$, i.e.

$$G(j\omega) = \frac{k}{j\omega(1 + j\omega T)}.$$

Log-magnitude plot:

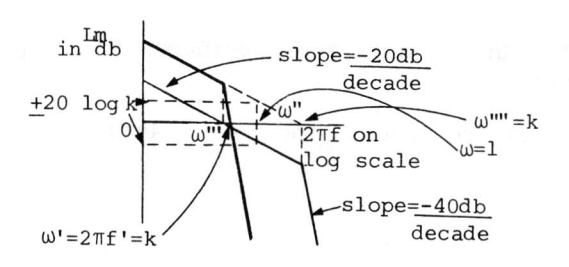

```
           Lm
        in db
                       ╱─ slope=-20db
                       ╱     decade
    ±20 log k┌ ─ ─ ─ ╱┐─ ─╱ω'''
           0 │      ╱ │ ╱          ω''''=k
             │ω''''╱──┴╱ 2πf on     ω=1
             │    ╱    ╲ log scale
             │   ╱      ╲
             │  ╱        ╲─ slope=-40db
   ω'=2πf'=k              ╲    decade
```

Fig. — curve for k>corner frequency
     — curve for k<corner frequency.

Facts:

A) For the low frequencies slope = -20 db/decade.

B) The gain k = static error coefficient for the ramp input.

Type 2 system:

This type of system is characterized by

$$G(s) = \frac{k}{s^2(1+sT)}.$$

The log-magnitude plot of $G(j\omega)$:

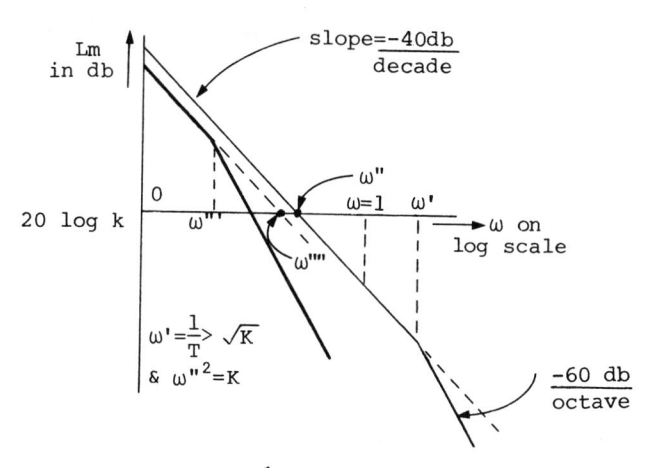

```
       Lm
     in db            ╱─ slope=-40db
                     ╱      decade
                    ╱
                   ╱          ω'''
              0   ╱       ω=1   ω'
  20 log k ───ω''''──┬─────┼────→ ω on
                 ╲ω''''    │       log scale
                   ╲       │
                    ╲      │
                     ╲     │
           ω'=1/T >√K╲     │
            & ω''²=K  ╲    │  -60 db
                       ╲   │  ─────
                        ╲  │  octave
```

Fig: - for $\omega''' = \frac{1}{T} < \sqrt{k}$ and $\omega''''^2 = k$

      - for $\omega' = \frac{1}{T} > \sqrt{k}$ and $\omega''^2 = k$

G-98

A) The gain k = static error coefficient for the parabolic input.

B) For low frequencies, the slope = -40 db/decade.

9.7 DIRECT POLAR PLOTS

DIRECT POLAR PLOTS: ILLUSTRATIVE EXAMPLES

Lag compensator:

A lag compensator is a simple series R-L circuit.

Its transfer function is:

$$G = \frac{R}{R + SL} \quad \text{i.e.} \quad G(j\omega) = \frac{R}{R + j\omega L}$$

$$G(j\omega) = \frac{1}{1 + j\omega \frac{L}{R}} = \frac{1}{1 + j\omega T}$$

where $T = \frac{L}{R}$ = time constant of the circuit.

Polar plot:

$$G(j\omega) = \frac{1}{1 + j\omega T}$$

For $\omega = 0$ radian,

$$G(j\omega) = \left| \frac{1}{1 + 0} \right| \underline{/\text{-}\tan^{-1} T\omega}$$

$$= 1 \; \underline{/0°} \quad \text{because } \tan^{-1} 0 = 0°$$

For $\omega = \infty$,

$$G(j\omega) = 0 \; \underline{/\text{-}90°}$$

The plot is shown below:

Magnitude of the transfer function

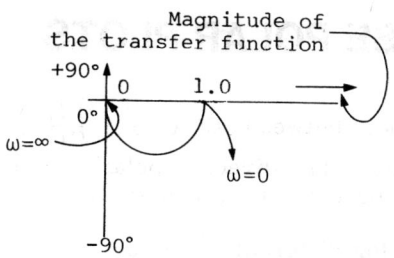

(Polar plot figure with axes at +90°, 0°, -90°, points ω=∞, ω=0, 1.0)

Note: Polar plot is a semicircle of diameter 1.

Lag-lead network: This is an R-C network.

$$G(s) = \frac{V_o(s)}{V_i(s)} = \frac{1 + (T_1 + T_2)s + T_1 T_2 s^2}{1 + (T_1 + T_2 + T_{12})s + T_1 T_2 s^2}$$

where $T_1 = R_2 C_2$, $T_2 = R_1 C_1$ and $T_{12} = R_2 C_1$.

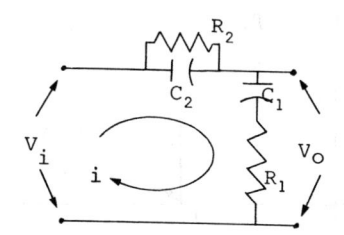

Fig. The lag-lead network.

The circuit acts as a lag network in the lower frequency range $0 - \omega_1$ and as a lead network when ω is beyond ω_1.

The polar plot:

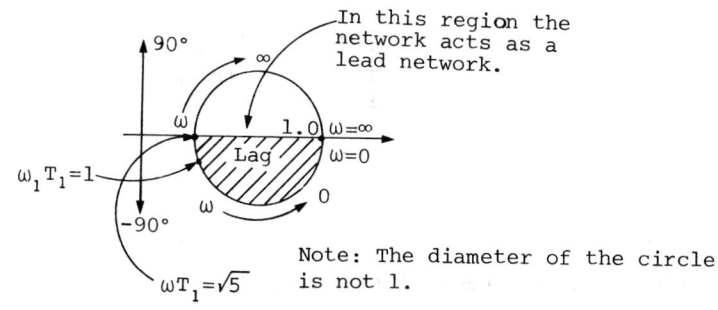

In this region the network acts as a lead network.

$\omega_1 T_1 = 1$

$\omega T_1 = \sqrt{5}$

Note: The diameter of the circle is not 1.

G-100

9.8 INVERSE POLAR PLOTS

It is a plot between $G^{-1}(j\omega) = \dfrac{\overline{E}(j\omega)}{C(j\omega)}$ and the angular frequency. The direct polar plots have some drawbacks when used for feedback systems.

Polar plots for different systems:

For a lag network:

The transfer function for the lag network at $\omega = 0$, $\overline{G}(j\omega) = 1$.

$$G(j\omega) = \dfrac{1}{1 + j\omega T}$$

hence,

$$\overline{G}^{-1}(j\omega) = 1 + j\omega T.$$

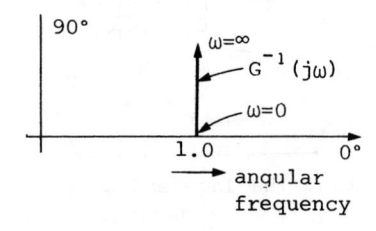

For a lead network:

$$G^{-1}(j\omega) = 1 + \dfrac{1}{j\omega T} \quad \text{at } \omega = 0,\ G^{-1}(j\omega) = 1.$$

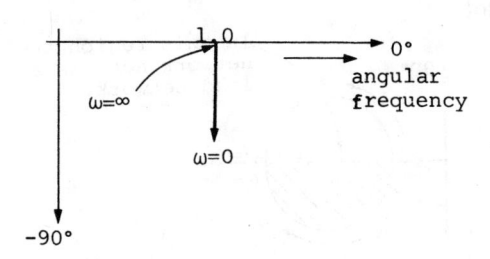

Type 0 and type 1 feedback systems:

$$G^{-1}(j\omega) = \frac{(1 + j\omega T_1)(1 + j\omega T_2)}{k} \quad \text{for type 0 system.}$$

$$= \frac{j\omega(1 + j\omega T')(1 + j\omega T'')(1 + j\omega T''')}{k}$$

for type 1 system.

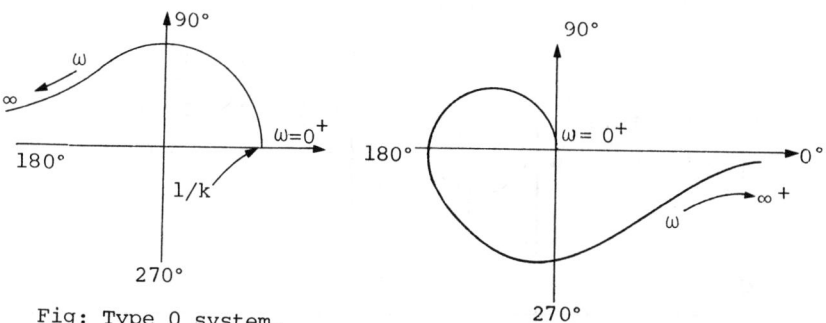

Fig: Type 0 system.

Fig. Type 1 system.

Characteristics:

A) When $\omega = \omega_1$, $\omega_1^2 T_1 T_2 = \omega_1^2 R_1 C_1 R_2 C_2 = 1$ so by proper choice of R_1, C_1, R_2 and C_2 this network can be made to act like a lead or a lag network.

B) Minimum value of the transfer function =

$$\frac{T_1 + T_2}{T_1 + T_2 + T_{12}} \quad \text{at } \omega = \omega_1 T_1.$$

From the polar plot, the following can be seen:

a) for frequencies below $\omega_1 T_1 = \sqrt{5.0}$, the circuit acts like an RL circuit, i.e. it is a lag compensator.

b) for frequencies greater than $\omega_1 T_1 = \sqrt{5.0}$, the circuit acts like an RC circuit. In this case it acts as a lead compensator.

Please note that this value is needed for designing the control system and for eliminating certain noisy frequencies.

C) Polar plots for different types of systems:

Table 9.1

Type of System	Forward transfer function of the feedback system.	Polar Plot	Comments/characteristics
0	$G(j\omega) = \dfrac{C(j\omega)}{E(j\omega)}$ $= \dfrac{k}{(1+j\omega T_1)(1+j\omega T_2)}$		i) The exact shape of the plot is determined by the time constants T_1 & T_2. ii) If the time constants T_1 & T_2 are nearly equal, the curve gets pulled more into the 3rd quadrant.
	$G(j\omega) = \dfrac{k(1+j\omega T')^2}{(1+j\omega T'')\times (1+j\omega T''')\times (1+j\omega T'''')^2}$		i) T'' & $T''' > T'$ and $T' > T''''$
1	$G(j\omega) = \dfrac{k}{j\omega(1+j\omega T_1)\times (1+j\omega T_2)(1+j\omega T_3)}$		i) $R = \lim_{\omega \to 0} \mathrm{Re}[G(j\omega)]$ $= -k(T_1+T_2+T_3)\ldots$ The true asymptote. ii) $\omega_y = \dfrac{1}{[T_1 T_2 + T_2 T_3 + T_3 T_1]^{\frac{1}{2}}}\ldots$ $\ldots$ can be obtained by the equation $I_w(G(j\omega)) = 0$ iii) R is always a direct function of the ramp error coefficient.
2	$G(j\omega) = \dfrac{k}{(j\omega)^2 \times (1+j\omega T_1)\times (1+j\omega T_2)}$		i) This polar plot represents an unstable feedback constant system.

D) $$\left[1 + \frac{2\zeta}{\omega_n}j2\pi f + \frac{-(2\pi f)^2}{(2\pi f_n)^2}\right]^{-1}$$

There are many curves for this factor depending on the value of ζ.

a) For $\zeta > 1$:

the function can be factored into two real factors. The log-magnitude curve is plotted for the individual factors and these curves are then combined..

b) For $\zeta < 1$:

In this case no factorizing is necessary.

$$\text{Lm[function]}^{-1} = -20\log \sqrt{\left(\frac{2\zeta\omega}{\omega_n}\right)^2 + \left(1 - \frac{\omega^2}{\omega_n^2}\right)^2}$$

$$\text{angle} = -\tan^{-1}\left[\frac{2\zeta\omega/\omega_n}{1 - \omega^2/\omega_n^2}\right]$$

Plot

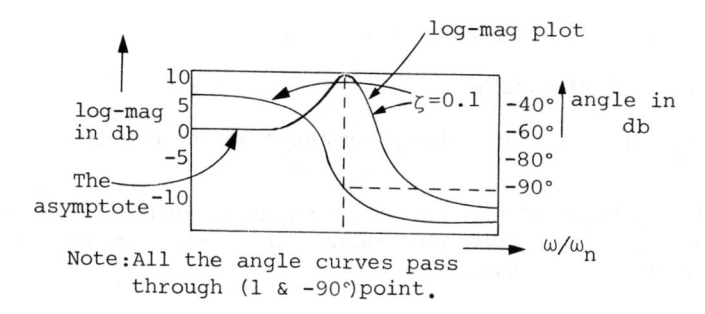

Note: All the angle curves pass through (1 and -90°) point.

FACTS

A) For very small values of ω, the asymptote is log-magnitude = 0, line as shown in the figure.

B) For higher values of ω, the high frequency asymptote has a slope of -40db per decade. The log-magnitude is approximately

$$-20\log \frac{\omega^2}{\omega_n^2} = -40\log \frac{\omega}{\omega_n}.$$

HOW TO PLOT AN INVERSE POLAR PLOT

A) First of all the system type, i.e. whether the system is type 0 or type 1, etc., is determined from the forward transfer function G(s).

B) Then $\lim_{\omega \to 0} G^{-1}(j\omega)$ is determined after substituting $s = j\omega$ in the transfer function and evaluating the value of

$G^{-1}(j\omega)$ by putting $\omega = 0$. This zero frequency point is always approached in the clockwise direction.

C) $\lim\limits_{\omega \to \infty} \overline{G}'(j\omega) = \infty \underline{/(a+b-c)90°}$ is determined. This gives the high-frequency end of the polar plot. In this expression, b is the number of zeros, ω is the number of poles and a is given as follows:

$$G^{-1}(s) = \frac{s^n(1 + T's)(1 + T''s)\ldots(1 + T_s^b)}{K(1 + T_1 s)\ldots(1 + T_c s)}$$

This is the most general form of $\overline{G}'(s)$.

D) Points of intersection:

$I_m[\overline{G}'(j\omega)] = 0$ this gives point on the negative real axis

Exact shape of the plot near negative real axis is very important and this particular point plays an important role in the system behavior.

$Re[G^{-1}(j\omega)] = 0$ on the imaginary axis.

E) If there are no time constants in the denominator, the curve is a smooth curve and the angle increases continuously as the angular frequency increases from 0 to ∞.

The presence of time constants results in dents or up and down (i.e. the angle will not vary continuously, i.e. it will not increase or decrease continuously) in the plot.

9.9 DEAD TIME

Some elements produce no output or may produce an output after a considerable time lag in response to the input signal to the control system. These are known as transport-lag elements; an analogous example in the electrical system is the D-type flip-flop.

$$V_{out}(t) = V_{in}(t - \gamma)u(t - \gamma)$$

G-105

where γ is called the dead time.

γ is represented by non-linear characteristics called dead-time characteristics given by the following curves.

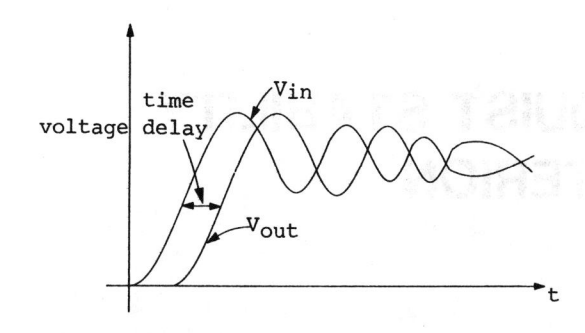

$$\bar{G}(j\omega) = e^{-j\omega\gamma} = 1 \underline{/-\omega\gamma}$$

is the transfer function of a dead time element.

Polar plot:

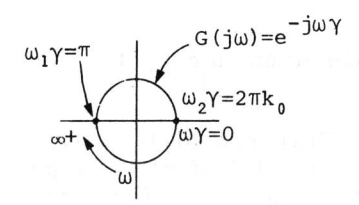

$k_0 = 1, 2, 3\ldots$ as ω increases.

CHAPTER 10

NYQUIST STABILITY CRITERION

10.1 DETERMINING AND ENHANCING THE SYSTEM'S STABILITY

Necessary condition for system's stability in terms of poles and zeros:

A) No zeros of T(s) should be present on the imaginary axis or in the right half of the s-plane as shown below. Note that the poles of T(s) are the poles of the open-loop transfer function.

B) There must be no zeros on the $j\omega$ axis; zeros present on the $j\omega$ axis result in a quasistable system and should there be any slight variation in the gain k, the system should become unstable.

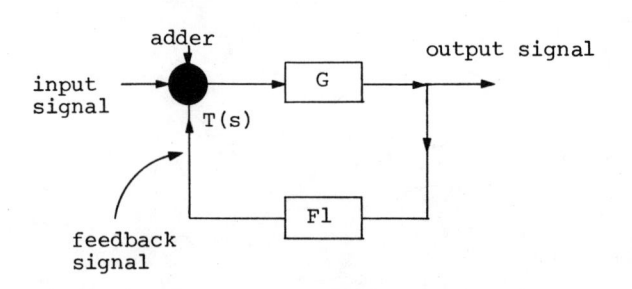

```
                    adder
                      ↓                    output signal
  input  ─────────→ ( + ) ────→[  G  ]────────┬──────→
  signal              │   T(s)                │
                      │                       │
                      └──────────[  F1  ]←────┘
                   ↑
              feedback
              signal
```

G-107

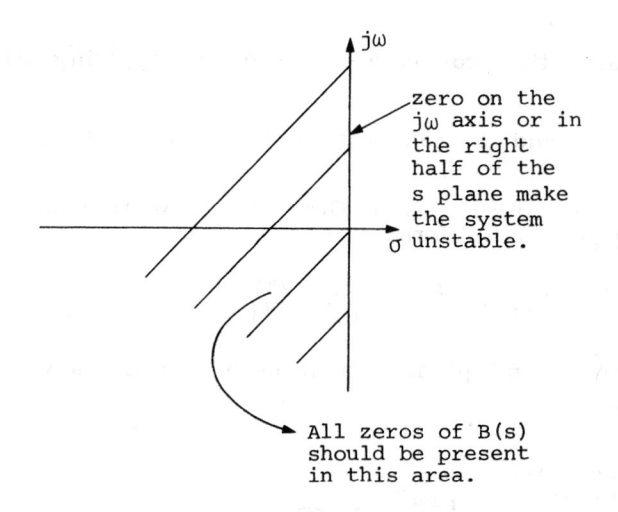

```
        ↑ jω
        │
        │    zero on the
        │    jω axis or in
        │    the right
        │    half of the
        │    s plane make
        │    the system
        │────→ unstable.
        │  σ
        │
        │
        │
        →  All zeros of B(s)
           should be present
           in this area.
```

Note the following about Nyquist plots:

A) Unlike the polar plots dealt with in the preceding chapter, Nyquist's plots are drawn for a range of frequencies from $-\infty$ to $+\infty$.

B) Nyquist's plots are plots of the open-loop transfer function $G(s)H(s)$.

C) The curve for negative frequencies is symmetrical to the curve for positive frequencies.

Nyquist's Criterion: A generalized statement

A) $$N = \frac{\text{Phase change of } 1 + G(s)H(s)}{2\pi}$$

$= P_{R.H.} - Z_{R.H.}$ Note that $T(s) = 1 + G(s)H(s)$

where,

$Z_{R.H.}$ = Number of zeros in the right half of the s-plane
$P_{R.H.}$ = Number of poles in the right half of the s-plane
N = The number of revolutions of $1 + G(s)H(s)$ about the origin.

Note: N may be positive, negative, or zero. Counterclockwise revolutions around T(s) are positive and clockwise revolutions are negative.

These are the conventions used in applying this criterion.

B) A pictorial and a mathematical approach: concept of revolutions

Suppose that $T(s) = 1 + G(s)H(s)$ is written in the following generalized form:

$$1 + G(j\omega)H(j\omega) = \frac{(s - z')(s - z'')}{(s - p')(s - p'')}$$

zeros and z' and p' are the poles of the characteristic equation.

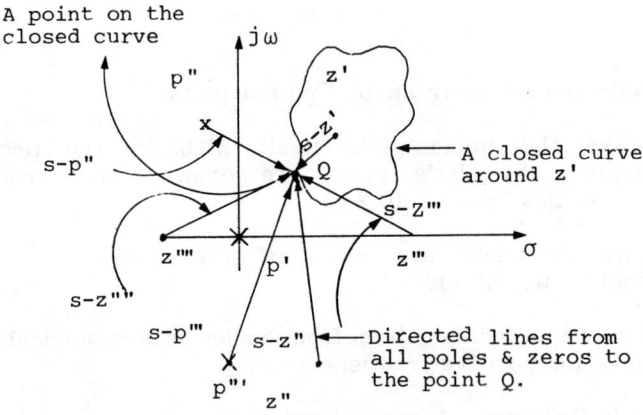

A point on the closed curve

s-p"

s-z"'"

A closed curve around z'

Directed lines from all poles & zeros to the point Q.

Note: The length of the directed lines are given by s-z'... & s-p'...

Fig. Pole-zero diagram for B(s)

Note that if the point Q rotates clockwise,

A) S-Z phasor turns through complete 360°.

B) Rest of the directed lines rotate through a net angle of 0°.

CONCLUSION

All the poles and zeroes which are outside the closed path, which itself lies in the right half of the s-plane,

G-109

give an individual contribution to the net angular rotation from 0° to T(s) as an arbitrary point moves around the closed path.

$$\begin{bmatrix}\text{The total number of}\\ \text{rotations of}\\ 1 + G(s)H(s)\end{bmatrix} = \begin{bmatrix}\text{Number of}\\ \text{poles en-}\\ \text{closed, P}\end{bmatrix} - \begin{bmatrix}\text{Number of}\\ \text{zeros en-}\\ \text{closed, Z}\end{bmatrix}$$

Note the rotations result due to clockwise movement of an arbitrary point on the closed curve which is drawn so as to enclose p poles and z zeros.

Stable system and the revolutions:

Table 10.1

Type of System	The Transfer Function $G(s)H(s)$ Of The System	Direct Polar Plot	Comments/Characteristics Important Points
Type 0	$\dfrac{K}{S(1+T's)(1+T''s)}$ i) No zeros ii) Two poles which are located on the left hand side		i) Nyquist criterion: $N=0$... since number of poles in the right half of S-plane ii) $P_{R.H.}=0$, hence $N=0=R_{R.H.}-Z_{R.H.}$ so the system is stable for all values of K.
Type 1	$\dfrac{K}{S(1+T's)(1+T''s)}$		i) If the gain K is increased beyond a certain value, then the system becomes unstable. ii) $N=-2=$ the rotations of $1=G(S)H(S)$ in the direction of increasing value of 0.
Type 2	$\dfrac{K(1+T_1S)}{S^2(1+T_2s)(1+T_3s)(1+T_4s)}$ $T_1>T_2$, T_3 and T_4 $P_{R.H.}=0$		i) $P_{R.H.}=0$, $N=0$ so $Z_{R.H.}=0$ hence the system is stable ii) If gain is increased, then the system can become unstable (because $-1+j0$ point is on the R.H. side of the intersection part of the curve and the -ve real axis).

A) For a stable automatic control system, the total number of revolutions of $1 + G(s)H(s)$ around the origin of the σ, $j\omega$ axis must be clockwise and

$$\begin{matrix}\text{Total number of}\\ \text{revolutions}\end{matrix} = \begin{matrix}\text{Number of poles that are}\\ \text{located in the right half of s-plane.}\end{matrix}$$

G-110

B) When there are no poles on the right-hand side of $j\omega$ axis, the necessary condition is

N_1 = The total number of revolutions about $(-1 + jo)$ point

    = 0.

10.2 INVERSE POLAR PLOTS: APPLICATION OF NYQUIST'S CRITERION

A MATHEMATICAL APPROACH

A) Clockwise revolutions of

$$T'(s) = \frac{1}{G(s)H(s)} + 1 = \text{Number of zeros on the right half of s-plane} = Z$$

B) Total number of anticlockwise revolutions of

$$\frac{1}{G(s)H(s)} + 1 \text{ due to poles} = \text{Number of poles located in the right half of s-plane} = P$$

C) N' = Total number of revolutions

$$\frac{1}{G(s)H(s)} + 1 \text{ about the origin of the s-plane} = P - Z$$

CRITERION FOR THE SYSTEM'S STABILITY

A) The net number of revolutions of $\frac{1}{G(s)H(s)} + 1$ about the origin point 0, must be counterclockwise and,

B) the number of revolutions should be equal to the number of poles that are located in the right half of the s-plane.

Note: The clockwise revolutions are treated as negative and the counterclockwise revolutions are treated as positive.

C) When there are no poles located in the right half of the s-plane, the stability criterion for the system is

$$\left[\begin{array}{c} \text{Total number of revolutions} \\ \text{about } -1 + \text{jo} \end{array} \right] = 0.$$

D) If the system's G(s)H(s) has some zeros in the right side, then the number of poles which may be present on the right half plane are found by using Routh's criterion after which point the criterion is applied.

CONCEPTS OF REVOLUTIONS OF T'(s)

Suppose that there are some poles and zeros of $T'(s) = 1 + \left[\dfrac{1}{G(s)H(s)} \right]$ which lie in the right half of the s-plane.

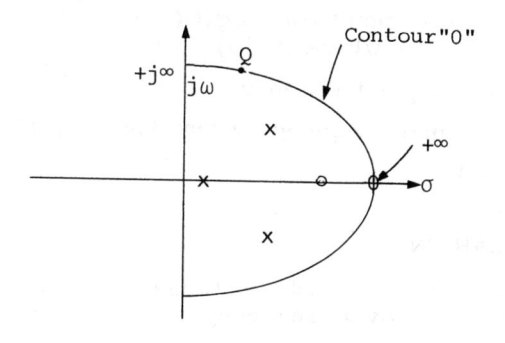

A closed contour O is drawn which can enclose all these poles and zeros. Such a contour should be semicircle of ∞ radius as shown.

Consider a point Q on this contour. Then, one revolution of T'(s) is defined as follows:

A) The point Q should go from +0 to +∞ on the imaginary jω axis,

B) then from +∞ to -∞ on the semicircle of ∞ radius,

C) then from -∞ to -0 on -jω axis.

G-112

10.3 PHASE MARGIN AND GAIN MARGIN: DEFINITIONS

The gain crossover:

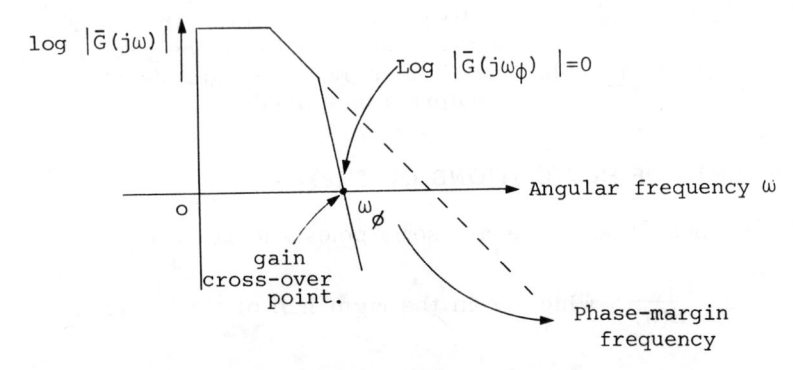

[Figure: plot of $\log |\bar{G}(j\omega)|$ versus angular frequency ω, showing $\text{Log} |\bar{G}(j\omega_\phi)| = 0$ at the gain cross-over point ω_ϕ, labeled as phase-margin frequency.]

It is the point on $\log |\bar{G}(j\omega)|$ versus ω curve where $\log |\bar{G}(j\omega)| = 0$ or $|\bar{G}(j\omega)| = 1$.

Phase margin frequency: ($\omega\phi$)
It is the angular frequency where $\log |\bar{G}(j\omega)| = 0$ or $|\bar{G}(j\omega)| = 1$.

PHASE MARGIN

Phase margin = 180° + (negative of the angle of $\bar{G}(j\omega)$ at the gain crossover frequency).

Phase crossover

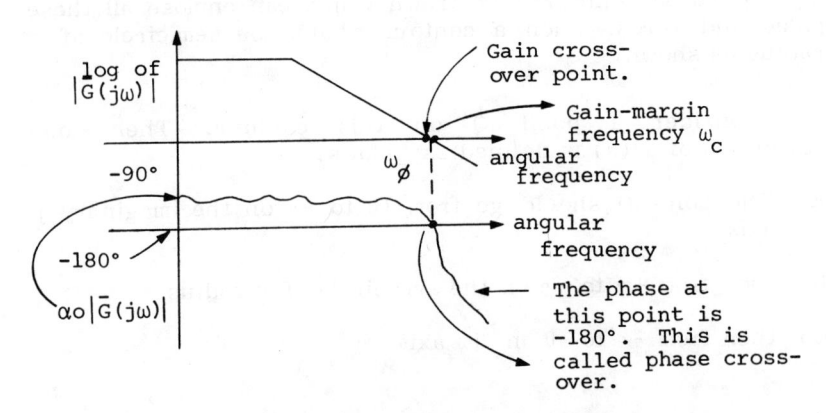

[Figure: plot showing log of $|\bar{G}(j\omega)|$ versus angular frequency, with gain cross-over point, gain-margin frequency ω_c, phase-margin frequency ω_ϕ, and phase axis with -90° and -180° markings. $\alpha o |\bar{G}(j\omega)|$. The phase at this point is -180°. This is called phase crossover.]

G-113

GAIN MARGIN

It is the extra gain which makes the system just unstable. It is given as:

$$\text{log magnitude of gain margin} = -\text{Lm}\overline{G}(j\omega_c).$$

10.4 SYSTEM STABILITY

A) For a stable system, the phase margin must be positive. A negative phase margin means an unstable system.

B) Gain margin expressed db should be a positive number for a stable system.

Negative gain margin in db implies an unstable system.

C) Phase margin gives the better estimate of damping ratio ζ than the gain margin.

Log magnitude-angle diagram, stability of the system log-magnitude-angle diagram

A) This is a plot between angle α of the torque for function $\overline{G}(j\omega)$ and its magnitude $|\overline{G}(j\omega)|$.

B) This is plotted from the following two plots:
   a) $\log|\overline{G}(j\omega)|$ versus ω and,
   b) angle α of $\overline{G}(j\omega)$ versus ω.

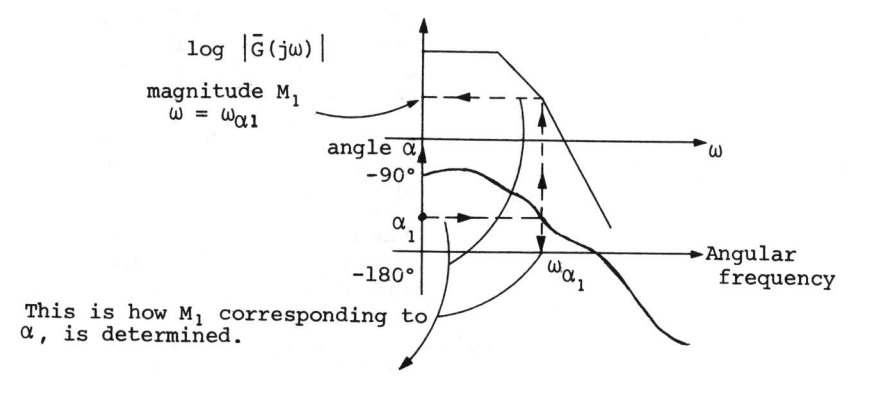

This is how M_1 corresponding to α_1 is determined.

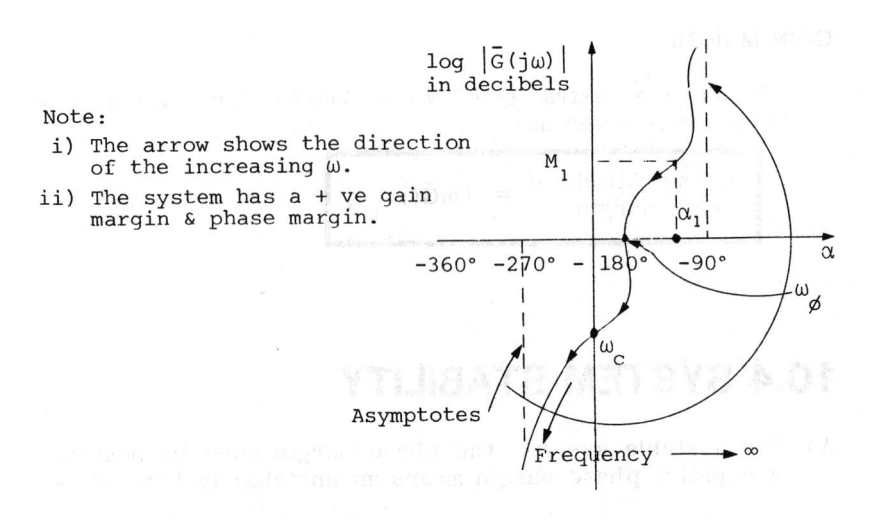

Note:
i) The arrow shows the direction of the increasing ω.
ii) The system has a + ve gain margin & phase margin.

10.5 STABILITY

Effects of changing the gain and the phase margin

Table 10.2

Value of gain k	Value of the phase margin	Value of the gain margin	Effect on system
–	positive	positive	The system is stable
changing	–	–	The log-magnitude curve is elevated or lowered without changing angle characteristics.
increased	–	–	The curve is raised and both gains and phase margins are decreased so the system becomes unstable. High values of gain k are necessary to reduce steady-state errors.
decreased	–	–	The system stability is increased because both the gain and phase margins are increased.

G-115

A simple rule to determine the stability.

A curve is drawn for $\bar{G}(j\omega)$ (i.e. the log magnitude-angle curve) for $\omega > 0$ and $\omega < \infty$.

The curve is traced in the direction of increasing value of the angular frequency.

The system is stable if the (0db, -180°) point is lying on the right-hand side of the curve.

```
         log |Ḡ(jω)|
            in db
                         Ḡ(jω)


                    0         angle
                    0 db & -180°

         direction of
         increasing ω
```

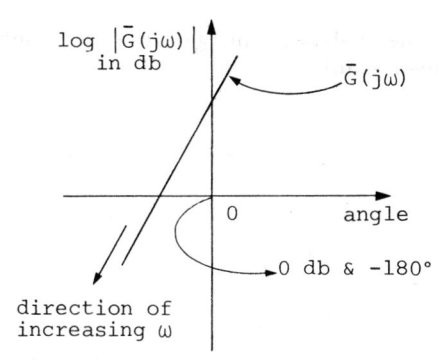

Fig. This is a stable system.

Conditionally stable system: This system is characterized by the curve shown below.

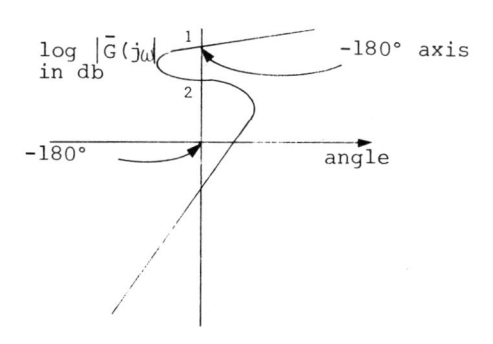

```
         log |Ḡ(jω)|  1    -180° axis
            in db
                     2
         -180°
                         angle
```

The characteristic is that the curve crosses $\log |\bar{G}(j\omega)|$ db axis at more than one point as shown.

G-116

10.6 EFFECT OF ADDING A POLE OR A ZERO: EFFECT ON THE POLAR PLOTS

A) If an additional zero is inforced in the pole-zero pattern of the system, it results into a counterclockwise rotation of the system's direct polar plot. So the system's stability is increased.

B) An extra pole rotates the polar plot clockwise, making the sytem less stable.

CHAPTER 11

PERFORMANCE EVALUATION OF A FEEDBACK CONTROL SYSTEM IN THE FREQUENCY-DOMAIN

11.1 PERFORMANCE EVALUATION USING DIRECT POLAR PLOT

This mathematical analysis of the system's direct polar plot gives us additional information regarding the steady-state error. This information is helpful in the design of compensating networks.

Feedback system: A block diagram.

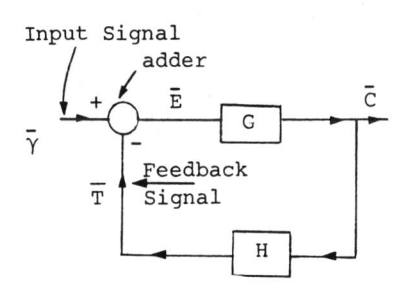

Analysis: Consider for the sake of analysis, H = 1 and let the following be the $\bar{G}(j\omega)\bar{H}(j\omega)$ curve.

G-118

A) The vector $\overline{AB}$ as shown in the figure gives

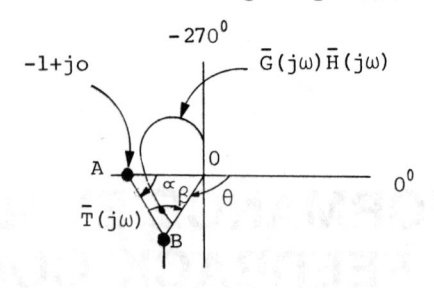

$$\overline{T}(j\omega) = 1 + \overline{G}\,\overline{H}$$

$$\overline{T}(j\omega) = \underbrace{|\overline{T}(j\omega)|}_{\text{length AB}} e^{j\alpha(\omega)}$$

B) $\dfrac{\overline{C}(j\omega)}{\overline{R}(j\omega)} = \dfrac{\overline{G}(j\omega)}{1 + \overline{G}(j\omega)}$ The closed-loop transfer function.

The vector $\overline{OB}$ in the figure is $\overline{G}(j\omega)$ because $\overline{H}(j\omega) = 1$. So,

$$\dfrac{\overline{C}(j\omega)}{\overline{R}(j\omega)} = \dfrac{\text{length of }\overline{OB}}{\text{length of }\overline{AB}} = \dfrac{OB}{AB} e^{j\beta} \quad \text{because } \theta - \alpha = \beta.$$

Note that both α and θ are negative and angle β is positive according to the convention followed.

C) $\dfrac{\overline{E}}{\overline{R}} = \dfrac{\text{Error signal phasor}}{\text{Input signal phasor}} = \dfrac{1}{AB\underline{/\alpha}}$.

Conclusion: The error between the input signal and the feedback signal is very very small if the length of the vector $\overline{AB}$ is very large. This is further explained as follows:

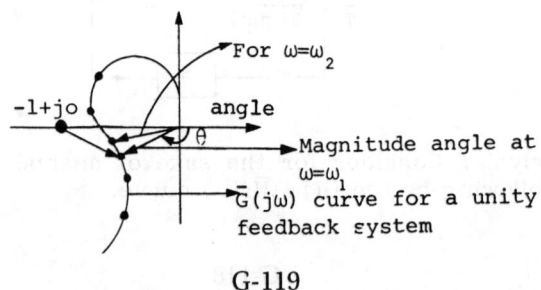

G-119

The frequency of operation ω_0 is chosen in such a way that the length of the vector will result in an error which is in the specified limit. This can be achieved by keeping ω = constant and by changing ζ, the damping ratio (this point is stressed in the next article).

11.2 RESONANT FREQUENCY AND THE MAXIMUM MAGNITUDE OF C/R OF A SECOND ORDER SYSTEM

SECOND-ORDER SYSTEM

Characteristics

A second order system is characterized by the following transfer functions:

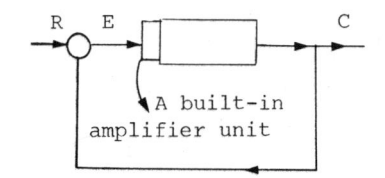

A built-in amplifier unit

$$\frac{\overline{C}(s)}{\overline{R}(s)} = \frac{\omega_n^2}{S^2 + 2\omega_n \zeta S + \omega_n^2}$$

where

$$\zeta = \frac{D}{2}(KA)^{-\frac{1}{2}}$$

$$\omega_n = \left(\frac{K}{A}\right)^{\frac{1}{2}}$$

and

$$\frac{\overline{C}(s)}{\overline{E}(s)} = \frac{K}{S(As+D)}$$

$$\frac{\overline{C}(j\omega_1)}{\overline{R}(j\omega_1)} = \text{Mag}(\omega_1) e^{j\beta(\omega_1)}.$$

G-120

This is applicable for $\omega = \omega_1$.

A) The maximum value of Mag is found from the following equation

$$\frac{d}{d\omega}(\text{Mag})^2 = 0.$$

B) The angular frequency at which $\text{Mag} = \text{Mag}_{max}$ is

$$\omega_m = \begin{pmatrix} \text{undamped natural} \\ \text{frequency} \end{pmatrix} [1-2(\text{damping ratio})^2]^{\frac{1}{2}}$$

$$\text{Mag}_{max} = \frac{1}{2\zeta\sqrt{1-\zeta^2}} \quad \text{and}$$

$$M_p = \frac{\text{Peak overshoot}}{\text{value of the step signal}} = 1 + e^{\frac{-\zeta\pi}{\sqrt{1-\zeta^2}}}$$

Plot: The following is a set of plots which highlights some of the facts regarding the frequency and line response. It is quite clear that for the smaller values of ζ, the values of Mag_{max} and M_p are larger.

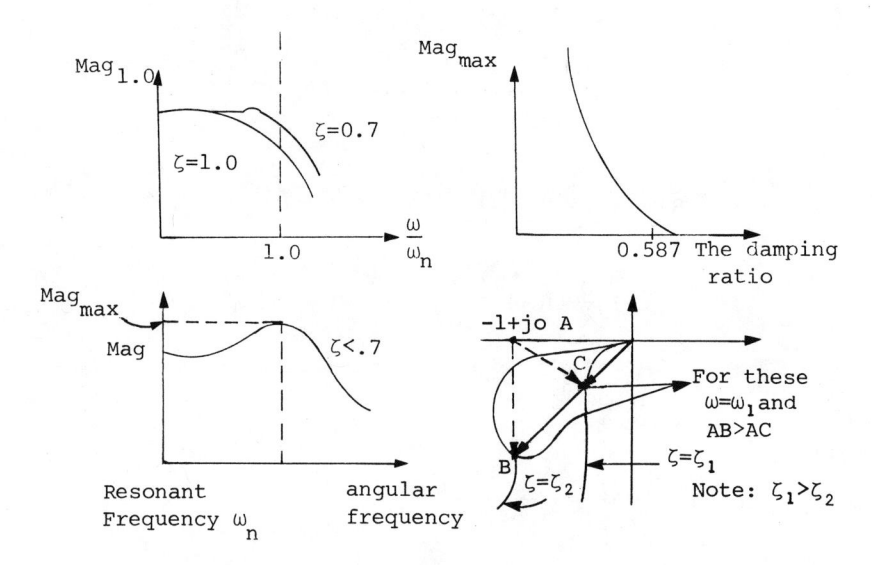

The steady-state or the error at $\omega = \omega_1$ can be altered by changing the damping constant. This is quite clear from the figure.

G-121

The time response can be made faster by increasing the resonance frequency which in turn depends on the damping ratio ζ. So, the lower the value of ζ, the higher the value of resonant frequency.

In order to make Mag_{max} higher, the $\bar{G}(j\omega)$ plot for this unity feedback system should be close to -1 + jo point.

A) Type 0 system: If the $\bar{G}(jo) = k$ point on the $\bar{G}(j\omega)\bar{H}(j\omega)$ curve is away from the origin, then for a step input there will be an exact steady-state time response. This exactness depends on how far this point is.

B) Type 1 system: Ramp input steady-state time response. In this case the $\omega \to o$ asymptote should be away from the imaginary axis of the plot.

11.3 PLOTTING MAXIMUM MAGNITUDE AND RESONANT FREQUENCY ON THE COMPLEX PLANE

These curves which represent $\text{Mag}(\omega)$ and $\beta(\omega)$ = constant values of the closed-loop transfer function are plotted on the direct polar plot of the system.

CONSTANT MAGNITUDE CIRCLES

Plotting procedures

A) First of all the $G(j\omega)H(j\omega)$ curve is plotted as shown in the following figure, and two vectors $\overline{OA}$ and $\overline{BA}$ are drawn.

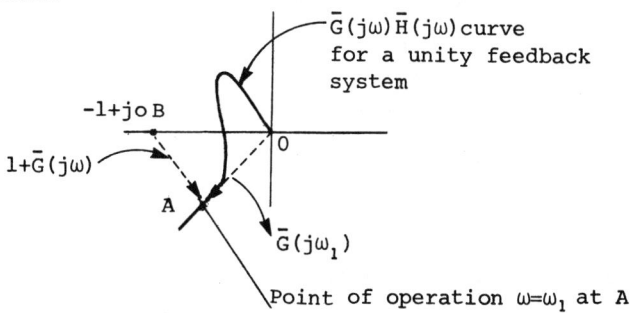

$\bar{G}(j\omega)\bar{H}(j\omega)$ curve for a unity feedback system

-1+joB

1+$\bar{G}(j\omega)$

$\bar{G}(j\omega_1)$

Point of operation $\omega=\omega_1$ at A

G-122

B) From this plot:

$$\frac{\overline{C}(j\omega_1)}{\overline{R}(j\omega_1)} = \frac{\text{output}}{\text{input}} = \frac{\overline{OA}}{\overline{BA}} = \text{Mag}(\omega_1) = \frac{|\overline{G}(j\omega_1)|}{|1 + \overline{G}(j\omega_1)|}.$$

C) Now if the curve $\overline{G}(j\omega)$ is plotted on an x-yj plane, then

$$\overline{G}(j\omega) = x + jy$$

and

$$\text{Mag}(\omega_1) = \frac{|x + jy|}{|1 + x + jy|}$$

so

$$\text{Mag}^2 = x^2 + y^2/(1-x)^2 + y^2.$$

This equation for Mg represents a circle which is shown as follows:

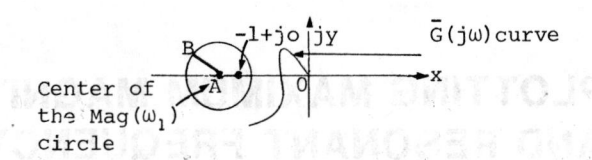

Note: $\quad OA = \dfrac{-\text{Mag}^2}{\text{Mag}^2 - 1}\quad$ and $\quad AB = \left|\dfrac{\text{Mag}}{\text{Mag}^2 - 1}\right|.$

Characteristics of Mag(ω_1) circle:

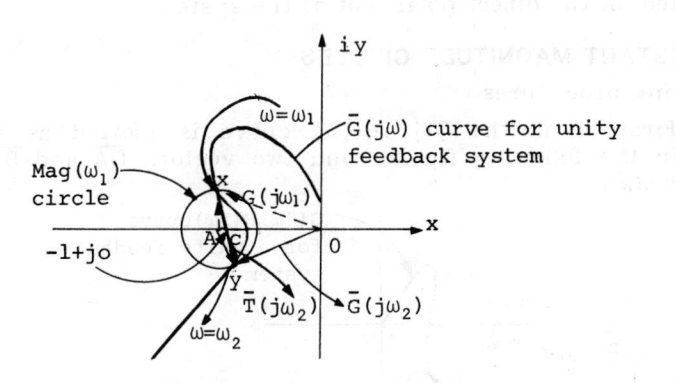

A)

For all points on Mag circle:

G-123

$$\frac{Oy}{Cy} = \frac{Ox}{Cx} = \text{Magnitude of } \overline{G}(j\omega) \text{ for } \omega = \omega_1.$$

B) If at x and y_1 the angular frequencies are ω_1 and ω_2 then

$$\left|\frac{\overline{G}(j\omega_1)}{\overline{B}(j\omega_1)}\right| = \left|\frac{\overline{G}(j\omega_2)}{\overline{B}(j\omega_2)}\right|$$

C)

[Figure: Nyquist-type plot showing circles of Mag_1, Mag_2, Mag_3 with $\overline{G}(j\omega)$, $\overline{G}(j\omega_3)$, $\overline{G}(j\omega_2)$, $\overline{G}(j\omega_1)$ labeled; $Mag_1 > Mag_2 > Mag_3$; point $-1+j0$ shown; Circle Mag_2 is tangent to $\overline{G}(j\omega)$.]

Circle Mag_2 is tangent to $\overline{G}(j\omega)$.

In this case for only the tangent point ω_2

$$\left|\frac{\overline{G}(j\omega_2)}{\overline{B}(j\omega_2)}\right| = Mag_2.$$

Locus of constant Mag curves:

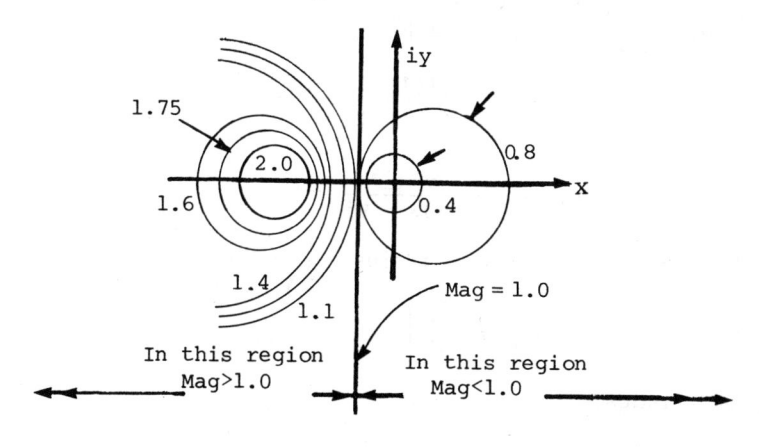

[Figure: constant magnitude circles labeled 1.75, 2.0, 1.6, 1.4, 1.1, 0.8, 0.4, Mag = 1.0; In this region Mag>1.0 / In this region Mag<1.0]

G-124

Constant angle contours: Plotting Procedure

A) $\dfrac{\overline{C}}{\overline{R}} = \text{Mag}(\omega)e^{i\beta(\omega)}$

$\phantom{\dfrac{\overline{C}}{\overline{R}}} = \dfrac{\overline{G}(j\omega)}{\overline{G}(j\omega) + 1}$

$\beta = \alpha - \theta$

$\overline{OB} = \overline{G}(j\omega_1)$ and $\overline{AB} = 1 + \overline{G}(j\omega_1) = \overline{B}(j\omega_1)$

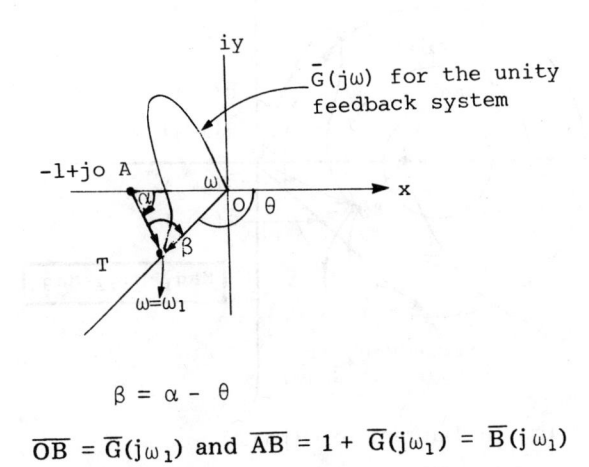

This is the closed-loop transfer function.

B) Now a constant angle contour is a curve for which the value of the angle β is always constant.

C) $\text{Mag}(\omega)e^{i\beta(\omega)} = \dfrac{x + iy}{(1+x) + iy}$

$\phantom{\text{Mag}(\omega)e^{i\beta(\omega)}} = \dfrac{(x+iy)[(1+x) - iy]}{(1+x^2)^2 - y^2}$

Hence

$\tan\beta = \dfrac{y}{(x+1)x + y^2} = N$

It represents a circle of radius $= \dfrac{1}{2}\sqrt{\dfrac{N^2 + 1}{N^2}}$

and its center is $\left(-\dfrac{1}{2}, \dfrac{1}{2N}\right)$.

Characteristics:

G-125

A)

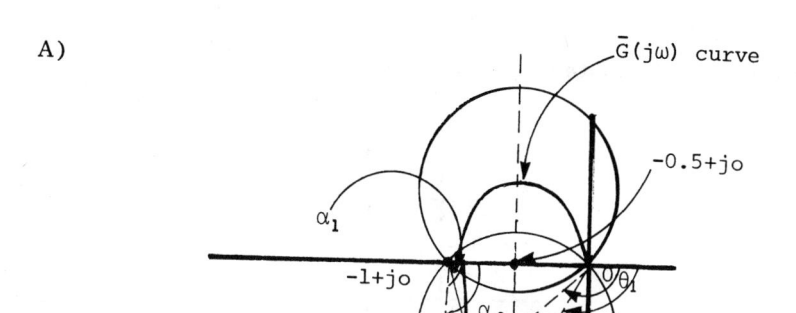

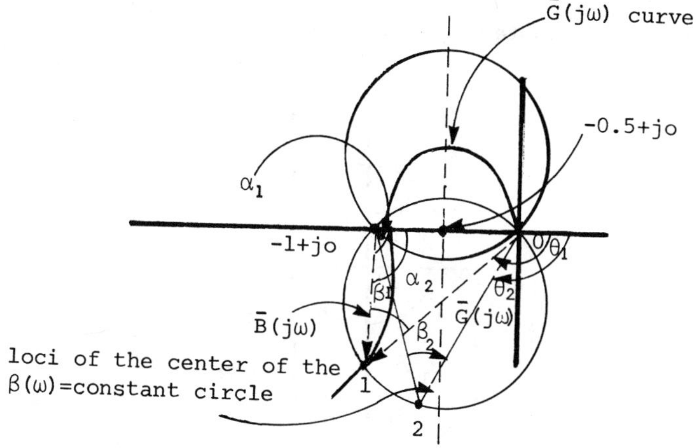

[Figure: $\bar{G}(j\omega)$ curve with loci of the center of the $\beta(\omega)$=constant circle, points labeled α_1, α_2, β_1, β_2, θ_1, θ_2, and reference points $-1+jo$, $-0.5+jo$, $\bar{B}(j\omega)$, $\bar{G}(j\omega)$]

Note: The circle always passes through $(-1+jo)$ and 0.

For points 1 and 2:

$$\boxed{\beta_1 = \beta_2 = [\text{value of the } \beta(\omega) = \text{constant}].}$$

B)

Note: (i) length a = length b if $\beta_1 = -\beta_2$

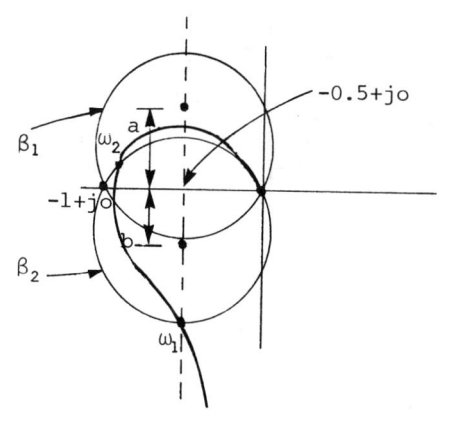

[Figure: circles with β_1, β_2, ω_1, ω_2, lengths a and b, reference points $-1+jo$, $-0.5+jo$]

$\beta_0 = \underline{/\bar{A}(j\omega_1)} - \underline{/\bar{B}(j\omega_1)} = 180° + \beta_2$.

This $\beta(\omega)$ at $\omega = \omega_1$

$\beta' = $ value of $\beta(\omega)$ at $\omega = \omega_2$ $= \beta_2$

C) Locus of constant β circles

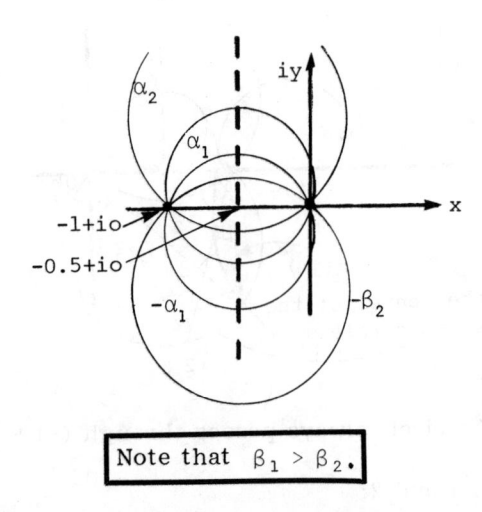

Note that $\beta_1 > \beta_2$.

D) Intersection with constant magnitude curve:

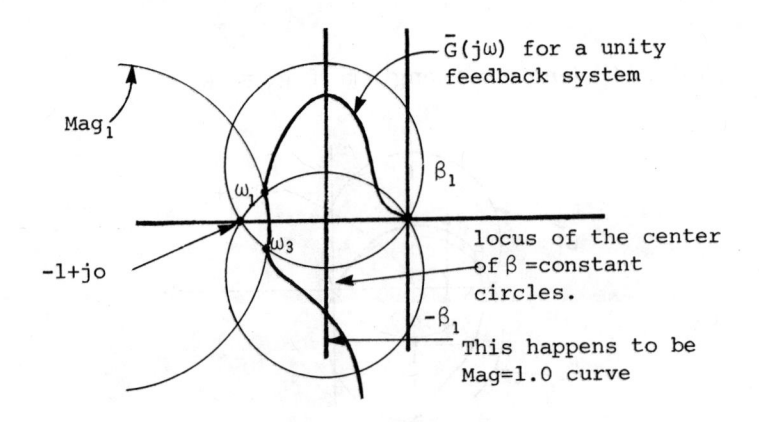

For $\omega = \omega_1$, $\quad \dfrac{\bar{C}(j\omega_1)}{\bar{R}(j\omega_1)} = \text{Mag}_1 e^{i(180° + \beta_1)}$.

For $\omega = \omega_3$, $\quad \dfrac{\bar{C}(j\omega_3)}{\bar{R}(j\omega_3)} = \text{Mag}_1 e^{i(180° - \beta_1)}$.

Gain adjustments using Mag circles:

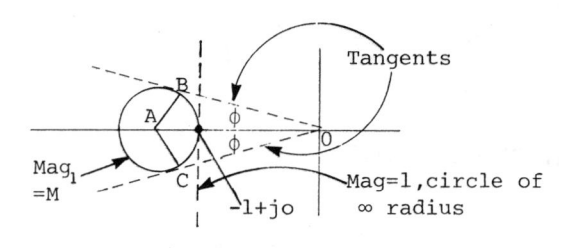

$$\sin \phi = \frac{AB}{AO} = \frac{M/(M^2-1)}{M^2/(M^2-1)} = \frac{1}{M}.$$

This adjustment procedure is given in the next article.

11.4 MAGNITUDE AND ANGLE CURVES IN THE INVERSE POLAR PLANE

MATHEMATICAL ANALYSIS

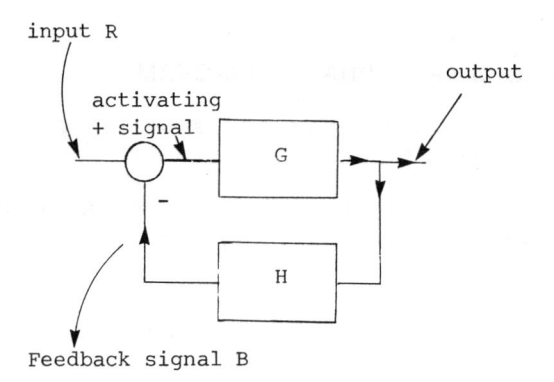

$$\frac{\overline{C}(j\omega)}{\overline{R}(j\omega)} = \overline{G}(j\omega) = |\overline{G}(j\omega)| \,\underline{/\theta(\omega)}$$
$$= \text{Mag}\,\underline{/\theta(\omega)}$$

$$\frac{\overline{R}(j\omega)}{\overline{C}(j\omega)} = \frac{1}{\text{Mag}} \,\underline{/-\theta(\omega)}$$

G-128

$$= \frac{1}{\overline{G}(j\omega)} + \overline{H}(j\omega)$$

PHASOR REPRESENTATION

Unity feedback system

[Phasor diagram with labels: $\frac{1}{G(j\omega_1)}$, $\frac{1}{\text{Mag}} \angle \theta(\omega)$, $\frac{1}{G(j\omega)}$, $\frac{1}{\text{Mag}} \angle \theta_1(\omega_1)$, constant $-\theta$ contour, $-1+j0$, $\frac{\overline{R}}{C}$, $\frac{\overline{R}}{C}(j\omega)$, $\frac{\overline{R}}{C} = \frac{1}{\overline{G}(j\omega)} + 1$, This is the $\overline{H}$ vector for unity feedback system, $\overline{H}(j\omega)$, $1+j0$]

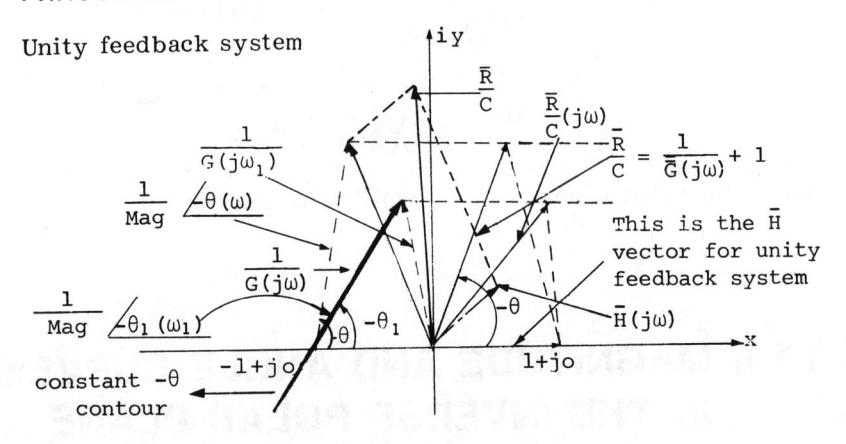

Note: The dot-dash-dot representation is for a nonunity feedback system.

CONCLUSIONS FROM PHASOR DIAGRAM

A) The curves of constant values of 1/Mag is a circle of radius 1/Mag with its center located at $-1 + j0$.

B) The contour for constant $-\theta$ is a radial line passing through $-1 + j0$ point, i.e. the center of the set of 1/Mag circles.

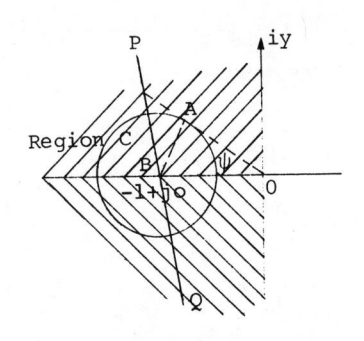

[Diagram showing Region C, points P, A, B, Q, $-1+j0$, iy axis, origin 0]

G-129

a) $\sin \psi = \dfrac{AB}{OB} = \dfrac{1/Mag}{1} = \dfrac{1}{Mag}$ compare this with that for direct

In region C and along the line PQ, the angle is $-\theta_1$ and in the D region, the angle is $-\theta_1 + 180°$.

CHARACTERISTICS

A)

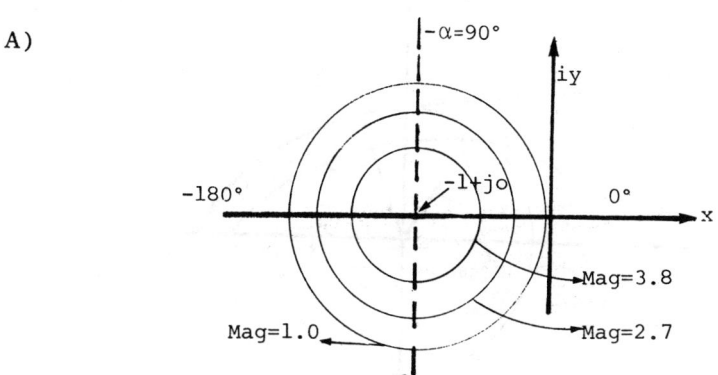

Fig: Inverse polar plot $-\theta = 270°$

Note: The Mag = 2.7 → $\dfrac{1}{Mag} = \dfrac{1}{2.7}$.

B) $\dfrac{1}{G(j\omega)}$ curve and Mag and θ curve:

$\dfrac{1}{G(j\omega)}$ curve and Mag and θ curve

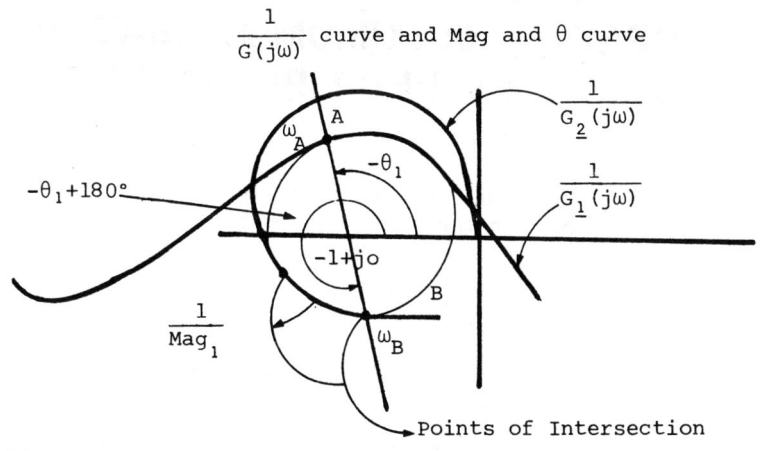

Points of Intersection

POINTS TO BE NOTED

A) At points A and B, the magnitude is the same but angles differ by 180°.

B) For large values of Mag_1, the circles are small and do not cross the $1/G_1(j\omega)$ curve.

C) Non-unity feedback case:

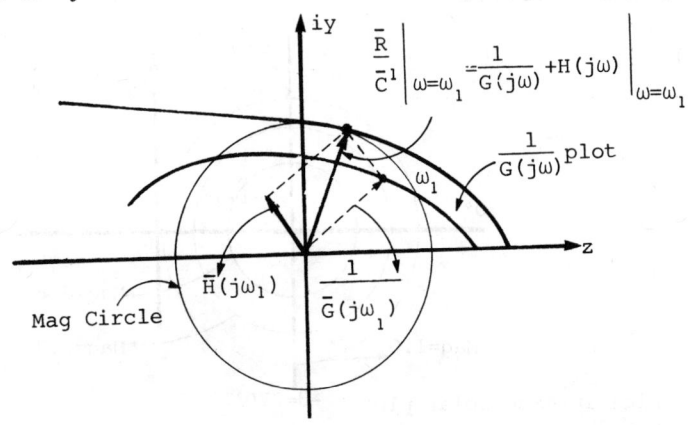

$$\left.\frac{\bar{R}}{\bar{C}}\right|_{\omega=\omega_1} = \left.\frac{1}{G(j\omega)} + H(j\omega)\right|_{\omega=\omega_1}$$

$\frac{1}{G(j\omega)}$ plot

Mag Circle $\bar{H}(j\omega_1)$ $\frac{1}{\bar{G}(j\omega_1)}$

Note: The center in this case is located at the origin.

11.5 GAIN ADJUSTMENT FOR A DESIRED MAXIMUM MAGNITUDE USING A DIRECT POLAR PLOT

Steps:

1: Express and plot the transfer function $G(j\omega)$ of the given system in the following form:

$$G(\omega) = \frac{K(1+j\omega T')(1+j\omega T'')\cdots}{(j\omega)^n(1+j\omega T_1)(1+j\omega T_2)\cdots}$$

2: Then the curve for the following modified form of $G(x)$ is plotted.

$$G(\omega)_{mod} = \frac{(1+j\omega T')(1+j\omega T'')\cdots}{(j\omega)^n(1+j\omega T_1)\cdots} = \frac{G(\omega)}{k}$$

G-131

3: Let the circle which is tangent to the $G(j\omega)$ curve be Mag_{max}. This is found out by trial and error.

4: A line is drawn whose slope is

$$\tan\psi = \tan\left[\sin^{-1}\frac{1}{Mag_1}\right].$$

5: Now a circle is found out by a trial and error method which is tangent to both $G(x)_{mod}$ and the line drawn in step number 4 is as follows.

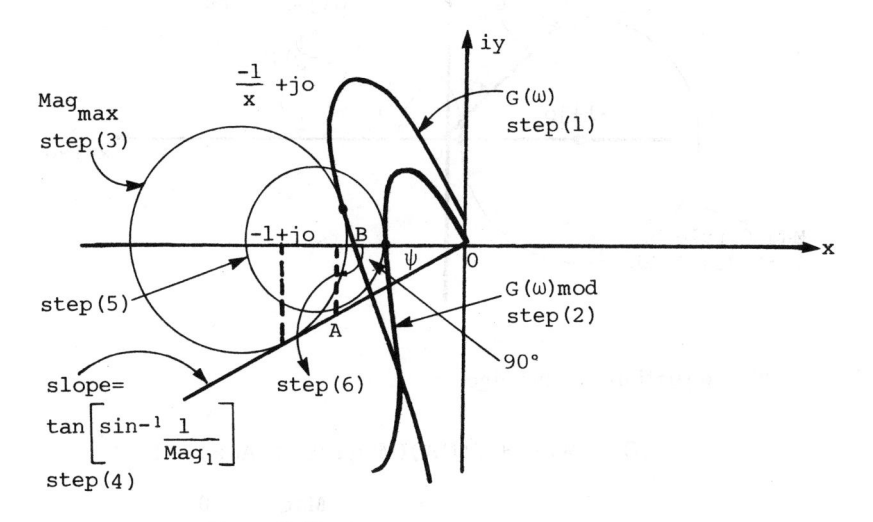

Please note the steps written beside each curve to understand the method.

6: Draw a line AB which is perpendicular to the x-axis.

7: The value of $K' = \frac{1}{OB}$, so the initial gain K has changed by a factor equal to $\frac{K'}{K}$ = length of line segment OB.

8: So the additional gain required to produce the given value of Mag_1 is given by $\frac{K'}{K}$ = length OB.

G-132

11.6 NICHOL'S CHART

These are the set of curves representing constant Mag and β on a log-magnitude versus angle plot.

MATHEMATICAL ANALYSIS

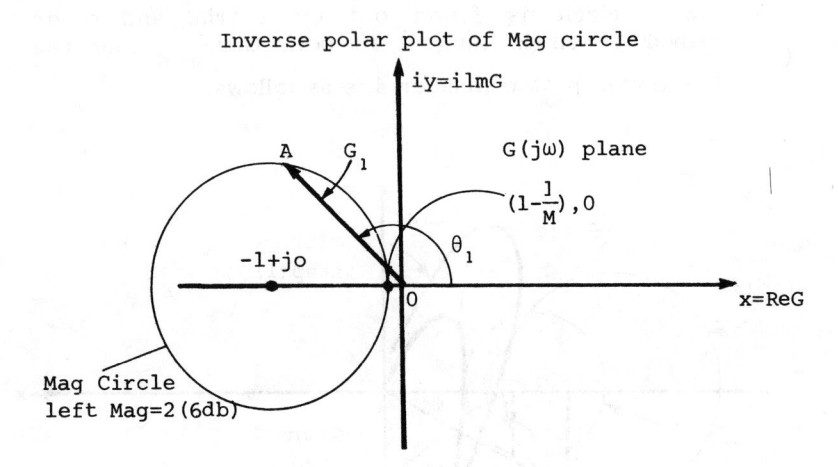

Inverse polar plot of Mag circle

$iy = ilmG$

$G(j\omega)$ plane

A G_1

$(1-\frac{1}{M}), 0$

θ_1

$-1+jo$

0

$x = ReG$

Mag Circle
left Mag=2 (6db)

The equation of the circle:

$$(OA)^2 Mag^2 + 2(OA)(Mag)^2 \cos(\angle Aox) - 1 + Mag^2 = 0.$$

$$|\overline{OA}| = -\cos\angle Aox \pm \sqrt{\cos^2(Aox) - \frac{Mag^2 - 1}{Mag^2}}$$

$$|\angle Aox| = \cos^{-1}\left[\frac{1 - Mag^2 - OA^2 Mag^2}{2(OA)(Mag)^2}\right].$$

MODIFICATIONS

A) A log magnification-angle diagram shows the relation between the magnitude of $G(j\omega)$ and angle.

B) An inverse plot depicts $G^{-1}(j\omega)$ of the system.

C) So following modifications are required:

a) $1/|\overline{OA}| = r$ and $-|\underline{/Aox}| = \theta$

b) The magnitude is represented in decibels.

Action Plot

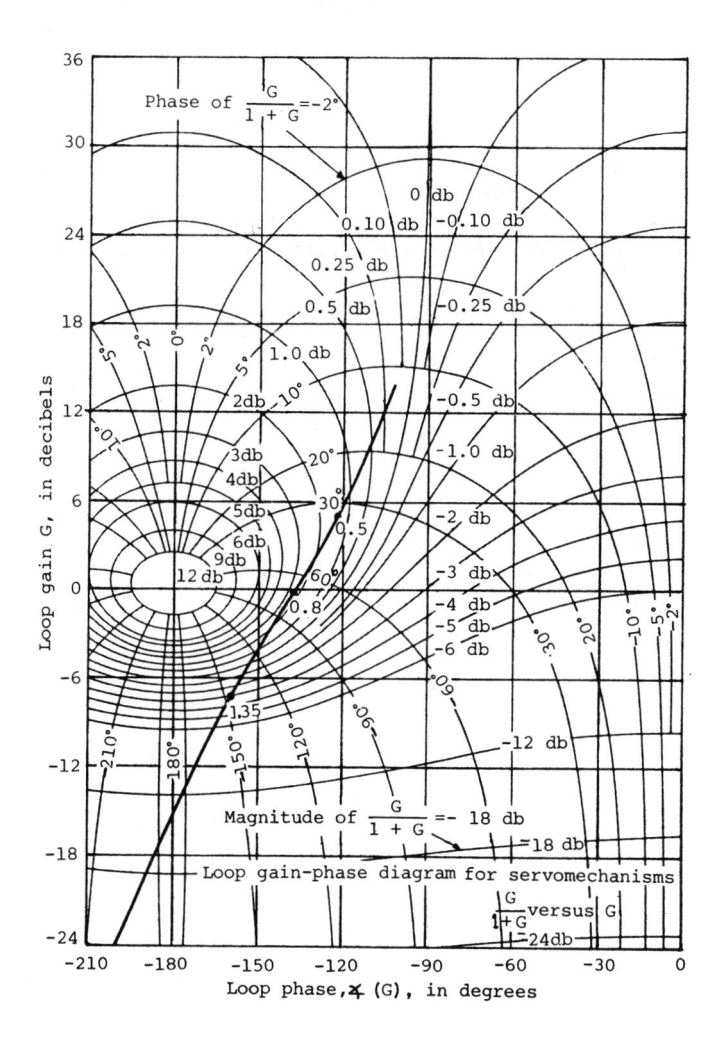

Loop gain-phase diagram for servomechanisms $\frac{G}{1+G}$ versus G

Loop phase, $\angle(G)$, in degrees

Loop gain, G, in decibels

Phase of $\frac{G}{1+G} = -2°$

Magnitude of $\frac{G}{1+G} = -18$ db

G-134

IMPORTANT POINTS OF NICHOL'S CHART

A) Constant Mag and angle curves repeat for every 360°.

B) There is symmetry at every 180° interval.

C) Mag = 1(0 db) curve is asymptotic to $\theta = -90°$ and $\theta = -270°$ line.

D) When Mag < -6db, the curves are negative because their magnitudes are negative as shown in the figure.

E) Mag > 1.0 curves are the closed curves and are bounded within -90° and -270°.

Note: Some of these points are shown by the corresponding number.

CHAPTER 12

SYSTEM STABILIZATION: USE OF COMPENSATING NETWORKS AND THE ROOT LOCUS

12.1 FUNCTION OF A COMPENSATING NETWORK

STABILIZED SYSTEM

Definition

The stabilized system must meet the following minimum requirements:

A) It ought to have a satisfactory transient response.

B) A large gain (k) to keep the steady-state error in the specified limit.

Compensating networks: location and function

Selection of location:

A) The selection depends on the type of the system and the modifications that are required.

B) Series cascading: The network is placed at a low-energy point so the dissipation is the lowest. An additional buffer amplifier is required to avoid the loading of the compensating network.

```
                          Series compensation
   input        ┌───┐  ┌───┐
   signal  +→(×)→│   │──│ G │──→ C
actuating   -   └───┘  └───┘
      signal              │
                          │ Parallel compensation
           ┌───┐  ┌───┐
           │   │──│ H │
           └───┘  └───┘
   feedback
              Compensating network with built-
              in buffer amplifier.
```

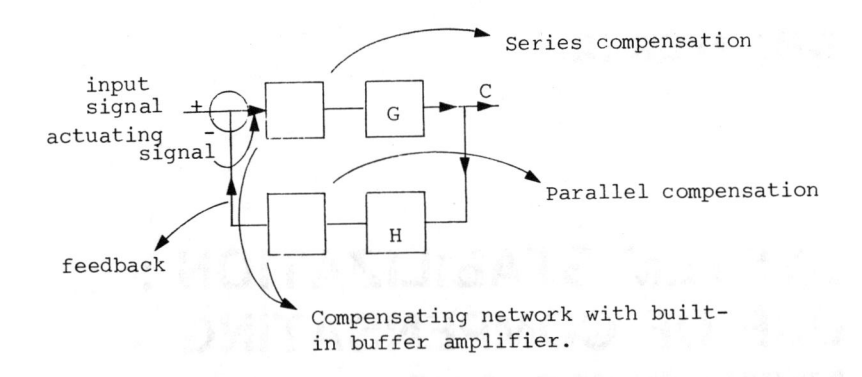

Function:

A) The compensation is done by introducing poles and zeros. Each additional pole increases the number of roots of the system by 1.

B) The compensating network effectively reshapes the root locus of the system.

12.2 TYPES OF COMPENSATIONS

12.2.1 PI (INTEGRAL AND PROPORTIONAL) CONTROL

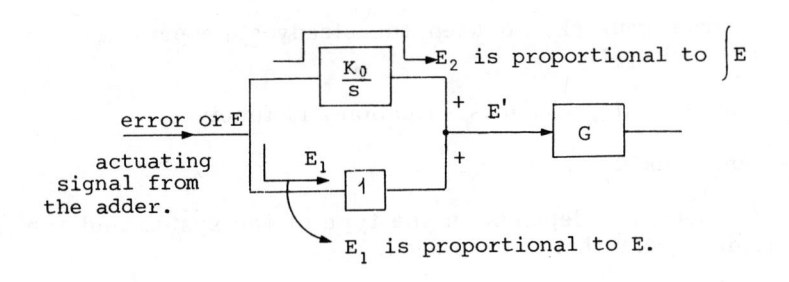

```
                    ┌─────┐
                    │ K₀/s│──→ E₂ is proportional to ∫E
   error or E       ├─────┤    +
   ───────────────→ │     │──→(+)→ E' →┌───┐
   actuating        │ E₁  │            │ G │
   signal from      ├─────┤    +       └───┘
   the adder.       │  1  │
                    └─────┘
                    E₁ is proportional to E.
```

Increase in type of system: The type of system is increased by a suitable network when 1_{ss} is very large, but the transient response is okay. This is done by adjusting the roots which are close to $j\omega$ axis (i.e. the dominant roots).

PI controller effectively does this. This controller works as follows:

Transfer function:

$$E' = E_1 + E_2 = (1 + \frac{k_0}{s})E$$

$$\boxed{G = \frac{E'}{E} = (1 + \frac{k_0}{s}).}$$

Pole-zero pattern:

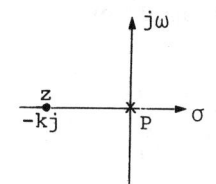

$$G = \frac{s + k_0}{s}.$$

Note: k_0 is selected very small.

Effects:

A) Effect due to the presence of pole P only: It shifts the root locus to the right, so the time response slows down.

B) Zero alone: The distance k_0 should be minimum so as to reduce the increase in response time.

C) The static error coefficient is ∞ because the type of system has increased.

D) $e_1(t)$ increases as long as there is an error signal $e(t)$.

12.2.2 LAG COMPENSATOR

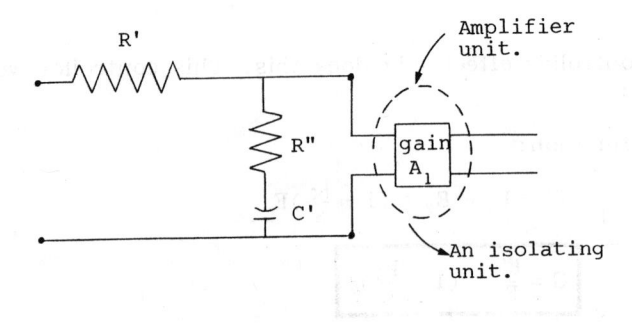

[Circuit diagram: R' resistor in series, R" resistor and C' capacitor in parallel branch, feeding into Amplifier unit (gain A_1) — An isolating unit.]

This is nothing more than an integral compensation.

Transfer function: (with amplifier)

$$G = A_1 \cdot \frac{(s + \frac{1}{T'})}{\left[s + \frac{1}{\beta T'}\right]\beta} \quad , \quad \beta > 1.$$

Steady-state accuracy:

A) $\displaystyle G_{or} = \frac{k' \prod_{i=1}^{n}(s - z_i)}{\prod_{j=1}^{n}(s - p_j)}$.

The general form of system's original open-loop transfer function.

B) $\displaystyle G' = G_{com} \times G_{or} = \frac{A_1 k'(s + \frac{1}{T'})}{\beta[(s + \frac{1}{\beta T'})]} \times \frac{\prod_{i=1}^{n}(s - z_i)}{\prod_{j=1}^{n}(s - p_j)}$.

$$\boxed{k'' = \begin{array}{l}\text{Static loop-sensitivity}\\ \text{for new root } S_1\end{array} = \frac{A_1 k'}{\beta}}$$

G-139

$$k_a = \prod_{i=1}^{n}(-z_i) / \prod_{j=1}^{n}(-p_j) k' \text{ for type 0}$$

$$k_b = \frac{\prod_{i=1}^{m}(-z_i)}{\prod_{j=1}^{n}(-p_j)} \beta k_0 .$$

Design Procedure

A) Pole $S_p = \frac{-1}{\beta T'}$ and zero $S_2 = \frac{-1}{T'}$ are placed very close by varying R', R" and C' (i.e. the root locus is slightly affected).

B) Then the angle contribution of the poles and zeros of this compensator is determined at the dominant root of the original closed-loop system.

C) If the angle contribution is < s, the new root locus will be changed slightly and the new pole s, will be slightly displaced from the uncompensated value S_0, the transient response will not change much, i.e.

$$k'' \approx k' \text{ and } k_b \approx \beta k_a .$$

D) The gain for new root s_1 increased by

$$\beta = \frac{\text{compensator zero}}{\text{compensator pole}} .$$

12.2.3 PROPORTIONAL PLUS DERIVATIVE (PD) COMPENSATOR

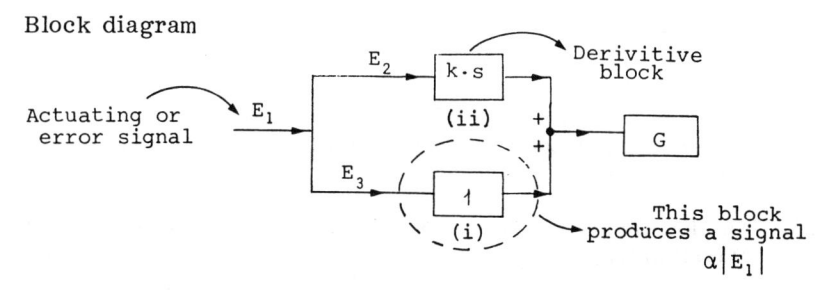

Block diagram

Function:

A) A zero is produced by block (i) and (ii).

B) Transfer function G = (1 + ks).

12.2.4 LEAD COMPENSATION

This is essentially a derivative compensator.

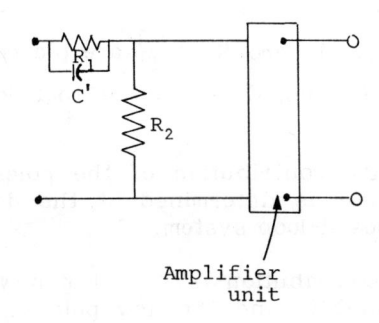

Transfer function

$$G = A'\beta \frac{1 + T's}{1 + \beta T's} = A' \frac{s - z_1}{s - p_1}$$

where $z_1 = -1/T'$ and $p_1 = -1/\beta T'$.

Pole-zero pattern:

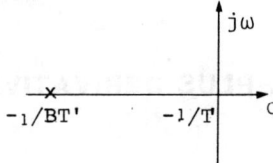

Note:

A) If β is very small then the location of the pole will be further away from the origin and it will have less effect on the root-locus.

B) The location of the zero is adjusted for the desired performance.

Characteristics:

A) The static loop sensitivity is proportional to the ratio

$$\frac{|s + 1/\beta T'|}{|s + 1/T'|} .$$

B) In order to increase the sensitivity, β must be small.

12.2.5 LEAD-LAG COMPENSATION

Advantages

A) A large increase in the gain, and

B) a large increase in the ω_n.

Transfer function:

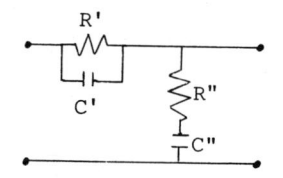

$$G = K_0 \frac{(s + \frac{1}{T'})(s + \frac{1}{T''})}{(s + \frac{1}{\beta T'})(s + \frac{\beta}{T''})}$$

$\beta > 1$ and $T' > T''$.

Relationship:

$$T' = R'C', \quad T'' = R''C''$$

$$\beta T' + \frac{T''}{\beta} = R'C' + R''C'' + R'C''.$$

G-142

Pole-zero pattern

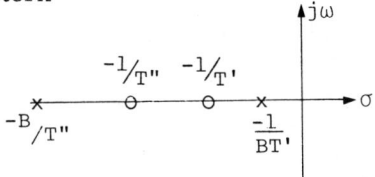

[Pole-zero plot showing poles at $-B/T''$ and $-1/BT'$ (×), zeros at $-1/T''$ and $-1/T'$ (○), on the σ–$j\omega$ plane]

Application:

$$\text{Uncompensated forward transfer function} = G_{un} = \frac{k_0}{(s+5)(s+1)s} .$$

Block-diagram of the system:

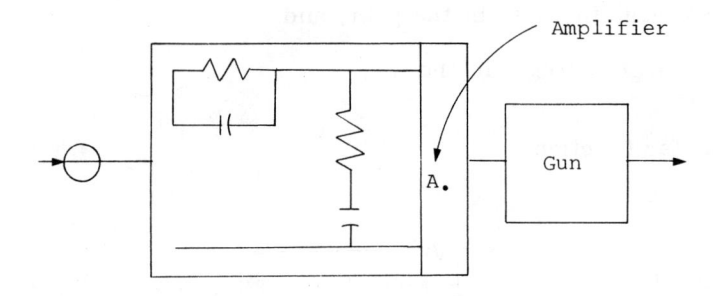

[Block diagram showing input summing junction, a block containing RC network labeled A. (Amplifier), feeding into a Gun block]

$$\text{New transfer } G_{NEW} = \frac{Ak_0(s + \frac{1}{T'})(s + \frac{1}{T''})}{(s+5)s(s+1)(s + \frac{1}{\beta T'})(s + \frac{\beta}{T''})} .$$

Selection of poles and zeros: Selecting β

The poles and zeros of this compensator are selected as follows:

A) Integral or lag element:
   a) $S_z = \frac{-1}{T'}$ and the pole $S_p = \frac{1}{\beta T'}$ are selected close to each other, with $\beta = 10$.
   b) To improve the gain, p and z are located closed to the origin and to the left.

G-143

B) Lead element:

$S_z = \frac{-1}{T''}$ is placed on the pole of the original system, so

$$\beta = 10 \text{ and,}$$

$$\frac{1}{T'} = \frac{1}{10}, \quad T'' = 1$$

$$\beta T' = 100 \text{ and } \frac{T''}{\beta} = 0.1 \text{ and}$$

$$G = \frac{k_0(s + \frac{1}{10})}{s(s + \frac{1}{100})(s + 5)(s + 10)}.$$

Root-locus:

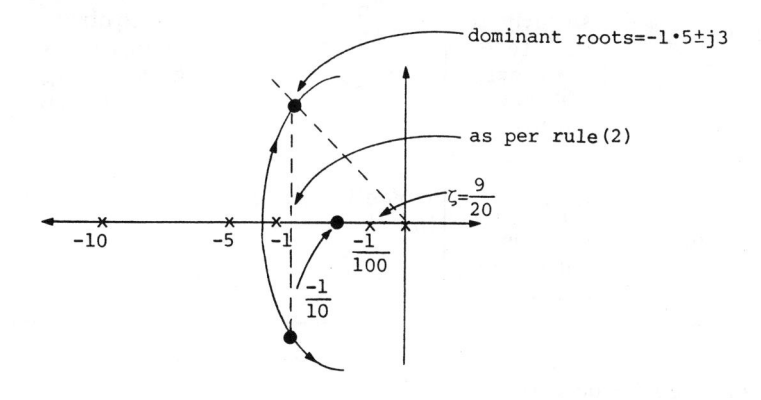

dominant roots = $-1 \cdot 5 \pm j3$

as per rule(2)

$\zeta = \frac{9}{20}$

K_1 = Static loop sensitivity at the dominant roots

$$= \frac{|S| \times |s + \frac{1}{100}| |s + 5| |s + \frac{1}{.1}|}{|s + \frac{1}{10}|}$$

$$= 143$$

ω_n = Undamped frequency $\cong 3$ rad/sec

Error coefficient for the ramp input $= \dfrac{k_1(\frac{1}{10})}{(\frac{1}{100})(s)(\frac{1}{0.1})} \cong 28'/\text{sec}.$

12.2.6 COMPARISON OF COMPENSATORS
Table 12.1

Type of compensator	Effect on Gain k	Effect on Undamped natural frequency ω_n	Comments
Lag	The gain increases by a factor $= \beta$, so the e_{ss} reduces considerably.	Decrease in ω_n so the settling time of the original system increases slightly	
Lead	k increases slightly so the steady-state error reduces slightly	Large increase in ω_n	An amplifier unit with gain A is required such that A > gain of original system.
Lag-lead	k_n increases considerably.	Large increase in ω_n	

Frequency response:

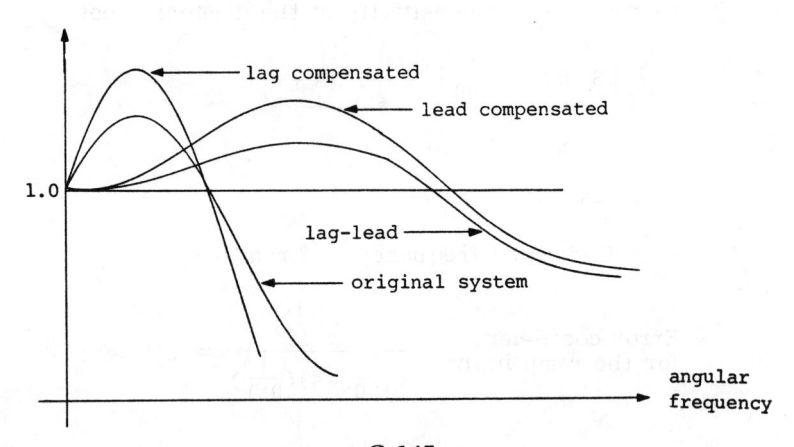

- lag compensated
- lead compensated
- lag-lead
- original system
- 1.0
- angular frequency

G-145

CHAPTER 13

FREQUENCY-RESPONSE PLOTS OF CASCADE COMPENSATED SYSTEMS

13.1 SELECTING A PROPER COMPENSATOR

A basic approach:

Let G_1 = The forward transfer function of the original feedback system.

Let G_2 = The desired transfer function, which means that the function will result in specified stability and e_{ss}.

Then $G = \dfrac{G_2}{G_1}$ is the equation from which the network of the compensator can be determined.

Characteristics of the fundamental compensators: The features of the basic types of compensators are given as follows:

Polar plots:

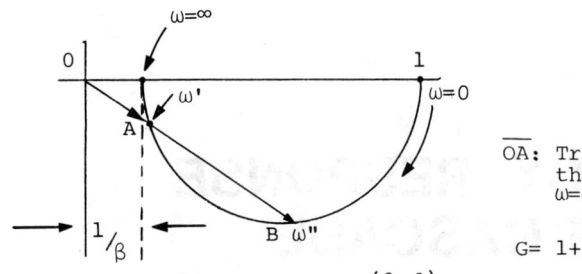

Fig. Lag compensator ($\beta > 1$)

$\overline{OA}$: Transfer function of the compensator for $\omega = \omega'$

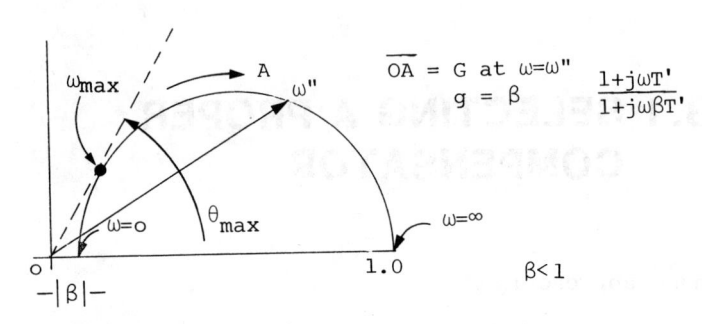

$$G = \frac{1+j\omega T'}{1+j\omega\beta T'}$$

Fig. Lead network

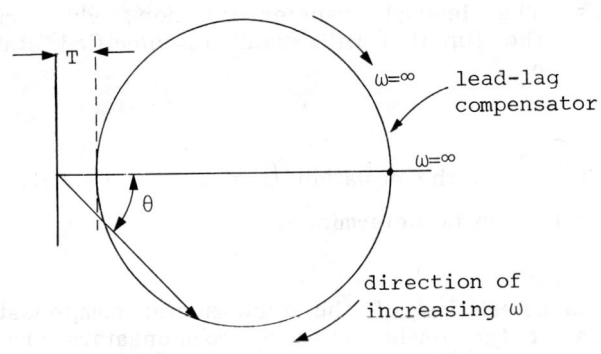

$\overline{OA} = G$ at $\omega = \omega''$
$g = \beta \quad \dfrac{1+j\omega T'}{1+j\omega\beta T'}$

$\beta < 1$

Fig. Lag-lead network

lead-lag compensator

direction of increasing ω

Lag-lead network:

$$T = \frac{T' + T''}{T' + T'' + T'''} \cdot$$

$$G = \frac{1 + j\omega(T' + T'') - \omega^2 T'T''}{1 + j\omega(T' + T'' + T''') - \omega^2 T'T''} \cdot$$

Table 13.1

Type of Compensator	Polarplot: Characteristics				Effects on the Original System Due to Addition of a Compensator	
	$\|G\| \omega = \omega_1$		$G(j\omega)$ increasing ω	Phasor $G(j\omega_1)$	Value of New ω_m	Over-all gain
	$\omega_1=0$	$\omega_1=\infty$	and/or increasing ω			
lag	1	$\frac{1}{\beta}$	angle decreases	angle varies from $0 - {}^-90°$	The phasor revolves clockwise	Lower	There is a very large increase. The gain is multiplied by β.
lead	β	1	increases	angle varies from $0 - {}^+90°$	The phasor revolves counter-clockwise	Higher	A small increase.

Mathematical approach:

$$G = |G| \angle \theta$$

The transfer function of a lag and lead compensator.

$$\theta = \tan^{-1} \omega T' - \tan^{-1} \omega\beta T'$$

$$= \tan^{-1}\left[\frac{\omega T' - \omega\beta T'}{1 + \omega^2 \beta T'^2}\right].$$

$$\tan \theta = \omega T' - \omega\beta T'/1 + \omega^2 \beta T'^2.$$

The maximum value of the angle θ at an angular frequency

$$\omega = 1/T'\sqrt{\beta} \quad \text{and} \quad \theta_{max} = \sin^{-1}\left[\frac{\frac{1}{\beta} - 1}{\frac{1}{\beta} + 1}\right] \cdot$$

For given values of β, θ and ω, the value of T' can be determined which is required for the compensator.

13.2 ANALYSIS OF A LAG NETWORK

When to use a lag network:

If the transient response of a feedback system is reasonably okay but the e_{ss} is very large, then the gain of the system is increased by introducing a lag network.

How to increase the gain of the system:

$$G_{com} = A_0 \frac{1 + T's}{1 + \beta T's}$$ The transfer function of the compensator.

Type 0 system: The e_{ss} can be reduced by increasing step error coefficient k.

$$k = \lim_{j\omega \to 0} G(j\omega) \quad \text{initial value}$$

$$k' = \lim_{j\omega \to 0} G_{com} G = A_0 k$$

So there must be a built-in amplifier of gain $A_0 = \frac{k'}{k}$.

Note: The change in the gain must not affect the transient response of the system.

Characteristics:

Note:

A) The phase margin at $\omega = \omega_{\theta_1}$ has been reduced.

B) The compensator is designed such that the log-magnitude β occurs at $\omega = \omega_{\theta_1}$.

C) The gain can now be increased so that the Lm plot have value of 0db at $\omega = \omega_{\theta_2}$.

G-149

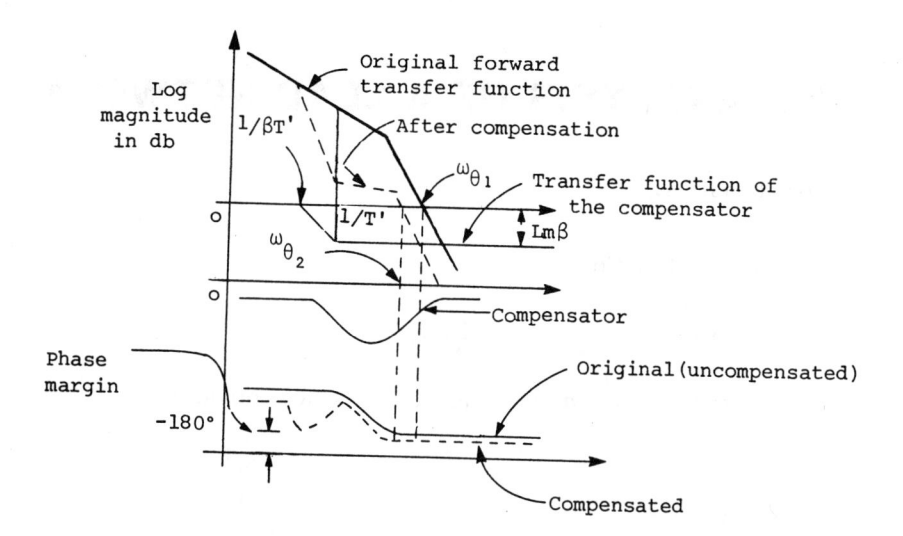

[Figure: Bode plot showing Log magnitude in db vs frequency, with labels: Original forward transfer function, After compensation, $1/\beta T'$, $1/T'$, ω_{θ_1}, ω_{θ_2}, Transfer function of the compensator, $Lm\beta$, Compensator, Phase margin, $-180°$, Original (uncompensated), Compensated]

Effects:

A) Gain adjustment for specified Mag_{max}.

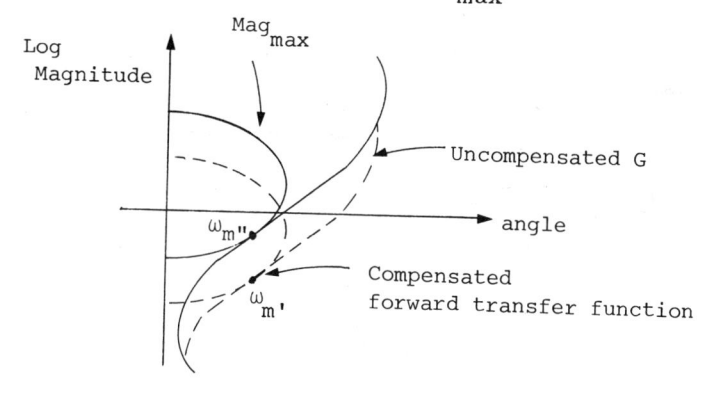

[Figure: Nichols chart showing Log Magnitude vs angle, with Mag_{max} contour, Uncompensated G, Compensated forward transfer function, ω_m'', ω_m']

    a) Increase in gain to obtain same Mag_{max} = The amount by which dotted curve must be raised to make it tangent to the Mag_{max} curve.

    b) Resonant frequency $\omega_{m'} < \omega_{m''}$.

       The reason is the negative angle introduced by the network.

B) The lag network acts as a low-pass filter.

13.3 ANALYSIS OF A LEAD NETWORK

Characteristics:

A) Transfer function:

$$G(j\omega) = \beta \, \frac{1 + j\omega T'}{1 + j\omega \beta T'} \cdot$$

Note that $\beta < 1$ and the amplifier is not considered.

B) Lm and angle:

$$\text{Lm}\,G(j\omega) = \text{Lm}\,\beta + \text{Lm}(1 + j\omega T') - \text{Lm}(1 + j\omega \beta T')$$

$$\underline{/G(j\omega)} = \underline{/1 + j\omega T'} - \underline{/1 + j\omega \beta T'}.$$

C) Plots:

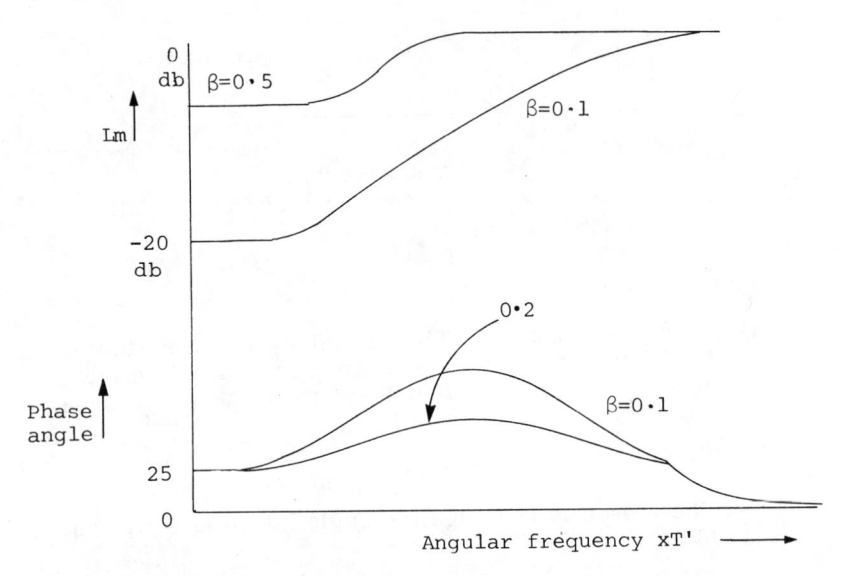

Application to a Feedback System: Curves and Important Facts

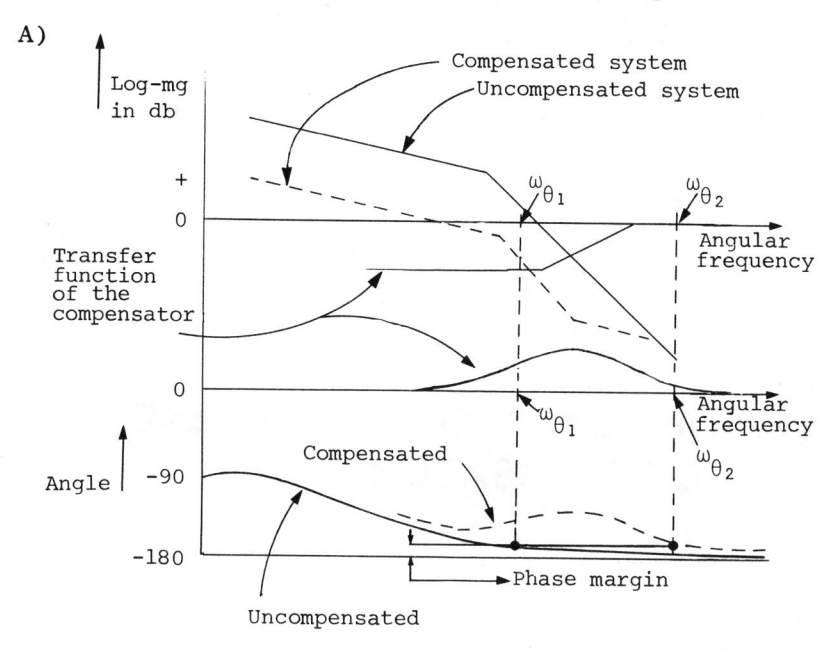

A)

[Figure: Bode plots showing log-magnitude and phase for compensated system, uncompensated system, and transfer function of the compensator, with phase-margin frequencies ω_{θ_1} and ω_{θ_2} indicated, and phase margin marked between -90 and -180 degrees.]

B) The lead network is a high-pass filter. For the frequencies below $\omega = 1/T'$, this lead network introduces an attenuation of value log-magnitude β.

C) The main function of the lead compensator is to increase ω_θ (the phase-margin frequency). By properly choosing T', the phase-margin frequency can be increased from ω_{θ_1} to ω_{θ_2}.

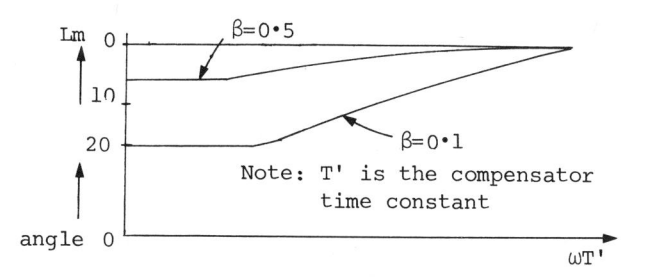

[Figure: Lm vs $\omega T'$ plot showing curves for $\beta = 0.5$ and $\beta = 0.1$. Note: T' is the compensator time constant]

This figure shows that for a constant β, if T' is made smaller, then the new gain of the system is very large.

D) Selection of T':

  a) Type 0 system:

  T' ≤ Second largest value of the time constant of the uncompensated system

  b) Type 1 or higher order:

  T' ≤ The largest time constant of the original forward transfer function

13.4 ANALYSIS OF A LAG-LEAD COMPENSATOR

Characteristics:

A) Transfer function:

$$G(j\omega) = \frac{A_0(1 + j\omega T')(1 + j\omega T'')}{(1 + j\omega \beta T')(1 + \frac{j\omega T''}{\beta})} \qquad \beta > 1 \text{ and } T' > T''$$

$$= A_0 \underbrace{\left[\frac{1 + j\omega T'}{1 + j\omega \beta T'}\right]}_{\text{lag effect}} \underbrace{\left[\frac{(1 + j\omega T'')}{(1 + \frac{j\omega T''}{\beta})}\right]}_{\text{lead effect}}$$

B) Plot:

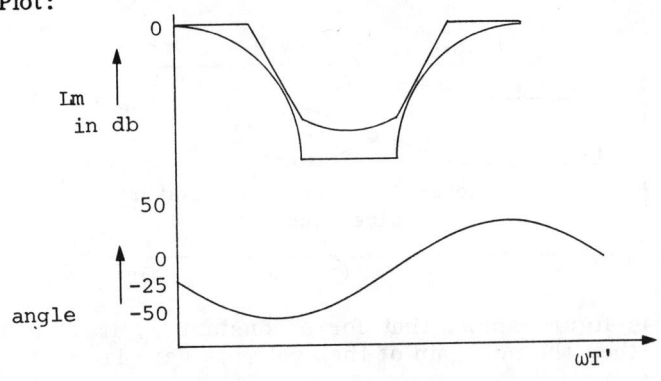

C) Uncompensated and compensated plots:

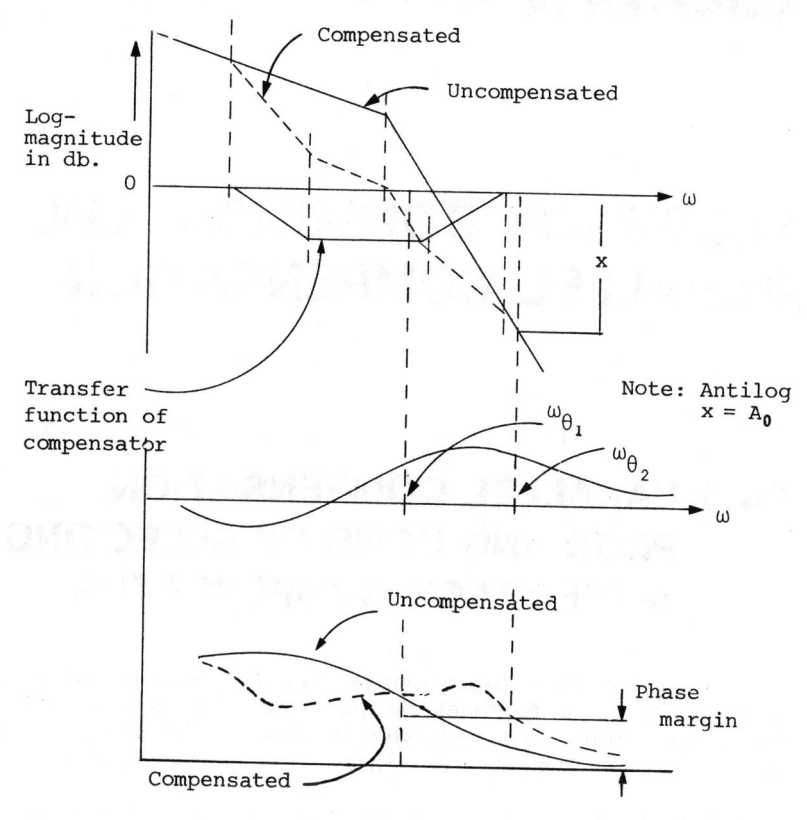

CHAPTER 14

FEEDBACK COMPENSATION: PARALLEL COMPENSATION

14.1 PARALLEL COMPENSATION: PROS AND CONS OF SELECTING A FEEDBACK COMPENSATOR

A faster time of response can be obtained with an introduction of a feedback compensator. The components of the compensator can be adjusted to obtain a small closed-loop time constant. Sometimes this time is an important factor.

Sometimes situations solely dictate the choice of a parallel compensation for greater stability.

The design procedure for a feedback compensator is more complicated than the series compensator.

Parameter variations: Effect on the overall stability

$\gamma_1 \longrightarrow \boxed{G_1} \longrightarrow C_1$

Forward transfer function

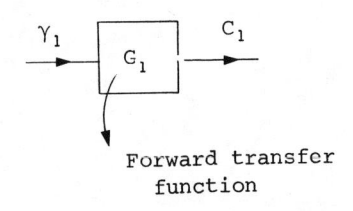

Fig.1 Open-loop system

CASE STUDY 1

Open-loop system:
$$C_1 = \gamma_1 G_1$$

$$dC_1 = \frac{dG_1}{G_1} C_1.$$

So a change in G can cause a corresponding change in the output, C_1.

CASE STUDY 2

$$C_2 = \frac{\gamma_2 G_2}{1 + G_2 H_2}.$$

$$dC_2 = \frac{\gamma_2 dG_2}{(1 + G_2 H_2)^2} = \left(\frac{1}{(1 + G_2 H_2)} \frac{dG_2}{G} \right) C_2.$$

Fig. 2

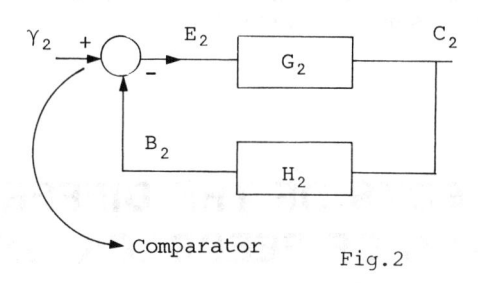

Thus when going from open-loop to closed-loop control, the effect of parameter changes upon the output is reduced by a factor of $1/(1 + GH)$. Note: In case of a series compensator, the value of $1 + G_2 H_2$ is high compared to $1 + G$ which results due to a parallel compensator. This is the plus point of a parallel compensator.

$$-dC_2 = \frac{\gamma_2 G_2^2 C_1 H_2}{(1 + G_2 H_2)^2} \qquad \text{Assuming } G_2 \text{ is fixed.}$$

$$\approx -C_2 \frac{dH_2}{H_2} \qquad \text{This is valid if } |G_2 H_2| > 1.$$

This also points out the advantages of a parallel compensator.

Sensitivity function of a system:

$$S = \frac{\Delta M}{\Delta \sigma} \Bigg|_{\text{for given variations in the parameter value.}}$$

Note:

A) M is the response and σ represents a set of parameters in the open-loop of the system.

B) $|S| \leq 1$.

C) $|S| \to 1$ as the system becomes more sensitive to the variations of a parameter.

D) Open-loop poles have very small effects on C (i.e. the output response) if and only if $S \leq 1$.

E) Parallel compensator makes a system less sensitive as compared to a series compensator.

14.2 EFFECTS OF THE DIFFERENT TYPES OF FEEDBACK ON THE SYSTEM'S TIME RESPONSE

A comparison

A) Specifications:

$$\text{input} \quad R(t) = \frac{u(t)}{0.1}.$$

B) $C_{ss} t = 1$; System with no feedback

$$C(s) = \frac{A}{S + 10} R(s)$$

$$= \frac{10A}{S(S + 10)}.$$

```
input  ┌─────────┐  output
──────▶│    A    │──────▶
       │  s + 10 │
       └─────────┘
```

The value of gain A can be obtained using the final value theorem.

$$C_{ss}(t) = \lim_{S \to 0} S(LS) = \frac{10A}{10} = 1.$$

So A = 1 and

$$C(s) = \frac{1}{(1 + \frac{1}{10}S)S}$$

so the system time constant is 1/10 second.

Original system with unity feedback:

$$C(s) = \frac{A_0}{\left[\frac{1}{10}s + 1 + \frac{A_0}{10}\right]s} \quad \text{because } R(s) = \frac{10}{S}.$$

Applying the final value theorem, $A_0 = 1.11$ the desired output is

$$C(s) = \frac{1}{s(1 + 0.09)s}.$$

System with series compensation:

```
r(t)=           ┌────────┐   ┌────────┐
1/.10 u(t) ───▶○───▶│  A₀    │──▶│ s+10   │─────┬──▶
               ▲    │(10+s)  │   │ s+100  │     │
               │    └────────┘   └────────┘     │
               │        G₁           G₂    Lead │
               │                         Compensator
               └─────────────────────────────────┘
```

$$\frac{C(s)}{R(s)} = \frac{G_1 G_2}{1 + G_1 G_2} = \frac{\frac{A_0}{s+10} \cdot \frac{s+10}{s+100}}{1 + \frac{A_0}{s+100}}$$

$$\Rightarrow \quad C(s) = \frac{10 A_0}{[(s + 100) + A_0]s} \quad \text{since } R(s) = \frac{10}{s}.$$

G-158

$$C(t)_{ss} = 1 = \lim_{s \to 0} SC(s) = \frac{10A_0}{A_0 + 100} \quad \text{from use of final value theorem.}$$

$\Rightarrow \quad A_0 + 100 = 10A_0$

$A_0 = 11.$

Hence,
$$C(s) = \frac{110}{S(S+11)} = \frac{1}{S\left(\frac{111}{110} + \frac{S}{110}\right)} = \frac{1}{S(1+0.009S)} \ .$$

Therefore the system's time constant is 0.009 second; note the improvement.

Original system with parallel compensation:

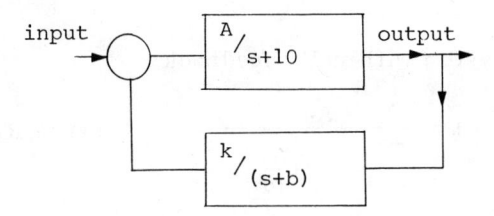

Let the closed-loop response be given by

$$C(s) = \frac{10}{s} \left[\frac{A(s+b)}{s^2 + (10+b)s + (10b + Ak)} \right]$$

$$C_{ss} = \lim_{S \to 0} SC(s) = \lim_{S \to 0} \left[\frac{10A(s+b)}{s^2 + (10+b)s + (10b + Ak)} \right]$$

$= 1.$

So simplifying this we get

$$1 = \frac{10Ab}{[10b + Ak]} \quad \text{i.e.} \quad 10b + Ak = 10Ab.$$

The system time constant in the previous case was .009. So the compensator can be designed for a time constant < .009. It is possible to get a faster time of response using a feedback compensator.

Location of a feedback compensator in the system:

A) The location of a feedback compensator totally depends on the type of system to be compensated.

B) There is always a unity feedback from the output to the input besides the feedback compensating loop. These two loops make the overall system complicated as compared to the system with a series compensator.

14.3 APPLICATION OF LOG-MAGNITUDE CURVE: FOR FEEDBACK COMPENSATION

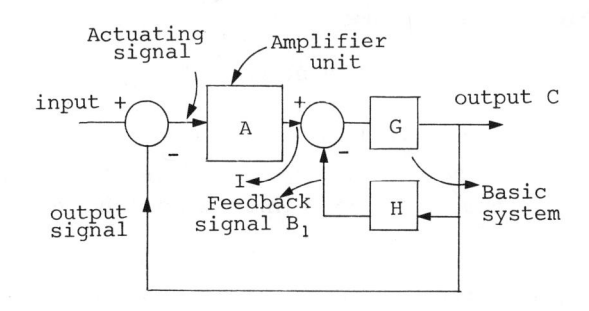

$$G_1 = \frac{C}{I} = \frac{G}{1 + GH} .$$

Note all of them are functions of $j\omega$ and G_1 is the transfer function of the system.

Let $|GH| \ll 1$, then $G_1 \approx G$.

Let $|GH| \gg 1$, then the following approximation is good

$$G_1 \approx \frac{1}{H} .$$

where G_1 and H are phasors and $j\omega$ dependent quantities.

Note: $|GH| \approx 1$ case is neglected.

Direct Application:

$$G = \frac{k'}{S(1 + ST')}.$$

Forward transfer function of a system whose $|H(j\omega)| = 1$ and angle $\angle H(j\omega) = 0°$.

A) Log-magnitude plot of GH:

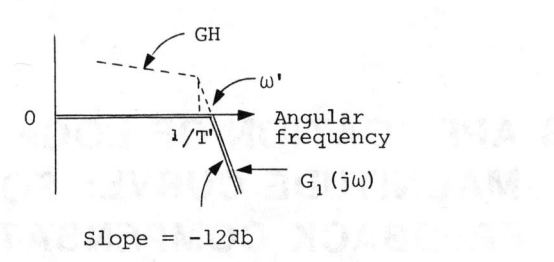

```
              GH
       ----↙
              ↘ ω'
    0 |————————→ Angular
        1/T'  ↘    frequency
              ↖— G₁(jω)

      Slope = -12db
```

B) Observations from the plot:

For $\omega < \omega'$, $|G_1(j\omega)|$ can be given by $\frac{1}{H(j\omega)}$.

For $\omega > \omega'$, $|G_1(j\omega)H(j\omega)| < 1$.

C) So $G_1(j\omega)$ is represented by a thick line as shown above.

D) From the above observations, it can be said that the $G_1(j\omega)$ should look like

$$G_1(j\omega) = \frac{1}{\left[\frac{2\omega j \zeta}{\omega'} + \frac{-\omega^2}{\omega'^2} + 1\right]}.$$

To obtain exact curve of $G_1(j\omega)$: Steps

A) The forward transfer function (overall) is expressed in the following form:

$$G_1(j\omega) = \frac{G''(j\omega)}{H(j\omega) + G''(j\omega)H(j\omega)}$$

where $G'' = G \cdot H$ and is a function of ω.

B) Next, the angle and log-magnitude of $G'' = H(j\omega)G_1(j\omega)$ are plotted on the Nichol's chart.

C) The points of intersection of $G''(j\omega)$ plot with constant magnitude and constant angle curves are determined.

D) These points are used to plot the curve $\dfrac{G''(j\omega)}{1+ G''(j\omega)}$.

E) $1/H(j\omega)$ curve is next plotted and it is combined to produce the entire $G_1(j\omega)$ curve.

CHAPTER 15

SYSTEM SIMULATIONS: USE OF ANALOG COMPUTERS

15.1 ANALOG COMPUTER: BASIC COMPONENTS

OPERATIONAL AMPLIFIER

Schematic diagram:

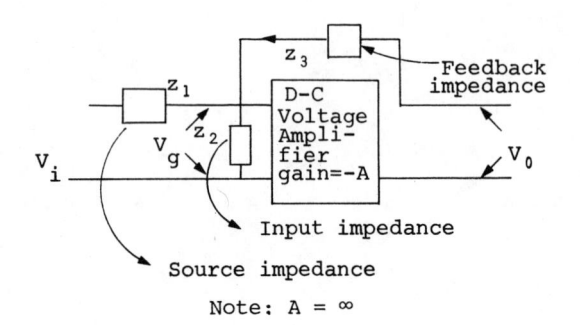

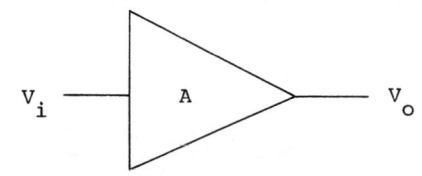

Note: $A = \infty$

Fig. Symbolic Representation

$$\frac{V_o}{V_i} = \frac{-Z_3}{Z_1} \left[\frac{1}{1 + \frac{1}{A}[1 + Z_3(Z_1+Z_2)/Z_1 Z_2]} \right].$$

For an ideal opamp, $Z_2 \gg Z_1$ and Z_3 and $A \to \infty$, so

$$\frac{V_o}{V_i} = \frac{-Z_3}{Z_1}$$

Opamp as a summer:

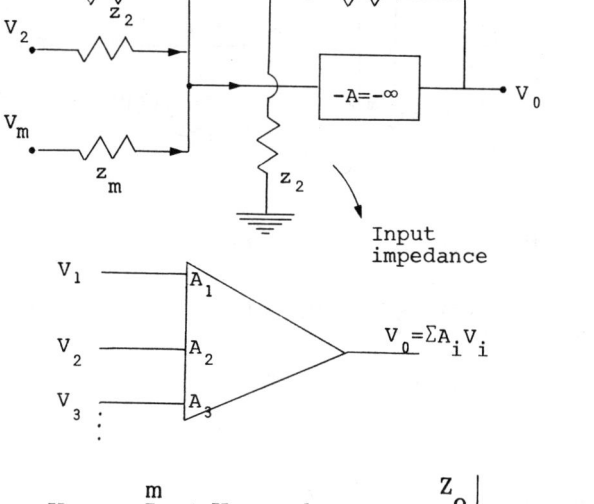

[Circuit diagram showing inputs V_1, V_2, V_m through impedances z_1, z_2, z_m to opamp with feedback z_0 and input impedance z_2, output V_0, gain $-A = -\infty$]

[Block symbol: inputs $V_1, V_2, V_3, \ldots$ with gains A_1, A_2, A_3, output $V_0 = \Sigma A_i V_i$]

$$V_o = -\sum_{i=1}^{m} A_i V_i \quad \text{where} \quad A_i = \frac{Z_o}{Z_i} \bigg|$$

i varies from 1 to m.

When the input and feedback impedances are pure resistors, this scheme is known as a summer.

Integrator:

$$\frac{V_o(s)}{V_i(s)} = \frac{-1}{RCs}$$

G-164

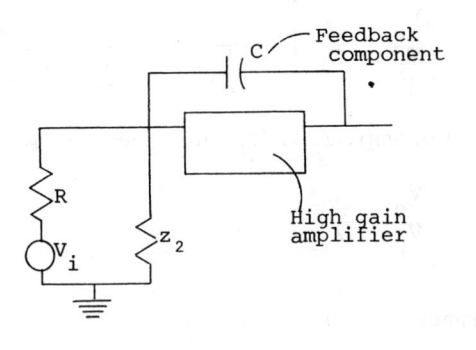

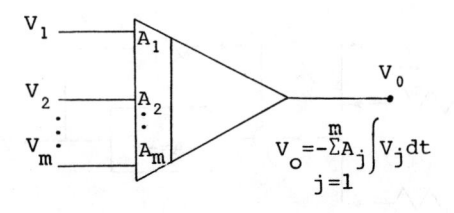

and in the time-domain this relation is

$$V_o(t) = \frac{-1}{RC} \int V_i(t)dt$$

$\frac{1}{RC}$ represents a change in gain.

Potentiometer:

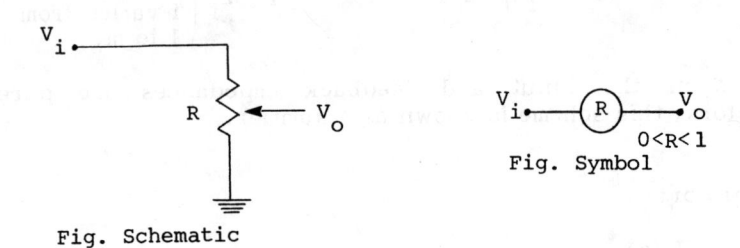

Fig. Schematic

Fig. Symbol

$0 < R < 1$

15.2 SIMULATIONS USING ANALOG COMPUTER

Second-order differential equation:

$$D^2x + \frac{a_1}{a_2}Dx + \frac{ax}{a_2} = f(t)/a_2 \quad \text{A second-order equation.}$$

This equation is solved for the highest derivative.

$$a_2 D^2 x = f(t) - ax - a_1 Dx$$

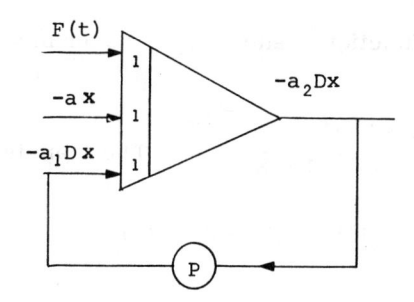

because

$$\int a_2 D^2 x = \int F(t) + \int -ax + \int -a_1 Dx$$

$$a_2 Dx = \int F(t) + \int -ax + \int -a_1 Dx.$$

(The output of integrator is $-a_2 Dx$ because of negative unity gain.)

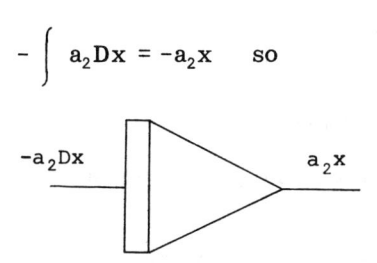

$$-\int a_2 Dx = -a_2 x \quad \text{so}$$

G-166

So the scheme for simulating the equation is:

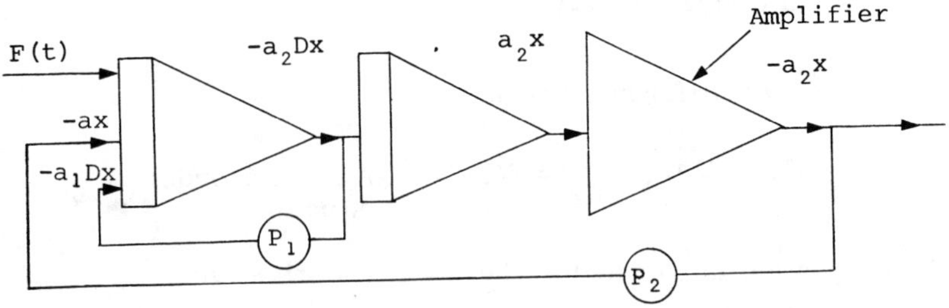

Amplifier
F(t)
-a2Dx
a2x
-a2x
-ax
-a1Dx
P1
P2

Simulation of a lag network:

Transfer function and the governing differential equation:

$$\frac{V_{out}(s)}{V_{in}(s)} = \frac{A}{1 + ST'} \qquad \text{The transfer function.}$$

$$(T'D + 1)V_{out}(t) = AV_{in}(t).$$

Computer simulation:

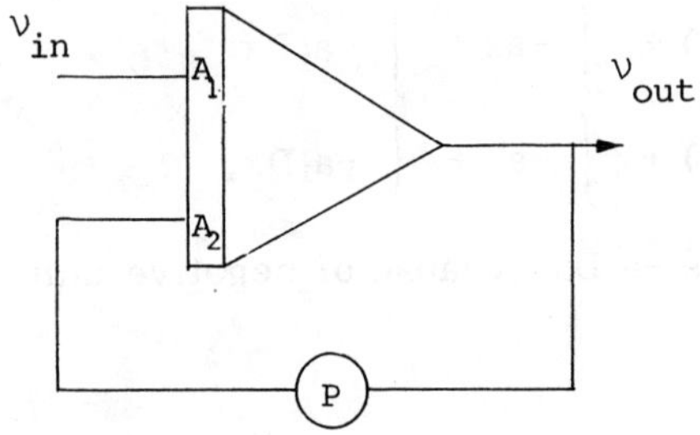

v in
A1
A2
v out
P

Note: i) $\frac{1}{PA_2} = T'$

ii) $\frac{-A_1}{PA_2} = A$

Simulation of a lead network:

Transfer function and the governing differential equation:

$$\frac{V_{out}(s)}{V_{in}(s)} = \frac{1 + T's}{1 + \beta T's} \qquad \text{Transfer function.}$$

$$V_{in} - V_{out} + T' Dv_{in} = \beta T' DV_{out} \quad \text{Differential equation.}$$

Mathematical manipulation:

$$\frac{V_{out}(s)}{V_{in}(s)} = \frac{1}{\beta}\left[\frac{\beta - 1}{1 + \beta T's} + 1\right].$$

Case #1: $\beta < 1$

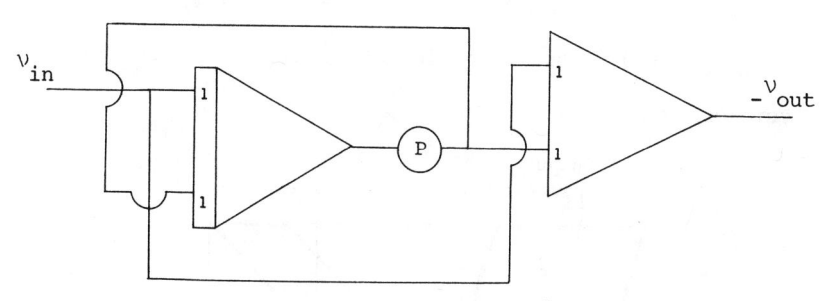

Note: P is set to $1/\beta T'$

Analog computer simulations of special curves:

Use of ideal diodes and relays

Saturation curve:

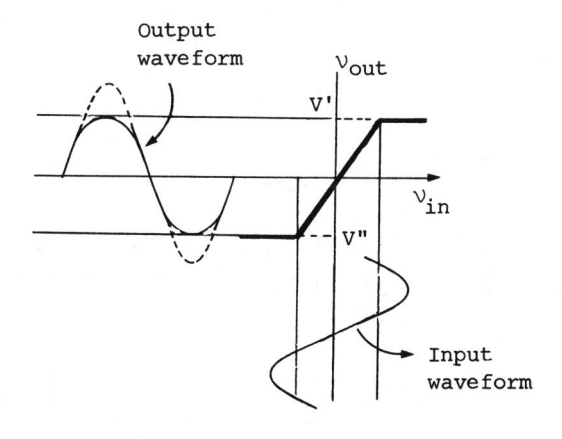

G-168

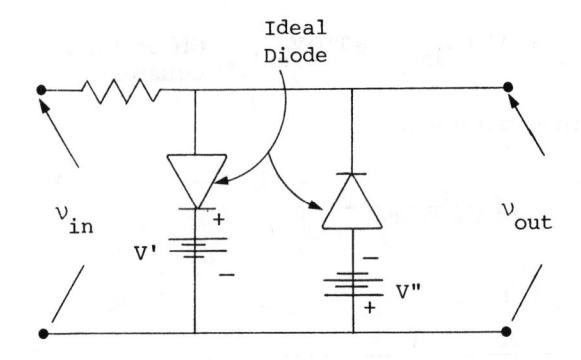

Fig. Shunt-type simulating circuit.

Hysteresis curve:

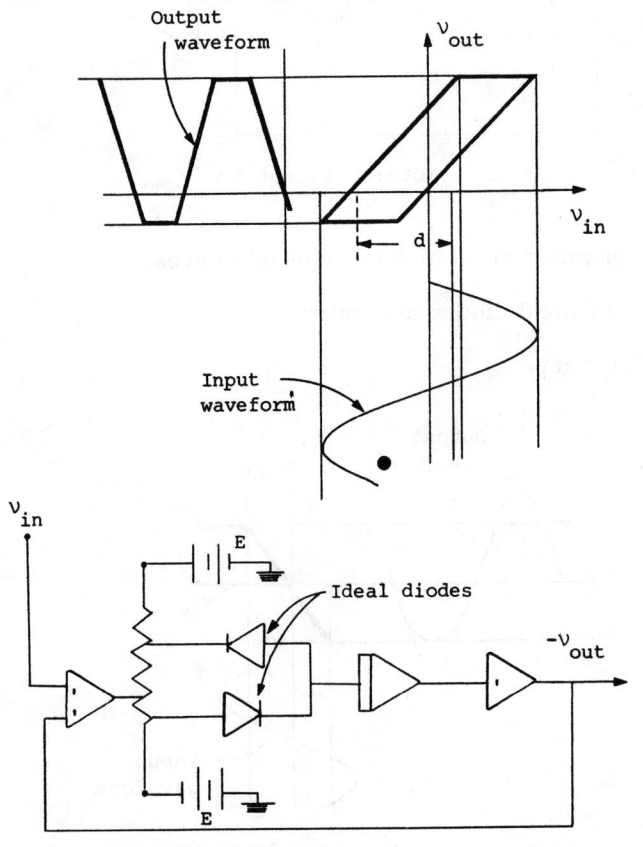

Fig. Back-lash circuit

G-169

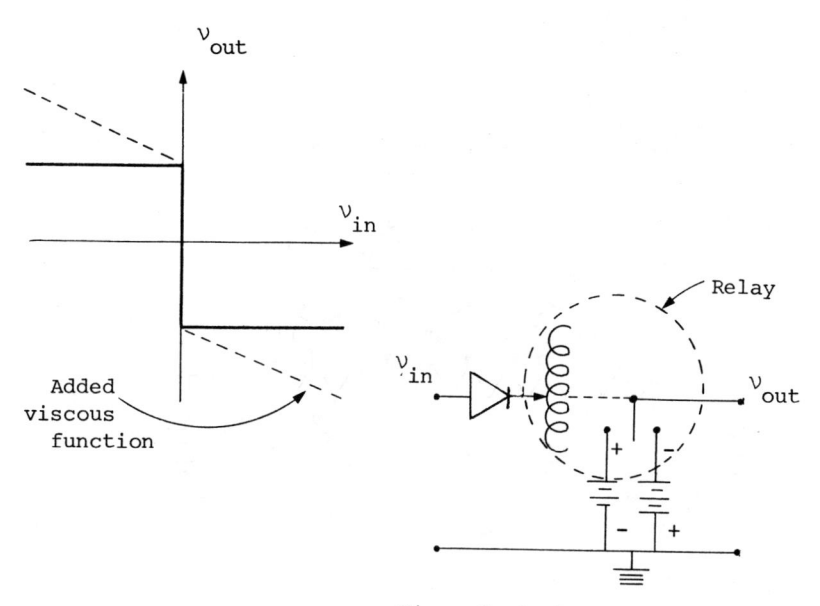

Fig.: Coulomb-friction circuit

Simulation of a continuous function:

The continuous function is represented as follows on a x-y plane:

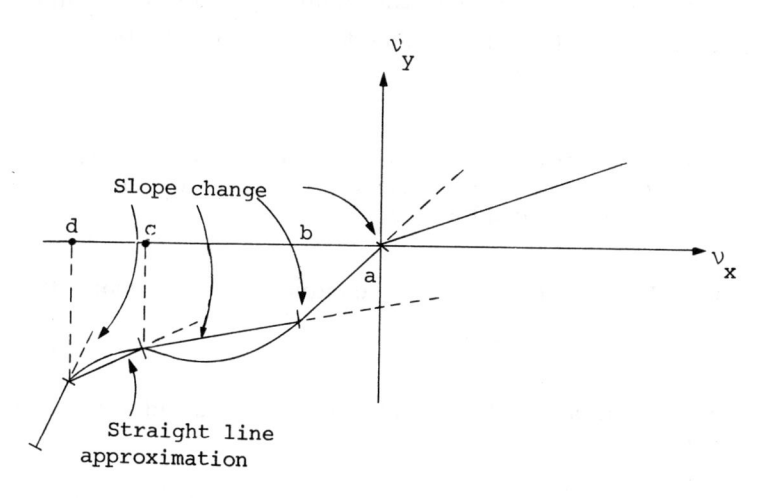

G-170

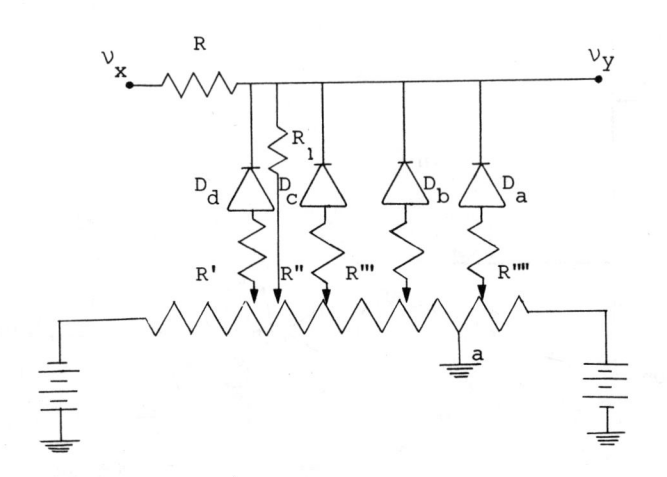

Procedure:

A) The number of ideal diodes required in the circuit is equal to number of changes in the slope.

B) When $V_x <$ d, then the diode D_d conducts and V_y depends on R and R'. For $V_x > d_1$ the diode enters in the cutoff region.

C) When the input voltage is in the range d-c, the slope from d to c depends solely on R and R_1 since all the diodes are non-conducting.

D) For $V_x >$ c, the diode D_c conducts and the slope from c to b depends on R/R".

E) Tap point a on the potentiometer is reached by trial and error.

G-171

15.3 APPLICATION OF ANALOG COMPUTERS FOR SYSTEM TUNING

Unity feedback system: Tuning

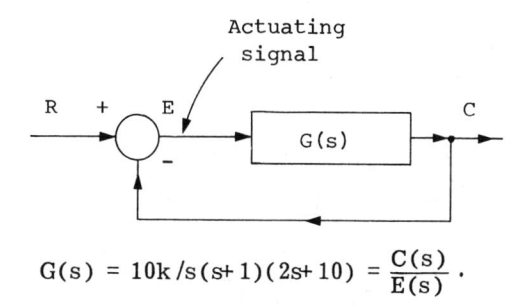

$$G(s) = 10k/s(s+1)(2s+10) = \frac{C(s)}{E(s)}.$$

So the corresponding differential equation is

$D^3C + 6D^2C + 5DC = 5ke$

$D^3C = 5ke - 6D^2 + 5DC.$

This is given by integrator #1.

The analog computer can be set up as follows:

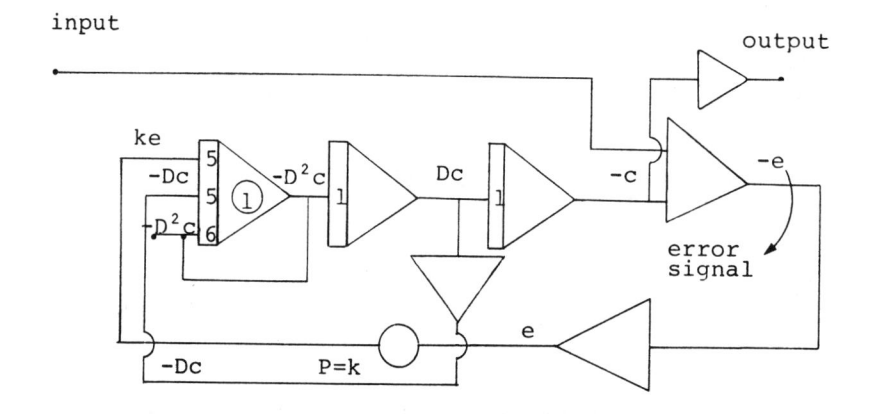

System response and adjusting potentiometer P:

The system gain k is adjusted by means of a potentiometer until the desired response is achieved.

Compensated unity feedback system:

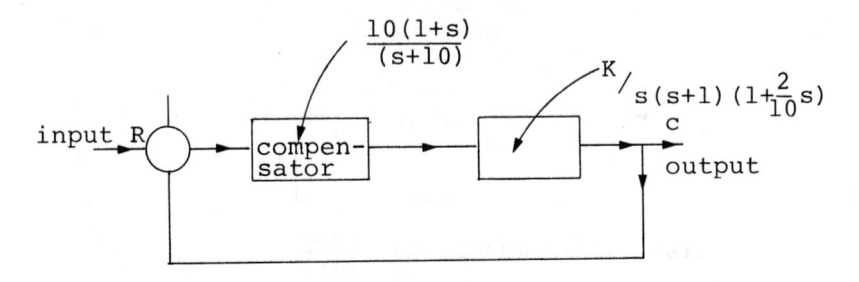

$$\frac{10(1+s)}{(s+10)}$$

input R → compensator → K → $\frac{1}{s(s+1)(1+\frac{2}{10}s)}$ → output C

The overall transfer function is

$$\frac{C(s)}{E(s)} = \frac{10k}{s(s+10)(1+\frac{2}{10}s)}$$

So the corresponding differential equation is

$$D^3C + 15D^2C + 50DC = 50ke$$

i.e. $D^3C = 50ke - 15D^2C - 50DC$

The computer set-up is:

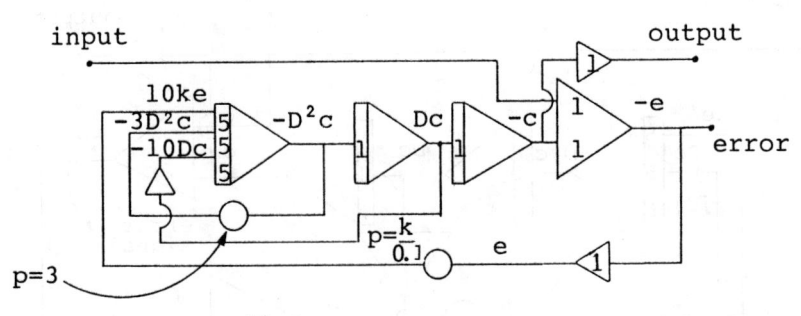

input → 10ke, -3D²c, -10Dc → -D²c → Dc → -c → -e → output / error

$p = \frac{k}{0.1}$, e, p=3

Time-response

Note: It can be seen that the settling time has been reduced considerably.

15.4 SETTING OF A CONTROL SYSTEM: CONTROLLER SETTING

ZIEGLER-NICHOLS METHOD
Salient features:

A) It is widely used for single-loop process control systems.

B) Procedure:

The loop is opened just after the controller and,

a small step input (Δs) is applied to the final control element.

The feedback signal is determined at the point where it is compared with the input signal.

The open-loop response to a small step signal is plotted and following quantities are determined from the curve.

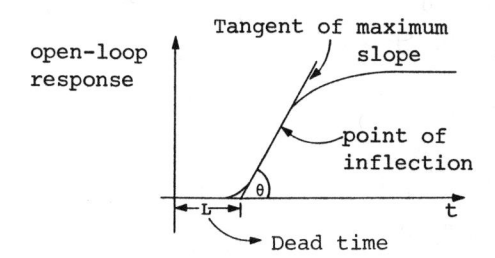

```
                     Tangent of maximum
    open-loop              slope
    response
                              
                              point of
                              inflection
                        θ
                    |-L-|                    t
                       Dead time
```

C) Calculations of the controller constants:

Proportional control

$$k_p = \Delta s / L \tan\theta$$

$$T_i = 3.33L \text{ minutes.}$$

PID control (proportional plus integral plus derivative control):

$$k_p = 1.2 \Delta s / L \tan\theta$$

$$T_i = 2L \text{ minutes}$$

$$T_d = 0.5L \text{ minutes.}$$

ULTIMATE-CYCLE METHOD

Following two parameters are required:

A) Ultimate gain k_u: It is the gain at which the closed-loop system just begins to oscillate.

B) P_u (ultimate period): It is the time period associated with frequency of oscillation ω_0.

Controller settings:

Proportional control

$$k_p = 0.5 k_u.$$

PI control

$$k_p = k_u/2.22$$
$$T_i = 0.83 P_u.$$

PID controller

$$k_p = 0.6 k_u$$
$$T_i = 0.5 P_u$$
$$T_d = 0.125 P_u.$$

Handbook of Mechanical Engineering

SECTION H
Mathematics for Engineers

CHAPTER 1

VECTORS, MATRICES, AND EQUATION SYSTEMS

Note: Vectors are indicated by **bold** type.

1.1 VECTORS IN THREE DIMENSIONS

FAMILIAR IDEAS

Vector **X** = (a, b, c) (See Figure 1.1)

Vector Components a, b, c

Equality of Vectors

Scalar

Zero Vector **0** = $(0, 0, 0)$

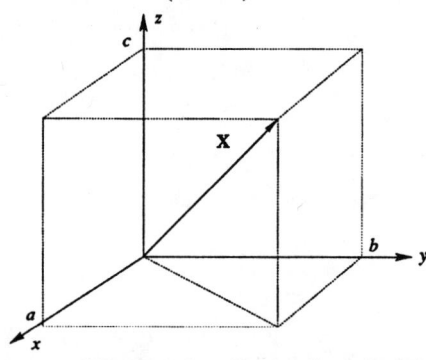

FIGURE 1.1 — Vector Components

H-1

DEFINITION 1.1 Norm (Length, Magnitude) of $X = (a, b, c)$

$|X| = \{a^2 + b^2 + c^2\}^{1/2}$.

DEFINITION 1.2 Unit Vector

A vector whose norm is 1, $|X| = 1$.

DEFINITION 1.3 Basis Vectors in x, y, z Space

$\mathbf{i} = (1, 0, 0), \mathbf{j} = (0, 1, 0), \mathbf{k} = (0, 0, 1)$.

REMARK 1.1 Representing X in Terms of Basis Vectors:

$X = (a, b, c) = a\mathbf{i}, b\mathbf{j} + c\mathbf{k}$

DEFINITION 1.4 Direction Cosines of $X = (a, b, c)$

The cosines of the angles α, β, Γ between X and the x, y, z axes (Figure 1.2).

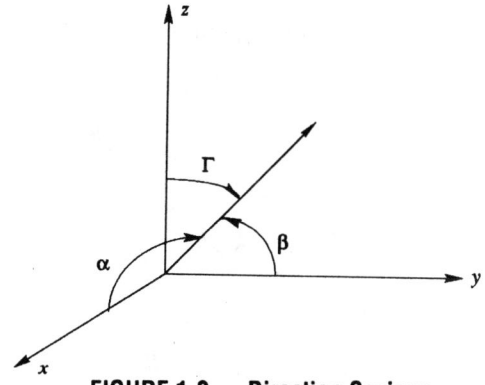

FIGURE 1.2 — Direction Cosines

THEOREM 1.1 $\Sigma = (\cos \alpha, \cos \beta, \cos \Gamma)$ is a Unit Vector

$1 = \cos^2\alpha + \cos^2\beta + \cos^2\Gamma$.

DEFINITION 1.5 Collinearity of X, Y

X, Y are parallel or have opposite directions.

DEFINITION 1.6 Sum of $X = (a, b, c)$, $Y = (A, B, C)$

$X + Y = (a + A, b + B, c + C)$.

DEFINITION 1.7 Product δX for δ = Scalar, X = Vector
$\delta X = (\delta a, \delta b, \delta c)$.

THEOREM 1.2 Properties of δX

1. $|\delta X| = |\delta| |X|$;
2. δX is parallel to X if $\delta > 0$,

    opposite to X if $\delta < 0$.

DEFINITION 1.8 Linear Combination of X, Y

$Z = AX + BY$, A, B = scalars.

REMARK 1.2 (Parallelogram Law)

$X + Y$ is diagonal of parallelogram of X, Y (Figure 1.3).

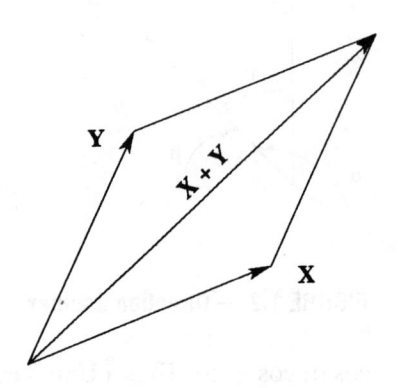

FIGURE 1.3 — Sum of Vectors

DEFINITION 1.9 Inner (Scalar) Product of $X = (a, b, c)$, $Y = (A, B, C)$

$(X, Y) = X \cdot Y = aA + bB + cC$.

THEOREM 1.3 Properties of $X \times Y$

For any X, Y and scalar δ,

$X \times Y = Y \times X$ (Commutative)

$(\delta X) \times Y = \delta (X \times Y)$ (Associative)

$|X \times Y| \le |X||Y|$ (Cauchy-Schwarz Inequality)

(= if and only if (iff) X and Y are collinear).

THEOREM 1.4 (Triangle Inequality) For any X, Y,

$|X + Y| \le |X| + |Y|$

$||X| - |Y|| \le |X - Y|$.

THEOREM 1.5 The Angle θ between X, Y (Figure 1.4)

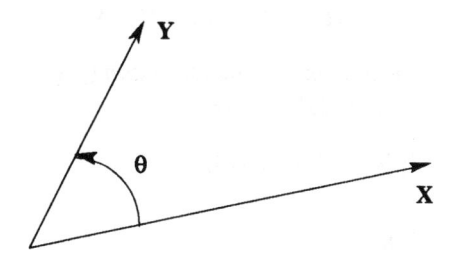

$\cos(\theta) = (X \times Y) / (|X||Y|)$.

FIGURE 1.4 — Angle Between Vectors

REMARK 1.3 A Unit Vector Parallel to X

For $X \ne 0$, $Y = X / |X|$ is parallel to X and $|Y| = 1$.

REMARK 1.4 Direction Cosines of X

$\cos \alpha = i \times (X / |X|)$, $\cos \beta = j \times (X / |X|)$,

$\cos \Gamma = k \times (X / |X|)$.

DEFINITION 1.10 X, Y Orthogonal

The angle between them is 90 or $\pi/2$.

> **REMARK 1.5** Condition for Orthogonality of **X, Y**
>
> $\mathbf{X} \times \mathbf{Y} = \mathbf{\theta}$.

DEFINITION 1.11 Orthonormal Vectors

Orthogonal unit vectors.

> **DEFINITION 1.12** Determinants of Second and Third Order
>
> $$\begin{vmatrix} a & b \\ c & d \end{vmatrix} = ad - bc \tag{1.1}$$
>
> $$\begin{vmatrix} a & b & c \\ d & e & f \\ g & h & k \end{vmatrix} = a\begin{vmatrix} e & f \\ h & k \end{vmatrix} - b\begin{vmatrix} d & f \\ g & k \end{vmatrix} + c\begin{vmatrix} d & e \\ g & h \end{vmatrix} \tag{1.2}$$

DEFINITION 1.13 Vector (Cross) Product $\mathbf{X} \times \mathbf{Y}$

Vector with *Right Hand Rule Direction* (Figure 1.5), norm equal to area of parallelogram defined by **X, Y**,

$$\mathbf{X} \times \mathbf{Y} = (|\mathbf{X}| \; |\mathbf{Y}| \sin(\theta)) \; \mathbf{N}.$$

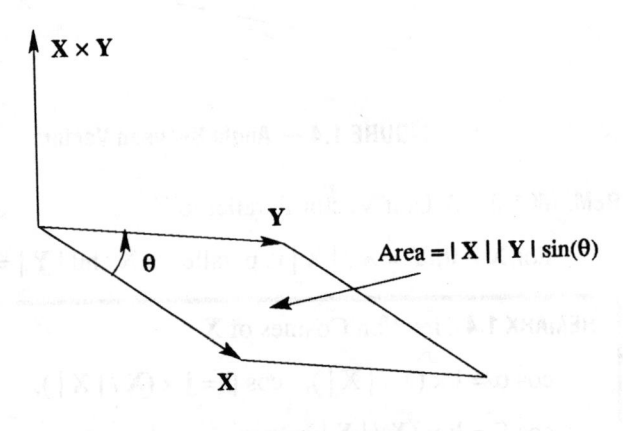

FIGURE 1.5 — Vector Product

THEOREM 1.6 Vector Products of the Basis Vectors

$$\mathbf{i} \times \mathbf{j} = \mathbf{k}, \quad \mathbf{j} \times \mathbf{i} = -\mathbf{k}, \quad \mathbf{j} \times \mathbf{k} = \mathbf{i},$$
$$\mathbf{k} \times \mathbf{j} = -\mathbf{i}, \quad \mathbf{k} \times \mathbf{i} = \mathbf{j}, \quad \mathbf{i} \times \mathbf{k} = -\mathbf{j}.$$

THEOREM 1.7 Vector Product $\mathbf{X} \times \mathbf{Y}$ of $\mathbf{X} = (a, b, c)$, $\mathbf{Y} = (d, e, f)$,

$$\mathbf{X} \times \mathbf{Y} = \begin{vmatrix} \mathbf{i} & \mathbf{j} & \mathbf{k} \\ a & b & c \\ d & e & f \end{vmatrix} = \mathbf{i}\begin{vmatrix} b & c \\ e & f \end{vmatrix} - \mathbf{j}\begin{vmatrix} a & c \\ d & f \end{vmatrix} + \mathbf{k}\begin{vmatrix} a & b \\ d & e \end{vmatrix}$$

THEOREM 1.8 Vector Product Properties

For any $\mathbf{X}, \mathbf{Y}, \mathbf{Z}, A$,

A. $(A\mathbf{X}) \times \mathbf{Y} = A(\mathbf{X} \times \mathbf{Y}) = \mathbf{X} \times (A\mathbf{Y})$;

B. $\mathbf{X} \times (\mathbf{Y} + \mathbf{Z}) = (\mathbf{X} \times \mathbf{Y}) + (\mathbf{X} \times \mathbf{Z})$ (Distributive);

C. $(\mathbf{X} + \mathbf{Y}) \times \mathbf{Z} = (\mathbf{X} \times \mathbf{Z}) + (\mathbf{Y} \times \mathbf{Z})$ (Distributive).

REMARK 1.6 Vector Product Non-Properties

Vector product is usually *not* commutative and *not* associative, $\mathbf{X} \times \mathbf{Y} \neq \mathbf{Y} \times \mathbf{X}$, $\mathbf{X} \times (\mathbf{Y} \times \mathbf{Z}) \neq (\mathbf{X} \times \mathbf{Y}) \times \mathbf{Z}$

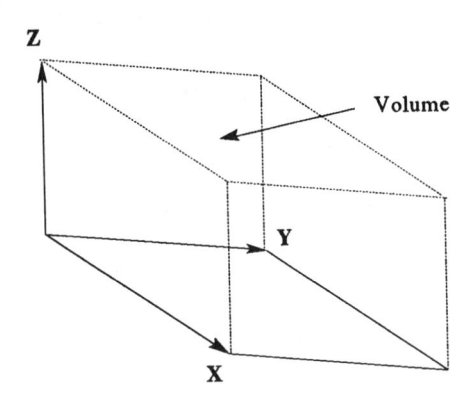

FIGURE 1.6 — {x y z} = Volume

DEFINITION 1.14 Scalar Triple Product $\{X\ Y\ Z\}$

$X \times (Y \times Z) = X, Y, Z$ parallelepiped volume (Figure 1.6).

1.2 VECTORS IN N-DIMENSIONS

DEFINITION 1.15 Norm (Magnitude, Length)

For $X = (x_1, x_2, ..., x_N)^{1/2}$

$|X| = \{x_1^2 + x_2^2 + ... + x_N^2\}^{1/2}$.

DEFINITION 1.16 Zero Vector

$0 = (0, 0, ..., 0)$.

DEFINITION 1.17 Unit Vector

Vector having norm 1.

DEFINITION 1.18 Unit Basis

$i_1 = (1, 0, ..., 0)$, $i_2 = (0, 1, 0, ..., 0)$

$i_{N-1} = (0, 0, ..., 1, 0)$, $i_N = (0, 0, ..., 1)$.

REMARK 1.7 Notation

$X = (x_1, x_2, ..., x_N)$ is often written as $X = \{x_j\}$.

DEFINITION 1.19 Sum of $X = \{x_j\}$, $Y = \{y_j\}$

$Z = X + Y = \{x_j + y_j\}$.

DEFINITION 1.20 Product δX of $X = \{x_j\}$, δ = Scalar

$\delta X = \{\delta x_j\}$.

REMARK 1.8 Representing $X = \{x_j\}$

$X = x_1 i_1 + x_2 i_2 + ... + x_N i_N$.

DEFINITION 1.21 **X, Y** Collinear

$\mathbf{Y} = a\mathbf{X}$, for some a.

DEFINITION 1.22   Scalar (Inner) Product

$\mathbf{X} \times \mathbf{Y} = x_1 y_1 + x_2 y_2 + \ldots + x_N y_N.$

THEOREM 1.9 Cauchy-Schwarz Inequality

$|\mathbf{X} \times \mathbf{Y}| \le |\mathbf{X}|\,|\mathbf{Y}|$

(= iff **X, Y** are collinear).

THEOREM 1.10 Triangle Inequality

$|\mathbf{X} + \mathbf{Y}| \le |\mathbf{X}| + |\mathbf{Y}|,$

$||\mathbf{X}| - |\mathbf{Y}|| \le |\mathbf{X} - \mathbf{Y}|$

(= iff **X, Y** are collinear).

DEFINITION 1.23   Angle θ Between **X, Y**

$\cos(\theta) = (\mathbf{X} \times \mathbf{Y}) / (|\mathbf{X}|\,|\mathbf{Y}|).$

DEFINITION 1.24   **X, Y** are

A. *Orthogonal* (Normal) if $\mathbf{X} \times \mathbf{Y} = 0$;

B. *Orthonormal* if they are orthogonal unit vectors;

C. *Parallel* if $\cos(\theta) = 1$.

REMARK 1.9   A Unit Vector Parallel to **X**

$\mathbf{Z} = \mathbf{X} / |\mathbf{X}|$ is parallel to **X**, $|\mathbf{Z}| = 1$.

DEFINITION 1.25 Direction Cosines of **X**

Components of $\mathbf{X} / |\mathbf{X}|$.

1.3 MATRICES AND SYSTEMS OF EQUATIONS

FAMILIAR IDEAS

Columns of a Matrix

Components (Elements) of a Matrix

$M \times N$ Matrix, M, N = Number of Rows, Columns

Rows of a Matrix

REMARK 1.10 Notation

$M \times N$ matrix $A = \{a_{ij}\}$ for i = row, j = column, and

$$A = \{a_{ij}\} = \begin{pmatrix} a_{11} & a_{12} & \cdots & a_{1N} \\ a_{21} & a_{22} & \cdots & a_{2N} \\ \cdots & & & \\ a_{M1} & a_{M2} & \cdots & a_{MN} \end{pmatrix}$$

DEFINITION 1.26 The $M \times N$ Matrix A is

I. a *Row Matrix* if it is a single row, $M = 1$;
II. a *Column Matrix* if it is a single column, $N = 1$;
III. a *Square Matrix of Order M* if $M = N$;
IV. a *Zero Matrix* 0 if all its elements are zero.

DEFINITION 1.27 Equality of $A = \{a_{ij}\}$ and $B = \{b_{ij}\}$

$$a_{ij} = b_{ij}, \quad i = 1, \ldots, M; \quad j = 1, \ldots, N.$$

DEFINITION 1.28 Diagonal of Square Matrix $A = \{a_{ij}\}$

The elements $a_{11}, a_{22}, \ldots, a_{NN}$; moreover, the $a_{ij}, i \neq j$ are the *Off-Diagonal Elements*.

DEFINITION 1.29 Diagonal Matrix

Off-diagonal elements are zero.

DEFINITION 1.30 Identity Matrix I of Order N

Diagonal matrix all of whose diagonal elements are 1.

DEFINITION 1.31 Sum of $M \times N$ Matrices A, B

$$C = A + B = \{a_{ij} + b_{ij}\}.$$

DEFINITION 1.32 Product αA, α = Scalar

$$\alpha A = \{\alpha \, a_{ij}\}.$$

DEFINITION 1.33 $-A = (-1)A$.

THEOREM 1.11 Properties of Arithmetic Operations

For any A, B, C and scalars α, β,

A. $A + B = B + A$ (Commutativity)

B. $(A + B) + C = A + (B + C)$ (Associativity)

C. $\alpha(A + B) = \alpha A + \alpha B$ (Distributivity)

D. $(\alpha + \beta)A = \alpha A + \beta A$ (Distributivity)

E. $A + 0 = A$

F. $A + (-A) = 0$

DEFINITION 1.34 Transpose A^T

Matrix obtained by interchanging rows and columns of A,

$$A^T = \{c_{ij}\}, \, c_{ij} = a_{ji}.$$

EXAMPLE 1.1 Transpose A^T

For $A = \begin{pmatrix} 3 & 2 & 1 \\ 4 & 5 & 6 \end{pmatrix}$, $A^T = \begin{pmatrix} 3 & 4 \\ 2 & 5 \\ 1 & 6 \end{pmatrix}$.

DEFINITION 1.35 Symmetric Matrix

$A = A^T$.

DEFINITION 1.36 Skew-Symmetric Matrix

$A^T = -A$.

DEFINITION 1.37 Products of Matrices

Let $A = \{a_{ij}\}$, $B = \{b_{jk}\}$ be $L \times M$ and $M \times N$ matrices. Their product $C = \{c_{ik}\} = AB$ is

$$c_{ik} = \sum_{j=1}^{M} a_{ij} b_{jk}$$

A is *Post-Multiplied* by B, B is *Pre-Multiplied* by A.

THEOREM 1.12 Properties of Matrix Multiplication

For any A, B, C and scalar α,

A. $(A + B)C = AC + BC$ (Distributivity)

B. $A(B + C) = AB + AC$ (Distributivity)

C. $A(BC) = (AB)C$ (Associativity)

D. $\alpha(AB) = (\alpha A)B = A(\alpha B)$

E. $IA = AI$ for I the identity matrix

EXAMPLE 1.2 Non-Properties of the Matrix Product

For

$$A = \begin{pmatrix} 1 & -1 \\ 1 & -1 \end{pmatrix}, \quad B = \begin{pmatrix} 1 & 1 \\ 1 & 1 \end{pmatrix}$$

$$AB = \begin{pmatrix} 1 & -1 \\ 1 & -1 \end{pmatrix} \begin{pmatrix} 1 & 1 \\ 1 & 1 \end{pmatrix} = \begin{pmatrix} 0 & 0 \\ 0 & 0 \end{pmatrix}$$

$$BA = \begin{pmatrix} 1 & 1 \\ 1 & 1 \end{pmatrix} \begin{pmatrix} 1 & -1 \\ 1 & -1 \end{pmatrix} = \begin{pmatrix} 2 & -2 \\ 2 & -2 \end{pmatrix}$$

Therefore,

A.   Matrix multiplication *may not be Commutative.*

B.   The product of non-zero matrices *may be zero.*

THEOREM 1.13 $(AB)^T = B^T A^T$.

REMARK 1.11   Vectors Are Matrices

An N-dimensional vector X is an $N \times 1$ column matrix. For an $M \times N$ matrix A, AX is an M-dimensional vector.

EXAMPLE 1.3

$$\begin{pmatrix} 1 & 2 & 3 \\ 4 & 5 & 6 \end{pmatrix} \begin{pmatrix} 3 \\ 2 \\ 1 \end{pmatrix} = \begin{pmatrix} 10 \\ 28 \end{pmatrix}$$

DEFINITION 1.38 Mappings of Linear Transformations

The $M \times N$ matrix A maps N-dimensional vectors **X** into M-dimensional vectors **Y** by premultiplication of **X**,

$$\mathbf{Y} = A\mathbf{X}.$$

Y is the *Image of* **X** under the mapping.

EXAMPLE 1.4   A Rotation

$$A = \begin{pmatrix} \cos(\theta) & -\sin(\theta) \\ \sin(\theta) & \cos(\theta) \end{pmatrix} \text{ maps } \mathbf{X} = \begin{pmatrix} x \\ y \end{pmatrix} \text{ onto}$$

$$\mathbf{Y} = \begin{pmatrix} u \\ v \end{pmatrix} = \begin{pmatrix} \cos(\theta) & -\sin(\theta) \\ \sin(\theta) & \cos(\theta) \end{pmatrix} \begin{pmatrix} x \\ y \end{pmatrix} = \begin{pmatrix} x\cos(\theta) & -y\sin(\theta) \\ x\sin(\theta) & +y\cos(\theta) \end{pmatrix}$$

Y is obtained by rotating **X** through the angle θ in the counterclockwise direction (Figure 1.7).

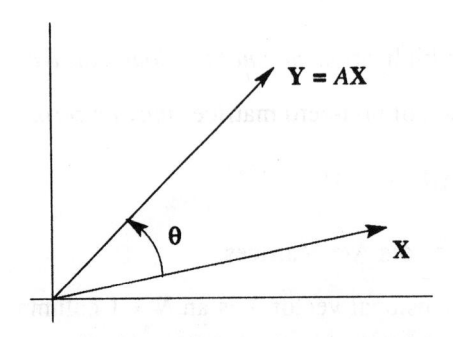

FIGURE 1.7 — Rotation by an Angle θ.

REMARK 1.12 A System of Two Equations in Two Unknowns

$$ax + by = A$$
$$cx + dy = B.$$

x, y are the *Unknowns*; *a, b, c, d* are the known *Coefficients*; *A,B* are the known *Right Hand Sides*. The system is *Homogeneous* if $A = B = 0$; if not, it is *Inhomogeneous*. The homogeneous system has the *Trivial Solution* $x = y = 0$. If the determinant

$$D = \begin{vmatrix} a & b \\ c & d \end{vmatrix} = ad - bc \neq 0,$$

then a unique *Solution* to the system is given by

$$x = (1/D)\begin{vmatrix} A & b \\ B & d \end{vmatrix} = (1/D)(Ad - Bb)$$

$$y = (1/D)\begin{vmatrix} a & A \\ c & B \end{vmatrix} = (1/D)(aB - cA)$$

If $D = 0$ then there are two possibilities:

A. $Ac = aB$: there are infinitely many solutions;

B. $Ac\ =\ aB$: there are no solutions.

DEFINITION 1.39 Inverse of $A = \{a_{ij}\}$

The $N \times N$ matrix A^{-1} obeying

$$A^{-1}A = AA^{-1} = I$$

for I the $N \times N$ identity matrix. A is *Nonsingular* if it has an inverse, and *Singular* if it does not.

THEOREM 1.14 Properties of the Inverse

If A, B are nonsingular then

A. $(A^{-1})^{-1} = A$

B. $(AB)^{-1} = B^{-1}A^{-1}$

DEFINITION 1.40 Determinant of Any Order

The *Determinant of First Order* $D = a_{11}$ is the value of a_{11}. Determinants of 2nd and 3rd order have been defined in (1.12). Assuming that determinants of order $N-1$ are defined, the *Determinant of N-th Order*

$$D = \begin{vmatrix} a_{11} & a_{12} & \cdots & a_{1N} \\ a_{21} & a_{22} & \cdots & a_{2N} \\ \cdots & & & \\ a_{N1} & a_{N2} & \cdots & a_{NN} \end{vmatrix}$$

is defined in terms of the *Cofactors* C_{ij} as

[ROW] $D = a_{i1}C_{i1} + a_{i2}C_{i2} + \ldots + a_{iN}C_{iN}$

[COLUMN] $D = a_{1j}C_{1j} + a_{2j}C_{2j} + \ldots + a_{Nj}C_{Nj}$

The *Cofactor* C_{ij} of a_{ij} is the value of the determinant obtained by removing the i-th row and j-th column of D, multiplied by $(-1)^{i+j}$.

DEFINITION 1.41 Determinant det(A) of a Matrix A

The determinant of the array of rows and columns of A.

THEOREM 1.15 Properties of Determinants

A. $\det(A^T) = \det(A)$;

B. $\det(A) = 0$ if any row or column of A has only zeros;

C. $\det(A)$ is preserved under addition of rows or columns;

D. $\det(AB) = \det(A)\det(B)$.

THEOREM 1.16 Inverse of a Nonsingular Matrix $A = \{a_{ij}\}$

$$A = (1/\det(A)) = \begin{pmatrix} C_{11} & C_{21} & \ldots & C_{N1} \\ C_{12} & C_{22} & \ldots & C_{N2} \\ \ldots & & & \\ C_{1N} & C_{2N} & \ldots & C_{NN} \end{pmatrix}$$

for C_{ij} the cofactor of a_{ij}.

THEOREM 1.17 Cramers Rule

If $D = \det(A) \neq 0$ for A the coefficient matrix of

$$a_{11}x_1 + a_{12}x_2 + \ldots + a_{1N}x_N = b_1$$
$$a_{21}x_1 + a_{22}x_2 + \ldots + a_{2N}x_N = b_2$$
$$\ldots$$
$$a_{N1}x_1 + a_{N2}x_2 + \ldots + a_{NN}x_N = b_N$$

then the system has a unique solution

$$x_1 = D_1/D, \quad x_2 = D_2/D, \ldots, x_N = D_N/D.$$

Here D_j is obtained by replacing the j-th column of D by the right-hand side values $b_1, b_2, \ldots, b_n$.

REMARK 1.13 How to Solve $AX = B$ for Nonsingular A

Premultiply the equation by A^{-1} giving $X = A^{-1}B$.

1.4 COMPLEX VECTORS AND MATRICES

FAMILIAR IDEAS

Complex Number, $z = x + iy$

Imaginary Unit, i, such that $i^2 = -1$

Complex Conjugate, $\bar{z} = x - iy$

Complex Vector, $\mathbf{Z} = \{z_j\} = (z_1, z_2, ..., z_N)$

DEFINITION 1.42 Complex Inner Product

For $Z = \{z_j\}$, $\mathbf{W} = \{w_j\}$,

$\mathbf{Z} \times \mathbf{W} = z_1 w_1 + z_2 w_2 + ... + z_N w_N$.

DEFINITION 1.43 Hermitian Matrix

$\bar{A}^T = A$.

REMARK 1.14 Real Valued Hermitian Matrix is Symmetric.

DEFINITION 1.44 Skew-Hermitian Matrix

$\bar{A}^T = -A$.

DEFINITION 1.45 Eigenvalues and Eigenvectors

μ, X satisfying $A\mathbf{X} = \mu\mathbf{X}$, $\mathbf{X} \neq 0$.

REMARK 1.15 Uniqueness of Eigenvector

Up to a multiplicative constant.

DEFINITION 1.46 Characteristic Equation

$$\det(A - \mu I) = 0. \qquad (1.4)$$

DEFINITION 1.47 Spectrum of A

The set of its eigenvalues.

DEFINITION 1.48 Unitary Matrix

$A^T = \mathbf{A}^{-1}$.

DEFINITION 1.49 Orthogonal Matrix

A real unitary matrix.

THEOREM 1.18 Some properties of Eigenvalues:

A. The eigenvalues of a Hermitian matrix are real;

B. The eigenvalues of a skew-Hermitian matrix are either zero or pure imaginary;

C. The eigenvalues of a unitary matrix have absolute value equal to one.

CHAPTER 2

ESSENTIALS OF CALCULUS

2.1 CALCULUS OF ONE VARIABLE

FAMILIAR IDEAS

Closed Interval (a,b): Set of all t with $a \leq t \leq b$

Number Sequence (a_n): Limit $\lim\limits_{n \to \infty} a_n$, convergence

Variable x, y: Independent, Dependent

Function $f(x)$: Bounded, Continuous, Uniformly Continuous, Domain of Definition, Range, Maximum, Minimum, Limit $\lim\limits_{x \to a} f(x)$

Function Sequence $\{f_j(x)\}$: Pointwise, Uniform Convergence

Infinite Series $\sum\limits_{n=0}^{\infty} a_n$: Limit, Absolute and Conditional Convergence, Convergence Tests

Infinite Series of Functions $\sum\limits_{n=0}^{\infty} f_n(x)$: Pointwise, Uniform Convergence

Integral $\int_a^b f(x)dx$: Definition as $\lim\limits_{n \to \infty} \sum\limits_{n=0}^{N} f(x_n) \Delta x$

H-18

Derivative $f'(x)$: Definition as $\lim_{h \to \infty} [f(x+h) - f(x)]/h$ of Sum, Product, Quotient of Functions Geometric Meaning as Slope of Curve $y = f(x)$

THEOREM 2.1 Mean Value Theorem

For f, f' continuous on $[a, b]$, for some c on $[a, b]$ (Figure 2.1)

$$f(b) - f(a) = f'(c)(b - a).$$

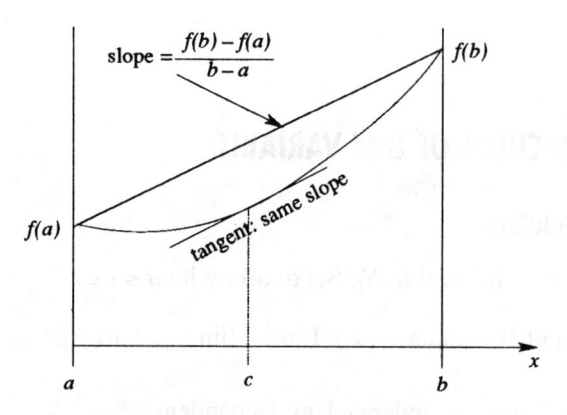

FIGURE 2.1 — Mean Value Theorem

DEFINITION 2.1 Differential $df[x, dx]$

$df[x, dx] = f'(x)dx$.

THEOREM 2.2 Justification for the Differential

If $f'(x)$ is continuous, then

$$f(x + dx) \approx f(x) + df[x, dx].$$

THEOREM 2.3 L'Hospitals Rule

If $f(x), g(x) \to 0$ as $x \to a$, then

$$\lim_{x \to a} f(x)/g(x) = \lim_{x \to a} f'(x)/g'(x).$$

THEOREM 2.4 The Taylor Series of $f(x)$ about $x = a$

$$f(x) = f(a) + f'(a)(x-a) + (1/2!)f''(a)(x-a)$$
$$+ \ldots + (1/n!)f^{(n)}(a)(x-a)^n + \ldots$$

DEFINITION 2.2 Antiderivative of f

$F(x)$ such that $F'(x) = f(x)$.

THEOREM 2.5 Fundamental Theorem of the Calculus

For continuous $f(x)$, $F'(t) = f(t)$ for

$$F(t) = \int_a^t f(x)dx.$$

THEOREM 2.6 Definite Integral

For any antiderivative $F(x)$ of $f(x)$,

$$\int_a^b f(x)dx = F(x)\Big|_{x=a}^{b} = F(b) - F(a).$$

REMARK 2.1 Elementary Functions (Figure 2.2)

A. Linear: $\quad y = ax + b$

B. Trigonometric: $\quad y = \cos(x), y = \sin(x),$

$\qquad y = \tan(x), \sec(x) = 1/\cos(x),$

$\qquad \text{cosec}(x) = 1/\sin(x)$

C. Exponential/Log: $\quad y = e^x,$

$\qquad y = \ln(x) = \int_1^x (1/t)\, dt$

D. Hyperbolic: $\quad \cosh(x) = \frac{1}{2}(e^x + e^{-x}),$

$\qquad \sinh(x) = \frac{1}{2}(e^x - e^{-x})$

$\qquad \tanh(x) = \sinh(x)/\cosh(x)$

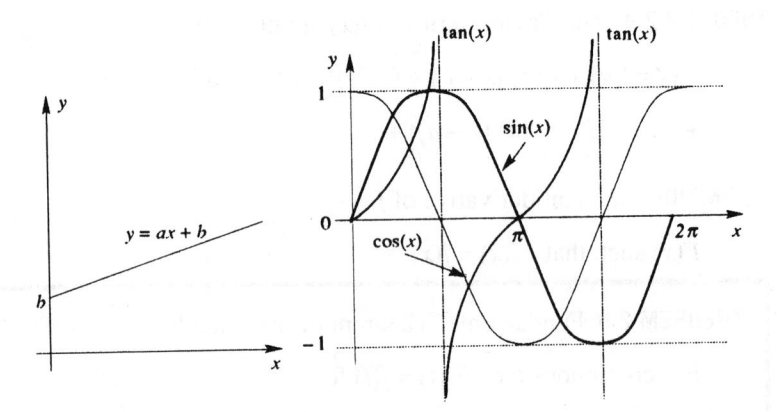

FIGURE 2.2 — Line and Trigonometric Functions

THEOREM 2.7 Basic Properties

A. For $x \approx 0$, $\sin(x) \approx x$, $\cos(x) \approx 1 - \frac{1}{2} x^2$

B. sin, cos, tan are periodic with period 2π

C. $\sin(x)/x \to 1$ for $x \to 0$

D. $\sin^2(x) + \cos^2(x) = 1$

E. $\sinh^2(x) + 1 = \cosh^2(x)$

F. $\sin(x + y) = \sin(x)\cos(y) + \cos(x)\sin(y)$

G. $1 + \tan^2(x) = \sec^2(x)$

DEFINITION 2.3 Compound Function

$z = H(x) = g(f(x))$ for $y = f(x)$, $z = g(y)$.

THEOREM 2.8 Chain Rule for Compound Function

If $H(x) = g(f(x))$, then $H'(x) = g'(f(x)) f'(x)$

DEFINITION 2.4 Inverse $x = g(y) = f^{-1}(y)$ of $y = f(x)$

Function such that $f(g(y)) = y$ and $g(f(x)) = x$.

THEOREM 2.9 Existence of the Inverse

If $f'(x) \neq 0$ then its inverse exists.

THEOREM 2.10 Derivative of the Inverse

$df^{-1}(y)/dy = 1/f'(x)$, for $y = f(x)$.

THEOREM 2.11 Properties of Definite Integral

A. $\int_a^b [Af(x) + Bg(x)]dx = A\int_a^b f(x)dx + B\int_a^b g(x)dx$

B. $(b-a) \min_{a \leq x \leq b} |f(x)| \leq \left| \int_a^b f(x)dx \right| \leq (b-a) \max_{a \leq x \leq b} |f(x)|$

C. Mean Value Theorem: For f continuous on $[a, b]$ there is some c for which

$$f(c)(b-a) = \int_a^b f(x)\,dx$$

D. Generalized Mean Value Theorem: For f, g continuous on $[a, b]$ and $g(x) \geq 0$ there is some c for which

$$\int_a^b f(x)g(x)\,dx = f(c)\int_a^b g(x)\,dx$$

REMARK 2.2 Improper Integrals — Typical Case

$$\int_0^1 \sqrt{x}\,dx = \lim_{\varepsilon \to 0} \int_\varepsilon^1 \sqrt{x}\,dx$$
$$= \lim_{\varepsilon \to 0} \{2 - 2\sqrt{\varepsilon}\} = 2$$

THEOREM 2.12 Techniques of Finding Integrals

A. Integration by Parts:

$$\int_a^b f(x)g'(x)dx = f(x)g(x)\Big|_{x=a}^b - \int_a^b g(x)f'(x)\,dx$$

H-22

B. Substitution: If $x = g(t)$, $g'(t) = 0$, then

$$\int_a^b f(x)dx = \int_{g^{-1}(a)}^{g^{-1}(b)} f(g(t))g'(t)dt.$$

2.2 VECTOR FUNCTIONS OF ONE VARIABLE

FAMILIAR IDEAS

Vector Function, $\mathbf{V}(t) = (x(t), y(t), z(t))$.

Calculus of Vector Functions of One Variable Continuity, Derivative

REMARK 2.3 Parametric Representation of a Curve

For $x(t), y(t), z(t)$ continuous on $[a, b]$, the vector function $\mathbf{V}(t) = (x(t), y(t), z(t))$ is a *Parametric Representation* of a curve C, written as $C : \mathbf{X} = \mathbf{V}(t)$ (Figure 2.3).

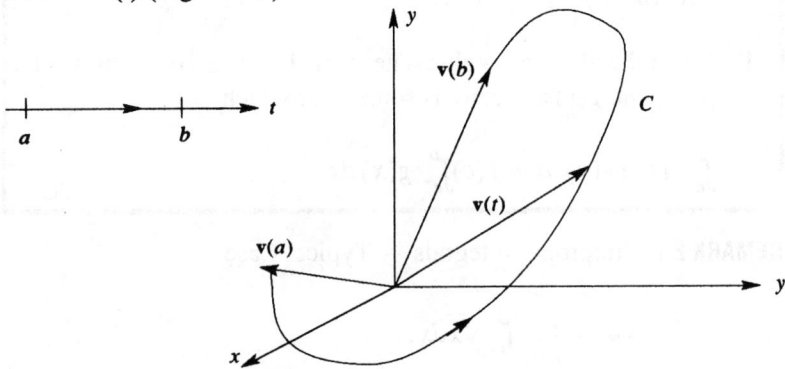

FIGURE 2.3 — Parametric Representation of C : $\mathbf{V} = \mathbf{V}(t)$

DEFINITION 2.5 Arc Length of C

If $x'(t), y'(t), z(t)$ are continuous, then the length of the arc of $C: \mathbf{X} = \mathbf{V}(t)$ from $\mathbf{V}(a)$ to $\mathbf{V}(t)$ for $a \le t \le b$ is a function $s(t)$ given by

$$s(t) = \int_a^t |\mathbf{V}'(\tau)|d\tau \qquad (2.1)$$

(See Figure 2.4.) If $\mathbf{V}'(t) \neq \mathbf{0}$, then s can replace t as the independent variable in the parametric representation.

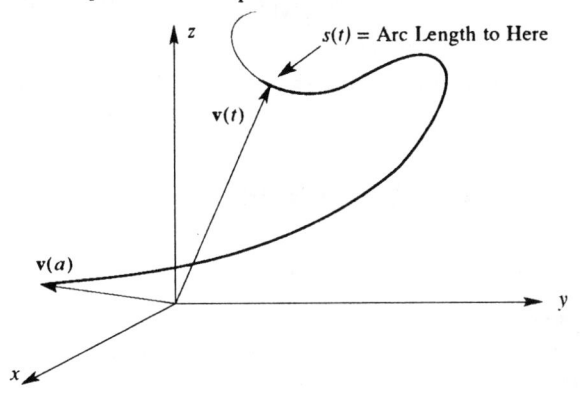

FIGURE 2.4 — Length of Arc $S(t)$ as Measured from $V(a)$

DEFINITION 2.6 Rectifiable Curve

A curve of finite length.

DEFINITION 2.7 Orientation

The direction of increasing arc length.

DEFINITION 2.8 Unit Tangent of C: $\mathbf{X} = \mathbf{V}(s)$ for Arc Length s:

$\mathbf{T}(s) = \mathbf{V}'(s)$

THEOREM 2.13 Frenet-Serret Formulas

$\mathbf{T}'(s) = \kappa(s)\mathbf{N}(s),$

$\mathbf{N}'(s) = -\kappa(s)\mathbf{T}(s) + \tau(s)\mathbf{B}(s),$

$\mathbf{B}'(s) = -\tau(s)\mathbf{N}(s),$

for $\mathbf{T}, \mathbf{N}, \mathbf{B}$ the *tangent, normal,* and *binormal* to C and $\kappa(s)$, $\tau(s)$ its *curvature* and *torsion*.

REMARK 2.4 Curves in Higher Dimensions

The parametric representation of a curve C in the space $(x, y, z, \ldots)$ is $\mathbf{V}(t) = (x(t), y(t), z(t), \ldots)$.

If $x'(t)$, $y'(t)$, $z'(t)$, ... are continuous, then the arc length from $V(a)$ to $V(t)$ is a function $s(t)$ again given by (2.1). If $V'(t) \neq 0$, then the arc length s can serve as the independent variable defining C in place of t. C is *rectifiable* if it has finite length. s can replace t in the representation of C. C is a *smooth curve* if $V'(x)$ is continuous.

2.3 FUNCTIONS OF TWO OR MORE INDEPENDENT VARIABLES

FAMILIAR IDEAS

Variables: $x, y, z, \ldots$

Function $z = f(x, y, z, \ldots)$: Domain, Range, Independent, Dependent Variables, Limits, Continuity

Partial Derivatives: $f_x(x, y, z, \ldots), \ldots$

Higher Order Derivatives: $f_{xx}(x, y, z, \ldots), \ldots$

Surface S: $z = f(x, y)$, Defined by $f(x, y)$ (Figure 2.5).

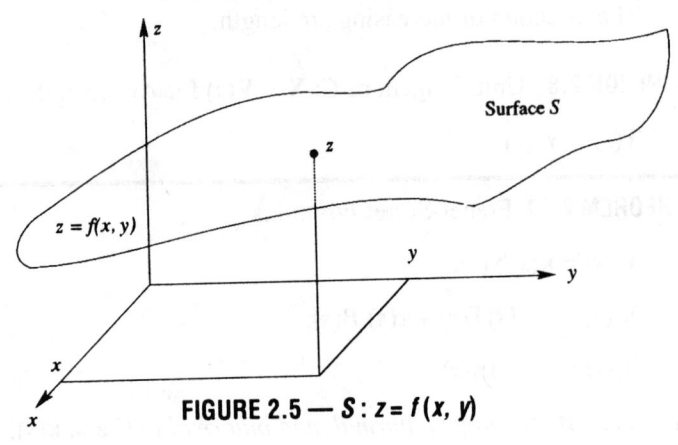

FIGURE 2.5 — $S: z = f(x, y)$

REMARK 2.5 Equation of a Plane

Plane normal to $N = (a, b, c)$, passing through $(0, 0, d/c)$ is

$$ax + by + cz = d.$$

DEFINITION 2.9 Differential $df[x, y, z, ..., dx, dy, dz, ...]$

$df[x, y, z, ..., dx, dy, dz, ...]$
$= f_x(x, y, z, ...) \, dx + f_y(x, y, z, ...) dy + ...$

REMARK 2.6 The Differential in Approximation

$f(x + dx, y + dy, z + dz, ...) \approx f(x, y, z, ...)$
$+ df[x, y, z, ..., dx, dy, dz, ...]$

THEOREM 2.14 Taylor Series for $f(x, y)$

$f(a + h, b + k) = f(a, b) + [hf_x(a, b) + kf_y(a, b)]$
$+ (1/2!)[h^2 f_{xx}(a, b) + 2hk f_{xy}(a, b) + k^2 f_{yy}(a, b)]$
$+ (1/3!)[h^3 f_{xxx}(a, b) + 3h^2 k f_{xxy}(a, b)$
$+ 3hk^2 f_{xyy}(a, b) + k^3 f_{yyy}(a, b)]$
$+ ...$

DEFINITION 2.10 Gradient $\nabla f(x, y, z, ...) = \text{grad.} \, f(x, y, z, ...)$

$\nabla f(x, y, z, ...) = (f_x(x, y, z, ...), f_y(x, y, z, ...), ...)$

DEFINITION 2.11 Directional Derivative

Derivative in the direction of the unit vector **U** is

$D_{\mathbf{u}} f = \mathbf{U} \cdot \nabla f.$

DEFINITION 2.12 Level Surfaces of $f(x, y, z, ...)$

Surfaces along which f is constant.

THEOREM 2.15 Direction of the Gradient

∇f is normal to the level surfaces of f and points in the direction of maximum rate of increase f.

DEFINITION 2.13 Differential $df[x, y, dx, dy]$

$df[x, y, dx, dy] = f_x(x, y) dx + f_y(x, y) dy$

THEOREM 2.16 Differential in Approximation

$$f(x + dx, y + dy) \approx f(x, y) + df[x, y, dx, dy]$$

DEFINITION 2.14 Compound Function

For $w = f(x, y, z, \ldots)$, $x = x(t)$, $y = y(t)$, $z = z(t)$, …,

$$w = g(t) = f(x(t), y(t), z(t), \ldots). \qquad (2.2)$$

THEOREM 2.17 Chain Rule for (2.2)

$$g'(t) = \nabla f \cdot \mathbf{V}',$$
$$\mathbf{V}'(t) = (x'(t), y'(t), z(t), \ldots).$$

DEFINITION 2.15 Compound Function

For $w = f(x, y, z, \ldots)$, $x = x(u, v)$, $y = y(u, v)$, $z = z(u, v)$, …,

$$g(u, v) = f(x(u, v), y(u, v), z(u, v), \ldots). \qquad (2.3)$$

THEOREM 2.18 Chain Rule for (2.3)

$$g_u(u, v) = f_x x_u + f_y y_u + f_z z_u + \ldots$$
$$g_v(u, v) = f_x x_v + f_y y_v + f_z z_v + \ldots$$

2.4 VECTOR FIELDS AND DIVERGENCE

DEFINITION 2.16 Vector Field

A vector function defined in the space $x, y, z, \ldots$,

$$\mathbf{V}(x, y, z, \ldots).$$

DEFINITION 2.17 Divergence $\nabla \cdot \mathbf{V} = \text{div.} V$

For

$$\mathbf{V}(x, y, z, \ldots) = (A(x, y, z, \ldots), B(x, y, z, \ldots), \ldots),$$
$$\nabla \cdot \mathbf{V} = A_x + B_y + C_z + \ldots.$$

DEFINITION 2.18 Curl of a 3-Dimensional Vector Field

For

$$\mathbf{V}(x, y, z) = (A(x, y, z), B(x, y, z), C(x, y, z)),$$

$$\text{curl. } \mathbf{V}(x,y,z) = \nabla \times \mathbf{V} = \begin{vmatrix} \mathbf{i} & \mathbf{j} & \mathbf{k} \\ \partial/\partial x & \partial/\partial y & \partial/\partial z \\ A & B & C \end{vmatrix}$$

$$= (C_y - B_z)\mathbf{i} + (A_z - C_x)\mathbf{j} + (B_x - A_y)\mathbf{k}$$

THEOREM 2.19 div(curl. f) = 0.

DEFINITION 2.19 LaPlacian $\Delta f(x, y, z, ...)$

$$\Delta f(x, y, z, ...) = \nabla^2 f(x, y, z, ...)$$
$$= f_{xx} + f_{yy} + f_{zz} + ...$$

2.5 THE DOUBLE INTEGRAL

FAMILIAR IDEAS (IN ANY NUMBER OF DIMENSIONS)

Point Set: A set of points (See Figure 2.6).

Connected Point Set: Any two points can be joined by a rectifiable curve.

Bounded Point Set: Can be contained in a sufficiently large circle (See Figure 2.7).

Closed Set: A set containing its boundary curve (See Figure 2.8).

Region: A closed, bounded, connected point set.

Non-Simply Connected Region: A region containing the interior of any circle contained in it (See Figure 2.9).

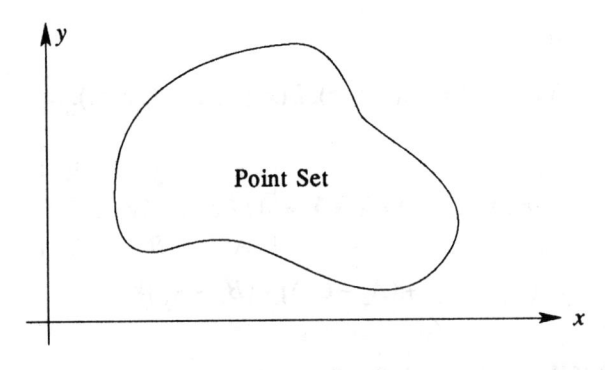

FIGURE 2.6 — A Point Set

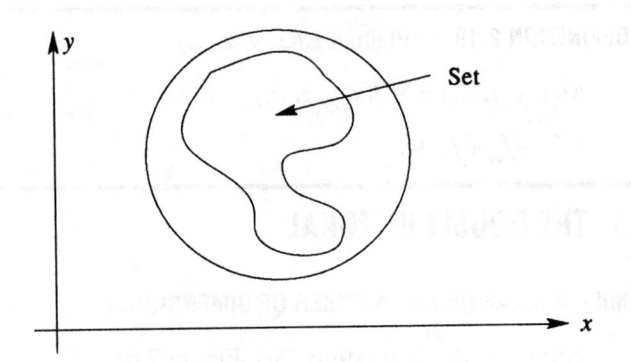

FIGURE 2.7 — Bounded Point Set

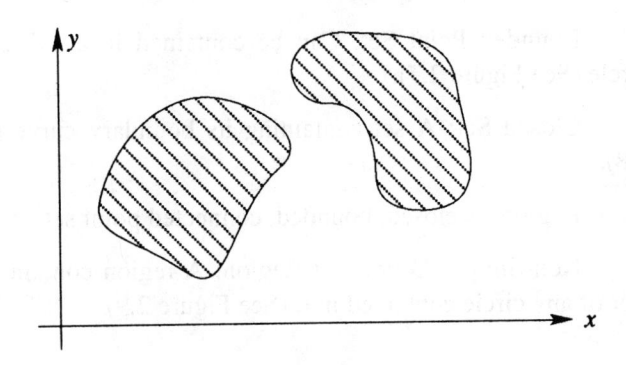

FIGURE 2.8 — Closed Set (Has 2 Parts)

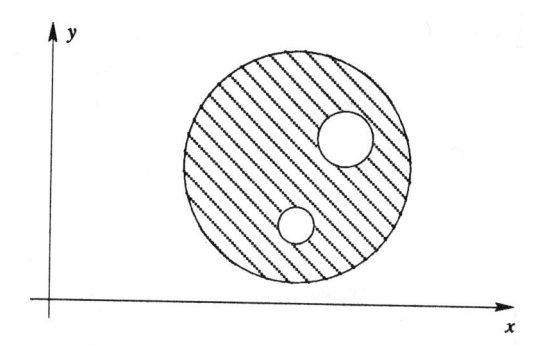

FIGURE 2.9 — A Non-Simply Connected Set

DEFINITION 2.20 The Integral of $f(x, y)$

For $f(x, y)$ continuous on a region D,

$$\iint_D f(x,y)dxdy = \lim_{N \to \infty} \sum_{i=1}^{N} f(x_i, y_j)\delta A$$

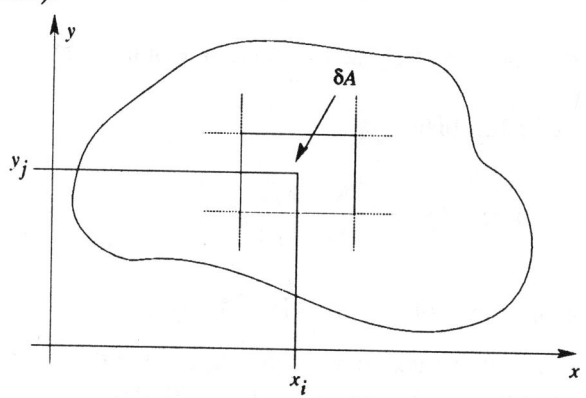

(Figure 2.10).

FIGURE 2.10 — Definition of Integral

DEFINITION 2.21 Iterated Integral

If D is the region between $y = y_1(x)$, $y = y_2(x)$, $y_1(x) \leq y_2(x)$, $a \leq x \leq b$ (see Figure 2.11), then

$$\iint_D f(x,y)dxdy = \int_a^b \left\{ \int_{y_1(x)}^{y_2(x)} f(x,y)dy \right\} dx$$

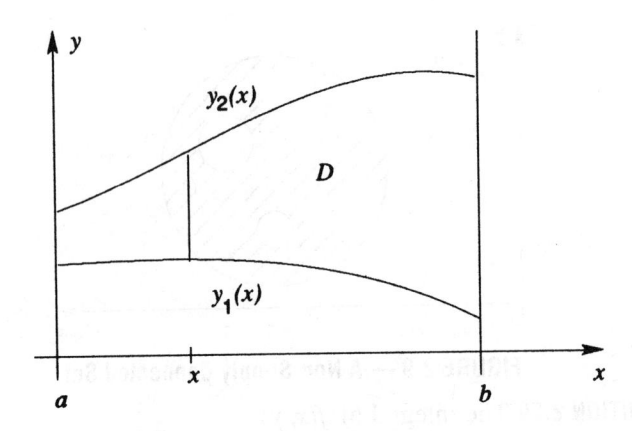

FIGURE 2.11 — Iterated Integral

REMARK 2.7 Mean Value Theorem

For continuous $f(x, y)$,

$$\iint_D f(x,y)dxdy = f(a,b)A(D),$$

where $A(D)$ is the area of D and (a, b) is some point of D.

DEFINITION 2.22 Jacobian

$$J = \partial(u,v)/\partial(x,y) = \begin{vmatrix} u_x & u_y \\ v_x & v_y \end{vmatrix} \quad (2.4)$$

THEOREM 2.20 Change of Variables in an Integral

If $u = u(x, y)$, $v = v(x, y)$ maps the region D of the x, y plane onto the region E of the u, v plane (see Figure 2.12), then

$$\iint_D f(x,y)dxdy = \iint_E f(x(u,v),y(u,v))Jdudv$$

for J the Jacobian (2.4).

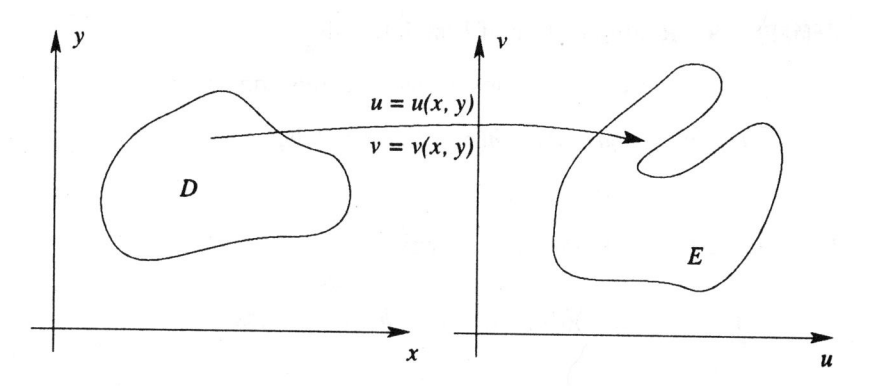

FIGURE 2.12 — Mapping of *D* onto *E*

2.6 LINE INTEGRALS

> **DEFINITION 2.23** Line Integral (Figure 2.13)
>
> $$\int_C f(x,y,z,\ldots)ds = \lim_{N \to \infty} \sum_{j=1}^{N} f(P_j)\delta s$$

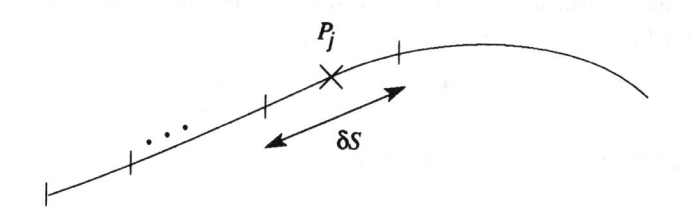

FIGURE 2.13 — Defining the Line Integral

> **REMARK 2.8** Properties of Line Integrals
>
> Properties of ordinary integrals hold for line integrals (Theorem 2.11).

H-32

REMARK 2.9 Representation of Line Integrals

Commonly, the integrands of line integrals appear as

$h(x, y, z, ...)dx/ds, \quad h(x, y, z, ...)dy/ds,$

$h(x, y, z, ...)dz/ds, ...$

for s the arc length. In this case we write

$$\int_C h(x,y,z,...)\{dx/ds\}ds = \int_C h(x,y,z,...)dx$$

DEFINITION 2.24 First-Order Differential Form

For given f, g, h, the expression

$fdx + gdy + hdz$

DEFINITION 2.25 Differential of $f(x, y, z)$

The function of the variables x, y, z, dx, dy, dz, defined as

$dF = F_x(x, y, z)dx + F_y(x, y, z)dy + F_z(x, y, z)dz$

DEFINITION 2.26 Exact Differential Form

The first-order differential form if it is the differential of a function F:

$f = F_x, g = F_y, h = F_z.$

THEOREM 2.21 Independence of Path

The line integral

$$\int_C (fdx + gdy + hdz) \qquad (2.5)$$

along a curve joining P, Q (see Figure 2.14) is independent of path iff the differential form $fdx + gdy + hdz$ is exact.

THEOREM 2.22 Independence of Path

The integral of (2.5) is independent of path iff

$$h_y = g_z, f_z = h_x, g_x = f_y$$

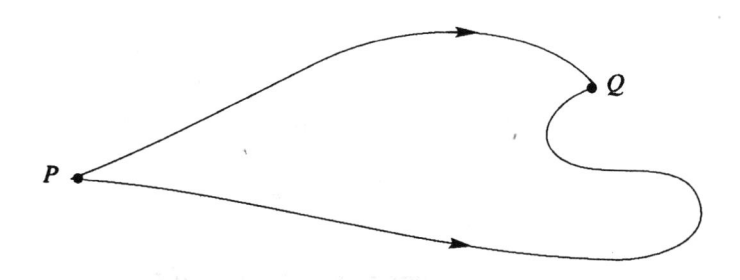

FIGURE 2.14 — Paths from P to Q

2.7 GREEN'S THEOREM

THEOREM 2.23 Green's Theorem:

$$\iint_D (f_x + g_y)dxdy = \int_C (fdy - gdx)$$

D is a region with boundary C, f, g, f_x, g_y are continuous in D, and C is smooth.

THEOREM 2.24 Area of a Region

The area $A(D)$ of the region bounded by the rectifiable, non-self-intersecting curve C (see Figure 2.15) is

$$A = \tfrac{1}{2}\int_C (xdy - ydx).$$

THEOREM 2.25 Integral of the Laplacian

$$\iint_D (f_{xx} + f_{yy})dxdy = \int_C (\partial f / \partial n)ds$$

for $\partial/\partial n$ the outer normal directional derivative to C.

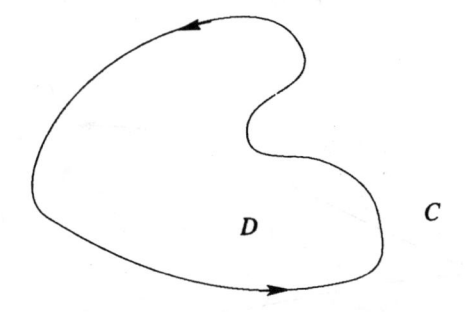

FIGURE 2.15 — The Area of D

2.8 SURFACES IN 3 DIMENSIONS

REMARK 2.10 Non-Parametric Representation of a Surface

$$S: F(x, y, z) = 0. \tag{2.6}$$

THEOREM 2.26 Unit Normal Vector to S of (2.6)

$$n = \nabla F / |\nabla F|.$$

REMARK 2.11 Parametric Representation of a Surface

$$\mathbf{V}(u, v) = (x(u, v), y(u, v), z(u, v)) \tag{2.7}$$

for (u, v) varying over a domain R of the u, v plane (see Figure 2.16); u, v are the *parameters* of the representation (2.7).

DEFINITION 2.27 Smooth Surface

A surface having a unique, continuous normal vector.

REMARK 2.12 A Curve on a Surface

Let S be given by (2.7) and let $C: u = u(t), v = v(t), a \le t \le b$ be a curve in R (see Figure 2.17). Then

$$\mathbf{V}[t] = (x(u(t), v(t)), y(u(t), v(t)), z(u(t), v(t)))$$

defines a curve C' on S.

H-35

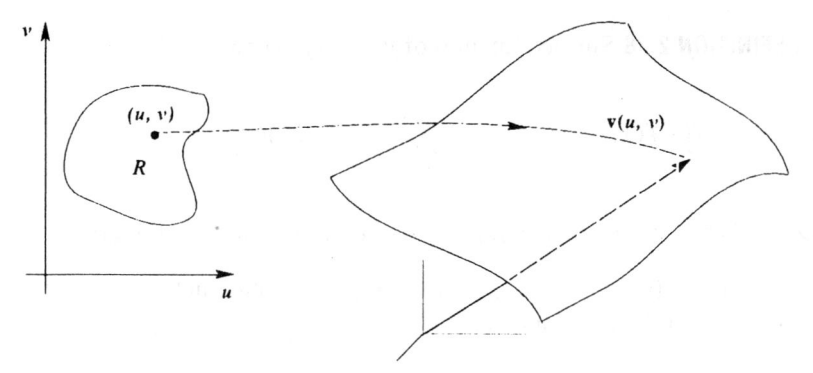

FIGURE 2.16 — Parametric Representation of Surface

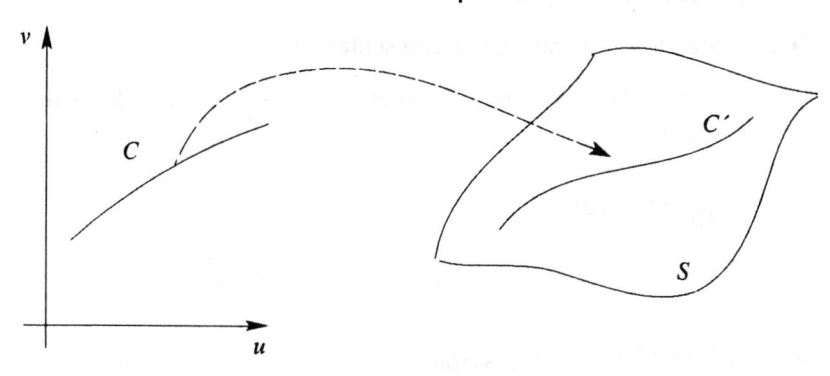

FIGURE 2.17 — A Curve on a Surface

THEOREM 2.27 The Length of a Curve on a Surface

The length of C' of Remark 2.12 is

$$L = \int_a^b \{Eu'^2 + 2Fu'v' + Gv'\}^{1/2} dt,$$

where E, F, G are the elements of the *First Fundamental Form of S*,

$$E = \mathbf{V}_u \times \mathbf{V}_u, F = \mathbf{V}_u \times \mathbf{V}_v, G = \mathbf{V}_v \times \mathbf{V}_v \tag{2.9}$$

THEOREM 2.28 Surface Area Integral

The area of the surface S of Remark 2.12 is

$$A(S) = \iint_D \{EG - F^2\}^{1/2} du dv.$$

DEFINITION 2.28 Surface Integral of $f(x, y, z)$ over S

$$\iint_S f(x,y,z)dA = \lim_{N\to\infty} \sum_{n=0}^{N} f(x_i, y_j, z_k)\delta A.$$

THEOREM 2.29 Parametric Representation of the Surface Integral

For $\mathbf{V}(u, v) = (x(u, v), y(u, v), z(u, v))$ the parametric representation of S on the u, v domain D

$$\iint_S f(x,y,z)dA = \iint_D f(x(u,v), y(u,v), z(u,v))\{EG - F^2\}^{1/2} du dv.$$

THEOREM 2.30 Non-Parametric Representation

For $z = g(x, y)$ the non-parametric representation of S on the (x, y) domain D

$$\iint_S f(x,y,z)dA$$
$$= \iint_D f(x,y,g(x,y))\{1 + (g_x(x))^2 + (g'(y))^2\}^{1/2} dxdy.$$

DEFINITION 2.29 Orientable Surface

There is a unique, continuous normal direction (Figure 2.18).

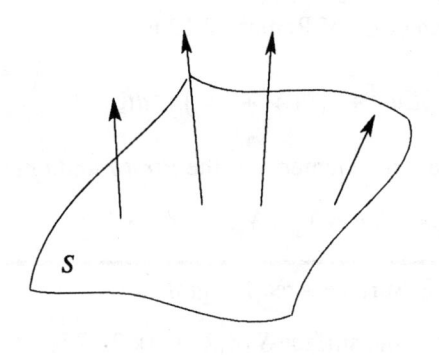

FIGURE 2.18 — Orientable Surface

REMARK 2.13 Integrals Over a Surface

For α, β, Γ the direction cosines of the unit normal to S,

$$\iint_S f(x,y,z)\alpha dA = \iint_S f(x,y,z)dydz$$
$$\iint_S f(x,y,z)\beta dA = \iint_S f(x,y,z)dzdx$$
$$\iint_S f(x,y,z)\Gamma dA = \iint_S f(x,y,z)dxdy$$

THEOREM 2.31 Stoke's Theorem

Let S be an oriented surface bounded by a smooth, closed, non-self-intersecting curve C (Figure 2.19). Let $\mathbf{V}(x, y, z)$ have continuous partial derivatives of first order in a region of x, y, z space containing S. Then

$$\iint_S (curl.\ \mathbf{V}) \cdot \mathbf{N} dA = \int_C \mathbf{V} \cdot \mathbf{T} ds,$$

where $\mathbf{T}$ is the unit tangent vector of C.

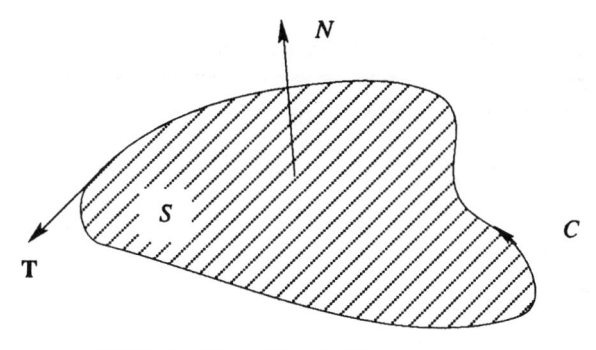

FIGURE 2.19 — Stoke's Theorem

2.9 VOLUME INTEGRALS

DEFINITION 2.30 Volume Integral of $f(x, y, z)$ (See Figure 2.20)

$$\iiint_D f(x,y,z)dV = \lim_{N \to \infty} \sum_{n=0}^{N} f(x_i, y_j, z_k)\delta V$$

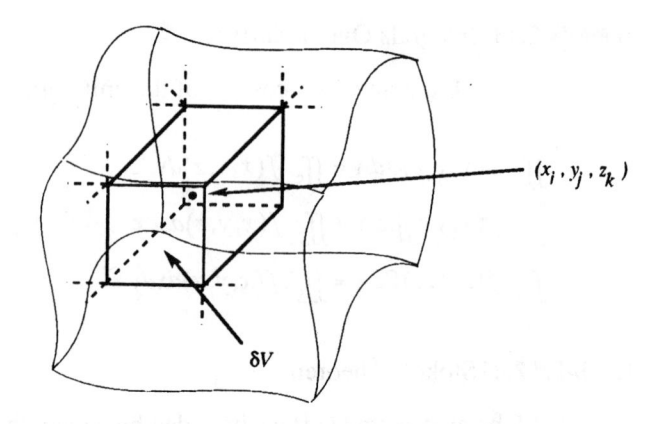

FIGURE 2.20 — Volume Integral

THEOREM 2.32 Divergence Theorem

If $\mathbf{V}(x, y, z)$ and its first partial derivatives are continuous in a region D bounded by an orientable surface S, then

$$\iiint_D \text{div. } \mathbf{V} dV = \iint_S \mathbf{V} \cdot \mathbf{N} \, dA \qquad (2.10)$$

for $\mathbf{N}$ the unit vector in the outer normal direction.

REMARK 2.14 Divergence Theorem

(2.10) can be restated as

$$\iiint_D (u_x + v_y + w_z)dxdydz = \iint_S (udydz + vdzdx + wdxdy).$$

REMARK 2.15 Integrating the Laplacian

For any f, g

$$\iiint_D [f\Delta g + (\nabla f)\cdot(\nabla g)]dV = \iint_S f(\partial g / \partial n)dA.$$

CHAPTER 3

COMPLEX FUNCTIONS

3.1 BASIC CONCEPTS

FAMILIAR IDEAS

Imaginary Unit i with $i^2 = -1$

Complex Numbers $z = x + iy$, x, y = Real Numbers

Real, Imaginary Parts of z: $x = Re(z)$, $y = Im(z)$

Complex Conjugate $\bar{z} = x - iy$

Absolute Value or Modulus $|z| = (x^2 + y^2)^{1/2}$

Arithmetic of Complex Numbers

ADDITION

$$(x_1 + iy_1) + (x_2 + iy_2) = (x_1 + x_2) + i(y_1 + y_2)$$

MULTIPLICATION

$$(x_1 + iy_1)(x_2 + iy_2) = (x_1 x_2 - y_1 y_2) + i(x_1 y_2 + x_2 y_1)$$

MULTIPLICATION BY COMPLEX CONJUGATE

$$(x + iy)(x - iy) = x^2 + y^2$$

DIVISION OF z_1 BY z_2

$z_1 / z_2 = (z_1 \bar{z}_2) / (z_2 \bar{z}_2)$

Complex Plane or Argand Diagram (See Figure 3.1)

   Parallelogram Law (See Figure 3.2)

   Triangle Inequality $|z_1 + z_2| \leq |z_1| + |z_2|$

   Polar Form $z = r \cos \theta + ir \sin \theta$ (See Figure 3.3)

   Euler's Formula $e^{i\theta} = \cos \theta + i \sin \theta$

   Exponential Form $z = re^{i\theta}$, $r = \text{mod } z$, $\theta = \arg z$

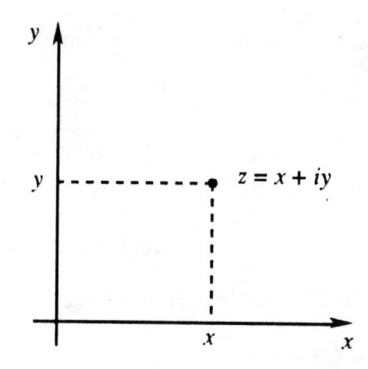

FIGURE 3.1 — Complex Plane

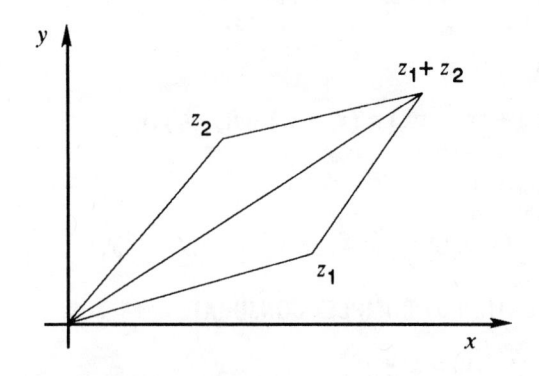

FIGURE 3.2 — Parallelogram Law

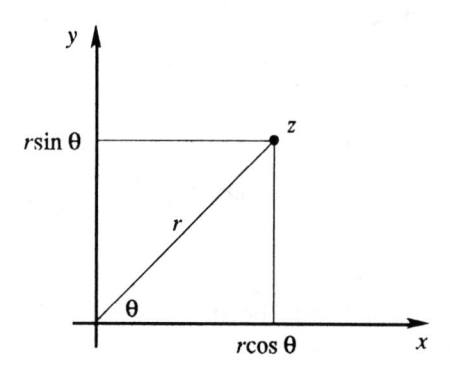

FIGURE 3.3 — Polar Form

THEOREM 3.1 DeMoivre's Formula

For $n = 0, \pm 1, \pm 2, \ldots$,

$(\cos \theta + i \sin \theta)^n = \cos(n\theta) + i \sin(n\theta)$

REMARK 3.1 The Circle of Radius R about $z = 0$

$z = z_1 + Re^{i\theta}$, $-\infty < \theta < \infty$

3.2 SETS IN THE COMPLEX PLANE

FAMILIAR IDEAS

Set, Point Set in the Plane

Subset

Interior, Boundary Points of a Set

Boundary of a Set

Bounded Set

DEFINITION 3.1 Open Set

A set not containing any of its boundary points.

DEFINITION 3.2 Closed Set

A set containing all of its boundary points.

DEFINITION 3.3 Closure of a Set

The set together with its boundary points.

DEFINITION 3.4 Connected Set

Any two of its points can be linked by a finite number of straight line segments in the set.

DEFINITION 3.5 Domain

An open connected set.

3.3 FUNCTIONS OF A COMPLEX VARIABLE

FAMILIAR IDEAS

    Complex Variable

    Complex Function $w = f(z)$

    Dependent, Independent Variables

    Domain of Definition of $f(z)$

    Range of $f(z)$

    Single-Valued Function

    Multi-Valued Function

DEFINITION 3.6 Polynomial of Order N

$$P(z) = \sum_{j=1}^{N} a_j z^j$$

DEFINITION 3.7 Rational Function

Ratio of polynomials $R(z) = P(z) / Q(z)$.

DEFINITION 3.8 Mapping (Transformation)

$w = f(z)$ (Figure 3.4).

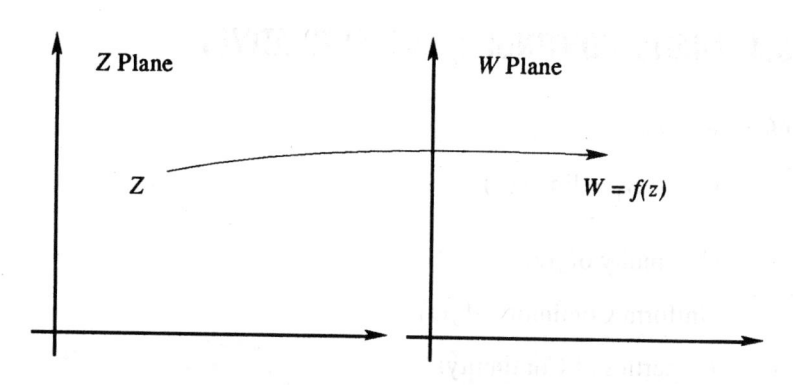

FIGURE 3.4 — Mapping $w = f(z)$

DEFINITION 3.9 Inverse

$f^{-1}(w)$ (Figure 3.5).

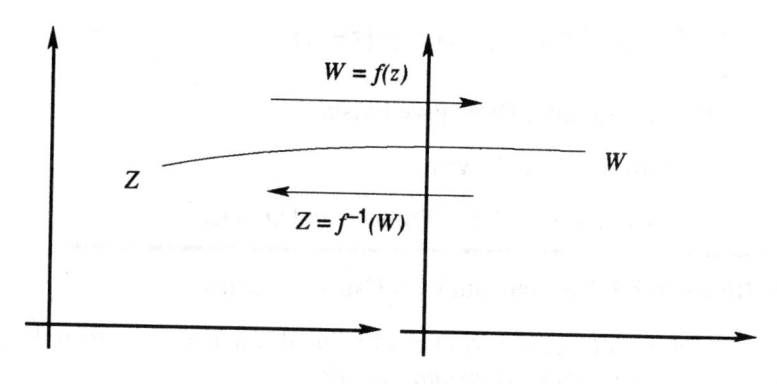

FIGURE 3.5 — Inverse Mapping

DEFINITION 3.10 Translation

$w = z + A$.

DEFINITION 3.11 Rotation

$w = ze^{i\theta}$, θ = Rotation Angle.

DEFINITION 3.12 Compound Function

$F(z) = g(f(z))$

3.4 LIMITS, CONTINUITY, AND DERIVATIVES

FAMILIAR IDEAS

Limit: $w_1 = \lim_{z \to z_1} f(z)$

Continuity of $f(z)$

Uniform Continuity of $f(z)$

Properties of Continuity:

    Sum, Product, and Quotient are Continuous

    Composite Function is Continuous

Derivative of $f(z)$ at $z = z_1$:

$$f'(z_1) = \lim_{z \to z_1} [f(z) - f(z_1)]/[z - z_1]$$

Differentiability: Derivative Exists

Properties of Derivatives:

    Derivatives of Sum, Product, and Quotient

THEOREM 3.2 Differentiability = Cauchy-Riemann

If $f(z) = u(x, y) + iv(x, y)$ is differentiable at $z = x + iy$, then u, v satisfy the *Cauchy-Riemann Equations*

$$u_x(x_1, y_1) = v_y(x_1, y_1),$$
$$u_y(x_1, y_1) = -v_x(x_1, y_1).$$

If u_x, u_y, v_x, v_y satisfy (3.1) then, $f = u + iv$ is differentiable.

THEOREM 3.3 Formula for the Derivative

$$f'(z) = u_x(x, y) + iv_x(x, y) = v_y(x, y) - iu_y(x, y).$$

EXAMPLE 3.1 Derivative of e^z

For $f(z) = e^z$, $f'(z) = e^z$.

DEFINITION 3.13 Analytic (or Holomorphic) at a Point

Differentiable in some circle about the point.

DEFINITION 3.14 Entire Function

Analytic at all points.

DEFINITION 3.15 Singular Point z_0

Not analytic at z_0 but at a point in every circle about it.

THEOREM 3.4 Analyticity of Sum and Product

The sum and product of analytic functions is analytic.

THEOREM 3.5 Analyticity of Quotient

The quotient $f(z) / g(z)$ of analytic functions is analytic wherever g does not vanish.

THEOREM 3.6 Analyticity of Composite Functions

If $w = f(z)$, $u = g(w)$ are analytic, then so is $G(z) = g(f(z))$ wherever it is defined.

3.5 HARMONIC FUNCTIONS

DEFINITION 3.16 Harmonic Function $A(x, y)$

Solves Laplace equation: $A = A_{xx}(x, y) + A_{yy}(x, y) = 0$

REMARK 3.2 Laplace Equation in Polar Coordinates

$r^2 A_{rr}(r, \theta) + r A_r(r, \theta) + A_{\theta\theta}(r, \theta) = 0.$

THEOREM 3.7 Cauchy-Riemann = Laplace

If $u(x, y)$, $v(x, y)$ satisfy the Cauchy-Riemann equations, then they are both harmonic.

THEOREM 3.8 Analytic Implies Harmonic

For $f(z) = u(x, y) + iv(x, y)$ analytic, u, v are harmonic.

DEFINITION 3.17 Harmonic Conjugates

Harmonic functions u, v satisfying Cauchy-Riemann equations.

3.6 THE ELEMENTARY COMPLEX FUNCTIONS

FAMILIAR IDEAS

Branches of the Function $y = \tan^{-1}(x)$ (See Figure 3.6)

Zeroes of $f(z)$: Points where f Vanishes

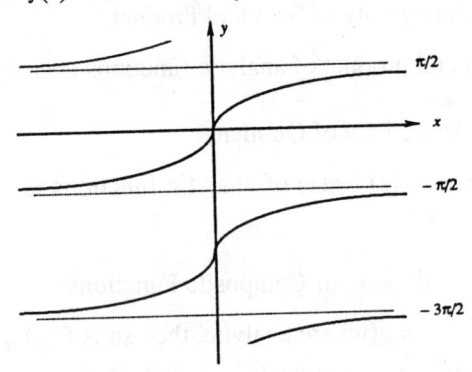

FIGURE 3.6 — Branches of tan⁻¹

DEFINITION 3.18 Branch of Multivalued Function $f(z)$

Any single-valued function $w = F(z)$, analytic in a domain in which it takes on one of the values at each point.

DEFINITION 3.19 Branch Cut

Boundary of the branch domain of a multivalued function.

EXAMPLE 3.2 $f(z) = z^{1/2}$

For $z = Re^{i\theta}$, $F(z) = \sqrt{R}e^{i\theta/2}$, $G(z) = -\sqrt{R}e^{i\theta/2}$ are branches if we choose the branch cut as the ray $\theta = 0$.

DEFINITION 3.20 $e^z = e^{x+iy} = e^x(\cos y + i \sin y)$

REMARK 3.3 Properties

A. entire with $(d/dz)e^z = e^z$

B. periodic with period $2\pi i$

C. never vanishes

D. $|e^z| = e^x$

DEFINITION 3.21 $\log(z)$

Inverse of e^z defined by $e^{\log(w)} = w$.

REMARK 3.4 $\log(z)$ is Multivalued

Since e^z is periodic,

$$w = \log(r) + i(\theta \pm 2n\pi), \ n = 0, 1, 2, \ldots, z = re^{i\theta}.$$

REMARK 3.5 Branches of $\log(z)$

A branch maps the z-plane onto any strip of height 2π in the w-plane; $\theta = \pi$ is a branch cut for the *Principal Branch* $\text{Log}(z)$ mapping z-plane onto strip $-\pi < v < \pi$.

THEOREM 3.9 Properties of Logarithm

A. $\log(z_1 z_2) = \log(z_1) + \log(z_2)$

B. $\log(z_1/z_2) = \log(z_1) - \log(z_2)$

C. $\log(z^\alpha) = \alpha \log(z), \ \alpha = 0$

D. $\log(z)$ is analytic and

E. $(d/dz) \log(z) = 1/z$

F. $e^{\log(z)} = z$

DEFINITION 3.22 Complex Powers

$z^\alpha = e^{\alpha \log(z)}$

DEFINITION 3.23 Trigonometric Functions

$$\sin(z) = [e^{iz} - e^{-iz}]/2i$$
$$\cos(z) = [e^{iz} - e^{-iz}]/2$$

THEOREM 3.10 Derivatives of Trigonometric Functions

$$(d/dz)\sin(z) = \cos(z), \quad (d/dz)\cos(z) = -\sin(z)$$

THEOREM 3.11 Properties

A. $\sin^2(z) + \cos^2(z) = 1$

B. $\sin(z + 2\pi) = \sin(z), \cos(z + 2\pi) = \cos(z)$

C. $\sin(-z) = -\sin(z), \cos(-z) = \cos(z)$

D. $\sin(z_1 + z_2) = \sin(z_1)\cos(z_2) + \cos(z_1)\sin(z_2)$

E. $\cos(z_1 + z_2) = \cos(z_1)\cos(z_2) - \sin(z_1)\sin(z_2)$

F. $\sin(2z) = 2\sin(z)\cos(z)$

G. $\cos(2z) = \cos^2(z) - \sin^2(z)$

H. $\sin(z + \pi/2) = \cos(z)$

I. $\cos(z) = 0$ for $z = \pi/2 \pm n\pi$, $n = 0, 1, 2, \ldots$

J. $\sin(z) = 0$ for $z = \pm n\pi$, $n = 0, 1, 2, \ldots$.

DEFINITION 3.24 Tangent

$$\tan(z) = \sin(z) / \cos(z)$$

DEFINITION 3.25 Hyperbolic Functions

$$\sinh(z) = (1/2)[e^z - e^{-z}]$$
$$\cosh(z) = (1/2)[e^z + e^{-z}]$$

REMARK 3.6 Properties of Hyperbolic Functions

A. $(d/dz)\sinh(z) = \cosh(z)$

B. $(d/dz)\cosh(z) = \sinh(z)$

C. $\sin(z) = -\sinh(iz)$, $\cos(z) = \cosh(iz)$

D. $\sinh(z) - i\sin(iz)$, $\cosh(z) = \cos(iz)$

E. $\cosh^2(z) - \sinh^2(z) = 1$

F. $\sinh(-z) = -\sinh(z)$, $\cosh(-z) = \cosh(z)$

G. $\sinh(z_1 + z_2) = \sinh(z_1)\cosh(z_2) + \cosh(z_1)\sinh(z_2)$

H. $\cosh(z_1 + z_2) = \cosh(z_1)\cosh(z_2) + \sinh(z_1)\sinh(z_2)$

I. $\sinh(z)$, $\cosh(z)$ are periodic with period $2\pi i$

J. $\cosh(z) = 0$ for $z = (1/2 \pm n)\pi i$, $n = 0, 1, 2, \ldots$

K. $\sinh(z) = 0$ for $z = n\pi i$, $n = 0, 1, 2, \ldots$

L. $\sinh(z) = \sinh(x)\cos(y) + i\cosh(x)\sin(y)$

M. $\cosh(z) = \cosh(x)\cos(y) + i\sinh(x)\sin(y)$

N. $|\sinh(z)|^2 = \sinh^2(x) + \sin^2(y)$

O. $|\cosh(z)|^2 = \sinh^2(x) + \cos^2(y)$.

REMARK 3.7 More on the Trigonometric Functions

A. $\sin(z) = \sin(x)\cosh(y) + i\cos(x)\sinh(y)$

B. $\cos(z) = \cos(x)\cosh(y) - i\sin(x)\sinh(y)$

C. $\sin(z)$, $\cos(z)$ are not bounded:

$|\sin(z)|^2 = \sin^2(x) + \sinh^2(y)$

$|\cos(z)|^2 = \cos^2(z) + \sinh^2(y)$

DEFINITION 3.26 Hyperbolic Tangent

$\tanh(z) = \sinh(z) / \cosh(z)$.

DEFINITION 3.27 Inverse Trigonometric Functions

$\sin^{-1}(z) = -i\log[iz + (1-z^2)^{1/2}]$,

$\cos^{-1}(z) = -i\log[z + i(1 - z^2)^{1/2}]$,

$\tan^{-1}(z) = (1/_{2i}) \log[(1 + iz) / (1 - iz)]$.

THEOREM 3.12 Derivatives of Inverse Functions

$(d / dz) \sin^{-1}(z) = 1 / (1 - z^2)^{1/2}$,

$(d / dz) \cos^{-1}(z) = -1 / (1 - z^2)^{1/2}$,

$(d / dz) \tan^{-1}(z) = 1 / (1 + z^2)$.

DEFINITION 3.28 Inverse Hyperbolic Functions

$\sinh^{-1}(z) = \log[z + (z^2 + 1)^{1/2}]$

$\cosh^{-1}(z) = \log[z + (z^2 - 1)^{1/2}]$

$\tanh^{-1}(z) = 1/_2 \log[(1 + z) / (1 - z)]$

3.7 INTEGRALS OF COMPLEX FUNCTIONS

FAMILIAR IDEAS

    Continuous Curve

    Rectifiable Curve

    Oriented (Directed) Curve

    Line Integral

    Simple (Jordan) Curve: Does Not Intersect Itself

    Simple (Jordan) Closed Curve: Intersects Only at Endpoints

    Positive (Negative) Orientation: Counterclockwise (Clockwise) Orientation (See Figure 3.7)

    Parametric Representation $z = z(t)$ of curve C

    Tangent to a Curve

    Smooth Curve: Has a Continuously Varying Tangent

    Contour: A Finite Number of Smooth Curves

Simply Connected Domain (See Figure 3.8)

Antiderivative F of f: $F' = f$

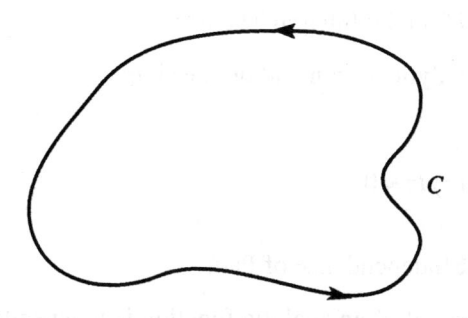

FIGURE 3.7 — Positive Orientation

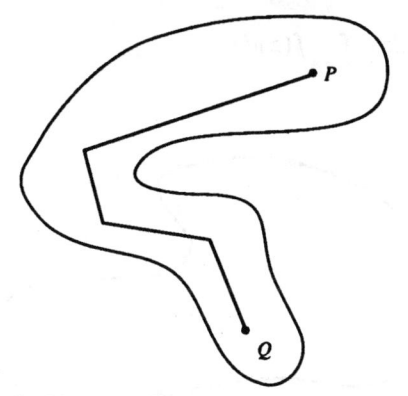

FIGURE 3.8 — Simply Connected Domain

THEOREM 3.13 Jordan Curve Theorem

A simple closed curve C divides the plane into three sets: points of C, interior to C, and exterior to C.

DEFINITION 3.29 Line Integral of $f(z)$ Along C

If $f = u + iv$ and C: $z = z(t)$, $a \leq t \leq b$, then

$I = \int_C f(z)dz$

$$= \int_a^b f[z(t)]z'(t)dt$$
$$= \int_a^b [(ux' - vy')dt + i\int_a^b [uy' + vx']dt.$$

THEOREM 3.14 Cauchy Integral Theorem

If f is analytic within and on the simple closed rectifiable curve C, then

$$\int_C f(z)dz = 0.$$

THEOREM 3.15 Independence of Path

The integral of an analytic function is independent of path (see Figure 3.9):

$$\int_{C_1} f(z)dz = \int_{C_2} f(z)dz.$$

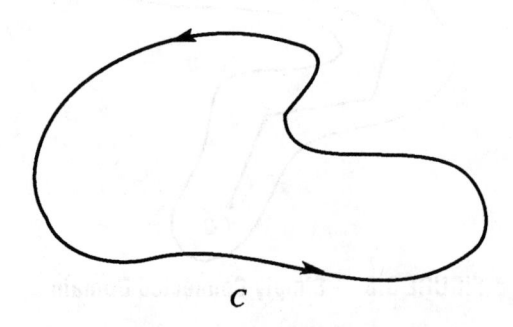

FIGURE 3.9 — Simple Closed Curve C

THEOREM 3.16 Fundamental Theorem of the Calculus

If $f(z) = F'(z)$, then for any curve C between P, Q in a domain of analyticity of f

$$\int_C f(z) = F(Q) - F(P).$$

THEOREM 3.17 Independence of Path Implies Antiderivative

If the line integrals of $f(z)$ are independent of path in a domain D then $f(z)$ has an antiderivative in D.

THEOREM 3.18 Analyticity Implies Antiderivative

If $f(z)$ is analytic, then it has an antiderivative $F(z)$.

THEOREM 3.19 Analyticity Implies Integral Formula

If $f(z)$ is analytic within a simple closed curve C, then for any z_0 within C

$$f(z_0) = (1/2\pi i)\int_C [f(s)/(s-z_0)]ds$$

THEOREM 3.20 Derivative in Terms of Boundary Data

$$f'(z_0) = (1/2\pi i)\int_C [f(s)/(s-z_0)^2]ds$$

THEOREM 3.21 Derivatives of Analytic Function are Analytic

$$f^{(n)}(z_0) = (n!/2\pi i)\int_C [f(s)/(s-z_0)^{n+1}]ds$$

THEOREM 3.22 Morera — Converse to Cauchy Integral Theorem

If $f(z)$ is continuous in D and its line integral around every closed curve C in D is zero, then $f(z)$ is analytic.

THEOREM 3.23 Maximum Modulus Theorem

If f is analytic in the closed curve C and M is its maximum modulus on C, then either $|f(z)| < M$ within C or $|f(z)| \equiv M$.

THEOREM 3.24 Cauchy Inequality

If C is the circle of radius R about z_0, $f(z)$ is analytic within and on C and $M(R)$ is its maximum modulus on C, then

$$|f^{(n)}(z_0)| \le n!M(R)/R^n.$$

THEOREM 3.25 Louville's Theorem

If $f(z)$ is entire and bounded everywhere in the complex plane, then it is constant.

3.8 NUMBER SEQUENCES AND SERIES

FAMILIAR IDEAS

Number Sequence $\{z_j\}$

Bounded Sequence

Convergence of a Sequence to a Limit: $\lim_{j \to \infty} z_j$

Null Sequence

Cauchy Sequence: $|z_j - z_k| \to 0, j, k \to \infty$

Infinite Series: $\sigma^* = \sum_{n=1}^{\infty} S_n$

Tests for Convergence of Series

Root Test, Integral Test, Ratio Test

EXAMPLE 3.3 Geometric Series

For $-1 < \alpha < 1$, $\sum_{n=0}^{\infty} \alpha^n = 1/[1-\alpha]$

DEFINITION 3.30 Absolute Convergence of a Series

$\sum_{n=1}^{\infty} |s_n|$ converges.

DEFINITION 3.31 Conditional Convergence

The series converges but not absolutely.

H-55

THEOREM 3.26 Absolute Convergence Implies Convergence

3.9 FUNCTION SEQUENCES AND SERIES

FAMILIAR IDEAS

    Function Sequence $\{f_n(z)\}$

    Pointwise Convergence

    Uniform Convergence

    Infinite Function Series $\sum_{n=0}^{\infty} f_n(z)$

    Power Series $\sum_{n=0}^{\infty} a_n(z - z_0)^n$

THEOREM 3.27 Limit of Uniformly Convergent Series is Continuous

THEOREM 3.28 Uniformly Convergent Series Can be Integrated Term by Term

$$\int \left[\sum_{n=0}^{\infty} f_n(z)\right] dz = \sum_{n=0}^{\infty} \int f_n(z) dz.$$

THEOREM 3.29 Convergent Power Series

A convergent power series can be integrated and differentiated term by term.

DEFINITION 3.32 Taylor Series

$$f(z) = \sum_{n=0}^{\infty} [f^{(n)}(z_o)/n!](z - z_o)^n.$$

DEFINITION 3.33 MacLaurin Series

    The Taylor Series for $z_0 = 0$.

DEFINITION 3.34 Laurent Expansion

For $f(z)$ analytic between circles C_0, C_1 (See Figure 3.10)

$$f(z) = \sum_{n=0}^{\infty} a_n(z-z_0)^n + \sum_{n=1}^{\infty} b_n/(z-z_0)^n;$$

b_1 is the *Residue* of f at z_0. The *Principal Part* is the latter sum of negative powers.

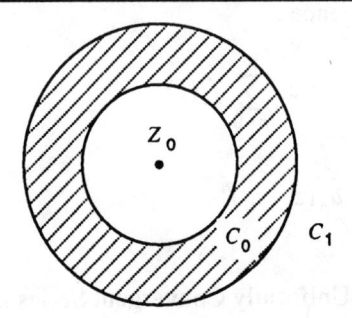

FIGURE 3.10 — Laurent Expansion

3.10 POLES AND RESIDUES

FAMILIAR IDEAS

Singular Point of $f(z)$

Isolated Singular Point z_0 of $f(z)$: The Only Singular Point in Some Circle About z_0

THEOREM 3.30 Value of Integral Around Closed Curve

If z_0 is an isolated singularity within C,

$$\int_C f(z)dz = 2\pi i b_1.$$

THEOREM 3.31 Integral Given by the Sum of the Residues

If f has finitely many isolated singularities in C, then its line integral is $2\pi i$ times the sum of its residues.

DEFINITION 3.35 Pole of Order N

The principal part has non-zero terms up to the power $-N$. For $N = 1$ we have a *Simple Pole*.

DEFINITION 3.36 Essential Singularity

The principal part has infinitely many terms.

THEOREM 3.32 Picard's Theorem

In any circle about an essential singularity $f(z)$ assumes every value except possibly for one, infinitely often.

3.11 ELEMENTARY MAPPINGS AND THE MOBIUS TRANSFORMATION

FAMILIAR IDEAS

Mapping by a Function $w = f(z)$

Domain D of f

Image S' of a Set S (See Figure 3.11)

Level Lines of $u(x, y)$

Expansion $(C > 1)$, Contraction $(C < 1)$, $f(z) = Cz$

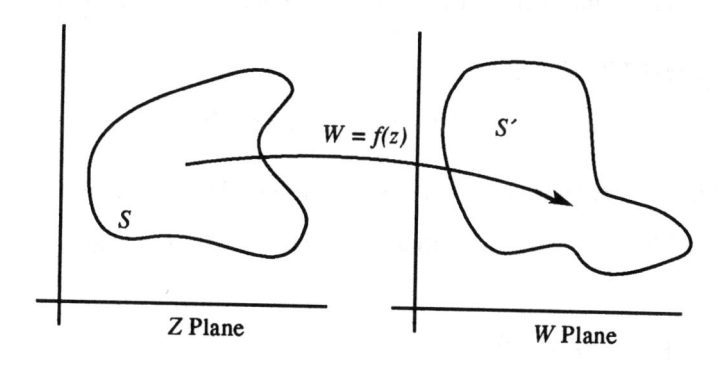

FIGURE 3.11 — Image S' of S

Rotations $w = f(z) = e^{i\theta}z$

Translations $w = fz = z + z_0$

Compound Mappings $w = f(z) = F(G(z))$

THEOREM 3.33 A Line is a Circle of Infinite Radius

THEOREM 3.34 Rotations, Contractions, Expansions, Translations Preserve rectangles, lines, and circles.

DEFINITION 3.37 Linear Transformation

$w = Az + B$

DEFINITION 3.38 Inversion

$w = 1/z$

THEOREM 3.35 Inversion Preserves Orientation

DEFINITION 3.39 Mobius = Bilinear = Linear Fractional Transformation

$w = f(z) = [az + b] / [cz + d]$.

THEOREM 3.36 Decomposition of Linear Fractional Transformation

The composition of inversion and linear transformation.

THEOREM 3.37 Mobius Transformation

Maps lines and circles into lines and circles.

THEOREM 3.38 Determining a Unique Mobius Transformation

There is a unique Mobius transformation mapping 3 distinct points into 3 distinct points.

THEOREM 3.39 Fixed Point z_0

$z_0 = f(z_0)$

THEOREM 3.40 Fixed Points of Mobius Transformation

There are at most two.

THEOREM 3.41 Composition of Mobius Transformations

Composition of two Mobius transforms is a Mobius transform.

3.12 CONFORMAL MAPPINGS AND HARMONIC FUNCTIONS

FAMILIAR IDEAS

Angle Between Two Smooth Arcs (See Figure 3.12)

Preservation of Angle (See Figure 3.13)

Preservation of Orientation or Sense (See Figure 3.14)

FIGURE 3.12 — Angle Between Arcs

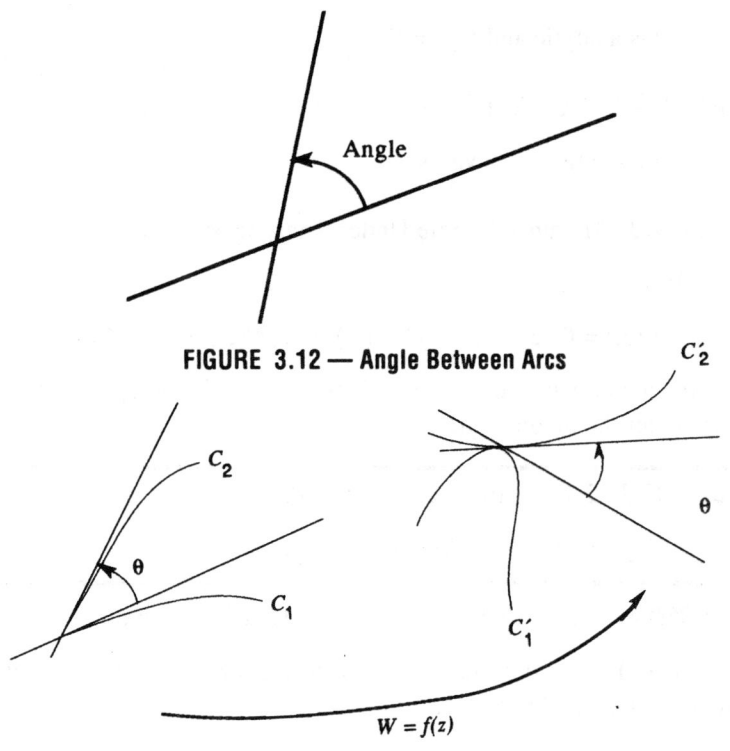

$W = f(z)$

FIGURE 3.13 — f Preserves Angle θ

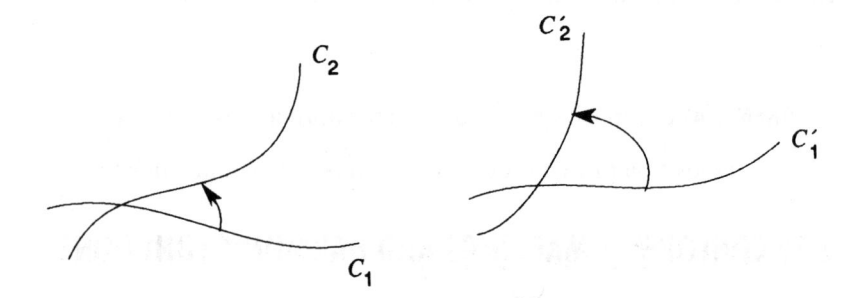

Figure 3.14 — f Preserves Orientation

THEOREM 3.42 Analytic Function Preserves Angle and Orientation

DEFINITION 3.40 Conformal Mapping

   f is analytic and $f'(z) = 0$.

DEFINITION 3.41 Critical Point

   f is analytic and f' vanishes.

THEOREM 3.43 Image of Angle Under Analytic Mapping

   If for $m = 2, 3, \ldots,$
   $$f'(z_0) = f''(z_0) = \ldots = f^{(m-1)}(z_0) = 0, f^{(m)}(z_0) \neq 0,$$
then the angle between any smooth curves meeting at z_0 is multiplied by m under the mapping $f(z)$.

DEFINITION 3.42 Harmonic Function $u(x, y)$

   Obeys the *Laplace Equation* $u_{xx} + u_{yy} = 0$.

THEOREM 3.44 Analytic Function Implies Cauchy-Riemann

   If $f(z) = u + iv$ is analytic, then u, v obey the Cauchy-Riemann equations and are harmonic.

DEFINITION 3.43 *u, v* are Harmonic Conjugates

u, v obey the Cauchy-Riemann equations.

THEOREM 3.45 Level Lines of Harmonic Conjugates are Perpendicular

---
DEFINITION 3.44 Poisson Integral Formula

For u harmonic in the circle $C: z = Re^{i\phi}$, $0 \leq \phi \leq 2\pi$

$$u(r,\theta) = (1/2\pi)\int_0^{2\pi} u(R,\phi)[(R^2-r^2)/(r^2+R^2-2rR\cos(\theta-\phi))]d\phi$$

THEOREM 3.46 Mean Value Property

$$u(z=0) = (1/2\pi)\int_0^{2\pi} u(R,\phi)d\phi.$$

DEFINITION 3.45 Dirichlet Problem

Find a harmonic function in domain *D* attaining given values on the boundary.

3.13 ZEROS AND SINGULAR POINTS OF ANALYTIC FUNCTIONS

FAMILIAR IDEAS

Zero of a Function $f(z)$: $f(z) = 0$

Zero of Order N: $f(z) = f'(z) = \ldots f^{N-1}(z) = 0, f^N(z) \neq 0$

THEOREM 3.47 Zeros and Poles

A zero of order *n* of $f(z)$ is a pole of order *n* of $1/f(z)$.

DEFINITION 3.46 Removable Singular Point

The Laurent expansion contains no negative powers.

THEOREM 3.48 The Numbers of Zeros and Poles

For f analytic in C with N zeros and P poles,

$$(1/2\pi i)\int [f'(z)/f(z)]dz = N - P.$$

THEOREM 3.49 Argument Principle

$$(1/2\pi)\, \Delta \mathrm{carg}(f) = N - P$$

THEOREM 3.50 Rouche's Theorem

For f, g analytic in C and $|f(z)| > |g(z)|$ on C, $f(z)$ and $f(z) + g(z)$ have the same number of zeros in C.

3.14 RIEMANN MAPPING THEOREM

DEFINITION 3.47 Conformal Equivalence of Domains D, D'

There exists a conformal mapping of D onto D'.

THEOREM 3.51 Riemann Mapping Theorem

For D founded by a simple closed Jordan curve C, there is an analytic function mapping D conformally onto $|w| < 1$.

CHAPTER 4

ORDINARY DIFFERENTIAL EQUATIONS

FAMILIAR IDEAS

Ordinary Differential Equation (ODE): An Equation Relating a Function $y(x)$ and Its Derivatives Up to Some Order N

Order of the ODE: The Highest Order of a Derivative in the Equation

Solution of the ODE: A Function Satisfying the Equation

Linear ODE: All Derivatives and $y(x)$ Itself Appear Linearly

4.1 ORDINARY DIFFERENTIAL EQUATIONS OF FIRST ORDER

FAMILIAR IDEAS

First Order ODE: $y'(x) = f(x, y(x))$

Initial Condition (IC): $y(x_0) = y_0$

Initial Value Problem (IVP): Find a Solution to the ODE Satisfying the IC

Separable ODE: Equation Can Be Written as $F(y)dy = G(x)dx$

DEFINITION 4.1 Exact ODE $A(x, y)dx + B(x, y)dy = 0$

There is a $U(x, y)$ such that $A = U_x, B = U_y$.

THEOREM 4.1 The Solution of an Exact Equation

The solution $y(x)$ obeys $U(x, y(x)) = $ constant.

THEOREM 4.2 When is $Adx + Bdy = 0$ Exact?

iff $A_y = B_x$.

DEFINITION 4.2 Integrating Factor $\Phi(x, y)$ for $Adx + Bdy = 0$

$\Phi(x, y) A(x, y)dx + \Phi(x, y)B(x, y)dy = 0$ is exact.

REMARK 4.1 Most General Linear First Order ODE

$$y'(x) + a(x)y(x) = b(x). \qquad (4.1)$$

DEFINITION 4.3 Homogeneity of (4.1)

$b(x) \equiv 0$.

DEFINITION 4.4 IVP for (4.1)

Find $y(x)$ satisfying equation (4.1) and the IC.

IC: $y(x_0) = y_0$

REMARK 4.2 Solution of the IVP

Multiply equation (4.1) by the integrating factor

$$\Phi(x) = \exp(\int_{x_0}^{x} a(s)ds)$$

giving us

$$\{\Phi(x)y(x)\}' = b(x)\Phi(x)$$

and integrate both sides from x_0 to x.

THEOREM 4.3 Solution to Homogeneous Equation $y' = c_0 y$

$y(x) = A \exp(c_0 x)$, $A = $ any constant

H-65

REMARK 4.3 IVP for the Most General ODE of First Order

Find $y(x)$ such that

ODE: $y'(x) = f(x, y(x))$

IC: $y(x_0) = y_0$.

> **REMARK 4.4** Equivalent Integral Equation Formulation
> $$y(x) = y_0 + \int_{x_0}^{x} f(s, y(s))ds \qquad (4.2)$$

DEFINITION 4.5 Picard Iteration for Equation (4.2)

A method for finding approximate

$Y_0(x) \equiv y_0$

$Y_{n+1}(x) = y_0 + \int_{x_0}^{x} f(s, Y_n(s))ds$

> **THEOREM 4.4** Fundamental Existence and Uniqueness Theorem
>
> If $f(x, y)$ and $f_y(x, y)$ are continuous, then the IVP has a unique solution $y = y(x)$ in some interval $\alpha \leq x \leq \beta$.

4.2 LINEAR ORDINARY DIFFERENTIAL EQUATION

FAMILIAR IDEAS

Second Order Linear ODE: $y''(x) + a(x)y'(x) + b(x)y(x) = c(x)$

Homogeneous Equation: $c(x) \equiv 0$

Coefficients: $a(x), b(x)$

Linear Independence (Dependence) of $y_1(x), y_2(x)$: Neither is a Constant Multiple of the Other

THEOREM 4.5 Superposition Principle

If u, v solve the homogeneous equation

$$y''(x) + a(x)y'(x) + b(x)y(x) = 0, \qquad (4.3)$$

then so does any linear combination $\alpha u(x) + \beta v(x)$.

REMARK 4.5 Homogeneous Equation with Constant Coefficients

To find the most general solution of

$$y''(x) + a_0 y'(x) + b_0 y(x) = 0$$

set $y = \exp(\Gamma x)$, giving the *Characteristic Equation (CE)*

$$\Gamma^2 + a_0 \Gamma + b_0 = 0.$$

If the CE has two distinct roots Γ_1, Γ_2, then

$$y(x) = A e^{\Gamma_1 x} + B e^{\Gamma_2 x}$$

If the CE has one double root Γ^*, then

$$y(x) = \{A + Bx\} e^{\Gamma^* x}.$$

REMARK 4.6 Complex Roots of the CE

If $\Gamma_1 = \alpha + i\beta$, $\Gamma_2 = \alpha - i\beta$, then

$$y(x) = e^{\alpha x} \{A \cos(\beta x) + B \sin(\beta x)\}.$$

DEFINITION 4.6 Solution with Non-Constant Coefficients

$$y(x) = A y_1(x) + B y_2(x)$$

for A, B constant, y_1, y_2 any linearly independent pair of solutions of equation (4.3).

DEFINITION 4.7 Wronskian of $y_1(x), y_2(x)$

$$W = \begin{vmatrix} y_1(x) & y_2(x) \\ y_1'(x) & y_2'(x) \end{vmatrix} = y_1(x) y_2'(x) - y_2(x) y_1'(x)$$

THEOREM 4.6 ODE for the Wronskian

For solutions y_1, y_2 of equation (4.3), $W'(x) = -a(x) W(x)$, or

$$W(x) = A\exp(-\int_{x_0}^{x} a(s)ds)$$

for some x_0 and constant A. $W(x) \equiv 0$ or is never zero.

> **THEOREM 4.7** Linear Dependence (Independence)
>
> Two solutions y_1, y_2 of equation (4.3) are linearly independent if their Wronskian is never zero. The solutions y_1, y_2 of equation (4.3) are dependent if their Wronskian vanishes identically.

DEFINITION 4.8 IVP for the Second Order Equation (4.3)

Find y(x) satisfying equation (4.3) and the initial conditions

$$y(x_0) = y_0, \quad y'(x_0) = y_0'$$

THEOREM 4.8 Existence Theorem for the IVP

For $a(x)$, $b(x)$ continuous in an interval about x_0 the IVP for equation (4.3) has a unique solution.

REMARK 4.7 *N*-th Order Linear Equation

The results for second order linear ODE's hold for equations of *N*-th order,

$$y^N(x) + a_1(x)y^{N-1}(x) + \ldots + a_N y(x) = 0 \qquad (4.4)$$

Linear Dependence (Independence) of functions $y_1, y_2, \ldots, y_N$ means that one of these functions can (cannot) be expressed as a linear combination of the others. Their *Wronskian* is the determinant

$$W = \begin{vmatrix} y_1 & y_2 & \ldots & y_N \\ y_1' & y_2' & \ldots & y_N' \\ \ldots & & & \\ y_1^N & y_2^N & \ldots & y_N^N \end{vmatrix}$$

For *N* solutions of equation (4.4) *W either vanishes identically or never, according as they are linearly dependent or independent.* Any solution of equation (4.4) can be written as a linear combination of N linearly independent solutions:

$$y(x) = A_1 y_1(x) + A_2 y_2(x) + \ldots + A_N y_N(x).$$

THEOREM 4.9 Solution of the Inhomogeneous ODE

$$y''(x) + a(x)y'(x) + b(x)y(x) = c(x) \quad (4.5)$$

The sum of any *particular solution* $Y_p(x)$ of equation (4.5) and the general solution of the homogeneous equation ($c(x) \equiv 0$) $Y_{gen}(x)$,

$$y(x) = Y_p(x) + Y_{gen}(x).$$

REMARK 4.8 Solving the IVP for Equation (4.5)

Find *any* particular solution y_p of (4.5) and the *general* solution of the homogeneous equation, add them,

$$y(x) = Y_p(x) + A y_1(x) + B y_2(x).$$

Then select A, B so that the initial conditions hold.

4.3 SYSTEMS OF FIRST ORDER ODE'S

DEFINITION 4.9 System of N First Order ODE's

$$y_1'(x) = F_1(x, y_1(x), y_2(x), \ldots, y_N(x))$$
$$y_2'(x) = F_2(x, y_1(x), y_2(x), \ldots, y_N(x)) \quad (4.6)$$
$$\ldots$$
$$y_N'(x) = F_N(x, y_1(x), y_2(x), \ldots, y_N(x))$$

DEFINITION 4.10 Initial Value Problem (IVP)

Find $y_1(x)$, $y_2(x)$, ..., $y_N(x)$ satisfying system (4.6) and assuming N given values at the initial point x_0.

THEOREM 4.10 Existence Theorem

If $F_j(x, y_1, y_2, \ldots, y_N)$ ($j = 1, \ldots, N$) and their partial derivatives are continuous then a unique solution exists in some interval about x_0.

DEFINITION 4.11 Autonomous ODE System

The system (4.6) if the F_j are independent of x:

$y_1'(x) = F_1(y_1(x), y_2(x), ..., y_N(x))$

$y_2'(x) = F_2(y_1(x), y_2(x), ..., y_N(x))$

...

$y_N'(x) = F_N(y_1(x), y_2(x), ..., y_N(x))$

DEFINITION 4.12 Trajectory

The curves in space given by $(y_1(x), y_2(x), ..., y_N(x))$.

DEFINITION 4.13 Phase Space

The space of $y_1, y_2, ..., y_N$ containing the trajectories.

DEFINITION 4.14 Critical Point

A point in phase space where $F_j(j = 1, ..., N) = 0$.

DEFINITION 4.15 Isolated Critical Point (ICP)

One having a neighborhood containing no others.

DEFINITION 4.16 Stable ICP

If for some x_0 a trajectory is sufficiently close to the ICP, then it will remain close to it permanently.

DEFINITION 4.17 Unstable ICP

One which is not stable.

DEFINITION 4.18 Attractive ICP

If for some x_0 a trajectory is sufficiently close to the ICP, then it will converge to it as $x \to \infty$.

4.4 METHODS FOR SOLVING ODE'S

FAMILIAR IDEAS

Taylor Series: Can Be Differentiated and Integrated Term by Term

Analytic Real-Valued Function at $x = x_0$: A Function Expandable in a Taylor Series about $x = x_0$

THEOREM 4.11 Analytic Solution to ODE

For $a(x)$, $b(x)$, $c(x)$ analytic at $x = x_0$ every solution of

$$y''(x) + a(x)y'(x) + b(x)y(x) = c(x)$$

is analytic and can be expanded in powers of $x - x_0$.

DEFINITION 4.19 Laplace Transform of $f(x)$

$$\mathcal{L}\{f\} = F(s) = \int_0^\infty e^{-sx} f(x)\,dx$$

THEOREM 4.12 Laplace Transform of N-th Derivative of $f(x)$

$$\mathcal{L}\{f^N\} = s^N \mathcal{L}\{f\} - s^{N-1}f(0) - s^{N-2}f'(0) - \ldots - f^{N-1}(0) \qquad (4.7)$$

DEFINITION 4.20 Convolution $h = f*g$ of $f(x)$, $g(x)$

$$h(x) = (f*g)(x) = \int_0^x f(y)g(x-y)\,dy$$

THEOREM 4.13 Laplace Transform of Convolution

$$\mathcal{L}\{f*g\} = \mathcal{L}\{f\}\,\mathcal{L}\{g\}.$$

REMARK 4.9 Laplace Transform Solution of IVP's

To solve in IVP for a *Linear, Inhomogeneous ODE with Constant Coefficients,* we apply the transform to the equation and use equation (4.7) to replace the transform of every derivative by a term involving only the transform of the solution. The values of the derivatives of the solution at the initial point (chosen as $x = 0$ in (4.7)) are known from the initial conditions. The equation is solved for the Laplace Transform of the solution which is then found using tables of Laplace Transforms.

4.5 BOUNDARY VALUE PROBLEMS

DEFINITION 4.21 Boundary Value Problem (BVP) for the ODE

$$y''(x) = F(x, y(x), y'(x)) \qquad (4.8)$$

For given h_a, k_a, h_b, k_b, m_a, m_b, find $y(x)$ satisfying (4.8) and the *Boundary Conditions*

$$h_a y(a) + k_a y'(a) = m_a \qquad (4.9)$$
$$h_b y(b) + k_b y'(b) = m_b$$

REMARK 4.10 Some BVP's Have No Solution

If $y''(x) + y(x) = 0$, $0 \leq x \leq \pi$, then

$$y(x) = A \cos(x) + B \sin(x).$$

For the boundary conditions $y(0) = 0$, $y(\pi) = 1$, the BVP has no solution since $\sin(0) = 0$ implies $A = 0$ while $\sin(\pi) = 0$ implies that $y(\pi)$ vanishes.

DEFINITION 4.22 Sturm Liouville Equation

$$(p(x)y'(x))' + (q(x) + \mu r(x))y(x) = 0. \qquad (4.10)$$

DEFINITION 4.23 Sturm Liouville Problem

Find constants μ and solutions to equation (4.10) not vanishing identically and satisfying the homogeneous BC obtained from system (4.9) by setting $m_a = m_b = 0$. Values of μ for which this can be done are *eigenvalues* while the solutions are *eigenfunctions*.

THEOREM 4.14 Orthogonality of Eigenfunctions $y_1(x)$, $y_2(x)$

$$\int_a^b r(x) y_1(x) y_2(x) dx = 0.$$

THEOREM 4.15 Eigenvalues are Real Numbers

If $r(x) = 0$, then the eigenvalues are real numbers.

CHAPTER 5

FOURIER ANALYSIS AND INTEGRAL TRANSFORMS

5.1 BASIC IDEAS

FAMILIAR IDEAS

Periodic Function: $f(t) = f(t + T)$, T = Period

Harmonic Vibration: $f(t) = A \sin[\omega t - \delta]$, $A \cos[\omega t - \delta]$, A = Amplitude, ω = Angular Frequency

Frequency: $f = \omega/2\pi$

Left, Right Limits $f(t_0 - 0)$, $f(t_0 + 0)$ of $f(t)$ for $t \to t_0$ (Figure 5.1)

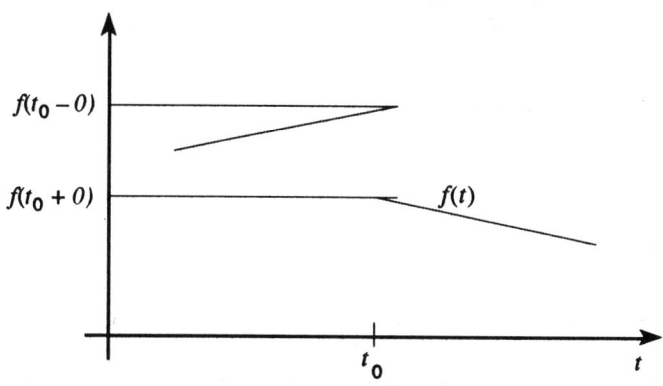

FIGURE 5.1 — Left and Right Hand Limits

Jump Discontinuity at a Point

Piecewise Continuous Function

Even, Odd Functions

DEFINITION 5.1 Mean Value at a Jump Discontinuity t_0

$$f_{mean}(t_0) = \tfrac{1}{2}\{f(t_0 + 0) + f(t_0 - 0)\}$$

DEFINITION 5.2 Phase Angle

The harmonic wave

$$f(t) = a\cos(\omega t) + b\sin(\omega t)$$

can be written as

$$f(t) = A\cos(\omega t - \delta),$$

for $\delta = \arctan(b/a)$ the *Phase Angle*, $A = [a^2 + b^2]^{1/2}$.

DEFINITION 5.3 Trigonometric Polynomial of Order N

A function

$$S_N(t) = \tfrac{1}{2}a_0 + \sum_{n=1}^{N}(a_n\cos(n\omega t) + b_n\sin(n\omega t))$$

$$= \tfrac{1}{2}a_0 + \sum_{n=1}^{N} A_n\cos(n\omega t - \delta_n)$$

DEFINITION 5.4 Superposition of Harmonic Waves

$$f(t) = \sum_{n=0}^{\infty} A_n\cos(\omega_n t - \delta_n),$$

with $0 < \omega_1 < \omega_2 < \ldots$.

DEFINITION 5.5 Fundamental and Higher Harmonics

$\cos(\omega_1 t - \delta_1)$, $\omega_n = n\omega_1$; ω_1, f_1 are the *Fundamental Angular Frequency* and *Fundamental Frequency*.

DEFINITION 5.6 Higher Harmonics

$\cos(\omega_2 t - \delta_2)$, ..., $\cos(\omega_n - \delta_n)$ are the second, third, ..., n-th harmonics, respectively; $\omega_2, \omega_3, ..., \omega_n$ are the second, third, ..., n-th harmonic angular frequencies. The $f_1, f_2, ...$ are the fundamental, second, ..., n-th harnomic frequencies.

DEFINITION 5.7 Beats

Rhythmic amplitude changes due to harmonic interactions.

THEOREM 5.1 Euler's Formula and Implications

$e^{iz} = \cos(z) + i\sin(z)$,

$e^{-iz} = \cos(z) - i\sin(z)$.

$\cos(z) = \frac{1}{2}(e^{iz} + e^{-iz})$,

$\sin(z) = -\frac{1}{2}i(e^{iz} - e^{-iz})$.

REMARK 5.1 Harmonic Wave Superpositions

$$f(t) = \sum_{n=-\infty}^{\infty} B_n \exp(i\omega_n t)$$

5.2 FOURIER SERIES

FAMILIAR IDEAS

Pointwise Convergence

Uniform Convergence

DEFINITION 5.8 Fourier Series of $f(t)$ on $-\pi < x < \pi$

$$f(t) = \tfrac{1}{2}a_0 + \sum_{n=1}^{\infty} [a_n \cos(nt) + b_n \sin(nt)],$$

with the *Fourier Coefficients*

$$a_0 = (1/2\pi)\int_{-\pi}^{\pi} f(s)ds,$$
$$a_n = (1/\pi)\int_{-\pi}^{\pi} f(s)\cos(ns)ds,$$
$$b_n = (1/\pi)\int_{-\pi}^{\pi} f(s)\sin(ns)ds.$$

DEFINITION 5.9 Dirichlet Conditions on $f(t)$ for Convergence

A. f is periodic;
B. at all but a finite number of points f is differentiable;
C. $f(t), f'(t)$ are piecewise continuous;
D. at points of discontinuity $f(t)$ is its mean value.

DEFINITION 5.10 Piecewise Smooth Function $f(t)$

f obeys the Dirichlet conditions.

THEOREM 5.2 Convergence Theorem for Fourier Series

If the Dirichlet conditions hold, then the Fourier Series converges at every point where f is continuous; at a jump convergence is to the mean value; convergence is uniform on every closed interval of continuity.

DEFINITION 5.11 Gibbs Phenomenon

"Overshooting" of Fourier polynomials at a jump.

THEOREM 5.3 Riemann-Lebesque Theorem

If the Dirichlet conditions hold, then

$$\lim_{n\to\infty}\int_a^b f(t)\cos(nt)dt = \lim_{n\to\infty}\int_a^b f(t)\sin(nt)dt = 0$$

THEOREM 5.4 Bessel's Inequality

$$\tfrac{1}{2}a_0^2 + \sum_{n=1}^{N}[a_n^2 + b_n^2] \le (1/\pi)\int_{-\pi}^{\pi}(f(t))^2 dt$$

THEOREM 5.5 Parseval's Identity

$$\tfrac{1}{2}a_0^2 + \sum_{n=1}^{N}[a_n^2 + b_n^2] = (1/\pi)\int_{-\pi}^{\pi}(f(t))^2\,dt$$

DEFINITION 5.12 Fourier Series in Complex Notation

$$f(t) = \sum_{n=-\infty}^{\infty} \alpha_n e^{int}$$

$$\alpha_n = (1/2\pi)\int_{-\pi}^{\pi} f(t)e^{-int}\,dt$$

THEOREM 5.6 Fourier Series for Even Functions

Can be written in terms of cosines.

THEOREM 5.7 Fourier Series for Odd Functions

Can be written in terms of sines.

DEFINITION 5.13 Multidimensional Fourier Series

$$f(s,t) = \sum_{m=0}^{\infty}\sum_{n=0}^{\infty} \gamma_{mn}\{a_{mn}\cos(ms)\cos(nt)$$

$+ b_{mn}\sin(ms)\cos(nt) + c_{mn}\cos(ms)\sin(nt)$
$+ d_{mn}\sin(ms)\sin(nt)\}$

$$\gamma_{mn} = \begin{cases} \tfrac{1}{4}, & \text{for } m = n = 0 \\ \tfrac{1}{2}, & \text{for } m > 0, n = 0 \text{ or } n > 0, m = 0 \\ 1, & \text{for } m, n > 0 \end{cases}$$

$a_{mn} = (1/\pi^2) \iint_D f(s,t)\cos(ms)\cos(nt)\,ds\,dt$

$b_{mn} = (1/\pi^2) \iint_D f(s,t)\sin(ms)\cos(nt)\,ds\,dt$

$$c_{mn} = (1/\pi^2) \iint_D f(s, t) \cos(ms) \sin(nt)\, dsdt$$

$$d_{mn} = (1/\pi^2) \iint_D f(s, t) \sin(ms) \sin(nt)\, dsdt$$

THEOREM 5.8 Termwise Integration of the Fourier Series

Fourier Series can be integrated term by term.

THEOREM 5.9 Termwise Differentiation of the Fourier Series

Fourier Series can be differentiated term by term.

5.3 FOURIER SERIES AND VECTOR SPACE CONCEPTS

FAMILIAR IDEAS

Vector in N Dimensions

Inner Product

Norm, Unit Vector

Cauchy-Schwarz Inequality

Angle Between Two Vectors

Triangle Inequality

Orthogonality of Vectors

Linear Combination of Vectors

DEFINITION 5.14 Vector Space

Vector collection containing the sum of any two and the product of any by a scalar.

THEOREM 5.10 Continuous Functions as Vectors

Piecewise continuous functions form a vector space.

DEFINITION 5.15 Inner Product of Continuous Functions on [a,b]

$$(f_1, f_2) = \int_a^b f_1(t) f_2(t) dt.$$

DEFINITION 5.16 Norm of $f(x)$

$$\|f\| = \{(f,f)\}^{1/2} = \left\{ \int_a^b f(t)^2 dt \right\}^{1/2}$$

THEOREM 5.11 Cauchy-Schwarz Inequality

$$|(f, g)| \le \|f\| \, \|g\|$$

DEFINITION 5.17 Angle Between f, g

$$\cos(\theta) = (f, g) / [\|f\| \, \|g\|]$$

DEFINITION 5.18 Orthogonality of f, g

$$(f, g) = 0.$$

THEOREM 5.12 Orthogonality of the Trigonometric Functions

For $a = -\pi$, $b = \pi$, $\cos(mx)$, $\sin(nx)$, $m,n = 0, 1, \ldots$ are orthogonal.

DEFINITION 5.19 Orthonormality

f, g are orthogonal and their norms are equal to one.

DEFINITION 5.20 Least-Square Distance Between f, g

$$d(f, g) = \|f - g\| \tag{5.1}$$

THEOREM 5.13 Fourier Polynomial is Best Approximation

The Fourier polynomial of order N is the closest linear combination of sin, cos functions in the sense of $\| \ \|$.

> **THEOREM 5.14** Generalized Fourier Series
>
> Let $\{0_n\}$ be an orthonormal family of functions. Their nearest linear combination to a function f in the sense of (5.1) is found by choosing the coefficients of the 0_n as the *Fourier Coefficients*
>
> $a_n = (f, 0_n)$.

THEOREM 5.15 Bessel's Inequality

$$\|f\|^2 \geq \sum_{n=1}^{N} \alpha_n^2.$$

DEFINITION 5.21 Completeness of the Orthonormal Family $\{0_n\}$

Every function with a finite norm is the limit of a sequence of linear combinations of the family elements.

THEOREM 5.16 Parseval's Identity

If the $\{0_n\}$ are complete, then Bessel's inequality becomes an equality for $N \to \infty$.

DEFINITION 5.22 Weighted Inner Products and Norms

The above hold if the inner product is redefined as

$$(f,g) = \int_a^b W(t)f(t)g(t)dt$$

with $W(t) > 0$ a given continuous function.

DEFINITION 5.23 Inner Product for Complex-Valued Functions

$$(f,g) = \int_a^b f(t)\overline{g(t)}dt$$

5.4 FOURIER TRANSFORMS

DEFINITION 5.24 Absolutely Integrable Function $f(t)$

$\int_{-\infty}^{\infty} |f(t)|dt$ exists.

THEOREM 5.17 Fourier Integral Theorem

$$f(t) = (1/\pi) \int_0^{\infty} d\gamma \int_{-\infty}^{\infty} f(\gamma) \cos[s(t-\gamma)]ds.$$

THEOREM 5.18 Equivalent Forms of the Fourier Integral Theorem

$$f(t) = (1/2\pi)\int_0^{\infty} \int_{-\infty}^{\infty} f(\gamma)e^{is(t-\gamma)}d\gamma ds$$

$$f(t) = (1/2\pi)\int_{-\infty}^{\infty} e^{ist}ds \int_0^{\infty} f(\gamma)e^{-is\gamma}d\gamma.$$

DEFINITION 5.25 Fourier Transform $F(\gamma)$ of $f(t)$

$$F(s) = \int_{-\infty}^{\infty} f(\gamma)\exp(-is\gamma)d\gamma.$$

We write $F = \Phi\{f\}$ or $F(\gamma) = \Phi\{f\}[\gamma]$.

THEOREM 5.19 Inverse $f = \Phi^{-1}\{F\}$ of the Fourier Transform

$$f(t) = (1/2\pi)\int_{-\infty}^{\infty} F(s)e^{ist}ds.$$

THEOREM 5.20 Fourier Transform of an Odd Function $f(t)$

$$F_s(\gamma) = \int_0^{\infty} f(t)\sin(t\gamma)dt.$$

THEOREM 5.21 Fourier Transform of an Even Function $f(t)$

$$F_c(\gamma) = \int_0^{\infty} f(t)\cos(t\gamma)dt.$$

THEOREM 5.22 Special Properties of the Fourier Transform

$$\Phi\{f(t - t^*)\} = \Phi(f)\exp(-i\gamma t^*)$$
$$\Phi^{-1}\{F(\gamma - \gamma^*)\} = f(t)\exp(i\gamma t^*)$$
$$\Phi\{f(\alpha t)\} = (1/|\alpha|)F(\alpha)$$
$$\Phi\{f(-t)\} = F(-\gamma)$$
$$\Phi\{\Phi\{f(t)\}\} = 2\pi f(-t)$$

THEOREM 5.23 Convolution Theorem

The Fourier Transform of the convolution of two functions is the product of their Fourier Transforms.

THEOREM 5.24 Parseval's Identities

Let $f(t)$, $g(t)$ have Fourier Transforms $F(\gamma)$, $G(\gamma)$. Then

$$\int_{-\infty}^{\infty} f(t)g(t)\,dt = (1/2\pi)\int_{-\infty}^{\infty} F(\gamma)G(\gamma)d\gamma,$$

$$\int_{-\infty}^{\infty} f(s)G(s)\,ds = \int_{-\infty}^{\infty} F(s)g(s)\,ds,$$

$$\int_{-\infty}^{\infty} f(t)^2 dt = (1/2\pi)\int_{-\infty}^{\infty} |F(\gamma)|^2 d\gamma.$$

THEOREM 5.25 Continuity of the Fourier Transform

If f is absolutely continuous, then its Fourier Transform $F(\gamma)$ is continuous for all γ and tends to zero as $|\gamma| \to \infty$.

THEOREM 5.26 The Transform of the Derivative

$$\Phi(f')[\gamma] = \gamma i \Phi(f).$$

THEOREM 5.27 The Transform of the Indefinite Integral

If

$$g(t) = \int_0^t f(s)\,ds,$$

then $\Phi(g)[\gamma] = (1/i\gamma) \Phi(f)[\gamma]$.

DEFINITION 5.26 The Fourier Transform of $f(s, t)$

$$F(\sigma,\gamma) = \int_{-\infty}^{\infty} \int_{-\infty}^{\infty} f(s,t) e^{-i(\sigma s + \gamma t)} ds dt.$$

THEOREM 5.28 Inverse Fourier Transform

$$f(s,t) = 1/(2\pi)^2 \int_{-\infty}^{\infty} \int_{-\infty}^{\infty} F(\sigma,\gamma) e^{i(s\sigma + t\gamma)} d\sigma d\gamma.$$

DEFINITION 5.27 Magnitude Spectrum of $f(t)$

Magnitude Spectrum $= |F(\gamma)|$.

DEFINITION 5.28 Phase Spectrum of $f(t)$

Phase Spectrum $= \arg(F(\gamma))$.

DEFINITION 5.29 Energy Content of a Function $f(t)$

$$\int_{-\infty}^{\infty} |f(t)|^2 dt.$$

THEOREM 5.29 Energy Content in Terms of Fourier Transform

Energy Content $= (1/2\pi) \int_{-\infty}^{\infty} |F(\gamma)|^2 d\gamma.$

DEFINITION 5.30 Cross-Correlation of f, g

$$R_{fg}(\gamma) = \int_{-\infty}^{\infty} f(t) g(t - \gamma) dt.$$

DEFINITION 5.31 Uncorrelated Functions

Their cross correlation is zero for all arguments.

DEFINITION 5.32 Auto-Correlation of $f(t)$

$$R_f(\gamma) = \int_{-\infty}^{\infty} f(t) f(t - \gamma) dt.$$

THEOREM 5.30 Fourier Transforms of Correlations

$\Phi\{R_{fg}\}\,[s] = \Phi\{f\}\,[s]\,\Phi\{g\}\,[-s]$

$\Phi\{R_f\}\,[s] = \Phi\{f\}\,[s]\,\Phi\{f\}\,[-s]$

THEOREM 5.31 Wiener-Khintchine Theorem

$|\Phi\{f\}\,[\gamma]|^2 = \Phi\{R_f\}\,[\gamma].$

5.5 SPECIAL FUNCTIONS

FAMILIAR IDEAS

    Fourier Series

    Fourier Coefficients

    Complete Family of Functions

    Sturm Liouville Problem

        Eigenvalues and Eigenfunctions

DEFINITION 5.33 Gamma Functions

$$\Gamma(p) = \int_0^\infty e^{-t} t^{p-1}\,dt, \quad \text{for } p > 0.$$

THEOREM 5.32 Gamma Function

$\Gamma(p + 1) = p!\quad \text{for } p = 1, 2, \ldots.$

DEFINITION 5.34 Bessel's Equation

$t^2 y'' + t y' + (t^2 - m^2) y = 0$, for m = constant.

THEOREM 5.33 Bessel Functions

For any m Bessel's Equation has a solution $J_m(t)$ having a finite limit as $t \to \infty$, and a solution $Y_m(t)$ which is unbounded as $t \to \infty$. J_m is a *Bessel Function of the First Kind of Order m*; Y_m is a *Bessel Function of the Second Kind of Order m, or Neumann Function.*

REMARK 5.2 Frobenius Method for Finding Bessel Functions

Look for a solution in the form

$$y = \sum_{n=0}^{\infty} a_n t^{n+\beta} ;$$

the unknown a_n are found by substitution.

THEOREM 5.34 Power Series for Bessel Functions of First Kind

$$J_m(t) = \sum_{n=1}^{\infty} \{(-1)^n (t/2)^{m+2n}\} / \{\Gamma(n+1)\Gamma(m+n+1)\}.$$

For $m = 1, 2, \ldots$

$$J_m(t) = \sum_{n=1}^{\infty} \{(-1)^n (t/2)^{2n+m}\} / \{n!(m+n)!\}.$$

For all m,

$$J_{-m}(t) = (-1)^m J_m(t).$$

THEOREM 5.35 Case of m Not an Integer

Then J_m and J_{-m} are linearly independent and the general solution of Bessel's equation is

$$y = a_1 J_m + a_2 J_{-m}.$$

THEOREM 5.36 Case of m an Integer

Then J_m and J_{-m} are linearly dependent and form an independent pair of solutions to Bessel's Equation, for which every solution can be represented as

$$y = a_1 J_m + a_2 Y_m.$$

THEOREM 5.37 Series Representation of the Neumann Function

For integer m,

$$Y_m(t) = (2/\pi) [\ln(t/2) + \gamma] J_m(t)$$

$$- (1/\pi) \sum_{k=0}^{m-1} (m - k - 1)! \, (t/2)^{2k-m} / k!$$

$$- (1/\pi) \sum_{k=0}^{\infty} (-1)^k \{W(k) + W(m + k)\} \, (t/2)^{2k+m} / [k!(m+k)!]$$

with $\gamma = 0.5772156 \ldots$ the *Euler's Constant* and

$$W(k) = 1 + 1/2 + 1/3 + \ldots + 1/k.$$

DEFINITION 5.35 Hankel Functions $H_m^{(1)}$, $H_m^{(2)}$

$$H_m^{(1)} = J_m + iY_m, \; H_m^{(2)} = J_m - iY_m.$$

THEOREM 5.38 Recurrence Relations for Bessel Functions

If C_m is J_m, Y_m, $H_m^{(1)}$ or $H_m^{(2)}$, then

$$C_{m+1}(t) = (2m/t) C_m(t) - C_{m-1}(t)$$

$$C_m' = \tfrac{1}{2}[C_{m-1} - C_{m+1}]$$

$$tC_m'(t) = mC_m(t) - tC_{m+1}(t)$$

$$tC_m'(t) = tC_{m-1}(t) - mC_m(t)$$

$$(d/dt) [t^m C_m(t)] = t^m C_{m-1}(t)$$

$$(d/dt) [t^{-m} C_m(t)] = - t^{-m} C_{m+1}(t)$$

THEOREM 5.39 Behavior for $t \approx 0$

$$J_m(t) \approx (\tfrac{1}{2}t)^m / m!$$

$$Y_0(t) \approx (2/\pi) \ln(t)$$

$$Y_m(t) \approx -(1/\pi) (m - 1)! \, (2/t)^m$$

THEOREM 5.40 Behavior for $t \to \infty$

$$J_m(t) \approx (2/\pi t)^{1/2} \cos\{t - \tfrac{1}{2} m\pi - \tfrac{1}{4}\pi\}$$

$$Y_m(t) \approx (2/\pi t)^{1/2}\sin\{t - \tfrac{1}{2}m\pi - \tfrac{1}{4}\pi\}$$

$$H_m^{(1)}(t) \approx (2/\pi t)^{1/2}\exp\{i(t - \tfrac{1}{2}m\pi - \tfrac{1}{4})\}$$

$$H_m^{(2)}(t) \approx (2/\pi t)^{1/2}\exp\{-i(t - \tfrac{1}{2}m\pi - \tfrac{1}{4})\}$$

DEFINITION 5.36 Bessel Inner Product $(f, g)_B$

$$(f, g)_B = \int_0^1 tf(t)g(t)dt.$$

DEFINITION 5.37 Bessel Norm

$$||f||_B^2 = \int_0^1 tf(t)^2 dt.$$

THEOREM 5.41 An Inner Product for Bessel Functions

For any α, β,

$$\int_0^1 tJ_m(\alpha t) J_m(\beta t)\, dt$$
$$= \{\alpha J_m(\beta)J_m'(\alpha) - \beta J_m(\alpha)J_m'(\beta)\} / \{\beta^2 - \alpha^2\}$$

THEOREM 5.42 Orthogonality of the Bessel Functions

If either or both of α, β are zeros of J_m, then

$$f(t) = J_m(\alpha t),\ g(t) = J_m(\beta t)$$

are orthogonal in the Bessel inner product.

THEOREM 5.43 Zeros of $J_m(t)$

Bessel functions have infinitely many positive zeros that can be arranged in increasing order. The first four zeros of J_2 are 5.14, 8.42, 11.62, and 14.80.

THEOREM 5.44 Fourier-Bessel Series for $f(t)$

For $\{\alpha_n\}$ the zeros of J_m and $m \geq -\tfrac{1}{2}$, let

$$f_n(t) = J_m(\alpha_n t).$$

Then

$$f(t) = \sum_{j=1}^{\infty} c_j f_n(t)$$

$$c_j = (f, f_j)_B / ||f_j||_B^2.$$

DEFINITION 5.38 Legendre's Differential Equation

$(1 - t^2)y'' - 2ty' + m(m + 1)y = 0$ for any m.

DEFINITION 5.39 Legendre Polynomials P_m

The solutions to Legendre's ODE for $m = 1, 2, \ldots$.

THEOREM 5.45 Orthogonality of the Legendre Polynomials

$$\int_{-1}^{1} P_m(t)P_n(t)dt = \begin{cases} 0, & \text{for } m \neq n \\ 2/(2n+1), & \text{for } m = n \end{cases}$$

THEOREM 5.46 Rodrigue's Formula for P_m

$$P_m(t) = [1/(2^m m!)] d^m / dt^m \{(t^2 - 1)^m\}.$$

THEOREM 5.47 Recurrence Formulas for P_m

$$P_{m+1}(t) = [(2m+1)/(m+1)]t P_m(t) - [m/(m+1)]P_{m-1}(t)$$

$$P_{m+1}'(t) = 2(m+1)P_m(t) + P_{m-1}'(t)$$

THEOREM 5.48 Legendre-Fourier Series for $f(t)$

$$f(t) = \sum_{m=1}^{\infty} \alpha_m P_m(t)$$

$\alpha_m = (f, P_m) / ||P_m||^2$

$||P_m||^2 = 2/(2m + 1)$.

CHAPTER 6

PARTIAL DIFFERENTIAL EQUATIONS (PDE'S)

6.1 FUNDAMENTAL IDEAS

FAMILIAR IDEAS

Domain R in 1, 2, 3 Space Dimensions (See Figure 6.1)

Boundary Γ of a Domain in 1, 2, 3 Space Dimensions

> **REMARK 6.1** Time and Space
>
> PDE's first arose in heat and mass transfer and wave propagation. Some processes are *Time-Dependent:* quantities of interest vary with time, such as changes of weather. Others are *Steady State*, not varying in time, such as a soap film spanning a wire frame (See Figure 6.2). Sometimes time is *Reversible*, with past and future interchangeable. Then the questions "what will a quantity be if it was given earlier" and "what was it if it is now some value" can both be answered. For other *Irreversible* processes this can't be done: there is a well defined sense of time that cannot be reversed, an example being the melting of an ice cube in hot coffee (See Figure 6.3).

REMARK 6.2 Number of Spatial Variables

We live in a three-dimensional world. However, the geometry

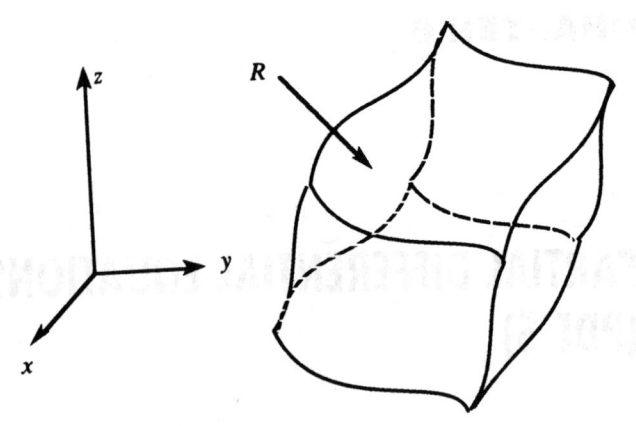

FIGURE 6.1 — Domain R

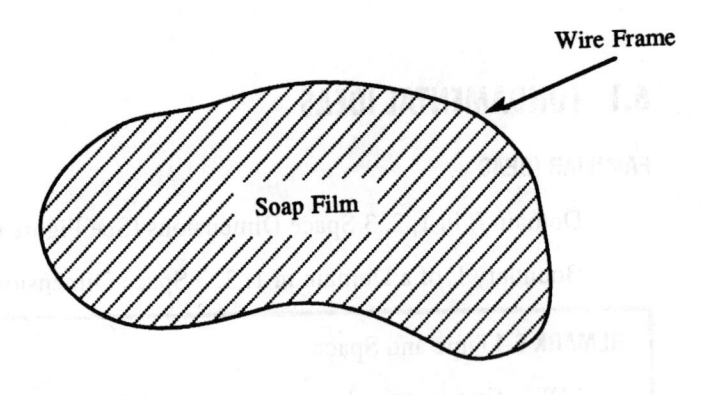

FIGURE 6.2 — Soap Film

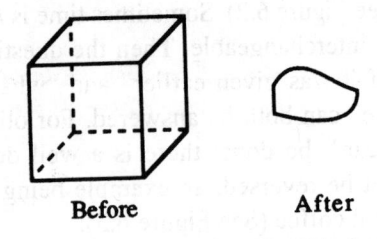

FIGURE 6.3 — Melting Ice Cube

of many processes permits us to represent quantities of interest in terms of functions of fewer than three spatial variables. A process may be axially symmetric, quantities of interest depending only on the depth along the cylinder axis and the distance from it (see Figure 6.4). Similarly, the radius of the cylinder may be so large that effectively the quantities of interest are independent of the distance from the axis, and depend only on the depth along it. Then the process is one dimensional.

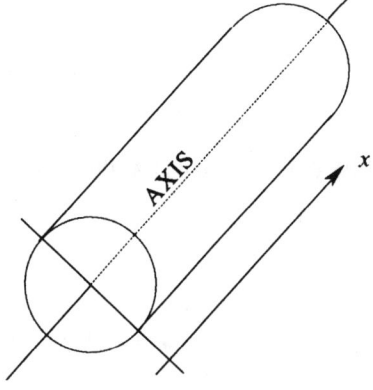

FIGURE 6.4 — Axially Symmetric Process

Note that a quantity may depend on space variables *and* other parameters of a process. We will then have more than three independent variables.

REMARK 6.3 Coordinate Systems

PDE's are represented in coordinate systems fitting the process being modeled. The most popular are the rectangular, cylindrical, and spherical, with the variables

Rectangular:	x, y, z	(See Figure 6.5)
Circular Cylindrical:	r, θ, z	(See Figure 6.6)
Spherical Polar:	$r, \theta, 0$	(See Figure 6.7)

Relations between them are given by:

Circular Cylindrical/Rectangular

$$x = r\cos(\theta), \quad y = r\sin(\theta), \quad z = z$$

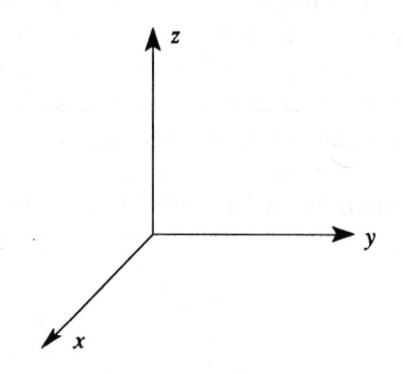

FIGURE 6.5 — Rectangular Coordinates

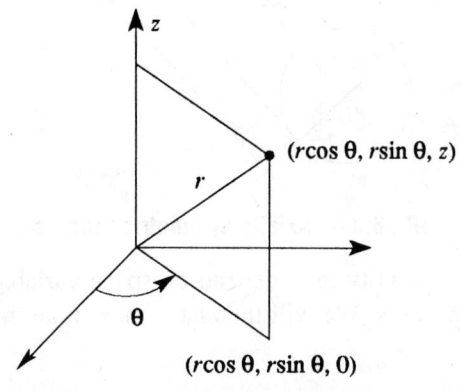

FIGURE 6.6 — Circular Cylindrical Coordinates

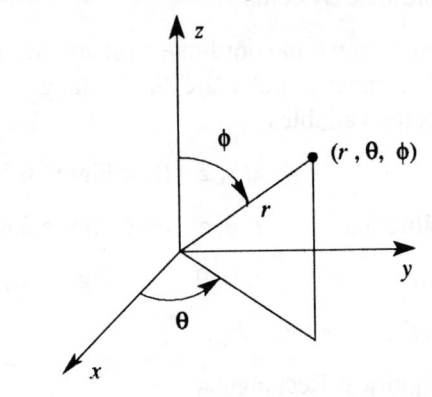

FIGURE 6.7 — Spherical Polar Coordinates

Spherical Polar/Rectangular

$$x = r\cos(\theta)\sin(\phi), \quad y = r\sin(\theta)\sin(\phi), \quad z = r\cos(\phi)$$

Where there is no dependence on z, the circular cylindrical reduce to the *Polar Coordinates*,

$$x = r\cos(\theta), \quad y = r\sin(\theta)$$

DEFINITION 6.1 PDE

A relation between a function $u(x, y, ...)$ and its partial derivatives with respect to $x, y, ...$ to various orders. We also say "PDE in u."

EXAMPLE 6.1 Three Classical PDE's

Three classical PDE's are

Heat Equation:	$u_t = \alpha u_{xx}$
Wave Equation:	$u_{tt} = c^2 u_{xx}$
Laplace Equation:	$u_{xx} + u_{yy} = 0$

DEFINITION 6.2 Qualities of a PDE in u

Solution of PDE: Any function obeying the PDE

Example: $u = e^x \cos(y)$ obeys $u_{xx} + u_{yy} = 0$

Linear PDE: Terms involving u are linear

Example: $u_t = \alpha u_{xx}$

Order of PDE: Order of highest order derivative

Example: $u_t = \alpha u_{xx}$, $u_{tt} = c^2 u_{xx}$ are 2nd order

Homogeneous PDE: No term that does not depend on u

Example: $u_{xx} + u_{yy} + 3u_x - 2u_y = 0$

Quasilinear PDE: Linear in highest order derivatives

Example: Minimal surface equation,

$$(1 + u_y^2)u_{xx} - 2u_x u_y u_{xy} + (1 + u_x^2)u_{yy} = 0$$

Nonlinear: Not linear

Example: $u_t + uu_x = 0$ and $u_t = \cos(u_x)$

Inhomogeneous: Not homogeneous

Example: Poisson's equation, $u_{xx} + u_{yy} = f$

DEFINITION 6.3 Superposition Principle

Any linear combination $au + bv$ of solutions u, v to a linear, homogeneous PDE is a solution.

DEFINITION 6.4 Types of $au_{xx} + bu_{xy} + cu_{yy} = 0$ (6.1)

Elliptic: $b^2 - 4ac < 0$

Hyperbolic: $b^2 - 4ac > 0$

Parabolic: $b^2 - 4ac = 0$

REMARK 6.4 Types of PDE's and Their Meaning

For hyperbolic and parabolic PDE's, one of the variables represents time while the other is space. These PDE's arise in time-dependent processes. In some cases, they are reversible; in others, they are not. Elliptic PDE's arise in steady-state processes. Each variable is spatial.

DEFINITION 6.5 Hadamard Well Posed Problem

A problem with a unique solution depending continuously on data.

REMARK 6.5 Well Posedness

A condition for the problem to be physically reasonable. In recent years, many *Non-Well-Posed* problems have become important in new areas of science and technology.

DEFINITION 6.6 Boundary Condition

A condition imposed on the PDE solution at the boundary.

DEFINITION 6.7 Initial Condition

A condition imposed on the PDE solution at an initial time.

DEFINITION 6.8 Separable Solution to a PDE

A solution that can be given as the product of functions of each of the variables.

DEFINITION 6.9 Similarity Solution to a PDE

A solution that can be given in terms of less variables.

6.2 THE LAPLACE EQUATION

DEFINITION 6.10 Laplace Equation in Rectangular Coordinates

$u_{xx} + u_{yy} = 0.$

REMARK 6.6 Type of Equation

Elliptic: $a = 1$, $b = 0$, $c = 1$, $b^2 - 4ac < 0$ in equation (6.1).

REMARK 6.7 Physical Meaning of Equation

Steady-state, equilibrium states.

REMARK 6.8 Equation in Higher Dimensions

$u_{xx} + u_{yy} + u_{zz} = 0.$

REMARK 6.9 Other Coordinate Systems

Circular Cylindrical:

$(1/r)(ru_r)_r + (1/r^2) u_{\theta\theta} + u_{zz} = 0$

Spherical Polar:

$$(1/r^2)(r^2 u_r)_r + (1/[r^2\sin(\phi)])(\sin(\phi)u_\phi)_\phi$$
$$+ (1/[r^2\sin^2(\phi)])u_{\theta\theta} = 0$$

Polar:

$$(1/r)(ru_r)_r + (1/r^2)u_{\theta\theta} = 0$$

REMARK 6.10 Properties of Equation

Order 2, Homogeneous, Linear

REMARK 6.11 Properties of Solutions

Maximum Principle: In a domain R (see Figure 6.8), the maximum and minimum of u are attained on the boundary Γ.

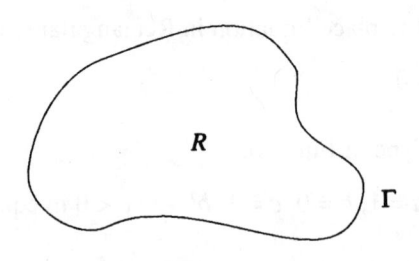

FIGURE 6.8 — Maximum Principle

REMARK 6.12 Form of Well Posed Problem

Dirichlet Problem: Find a solution u in R (see Figure 6.9) attaining given values on Γ.

EXAMPLE 6.2 A Well Posed Problem and Its Solution

For $f(\theta)$ continuous on $[0, 2\pi]$ ($f(0) = f(2\pi)$) find $u(r, \theta)$ obeying Laplace's equation in $0 \le r \le 1$ (see Figure 6.10) and B.C.

$$u(1,\theta) = f(\theta).$$

FIGURE 6.9 — Well Posed Problem

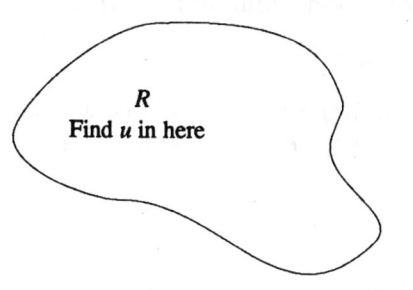

Γ: boundary conditions

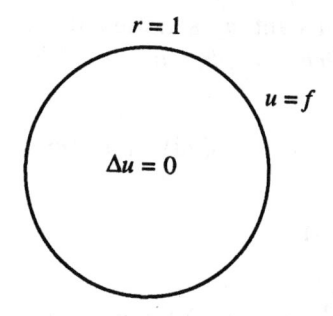

FIGURE 6.10 — Well Posed Problem in the Unit Circle

Separation of Variables for $u(r, \theta) = A(r)B(\theta)$ gives

$$(1/r)(rA'(r))'B(\theta) + (1/r^2)B''(\theta)A(r) = 0$$

or

$$[r^2 A''(r) + rA'(r)]/A(r) = -B''(\theta)/B(\theta). \qquad (6.2)$$

But the left and right sides of this equation are functions of independent variables. Thus, a change in θ must not induce a change in the left hand side and vice versa, or both sides must be constants. For $n = 1, 2, \ldots,$ let this value be n^2. Equating the left hand side to n^2 yields

$$r^2 A''(r) + rA'(r) - n^2 A(r) = 0.$$

Seeking a solution $A(r)$ of the form $A(r) = r^\delta$ yields

$$\delta(\delta - 1) + \delta - n^2 = 0$$

or $\delta = \pm n$. A solution $A(r) = r^{-n}$ would not be defined at $r = 0$ and is unacceptable. Hence, $A(r) = r^n$. The right hand side of equation (6.2) gives the ODE

$$B''(\theta) + n^2 B(\theta) = 0$$

having the general solution

$$B(\theta) = a_n \cos(\theta) + b_n \sin(\theta).$$

Thus, for any $n = 1, 2, \ldots$, the separable solution u is

$$u(r, \theta) = r^n(a_n \cos(n\theta) + b_n \sin(n\theta)).$$

Similarly, any constant a_0 satisfies the Laplace equation. By superposition the *General Solution* is

$$u(r,\theta) = \sum_{n=0}^{\infty} r^n(a_n \cos(n\theta) + b_n \sin(n\theta)).$$

At the boundary $r = 1$,

$$u(1,\theta) = \sum_{n=0}^{\infty} (a_n \cos(n\theta) + b_n \sin(n\theta)).$$

Hence, the boundary condition becomes

$$f(\theta) = \sum_{n=0}^{\infty} (a_n \cos(n\theta) + b_n \sin(n\theta)),$$

which is met by choosing the coefficients a_n, b_n as the Fourier coefficients of $f(\theta)$, resulting in the solution.

REMARK 6.13 Related Equations

The steady-state with a heat source is Poisson's PDE

$$u_{xx} + u_{yy} = f(x, y).$$

REMARK 6.14 Relation to Complex Functions

For analytic $f = u + iv$, u, v obey the Cauchy-Riemann equations

$$u_x = v_y, \quad u_y = -v_x.$$

Differentiating the first equation with respect to x, the second with respect to y, and adding gives the Laplace PDE for u. Similarly, we obtain the Laplace equation for v.

There is an intimate relation between analyticity and solving the Laplace equation.

DEFINITION 6.11 Harmonic Functions

Solutions of Laplace's equation.

6.3 THE HEAT EQUATION

DEFINITION 6.12 Heat Equation in Rectangular Coordinates

$$u_y = \alpha u_{xx}, \quad \alpha > 0. \tag{6.3}$$

REMARK 6.15 Type of Equation

Parabolic: $a = \alpha$, $b = 0$, $c = 0$, $b^2 - 4ac = 0$ in equation (6.1).

REMARK 6.16 Representation of Variables

The heat equation is parabolic; one of the variables is "time;" the equation models a "marching" process with time y. We usually write "t" in place of y to stress that it is not a spatial variable, giving us

$$u_t = \alpha u_{xx}.$$

REMARK 6.17 Physical Meaning

The heat equation gives the temperature rate of change in terms of the present value of T_{xx}. It is a tool for predicting future temperature from its history.

EXAMPLE 6.3 Heat Transfer in a Rod

Insulate a cylindrical metal rod along its length, leaving its faces (*A, B* in Figure 6.11) exposed. At any of its points the temperature is constant on any cross section and depends only on the distance from the faces. Placing the *x*-axis along the rod length, the rod temperature is $u(x, t)$. If there are no internal heat sources and the rod physical properties are constant, then u obeys (6.3).

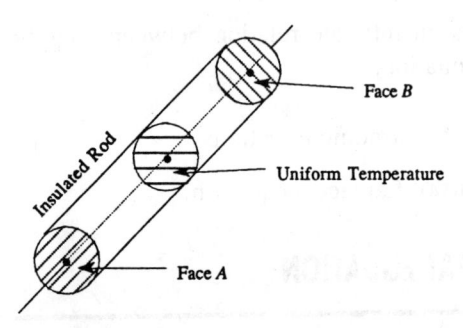

FIGURE 6.11 — Insulated Rod

REMARK 6.18 Equation in Higher Dimensions

$$u_t = \alpha[u_{xx} + u_{yy} + u_{zz}].$$

REMARK 6.19 Other Coordinate Systems

Circular Cylindrical

$$u_t = \alpha[(1/r)(ru_r)_r + (1/r^2)u_{\theta\theta} + u_{zz}]$$

Spherical Polar

$$u_t = \alpha[(1/r^2)(r^2 u_r)_r + (1/[r^2\sin(\phi)])(\sin(\phi)u_\phi)_\phi + (1/[r^2\sin^2(\phi)])u_{\theta\theta}]$$

Polar

$$u_t = \alpha[(1/r)(ru_r)_r + (1/r^2)u_{\theta\theta}]$$

REMARK 6.20 Properties of Equation (6.3)

Order 2, Homogeneous, Linear

REMARK 6.21 Properties of Solutions

Maximum Principle: The maximum and minimum values of any solution to the heat equation in the rectangle

$$R: 0 \leq t \leq t^*, 0 \leq x \leq a$$

of the x, t plane (see Figure 6.12) are achieved either at the time $t = 0$ or at the boundaries $x = 0, a$.

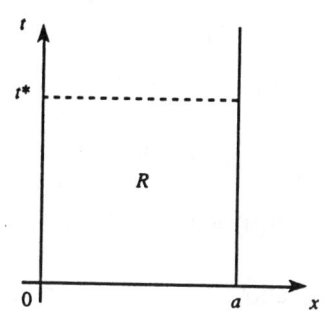

FIGURE 6.12 — Maximum Principle

REMARK 6.22 Form of Well Posed Problem

Solve equation (6.3) on $0 < x, t < \infty$ for the initial condition

$$u(x, 0) = f(x)$$

and the boundary condition

$$u(0, t) = g(t)$$

(Figure 6.13).

$u(0, t) = g(t)$

$u_t = \alpha u_{xx}$

$u(x, 0) = f(x)$

FIGURE 6.13 — Problem for Heat Equation

EXAMPLE 6.4 A Well Posed Problem and Its Solution

PDE $u_t(x, t) = \alpha u_{xx}(x, t)$

IC $u(x,0) = f_0$

BC $u(0, t) = g_0$

$u(x, t)$ is a similarity solution

$$u(x, t) = g_0 + (f_0 - g_0)\,\mathrm{erf}(x/[2\sqrt{\alpha t}\,]).$$

Here is the error function erf is defined by

$$\mathrm{erf}(s) = (2/\sqrt{\pi})\int_0^s \exp(-w^2)\,dw$$

satisfying the conditions: $\mathrm{erf}(0) = 0$, $\mathrm{erf}(\infty) = 1$.

REMARK 6.23 Related Equations

Laplace Equation: $u_{xx} + u_{yy} + u_{zz} = 0$

Steady-state form of the heat equation

Heat Source: $u_t = \alpha u_{xx} + W$

Nonlinear Case: For ρ, c the density and specific heat (taken as constants) and $k(u)$ the (temperature dependent) thermal conductivity, the heat transfer model is the nonlinear heat equation

$$\rho c\, u_t = (k(u)u_x)x$$

6.4 THE FIRST ORDER WAVE EQUATION

DEFINITION 6.13 In Rectangular Coordinates

$$u_t + c u_x = 0, \quad c = \text{given} \tag{6.4}$$

REMARK 6.24 Physical Meaning

Sound is a pressure wave moving through air causing the air density to vary with position x and time t; the wave moves at the sound speed c. If the direction of motion is that of increasing x then $c > 0$; if not, then $c < 0$. In either case, u will approximately obey the PDE (6.4).

REMARK 6.25 Equation in Higher Dimensions

In three space dimensions,

$$u_t + c_1 u_x + c_2 u_y + c_3 u_z = 0.$$

REMARK 6.26 Other Coordinate Systems

Circular Cylindrical

$$u_t + c_1 u_r + (c_2/r)u_\theta + c_3 u_z = 0$$

Spherical Polar

$$u_t + c_1 u_r + (c_2/[r\sin(\phi)])u_\theta + (c_3/r)u_\phi = 0$$

Polar

$$u_t + c_1 u_r + (c_2/r)u_\theta = 0$$

REMARK 6.27 Properties of the PDE

Order 1, Homogeneous, Linear

REMARK 6.28 Properties of Solutions

$$u(x, t) = f(x - ct),$$

constant on all lines (Figure 6.14) $L: x = ct + x_0$.

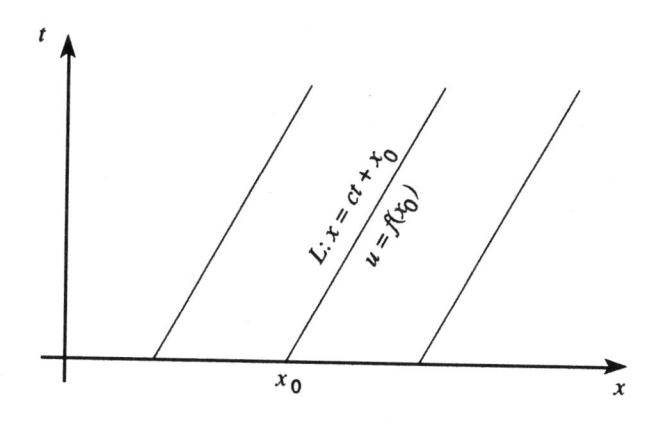

FIGURE 6.14 — *f(x − ct)* Constant on *L*

DEFINITION 6.14 Characteristics of Equation (6.4)

The lines L of Remark 6.28.

REMARK 6.29 Form of Well Posed Problem

An initial condition and a single boundary condition at the upstream boundary.

EXAMPLE 6.5 A Well Posed Problem and Its Solution

PDE $u_t + u_x = 0$, ($c = 1$ in equation (6.4))

IC $u(x,0) = 0$, $x > 0$

BC $u(0, t) = 1$, $t > 0$

The solution is constant along all lines

$x = t + \text{const.}$

$u = 1$ for $0 < x < t$ and $u = 0$ beyond (Figure 6.15).

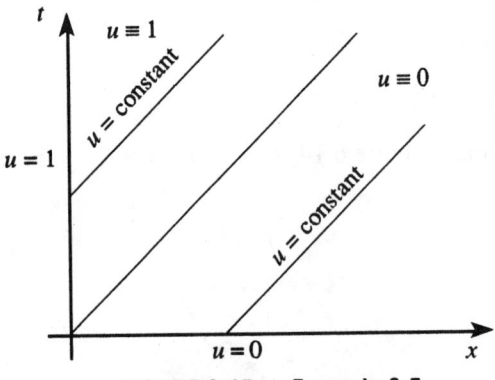

FIGURE 6.15 — Example 6.5

EXAMPLE 6.6 A Non-Well Posed Problem

For $u_t + u_x = 0$ signals move to the right; we cannot give initial conditions to the left of a boundary where we impose a condition (see Figure 6.16) since this would result in the impossibility of a line of constancy of the solution along which the solution has one value, meeting the boundary where a different value is imposed.

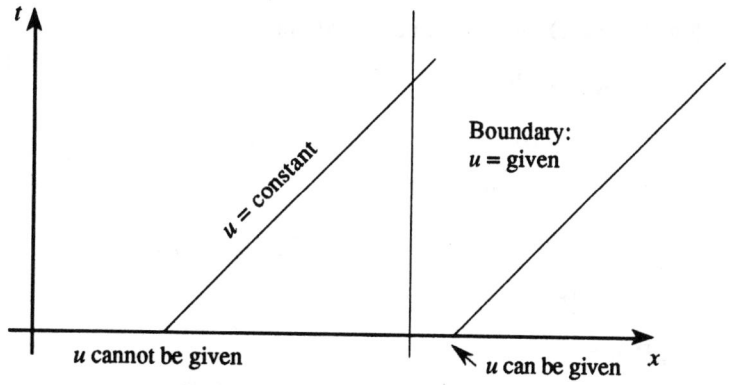

FIGURE 6.16 — Non-Well Posed Problem

6.5 THE WAVE EQUATION

DEFINITION 6.15 Wave Equation in Rectangular Coordinates

$u_{yy} = c^2 u_{xx}$, c = given

REMARK 6.30 Type of Equation

Hyperbolic: $a = c^2$, $b = 0$, $c = -1$, $b^2 - 4ac > 0$ in equation (6.1).

REMARK 6.31 Representation of Variables

The PDE is hyperbolic: one of the variables is time, and this models a "marching" process. The timelike variable is y. Accordingly, we write t in its place and have

$$u_{tt} = c^2 u_{xx}. \tag{6.5}$$

REMARK 6.32 Physical Meaning

The PDE describes wave propagation and vibrations. For example, if one subjects a string to vibrations $u(x, t)$ represents the time-dependent amplitude at the point x, time t. c is the wave speed.

REMARK 6.33 Equation in Higher Dimensions

$$u_{tt} = c^2[u_{xx} + u_{yy} + u_{zz}]$$

REMARK 6.34 Other Coordinate Systems

Circular Cylindrical

$$u_{tt} = c^2[(1/r)(ru_r)_r + (1/r^2)u_{\theta\theta} + u_{zz}]$$

Spherical Polar

$$u_{tt} = c^2[(1/r^2)(r^2 u_r)_r + (1/[r^2\sin(\phi)])(\sin(\phi)u_\phi)_\phi$$
$$+ (1/[r^2\sin^2(\phi)])u_{\theta\theta}]$$

Polar

$$u_{tt} = c^2[(1/r)(ru_r)_r + (1/r^2)u_{\theta\theta}]$$

REMARK 6.35 Properties of Wave Equation

Order 2, Homogeneous, Linear

REMARK 6.36 Properties of Solutions

General solution can be expressed as

$$u(x, t) = A(x - ct) + B(x + ct),$$

that is, the superposition of two waves moving at speed c, in opposite directions. The solution at any x_0 and time $t_0 > 0$ depends on the "past" within the *Domain of Dependence* bounded by the lines of slopes $\pm 1/c$ passing through the point (see Figure 6.17). Similarly, the solution

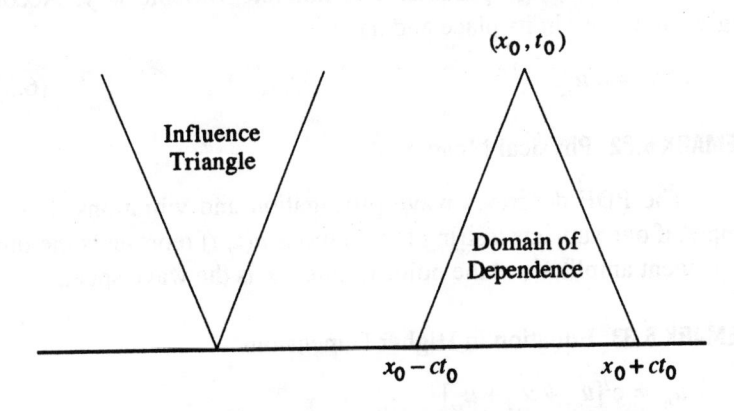

FIGURE 6.17 — Domains of Dependence and Influence

value at any initial time t_1 will effect the future only within the *Influence Triangle* bounded by the two lines of slope $\pm 1/c$ passing through this point.

DEFINITION 6.16 Characteristics

The families of lines of slope $\pm 1/c$ covering the plane.

REMARK 6.37 Form of Well Posed Problem

Consists of the PDE plus *two* initial conditions specifying u, u_t initially, and a single boundary condition at each boundary line. An example is the initial-value problem:

Find $u(x, t)$ satisfying equation (6.5) on $[a, b]$ and obeying the initial conditions

$$u(x, 0) = f(x), \quad u_t(x, 0) = g(x).$$

and the boundary conditions

$$u(a, t) = \alpha(t), \quad u(b, t) = \beta(t), t > 0,$$

for f, g, α, β given functions.

EXAMPLE 6.7 A Well Posed Problem and Its Solution

Solve the wave equation on the infinite line $-\infty < x < \infty$ for the initial conditions

$$u(x, 0) = f(x), \quad u_t(x, 0) = g(x).$$

The solution is given by D'Alembert's formula

$$u(x, t) = \tfrac{1}{2}[f(x - ct) + f(x + ct)] + (1/2c) \int_{x-ct}^{x+ct} g(s)ds.$$

REMARK 6.38 Related Equations

For many cases it is known that a vibrating system has a known angular frequency ω. For two space dimensions the solution will be of the form

$$u(x, y, t) = e^{-i\omega t} U(x, y).$$

Substitution into the wave equation

$$u_{tt} = c^2(u_{xx} + u_{yy})$$

yields the differential equation for $U(x, y)$,

$$U_{xx} + U_{yy} + (\omega/c)^2 U = 0.$$

This is the *Helmholt* or *Reduced Wave Equation*.

CHAPTER 7

CALCULUS OF VARIATIONS

Calculus of Variations is concerned with finding extreme values (maxima, minima). We examine this subject initially for functions and then for functions of functions or "functionals."

FAMILIAR IDEAS

Connected Set: A Set Each Pair of Whose Points Can Be Connected by Finitely Many Broken Lines (See Figure 7.1)

Boundary Point: Any Circle About the Point Contains Points Belonging to the Set and Points Not Belonging to It

Closed Domain: A Domain Containing All Its Boundary Points

Relative Minimum of $f(x)$: x^* such that $f(x) \geq f(x^*)$ for All $x \approx x^*$

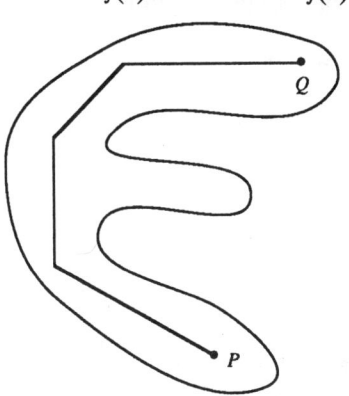

FIGURE 7.1 — Connected Set

Relative Maximum: x^* such that $f(x) \le f(x^*)$ for All $x \approx x^*$

Relative Extremum: Relative Maximum or Minimum

Stationary Point of $f(x)$: x^* for which $f'(x^*) = 0$

Absolute Maximum (Minimum): x^* such that $f(x) < (>) f(x^*)$ for All $x = x^*$

Absolute Extremum: Absolute Maximum or Minimum

Relative Extremum is a Stationary Point

If x^* is a Stationary Point and $f'' < (>) 0$ then x^* is a Relative Maximum (Minimum)

7.1 BASIC THEORY OF MAXIMA AND MINIMA

PROBLEM 7.1 Fundamental Problem of Extrema of Functions

Find values $X = (x_1, ..., x_N)$ where a function $f(X)$ continuous in a closed domain of N dimensional space has relative and absolute extrema.

THEOREM 7.1 Weierstrass Theorem

A continuous function in a closed domain has a largest and smallest value either in the interior or on the boundary.

DEFINITION 7.1 Distance Between Points

For $X = (x_1, ..., x_n), X^* = (x_1^*, ..., x_n^*)$,

$$|X - X^*| = (\sum_{i=1}^{n} (x_i - x_i^*)^2)^{1/2}$$

DEFINITION 7.2 Spherical Neighborhood of X^*

The points X for which $|X - X^*| < \varepsilon$ for some ε.

DEFINITION 7.3 Relative Minimum X^*

$f(X^*) \le f(X)$ for all X in some spherical neighborhood.

DEFINITION 7.4 Relative Maximum X^*

$f(X^*) \geq f(X)$ for all X in some spherical neighborhood.

DEFINITION 7.5 Relative Extremum

Relative maximum or minimum.

DEFINITION 7.6 Absolute Minimum/Maximum X^* in R

$f(X^*)$ is least/greatest over all R.

EXAMPLE 7.1 Absolute Maximum and Minimum (Figure 7.2)

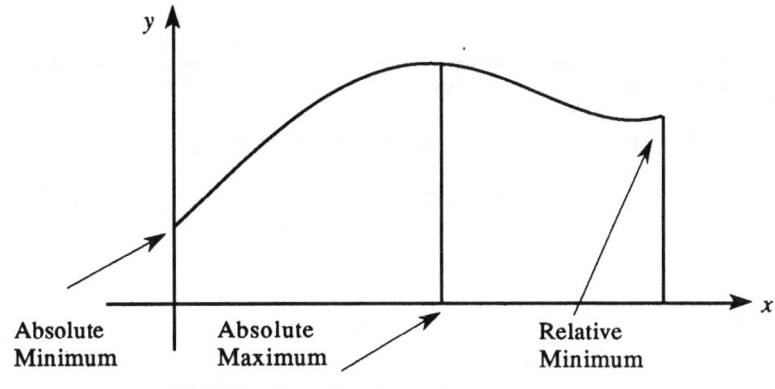

FIGURE 7.2 — Absolute Minima/Maxima

DEFINITION 7.7 Gradient of $f(X)$

$\nabla f(X) = (\partial f/\partial x_1, \ldots, \partial f/\partial x_n)$.

DEFINITION 7.8 Stationary Point X^* of $f(X)$

$\nabla f(X^*) = 0$.

REMARK 7.1 Stationary Point Need Not Be Extremum

$x = y = 0$ is stationary but not extremum for $f(x, y) = xy$.

REMARK 7.2 Constraints

The problem of finding minima or maxima is often constrained by the condition that the extremizing point X^* satisfy one or more auxiliary conditions.

DEFINITION 7.9 Constrained Extremizing Problem

Among all points X for which

$$g_j(X) = 0, \; j = 1, 2, \ldots, m$$

find a point X^* for which $f(X)$ has an extremum.

EXAMPLE 7.2 A Constrained Extremizing Problem

Among all bodies with given surface area the sphere has the largest volume.

REMARK 7.3 Lagrange Multipliers for Constrained Extrema

An extremum X^* of $f(x)$ is a stationary point for

$$F(X, \Gamma) = f(X) + \Gamma_1 g_1(X) + \ldots + \Gamma_n g_n(X):$$

$\partial F/\partial x_1 = 0$

$\partial F/\partial x_2 = 0$

...

$\partial F/\partial x_n = 0$

$\partial F/\partial \Gamma_1 = 0$

...

$\partial F/\partial \Gamma_n = 0$

There are $2n$ equations for the $2n$ values we seek.

EXAMPLE 7.3 The Right Cylinder of Given Area and Maximum Volume

Find the right cylinder of largest volume with given total surface area A_0.

Let r, h be the (unknown) radius and height of the cylinder. Its volume is $V = \pi r^2 h$ while the surface area constraint is $g(r, h) = 2\pi r^2 + 2\pi rh - A_0 = 0$. One way of solving this problem is to solve for h in terms of r from the constraint, substitute the result and then minimize V. We thus obtain

$$h = [A_0 - 2\pi r^2] / [2\pi r]$$
$$V = \pi r^2 [A_0 - 2\pi r^2] / [2\pi r]$$
$$= \tfrac{1}{2} r [A_0 - 2\pi r^2]$$

Differentiation of V with respect to r and setting the result equal to zero yields

$$r = [A_0 / 6\pi]^{1/2}, \; y = [2A_0 / 3\pi]^{1/2}.$$

The same result is obtained using a single Lagrange multiplier by seeking the stationary point of

$$F(r, h, \Gamma) = \pi r^2 h + \Gamma(2\pi r^2 + 2\pi rh - A_0).$$

Differentiating with respect to r, h, Γ and setting the results equal to zero yields the equations

$$F_r = 2\pi rh + \Gamma(4\pi r + 2\pi h) = 0,$$
$$F_\Gamma = 2\pi r^2 + 2\pi r\Gamma - A_0,$$
$$F_h = \pi r^2 + 2\pi r\Gamma = 0.$$

The third equation yields $\Gamma = -r/2$. The first then gives $r = h/2$ which, via the second, yields the earlier result.

REMARK 7.4 When Are Lagrange Multipliers Useful?

When elimination of variables becomes cumbersome or impossibly complex.

THEOREM 7.2 Direction of Gradient

That of the maximum rate of increase of the function. The negative of the gradient is in the direction of maximum decrease.

> **REMARK 7.5** Gradient Method ≡ Method of Steepest Descent
>
> Starting at any point we move through small steps, each in the direction opposite that of the current gradient, until reaching a relative minimum. Starting at X_0 we set
>
> $$X_1 = X_0 - t_1 \nabla f(X_0)$$
> $$X_2 = X_1 - t_2 \nabla f(X_1)$$
> ...
>
> for $t_1, t_2, \ldots$ sufficiently small.

REMARK 7.6 Shortcomings of the Gradient Method

The gradient method has two major limitations. The first is that the gradient may not exist, or be difficult to evaluate. A simple way to deal with this case is to replace the derivatives by finite differences to indicate a direction to be followed. The second limitation is that the method will lead to the nearest relative extremum instead of to the absolute extremum desired. Methods exist for "leaping" from a relative extremum in order to proceed to others. "Simulated Annealing" is one method, based on the use of tools of probability and giving the absolute extremum.

REMARK 7.7 Maximum Search

The gradient method finds relative maxima by moving in the direction of the gradient.

7.2 THE SIMPLEST PROBLEM OF VARIATIONAL CALCULUS

DEFINITION 7.10 Admissible Functions

For a given $f(x, y, z)$, $y = y(x)$ is *Admissible on* $[a, b]$ if

i) $y(x), y'(x), y''(x)$ are continuous on $[a, b]$;

ii) $f(x, y(x), y'(x))$ is defined and continuous on $a \leq x \leq b$;

iii) $y(a) = y_0$, $y(b) = y_1$, $y_0, y_1 =$ given.

DEFINITION 7.11 The Fundamental Problem

Let Ω be the collection of all admissible functions $y(x)$. Find $y = Y(x)$ in Ω giving the least value of

$$F[f] = \int_a^b f(x, y(x), y'(x))dx.$$

REMARK 7.8 Functionals

A *Functional* is a rule assigning a unique real number to each function $y(x)$ in Ω. $F[f]$ is a function over Ω.

REMARK 7.9 Notation

The third variable of f is the derivative y'. We write the partial derivative of f with respect to it as $f_{y'}$.

DEFINITION 7.12 First Variation $\delta F[f]$

$$\delta F[y] = \int_a^b \{\phi(x)f_y + \phi'(x)f_{y'}\}dx$$

for $\phi(x)$ vanishing at $x = a, b$.

DEFINITION 7.13 Stationary Function

A function $y(x)$ with vanishing first variation, $\delta F[y]$.

THEOREM 7.3 Euler Equation

If $Y(x)$ minimizes $F[y]$, then $Y(x)$ obeys the Euler ODE

$$F_y(x, Y(x), Y'(x)) = (d/dx)F_{y'}(x, Y(x), Y'(x)).$$

REMARK 7.10 On Euler's Equation

Euler's equation is a 2nd order ODE, for which the boundary conditions for admissibility at $x = a, b$ are suitable.

EXAMPLE 7.4 A Functional and Euler Equation

Let Ω be the set of all $y(x)$ with y', y'' continuous, and $y(0) = 0$, $y(1) = 1$. For the functional

$$F[y] = \int_0^1 [y(x)^2 + y'(x)^2]dx$$

$$f(x, y, z) = y^2 + z^2,$$

$$f_y = 2y, \quad f_{y'} = 2y'$$

and the Euler equation becomes

$$2y(x) = (d/dx)(2y'(x)) = 2y''(x)$$

or

$$y''(x) = y(x).$$

This equation has the general solution

$$y(x) = Ae^x + Be^{-x}.$$

Substitution into the boundary conditions yields

$$A + B = 0, \quad Ae + B/e = 1$$

or $A = 1/[e - 1/e]$, $B = -1/[e - 1/e]$. Hence, the solution to the minimum problem over Ω is

$$y(x) = (e^x - e^{-x})/[e - 1/e]$$
$$= \sinh(x)/\sinh(1).$$

THEOREM 7.4 Fundamental Theorem of the Calculus of Variations

If $w(x)$ is any continuous function on $[a, b]$ such that

$$\int_a^b w(x)\phi(x)dx = 0$$

for all continuous $\phi(x)$ vanishing at $x = a, b$, then $w(x) \equiv 0$.

DEFINITION 7.14 A Functional of Many Functions

$$F[y_1, y_2, ..., y_N] = \int_a^b f(x, y_1(x), ..., y_N(x), y_1'(x), ..., y_N'(x))dx.$$

DEFINITION 7.15 Admissible Vector Function

A vector function $\mathbf{Y}(x) = (y_1, y_2, ..., y_N)$ defined on $[a, b]$, as-

H-117

suming given values at $x = a, b$, for which the integrand $F[y_1, y_2, ..., y_N]$ is defined.

DEFINITION 7.16 Minimum Problem for F

Among all admissible functions in Ω find one minimizing F.

THEOREM 7.5 Euler's Equation for F

If $Y^*(x)$ minimizes F, then it satisfies the system of N Euler equations

$$\partial F/\partial y_i = (d/dx)\partial F/\partial y_i', \ i = 1, ..., N.$$

DEFINITION 7.17 Functionals in Two Dimensions

Let R be a closed domain in the x, y plane and $f(x, y, u, v, w)$ be given. For any $u(x, y)$ define the functional $F[u]$ by

$$F[u] = \iint_R f(x, y, u(x, y), u_x(x, y), u_y(x, y)) \, dxdy;$$

$u(x, y)$ is *admissible* if $F[u]$ is defined and u takes on given values at the boundary Γ of R. Ω is the set of all admissible functions.

DEFINITION 7.18 Minimum Problem for $F[u]$

Find $u^*(x, y)$ in Ω minimizing $F[u]$.

THEOREM 7.6 Euler's Equation for $F[u]$

$$F_u(x, y, u(x, y), u_x(x, y), u_y(x, y))$$
$$= (\partial/\partial x)\, F_{u_x}(x, y, u(x, y), u_x(x, y), u_y(x, y))$$
$$+ (\partial/\partial x)\, F_{u_y}(x, y, u(x, y), u_x(x, y), u_y(x, y))$$

EXAMPLE 7.5 Poisson's Equation as Euler's Equation

For given $g(x, y)$,

$$F(u) = \iint_R \{u_x^2 + u_y^2 + 2ug\} \, dxdy$$

Euler Equation: $g = u_{xx} + u_{yy}$.

7.3 SOME CLASSICAL PROBLEMS

Some classical Calculus of Variations Problems are

PROBLEM 7.2 Steiner's Problem (see Figure 7.3)

Find a road system of least total length joining N cities.

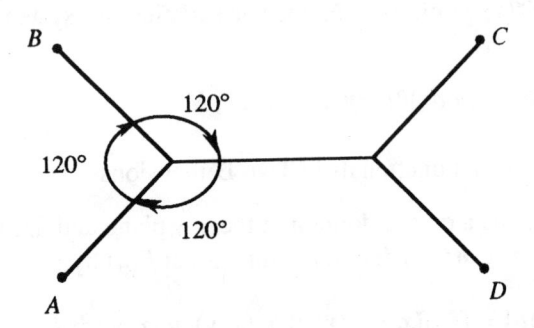

FIGURE 7.3 — Cities at A, B, C, D

PROBLEM 7.3 Brachistrochrone Problem

Find the path of least travel time for a particle, sliding without friction but under the influence of gravity, from a point A to a lower point B.

PROBLEM 7.4 Catenary Problem

What is the curve assumed by a string of given length with fixed end points, subject to the influence of gravity? The minimum problem is that of making the center of gravity of the string as low as possible.

PROBLEM 7.5 Plateau Problem

Find a surface of least area spanning a given closed wire frame in three dimensions.

PROBLEM 7.6 Hamilton's Principle

In progressing from a time t_0 to a time t_1, a mechanical system whose properties are determined by the kinetic energy T and the potential energy U will proceed in such a way as to make the integral

$$J = \int_{t_0}^{t_1} (T-U)dt$$

stationary over all possible motions.

7.4 CONTROL

DEFINITION 7.19 Control Process

A process governed by differential equations (ODE or PDE) including a forcing or *Control* term to be used to produce a desired performance for the system being modeled.

EXAMPLE 7.6 Temperature Control

Find a source $W(x, t)$ providing a desired temperature history for a rod subject to the conditions:

$T_t = \alpha T_{xx} + W(x, t)$	(PDE)
$T(a, t) = T(b, t) = 0$	(Boundary Conditions)
$T(x, 0) = f(x)$	(Initial Conditions)

EXAMPLE 7.7 A Least Time Control Problem

Find $u(t)$ such that $x''(t) = u(t)$, $x(0) = x'(0) = 0$,

$x(t^*) = 1$, $x'(t^*) = 0$, $-1 \leq u(t) \leq 1$ and

t^* is least.

This problem is one of finding the acceleration (positive or negative) minimizing the travel time from 0 to 1, and subject to the constraint of never exceeding 1 in absolute value. Moreover, the body is to begin and end its trip at rest.

This is a variational problem which cannot be handled via the Euler equation approach. The solution $u(t)$ is a *Bang-Bang* control, jumping between the constraining values ± 1. On some time interval $0 \le t \le t_{sw}$, $u(t) = 1$; it then switches to the value -1 on $t_{sw} \le t \le t^*$. The relation between t^* and t_{sw} is found by solving the problem

$$x''(t) = 1, x(0) = 0, x'(0) = 0$$

yielding the solution

$$x(t) = \tfrac{1}{2} t^2$$

with the velocity

$$v(t) = t.$$

This is the solution up to the (unknown) time t_{sw}. Beyond $t_{sw} x(t)$ is found via

$$x''(t) = -1, x(t^*) = 1, x'(t^*) = 0.$$

The solution to this problem is

$$x(t) = 1 - \tfrac{1}{2}(t - t^*)^2$$

with the velocity

$$v(t) = -(t - t^*).$$

At the time t_{sw} we require that $x(t)$ be continuous. Thus,

$$\tfrac{1}{2} t_{sw}^2 = 1 - \tfrac{1}{2}(t_{sw} - t^*)^2.$$

Solving for t^* in terms of t_{sw} yields

$$t^* = t_{sw} + [2 - t_{sw}^2]^{1/2}$$

From the calculus we find that t^* attains its least value at $t_{sw} = 1$. In turn $t^* = 2$ and $u(t)$ is found as the bang-bang control

$$u(t) = \begin{cases} 1, & 0 \le t \le 1 \\ -1, & 1 \le t \le 2. \end{cases}$$

7.5 DYNAMIC PROGRAMMING

DEFINITION 7.20 State of a System

A description of the system in terms of one or more quantities. Let the possible states of the system be denoted by S.

DEFINITION 7.21 Controller of a System

One or more functions representing our means for controlling the system. This can be represented as a function $u(t)$.

REMARK 7.11 On Dynamic Programming

Dynamic programming is a method for determining how to move a system from an initial state S_{init} at some starting time $t = a$ to a final state S_{end} at time $t = b$ at least total "cost."

DEFINITION 7.22 Cost of a Process

The cost of the system evolving from a state S_i to the state S_j over a time period $[a,b]$. It depends on the states, the times, and the controller and is written as $C[a, b, S_i, S_j, u]$.

DEFINITION 7.23 Least Cost of a Process

Given an initial state S_i and a final state S_j assume there is a controller U minimizing the cost

$$C[t_i, t_j, S_i, S_j, u]$$

over $[a, b]$. This least cost is denoted by $C^*[a, b, S_i, S_j]$.

THEOREM 7.7 Principle of Dynamic Programming

Let $a < c < b$ be any three times. Assume that we have found a least cost controller for moving from the initial state S_{init} to *any* possible state S at time c. Then the least cost process of going from S_{init} at time $t = a$ to S_{end} at time $t = b$ is found by minimizing the cost

$$\min. \{C^*[a, c, S_{init}, S] + C[c, b, S, S_{end}, u]\} \text{ all } u \text{ on } [c, b].$$

We can use this to incrementally find the least path in discrete time steps over the interval [a, b] (Figure 7.4).

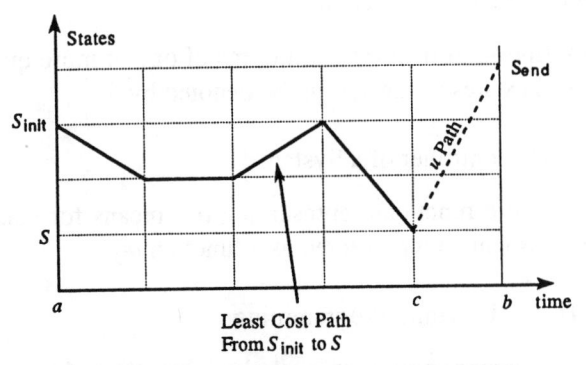

FIGURE 7.4 — Dynamic Programming Setting

7.6 LINEAR PROGRAMMING

The linear programming (LP) problem is best illustrated by the following simple problem:

EXAMPLE 7.8 An LP Problem

Maximize the function

$$y = -ax + b, \quad a, b > 0$$

subject to the conditions

$$y \leq -Ax + B, \quad A, B > 0 \tag{7.1}$$

$$x, y \geq 0.$$

There are four possibilities as seen when we compare the lines

$$L_1: y = -ax + b, L_2: y = -Ax + B$$

(see Figure 7.5). Let us refer to those points (x, y) lying below L_2 and on L_1 as *feasible*. In the first case, every point on L_1 lies below L_2 and the constraint (7.1) is not limiting. In case II, it is, and the solution is at the intersection of the two lines. For case III, the constraint is again not limiting. However, in case IV, there are no feasible points and no solution.

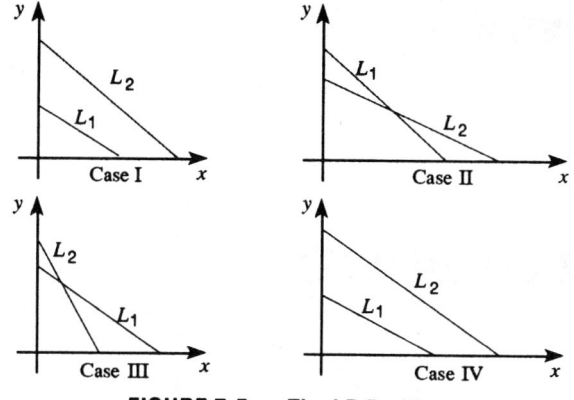

FIGURE 7.5 — The LP Problem

REMARK 7.12 Simplex Method

The solution will always be among the vertices of the region bounded by the lines and axes. This fact is the basis for the *Simplex Method* which is used for solving large-scale programming problems. The basic idea of the simplex method is to systematically move through vertices until reaching the solution vertex.

DEFINITION 7.24 General Linear Programming Problem

Maximize

$$f = c_1 x_1 + \ldots + c_n x_n$$

subject to the constraints

$$a_{11} x_1 + \ldots + a_{1n} x_n = b_1$$
$$a_{21} x_1 + \ldots + a_{2n} x_n = b_2$$
$$\ldots$$
$$a_{m1} x_1 + \ldots + a_{mn} x_n = b_m$$
$$x_j \geq 0, \quad j = 1, \ldots, n.$$

CHAPTER 8

NUMERICAL METHODS

8.1 SOLUTION OF EQUATIONS

DEFINITION 8.1 Roundoff Error

The error in replacing a number having infinitely many decimal places by one with a finite number of decimal places, as for example, $\pi \approx 3.14159$ and $\sqrt{2} \approx 1.414$.

DEFINITION 8.2 Approximation

A value x_{approx} that is to be used in place of the true value x_{true} of a quantity.

DEFINITION 8.3 Absolute Error of Approximation

$$\varepsilon_{abs} = |x_{true} - x_{approx}|.$$

DEFINITION 8.4 Relative Error of Approximation

$$\varepsilon_{rel} = e_{abs} / |x_{true}|.$$

DEFINITION 8.5 Fixed Point of a Function $\phi(x)$

An x^* for which

$$x^* = \phi(x^*)$$

(see Figure 8.1).

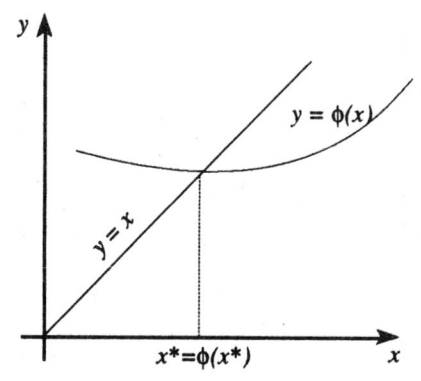

FIGURE 8.1 — Fixed Point of $\phi(x)$

METHOD 8.1   Iteration Method for Finding the Fixed Point

For some *Initial Iterate* x_0 we use the *Iteration Scheme*

$$x_1 = \phi(x_0), x_2 = \phi(x_1), \ldots, x_{n+1} = \phi(x_n) \ldots$$

THEOREM 8.1   Convergence of Iteration Scheme

If ϕ has a continuous derivative and $|\phi'(x)| \le \alpha < 1$, then the iteration scheme $x_{n+1} = \phi(x_n)$ converges.

METHOD 8.2   Bisection Method for Solving $f(x) = 0$

Let a, b be approximations to a root of $f(x) = 0$ such that f differs in sign at these points, $f(a) f(b) < 0$; then we choose the next approximation to be the average of a, b, $c = \frac{1}{2}(a + b)$, examine the sign of $f(c)$ and select that one of the values a, b at which f differs in sign from $f(c)$ to serve with c as the new pair of values to be averaged. (See Figure 8.2.)

METHOD 8.3   Newton-Raphson Method

If f, f' are continuous and f' is not zero at the root of $f(x) = 0$, then for an initial iterate x_0 close to the desired root, we compute the iterations

$$x_{n+1} = x_n - f(x_n)/f'(x_n), \quad n = 0, 1, 2, \ldots.$$

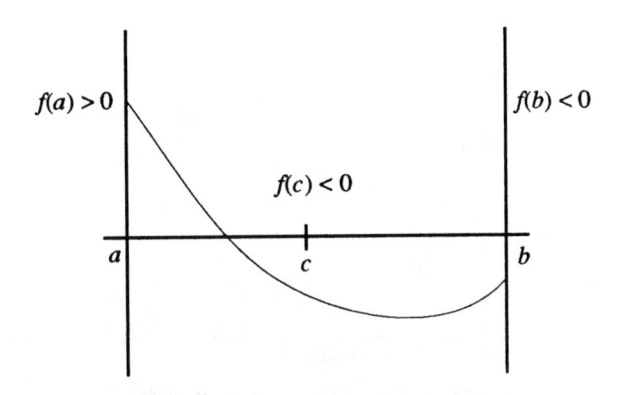

FIGURE 8.2 — Bisection Method

METHOD 8.4 Rule of False Position

If the derivative of f needed in the Newton-Raphson method is difficult to compute, then it can be replaced by a finite difference giving us

$$x_{n+1} = x_n - f(x_n) / \{[f(x_n) - f(x_{n-1})] / [x_n - x_{n-1}]\}$$

8.2 FUNCTION APPROXIMATION

DEFINITION 8.6 Interpolation Polynomial $\phi(x)$

Find a polynomial $\phi(x)$ of degree $\leq N$ assuming given values $f_0, f_1, \ldots, f_N$ of the function $f(x)$ at $N+1$ distinct points $x_0, x_1, \ldots, x_N$:

$$\phi(x_0) = f_0, \phi(x_1) = f_1, \ldots, \phi(x_n) = f_N.$$

x_i, $i = 0, N$ are the *Interpolation Points*.

THEOREM 8.2 Uniqueness of Interpolation Polynomial

The interpolation polynomial is unique.

THEOREM 8.3 Error of Interpolation Polynomial

For any x there is a point Γ such that

$$f(x) - \phi(x) = (x - x_0)(x - x_1) \ldots (x - x_N) f^{N+1}(\Gamma) / (N+1)!$$

DEFINITION 8.7 Lagrange Interpolation Polynomial

For any $k = 0, 1, 2, \ldots, N$, the *Exclusive Product* P_k is the product of all terms $(x - x_i)$ *except for* $(x - x_k)$

$$P_0(x) = (x - x_1)(x - x_2) \ldots (x - x_N)$$
$$P_1(x) = (x - x_0)(x - x_2) \ldots (x - x_N)$$
$$\ldots$$
$$P_N(x) = (x - x_0)(x - x_1) \ldots (x - x_{N-1}).$$

For

$$Q_k(x) = P_k(x) / P_k(x_k),$$
$$Q_k(x_j) = 0 \text{ for } j = k, = 1 \text{ for } j = k.$$

The *Lagrange Interpolation Polynomial* is

$$\phi(x) = \sum_{j=0}^{N} f_j Q_j(x).$$

REMARK 8.1 Other Approaches

As the order of the interpolation polynomial grows larger, the approximation may become poorer. One alternative is the *Method of Least Squares*, defined as:

DEFINITION 8.8 Least Squares Problem

Let $f(x)$ be continuous on $[a, b]$. Find the "closest" line

$$\phi(x) = Ax + B$$

to f in the sense that it minimizes the integral

$$\delta(f, \phi) = \{\int_a^b [\phi(x) - f(x)]^2 \, dx\}^{1/2}$$

METHOD 8.5 Solution of the Least Squares Problem

Minimizing $\delta(f, \phi)$ over all polynomials ϕ is equivalent to minimizing its square, considered as a function of A, B:

$$F(A, B) = \int_a^b [Ax + B - f(x)]^2 \, dx.$$

Setting the first derivatives equal to zero yields

$$\alpha_{11} A + \alpha_{12} B = \beta_1$$
$$\alpha_{21} A + \alpha_{22} B = \beta_2$$

$$\alpha_{11} = \int_a^b x\,dx = \tfrac{1}{2}\{b^2 - a^2\}$$

$$\alpha_{12} = \int_a^b dx = b - a$$

$$\alpha_{21} = \int_a^b x^2\,dx = (\tfrac{1}{3})\{b^3 - a^3\}$$

$$\alpha_{22} = \alpha_{11}$$

$$\beta_1 = \int_a^b f(x)\,dx, \quad \beta_2 = \int_a^b xf(x)\,dx.$$

The solution is then

$$A = [\beta_1 \alpha_{22} - \beta_2 \alpha_{12}]/D, \; B = [\beta_2 \alpha_{11} - \beta_1 \alpha_{21}]/D,$$

where $D = \alpha_{11}\alpha_{22} - \alpha_{12}\alpha_{21}$.

REMARK 8.2 Least Squares for Data Points

To approximate data points (x_j, f_j) by a line in the least squares sense we use the "distance"

$$\delta(f, \phi) = \{\sum_{j=1}^N [Ax_j + B - f_j]^2\}^{1/2}.$$

REMARK 8.3 Regression Line of Statistics

The closest least squares line is the *Regression Line*.

REMARK 8.4 Fourier Series and Least Squares

The Fourier polynomial is the nearest trigonometric polynomial in the least squares sense,

$$\phi(x) = a_0 + \sum_{n=1}^{N} \{a_n \cos(nx) + b_n \sin(nx)\}.$$

8.3 NUMERICAL INTEGRATION

We now turn to the evaluation of the definite integral

$$I = \int_a^b f(x)dx.$$

To do this we divide $[a, b]$ into N equal subintervals of length $h = (b - a)/N$ via the *Mesh Points* $x_0 = a, x_1 = a + h, ..., x_N = a + Nh = b$. I is the sum of the integrals over each of these subintervals:

$$I = \sum_{n=1}^{N} I_n$$

$$I_n = \int_{x_{n-1}}^{x_n} f(x)dx.$$

The simplest methods are the *Trapezoid* and *Simpson's Rule*. In the first, we replace $f(x)$ on each subinterval by the line joining the endpoints of its curve (see Figure 8.3). Simpson's Rule replaces f by an interpolating parabola at the endpoints and center of each subinterval (see Figure 8.4).

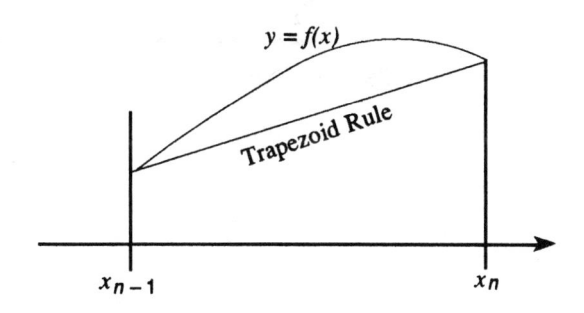

Figure 8.3 — Trapezoid Rule

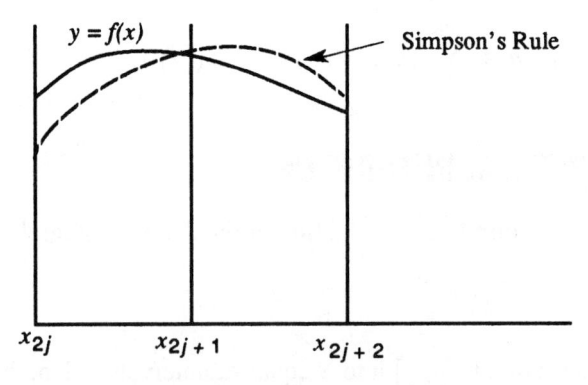

FIGURE 8.4 — Simpson's Rule

DEFINITION 8.9 Trapezoid Approximation for I

For $f_0 = f(x_0), \ldots, f_N = f(x_N)$ the Trapezoid Rule is

$$T_N = \tfrac{1}{2}h\,(f_0 + f_N) + h \sum_{n=1}^{N-1} f_n.$$

THEOREM 8.4 Error of Trapezoid Approximation

If f, f', f'' are continuous on $[a, b]$ with $|f''(x)| \leq M_2$,

$$|I - T_N| \leq (1/12)h^2 M_2 (b - a).$$

REMARK 8.5 Trapezoid Rule Has Second Order Accuracy

The error is proportional to h_2.

DEFINITION 8.10 Simpson's Rule Approximation for I

Suppose the number N of subdivisions of $[a, b]$ is even:

$N = 2M$.

Simpson's rule S_N is

$$S_N = (h/3)(f_0 + f_{2m}) + (4h/3)(f_1 + f_3 + \ldots + f_{2m-1})$$
$$+ (2h/3)(f_2 + f_4 + \ldots + f_{2m-2}).$$

THEOREM 8.5 Error of Simpson Approximation

For $f, \ldots, f^4$ continuous with $|f^{(4)}(x)| \leq M_4$,

$$|I - S_N| \leq (1/180)h^4 M_4 (b - a).$$

REMARK 8.6 Simpson's Rule Has 4-th Order Accuracy

REMARK 8.7 Monte-Carlo Integration

A different numerical integration approach is based on a "shooting" exercise using random numbers. For simplicity assume that the integral I is on the unit interval $[0, 1]$, with $0 \leq f(x) \leq 1$. Suppose that we randomly "shoot" at the square $0 \leq x, y \leq 1$. The chance (probability) of hitting a point below the graph of $y = f(x)$ is equal to the area under this graph, or the integral of $f(x)$ (see Figure 8.5). But in M random selections of points in the square, if N is the number of hits below the graph then $N/M \to I$ as $M, N \to \infty$. Thus, for M sufficiently large N/M is an approximation to I. The implementation of such a "Monte-Carlo method" on a computer (or even by hand) is straightforward.

8.4 NUMERICAL LINEAR ALGEBRA

DEFINITION 8.11 System of Linear Equations

$$a_{11}x_1 + \ldots + a_{1N}x_N = b_1$$
$$a_{21}x_1 + \ldots + a_{2N}x_N = b_2$$
$$\ldots$$
$$a_{N1}x_1 + \ldots + a_{NN}x_N = b_N \qquad (8.1)$$

The *Coefficients* $\{a_{ij}\}$ and *Right Hand Side* $\{b_j\}$ are given. The system is $N \times N$ and the $\{x_i\}$ form its *solution*.

REMARK 8.8 Cramer's Rule

Using Cramer's rule for large N is impractical because of the large number of multiplications and roundoff errors.

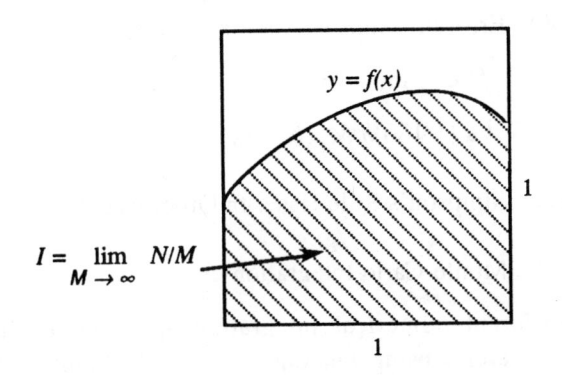

FIGURE 8.5 — Monte-Carlo Method

METHOD 8.6 Gauss Elimination for Solving a System

We successively eliminate unknown variables by an ordered subtraction of equations from each other. We illustrate this approach for the system of order 3.

i) $x_1 + 4x_2 + 2x_3 = -2$
ii) $5x_1 + x_2 - 2x_3 = 1$
iii) $-x_1 - 2x_2 + 6x_3 = 0$

Multiply i) by 5 and subtract it from ii); similarly add the original i) to iii); we obtain the system

i) $x_1 + 4x_2 + 2x_3 = -2$
ii) $-19x_2 - 12x_3 = 11$
iii) $2x_2 + 8x_3 = -2$

x_1 appears only in i) while ii), iii) is a system of two equations in two unknowns. Now multiply ii) by 2/19 and add the result to iii), yielding

iii) $(-24/19 + 8)x_3 = -2 + 22/19$.

This is now solved for x_3:

$$x_3 = -1/8.$$

From ii)

$$x_2 = -(1/19)(11 + 12x_3)$$
$$= -1/2.$$

while from i),

$$x_1 = -2 - 4x_2 - 2x_3 = 1/4.$$

DEFINITION 8.12 Pivotal Coefficient

The coefficient of the first term reached in each equation. In the above example, these pivots are 1 and -19.

REMARK 8.9 The Case of a Small Pivotal Coefficient

Suppose the coefficient of x_2 in ii) is a small number such as -0.00019. Elimination then requires multiplication of the equation by $2/0.00019$, introducing the possibility of a large roundoff error in the final equation for x_3. For this reason *Pivoting Methods* are introduced. In these we choose the *largest* leading coefficient as the next pivot. Thus, if the coefficient of x_2 is indeed -0.00019, we would consider equations ii), iii) as

ii) $\quad -12x_3 - 0.00019x_2 = 11$

iii) $\quad 8x_3 + \quad 2x_2 = -2$

so that the next pivot is -12 instead.

DEFINITION 8.13 Upper Triangular Matrix

$$A = \begin{pmatrix} a_{11} & a_{12} & \cdots & a_{1N} \\ a_{21} & a_{22} & \cdots & a_{2N} \\ \cdots & & & \\ a_{N1} & a_{N2} & \cdots & a_{NN} \end{pmatrix}$$

is upper triangular if terms below the diagonal are zero:

$$a_{ij} = 0, \text{ for } i > j.$$

DEFINITION 8.14 Lower Triangular Matrix

All the terms above the diagonal are zero:

$a_{ij} = 0$, for $i < j$.

THEOREM 8.6 Cholesky Decomposition

Under certain conditions A can be written as

$$A = LU$$

for L, U lower and upper triangular matrices. L and U are uniquely determined if we specify the diagonal elements of one of them. If A is symmetric ($a_{ij} = a_{ji}$) then L is the transpose of U and vice versa.

METHOD 8.7 Cholesky's Method

Solving system (8.1) is equivalent to solving the vector equation

$$A\mathbf{X} = \mathbf{B}.$$

Using the Cholesky decomposition this becomes

$$LU\mathbf{X} = \mathbf{B}$$

and premultiplying by the inverse L^{-1} of L,

$$U\mathbf{X} = L^{-1}\mathbf{B}.$$

But this is the form obtained by Gauss elimination, which can be solved for each variable in turn.

REMARK 8.10 Matrix Inversion

Finding the inverse $C = A^{-1}$ of a square, nonsingular matrix A, is equivalent to finding C such that

$$AC = I$$

for I the identity matrix. Let $\mathbf{Y}_1, \mathbf{Y}_2, \ldots, \mathbf{Y}_N$ be the vectors formed from the first, second, ..., N-th columns of C. Define the vectors $\mathbf{i}_1, \mathbf{i}_2, \ldots, \mathbf{i}_N$ as those formed from the corresponding columns of the identity matrix I. Then the matrix equation (8.2) can be regarded as N linear systems of equations with the same coefficient matrix. Each can be solved by Gauss Elimination to yield the columns of C as their solution.

DEFINITION 8.15 Sparse System of Equations

Systems for which the coefficient matrix is of large order but with few non-zero terms. Sparse systems arise frequently in engineering applications.

REMARK 8.11 Elimination Versus Iterative Methods

Elimination methods such as that of Gauss require handling and storage of the system coefficients, whether the system is sparse or not. For this reason, it is often preferable to use iterative methods for solving such systems.

REMARK 8.12 Iterative Methods for a System of Equations

In iterative methods we represent the system in the form of the fixed point equation $x = \phi(x)$ examined earlier.

METHOD 8.8 Jacobi Iteration

The system $A\mathbf{X} = \mathbf{B}$ can be written

$$\mathbf{X} = (I - A)\mathbf{X} + \mathbf{B}.$$

Starting with an initial iterate $\mathbf{X}_0$, we define the iteration sequence $\{\mathbf{X}_n\}$ by

$$\mathbf{X}_{n+1} = (I - A)\mathbf{X}_n + \mathbf{B}.$$

METHOD 8.9 Gauss-Seidel Iteration

In practice we find the coefficients of $\mathbf{X}_{n+1}$ in turn, starting from the first. The latest known approximation to each is not used until the next complete cycle. The Gauss-Seidel method uses the latest known values of each element at each step.

REMARK 8.13 On Iteration Methods

A key extension of the Gauss-Seidel method is the SOR (Successive Over-Relaxation) method, in which convergence is accelerated by using information about the error of the previous iteration.

Recently, with the growing popularity of *Parallel Computers*, it has become clear that the choice of a numerical method must take into account the architecture of the computer being used. Thus, Jacobi's method, which is poor for a serial computer, is readily suitable to a parallel computer.

REMARK 8.14 Ill-Conditioned Systems

Corresponding to the concept of non-well-posedness is that of ill conditioning of systems of equations. A system is ill conditioned if small errors in the coefficients or introduced in the solution process have a large effect on the solution. This corresponds to the case of a determinant whose value is close to zero.

DEFINITION 8.16 Eigenvalue of a Matrix

A number Γ for which $\det(A - \Gamma I) = 0$ or equivalently, there is a vector (*eigenvector*) $\mathbf{v}$ such that $A\mathbf{v} = \Gamma \mathbf{v}$.

THEOREM 8.7 Eigenvalues of a Symmetric Matrix

The eigenvalues of a symmetric matrix are real numbers.

THEOREM 8.8 Gershgorin's Theorem

Let Γ be any eigenvalue of A. Then for some j,

$$|a_{jj} - \Gamma| \leq |a_{j1}| + |a_{j2}| + \ldots + |a_{j,\,j-1}| +$$
$$|a_{j,\,j+1}| + \ldots + |a_{jN}|$$

THEOREM 8.9 Rayleigh Quotient for Symmetric Matrix A

The *largest eigenvalue* is given by the Rayleigh Quotient

$$\Gamma = \max_{\mathbf{X}} \{(A\mathbf{X}, \mathbf{X}) / (\mathbf{X}, \mathbf{X})\}.$$

Here $(\mathbf{X}, \mathbf{Y})$ is the inner product of $\mathbf{X}$ and $\mathbf{Y}$. Furthermore, suppose that $\mathbf{X}^*$ is the maximizing vector. The *second largest eig-*

envalue μ is given by

$$\mu = \max_{X \perp X^*} \{(AX,X)/(X,X)\}.$$

8.5 SOLVING ORDINARY DIFFERENTIAL EQUATIONS

REMARK 8.15 Initial Value Problem

Find $y(x)$ for $x > a$ satisfying the initial condition $y(a) = y_0$ and the ODE

$$y'(x) = f(x, y(x))$$

REMARK 8.16 Definition of a Mesh

For $h > 0$ a (small) *Mesh Length* define *Mesh Points*

$$x_0 = a, x_1 = a + h, \ldots, a_j = a + jh.$$

METHOD 8.10 Euler's Method

At any point x_j

$$y'(x_j) \approx [y(x_{j+1}) - y(x_j)] / h.$$

Hence, the differential equation yields

$$[y(x_{j+1}) - y(x_j)] / h \approx f(x_j, y(x_j))$$

or

$$y(x_{j+1}) \approx y(x_j) + hf(x_j, y(x_j)).$$

Let Y_j be the (desired) approximation to $y(x_j)$. In Euler's method Y_j is found by replacing "$\approx$" by =:

$$Y_0 = y_0, Y_1 = Y_0 + hf(x_0, Y_0) \ldots$$
$$Y_n = Y_{n-1} + hf(x_{n-1}, Y_{n-1}).$$

EXAMPLE 8.1 An Initial Value Problem

$$y' = y, y(0) = 1.$$

Euler's method takes the form

$$Y_n = (1 + h)Y_{n-1} = \ldots = (1 + h)^n.$$

Let x^* be any point, and let h, n be so related that

$$hn = x^* = x_n.$$

Then

$$Y_n = (1 + x^*/n)^n$$

and this expression converges to x^* as $n \to \infty$.

REMARK 8.17 Accuracy of Euler's Method

The error of Euler's method is proportional to h. This can be improved, most notably by the *Romberg Method*, combining approximations obtained for various h.

A collection of methods due to Runge-Kutta is the standard computer approach to initial value problems for ODE's.

METHOD 8.11 Runge-Kutta Method with Error of Order h^4

To go from Y_n to Y_{n+1} we define

$A = hf(x_n, Y_n)$

$B = hf(x_n + \frac{1}{2}h, Y_n + \frac{1}{2}A)$

$C = hf(x_n + \frac{1}{2}h, Y_n + \frac{1}{2}B)$

$D = hf(x_n + h, Y_n + C)$

$Y_{n+1} = Y_n + (1/6)[A + 2B + 2C + D]$.

8.6 SOLVING PARTIAL DIFFERENTIAL EQUATIONS

THEOREM 8.10 Approximations to the First Derivative

$$f'(x) \approx (1/h)\{f(x + h) - f(x)\}$$

and

$$f'(x) \approx (1/h) \{f(x) - f(x - h)\}.$$

These approximations are, respectively, the *Forward* and *Backward* difference approximations to the derivatives. Each is in error proportional to h.

THEOREM 8.11 An Approximation to the Second Derivative

$$f''(x) \approx (1/h^2) \{f(x + h) - 2f(x) + f(x - h)\}$$

This is the *Centered Difference Approximation* to the second derivative. Its error is proportional to h^2.

THEOREM 8.12 An Approximation to the Heat Equation

For h, k positive (small) values, the heat equation

$$T_t(x, t) = \alpha T_{xx}(x, t)$$

can be approximated by

$$(1/k) [T(x, t + k) - T(x, t)]$$
$$\approx (\alpha/h^2) [T(x + h, t) - 2T(x, t) + T(x - h, t)]. \quad (8.3)$$

Here T_t has been replaced by a forward difference in t while for T_{xx} we have written its centered difference approximation.

METHOD 8.12 A Numerical Scheme for the Heat Equation

For given *Mesh Widths* h, k, introduce the two-dimensional mesh (x_i, t_j) (see Figure 8.6). Let T_{ij} denote the approximation to $T(x_i, t_j)$. Then for each i, j the approximation relation (8.3) leads to the equation

$$(1/k)T_{i,j+1} - T_{ij} = (\alpha/h^2) [T_{i+1,j} - 2T_{ij} + T_{i-1,j}]$$

and hence to the numerical scheme

$$T_{i,j+1} = T_{ij} + (\alpha k/h^2) [T_{i+1,j} - 2T_{ij} + T_{i-1,j}] \quad (8.4)$$

This is a *Marching Scheme* enabling us to move to later j values, knowing the solution for earlier j. It is an *Explicit Scheme* since it represents a single later value of temperature in terms of earlier values. The scheme is suitable for solving initial-boundary value problems over a finite region (see Figure 8.7).

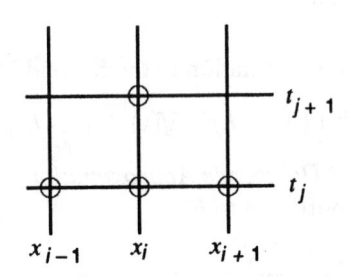

FIGURE 8.6 — Mesh for Heat Equation

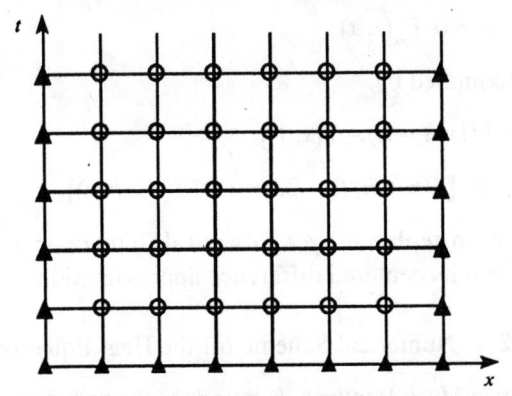

FIGURE 8.7 — Marching Simulation: △ = Data Points, O = Calculated

REMARK 8.18 Instability of the Explicit Scheme

If the ratio

$$= \alpha k / h^2$$

is greater than 1/2 the scheme (equation 8.4) is *Unstable*, meaning that it responds exponentially to small changes in data. To use the scheme one must select h, k such that

> Stability Condition $\alpha k / h^2 \leq 1/2$.

H-141

In this case the approximations converge to the solution of the PDE as h, k tend to zero.

REMARK 8.19 Shortcoming of the Explicit Scheme

The stability condition is a strong limitation on the size of the time step k of the explicit scheme. One wants the spatial mesh size h to be small in order that the centered difference scheme accurately approximate the second derivative T_{xx}. Hence, k would have to be yet smaller, since it would be bounded by h^2. For this reason, it is often preferable to seek a scheme which is not limited by a stability condition.

METHOD 8.13 Implicit Difference Scheme for the Heat Equation

This scheme is found by replacing the time derivative T_t by a *Backward Difference*, giving us

$$(1/k)T_{ij} - T_{i,j-1} = (\alpha/h^2)[T_{i+1,j} - 2T_{ij} + T_{i-1,j}] \quad (8.5)$$

Now assuming that we are moving forward in time, there are three unknowns in the equation: T_{ij}, $T_{i\pm1,j}$. When solving the heat equation with boundary conditions on an interval there is one equation for each mesh point. Hence, the resulting linear system of equations for the unknowns at the time step j can be solved for these unknowns. The methods used include Gauss Elimination and iteration. The system is sparse. In addition, the scheme is *Unconditionally Stable* for all $h, k > 0$.

REMARK 8.20 Other Schemes

Other schemes for the heat equation include the *Crank-Nicholson* scheme, balancing the implicit scheme of equation (8.5) and the explicit scheme of equation (8.4). Another scheme of interest is that of *Dufort-Frankel*, which is explicit, unconditionally stable, but may not correctly model the heat transfer process.

REMARK 8.21 Laplace's Equation

For the Laplace equation $u_{xx} + u_{yy} = 0$ in a domain of the x, y plane, let h be any mesh width and introduce points (x_i, y_j) (see Figure

8.8). Replacing the derivatives by centered differences at any point x, y yields

$$(1/h^2) [u(x + h, y) - 2u(x, y) + u(x - h, y)]$$
$$+ (1/h^2) [u(x, y + h) - 2u(x, y) + u(x, y - h)] \approx 0.$$

For u_{ij} the desired approximation to $u(x_i, y_j)$

$$(1/h^2) [U_{i+1,j} - 2U_{ij} + U_{i-1,j}]$$
$$+ (1/h^2) [U_{i,j+1} - 2U_{ij} + U_{i,j-1}] = 0$$

or

$$U_{ij} = (1/4) [U_{i+1,j} + U_{i-1,j} + U_{i,j+1} + U_{i,j-1}]$$

that is, each value is the average of its neighboring values. Again we have an equation for each point interior to the domain. Assigning boundary values uniquely determines the solution at every mesh point. The system can be solved by iteration.

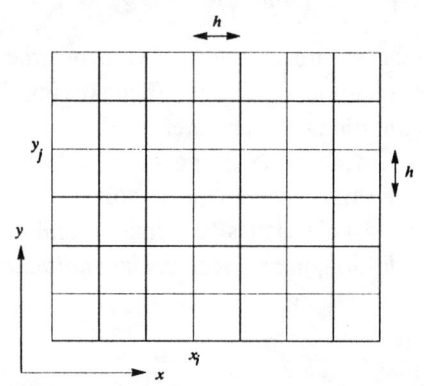

FIGURE 8.8 — Mesh for Laplace Equation

METHOD 8.14 Ritz Method and Finite Elements

An alternate approach to solving the Laplace equation in a domain R is based on the fact that it is the Euler equation for the functional

$$I[u] = \iint_R [u_x^2 + u_y^2] \, dxdy.$$

Suppose that any function on R can be expanded as a linear combination of *Basis Functions* $\{\phi_j\}$. Consider a finite linear combination of these functions

$$\sum_{n=1}^{N} a_j \phi_j(x, y)$$

and seek the coefficients minimizing the functional

$$F(a_1, a_2, ..., a_N) = I\,[\,\sum_{n=1}^{N} a_j \phi_j(x, y)].$$

Their determination rests merely on differentiation of F with respect to each one of them and solving the resulting equation system. This approach constitutes the *Rayleigh-Ritz-Galerkin* method and is the foundation of *Finite Element Methods*.

REMARK 8.22 The Wave Equation

For the wave equation $u_{tt} = c^2 u_{xx}$, we replace both second order derivatives by centered differences. As in the case of the heat equation, there is a stability condition

$ck/h \le 1.$

Unlike the heat equation this condition is not overly restrictive. Moreover, attempts to use implicit approaches must be treated with caution because of the phenomenon of *Overstability* or artificial damping of wave motion that is not actually damped.

CHAPTER 9

STATISTICS AND PROBABILITY

9.1 ON STATISTICS AND PROBABILITY

REMARK 9.1 On Statistics and Probability

The fall of dice is a "random event": the rules governing their flight are so complex that to model them is essentially fruitless, but over many throws the outcomes become statistically predictable. If a die is fair then on the average each of its faces appears once in six throws.

Many phenomena encountered in engineering practice are statistical in the same way as the throw of dice. Significant examples include defective items on a production line, distribution of materials in mixtures, and the anticipated failures in communications networks.

DEFINITION 9.1 Experiment

A process carried out following a well-defined set of rules that can be repeated arbitrarily often and whose results cannot be predicted.

DEFINITION 9.2 Event

A possible outcome of an experiment.

DEFINITION 9.3 Sample Space S

The set of all possible outcomes of the experiment.

DEFINITION 9.4 Complement E^- of an Event E

The event "E does not happen."

DEFINITION 9.5 Venn Diagram

The graphical representation of a sample space and its events.

DEFINITION 9.6 Union $A \cup B$ of Events A, B

The event "either A or B or both occur."

DEFINITION 9.7 Intersection $A \cap B$ of Events A, B

The event "both A and B occur."

DEFINITION 9.8 Impossible Event ϕ

An event that can never occur.

DEFINITION 9.9 Disjoint Events A, B

Events that cannot both be true: $A \cap B = \phi$.

DEFINITION 9.10 Probability Function S

An assignment $P(E)$ of a real number to every element in the sample space S, such that:

i) for all sets E of S, $0 \leq P(E) \leq 1$,

ii) $P(S) = 1$,

iii) If $A \cap B = \phi$, then $P(A \cup B) = P(A) + P(B)$.

THEOREM 9.1 For any events A, B (see Figure 9.1)

$$P(A \cup B) = P(A) + P(B) - P(A \cap B).$$

THEOREM 9.2 For any event A,

$$P(A^-) = 1 - P(A).$$

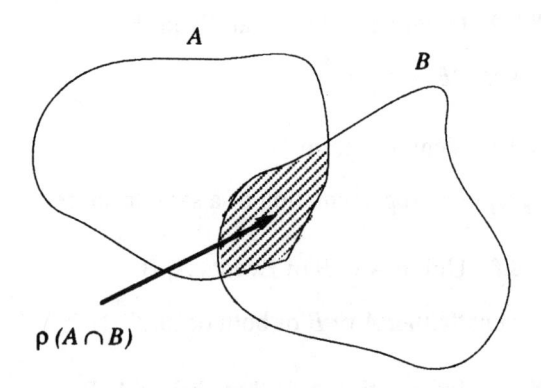

FIGURE 9.1 — Theorem 9.1

$p(A \cap B)$

DEFINITION 9.11 Conditional Probability $P(A \mid B)$

The probability that A occurs given that B occurs, given by

$P(A \mid B) = P(A \cap B) / P(B)$.

DEFINITION 9.12 Independent Events A, B

Events satisfying $P(A \mid B) = P(A)$ or equivalently

$P(A \cap B) = P(A) P(B)$.

THEOREM 9.3 Bayes Theorem

For events A, B,

$P(A \mid B) = P(B \mid A) [P(A) / P(B)]$.

DEFINITION 9.13 Factorial $N!$

$N! = 1 \times 2 \times 3 \times \ldots \times N$,

$0! = 1$.

THEOREM 9.4 Stirling's Formula

For large N,

$N! \approx (2\pi N)^{1/2} (N/e)^N$.

DEFINITION 9.14 Binomial Coefficient

For $N \geq K$,

$$\binom{N}{K} = \frac{N!}{K!(N-K)!}$$

DEFINITION 9.15 Permutation of N Objects

An arrangement of the objects in some order.

THEOREM 9.5 Number of Permutations

The number of permutations of N objects is $N!$.

DEFINITION 9.16 Combination of K Out of N Objects

A selection of K of the N objects without regard to order and with no repetition.

THEOREM 9.6 Number of Combinations

The number of combinations of K objects from a collection of N objects is

$$\binom{N}{K} = \frac{N!}{K!(N-K)!}$$

DEFINITION 9.17 Random Variable X

A real valued variable whose value is determined by the outcome of an experiment.

DEFINITION 9.18 Discrete Random Variable X

A random variable assuming only discrete values (e.g., 1, 2, 3, ...).

DEFINITION 9.19 Continuous Random Variable X

A random variable assuming any values over an interval of the real line.

DEFINITION 9.20 Sample Space for Random Variable X

The collection of events determining the values of X.

DEFINITION 9.21 Probability Distribution Function for X

The function
$$F(x) = P(X \leq x).$$

DEFINITION 9.22 Density Function $f(x)$ of a Discrete Random Variable

If X assumes the values $x, y, \ldots$ then
$$f(x) = P(X = x).$$

EXAMPLE 9.1 A Density Function for Dice Throws

Let X assume the values 1, 2, 3, 4, 5, 6 according to the outcome of throwing a fair die. Then
$$f(1) = f(2) = \ldots = f(6) = 1/6$$

DEFINITION 9.23 $f(x)$ for a Continuous Random Variable
$$f(x) = F'(x).$$

THEOREM 9.7 Properties of the Distribution Function
$$F(-\infty) = 0, \quad F(\infty) = 1, F(x) \text{ is increasing}$$

EXAMPLE 9.2 Normal Distribution Function
$$F(x) = (1/\sqrt{2\pi}) \int_{-\infty}^{x} \exp(-\tfrac{1}{2} s^2) ds$$

is the distribution function for a *Normally Distributed* random variable (see Figure 9.2). Its density function is
$$f(x) = F'(x) = (1/\sqrt{2\pi}) \exp(-x^2/2).$$

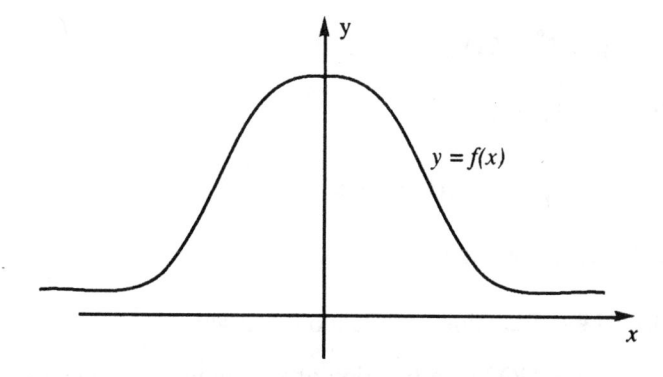

FIGURE 9.2 — Normal Distribution Density

REMARK 9.2 Mean and Standard Deviation

The mean value characterizes the anticipated or expected or average value of a random variable. On the other hand, the standard deviation gives us the spread or deviation of the random variable about its mean value. We define these now for both the discrete and continuous random variable X.

DEFINITION 9.24 Discrete Mean, Variance, and Standard Deviation

Let X assume the values $\{x_j\}$ and have density function $f(x_j)$. Its *Mean* or *Expected Value* $\mu = E[X]$ is

$$\mu = \sum_j x_j f(x_j);$$

its variance is

$$\sigma^2 = \sum_j (x_j - \mu)^2 f(x_j)$$

and its standard deviation is σ.

DEFINITION 9.25 Continuous Mean, Variance, and Standard Deviation

If X is a continuous random variable with density function $f(x)$, then the mean or expected value $\mu = E(X)$ is

$$\mu = \int_{-\infty}^{\infty} xf(x)\,dx$$

while its variance is

$$\sigma^2 = \int_{-\infty}^{\infty} (x-\mu)^2 f(x)\,dx.$$

The standard deviation is again σ.

THEOREM 9.8 Function of a Random Variable

Let $Y = G(X)$ be a function of the random variable X. Then Y is a random variable with the expected value

$$E(G(X)) = \sum_j G(x_j) f(x_j)$$

for the discrete case, while the continuous case

$$E(G(X)) = \int_{-\infty}^{\infty} G(x) f(x)\,dx.$$

DEFINITION 9.26 Binomial Distribution

In a coin toss there is a probability p of observing "heads" and q of observing "tails," $p + q = 1$. Let x be the random variable defined as the number of heads in N coin tosses. Then X is a discrete random variable. Its density function is

$$f(x) = \binom{N}{x} p^x q^{N-x}.$$

The mean value of X is $\mu = Np$, while its variance is $\sigma = (Npq)^{1/2}$.

DEFINITION 9.27 Poisson Distribution

The Poisson random variable is associated with transportation, queues, and service lines. It is a discrete random variable with the density function

$$f(x) = (\mu^x / x!) e^{-\mu}. \quad x = 0, 1, 2, \ldots$$

H-151

DEFINITION 9.28 Normal Distribution

The most commonly used distribution is the normal distribution. For mean μ and standard deviation σ, its density function is

$$f(x) = (1/[\sigma\sqrt{2\pi}]) \exp\{-\tfrac{1}{2}[(x-\mu)/\sigma]^2\}.$$

THEOREM 9.9 Law of Large Numbers

As $N \to \infty$ the binomial distributed random variable X tends to a normally distributed random variable.

DEFINITION 9.29 Parameter Estimators

If the values $x_1, x_2, \ldots, x_N$ of a random variable X are obtained from a series of experiments then estimates of the mean value and variance of X are

$$\mu_{est} = (1/N) \sum_{j=1}^{N} x_j$$

$$\sigma_{est}^2 = (1/[N-1]) \sum_{j=1}^{N} (x_j - \mu_{est})^2$$

REMARK 9.3 Applying Statistical Methods

One of the key engineering applications of statistics and probability is to the question of whether a sample taken truly represents the state of the general population of interest. Thus, if we wish to know if a production line is operating properly, we will routinely select a small sample of items and test them. If many are defective does this mean that we must halt production? What if the line is working properly and we just had "bad luck" in our random sample? On the other hand, we might have selected a sample of which very few are defective. Does this mean that the line is working well, or did we just have "bad luck" again, this time on the optimistic side? In any case, we assume that defective items follow a particular probability distribution, and we use the mean and standard deviation as tools for determining the probability that our line is/is not working well as a reflection of our sample.

Handbook of Mechanical Engineering

SECTION I
Physics

CHAPTER 1

VECTORS AND SCALARS

1.1 BASIC DEFINITIONS OF VECTORS AND SCALARS

a) A vector is a quantity that has both magnitude and direction.

   i) Some typical vector quantities are: displacement, velocity, force, acceleration, momentum, electric field strength and magnetic field strength.

b) A scalar is a quantity that has magnitude but no direction.

   i) Some typical scalar quantities are: mass, length, time, density, energy and temperature.

1.2 ADDITION OF VECTORS ($\bar{a} + \bar{b}$) –GEOMETRIC METHODS

a) Triangle Method (Head-to-Tail Method)

I-1

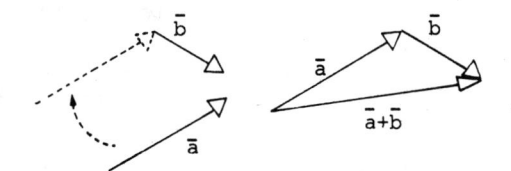

(i) Attach the head of $\bar{a}$ to the tail of $\bar{b}$.

(ii) By connecting the head of $\bar{a}$ to the tail of $\bar{b}$, the vector $\bar{a}+\bar{b}$ is defined.

Fig. 1.1 TRIANGLE METHOD OF ADDING VECTORS

b) Parallelogram Method (Tail-to-Tail Method)

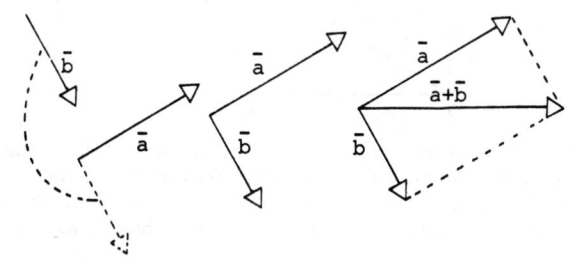

(i) Join the tails of the two vectors.

(ii) Construct a parallelogram having $\bar{a}$ and $\bar{b}$ as two of it's sides. The long diagonal of the parallelogram represents the vector $\bar{a}+\bar{b}$.

Fig. 1.2 THE PARALLELOGRAM METHOD OF ADDING VECTORS

1.3 SUBTRACTION OF VECTORS

a) The subtraction of a vector is defined as the addition of the corresponding negative vector. Therefore, the vector $\vec{P} - \vec{F}$ is obtained by adding the vector $(-\vec{F})$ to the vector $\vec{P}$, i.e., $\vec{P} + (-\vec{F})$. (See Fig. 1.3)

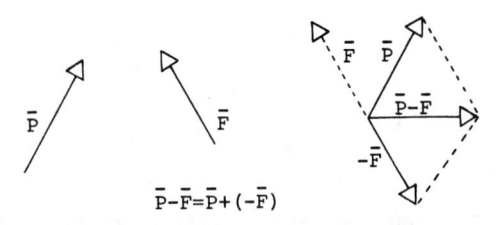

Fig. 1.3 THE SUBTRACTION OF A VECTOR

1.4 THE COMPONENTS OF A VECTOR

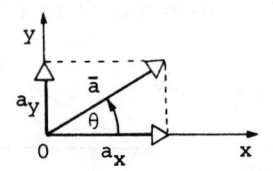

a_x and a_y are the components of a vector $\bar{a}$. The angle θ is measured counterclockwise from the positive x-axis. The components are formed when we draw perpendicular lines to the chosen axes.

Fig. 1.4 THE FORMATION OF VECTOR COMPONENTS ON THE POSITIVE X-Y AXIS

a) The components of a vector are given by

$$A_x = A \cos \theta$$
$$A_y = A \sin \theta$$

i) A component is equal to the product of the magnitude of vector A and cosine of the angle between the positive axis and the vector.

b) The magnitude can be expressed in terms of the components.

$$A = \sqrt{A_x^2 + A_y^2}$$

i) For the angle θ,

I-3

$$\boxed{\tan \theta = \frac{A_y}{A_x}}$$

c) A vector $\vec{F}$ can be written in terms of its components F_x and F_y

$$\boxed{\vec{F} = \vec{i}F_x + \vec{j}F_y}$$

where $\vec{i}$ and $\vec{j}$ represent perpendicular unit vectors (magnitude = 1) along the x- and y-axis.

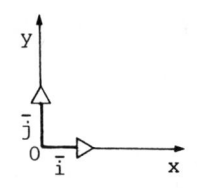

(a) The unit vectors $\bar{i}$ and $\bar{j}$ of the two-dimensional rectangular coordinate system.

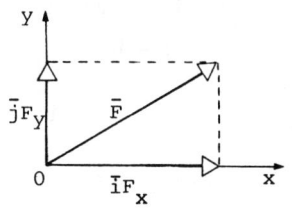

(b) The components F_x and F_y.

Fig. 1.5 VECTOR COMPONENTS AND THE UNIT VECTOR

1.5 THE UNIT VECTOR

a) A Unit Vector in the direction of a vector $\vec{a}$ is given by

$$\boxed{\vec{u} = \frac{\vec{a}}{|\vec{a}|} = \frac{\vec{a}}{a} = \left(\frac{a_x}{a}\vec{i} + \frac{a_y}{a}\vec{j} \right)}$$

1.6 ADDING VECTORS ANALYTICALLY

a) Analytical addition involves adding the components of the individual vectors to produce the sum, expressed in terms of its components.

b) To find $\vec{a} + \vec{b} = \vec{c}$ analytically:

　i) Resolve $\vec{a}$ in terms of its components:
$$\vec{a} = \vec{i}a_x + \vec{j}a_y$$

　ii) Resolve $\vec{b}$ in terms of its components:
$$\vec{b} = \vec{i}b_x + \vec{j}b_y$$

　iii) The components of $\vec{c}$ equal the sum of the corresponding components of $\vec{a}$ and $\vec{b}$:
$$\vec{c} = \vec{i}(a_x + b_x) + \vec{j}(a_y + b_y)$$
　and
$$\vec{c} = \vec{i}c_x + \vec{j}c_y$$

and the magnitude $|c| = \sqrt{c_x^2 + c_y^2}$

with θ given by $\tan \theta = \dfrac{c_y}{c_x}$

1.7 MULTIPLICATION OF VECTORS

a) Multiplication of a vector by a scalar

　i) The product of a vector $\vec{a}$ and a scalar k, written as $k\vec{a}$, is a new vector whose magnitude is k times the magnitude of $\vec{a}$; if k is positive, the new vector has the same direction as $\vec{a}$; if k is negative, the new vector has a direction opposite that of $\vec{a}$.

b) The Scalar Product (Dot Product)

　i) The Dot Product of two vectors yields a scalar:

I-5

$$\vec{a} \cdot \vec{b} = ab \cos \theta$$

c) The Vector Product (Cross Product)

  i) The Cross Product of two vectors yields a vector:

$$\vec{a} \times \vec{b} = \vec{c}$$
$$\text{and}$$
$$|\vec{c}| = ab \sin \theta$$

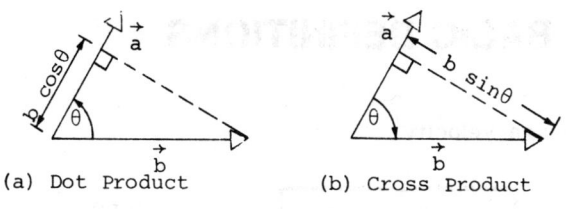

(a) Dot Product   (b) Cross Product

Fig. 1.6 VECTOR MULTIPLICATION

ii) The direction of the Vector Product $\vec{a} \times \vec{b} = \vec{c}$ is given by the "Right-Hand Rule":

  1) With $\vec{a}$ and $\vec{b}$ tail-to-tail, draw the angle θ from $\vec{a}$ to $\vec{b}$.

  2) With your right hand, curl your fingers in the direction of the angle drawn. The extended thumb points in the direction of $\vec{c}$.

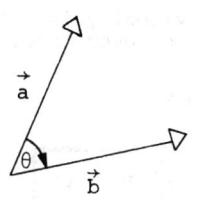

Fig. 1.7 The direction of the vector product, $\vec{c} = \vec{a} \times \vec{b}$ ($|\vec{c}|$=ab sinθ), is into the page.

CHAPTER 2

ONE DIMENSIONAL MOTION

2.1 BASIC DEFINITIONS

a) Average Velocity

$$v = \frac{\Delta x}{\Delta t} = \frac{x_2 - x_1}{t_2 - t_1}$$ units: $\frac{\text{meters}}{\text{sec}}$

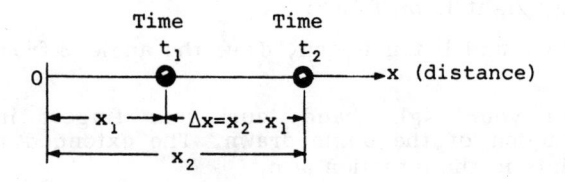

Fig. 2.1 Average velocity is a function of distance over time.

b) Instantaneous Velocity

$$v = \lim_{\Delta t \to 0} \frac{\Delta x}{\Delta t} = \frac{dx}{dt} = v(t)$$ units: $\frac{\text{meters}}{\text{sec}}$

c) Average Acceleration

$$a = \frac{\Delta v}{\Delta t} = \frac{v_2 - v_1}{t_2 - t_1}$$ units: $\frac{\frac{\text{meters}}{\text{sec}}}{\text{sec}} = \frac{\text{meters}}{\text{sec}^2}$

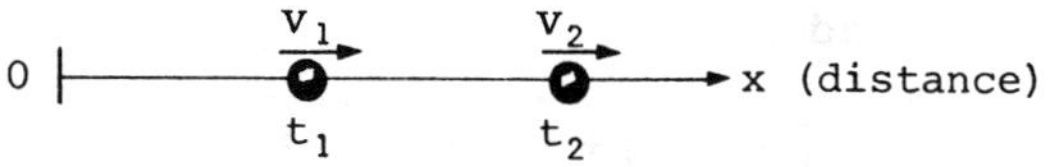

```
    v₁         v₂
0 |──●─────────●─────→ x (distance)
    t₁         t₂
```

Fig. 2.2 Average acceleration is a function of velocity over time.

v_1 = Initial Velocity

v_2 = Final Velocity

d) Instantaneous Acceleration

$$\boxed{a = \lim_{\Delta t \to 0} \frac{\Delta v}{\Delta t} = \frac{dv}{dt} = a(t)}\qquad \text{units:}\ \frac{\text{meters}}{\text{sec}^2}$$

2.2 MOTION WITH CONSTANT ACCELERATION

a = constant

a) The velocity, $v\ \left(\frac{\text{meters}}{\text{sec}}\right)$

i) In terms of a and t,

$$\boxed{v = v_0 + at}$$

where $v_0 = v(0)$ is the velocity at time $t = 0$ (the initial velocity).

ii) In terms of a and x,

$$\boxed{v^2 = v_0^2 + 2a(x-x_0)}$$

or

$$v = \sqrt{v_0^2 + 2a(x-x_0)}$$

where $x_0 = x(0)$ is the position at time $t = 0$, and v_0 is the velocity at time $t = 0$.

b) The position, x (meters)

i) In terms of a and t,

$$x = x_0 + v_0 t + \tfrac{1}{2}at^2$$

ii) In terms of v and t,

$$x = x_0 + \tfrac{1}{2}(v_0 + v)t$$

Acceleration — $a = \text{constant}$

Velocity — $v = v_0 + at$, slope = a

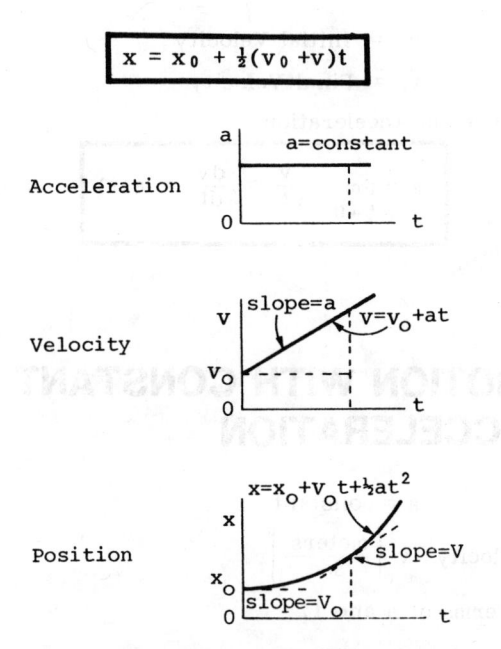

Position — $x = x_0 + v_0 t + \tfrac{1}{2}at^2$, slope = v, initial slope = v_0

Fig. 2.3 Graphs of one-dimensional motion with constant acceleration.

2.3 FREELY FALLING BODIES

$$a_y = \text{constant} = -g$$

$g = 32.2 \text{ ft/sec}^2$
$= 9.81 \text{ m/sec}^2$

The equations of motions are the same as for other systems of constant acceleration. One must note, however, that a falling object has a negative velocity and a negative acceleration (-g), with respect to a positive height, y.

CHAPTER 3

PLANE MOTION

3.1 DISPLACEMENT, VELOCITY AND ACCELERATION IN GENERAL PLANAR MOTION

a) The Displacement (Position) Vector

$$\bar{r} = \bar{i}x + \bar{j}y$$

b) The Velocity Vector

$$\bar{v} = \frac{d\bar{r}}{dt} = \bar{i}v_x + \bar{j}v_y$$

c) The Acceleration Vector

$$\bar{a} = \frac{d\bar{v}}{dt} = \bar{i}a_x + \bar{j}a_y$$

The vectors $\bar{r}$, $\bar{v}$, $\bar{a}$ and their components are shown in Fig. 3.1.

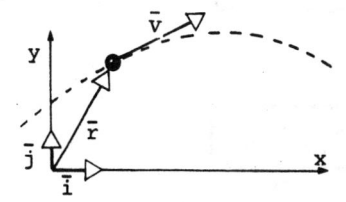

(a) A particle moves in a curved path in the xy plane. At time t the position vector is $\bar{r}$ and its velocity is $\bar{v}$.

I-10

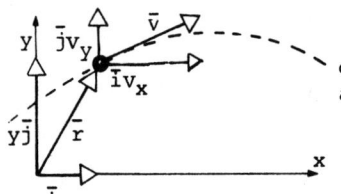

(b) The components of vectors $\bar{r}$ and $\bar{v}$, at time t.

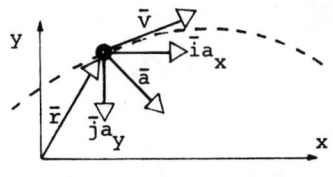

(c) The particle experiences an acceleration $\bar{a}$ at time t. The components are a_x and a_y. Notice that a_y is negative.

Fig. 3.1 A PARTICLE TRAVELING IN PLANE MOTION

3.2 MOTION IN A PLANE WITH CONSTANT ACCELERATION

The Equations of Motion

a) $\bar{a}$ = constant

b) a_x = constant $\qquad a_y$ = constant

The Velocity Vector, $\bar{v}$

c) $\bar{v} = \bar{i} v_x + \bar{j} v_y$

   i) The components of velocity in the x- and y-directions are:

$$\boxed{v_x = v_{x_0} + a_x t}$$

and

$$\boxed{v_y = v_{y_0} + a_y t}$$

I-11

ii) In terms of position and acceleration,

$$\boxed{v_x^2 = v_{x_0}^2 + 2a_x(x-x_0)}$$

and

$$\boxed{v_y^2 = v_{y_0}^2 + 2a_y(y-y_0)}$$

The velocity vector, $\bar{v}$

where
$$\boxed{\bar{v} = \bar{v}_0 + \bar{a}t}$$

and
$$\bar{v}_0 = \bar{i}v_{0_x} + \bar{j}v_{0_y} \quad \text{(the initial velocity vector)}$$
$$\bar{a} = \bar{i}a_x + \bar{j}a_y$$

The Position Vector, $\bar{r}$

$$\boxed{\bar{r} = x\bar{i} + y\bar{j}}$$

where
$$x = x_0 + \tfrac{1}{2}(v_{x_0} + v_x)t$$

or
$$x = x_0 + v_{x_0}t + \tfrac{1}{2}a_x t^2$$

and
$$y = y_0 + \tfrac{1}{2}(v_{y_0} + v_y)t$$

or
$$y = y_0 + v_{y_0}t + \tfrac{1}{2}a_y t^2.$$

The position vector may also be written as

$$\boxed{\bar{r} = \bar{r}_0 + \bar{v}_0 t + \tfrac{1}{2}\bar{a}t^2}$$

3.3 PROJECTILE MOTION

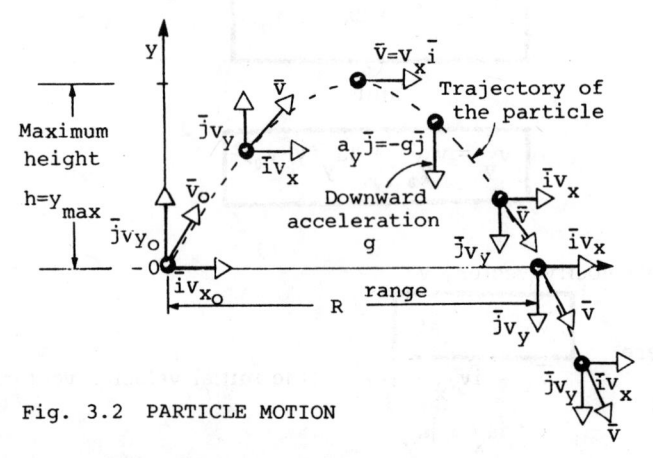

Fig. 3.2 PARTICLE MOTION

Acceleration is constant

$$a_x = 0 \qquad a_y = -g$$

$$g = 9.81 \text{ m/sec}^2 = 32.2 \text{ ft/sec}^2$$

The Initial conditions:

$$x_0 = 0 \qquad y_0 = 0$$

$$\bar{v}_0 = \bar{i}v_{x_0} + \bar{j}v_{y_0}$$

$$v_{x_0} = v_0 \cos\theta_0 \qquad v_{y_0} = v_0 \sin\theta_0$$

a) The velocity

The magnitude of velocity at any instant t is

$$v = \sqrt{v_x^2 + v_y^2}$$

The horizontal component of velocity is constant throughout the projectile motion since there is no horizontal acceleration:

$$\boxed{v_{x_0} = v_0 \cos\theta_0 = \text{constant}}$$

The vertical component of velocity is

$$v_y = v_{y_0} - gt$$

or

$$\boxed{v_y = (v_0 \sin\theta_0) - gt}$$

since the negative acceleration of gravity acts vertically.

b) The Position

The Horizontal Position Component is

$$x = v_x t$$

or

$$\boxed{x = (v_0 \cos\theta_0)t}$$

The Vertical Position Component is

$$y = v_{y_0} t - \tfrac{1}{2}gt^2$$

or

$$\boxed{y = (v_0 \sin\theta_0)t - \tfrac{1}{2}gt^2}$$

or, as a function of the horizontal position component,

$$\boxed{y = (\tan\theta_0)x - \frac{g}{2(v_0 \cos\theta_0)^2} x^2}$$

This is the trajectory of the particle. The angle made by the velocity vector with the horizontal at any instant is

$$\boxed{\text{Tan } \theta = \frac{v_y}{v_x}}$$

or

$$\theta = \text{Tan}^{-1} \frac{v_y}{v_x}$$

i) The Range

In terms of v_0 and θ_0,

$$\boxed{R = \frac{v_0^2}{g} \sin 2\theta_0}$$

In terms of time,

$$\boxed{R = v_{x_0} t_2 = (v_0 \cos \theta_0) t_2}$$

Here, t_2 is the time required for the particle to traverse the full range.

The maximum range is obtained when the initial angle of flight is $\theta_0 = 45° = \frac{\pi}{4}$ rad:

$$\boxed{R_{max} = \frac{v_0^2}{g}}$$

ii) The Maximum Height (Elevation)

In terms of v_0 and θ_0,

$$\boxed{y_{max} = \frac{v_0 \sin \theta_0}{2g} = \frac{v_{0_y}}{2g}}$$

The time t_1 required to attain maximum height is

$$\boxed{t_1 = \frac{v_{0_y}}{g} = \tfrac{1}{2} t_2}$$

I-15

3.4 UNIFORM CIRCULAR MOTION

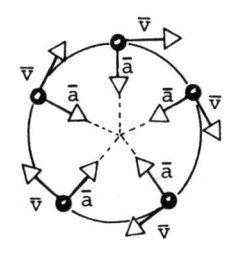

Fig. 3.3 A PARTICLE IN UNIFORM CIRCULAR MOTION
Notice that the acceleration vector is always directed toward the center of the circle and thus perpendicular to $\bar{v}$.

The magnitude of velocity is constant

$$v = \text{constant}$$

a) Centripetal Acceleration

$$a = \frac{v^2}{r}$$

a = Acceleration
v = Tangential Component of Velocity
r = Radius of the Path

b) For uniform circular motion, a can also be written as

$$a = \frac{4\pi^2 r}{T^2}$$

where T, the period or time for one revolution, is given by

$$T = \frac{2\pi r}{v}$$

c) The tangential component of the acceleration is the rate at which the particle speed changes:

$$a_T = \lim_{\Delta t \to 0} \frac{\Delta v}{\Delta t} = \frac{dv}{dt}$$

CHAPTER 4

DYNAMICS OF A PARTICLE

4.1 NEWTON'S SECOND LAW

a) $\boxed{\bar{F} = m\bar{a}}$ → units: $\left(kg \cdot \dfrac{meters}{sec^2}\right)$ = Newtons

When a body of mass m is acted upon by a force $\bar{F}$, the force $\bar{F}$ and the acceleration of the body are related by the above equation.

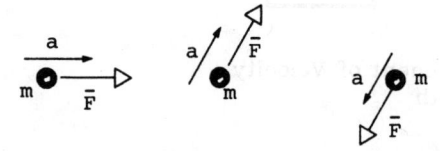

Fig. 4.1 The acceleration is in the direction of the applied force.

For several forces:

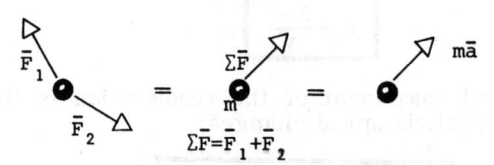

$$\boxed{\Sigma\bar{F} = m\bar{a}}$$

$\Sigma\bar{F} = \bar{F}_1 + \bar{F}_2$

Fig. 4.2 Newton's second law also holds for several applied forces.

4.2 NEWTON'S THIRD LAW

To every action there is always an equal, opposing reaction.

If a body A exerts a force F on body B, then body B exerts an equal and oposite force -F on body A.

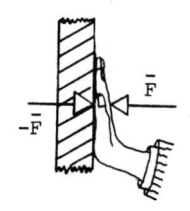

(a) A wall pushes you with the same force with which you push it.

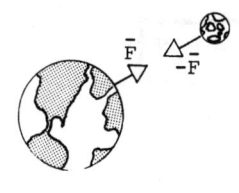

(b) The gravitational force that the earth exerts on the moon is equal to and opposite to the force that the moon exerts on the earth.

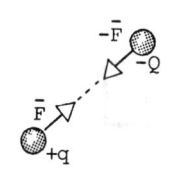

(c) The electrostatic force that the +q charge exerts on the -Q charge is equal to and opposite to the force that the -Q charge exerts on the +q charge.

Fig. 4.3 EXAMPLES OF NEWTON'S THIRD LAW.

4.3 MASS AND WEIGHT

a) Mass → units: (Kilograms (kg))

For a given body, the ratio of the magnitude of the force to that of the acceleration is a constant and is called it's mass:

$$m = \frac{F}{a} = \text{constant} \quad \text{(for a given body)}.$$

b) **Weight** → units: (Newtons)

The weight of a body is the gravitational force exerted on the body by the earth and is given by the product of the mass and the gravitational acceleration.

$$\boxed{W = mg}$$

4.4 FRICTION

For Impending Motion,

$$\text{Frictional Force} = \boxed{F_s = \mu_s N}$$

μ_s = Coefficient of Static Friction
N = Normal Force

For a body already in motion,

$$\boxed{F_k = \mu_k N}$$

μ_k = Coefficient of kinetic friction

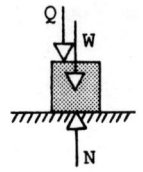

(a) $Q_x=0$. No friction exists.
N=Normal force
$\quad =Q+W$
F=Friction force
$\quad$=zero

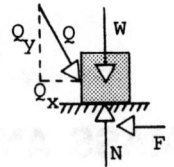

(b) No motion.
$F=-Q_x$
$F<\mu_s N$
$N=Q_y+W$

I-19

(c) Motion impending
$F = F_{max} = \mu_s N$
$-Q_x = F$
$N = Q_y + W$

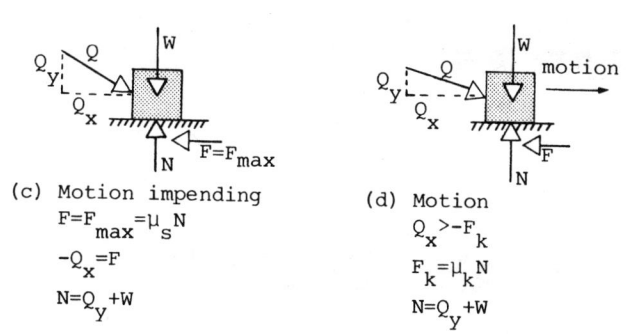

(d) Motion
$Q_x > -F_k$
$F_k = \mu_k N$
$N = Q_y + W$

Fig. 4.4 Simple cases involving friction.

4.5 THE DYNAMICS OF UNIFORM CIRCULAR MOTION— CENTRIPETAL FORCE

The force acting on a body of mass m undergoing uniform circular motion is the centripetal force, given by

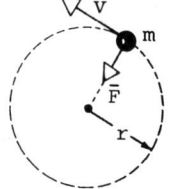

$$F = ma = m\frac{v^2}{r}$$

(a) A body of mass m traveling in uniform circular motion. The body is pulled toward the center of the circle of radius r with a force $F = \frac{mv^2}{r}$.

Fig. 4.5 CENTRIPETAL FORCE OF A PARTICLE.

Here, v is the magnitude of velocity, which is constant, and r is the radius of the circle.

I-20

CHAPTER 5

WORK AND ENERGY

5.1 WORK DONE BY A CONSTANT FORCE

$\overline{F}$ = constant acting on a body at an angle ϕ in the direction of motion

d = displacement of a body

The work done by the force $\overline{F}$ on the body is

$$W = (F\cos\phi)d$$

F = magnitude of the force $\overline{F}$

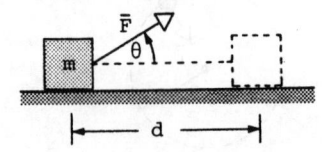

Fig. 5.1 Work done on a block, m

The work may also be written as:

$$W = \overline{F} \cdot \overline{d}$$ units: $\left(\frac{kg \cdot m}{sec^2} \cdot m \right)$ = joules

where $\overline{d}$ = Displacement Vector.

I-21

Note: Work can only be done by the component of force in the direction of displacement.

5.2 WORK DONE BY A VARYING FORCE

$\bar{F}$ = force acting on particle
$\bar{r}$ = displacement vector of path of particle

The work done on a particle as it moves from a point a to a point b is

$$W = \int_a^b \bar{F} \cdot d\bar{r} = \int_a^b F \cos\phi \, dr$$

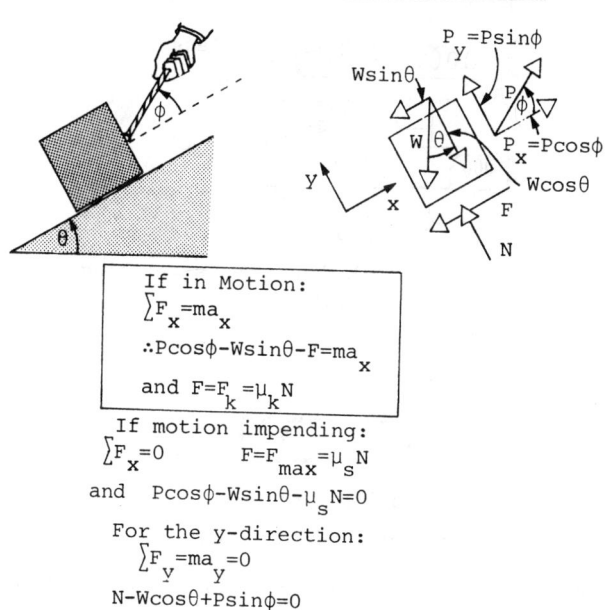

If in Motion:
$\sum F_x = ma_x$
$\therefore P\cos\phi - W\sin\theta - F = ma_x$
and $F = F_k = \mu_k N$

If motion impending:
$\sum F_x = 0 \qquad F = F_{max} = \mu_s N$
and $P\cos\phi - W\sin\theta - \mu_s N = 0$

For the y-direction:
$\sum F_y = ma_y = 0$
$N - W\cos\theta + P\sin\phi = 0$

Fig. 5.2 Work done on a block on an inclined plane.

I-22

5.3 WORK DONE BY A VARYING FORCE-ONE DIMENSIONAL CASE

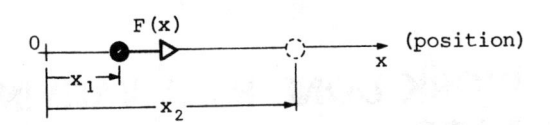

$F(x)$ = The force acting on the particle - a function of position - acts along the x-direction.

Total work done in going from x_1 to x_2

$$W_{12} = \int_{x_1}^{x_2} F(x)\,dx$$

Work done by a spring.

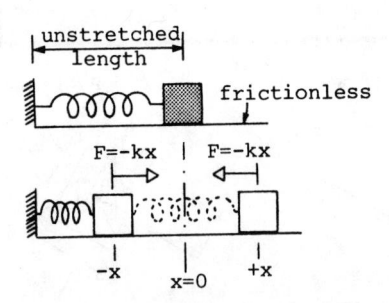

Fig. 5.3 Work done when stretching a spring.

Force of the spring

$$\boxed{F(x) = -kx}$$

k = spring constant.

Applied force = $F' = kx$

Work done by the applied force in going from x_1 to x_2:

$$W_{12} = \tfrac{1}{2}kx_2^2 - \tfrac{1}{2}kx_1^2.$$

I-23

If $x_1 = 0$, then

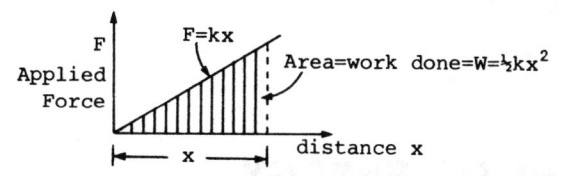

$$W = \tfrac{1}{2}kx_2^2$$

```
       F       F=kx
 Applied    ╱│ Area=work done=W=½kx²
   Force  ╱▓▓│
        ╱▓▓▓▓│
      ╱▓▓▓▓▓▓│
    |←— x —→|  distance x
```

Fig. 5.4 Area under the force curve is the work done in stretching a spring a distance x.

5.4 POWER

Power (P) is defined as the rate at which work is performed.

a) Average Power

$$\overline{P} = \frac{W}{t} \quad \rightarrow \quad \text{units } \frac{\text{joules}}{\text{sec}} = \text{watts}$$

W = work done during an interval of time t

b) Instantaneous Power

$$P = \frac{dW}{dt} \quad \rightarrow \quad \text{units } \frac{\text{joules}}{\text{sec}} = \text{watts}$$

or

$$P = \overline{F} \cdot \overline{v}$$

here,

$\bar{F}$ = force acting on particle

$\bar{v}$ = velocity vector

If the force is in the direction of the velocity, then

$P = Fv.$

If the power is constant, then

$W = Pt.$

5.5 KINETIC ENERGY

$\boxed{K = \tfrac{1}{2}mv^2}$ → units: $\dfrac{kg \cdot m^2}{sec^2}$ = joules

m = mass of object

v = velocity of object

5.6 WORK-ENERGY THEOREM

The work done by the resultant external force acting on a particle is equal to the change in kinetic energy of the particle.

$\boxed{W = K_2 - K_1 = \tfrac{1}{2}mv_2^2 - \tfrac{1}{2}mv_1^2 = \Delta K}$

CHAPTER 6

CONSERVATION OF ENERGY

6.1 CONSERVATIVE FORCES

A force, or force field, is said to be conservative if any one (and hence all) of the three following properties are satisfied:

a) The work done by the force on a particle that moves through any round trip is zero.

b) The work done by the force on a particle is independent of the path followed.

c) The kinetic energy of a particle subject to a force returns to its initial value after any round trip.

A force is said to be nonconservative, or dissipative, if any one of the above three conditions are not met.

The force of friction is an example of a nonconservative force.

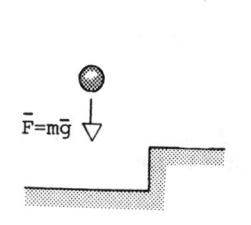

$\vec{F}=m\vec{g}$

$F=\dfrac{GmM}{r^2}$

(a) The gravitational force near the earth's surface is an example of a conservative force.

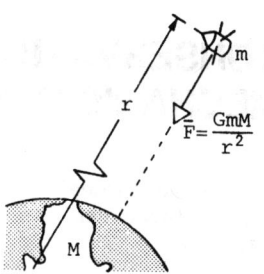

(b) The gravitational force that the earth exerts on a satellite is an example of a conservative force.

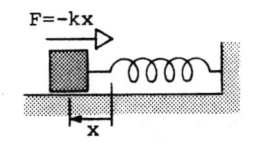

(c) The elastic force F that the spring exerts on an attached mass is an example of a conservative force.

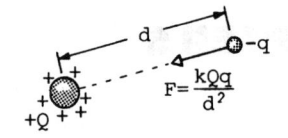

(d) The electrostatic force that is exerted on the -q charge by the +Q charge is an example of a conservative force.

Fig. 6.1 EXAMPLES OF CONSERVATIVE FORCES.

6.2 POTENTIAL ENERGY

A conservative force field has an associated potential energy function U, such that the change in potential energy ΔU, is equal to the negative of the work done.

$$\Delta U = -W$$

W = work done by the conservative force.

6.3 CONSERVATION OF MECHANICAL ENERGY

The total mechanical energy E of a particle subject to a conservative force remains constant throughout the motion of the particle.

$$K + U = \tfrac{1}{2}mv^2 + U = E = \text{constant}$$

K = kinetic energy = $\frac{1}{2}mv^2$

U = potential energy

$K_1 + U_1 = K_2 + U_2$

The kinetic plus potential energy at station 1 is equal to the kinetic plus potential energy at station 2.

These results hold true only if no work is done by nonconservative forces.

6.4 ONE-DIMENSIONAL CONSERVATIVE SYSTEMS

a) The Gravitational System

Force: $W = mg$

Potential Energy: $\boxed{U(y) = mgy}$

$y = 0$ at surface of earth.

Conservation of mechanical energy for Gravitation

$$\boxed{\tfrac{1}{2}mv_1^2 + mgh_1 = \tfrac{1}{2}mv_2^2 + mgh_2}$$

b) The Spring

Force: $F = -kx$

Potential Energy: $\boxed{U(x) = \tfrac{1}{2}kx^2}$

$x_0 = 0$ is the position of the end of the spring when unstretched.

Conservation of Mechanical Energy for a Spring

$$\boxed{\tfrac{1}{2}mv_1^2 + \tfrac{1}{2}kx_1^2 = \tfrac{1}{2}mv_2^2 + \tfrac{1}{2}kx_2^2}$$

6.5 NON-CONSERVATIVE FORCES

The work done by friction (nonconservative force) is equal to the change in internal energy:

$$W_{fr} = \Delta E = -\Delta U_{int}$$

$\Delta E = E_f - E_i =$
final mechanical energy - initial mechanical energy
ΔU_{int} = change in internal (or thermal) energy

Conservation of Energy for a system acted upon by conservative and frictional forces

$$\Delta E + \Delta U_{int} = \Delta K + \Delta U + \Delta U_{int} = 0$$

The sum of the mechanical and internal energy of a system acted upon only by conservative and frictional (nonconservative) forces remains constant.

6.6 THE CONSERVATION OF ENERGY

The total energy - kinetic plus potential plus internal plus all other forms of energy - must remain constant, i.e., energy may be transformed from one kind to another, but it cannot be created or destroyed.

$\Delta K + \Sigma \Delta U + \Delta U_{int}$ + (all changes in other forms of energy) = 0

6.7 THE RELATIONSHIP BETWEEN MASS AND ENERGY

a) Redefinition of mass, m

$$m = \frac{m_0}{\sqrt{1-\left(\frac{v}{c}\right)^2}}$$

m_0 = rest mass
c = speed of light
   = 3×10^8 m/sec

The mass of a particle is not constant but instead varies with the v, where v is the velocity relative to the observer.

b) Kinetic Energy

$$K = (m-m_0)c^2$$

c) The total energy, E

$$K + \Sigma U + U_{int} + \ldots + m_0 c^2 = mc^2 = E_{tot}$$

E = Total Energy = constant

The law of conservation of mass-energy

$$\Delta K + \Sigma \Delta U + \Delta U_{int} + \ldots + \Delta m_0 c^2 = 0$$

d) Mass-energy equivalence formula

$$E = -\Delta m_0 c^2$$

E = Total Energy
Δm_0 = Change in Rest Mass
c = Speed of Light

CHAPTER 7

THE DYNAMICS OF SYSTEMS OF PARTICLES

7.1 CENTER OF MASS OF A SYSTEM OF PARTICLES

a) The x-coordinate

i) $x_c = \dfrac{m_1 x_1 + m_2 x_2 + \ldots + m_n x_n}{m_1 + m_2 + \ldots + m_n} = \dfrac{\sum_{i=1}^{n} m_i x_i}{\sum_{i=1}^{n} m_i}$

$$\boxed{x_c = \dfrac{\sum_{i=1}^{n} m_i x_i}{M}}$$

b) The y-coordinate

i) $y_c = \dfrac{m_1 y_1 + m_2 y_2 + \ldots + m_n y_n}{\sum m_i} = \boxed{\dfrac{\sum_{i=1}^{n} m_i y_i}{M} = y_c}$

The z-coordinate (for three dimensions)

$$\boxed{z_c = \dfrac{\sum_{i=1}^{n} m_i z_i}{M}}$$

Here

$$M = \text{total mass} = \sum_{i=1}^{n} m_i$$

n = number of particles in the system
x_i = x-coordinate of ith particle
y_i = y-coordinate of ith particle
z_i = z-coordinate of ith particle

In vector notation

$$\bar{r}_c = \bar{i}\, x_c + \bar{j}\, y_c + \bar{k}\, z_c$$

7.1.1 CENTER OF MASS IN VECTOR NOTATION

Center of Mass in Vector Notation

$$\boxed{\bar{r}_c = \frac{1}{M} \sum_{i=1}^{n} m_i\, \bar{r}_i}$$

7.2 MOTION OF THE CENTER OF MASS

Newton's second law for a system of particles

The center of mass of a system of particles moves as if all the mass of the system were concentrated at the mass center and all the external forces were acting at that point.

$$\boxed{\bar{F}_{ext} = M\, \bar{a}_c}$$

$\bar{a}_c$ = acceleration of the center of mass
$\bar{F}_{ext}$ = the sum of the external forces
M = the total mass of the system

7.3 WORK-ENERGY THEOREM FOR A SYSTEM OF PARTICLES

$$W_c = \Delta K_c = K_{c_f} - K_{c_i} = \tfrac{1}{2}MV_{c_f}^2 - \tfrac{1}{2}MV_{c_i}^2$$

W_c = Center-of-mass work

K_{c_i}, K_{c_f} = Initial and Final kinetic energy associated with the center of mass.

V_{c_i}, V_{c_f} = Initial and Final velocity of the center of mass

7.4 LINEAR MOMENTUM OF A PARTICLE

$$\boxed{\bar{p} = m\bar{v}} \rightarrow \text{units: } \frac{\text{kg} \cdot \text{m}}{\text{sec}}$$

$\bar{p}$ = linear momentum of particle
m = mass of particle
$\bar{v}$ = velocity of particle

7.4.1 NEWTON'S SECOND LAW

$$\boxed{\bar{F} = \frac{d\bar{p}}{dt} = \frac{d(m\bar{v})}{dt}}$$

where $\bar{F}$ is the net force on the particle.

7.5 LINEAR MOMENTUM OF A SYSTEM OF PARTICLES

a) Total Linear Momentum

$$\bar{P} = \sum_{i=1}^{n} \bar{p}_i = \bar{p}_1 + \bar{p}_2 + \ldots + \bar{p}_n = m_1\bar{v}_1 + m_2\bar{v}_2 + \ldots + m_n\bar{v}_n$$

$\bar{P}$ = total linear momentum of system

$\bar{p}_i, m_i, \vec{v}_i,$ = linear momentum, mass, and velocity of ith particle, respectively.

b) Newton's Second Law for a System of Particles (Momentum Form.)

$$\bar{F}_{ext} = \frac{d\bar{P}}{dt}$$

$\bar{F}_{ext}$ = sum of all external forces

7.6 CONSERVATION OF LINEAR MOMENTUM

If the sum of the external forces acting on a system is zero, the total linear momentum of the system remains unchanged.

$$\bar{P} = \text{constant}$$

For a system of particles:

$$\bar{p}_1 + \bar{p}_2 + \ldots + \bar{p}_n = \bar{P} = \text{constant}$$

or

$$m_1 \bar{v}_1 + m_2 \bar{v}_2 + \ldots + m_n \bar{v}_n = \text{constant}$$

and the total momentum of the system as a result of collisions is constant

$$\bar{p}_1 + \bar{p}_2 + ... + \bar{p}_n = \bar{P} = \text{constant}$$

7.7 ELASTIC AND INELASTIC COLLISIONS IN ONE DIMENSION

When kinetic energy is conserved, the collision is Elastic. Otherwise, the collision is said to be Inelastic.

a) For an Elastic collision,

$$\tfrac{1}{2}m_1 v_{1i}^2 + \tfrac{1}{2}m_2 v_{2i}^2 = \tfrac{1}{2}m_1 v_{1f}^2 + \tfrac{1}{2}m_2 v_{2f}^2$$

b) For an inelastic collision, some kinetic energy is transformed into internal energy. However, linear momentum is still conserved. If the two bodies stick and travel together with a common final velocity after collision, it is said to be completely inelastic. From conservation of momentum, we have

$$m_1 v_{1i} + m_2 v_{2i} = (m_1 + m_2) v_f$$

7.8 COLLISIONS IN TWO AND THREE DIMENSIONS

Since momentum is linearly conserved, the resultant components must be found and then the conservation laws applied in each direction.

a) The x-component

I-35

i) $\boxed{m_1 v_{1_i} = m_1 v_{1_f} \cos \theta_1 + m_2 v_{2_f} \cos \theta_2}$

b) The y-component

i) $\boxed{m_2 v_{2_i} = m_1 v_{1_f} \sin \theta_1 + m_2 v_{2_f} \sin \theta_2}$

> where θ_1 = the angle of deflection, after the collision, of mass m_1
>
> θ_2 = the angle of deflection, after the collision, of mass m_2

c) For three dimensions, there would be an added z-component and an added angle, θ_3.

For the above cases, "i" denotes initial value; "f" denotes final value.

CHAPTER 8

ROTATIONAL KINEMATICS

8.1 ROTATIONAL MOTION – THE VARIABLES

(a) Average Angular Velocity

$$\bar{\omega} = \frac{\Delta \theta}{\Delta t} = \frac{\theta_2 - \theta_1}{t_2 - t_1}$$ → units: $\frac{\text{Radians}}{\text{Second}}$

$\bar{\omega}$ = Angular Velocity
θ = Angular Displacement
t = Time

(b) Instantaneous Angular Velocity

$$\omega = \frac{d\theta}{dt}$$ → units: $\frac{\text{Radians}}{\text{Second}}$

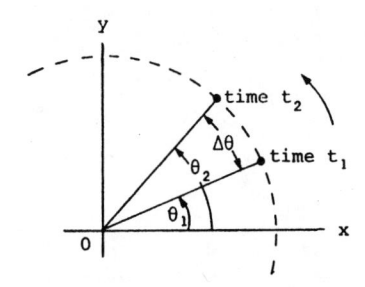

Fig. 8.1 Angular Velocity of a particle.

I-37

(c) Average Angular Acceleration

$$\boxed{\bar{\alpha} = \frac{\omega_2 - \omega_1}{t_2 - t_1} = \frac{\Delta\omega}{\Delta t}} = \frac{\text{Radians}}{\frac{\text{Second}}{\text{Second}}} = \frac{\text{Radians}}{\text{Sec}^2}$$

$\bar{\alpha}$ = Angular Acceleration

ω_1, ω_2 = Instantaneous Angular Velocities at times t_1 and t_2.

(d) Instantaneous Angular Acceleration

$$\boxed{\alpha = \frac{d\omega}{dt}} \rightarrow \text{units:} \ \frac{\text{Rads}}{\text{Sec}^2}$$

8.2 ROTATIONAL MOTION WITH CONSTANT ANGULAR ACCELERATION

Similarity Table

Rotational Motion	Linear Motion Equivalent
α = constant	a = constant
$\omega = \omega_0 + \alpha t$	$v = v_0 + at$
$\theta = \frac{\omega_0 + \omega}{2} t$	$x = \frac{v_0 + v}{2} t$
$\theta = \omega_0 t + \frac{1}{2}\alpha t^2$	$x = v_0 t + \frac{1}{2}at^2$
$\omega^2 = \omega_0^2 + 2\alpha\theta$	$v^2 = v_0^2 + 2ax$
θ_0, θ = initial and final angular displacements		
ω_0, ω = initial and final angular velocities		

I-38

8.3 RELATION BETWEEN LINEAR AND ANGULAR KINEMATICS FOR A PARTICLE IN CIRCULAR MOTION

P is a particle
S = arc length
θ = angle
r = radius

a_T = tangential acceleration
a_n = normal acceleration
a = total accelaration

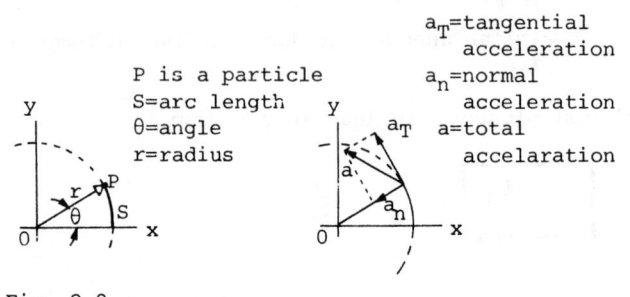

Fig. 8.2

(a) Velocity - Angular Velocity

$$v = \omega r$$

v = Linear Velocity
ω = Angular Velocity
r = Radius of Path of the Particle

(b) Acceleration - Angular Acceleration

$$a_T = \alpha r$$ a_T = Tangential Component of Acceleration

$$a_n = \frac{v^2}{r} = \omega^2 r$$ a_n = Normal Component of Acceleration

Note that $\omega^2 r = \frac{v^2}{r}$ = centripetal acceleration in uniform circular motion.

CHAPTER 9

ROTATIONAL DYNAMICS

9.1 TORQUE

a) Vector Equation

$$\tau = \bar{r} \times \bar{F}$$

This indicates that the torque is formed by the component of force perpendicular to $\bar{r}$.

b) Scalar Equation

$$\tau = rF \sin\theta$$

τ = Torque
r = Radius
F = Applied Force
θ = Angle Formed by $\bar{r}$ and $\bar{F}$

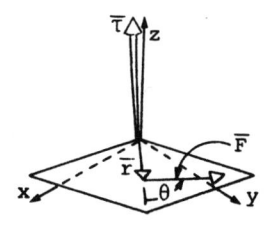

Fig. 9.1 Torque

I-40

9.2 ANGULAR MOMENTUM

a) Vector Equation

$$\overline{\ell} = \overline{r} \times \overline{p}$$

b) Scalar Equation

$$\ell = rp\sin\theta$$

ℓ = Angular Momentum
r = Radius
P = Linear Momentum
θ = Angle Formed by $\overline{r}$ and $\overline{p}$

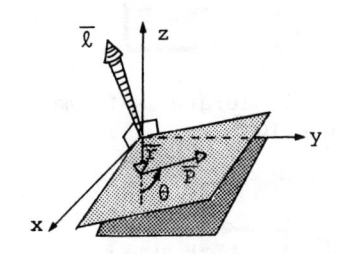

Fig. 9.2 Angular momentum

c) Relation Between Torque and Angular Momentum

$$\overline{\tau} = \frac{d\overline{\ell}}{dt}$$

This vector equation is equivalent to three scalar equations:

$$\tau_x = \frac{d\ell_x}{dt}, \quad \tau_y = \frac{d\ell_y}{dt}, \quad \tau_z = \frac{d\ell_z}{dt}$$

x-component   y-component   z-component

9.3 KINETIC ENERGY OF ROTATION AND ROTATIONAL INERTIA

a) Kinetic Energy

$$K = \tfrac{1}{2} m r^2 \omega^2$$

since $r^2 \omega^2 = v^2$

K = kinetic Energy
m = mass
r = radius
ω = angular velocity

b) Total Kinetic Energy

$$K = \tfrac{1}{2}(m_1 r_1^2 + m_2 r_2^2 + \ldots + m_n r_n^2)\omega^2$$

$$K = \tfrac{1}{2}(\Sigma\, m_n r_n^2)\omega^2$$

c) Moment of Inertia

$$I = \Sigma\, m_n r_n^2$$

The equation for kinetic energy becomes,

$$K = \tfrac{1}{2} I \omega^2$$

I-42

9.4 SOME ROTATIONAL INERTIAS

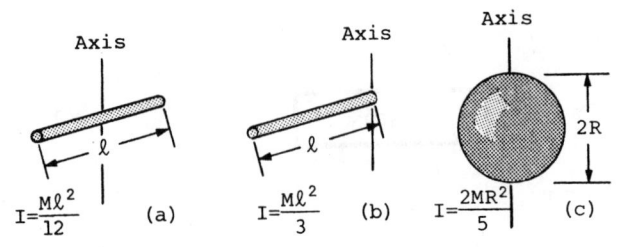

(a) $I = \dfrac{M\ell^2}{12}$

(b) $I = \dfrac{M\ell^2}{3}$

(c) $I = \dfrac{2MR^2}{5}$

Fig. 9.3 The rotational inertia for (a) a thin rod about an axis through its center, perpendicular to the length, (b) a thin rod about an axis through one end, perpendicular to the length, and (c) a solid sphere about any diameter.

9.5 ROTATIONAL DYNAMICS OF A RIGID BODY

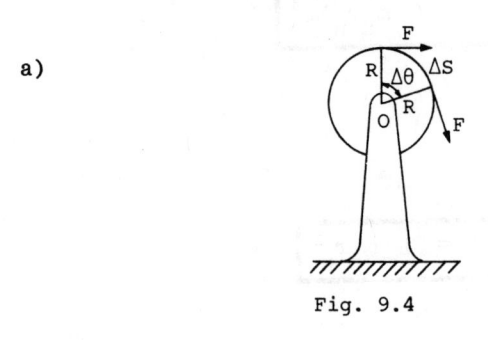

a)

Fig. 9.4

$$\Delta w = \tau \Delta \theta$$

Units - Joules

w - work done

τ - FR is the torque, τ, due to the force F.

b) If torque is constant while angle changes by a finite amount from θ_1, θ_2,

$$w = \tau(\theta_2 - \theta_1)$$

c) Power, P

$$P = \tau \omega$$

ω - angular velocity

d) Angular Momentum, L

$$L = I\omega$$

e) Torque, τ

$$\tau = I\alpha$$

9.6 ROLLING BODIES

a) Rotational Inertia

1) Rolling Cylinder or Disk

$$I_p = I_{cm} + MR^2$$

I_p = Rotational Inertia About Axis Through P

I_{cm} = Rotational Inertia About Axis Through C

M = Mass

R = Radius

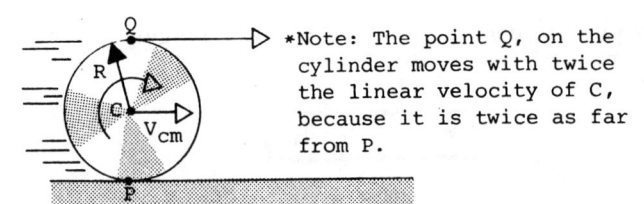

*Note: The point Q, on the cylinder moves with twice the linear velocity of C, because it is twice as far from P.

Fig. 9.5 Rolling cylinder with mass, M, and velocity of center of mass, V_{cm}.

b) Kinetic Energy

$$K = \tfrac{1}{2} I_p \omega^2$$

or

$$K = \tfrac{1}{2} I_{cm} \omega^2 + \tfrac{1}{2} MR^2 \omega^2$$

or

$$K = \tfrac{1}{2} I_{cm} \omega^2 + \tfrac{1}{2} M v_{cm}^2$$

9.7 CONSERVATION OF ANGULAR MOMENTUM

When no external force is acting on a system,

$$I\omega = I_0 \omega_0 = \text{Constant}$$

angular momentum is conserved.

CHAPTER 10

HARMONIC MOTION

10.1 OSCILLATIONS - SIMPLE HARMONIC MOTION (SHM)

a) Equations of Motion - Simple Oscillation

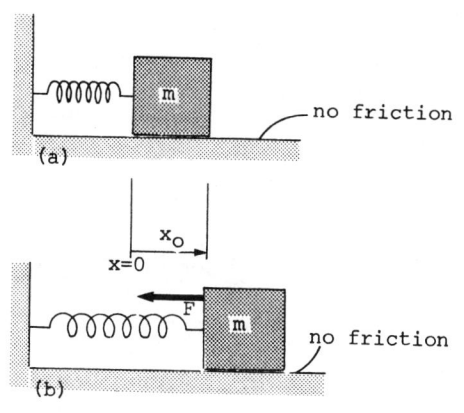

Fig. 10.1 Oscillation of a spring with mass

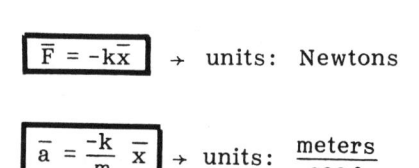

$$\overline{F} = -k\overline{x}$$ → units: Newtons

$$\overline{a} = \frac{-k}{m}\overline{x}$$ → units: $\frac{meters}{sec^2}$

I-46

F = Restoring Force
k = Spring Constant
x = Displacement
a = Acceleration
m = Mass

b) Equations of Motion - The Variables for SHM

The Period of Motion

$$\boxed{T = \frac{2\pi}{\omega}} = \boxed{2\pi\sqrt{\frac{m}{k}}} \rightarrow \text{units: Seconds}$$

The Frequency of Motion

$$\boxed{\nu = \frac{1}{T}} = \frac{\omega}{2\pi} = \frac{1}{2\pi}\sqrt{\frac{k}{m}} \rightarrow \text{units: } \frac{1}{\text{seconds}}$$

The Angular Frequency of Motion

$$\boxed{\omega = 2\pi\nu} = \frac{2\pi}{T} = \sqrt{\frac{k}{m}} \rightarrow \text{units: } \frac{\text{Rads}}{\text{sec}}$$

Frequency in radians/second

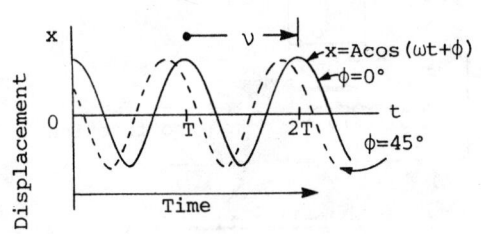

Fig. 10.2 Simple harmonic oscillation

c) Differential Equations of Motion

Displacement: $\boxed{\bar{x} = A\cos(\omega t + \phi)}$

Velocity: $\frac{dx}{dt} = \boxed{\bar{v} = -\omega A \sin(\omega t + \phi)}$

Acceleration: $\dfrac{d^2x}{dt^2} = \bar{a} = -\omega^2 A \cos(\omega t + \phi)$

x = Displacement
v = Velocity (Linear)
a = Acceleration (Linear)
A = Amplitude
ω = Angular Velocity
t = Time
φ = Phase Angle

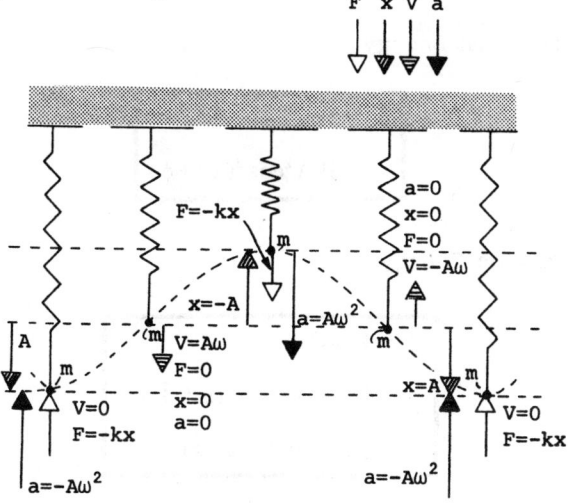

Fig. 10.3 Simple harmonic oscillation with a constant applied force

For maximum velocity,

$$v_{max} = \omega A$$

For maximum acceleration,

$$a_{max} = \omega^2 A$$

Example of Displacement equation:

$$x = \underbrace{0.53}_{\text{amplitude}} \cos(\underbrace{9.3t}_{\substack{\text{Angu-}\\\text{lar fre-}\\\text{quency}}} + \underbrace{46}_{\substack{\text{Phase}\\\text{Angle}}})$$

10.2 ENERGY CONSIDERATIONS OF SHM

a) Potential Energy, U

$$\boxed{\begin{aligned} U &= \tfrac{1}{2}kx^2 \\ &= \tfrac{1}{2}kA^2\cos^2(\omega t + \phi) \end{aligned}}$$

b) Kinetic Energy, K

$$\boxed{\begin{aligned} K &= \tfrac{1}{2}mv^2 \\ &= \tfrac{1}{2}m\omega^2 A^2 \sin^2(\omega t + \phi) \\ &= \tfrac{1}{2}kA^2 \sin^2(\omega t + \phi) \end{aligned}}$$

c) Total Energy, T

$$T = U + K$$
$$\boxed{T = \tfrac{1}{2}kA^2}$$

d) Velocity, $\bar{v}$

$$\boxed{\bar{v} = \frac{dx}{dt} = \pm\sqrt{\frac{k}{m}}(A^2 - x^2)}$$

I-49

10.3 PENDULUMS

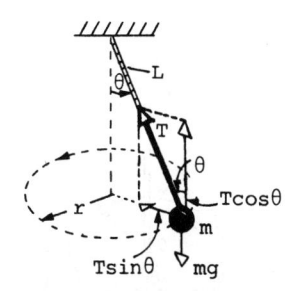

Fig. 10.4 The Conical Pendulum.
A mass revolves in a horizontal circle of radius r. The velocity is v of constant magnitude. The centripetal force is $F = T\sin\theta = m\dfrac{v^2}{r}$.

a) Simple Pendulum

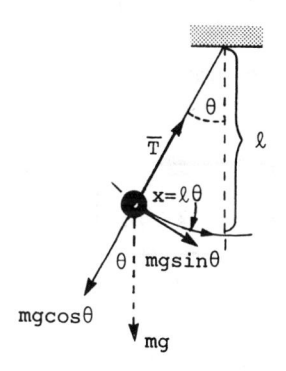

Fig. 10.5

i) Force, F

$$\overline{F} = -mg\theta = \dfrac{-mg}{L}x$$

ii) Period T

$$T = 2\pi\sqrt{\dfrac{L}{g}}$$

I-50

L → Length of Supporting Cord

b) Torsional Pendulum

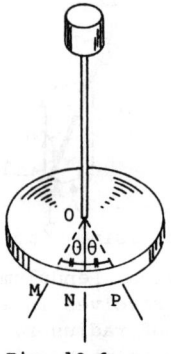

Fig. 10.6

i) Torque, τ

$$\tau = -k\theta$$

k = Torsional Constant

ii) Period T

$$T = 2\pi\sqrt{\frac{I}{k}}$$

10.4 SIMPLE HARMONIC MOTION AND UNIFORM CIRCULAR MOTION

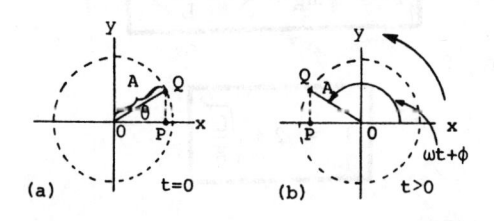

(a) t=0 (b) t>0

I-51

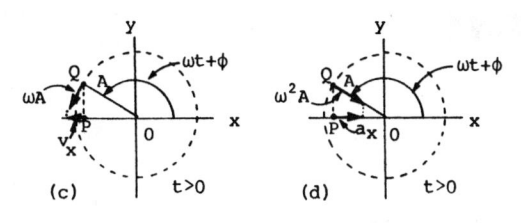

Fig. 10.7

a) Equations of Motion

Radius, r

$$r = \sqrt{x^2 + y^2} = A$$

Velocity, v

$$v = \sqrt{v_x^2 + v_y^2} = \omega A$$

Acceleration, a

$$a = \sqrt{a_x^2 + a_y^2} = \omega^2 A$$

CHAPTER 11

SOUND WAVES

11.1 SPEED OF SOUND

a) Wave Speed in Fluid*

$$v = \sqrt{\frac{B}{\rho_0}}$$ → units: $\frac{\text{meters}}{\text{seconds}}$

v = Speed of Sound in a Fluid
B = Modulus of Elasticity
ρ_0 = Density of medium

*Note: Gas is also a fluid, yet a more precise equation may be given, as seen below.

b) Wave Speed in a Gas

$$v_g = \sqrt{\frac{\gamma p_0}{\rho_0}}$$ → units: $\frac{\text{meters}}{\text{seconds}}$

(Ideal gas)

v_g = Speed of Sound
γ = Ratio of Specific Heats for a Gas
p_0 = undisturbed pressure
ρ_0 = Density of Medium

11.2 INTENSITY OF SOUND

a) Average Intensity

$$I = \frac{1}{2} \frac{P_m^2}{\sqrt{B\rho_0}}$$

I = Average Intensity

P_m = Pressure Amplitude

B = Bulk Modulus of Elasticity

ρ_0 = Density of Medium

11.3 ALLOWED FREQUENCIES OF A STRING, FIXED AT BOTH ENDS, OR AN ORGAN PIPE, OPEN AT BOTH ENDS

$$\nu_n = \frac{n}{2\ell} v, \quad n = 1, 2, 3, \ldots$$

ν_n = Frequency

ℓ = Length of String or Pipe

v = Speed of Wave

n = Number of Modes

11.4 THE BEAT EQUATION FOR PARTICLE DISPLACEMENT

$$y = \left[2y_m \cos 2\pi \left(\frac{\nu_1 - \nu_2}{2}\right) t\right] \cos 2\pi \left(\frac{\nu_1 + \nu_2}{2}\right) t$$

y = Displacement
y_m = Amplitude
ν_1 = Frequency of Wave 1
ν_2 = Frequency of Wave 2
t = Time

11.5 THE BEAT FREQUENCY

$$\nu_B = \nu_1 - \nu_2$$

where ν_B = Beat Frequency

11.6 THE DOPPLER EFFECT

a) Source at Rest

$$\nu' = \nu \left(\frac{v \pm v_0}{v}\right)$$

I-55

ν' = Observed (Heard) Frequency

ν = Actual Frequency

v = Speed of Sound in a Medium

v_0 = Speed of Observer

Note: The ± signs denote when the motion of sound is toward or away from the source; it's + when the motion is toward the source, - when the motion is away from the source.

b) Observer at Rest

$$\nu' = \nu \left(\frac{v}{v \pm v_s} \right)$$

where v_s is the speed of the source.

c) The General Doppler Equation

$$\nu' = \nu \left(\frac{v \pm v_0}{v \pm v_s} \right)$$

11.7 HALF ANGLE OF A SHOCK WAVE

$$\sin \theta = \frac{v}{v_s}$$

θ = Half Angle of Mach Cone

v = Speed of Sound in a Medium

v_s = Speed of Source

I-56

CHAPTER 12

GRAVITATION

12.1 NEWTON'S LAW OF GRAVITATION

$$\boxed{F = G \frac{m_1 m_2}{r^2}} \rightarrow \text{units: Newtons}$$

F = Force of Gravity
G = Gravitational Constant
m = Mass
r = Distance Between m_1 and m_2
G = 6.6726 × 10^{-11} m³/kg·s²

12.2 THE MOTION OF PLANETS AND SATELLITES–KEPLER'S LAWS OF PLANETARY MOTION

a) Law of Orbits - All planets move in elliptical orbits having the sun as one focus.

Fig. 12.1

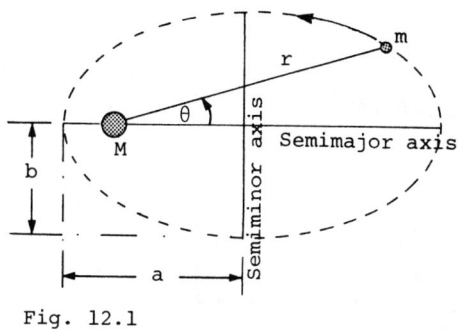

i) Equation of Orbit

$$\frac{1}{r} = \frac{1}{b^2}(a + \sqrt{a^2 - b^2}\cos\theta)$$

  r = Distance between Masses
  b = Semi-minor Axis
  a = Semi-major Axis
  θ = Angle Made with Semi-major Axis

b) Law of Areas - A line joining any planet to the sun sweeps out equal areas in equal times.

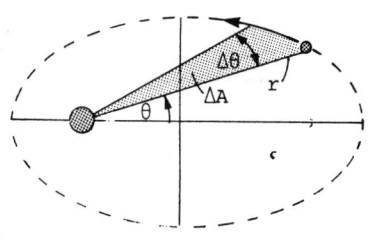

Fig. 12.2

i) Rate of Area Change

$$\frac{dA}{dt} = \frac{1}{2}r^2\omega$$

I-58

A = Area
r = Distance
ω = Angular Velocity

c) Law of Periods

$$T^2 = \left(\frac{4\pi^2}{GM}\right) a^3$$

T = Period
G = Gravitational Constant
M = Mass
a = Semi-major Axis

12.3 ANGULAR MOMENTUM OF ORBITING PLANETS

a) Angular Momentum

$$\ell = mr^2\omega$$

ℓ = Angular Momentum
m = Mass
ω = Angular Velocity

b) Rate of Area Change

$$\frac{dA}{dt} = \ell/2m$$

A = Area
ℓ = Angular Momentum
m = Mass

12.4 GRAVITATIONAL POTENTIAL ENERGY, U

$U = -W_\infty$ = (Work done by gravity on the particle as it moves from infinity to a distance r from the earth's center)

$$\boxed{U = \frac{-GMm}{r}}$$

G = Gravitational Constant

M = Mass of Denser Particle

m = Mass of Orbiting Particle

r = Distance

12.5 POTENTIAL ENERGY FOR A SYSTEM OF PARTICLES

For 3 particles:

$$U = -\left(\frac{Gm_1 m_2}{r_{12}} + \frac{Gm_1 m_3}{r_{13}} + \frac{Gm_2 m_3}{r_{23}} \right)$$

For 4 particles, there are 6 different gravitational potentials.

For 5, there are 10.

For n particles, there are $\frac{1}{2}(n-1)n$ individual gravitational potentials.

I-60

12.6 KINETIC ENERGY

$$\boxed{\begin{array}{l} K = \tfrac{1}{2} m \omega^2 r_0^2 \\ K = \dfrac{\;}{r_0} \end{array}}$$

12.7 TOTAL ENERGY

$$T = K + U$$
$$T = \tfrac{1}{2} \frac{GMm}{r_0} - \frac{GMm}{r_0}$$

$$\boxed{T = \frac{-GMm}{2r_0}}$$

CHAPTER 13

EQUILIBRIUM OF RIGID BODIES

13.1 RIGID BODIES IN STATIC EQUILIBRIUM

a) Sum of the forces

$$\sum_n \overline{F} = (\overline{F}_1 + \overline{F}_2 + \ldots + \overline{F}_n) = 0$$

b) Sum of the Torques

$$\sum_n \overline{\tau} = (\overline{\tau}_1 + \overline{\tau}_2 + \ldots + \overline{\tau}_n) = 0$$

13.2 FREE BODY DIAGRAMS

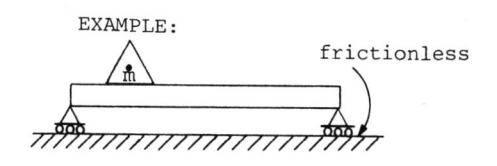

EXAMPLE:

frictionless

Free-body diagram

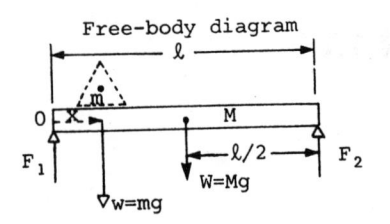

Fig. 13.1

$\Sigma F_x = 0$

$\Sigma F_y = F_1 - w - W + F_2 = 0$

$\Sigma \tau_0 = F_1(0) - w(x) - W(\ell/2) + F_2(\ell) = 0$

Note: In choosing a reference point about which all torques are measured, one may choose any point in the system.

CHAPTER 14

FLUID STATICS

14.1 PRESSURE IN A STATIC FLUID

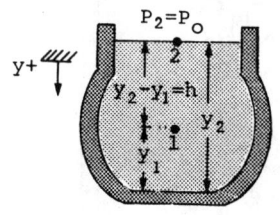

Fig. 14.1

$$\boxed{p = p_0 + \rho gh} \rightarrow \text{units: } \frac{\text{Newtons}}{\text{Meters}^2} = \frac{\text{Force}}{\text{area}}$$

p = Pressure at point 1 at any level

p_0 = Surface pressure, i.e., atmospheric pressure

ρ = Density

g = Gravitational Constant

h = Distance from surface

14.2 ARCHIMEDES' PRINCIPLE (BUOYANCY)

a) Buoyancy Force, F_B

$$F_B = W_{df}$$

F_B = Buoyancy Force
W_{df} = Weight of Displaced Fluid

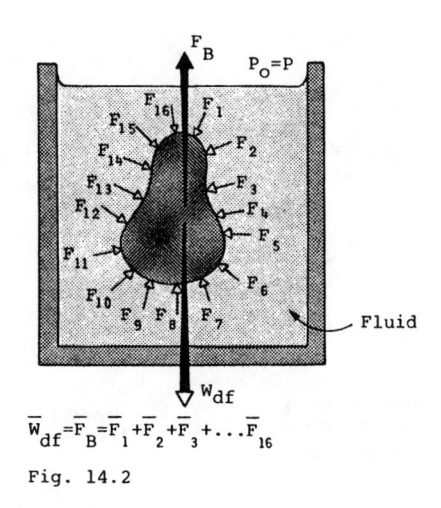

$$\overline{W}_{df} = \overline{F}_B = \overline{F}_1 + \overline{F}_2 + \overline{F}_3 + \ldots \overline{F}_{16}$$

Fig. 14.2

14.3 MEASUREMENT OF PRESSURES

a) Atmospheric Pressure, P_0 (Mercury)

$$\boxed{P_0 = \rho g h}$$

ρ = Density of Mercury

g = Acceleration Due to Gravity

h = Height Change of the Mercury

b) Gauge Pressure

$$\boxed{P - P_0 = \text{Gauge Pressure}}$$

CHAPTER 15

FLUID DYNAMICS

15.1 EQUATION OF CONTINUITY

a) Steady Flow

$$\rho_1 A_1 v_1 = \rho_2 A_2 v_2 = \text{constant}$$

ρ = Density
A = Cross-sectional Area
v = Velocity

b) Steady Incompressible Flow ($\rho_1 = \rho_2$)

$$A_1 v_1 = A_2 v_2 = \text{constant}$$

15.2 BERNOULLI'S EQUATION

a) Steady Flow

$$p_1 + \tfrac{1}{2}\rho v_1^2 + \rho g y_1 = p_2 + \tfrac{1}{2}\rho v_2^2 + \rho g y_2 = \text{constant}$$

where $p + \rho g y$ = static pressure
$\frac{1}{2} \rho v^2$ = dynamic pressure

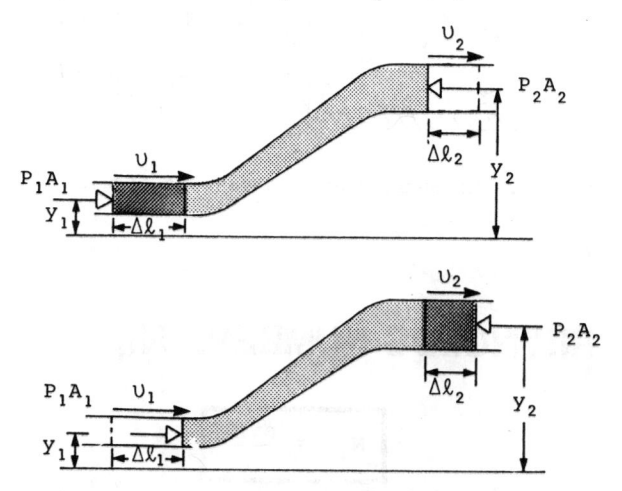

Fig. 15.1 A portion of fluid (shaded area) travels through a section of pipe with a changing cross-sectional area.

15.3 VISCOSITY

$$\boxed{F = \eta \frac{Av}{L}} \rightarrow \text{units: } kg/m \cdot s^2$$

η = Coefficient of Viscosity
F = The Forces that tend to drag the liquid left and right
A = Area of the liquid over which the forces are applied
L = Transverse Dimension
v = Velocity

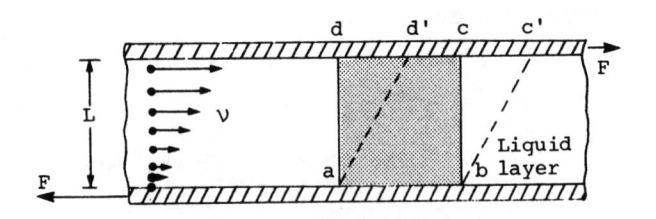

Fig. 15.2 Laminar flow of a viscous liquid

15.4 REYNOLDS NUMBER: N_R

$$N_R = \frac{\rho v D}{\eta}$$

ρ = Density of the fluid
v = Average velocity
D = Diameter of the pipe
η = Coefficient of viscosity

CHAPTER 16

TEMPERATURE

16.1 THE KELVIN, CELSIUS AND FAHRENHEIT SCALES

a) Absolute (Kelvin) Temperature

T_{tr} = 273.16K

T_{tr} = Temperature at which H_2O (water) can coexist as a vapor, a liquid and a solid (triple point)

K = Unit of Temperature Interval

b) Celsius Temperature

$$t_c = (T - 273.15)\ ^0C$$

t_c = Celsius Temperature
T = Kelvin Temperature

c) Fahrenheit Temperature

$$t_F = (9/5)t_c + 32°F$$

t_F = Fahrenheit Temperature

I-70

t_c = Celsius Temperature

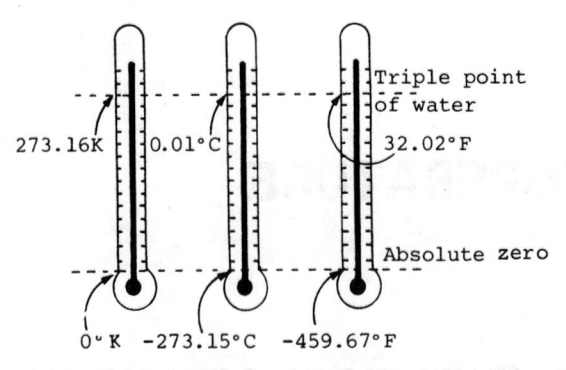

273.16K   0.01°C   Triple point of water / 32.02°F

0°K   -273.15°C   -459.67°F   Absolute zero

Fig. 16.1

16.2 COEFFICIENTS OF THERMAL EXPANSION

a) Coefficient of Linear Expansion

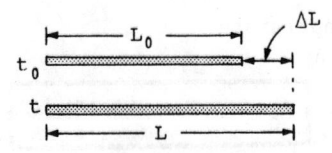

Fig. 16.2

$$\alpha = \frac{\Delta L}{L_0} \frac{1}{\Delta t}$$

Replacing ΔL by $L-L_0$ and solving for L.

$$L = L_0(1 + \alpha \Delta t)$$

α = Coefficient of linear thermal expansion
ΔL = Change in length
L_0 = Length at some reference temperature
Δt = Change in temperature

b) Surface Expansion

$$\boxed{A = A_0(1 + \gamma \Delta t)} \rightarrow \text{units: Meters}^2$$

$$\boxed{\gamma = 2\alpha}$$

γ = Coefficient of surface expansion
A = Area at temperature t
A_0 = Area at t_0
Δt = Change in temperature

c) Volume Expansion

$$\boxed{V = V_0(1 + \beta \Delta t)} \rightarrow \text{units: Meters}^3$$

$$\boxed{\beta = 3\alpha}$$

β = Coefficient of volume expansion
V = Volume at temperature t
V_0 = Volume at t_0
Δt = Change in temperature

CHAPTER 17

HEAT AND THE FIRST LAW OF THERMODYNAMICS

17.1 QUANTITY OF HEAT AND SPECIFIC HEAT

a) Heat Capacity

$$\boxed{C = \frac{Q}{\Delta T}} \text{ units: } \frac{\text{Kilocalories}}{\text{Temperature}} \text{ (Celsius)}$$

C = Heat Capacity
Q = Heat Applied
ΔT = Change in Temperature

b) Specific Heat

$$\boxed{c = \frac{Q}{m \Delta T}} \text{ units: cal/gm} \cdot {}^\circ C$$

c = Specific Heat
Q = Heat Applied
m = Mass
ΔT = Change in Temperature

I-73

17.2 HEAT CONDUCTION

$$H = -kA \frac{dT}{dx}$$

H = Heat Flow Rate
k = Thermal Conductivity
A = Cross-sectional Area
dT = Temperature Difference
dx = Thickness
$\frac{dT}{dx}$ = Thermal Gradient

For a rod of constant cross-sectional area

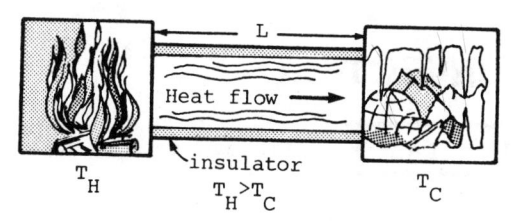

Fig. 17.1

$$H = kA \frac{T_h - T_c}{L}$$

17.3 THERMAL RESISTANCE, R

$$R = \frac{L}{k}$$

a) Heat Conduction

$$\boxed{H = A \frac{T_h - T_c}{R}}$$

L = Length of Rod
H = Heat Flow Rate
A = Cross-sectional Area
R = Thermal Resistance
T_h = Higher Temperature

T_c = Lower Temperature

17.4 CONVECTION

$$\boxed{H = LA\Delta t}$$ units: cal/sec. cm² deg.

L = Convection Coefficient
H = Heat Convection Current
A = Surface Area
Δt = Change in Temperature

17.5 THERMAL RADIATION

$$\boxed{*R = e\sigma T^4}$$

R = Energy radiated per unit area
σ = Stefan's constant, 5.6703×10^{-8} w/m²°K⁴

T = Kelvin temperature
e = Emissivity of the surface
*Also known as Stefan's law.

17.6 HEAT AND WORK

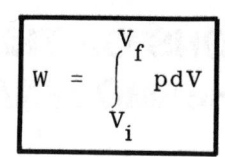

a)

$$W = \int_{V_i}^{V_f} p\, dV$$

W = Work Done
V_i = Initial Volume
V_f = Final Volume
p = Pressure
dV = Differential Change in Volume

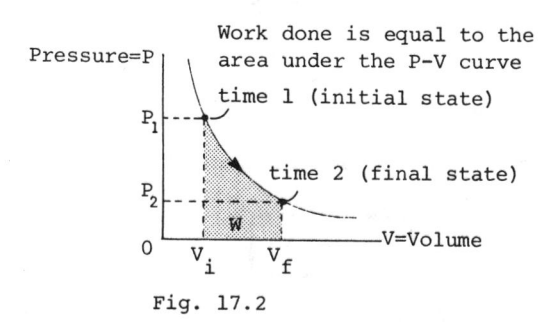

Work done is equal to the area under the P-V curve

Fig. 17.2

17.7 THE FIRST LAW OF THERMODYNAMICS

$$Q = u_f - u_i + W$$

Q = Heat Absorbed
$u_f - u_i$ = Change in Internal Energy
W = Work Done

17.8 APPLICATIONS OF THE FIRST LAW OF THERMODYNAMICS

a) Adiabatic Process

$$U_f - U_i = -W$$

U_f = Final Internal Energy
U_i = Initial Internal Energy
W = Work Done

b) Isobaric (Constant Pressure) Process

$$W = p(V_f - V_i)$$

p = Pressure
V_f = Final Volume
V_i = Initial Volume

c) Isochoric Process (Volume Unchanged)

$$Q = U_f - U_i$$

Q = Heat Absorbed

I-77

CHAPTER 18

KINETIC THEORY OF GASES

18.1 IDEAL GAS EQUATION

a)
$$pV = \mu RT$$

p = Pressure $\left(\dfrac{\text{Newtons}}{\text{Meters}^2}\right)$
V = Volume (Meters^3)
μ = Number of Moles
R = Universal Gas constant
T = Temperature (Kelvin)

For Ideal Gas: $R = 8.314 \text{ J/mol} \cdot {}^0\text{K} = 1.986 \text{ cal/mol} \cdot {}^0\text{K}$

b)
$$\dfrac{p_i V_i}{T_i} = \dfrac{p_f V_f}{T_f} = R = \text{a constant}$$

(For a fixed mass of gas: a closed system)

"i" denotes initial value

"f" denotes final value

18.2 PRESSURE AND MOLECULAR SPEED OF AN IDEAL GAS

a)
$$p = 1/3\, \rho v^2$$

p = Pressure
ρ = Density
v = Speed

b)
$$v_r = \sqrt{\frac{3p}{\rho}}$$

v_r = Root-Mean-Square Speed
p = Pressure
ρ = Density

18.3 KINETIC ENERGY OF AN IDEAL GAS

a)
$$\tfrac{1}{2} M \bar{v}_r^2 = \tfrac{3}{2} RT = \text{Kinetic Energy}$$

M = Molecular Weight
v_r = Root-Mean-Square Speed
R = Ideal Gas Constant
T = Temperature

b)
$$\tfrac{1}{2} m v_r^2 = \tfrac{3}{2} kT$$

m = Molecular Mass
v_r = Root-Mean-Square Speed
k = Boltzmann Constant
T = Temperature
k = 1.381 × 10⁻²³ J/Molecule°K

c) Two Diffused Gases:

$$\frac{v_{r_1}}{v_{r_2}} = \sqrt{\frac{m_2}{m_1}}$$

(For two different gases in a closed system, during a diffusion process)

v_r = Root-Mean-Square Speed

m = Mass

18.4 INTERNAL ENERGY OF AN IDEAL MONATOMIC GAS

$$U = 3/2 \; \mu RT$$

U = Internal Energy
μ = Number of Moles
R = Universal Gas Constant
T = Temperature

18.5 SPECIFIC HEATS OF AN IDEAL GAS

a) Ideal Monatomic Gas

i)
$$C_v = \frac{3}{2} R$$

C_v = Molar Heat Capacity at Constant Volume

R = Universal Gas Constant

ii)
$$\gamma = \frac{C_p}{C_v} \cong \frac{5}{3} \cong 1.67$$

γ = Ratio of Specific Heats
C_p = Molar Heat Capacity at Constant Pressure
C_v = Molar Heat Capacity at Constant Volume

b) Ideal Diatomic Gas

$$\gamma = \frac{C_p}{C_v} \cong \frac{7}{5} \cong 1.40$$

c) Ideal Polyatomic Gas

$$\gamma = \frac{C_p}{C_v} \cong \frac{4}{3} \cong 1.33$$

d) Universal Gas Constant, R

$$R = C_p - C_v$$

18.6 MEAN FREE PATH

$$\ell = \frac{1}{\sqrt{2}\,\pi\, n_v\, d^2}$$

ℓ = Mean Free Path
n_v = Number of Molecules per Unit Volume
d = Molecular Diameter

I-81

CHAPTER 19

ENTROPY AND THE SECOND LAW OF THERMODYNAMICS

19.1 EFFICIENCY OF A HEAT ENGINE

a)
$$e = \frac{W}{Q_h} = 1 - \frac{Q_c}{Q_h}$$

e = Efficiency
W = Work Done by Engine
Q_c = Heat (Lower)
Q_h = Heat (Higher)

b)
$$e = 1 - \frac{T_l}{T_h}$$

e = Efficiency
T_l = Temperature (Lower)
T_h = Temperature (Higher)

19.2 ENTROPY, S

a) For a Closed Cycle,

$$\oint dS = 0$$

dS = Rate of Entropy Change

b) Change Between Two States,

$$\int_a^b dS = \int_a^b \frac{dQ}{T}$$ Units: Joules/Kelvin
i.e., J/K

dS = Rate of Entropy Change
dQ = Rate of Heat Change
T = Temperature

c) For Free Expansion,

$$S_f - S_i = \int_i^f \frac{dQ}{T} = \mu R \ln\left(\frac{V_f}{V_i}\right)$$

S_f = Final Entropy

S_i = Initial Entropy

dQ = Rate of Heat Change

T = Temperature

μ = Number of Moles

R = Ideal Gas Constant

V_f = Final Volume

V_i = Initial Volume

I-83

Handbook of Mechanical Engineering

SECTION J

Index

INDEX

Absolute acceleration, A-164
Absolute pressure, C-16
Absolute velocity, A-161
Absorptivity, E-110
Acceleration, I-10
   absolute, A-164
   angular, I-38
   centripetal, A-94
   relative, A-164
   tangential, A-94
Accuracy, A-2
   instantaneous, A-85
Activity, D-81
   coefficient, D-81
Actual cycle, D-65
Adiabatic:
   bulk modulus, D-60
   compressibility, D-59—D-50
   flame temperature, D-85
   flow, C-52
   process, D-2, I-77
   saturation process, D-79
Air-standard power cycle, D-67
Aluminum, column design, B-144
Amagat-Leduc rule of partial volume, D-76
Amplitude, A-222,B-172
Analog computer, G-163—G-174
Analysis:
   dimensional, C-67—C-69
   heat conduction, E-22
   of a frame, A-48
   of a lag network, G-149
   of a lag-lead network compensator, G-153
   of a lead network, G-151
Angle:
   of friction, A-57
   of rotation, B-103, B-106
   of twist, B-178, B-184
Angular:
   acceleration, I-37
   curves, G-128
   kinematics, I-39
   momentum, A-107, A-147, A-149, A-174, A-188, I-41, I-59
   momentum of orbiting planets, I-59
   of departure, G-76
   of a shock wave, I-56
   velocity, A-99
Aphelion, A-111
Applied frequency, F-57
Apogee, A-111
Apside, A-116
   angle, A-116

Area:
   composite, B-49
   composite body, A-38, A-69
Archimedes' principle, I-65
Armature control, G-11
Asymptote, G-96
   of root loci, G-73
Atmosphere, C-15
Availability, D-40
Availability functions, D-38—D-40
Axial force, B-1—B-11
   beams, B-4, B-7
   diagram, B-4, B-7
   stress, B-216
Axle friction, A-60
Axis:
   motion about, A-199
   of a wrench, A-29
   principle, A-72, A-192

Balanced:
   dynamically, A-195
   statically, A-192
Basis, H-6
Bayes theorem. H-147
Beams, F-63
   and cables, A-49
   axial force, B-4, B-7
   column, B-169
   deflection of, B-91—B-108
   design criteria, B-73
   design procedures, B-76
   design of, B-73—B-89
   diagrams and formulas, B-78
   distributed load, A-43
   economy in design, B-77
   plastic analysis of, B-77
   reinforced concrete, B-151
   shear center, B-69
   shear stresses, B-68
   statically indeterminate, B-110—B-127
   stresses in, B-64—B-69
   supported, B-7
Bearings:
   failure, B-187
   thrust, A-61
Beat:
   equation, I-55
   frequency, I-55
Beattie—Bridgeman equation, D-57
Behavior of Laplace transforms, F-6
Belt friction, A-63
Bending:
   and torsion, B-223
   energy, B-207
   stress, B-216
Bending moment, B-1—B-11,
   beam, B-7
   diagram, B-4

J-1

Bending stress distribution, B-64
Benedict-Webb-Rubin equation, D-57
Bernoulli equation of fluid motion, C-39, I-67
Bessel function, F-27, H-85
   inequality, H-77
   modified, F-28
Beta function, F-36
Binomial distribution, H-151
Bisector method, H-126
Black surface, E-111, E-124
Blade passages, D-93—D-104
   flow, D-101
Blasius solution, E-52
Block diagram, G-51—G-55
Blower, C-135
Bode plot, G-89—G-105
Body:
   rigid, A-19, A-30
   composite, A-38, A-69
Boiling, E-127—E-133
   phenomenon, E-127
   pool, E-127
Bolted joints, B-185—B-195
   eccentrically loaded, B-218
Boltzmann's constant, D-43
Boundary conditions, E-34—E-35, F-100
Boundary layer:
   equations, E-84
   theory, C-89—C-99
   thickness, C-89—C-90, E-88
Boundary value problem, F-98—F-103, H-72
Brayton cycle, D-72
   with regeneration, D-73
Brittle, B-28
   fracture conditions, B-162
Bulk temperature, E-73
Buoyancy, C-20, I-65

Cables, A-49
   analytical relationship, A-51
   concentrated load, A-52
   distributed load, A-53
   parabolic, A-55
Calculus, H-18—H-38
   of one variable, H-18
Cantilever, deflection, B-106, B-108
Carnot cycle, D-31, D-67
Cartesian coordinates, E-5
Cascade compensated systems, G-146—G-153
Castigliano's theorem, B-202
Cauchy problem, F-101
Cauchy-Schwarz inequality, H-8
Cavitation, C-133, C-137
Celsius scale, I-70

Center:
   of gravity, A-36, A-232, B-35—B-55
   of mass, A-170, I-31, I-32
   of mass, motion, I-32
   of oscillation, A-232
   of rotation, instantaneous, A-162
   of suspension, A-232
Centrifugal pumps, C-137
Centripetal:
   acceleration, A-94
   force, I-20
Centroid, A-36, A-37, B-35—B-55
   by integration, A-40
   of integration, B-39
Centroidal motion, A-87
Chain rule, H-27
Changes of scale property, F-8
   Laplace transform, F-8
   inverse Laplace transform, F-20
Channel, flow, C-116—C-124
Characteristic:
   equation, G-53
   function, G-33
Chemical equilibrium, D-89—D-92,
   chemical potential, D-89
   requirements, D-89
Chemical potential, D-87
Chemical reaction, D-82—D-87
Cholesky decomposition, H-135
Circle, Mohr's A-74
Circular:
   orbits, A-115
   permutations, A-9
   pipes, E-59
   shaft, B-178, B-180
Circulation, C-102, C-112
Clapeyron equation, D-54
Clausius:
   inequality, D-34
   statement, D-30
Closed system, D-1
Coefficient:
   activity, D-81
   convective heat transfer, E-2
   definitions, C-139
   diffuser, D-100
   Joule-Thompson, D-25
   nozzle, D-100
   of contractions, C-139
   of discharge, C-140, D-100
   of linear expansion, D-59
   of restitution, A-135
   of thermal expansion, I-71
   of velocity, C-140
   skin friction, E-53
   velocity, D-101
Colburn's analogy, E-69
Collinearity, H-2

Collision, I-35
   elastic, I-35
Column, B-132—B-145
   aluminum, design, B-144
   beam, design of, B-169
   definition, B-132
   design, B-143
   eccentricity loaded, B-145
   long, Euler's formula, B-136
   short, with eccentric load, B-134
   steel, design, B-143
   structural steel formula, B-141
   timber, design, B-144
Combined stresses, B-216—B-223
   axial and bending, B-216
Combustion process, D-82
Compensating network, G-55, G-136—G-145
Compensators:
   comparison of, G-145
   feedback, G-155—G-160
   lag, G-139
   lead, G-141
   lead-lag, G-142, G-153
   parallel, G-155—G-160
   PI control, G-137
   proportional derivative, G-140
   selecting, G-146
Complex:
   function, H-40—H-64
   integral, H-51
   integration theorem, G-32
   plane G-122, H-42
   variable, H-43
   vector, H-16
Composite:
   areas, A-69, B-49
   bodies, A-38, A-69, B-37
   member, in parallel, B-147
   member, in series, B-148
   structure, B-147—B-151
Compressibility, D-43
   adiabatic, D-59
   isothermic, D-59
Compressible flow, C-11, C-41—C-45
   one-dimensional, C-47—C-63
   internal, C-143
Compression:
   combined shear, B-219
   ratio, D-70
   shear, B-220
   shear stress, B-219
Computer, analog, G-163—G-174
Concrete, reinforced, beams B-151
Condensation, E-127—E-133
   film, E-133
Condenser, D-66

Conduction E-1, I-74,
   heat, steady-state, one-dimensional, E-5—E-19
   heat, steady-state, two-dimensional, E-21—E-28
   heat, unsteady-state, one-dimensional, E-32
Conformal mapping, H-60
Connections, A-31
Conservation:
   of angular momentum, A-107, A-130, A-186, I-45
   of energy, A-127, A-152, A-184, A-236
   of linear momentum, I-34
   of mass, D-23
      control volume, D-93
   of mechanical energy, I-27
   of momentum A-151, I-26—I-30
Conservative force, A-122, I-26
Constant-pressure specific heat, D-22
Constant-volume specific heat, D-22
Constrained motion, A-178
Continuity, H-45
   equation, C-33, E-47, I-67
Continuous function, piecewise, F-4
Continuum, fluid, C-5
Contraction, C-83
Control, H-120
   armature, G-11
   field, G-12
   ratio, G-53
   system, G-174
Control volume, C-1, C-23—C-26, D-23, D-24
   conservation of mass, D-93
   momentum equation, D-93
Convection, E-2, E-34, E-35, I-75
   forced, E-73—E-83
   free, E-84—E-94
   mixed, E-94
Convective heat transfer coefficient, E-2
Converging:
   nozzle, D-97
   /diverging nozzle, D-97—D-99
Convolution:
   property, F-22—F-23
   theorem, F-89, H-83
Corner frequency, G-96
Correlation, forced convection, E-73—E-83
Corresponding states, law of, D-59
Cosine:
   direction, H-1
   Fourier formula, F-88
   integral, F-31
Coulomb's law, A-113, I-84
Couple:
   addition, A-24
   moment of, A-24
   resolution of forces, A-25

Cover plate, B-186
Cramer's rule, H-15, H-133
Critical:
    damping coefficient, A-240
    depth, C-120
    point, D-8
    point of a material, B-29
Cross-flow heat exchanger, E-98
Cross-section, fin E-16—E-17
Curl, H-28
Curvilinear motion, A-85
Cutoff ratio, D-70
Cycle, D-20
    Brayton, D-72—D-73
    Carnot, D-31, D-67
    Diesel, D-69
    Dual, D-70
    Ericsson, D-71
    gas refrigeration, D-74
    ideal, D-65
    Otto, D-68
    Rankine, D-61
    regenerative, D-64
    Stirling, D-71
    reheat, D-63
    vapor power, D-61
Cylinder:
    concentric, E-121
    flow around, C-112
    flow inside, E-76, E-78—E-79
    heat conduction, E-25
    horizontal, E-91
    temperature, multilayer, E-10
    temperature, single, E-9
    temperature, specified, E-12
Cylindrical:
    coordinates, A-97, E-6

D'Alembert's principle, A-148, A-177
Dalton's rule of partial pressure, D-75
Damped:
    forced vibration, A-241
    free vibration, A-239
    motion, A-249
    vibration, A-222
Dead load, B-163
Dead time, G-105
Decade, G-93
Decibel, G-92
Deflection:
    and elastic curve, B-9
    and slope, B-101
    determination of, B-94
    formula, A-53
    of axially loaded members, B-25
    of beams, B-91—B-108
    of free end of cantilever, B-106, B-108
De Moivre's formula, H-42

Density, D-5
    field, C-5
Derivative, H-45
    inverse Laplace transform, F-20
    Laplace transform, F-10
    of a vector, A-9, A-10
    of Laplace transform, F-12
Design:
    column, B-143—B-144
    criteria, B-73
    economy, B-77
    failure criteria, B-158—B-171
    for brittle fracture conditions, B-162
    for dynamic strength, B-163
    for elastic strength, B-159
    for plastic strength, B-162
    for static strength, B-159
    for torsional strength, B-167
    of beam columns, B-169
    of beams, B-73—B-89
    procedure, B-76
Determinant, H-15
Dew point temperature, D-78
Diagonal matrix, H-9
Diesel cycle, D-69
Dietrici equation, D-57
Difference equation, F-91—F-96
Differential analysis of fluid motion, C-33—C-39
Differential equation, H-64—H-72
    constant coefficients, F-42
    Laplace transform, F-68, F-70, G-32, G-41
    ordinary linear, F-42—F-63, H-66, H-138
    systems of, F-48, H-69
    systems response, G-16—G-28
    variable coefficients, F-46
Diffuser coefficient, D-100
Diffusivity, E-2
Dimensional analysis, C-67—C-69
Dimensions, D-6
Dirac delta function, F-35
Direct method, F-66
Direct polar plot, G-99, G-118, G-131
Dirichlet problem, F-100, H-62
    conditions, F-83
Disk friction, A-61
Displacement, I-10
    finite, A-80
    particle, I-55
Distance, A-15
Distributed loading, B-3
Distribution:
    binomial, H-151
    normal, H-152
    Poisson, H-151
Divergence, H-27

Doppler effect, I-55
Double integral, H-28
Double pipe heat exchanger, E-97
Doublet, C-110
Drag force, C-96
   coefficient, C-99
Drift, A-203
Dry friction, A-56, A-239
Dual cycle, D-70
Duct:
   constant area, C-52, C-55
   flow in, C-76
   non-circular, C-88
Ductile, B-28
Dynamic(s), A-188—A-210, I-17—I-20
   balance, A-192
   effects, A-202
   fluid, I-67—I-69
   load, B-163
   programming, H-122
   rotational, I-40—I-45
   similarity, C-68
   strength, B-163
   systems of particles, I-31—I-35
   torsional strength, B-169
   viscosity, D-3

Earth, rotational effects, A-200
Eccentric impact, A-186
Eccentric load:
   axial, B-217
   bolt joints, B-218
   column, B-145
   short column, B-134
   stresses, B-217
Eccentricity, A-109
Eckert number, E-48
Economy, in beam design, B-77
Edge distance, B-186
Effectiveness, E-103
Efficiency, A-121, A-185
   blade, D-103
   fin, E-19
   heat engine, D-27, I-82
   mechanical, A-80, A-121
   nozzle, D-100
   of a bolted joint, B-187
   thermal, D-68
Eigenfunctions, H-72
Eigenvalues, H-72, H-137
Elastic:
   collision, I-35
   curve and deflection, B-91
   design, B-159
   limit, B-30
   strain energy, B-23
   strength, B-163, B-167
Elasticity, B-30

Elementary functions, Laplace transform, F-3
Ellipsoid:
   momental, A-193
   of inertia, A-74
Emissivity, E-2, E-112
Endothermic reaction, D-83
Energy, A-118, A-145, D-19—D-26, I-21—I-25
   and momentum, A-118, A-181—A-186
   bending, B-207
   conservation of, A-127, A-152, A-184, A-236, I-26—I-30
   content, H-84
   elastic strain, B-23
   equation, E-47
   kinetic, A-151, A-183, A-196, I-25, I-42, I-61
   methods, B-197—B-214
   of a system, D-20
   potential, A-82, A-123, A-243, I-60
   shear-strain, B-214
   specific, C-119
   torsion, B-211
   total, I-61
   work principle, A-120, A-181
Enthalpy, C-42, D-22, D-56
   chart, real gases, D-50
   gas-vapor mixture, D-79
   of formation, D-83
   of reaction, D-83
Entropy, C-42, D-27—D-37, D-58, I-82
   absolute, D-85
   change during irreversible process, D-35
   chart, real gases, D-52
   gas-vapor mixture, D-49
   principle of increase, D-34
Entry length, E-60
Equation:
   Beattie—Bridgeman, D-57
   Benedict-Webb-Rubin, D-57
   Bernoulli's, C-39, I-67
   Clapeyron, D-54
   continuity, C-33, E-47, I-67
   Dietrici, D-57
   difference, F-91—F-96
   energy, E-47
   Euler's, C-38, E-48
   fluid statics, C-13
   Hamilton's, A-219
   heat conduction in one-dimension, E-5
   ideal gas, I-78
   integral momentum, C-91
   isentropic flow, C-47
   Lagrange's, A-214, A-220
   Laguerre, F-46
   momentum, C-35, E-47
   Navier-Stokes, E-47

J-5

of a beam, B-92
of motion, A-173, A-198
of state, C-41, D-56, D-59
systems, H-1—H-16
van der Waal's, D-56
Equilibrium:
  between two phases, D-91
  chemical, D-89—D-92
  constant, D-90
  dynamic, A-101
  multicomponent system, D-91
  mutual, D-3
  of a particle, A-18
  rigid body, A-30, A-33—A-35, I-62
  solid body, B-1
  static, A-170
  thermodynamic, D-3
Equipollent, A-26
Equivalent static load, B-163
Erf (x), F-96
Ericsson cycle, D-71
Error:
  evaluation, G-60, H-125—H-127
  function, F-28—F-31
  series, G-61
Escape velocity, A-111
Euler:
  constant, H-87
  equation of fluid motion, C-38
  equation of motion, A-198, E-48, H-116
  formula for long columns, B-136
  method, H-138—H-139
Even:
  extension, Fourier transform, F-80
  function, F-77
Existence, H-66
  of Laplace transform, F-6
Exothermic reaction, D-83
Expansion, C-83
  volume, D-59
Explicit scheme, H-140
Exponential:
  integral, F-31
  order function, F-5
Extensive property, D-2
External flow, C-71

Fahrenheit scale, I-70
Failure:
  criteria in design, B-158—B-171
  types in joints, B-186
Fanno line, C-54
Fatigue, strength, B-171
Feedback:
  compensation, G-155—G-160
  control, G-118—G-133
  system, G-59
Feedwater heater, D-64

Field:
  control, G-12
  density, C-5
  inverse-square repulsive, A-113
  velocity, C-5
Film:
  boiling, E-131
  condensation, E-133
  temperature, E-70
Fin, E-15
  efficiency, E-19
Final value theorem, F-65
Finite element method, H-144
First law analysis of reacting systems, D-84
First law of thermodynamics, D-20—D-26, I-73—I-77
  change of state, D-21
  cycle, D-20
  control volume, D-24
First shift property, F-8
  inverse Laplace transform, F-18
  Laplace transform, F-8
First translation property, F-18
Fitting, C-85
Fixed supports, B-3
Flexure formula, B-65
Flow:
  adiabatic, C-52
  around immersed bodies, C-96
  blade passage, D-93—D-104
  compressible, C-11, C-41—C-45
    one-dimensional, C-47—C-63
  cylinder, E-78—E-79
  cylindrical pipes, E-76
  external, C-71
  gradient, C-93
  graph algebra, G-56
  ideal, C-102—C-112
  incompressible, C-11, C-71—C-86, C-104
  internal, C-71
  irrotational, C-11, C-105
  isentropic, C-47—C-51, D-95
  laminar, C-11, C-71—C-72, C-74, E-46, E-74, E-76
  measurement, C-139—C-143
  meter, C-140, C-143
  non-uniform, C-122
  nozzle, C-140, D-93—D-104
  one-dimensional, C-6
  open channel, C-116—C-124
  over a flat plate, C-92, E-73
  pipe, C-86
  regime, E-95
  separation, C-95
  simple, C-107, C-112
  steady uniform, C-116
  tube, E-80

turbulent, C-11, C-71, E-46, E-74, E-77
two-dimensional, C-105
viscous, E-45
Fluid:
   confined, E-92
   definition, C-1
   dynamics, I-67—I-69
   ideal flow, C-102—C-112
   mechanics, C-1, E-45—E-48
   motion, C-11, C-22, I-64—I-65
   Newtonian viscosity, C-9
   pressure, I-64
   properties, C-5
   statics, C-13—C-22, I-64—I-65
Force:
   axial, B-1—B-11
   central, A-106
   central fields, A-106
   centrifugal, A-188
   centripetal, I-20
   concurrent, A-27
   conservative, A-122, I-26
   coplanar, A-27
   definition, A-14
   hydrostatic, C-17
   impulsive, A-217
   internal, A-44
   moment about a particle, A-20
   non-conservative, I-29
   on a particle, A-13, A-19
   reduction of, A-26
   resolution of, A-25
   restoring, F-52
   shear, B-1—B-11
   systems, A-147
   total resultant, A-28
   transverse, A-188
   vector, B-1
   velocity dependent, A-102
   work of, A-78, A-118, I-21—I-23
Forced:
   boiling, E-127
   convection, E-73—E-83
   vibration, A-222, A-236
      damped, A-241
Forcing frequency, A-237
Forward transfer function, G-53
Fouling factor, E-100
Fourier formula:
   cosine, F-88
   sine, F-88
Fourier integral transform, F-86
Fourier series, F-77, H-76—H-79
   cosine, F-79
   sine, F-79
Fourier transform, F-77—F-89, H-79
   analysis, H-74—H-85
   integral, F-86, G-89

Frame, A-47
   analysis of A-48
Fredholm integral equation, F-91
Free:
   body diagram, A-15, A-30
   convection, E-84—E-94, E-129
      confined fluid, E-92
      correlation, E-90
      laminar, E-85
   falling bodies, I-7—I-9
   vibration, A-222, A-234
      damped, A-239
Frequency, A-222, A-253, B-171, E-109
   applied, F-57
   domain, analysis, G-89—G-105
   domain, feedback control, G-118—G-133
   forcing, A-237
   natural, F-57
   resonant, G-120
   response, G-38, G-78, G-85, G-89—G-90
      plots, G-146—G-153
Friction, A-56, I-19
   angle of, A-57
   axle, A-60
   belt, A-63
   disk, A-61
   dry, A-56, A-239
   factor, C-80
   in duct, C-52
   kinetic, A-57—A-58
   static, A-56—A-57
Frobenius method, H-86
Fugacity, D-80
Function, F-25—F-38
   analytic, H-62
   approximation, H-127
   compensating network, G-136
   complex, H-40—H-64
   continuous, F-4
   Dirac delta, F-35
   elementary, Laplace, F-3
   exponential order, F-4—F-5
   harmonic, H-46, H-60
   hyperbolic, H-49
   null, F-16
   of two variables, F-98
   parabolic, G-19
   periodic, F-13
   piecewise smooth, F-81
   ramp, G-19
   rational, F-72
   sequence, H-56
   series, H-56
   square-wave, F-32
Fusion curve, D-9

J-7

Gauge, B-186
    pressure, C-16
Gain:
    adjustment, G-128
    curve, G-97
    margin, G-113
Gamma function, F-25, F-26, H-85
Gas, D-42—D-53
    heat exchange, E-124—E-125
    ideal, I-78
    kinetic theory of, I-78—I-81
    mixtures, D-77
    perfect (ideal), D-45
    radiation, E-123
    real, D-49—D-52
    refrigeration cycle, D-74
    semi-perfect, D-44
    universal constant, D-42
Gauss:
    elimination, H-133
General harmonic oscillator, A-224
Generating function, F-29
Geometric:
    sections, properties of, B-55—B-63
    similarity, C-68
Gershgoin theorem, H-137
Gibb's:
    function, D-87
    phase rule, D-92
    phenomenon, H-77
Gordan-Rankine formula, B-141
Gradient, H-26
    method, H-115
Graetz number, E-96
Grant's rule, G-78
Grashof number, E-49, E-86
Gravitation, I-57—I-61
Gravitational:
    field, intensity, A-125
    potential, A-125
Gravity:
    center of, A-36
    force, A-82
Gray surface, E-115, E-125
Green's theorem, H-34
Gyration, radius of A-67, B-46
Gyroscope, A-204

Hamilton:
    equation, A-218
    variation principle, A-218
Hankel function, H-87
Harmonic:
    function, H-46, H-60
    motion, I-46—I-51
        simple, A-223
    oscillator, A-223
Hazony and Riley rule, G-35
Head loss, C-80

Heat, D-14—D-18, I-73—I-77
    capacity, I-73
    conduction, E-5—E-19, I-74
        one-dimensional, E-5—E-19
        two-dimensional, E-21—E-28
        unsteady-state, E-30—E-42
    duct, C-55
    engine, D-27, I-82
    equation, H-100
    exchanger, E-97—E-106
        parallel flow, E-105—E106
        counterflow, E-106
    flow lines, E-27
    generation, E-13—E-14
    of fusion, D-10
    of vaporization, D-10
    transfer, E-1—E-4
        coefficient, E-125
        correlation, E-129
        high speed flow, E-83
        liquid-metal, E-82
    units, D-18
Heaviside:
    expansion formula, F-76
    function, F-31
Helmholt wave equation, H-109
Hermitian matrix, H-16
Hodograph, A-87
Holonomic, A-212
    constraint, A-220
Homogeneous equation, H-65
Hooke's law of isotropic materials, B-21
H-S diagram, D-11
Hydraulic:
    efficiency, C-127
    jump, C-121
    transmission, G-9
Hydrostatic:
    forces, C-17
    loading, B-4
Hyperbolic function, H-49
Hysteresis curve, G-169

Ideal cycle, D-65
Ideal fluid flow, C-102, C-112
Ideal gas, D-45, I-78—I-81
    diagrams, D-47
    equation, I-78
    isentropic flow, C-50, D-96
    kinetic energy, I-79
    properties, D-45
    speed, I-79
Immersed body, flow, C-96
Impact, A-134
    central, A-134
    eccentric, A-134, A-186
    factor, B-163
    load, B-163
    phenomena, B-200
    solution methods, A-139

Implicit scheme, H-141
Impulse:
   and momentum, A-132, A-152, A-185, A-195
   function, G-41—G-42
   motion, A-133
   stage, D-104
   turbines, D-103
Impulsive force, A-133, A-217
Incompressible flow, C-11, C-71—C-86, C-104
   internal, C-140
Index, Bessel function, F-27
Inelastic collision, I-35
Inequality of Clausius, D-34
Inertia:
   and principle axes, A-72, B-52
   ellipsoid of, A-74
   forces, A-165
   moment of, A-64, A-183, B-35—B-55
   polar moment, A-66, B-178
   product of, A-70, B-50
   rotational, I-42
Initial value theorem, F-65, G-32
   problem, F-42
Instantaneous:
   acceleration, A-85
   velocity, A-85
Integral:
   complex function, H-51
   cosine, F-31
   double, H-28
   equation, F-91—F-96
      convolution type, F-92
   exponential, F-31
   improper, H-22
   line, H-32
   sine, F-31
   solution, E-88
   transform, F-1, H-74—H-85
      inverse Laplace, F-21
      Laplace, F-1
   volume, H-38
Integration:
   centroid, A-40
   method, beam, B-111
   method, beam deflection, B-101
   method, moment of inertia, A-64
   numerical, H-130
Intensity:
   of radiation, E-115
   of sound, I-54
Intensive properties, D-2
Internal:
   energy, C-41, D-19, D-56
   flow, C-71
   force, A-44
   resisting moment, B-6
Interpolation, H-127

Inverse:
   Laplace transform, F-15—F-23, F-65—F-76, G-33
   convolution, F-22
   definition, F-15
   properties, F-18
   polar plane, G-128
   polar plots, G-101, G-111
   square repulse field, A-113, A-131
   value theorem, F-65
Inversion theorem, Fourier transform, F-87
Irradiation, E-119
Irreversible process, D-2
   entropy change, D-35
Irrotational flow, C-11, C-105
Isentropic:
   compression ratio, D-68
   flow, C-47—C-51
      converging nozzle, C-51
   equation, C-47
   ideal gas, C-50, D-96
   properties, C-49
   pressure ratio, D-68
   stagnation, C-45, D-94
Isobaric process, D-2
Isolated system, D-1
Isometric process, D-2
Isothermal:
   compressibility, D-59—D-60
   process, D-2
Isotropic, materials Hook's law, B-21

Jacobian, H-31
Joint, B-185—B-195
   bolted, B-185, B-218
   method of, A-45
   riveted, B-187
   welded, B-190
Jordan curve theorem, H-52
Joule-Thompson coefficient, D-24

Kelvin-Planck statement, D-30
Kelvin scale, I-70
Kepler:
   laws, A-112
   laws of planetary motion, I-57
Kernel, F-1
Kinematics, A-85—A-99
   of rigid bodies, A-157—A-169
   rotational, I-37—I-39
   similarity, C-68
   viscosity, E-3
Kinetic energy, A-120, A-151, A-196, D-19, I-61
   of an ideal gas, I-79
   of rotation, I-42
   plane motion, A-183
Kinetics, A-100—A-115
Kinetic theory of gases, I-78—I-81

Kirchoff's:
  identity, E-112
  laws, F-62

Lag:
  compensator, G-139
  lead network, G-153
  network, G-149
Lagrangian:
  equation, A-214, A-220, A-245, G-14
  function, A-245
  mechanics, A-212—A-220
  multiplication, H-114
Laguerre equation, F-46
Laminar flow, C-11, C-71, E-46
  flat plate, E-74
  infinite parallel plate, C-72
  pipe, C-74, E-74
Laminar free convection, E-85
Laplace:
  equation, E-28, H-96
  inverse transform, F-15—F-23
  transform, F-1—F-23, F-65—F-76, F-91—F-96, G-29—G-41
Laplacian, H-28
Latent heat:
  of fusion, D-10
  of vaporization, D-10
Lead, A-60
  compensator, G-141
  lag compensator, G-142
  network, G-142
Least squares method, H-128
Legendre:
  equation, H-89
  polynomials, H-89
Length, characteristic, E-91
Lerch's theorem, F-16
L'Hopital's rule, H-19
Lift force, C-96
Limits, H-45
Line:
  composite, A-39
  integral, H-32
Linear:
  algebra, H-132
  dependence, H-68
  independence, H-68
  momentum, A-147, I-33—I-34
    conservation of, I-34
  programming, H-123
Linearity property, F-7
  inverse Laplace transform, F-18
  Laplace transform, F-7
Link support, B-2
Liquid-metal heat transfer, E-82
Live load, B-163

Load:
  beams, static conditions, B-78
  cables, A-49, A-52—A-53
  concentrated moving, B-89
  dead, B-163
  deformation relationships, B-197
  distributed on beam, A-43
  dynamic, B-163
  equivalent static, B-163
  function, A-28—A-29
  impact, B-163
  live, B-163
  special conditions, A-45
  terms, B-124
Loaded member, stress, B-15—B-16
Loading:
  distributed, B-3
  hydrostatic, B-4
Logarithmic plots, G-92
Log-magnitude:
  curve, G-97, G-160
  diagram, G-94
Log-mean temperature difference, E-100
Long column, Euler's formula, B-136
Longitudinal strain, B-20
Losses:
  piping, D-65
  pump, D-66
  turbine, D-66
Lumped system, E-30

Mach:
  cone, C-44
  number, D-43—D-94
Machine, A-48
  perpetual motion, D-30
MacLaurin series, H-56
Magnification factor, A-238, A-292
Magnitude:
  curve, G-128
  maximum, G-131
  spectrum, H-84
Manometer, C-16
Mapping:
  conformal, H-60
  elementary, H-58
  Riemann, H-63
Marching scheme, H-140
Mason's rule, G-56
Mass, I-18
  center angular momentum, A-149
  center, motion of A-148
  control volume, D-93
  fraction, D-75
  variable, system A-155
Material:
  properties, B-28
  strength, B-32
  ultimate strength, B-29

Matrix, H-1—H-6
  algebra, G-44
  complex, H-16
  inversion, H-136—H-138
  transformation, A-11
Maxima, theory, H-111
Maxwell relations, D-54
Mean, H-150
  free path, I-81
  square error, F-84
  value theorem, H-19
Measurement:
  flow, C-139—C-143
  of pressure, I-65
Mechanical:
  efficiency, A-80, A-121
  translation system, G-3
  vibration, A-222—A-252
Mechanics, A-1
  fluid, C-1
  Lagrangian, A-212—A-220
  space, A-108, A-129
Melting curve, D-9
Method,
  analytic, E-21
  energy, B-197—B-214
  graphical, E-27
  numerical, E-28
  of virtual work, A-78
Minima, theory, H-111
Mixed convection, E-94
Mixture, D-75—D-81
  enthalpy, D-79
  involving gases and vapor, D-77
  reactive gases, D-90
Mobius transformation, H-58
Modulus of resilience, B-163
Mohr's circle, A-74
  for stress, B-221
Moisture, D-12
Mole fraction, D-75
Moment:
  about a given axis, A-22
  area method, A-92, B-101, B-118
  bending, A-50, B-1—B-11
  diagrams, B-11
  first, A-37
  integration method, A-64
  of a couple, A-24
  of force about a particle, A-20
  of inertia, A-64, B-35—B-55
  polar, of inertia, A-66, B-45, B-178
  shear, A-50
  vector, B-1
Momental ellipsoid, A-193
Momentum:
  and energy, A-118, A-181—A-186
  and impulse, A-132, A-152, A-185, A-195

angular, A-105, A-147, A-174, A-186, A-188, I-41
angular, about mass center, A-149
conservation of, A-151
equation, D-93, E-47
generalized, A-217
linear, A-100, A-147
Monatomic gas, I-80
Moody diagram, C-81
Motion, I-7—I-9
  about a fixed axis, A-199
  about a fixed point, A-169, A-199
  absolute, A-90
  centroidal, A-87
  charged particle, A-143, A-145
  circular, I-39
  constrained, A-178
  curvilinear, A-85, A-90
  dependent, A-90
  equations of, A-101, A-173, A-198
  fluid, C-11
  impulse, A-133
  of a gyroscope, A-204
  of a particle, A-87
  of mass center, A-148
  of planets, I-57
  plane, I-10—I-16
  projectile, A-140, I-13
  rectilinear, A-85, A-89
  relative, A-89
  rigid body, A-197, C-22
  simple harmonic, A-223, I-46—I-51
  torque-free, A-210
  uniform, A-89, I-16
Multicomponent system, D-91
Multiforce member, A-47
Multiphase system, D-91
Mutual inductance, I-123

Natural frequency, F-57
Navier-Stokes equation, E-47
Network:
  lag, G-149
  lead, G-151
Neumann function, H-86
Newton:
  fluid viscosity, C-9
  law of cooling, E-2
  law of gravitation, A-1, I-57
  law of viscosity, E-3
  laws, A-1, A-13, A-100, C-106
  Raphson method, H-126
  three fundamental laws of motion, A-1, I-17—I-18
Nichol's chart, G-133
Node method, G-2
Non-circular ducts, C-88
Non-inertial reference system, A-165
Non-singular matrix, H-14

J-11

Non-uniform flow, C-122
Norm, H-2
Normal:
    distribution, H-152
    stress, B-13, B-15
Nozzle:
    coefficient, D-100
    converging-diverging, C-62, D-97-D-99
    efficiency, D-100
    flow, D-93,D-104
    isentropic flow, C-51
Nucleate boiling, E-129
Null:
    function, F-16
    vector, A-5
Number:
    sequences, H-55
    series, H-55
Numerical methods, H-125—H-139
    integration, H-130
    linear algebra, H-132
Nusselt number, E-49, E-73
Nyquist stability, G-107—G-117

Octave, G-93
Odd:
    extension, F-80
    function, F-77
Offset method, B-29
Open-loop transfer function, G-53
Open system, D-1, D-23
Operational amplifier, G-163
Orbit, I-58
    nearly circular, A-115
Order, Bessel function, F-27
Ordinary differential equations, F-42—F-63, H-64—H-72
    linear, F-4—F-63
Orifice plate, C-141
Orthogonal vectors, H-5
Orthonormal, H-5
Oscillation, I-46
Oscillator:
    general harmonic, A-224
    non-isotropic, A-225
    three-dimensional, A-225
    two-dimensional, A-224
Otto cycle, D-68
Overall heat transfer coefficient, E-99

Pappus—Guldinus theorem, A-42, B-41
Parabolic:
    cable, A-55
    function, G-19
Parallel:
    axis theorem, A-68, B-47
    compensation, G-155—G-160
    composite member, B-147
    plate, C-72

Parallelogram law for vector addition, A-1, A-4, H-3, I-2
Parseval's theorem, F-84, F-86
    identity, F-85, H-78
Partial:
    differential equation, F-98—F-99, H-90—H-106, H-139
    second order, F-99
    fraction method, F-72
    pressure, Dalton's rule, D-75
    volume, D-76
Particle, A-1, A-85—A-155
    dynamic, I-17—I-20
    energy and momentum, A-118
    motion of, A-87, A-143, A-145
Pathline, C-6
Peclet number, E-74
Pendulum, I-50
    physical, A-232
    simple, A-226, F-58
    spherical, A-228
    torsion, A-231
Perfect gas, D-45
    diagrams, D-47
    expression, D-76
    properties, D-45
Perihelion, A-111
Period, A-112, A-222, B-171, I-51
    of damped vibration, A-241
Periodic function, F-13, H-74
    extension, F-81
Permanent deformation, B-30
Permittivity constant, I-84
Permutation, H-148
Perpendicular axis theorem, A-67, B-47
Perpetual motion machine, D-30
P—H diagram, D-11
Phase:
    angle, H-75
    diagram, G-94
    equilibrium, D-91
    margin, G-113
    spectrum, H-84
Phasor, G-119
    representation, G-129
    diagram, G-129
Phenomena, impact, B-200
Physical:
    pendulum, A-232
    similarity, C-67—C-69
    types, C-68
PI control compensation, G-137
Piecewise smooth function, F-81
Pinned supports, B-3
Pipe:
    circular, E-70, E-78
    non-circular, E-60, E-71, E-79
Pipe flow, C-76, C-86, E-76
    laminar, C-74, E-76
    turbulent, E-77

Piping losses, D-65
Pitch, A-60,
   of a joint, B-186
   of a screw, A-59
   of a wrench, A-28
Planck's constant, E-109
Plane:
   complex, H-42
   infinite, E-121
   motion, A-157, A-170—A-178, I-10—I-16
      kinetic energy, A-183
   wall, E-7
Planet, I-59
Plastic, B-30
   analysis of beams, B-77, B-127
   strength, B-162, B-166, B-168
Plate:
   flat flow, C-92, E-74
   heat conduction in, E-22
   horizontal, E-91—E-92
   thin, A-76
   turbulent flow, E-82
   vertical, E-93, E-133
Plot:
   direct polar, G-99
   frequency response, G-90
   inverse polar, G-101
   logarithm, G-92
   polar, G-92
Polhausen solution, E-55
Point:
   motion about fixed, A-199
Poisson:
   distribution, H-151
   integral formula, H-62
   ratio, B-23
Polar:
   coordinates, A-95, C-106, E-26
   moment of inertia, A-66, B-45, B-177
   plots, G-89—G-105, G-117
Pole, G-117
   zero diagram, G-38, G-80—G-87
Polygon rule, A-4
Polytropic process, D-47
Pool boiling, E-127
Position, A-85
Potential:
   chemical, D-87, D-89
   gravitation, A-125
   lines, C-105
Potential energy, A-82, A-123, D-20, I-27
   function, A-243
   gravitation, I-60
Power, A-121, A-184, I-24
   cycle, D-61—D-74
   series method, F-66, G-49
Prandtl number, E-49, E-86

Precession, A-207
   steady, A-208
Pressure, D-4, I-79
   absolute, C-16
   gage, C-16
   gradient flow, C-93, C-95
   in static fluid, I-64—I-65
   partial, Dalton's rule, D-75
   units, D-5
   vessels, thin walled, B-194
Principle:
   axes, B-53
   axes and inertia, A-72—A-73, A-192, B-52
   axes of inertia, A-190
   Hamilton's variation, A-218
   moments of inertia, A-73
   of the increase of entropy, D-34
   of transmissibility, A-1, A-19
   of virtual work, A-78
   of work and energy, A-120, A-181
Probability, H-145
Product:
   inner, H-81
   of inertia, A-70, B-50
   scalar, A-5
   triple, A-8
   vector, A-7, H-6—H-9
Programming:
   dynamic, H-122
   linear, H-123
Projectile motion, I-13
Properties:
   fluid, C-5
   geometric sections, B-55
   ideal gas, D-45
   isentropic flow, C-49
   isentropic stagnation, C-45, D-94
   material, B-28
   of inverse Laplace transform, F-18—F-23
   of Laplace transform, F-7—F-9
   of root loci, G-69
   perfect gas, D-45
   pure substances, D-7—D-12
   radiation E-110
   semi-perfect gas, D-44
   thermodynamic, D-11
Proportional:
   limit, B-28
   plus derivative compensator, G-140
Pump, C-135
   centrifugal, C-137
   losses, D-66
Pure substance, D-7—D-12
P-T program, D-8
P-V diagram, D-10
P-V-T behavior, D-7

Quality, D-12
Quasi-static process, D-2

Radial flow, pumps and blowers, C-135
Radiation, E-2
   and convection, E-125
   between two gray surfaces, E-121
   gas, E-123
   heat transfer, E-109—E-125
   shape factor, E-116
   shields, E-122
Radiosity, E-119
Radius of gyration, A-67, B-46
Ramp function, G-19
Range, I-15
Rankine cycle, D-61
   with super heater, D-62
Rational function, F-72
Rayleigh:
   line, C-57
   number, E-49, E-89
   quotient, H-137
   Ritz-Galerkin method, H-144
Reaction:
   chemical, D-82—D-87
   enthalpy, D-83
   stage, D-103
Real gas, D-49
   enthalpy chart, D-50
   entropy chart, D-52
Rectangular:
   block, E-92
   component method, A-5, A-16—A-17
   shaft, angle, B-181
   shaft, torque, B-181
Rectilinear motion, A-85
Reflectivity, E-110
Refrigeration cycle, D-61—D-74
   vapor, D-66
Refrigerator, D-28
Regenerative cycle, D-64
Regression of data, H-129
Reheat cycle, D-63
Reinforced concrete beams, B-151
Relative:
   acceleration, A-164
   velocity, A-161
Residual volume, D-53
Residue, H-57
Resistance:
   air, A-140—A-141
   rolling, A-62
Resonance, F-56
Resonant frequency, G-120, G-122
Reversible:
   processed, D-2
   steady-state steady flow process, D-37
   work, D-38
Revolution, G-112

Reynolds:
   analogy, E-68
   number, E-46, E-49, D-74, I-69
   transport equation, C-23
Riemann:
   Lebesque theorem, H-77
   mapping theorem, H-63
Right hand rule, H-5
Rigid body, A-1, A-19
   energy and momentum, A-181—A-186
   equilibria, A-30, I-62
   fluid motion, C-22
   kinematics, A-157—A-169
   motion, A-197
   plane motion, A-170—A-178
   rotational dynamics, I-43
Ring, thin, gravitational potential, A-126
Riveted joints, B-185—B-195
Roller, A-31
   support, B-2
Rolling resistance, A-62
   body, A-176, I-44
Romberg method, H-139
Root locus, G-67—G-78, G-136—G-145
Roots:
   complex, G-21
   real, G-20
Rotation, A-158
   about a fixed axis, A-157
   center of, A-162
   centroidal, A-175
   non-centroidal, A-175
Rotational:
   dynamics, I-40—I-45
   inertia, I-42
   kinematics, I-37—I-39
   motion, I-37—I-38
Rouche's theorem, H-63
Routh's stability criterion, G-39
   array, G-76

Saddle points, G-74
Safety factor, B-159
Saturation process, adiabatic, D-79
Scalar, A-3, I-1—I-5
   product, A-5
Screw, square threaded, A-59
Seam:
   circumferential, B-195
   longitudinal, B-194
Second law:
   analysis of reacting systems, D-86
   of thermodynamics, D-27—D-37, I-82
Second order system, G-120
Second shift property, F-9
   inverse Laplace transform, F-19
   Laplace transform, F-9
Sections, method of, A-47
Semi-perfect gas, D-44
   properties, D-45

Sensitivity function, G-157
Separation of flow, C-95
Sequence:
    function, H-56
    number, H-55
Series:
    cascading, G-136
    composite member, B-148
    Fourier, H-74
    function, H-56
    MacLaurin, H-56
    number, H-55
    Taylor, H-20
Servo characteristics, G-58—G-63
Servomotor, G-10
Shape factor, radiation, E-116
Shear, B-5
    beams and beam loadings, B-7
    center, B-70
    combined, B-219
    diagrams, B-4, B-9
    failure, B-186
    flow, B-182
    force, B-1—B-11, B-74
    strain, B-20—B-21
    strain energy, B-214
    stress, B-219, C-77
    tension, B-220
Shearing stress, B-13, B-15
Shell and tube heat exchanger, E-97
Shield, radiation, E-122
Shock:
    normal, C-58, D-98
    oblique, C-63
Short column with eccentric load, B-134
Signal flow graph, G-51—G-55
Similarity, physical, C-67—C-69
    solution, E-86
    turbomachine, C-126
    types, C-68
Simple:
    flow, C-107
    flow superposition, C-112
    harmonic motion, A-223
    pendulum, A-226, F-58
Simpson's rule, H-130
Simulation, system G-163—G-174
Sine:
    Fourier formula, F-88
    integral, F-31
Singular:
    matrix, H-14
    points, H-62
Sink, C-108
Skew, H-6—H-17
Skin friction coefficient, C-93, E-53

Solid:
    body, equilibrium, B-1
    non-circular member, B-181
Solutions, D-75—D-81
Sound:
    intensity, I-54
    speed, D-94, I-53
    wave, C-43, I-53—I-56
Sources, C-108
Space:
    mechanics, A-108, A-129
    truss, C-46
Spatial dynamics, A-188—A-210
Specific:
    energy, C-119
    heat, C-43, D-56, I-73
        constant pressure, D-22
        constant volume, D-22
        ideal gas, I-80
    humidity, D-78
    speed, C-128
    volume, D-5
    weight, A-37
Spectrum of electromagnetic radiation, E-109
Speed, A-87
    molecular, I-79
    of sound, C-43, D-94, I-53
Sphere, E-92
Spherical:
    coordinate, A-98, E-7
    pendulum, A-228
    shell, gravitational potential, A-125
Spring:
    constant, F-52
    vibration, F-51
    work, A-81—A-82, A-119
Square-wave function, F-32
Stability, A-115, G-115, H-141
    and buoyancy, C-20
    Nyquist, G-107—G-117
    of virtual work, A-83
    system, G-136—G-145
    vibration, A-242
Standard deviation, H-150
Stanton number, E-74
State, D-9
    change of, D-21
    of a pure substance, D-7—D-12
Static equilibrium, A-170, I-62
    and balance, A-192
    effects, A-200
    loop sensitivity, G-67, G-139
    strength, B-159
Statically indeterminate beam, B-110, B-127
Statics, fluid, I-64—I-65
Statistics, H-145

Steady:
  precessation, A-207
  state heat conduction, one dimension, E-5—E-19
  state heat conduction, two dimensions, E-21—E-28
  state steady-flow process(SSSF), D-24, D-36
  state response, G-16
Steel:
  column, design, B-143
  column, structural formula, B-141
Stefan—Boltzmann law of thermal radiation, E-2, E-111
Stirling cycle, D-71
Stokes' theorem, C-103, H-38
Strain, B-19—B-26
  definition, B-19
  elastic energy, B-23
  energy, B-199
  longitudinal, B-20
  shear, B-20—B-21, B-214
  stress diagram, B-30
  stress relationship, B-28—B-30
  tensor, B-21
  thermal, B-23
Streakline, C-7
Stream function, C-104
Streamline, C-7, C-101
Stream tube, C-7
Strength:
  dynamic, B-163
  elastic, B-159, B-163, B-167
  fatigue, B-171
  of materials, B-32
  plastic, B-162, B-166, B-168
  static, B-159
  torsional, B-169
Stress, B-12—B-16
  axially loaded member, A-15, B-16
  bearing, B-15
  bending distribution, B-64
  combined, B-216—B-223
  concentration, B-26
  definition at a point, B-12—B-13
  eccentric loading, B-217
  failure in joints, B-186
  field, C-8
  in beams, B-64—B-69
  Mohr's circle, B-221
  normal, B-13, B-15
  on longitudinal seam, B-194
  on transverse, B-195
  shearing, B-13, B-15, B-68
  strain diagram, B-30
  strain energy, B-199
  strain relationship, B-28—B-32
  tensor, B-14
  true, B-28

Structural steel column formula, B-141
Structure, A-44
  composite, B-147—B-151
  multiforce, A-47
  three dimensional, A-32
  two dimensional, A-31
Struts, B-132
Sturm-Liouville equation, H-72
Sublimation curve, D-9
Superposition:
  method, B-115
  simple flow, C-112
  principle, H-66
Support, A-31
  and loadings, B-2
  fixed, B-3
  pinned, B-3
  roller and link, B-2
Supported beam, B-7, B-101, B-105
Surface:
  black, E-111
  flat, E-89—E-90
  gray, E-115, E-121
  in three dimensions, H-35
  non-black, E-119
  of revolution A-42, B-41
  submerged, C-17
  temperature, E-7—E-12
  waves, C-119
System, C-1
  analysis, G-89—G-105
  cascade compensated, G-146—G-153
  closed, D-1
  compressible, work, D-15
  continuous, A-153
  dynamic, I-31—I-35
  energy of, D-20
  feedback control, G-118—G-133
  isolated, D-1
  modeling, G-1—G-14
  multicomponent, D-91
  multi-dimension, E-40
  multiphase, D-91
  non-inertial reference, A-165
  of equations, H-1—H-16
  open, D-1
  reacting, D-84, D-86
  response, G-16—G-28
  second order, G-120
  simulations, G-163—G-174
  stability, G-114
  stabilization, G-136—G-145
  thermodynamic, D-1
  tuning, G-172
  variable mass, A-155
  vibrating, A-245

Tangent, H-49
Tangential acceleration, A-94

Taylor series, H-20
Temperature, I-70—I-71
   adiabatic flame, D-85
   bulk, E-73
   dew point, D-78
   dry bulb, D-78
   film, E-73
   log-mean, E-100
   mean, E-72
   saturation, D-8
   scale, D-32
   surface, E-7—E-10
   wet bulb, D-78
Tension:
   combined shear, B-219
   failure, B-187
   shear, B-220
   shear stress, B-219
Tensor:
   strain, B-21
   stress, B-14
Terminal velocity, A-102
Theorem:
   Bayes, H-147
   Buckingham Pi, C-69
   Castigliano, B-202
   convolution, F-89
   final value, F-65
   Green's, H-34
   initial value, F-65
   inversion, F-84
   Lerch, F-16
   Liouville, H-55
   mean value, H-19
   Pappus—Guldinus, A-42, B-41
   parallel axis, A-68, B-47
   Parseval, F-84, F-86
   perpendicular axis, A-67, B-47
   Riemann mapping, H-63
   Stokes', C-103, H-38
   three moment, B-121
   Varignon, A-22
Thermal:
   conductivity, E-1
   diffusivity, E-2
   efficiency, D-68
   expansion, I-71
   radiation, E-111, I-75
   resistance, I-74
   strain, B-23
   systems, G-7
Thermodynamics, C-41
   equilibrium, D-3
   first law, D-19—D-26, I-73—I-77
   properties, D-11
   relations, D-54—D-59
   second law, D-27—D-37, I-82
   systems, D-1
   temperature, D-32

   third law, D-85
   work, D-14, D-16
   zeroth law, D-4
Thin:
   plate, A-76
      circular, A-77
      rectangular, A-77
   walled pressure vessel, B-194
Third law of thermodynamics, D-85
Three dimension:
   surfaces, H-35
   vectors, H-1
Three moment theorem, B-121
Timber, column design, B-144
Time:
   constant, G-23, G-28
   domain analysis, G-58—G-63
   response, G-85, G-89, G-157
Torque, B-176, B-182, I-40
   circular shaft, B-180
   free motion, A-210
   rectangular shaft, B-181
Torsion, B-176—B-184
   and bending, B-223
   angle of twist, B-181
   energy, B-211
   formula, B-177
   pendulum, A-231
Torsional strength:
   design, B-67
   dynamic, B-169
Total:
   hemispheric emissivity, E-112
   internal reflection, I-138
   resultant force, A-28
Toughness, B-166
Transfer:
   function, G-51—G-55
   units, E-106
Transform:
   Fourier, F-77—F-89
   integral, F-1
   Laplace, F-1—F-13
Transformation:
   matrix, A-11
   Mobius, H-58
Transient response, G-20, G-80
Transition, D-9
Translation, A-157
Transmissivity, E-110
Transpose, H-10
Trapezoid rule, H-130
Triangle method, I-1
Triangular inequality, H-4
Trigonometric series, F-77
Triple:
   point, D-9
   product, A-8
Throttling process, D-25

J-17

True stress, B-28
Trusses, A-44
   space, A-46
T-S diagram, D-10
Tube, flow, E-76, E-80, E-82
Turbine, C-130
   losses, D-65
   reaction stages, D-103
Turbomachinery, C-126—C-137
Turbulent:
   boundary layer, C-94
   flow, C-11, C-71, E-46, E-75, E-77, E-82, E-89
Turning point, A-128
T-V program, D-9
Twist, B-179
   angle of, B-178, B-184
   circular shaft, B-180
   rectangular shaft, B-181

Ultimate:
   cycle method, G-175
   gain, G-175
Undamped vibration, A-222
Uniform:
   circular motion, I-51
   convergence, F-82—F-83
   flow, C-107
   state uniform flow process (USUF), D-26, D-36
Uniqueness, H-66
   of Laplace transform, F-16
Unit:
   impulse function, F-35
   step function, F-32, F-35
   vector, A-10, A-14, H-1, I-4
Units, A-2, C-2, D-6
   heat and work, D-18
   heat transfer, E-4, E-106
   pressure, D-5
Unity feedback, G-63
Universal gas constant, D-42
Unsteady-state heat conduction, E-30—E-42

Valves, C-85
van der Waal equation, D-56
Vapor:
   enthalpy, D-79
   entropy, D-79
   mixture, D-77
   power cycle, D-61
   refrigeration cycle, D-66
Vaporization:
   curve, D-9
   heat of, D-10
Varignon's theorem, A-22
Vector, A-3, H-1—H-16, I-1—I-5
   addition, A-4

complex, H-16
fields, H-27
force, B-1
functions, A-9, H-23
in N-dimensions, H-7
in three dimensions, H-1
moment, B-1
product, A-7
space, H-79
unit, A-10
Velocity, I-10
   absolute, A-161
   angular, A-99
   areal, A-108
   escape, A-111
   fictitious, E-88
   field, C-5
   instantaneous, A-85
   horizontal, C-117
   potential, C-103
   profile, C-76
   relative, A-161
   terminal, A-102
   vertical, C-118
Venturi:
   flame, C-125
   meter, C-142
Vessel, thin walled pressure, B-194
Vibration, B-171
   damped, F-53
   damped forced, A-241
   damped free, A-239
   forced, A-236
   free, A-234, F-54
   mechanical, A-222—A-252
   of a loaded string, A-250
   of a spring, F-51
   undamped, F-54
Virtual work, A-78
   principle of, A-79
Viscosity, E-3, I-68
   dynamic, E-3
   kinematic, E-3
   Newtonian fluid, C-9
Viscous flow, C-71—C-86, E-45
Volterra integral equation, F-91
Volume, D-60
   composite, A-39
   control, D-23
   expansion, D-59
   integral, H-38
   of revolution, A-43, B-42
   partial, D-76
   residual, D-53
Von Karman's analogy, E-69
Vortex, C-109

Wall:
    multilayer, E-8
    plane, E-7, E-11, E-13
Wave, A-252
    equation, H-103—H-106, H-144
    sound, C-43, I-53—I-56
    surface, C-119
Wavelength, E-109
Wedge, A-58
Weight, A-37, A-81, I-18
Weir, C-143
Weld:
    butt, B-191
    fillet, B-190
Welted joint, B-185—B-195
Wien's displacement law, E-113
Work, D-14—D-18, I-21—I-25, I-76
    definition, D14
    energy principle, A-120, A-181
    energy theorem, I-33

of a force, A-78, A-118, I-21—I-23
of a gravitational force, A-120
of a moment, A-79
units, D-18
virtual, method of, A-78
virtual, principle of, A-79
Wrench, pitch, A-28
Wronskian, H-67

Yield point, B-29
Young's:
    modulus, B-22

Zero, G-117, H-62
    force member, A-45
Zeroth law of thermodynamics, D-4
Ziegler-Nichols method, G-174
Z-transform, G-44—G-47

J-19

HANDBOOK of MATHEMATICAL, SCIENTIFIC, & ENGINEERING FORMULAS, FUNCTIONS, GRAPHS, TABLES, & TRANSFORMS

RESEARCH & EDUCATION ASSOCIATION

A particularly useful reference for those in math, science, engineering and other technical fields. Includes the most-often used formulas, tables, transforms, functions, and graphs which are needed as tools in solving problems. The entire field of special functions is also covered. A large amount of scientific data which is often of interest to scientists and engineers has been included.

Available at your local bookstore or order directly from us by sending in coupon below.

RESEARCH & EDUCATION ASSOCIATION
61 Ethel Road W., Piscataway, New Jersey 08854
Phone: (732) 819-8880   website: www.rea.com

VISA   MasterCard

☐ Payment enclosed
☐ Visa   ☐ MasterCard

Charge Card Number

Expiration Date: _____ / _____
                      Mo   Yr

Please ship **"Math Handbook"** @ $34.95 plus $4.00 for shipping.

Name _____

Address _____

City _____ State _____ Zip _____

The HANDBOOK of CHEMICAL ENGINEERING

Staff of Research and Education Association

Available at your local bookstore or order directly from us by sending in coupon below.

RESEARCH & EDUCATION ASSOCIATION
61 Ethel Road W., Piscataway, New Jersey 08854
Phone: (732) 819-8880   website: www.rea.com

☐ Payment enclosed
☐ Visa   ☐ MasterCard

Charge Card Number

Expiration Date: _____ / _____
                    Mo   Yr

Please ship the **"The Handbook of Chemical Engineering"** @ $38.95 plus $4.00 for shipping.

Name _____

Address _____

City _____ State _____ Zip _____

The HANDBOOK of ELECTRICAL ENGINEERING

R̃EA

Staff of Research and Education Association

Available at your local bookstore or order directly from us by sending in coupon below.

RESEARCH & EDUCATION ASSOCIATION
61 Ethel Road W., Piscataway, New Jersey 08854
Phone: (732) 819-8880   website: www.rea.com

VISA   *MasterCard*

Charge Card Number

☐ Payment enclosed
☐ Visa ☐ MasterCard

Expiration Date: _____ / _____
                    Mo   Yr

Please ship **"The Handbook of Electrical Engineering"** @ $38.95 plus $4.00 for shipping.

Name _____

Address _____

City _____ State _____ Zip _____

MAXnotes®
REA's Literature Study Guides

MAXnotes® are student-friendly. They offer a fresh look at masterpieces of literature, presented in a lively and interesting fashion. **MAXnotes®** offer the essentials of what you should know about the work, including outlines, explanations and discussions of the plot, character lists, analyses, and historical context. **MAXnotes®** are designed to help you think independently about literary works by raising various issues and thought-provoking ideas and questions. Written by literary experts who currently teach the subject, **MAXnotes®** enhance your understanding and enjoyment of the work.

Available **MAXnotes®** include the following:

Absalom, Absalom!	Henry IV, Part I	Othello
The Aeneid of Virgil	Henry V	Paradise
Animal Farm	The House on Mango Street	Paradise Lost
Antony and Cleopatra	Huckleberry Finn	A Passage to India
As I Lay Dying	I Know Why the Caged	Plato's Republic
As You Like It	Bird Sings	Portrait of a Lady
The Autobiography of	The Iliad	A Portrait of the Artist
Malcolm X	Invisible Man	as a Young Man
The Awakening	Jane Eyre	Pride and Prejudice
Beloved	Jazz	A Raisin in the Sun
Beowulf	The Joy Luck Club	Richard II
Billy Budd	Jude the Obscure	Romeo and Juliet
The Bluest Eye, A Novel	Julius Caesar	The Scarlet Letter
Brave New World	King Lear	Sir Gawain and the
The Canterbury Tales	Leaves of Grass	Green Knight
The Catcher in the Rye	Les Misérables	Slaughterhouse-Five
The Color Purple	Lord of the Flies	Song of Solomon
The Crucible	Macbeth	The Sound and the Fury
Death in Venice	The Merchant of Venice	The Stranger
Death of a Salesman	Metamorphoses of Ovid	Sula
Dickens Dictionary	Metamorphosis	The Sun Also Rises
The Divine Comedy I: Inferno	Middlemarch	A Tale of Two Cities
Dubliners	A Midsummer Night's Dream	The Taming of the Shrew
The Edible Woman	Moby-Dick	Tar Baby
Emma	Moll Flanders	The Tempest
Euripides' Medea & Electra	Mrs. Dalloway	Tess of the D'Urbervilles
Frankenstein	Much Ado About Nothing	Their Eyes Were Watching God
Gone with the Wind	Mules and Men	Things Fall Apart
The Grapes of Wrath	My Antonia	To Kill a Mockingbird
Great Expectations	Native Son	To the Lighthouse
The Great Gatsby	1984	Twelfth Night
Gulliver's Travels	The Odyssey	Uncle Tom's Cabin
Handmaid's Tale	Oedipus Trilogy	Waiting for Godot
Hamlet	Of Mice and Men	Wuthering Heights
Hard Times	On the Road	Guide to Literary Terms
Heart of Darkness		

RESEARCH & EDUCATION ASSOCIATION
61 Ethel Road W. • Piscataway, New Jersey 08854
Phone: (732) 819-8880   website: www.rea.com

Please send me more information about **MAXnotes®**.

Name _____

Address _____

City _____ State _____ Zip _____

REA's
HANDBOOK OF
ENGLISH
Grammar, Style, and Writing

All the essentials you need to know are contained in this simple and practical book

Learn quickly and easily
- Rules and exceptions in grammar
- Spelling and proper punctuation
- Common errors in sentence structure
- 2,000 examples of correct usage
- Effective writing skills

Complete practice exercises with answers follow each chapter

Research & Education Association

Available at your local bookstore or order directly from us by sending in coupon below.

RESEARCH & EDUCATION ASSOCIATION
61 Ethel Road W., Piscataway, New Jersey 08854
Phone: (732) 819-8880   website: www.rea.com

VISA   **MasterCard**

Charge Card Number

☐ Payment enclosed
☐ Visa   ☐ MasterCard

Expiration Date: ____ / ____
                    Mo   Yr

Please ship **"Handbook of English"** @ $22.95 plus $4.00 for shipping.

Name _____

Address _____

City _____ State _____ Zip _____

The Best Test Preparation & Review Course

FE/EIT

Fundamentals of Engineering / Engineer-in-Training
– AM Exam –

The only book with

3 Full-Length 4-Hour Exams

plus **12 Comprehensive Reviews**

Completely up-to-date based on official exam questions

- Solutions to every test question
- Prepared by test experts in each field
- Comprehensive reviews with sample problems
- Test-taking tips *and* strategies to help *you* achieve a *TOP SCORE*

Chemical • Mechanical • Electrical • Civil

Prepares you for Certification as an Engineer-In-Training, a requirement before receiving the Professional Engineering License (PE)

REA *Research & Education Association*

Available at your local bookstore or order directly from us by sending in coupon below.

RESEARCH & EDUCATION ASSOCIATION
61 Ethel Road W., Piscataway, New Jersey 08854
Phone: (732) 819-8880   website: www.rea.com

VISA *MasterCard*

Charge Card Number

☐ Payment enclosed
☐ Visa   ☐ MasterCard

Expiration Date: ____ / ____
                   Mo   Yr

Please ship REA's **"FE/EIT AM Exam"** @ $49.95 plus $4.00 for shipping.

Name _____
Address _____
City _____ State _____ Zip _____

Available at your local bookstore or order directly from us by sending in coupon below.

RESEARCH & EDUCATION ASSOCIATION
61 Ethel Road W., Piscataway, New Jersey 08854
Phone: (732) 819-8880   website: www.rea.com

☐ Payment enclosed
☐ Visa   ☐ MasterCard

Charge Card Number

Expiration Date: ____ / ____
                    Mo   Yr

Please ship REA's **"FE/EIT PM Exam"** @ $41.95 plus $4.00 for shipping.

Name _____

Address _____

City _____ State _____ Zip _____

Research & Education Association

The Best Test Preparation & Review Course

FE/EIT

Fundamentals of Engineering / Engineer-in-Training

PM Exam Mechanical Engineering

The only book with **2** Full-Length 4-Hour Exams *plus* Comprehensive Reviews

- Solutions to every test question
- Prepared by test experts in each field
- Comprehensive reviews with sample problems
- Test-taking tips *and* strategies to help *you* achieve a *TOP SCORE*

Completely up-to-date based on official exam questions

Prepares you for Certification as an Engineer-In-Training, a requirement before receiving the Professional Engineering License (PE)

Available at your local bookstore or order directly from us by sending in coupon below.

RESEARCH & EDUCATION ASSOCIATION
61 Ethel Road W., Piscataway, New Jersey 08854
Phone: (732) 819-8880   website: www.rea.com

VISA   **MasterCard**

Charge Card Number

☐ Payment enclosed
☐ Visa   ☐ MasterCard

Expiration Date: _____ / _____
                      Mo   Yr

Please ship REA's **"FE/EIT PM Mechanical Engineering Exam"** @ $39.95 plus $4.00 for shipping.

Name _____

Address _____

City _____ State _____ Zip _____

REA's Test Preps
The Best in Test Preparation

- REA "Test Preps" are **far more** comprehensive than any other test preparation series
- Each book contains up to **eight** full-length practice tests based on the most recent exams
- **Every** type of question likely to be given on the exams is included
- Answers are accompanied by **full** and **detailed** explanations

REA publishes over 60 Test Preparation volumes in several series. They include:

Advanced Placement Exams (APs)
Biology
Calculus AB & Calculus BC
Chemistry
Computer Science
English Language & Composition
English Literature & Composition
European History
Government & Politics
Physics
Psychology
Spanish Language
Statistics
United States History

College-Level Examination Program (CLEP)
Analyzing and Interpreting Literature
College Algebra
Freshman College Composition
General Examinations
General Examinations Review
History of the United States I
Human Growth and Development
Introductory Sociology
Principles of Marketing
Spanish

SAT II: Subject Tests
Biology E/M
Chemistry
English Language Proficiency Test
French
German
Literature

SAT II: Subject Tests (cont'd)
Mathematics Level IC, IIC
Physics
Spanish
United States History
Writing

Graduate Record Exams (GREs)
Biology
Chemistry
Computer Science
General
Literature in English
Mathematics
Physics
Psychology

ACT - ACT Assessment

ASVAB - Armed Services Vocational Aptitude Battery

CBEST - California Basic Educational Skills Test

CDL - Commercial Driver License Exam

CLAST - College-Level Academic Skills Test

ELM - Entry Level Mathematics

ExCET - Exam for the Certification of Educators in Texas

FE (EIT) - Fundamentals of Engineering Exam

FE Review - Fundamentals of Engineering Review

GED - High School Equivalency Diploma Exam (U.S. & Canadian editions)

GMAT - Graduate Management Admission Test

LSAT - Law School Admission Test

MAT - Miller Analogies Test

MCAT - Medical College Admission Test

MTEL - Massachusetts Tests for Educator Licensure

MSAT - Multiple Subjects Assessment for Teachers

NJ HSPA - New Jersey High School Proficiency Assessment

PLT - Principles of Learning & Teaching Tests

PPST - Pre-Professional Skills Tests

PSAT - Preliminary Scholastic Assessment Test

SAT I - Reasoning Test

SAT I - Quick Study & Review

TASP - Texas Academic Skills Program

TOEFL - Test of English as a Foreign Language

TOEIC - Test of English for International Communication

RESEARCH & EDUCATION ASSOCIATION
61 Ethel Road W. • Piscataway, New Jersey 08854
Phone: (732) 819-8880   website: www.rea.com

Please send me more information about your Test Prep books

Name _____

Address _____

City _____ State _____ Zip _____

REA's Problem Solvers

The "PROBLEM SOLVERS" are comprehensive supplemental textbooks designed to save time in finding solutions to problems. Each "PROBLEM SOLVER" is the first of its kind ever produced in its field. It is the product of a massive effort to illustrate almost any imaginable problem in exceptional depth, detail, and clarity. Each problem is worked out in detail with a step-by-step solution, and the problems are arranged in order of complexity from elementary to advanced. Each book is fully indexed for locating problems rapidly.

ACCOUNTING
ADVANCED CALCULUS
ALGEBRA & TRIGONOMETRY
AUTOMATIC CONTROL SYSTEMS/ROBOTICS
BIOLOGY
BUSINESS, ACCOUNTING, & FINANCE
CALCULUS
CHEMISTRY
COMPLEX VARIABLES
DIFFERENTIAL EQUATIONS
ECONOMICS
ELECTRICAL MACHINES
ELECTRIC CIRCUITS
ELECTROMAGNETICS
ELECTRONIC COMMUNICATIONS
ELECTRONICS
FINITE & DISCRETE MATH
FLUID MECHANICS/DYNAMICS
GENETICS
GEOMETRY
HEAT TRANSFER
LINEAR ALGEBRA
MACHINE DESIGN
MATHEMATICS for ENGINEERS
MECHANICS
NUMERICAL ANALYSIS
OPERATIONS RESEARCH
OPTICS
ORGANIC CHEMISTRY
PHYSICAL CHEMISTRY
PHYSICS
PRE-CALCULUS
PROBABILITY
PSYCHOLOGY
STATISTICS
STRENGTH OF MATERIALS & MECHANICS OF SOLIDS
TECHNICAL DESIGN GRAPHICS
THERMODYNAMICS
TOPOLOGY
TRANSPORT PHENOMENA
VECTOR ANALYSIS

If you would like more information about any of these books, complete the coupon below and return it to us or visit your local bookstore.

RESEARCH & EDUCATION ASSOCIATION
61 Ethel Road W. • Piscataway, New Jersey 08854
Phone: (732) 819-8880   website: www.rea.com

Please send me more information about your Problem Solver books

Name _____

Address _____

City _____ State _____ Zip _____